Urban Operations

Plus

Counterinsurgency

Department of Defense

Pentagon
Publishing

Field Manual
No. 3-06

Headquarters
Department of the Army
Washington, DC, 26 October 2006

Urban Operations

Contents

Distribution Restriction: Approved for public release; distribution is unlimited.

***This publication supersedes FM 3-06, 1 June 2003.**

Field Manual
No. 3-06

Headquarters
Department of the Army
Washington, DC, 26 October 2006

Urban Operations

Contents

***This publication supersedes FM 3-06, 1 June 2003.**

Figures

Historical Vignettes

Preface

Doctrine provides a military organization with a common philosophy, a language, a purpose, and unity of effort. Rather than establishing a set of hard and fast rules, the objective of doctrine is to foster initiative and creative thinking. To this end, FM 3-06 discusses major Army operations in an urban environment. This environment, consisting of complex terrain, a concentrated population, and an infrastructure of systems, is an operational environment in which Army forces will operate. In the future, it may be the predominant operational environment. Each urban operation is unique and will differ because of the multitude of combinations presented by the threat, the urban area itself, the major operation of which it may be part (or the focus), and the fluidity of societal and geopolitical considerations. Therefore, there will always exist an innate tension between Army doctrine, the actual context of the urban operation, and future realities. Commanders must strike the proper balance between maintaining the capability to respond to current threats and preparing for future challenges.

PURPOSE

This manual provides the analytical tools for evaluating an urban operation to determine if the operation is necessary for overall mission success. It also provides the means to understanding and determining the impacts of the urban environment on military operations and provides information on managing, taking advantage of, and mitigating the effects of those impacts as appropriate. As such, this manual demonstrates how to apply the doctrinal principles in FM 3-0 to this unique environment.

SCOPE

Chapter 1 introduces theoretical and historical perspectives of urban operations that serve as the underlying basis for the rest of the manual. While this manual has incorporated lessons learned from recent and ongoing operations, it has deliberately taken a broad and varied historical perspective in order to remain relevant to future threats and circumstance differing from those that the Army currently faces. Chapter 2 discusses the characteristics of urban centers and populations as well as their impact on operations. It is unlikely that Army forces will ever operate in a benign urban environment; therefore, Chapter 3 discusses the varied nature of potential urban threats. Chapter 4 describes the effects of the urban environment on warfighting functions and tactics. An understanding of the complexities of the urban environment, the nature of the enemy as an adaptive, learning opponent, and the effects of the environment on warfighting capabilities and skills is essential to sound decision making. Chapters 5 and 6 discuss the potential costs of urban operations and risk reduction measures that the commander and his staff must consider early in their planning. These chapters also outline an urban operational framework and specific urban considerations that create the foundations necessary for successfully applying operational doctrine to an urban environment.

The second part of the manual (Chapters 7 – 10) discusses how urban operations are conducted and resourced. Urban operations include major offensive and defensive operations in urban environments as well as stability or civil support operations ranging from peace operations and counterterrorism to disaster relief and humanitarian assistance. For the different types of operations—offense, defense, and stability or civil support—the purpose, characteristics, organization, and considerations are discussed. However, commanders consider that most urban operations will normally involve the simultaneous execution of offense and defense with stability or civil support (although proportional emphasis will shift over time) and plan accordingly.

APPLICABILITY

This manual is intended for Army commanders and their staffs at the brigade level and above. It addresses full spectrum operations that Army units will execute in urban settings. However, users should also consult JP 3-06 for specific joint information. Additionally, users should be familiar with FM 3-06.1, FM 3-06.11, TC 90-1, and urban operations chapters, appendices, or sections found in other infantry, armor, combined arms, and proponent field manuals for the tactics, techniques, and procedures (TTP) and appropriate proponent information necessary to conduct tactical urban operations at the brigade level and below. This publication applies to the Active Army, the Army National Guard (ARNG)/Army National Guard of the United States (ARNGUS), and United States Army Reserve unless otherwise stated.

ADMINISTRATIVE INSTRUCTIONS

When this FM 3-06 was approved, doctrine was incomplete for incorporating stability operations and civil support operations in place of stability operations and support operations. This manual incorporates stability and reconstruction and civil support operations doctrine found in FM 1 and FMI 5-0.1; it will be revised to reflect the future updates of FM 3-0 and FM 3-07 and, when published, FM 3-28. (Note: Stability and reconstruction operations has been redesignated stability operations to comply with Department of Defense Directive 3000.05.)

Chapter 2 defines "city" according to a population size. However, in historical vignettes and accounts, the term "city" may be applied in its common usage without specific regard to size to maintain conformity with most other historical reports.

In this manual, the term "threat" is applied broadly to include an enemy force (conventional or unconventional), an armed belligerent in a peace operation, antagonistic or unfriendly elements of the civilian population, or some other hazardous condition in the urban environment that negatively influences mission accomplishment.

The term military operations on urban terrain (MOUT) is replaced by urban operations (UO). MOUT is an acronym from FM 90-10, which is superseded by this manual.

Otherwise, the glossary lists most terms used in FM 3-06 that have joint or Army definitions. Where Army and joint definitions are different, (Army) follows the term. Definitions for which FM 3-06 is the proponent manual (the authority) are marked with an asterisk (*). The proponent or amplifying manual for other terms is listed in parentheses after the definition.

The manual attempts to incorporate a broad range of historical vignettes into each chapter where the account supports the doctrinal line of reasoning. Two historical vignettes, however, were included as appendices (A and C) because of their longer lengths.

Unless this publication states otherwise, masculine nouns or pronouns do not refer exclusively to men.

This publication contains copyrighted material.

The proponent for this publication is the United States Army Training and Doctrine Command. Send written comments and recommended changes on DA Form 2028 (Recommended Changes to Publications and Blank Forms) directly to Commander, U.S. Army Combined Arms Center and Fort Leavenworth, Combined Arms Doctrine Directorate, ATTN: ATZL-CD, (FM 3-06), 201 Reynolds Avenue, Fort Leavenworth, KS 66027-2337. Send comments and recommendations by e-mail to webb-cadd@leavenworth.army.mil. Follow the DA Form 2028 format or submit an electronic DA Form 2028.

This page intentionally left blank.

This page intentionally left blank.

Chapter 1

The Urban Outlook

Today's security environment demands more from Army leaders than ever before. Army leaders must not only be able to lead Soldiers but also influence other people. They must be able to work with members of other Services and governmental agencies. They must win the willing cooperation of multinational partners, both military and civilian. But ultimately, the Army demands self-aware and adaptive leaders who can compel enemies to surrender in war and master the circumstances facing them in peace. Victory and success depend on the effectiveness of these leaders' organizations. Developing effective organizations requires hard, realistic, and relevant training.

FM 1

Given the prevalence of large cities throughout the world, Army forces will likely be required to conduct operations in, around, and over large urban areas. These operations will normally be in support of a joint force commander (JFC) conducting military operations pursuant to U.S. national security policy. This manual is designed to facilitate the planning and conduct of the full range and spectrum of land operations in a complex urban environment. Each urban environment and urban operation is unique; prescribing a specific doctrinal "solution" for each situation is impossible. Instead, this manual provides a framework to commanders and their staffs for understanding the urban environment, for analyzing and deciding whether or not to initiate urban operations (UO), and for applying operational doctrine to this complex environment. It also provides a broad base of historical vignettes and examples to help develop a refined analytical perspective and stimulate thought. The manual also includes some planning points and tactics and techniques to assist in preparing for and conducting UO. This information provides a foundation for approaching major UO. Combined with other joint and Army doctrine, this information also will help commanders and their staffs learn to adapt and succeed in this challenging environment.

THE PROSPECT OF URBAN OPERATIONS

1-1. The world is undergoing massive urbanization. Although exceptions exist, an overall trend of migration from rural to urban areas is occurring throughout the globe. (Australia, one of the world's most urbanized countries, is actually becoming less urbanized.) This trend is especially evident in developing nations. Combined with the exponential growth of the global population in the last quarter century, this migration has created massive urban areas that hold the centers of population, government, and economics in their respective regions. In many cases, rapid urbanization has overburdened already weak infrastructures, scarce resources, and a fragile economic base. As urbanization has changed the demographic landscape, potential enemies recognize the inherent danger and complexity of this environment to the attacker, and may view it as their best chance to negate the technological and firepower advantages of modernized opponents. Given the global population trends and the likely strategies and tactics of future threats, Army forces will likely conduct operations in, around, and over urban areas—not as a matter of fate, but as a deliberate choice linked to national security objectives and strategy, and at a time, place, and method of the commander's choosing.

Army Urban Operations

Army forces conduct UO either as one component of a larger operation or as a single operation focused totally on a specific urban environment. Major Army UO are often part of a joint and multinational effort requiring rigorous interagency and civil-military coordination that typically includes the full spectrum of military operations. Commanders of Army major operations must determine if UO are essential to mission accomplishment. If so, commanders must carefully integrate the operations into campaign planning to support the operational objectives of the JFC.

Army leaders conducting UO must—

- Understand the urban environment to determine decisive points.
- Shape the operation to set the conditions for success.
- Precisely mass the effects of combat power to thoroughly engage the decisive points that lead to centers of gravity.
- Continually consolidate gains essential to the retention of the initiative.
- Transition the urban area to the control of another force or agency or, ultimately, back to legitimate and functioning civilian control.

AN URBAN PERSPECTIVE

1-2. As a subset of all Army operations, UO are operations focused on an urban environment. UO include full spectrum operations—offensive, defensive, and stability or civil support—that may be executed, either sequentially or (more likely) simultaneously, during the conduct of a single urban operation (see figure 1-1). UO may be the sole mission of the commander or one of several tasks nested in a larger operation. Regardless of the types of operations conducted or whether the urban area is the single focus of the operation or only one component of a larger operation or campaign, the complex urban environment significantly affects the overall conduct of the mission.

1-3. When conceptualizing urban operations, commanders must understand two important terms: urban area and urban environment. The first is a subset of the second. An ***urban area* is a topographical complex where man-made construction or high population density is the dominant feature**. Focusing on urban areas means concentrating on the physical aspects of the area and their effects on weapons, equipment, line-of-sight, and tactics, techniques, and procedures. The ***urban environment* includes the physical aspects of the urban area as well as the complex and dynamic interaction and relationships between its key components—the terrain (natural and man-made), the society, and the supporting infrastructure—as an overlapping and interdependent system of systems.**

1-4. Importantly, commanders must also understand and consider that critical elements of the infrastructure may lie far beyond the area's physical confines. For example, the generating source providing power to the urban energy system is part of that system but may be located well outside of the urban area. Similarly, effects of the interaction between components of the infrastructure, located both inside and outside the urban area, extend well into smaller, neighboring urban areas and surrounding rural areas and often form their political, economic, and cultural focus. Understanding the total urban environment is essential to planning and conducting full spectrum urban operations.

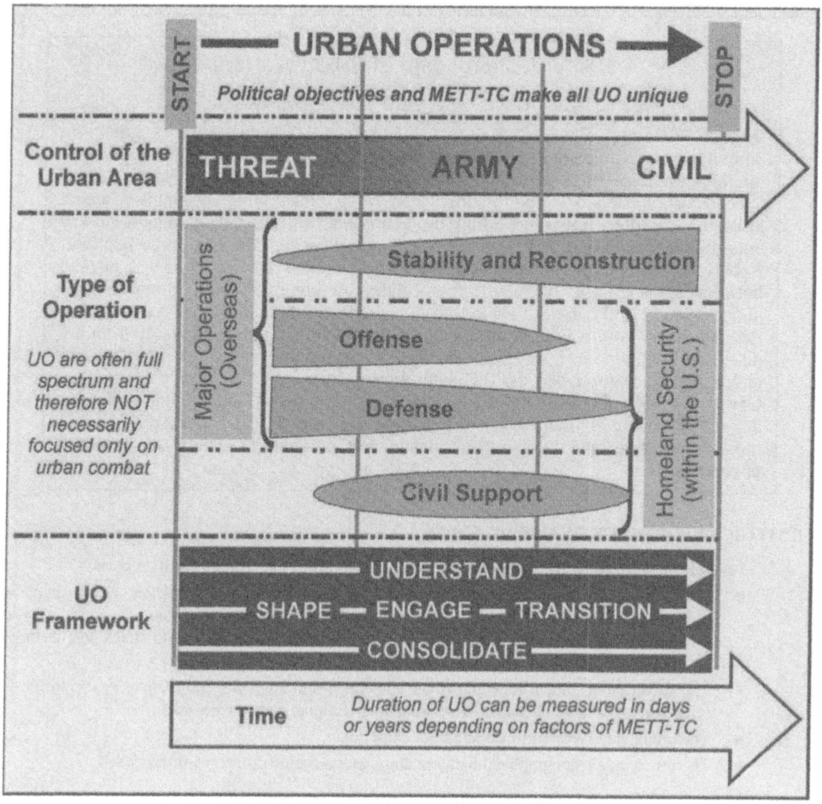

Figure 1-1. Full spectrum urban operations

HISTORICAL SIGNIFICANCE OF URBAN AREAS IN WARFARE

1-5. Urban areas always have been central to, or have significantly influenced, military operations. One of the first urban-centered battles was the siege of Troy at the beginning of Greek history. Moreover, much of the history of early Greece revolved around wars between its city-states or with Persia and centered on the conquest, siege, or blockade of cities. Five hundred years later, the Roman Empire replaced Greece as the dominant world power; although, urban areas remained central to the Roman method of warfare. Even Rome's history can be viewed as a microcosm of urban warfare over the past two thousand years. Though military operations within the physical confines of many of these historic urban areas were not the norm, the focus of these operations was their conquest or control.

1-6. Although Rome last saw combat in 1944, urban areas have been no less prominent in warfare since that time. Seoul in Korea, Beirut in Lebanon, Panama City in Panama, Grozny in Chechnya, Sarajevo in Bosnia-Herzegovina, Kabul in Afghanistan, and Baghdad in Iraq have been centers of conflict in the last 50 years. Urban areas, now more pervasive than ever before, will continue to be essential to successful operational and strategic warfighting. Today, armies cannot expect to execute major military operations without the influence of the urban environments within their area of operations.

Rome: A Microcosm of Urban Warfare

During two millennia, Rome has been the center of at least 12 battles. The Gauls lay siege to Rome first in 387 BC. That first siege lasted six months and ended after the barbarians burnt much of the city. The surviving patrician families paid a ransom for the withdrawal of Brennus' army. From 408 to 410 AD, the Goth leader, Alaric, successfully besieged Rome no less than three times. The Byzantine General Belisarius captured Rome twice from the Goths and withstood siege inside the city once between 536 and 549. Five hundred years later in 1084, Norman adventurer Robert Guiscard captured medieval Rome and sacked the city during a dispute between the Pope and the Holy Roman Empire. Forces of the Holy Roman Empire again stormed and captured the city to punish the Pope in 1527. During the Italian Revolution in 1849, a French army supporting the Pope captured the city from the Italian revolutionary army under Garibaldi. In 1944, the last military action took place in and around Rome when the U.S. Fifth Army captured the city from the retreating German army. Rome's turbulent history—fought over ethnic and religious differences, prestige, and military necessity—demonstrates the importance of urban areas in warfare and the various causes and combatants within this complex environment.

Strategic Importance of Urban Areas

1-7. Several reasons have attracted (and continue to attract) armies to combat in urban areas:

- A military force chooses to position itself in an urban area to capitalize on the perceived defensive advantages offered by the environment. In contrast, an opposing force, by analyzing the factors of the situation, determines that it must enter the urban area to attack and destroy its enemy (or devote essential combat power to their isolation).
- The urban environment's people (their allegiance and support), infrastructure, capabilities, or other resources have or can be of significant operational or strategic value.
- The urban area has significant symbolic importance.
- The urban area's geographical location dominates a region or avenue of approach.

1-8. Russia's 1994 experience in Chechnya illustrates an increasingly important motivation for conducting urban operations. The Chechen rebels, after failing to engage Russian forces outside the city, chose to turn Grozny into the main battlefield. Leaders of the defeated Chechen conventional forces recognized that fighting in the urban area provided them their best chance for success. The complexities of urban combat and the perceived advantages of defending an urban area mitigated their numerical and technological inferiority. The urban area provided the Chechens protection from fires, resources, interior lines, and covered and concealed positions and movement. Given such advantages offered by the environment, smaller or less-sophisticated military forces have similarly chosen to fight in urban areas either as a deliberate strategy or to escape certain destruction in open terrain.

1-9. Such advantages of operating in an urban environment also prompt forces to conduct an urban operation to facilitate a larger campaign plan and decisive battle in another location. The urban operation can focus the enemy on the urban area and allow other forces to conduct operations elsewhere. From a defensive perspective, an urban defense may gain time and space to reorganize forces in new defensive positions, to divert enemy forces from other critical tasks, or to prepare to conduct offensive operations. To some extent, these reasons motivated Soviet forces defending Leningrad and Stalingrad from the Germans in World War II. The stubborn defense permitted the Soviets to reorganize for later offensive operations. From an offensive perspective, an attack on an urban area may be a shaping operation used to divert enemy resources from the decisive operation that will follow.

1-10. Armies also fight in an urban area to obtain some critical feature or resource in the area, such as a port facility. The desire to control an important seaport and access to the Persian Gulf largely motivated the Iranian and Iraqi struggle for Basra in the 1980s. Earlier, in 1944, British forces fought German units in Arnhem for control of the Rhine River Bridge. Other key infrastructure of the urban environment may have operational or strategic significance and can compel military forces to attack or defend the area. As urban areas account for an increasing share of a country's national income, often generating over 50 percent of gross national product, the strategic implications for their control or influence become even greater.

1-11. Urban areas are often located on terrain that dominates a region or an avenue of approach. In these cases, offensive armies capture these areas to proceed with security to another objective. Conversely, defensive forces commonly defend the area to deny the area of operations. To illustrate, Cassino, Italy during World War II stood astride the critical highway approach up the Liri valley to Rome. The allies had to attack and capture the monastery to facilitate the allied offensive north. Cassino's location made bypassing virtually impossible. Likewise, in the early 1980's, Israeli army urban operations in Beirut were a result of its strategic location near the Israeli security zone; various Arab insurgent and terrorist groups used Beirut as a base for attacks against Israel. Beirut evolved as the major base of the Palestine Liberation Organization, a major opponent of Israel. Beirut's location made it a security threat to Israel and thus compelled several major Israeli operations in the urban area (see Appendix A).

1-12. Another reason for engaging in urban operations is the symbolic—historical, cultural, political, and even economic—importance of many urban areas. Often, capital cities—such as Rome, Paris, Seoul, Berlin, and Baghdad—are identified as the strategic centers of gravity of their respective nations. Possessing or threatening these urban areas may impact directly on the outcome of a conflict. The objective of Germany's wars with France in 1870 and 1914 was ultimately Paris. Napoleon's 1812 campaign had as its objective Moscow, as did Hitler's 1941 offensive into Russia. The objective of the Soviet 1945 offensive was Berlin, and the North Vietnamese 1975 offensive had as its objective the South's capital of Saigon. Still, history also reminds us that commanders must assess the sustainability and decisiveness of operations directed toward these "prestige" objectives. For example, in 1812, Napoleon captured Moscow but had to evacuate it within 30 days. He lacked supplies and shelter, failed to destroy the Russian Army, and failed to defeat the political will of the Czar and the people. Similarly, the North Korean occupation of Seoul during the Korean War was equally indecisive.

U.S. Army's Experience in Urban Operations

1-13. The U.S. Army has a varied history of conducting urban operations. The American Revolution saw the Army conduct several urban operations. These operations included the unsuccessful defense of New York, the successful attack on Trenton, and the decisive siege and attack on British forces at Yorktown. The Mexican War also had a successful assault on the fortified city of Monterey and the decisive siege of Mexico City. During the American Civil War, the armies, in the tradition of Napoleonic maneuver warfare, avoided urban areas and fought in the open. However, the opposing armies frequently made urban areas their objective because of their importance as railheads. Success in the siege of several key urban areas—Vicksburg, Atlanta, and Petersburg—contributed to the Northern victory.

1-14. Following the Civil War, the U.S. Army faced no large-scale urban combat for several generations. The Indian Wars, the Spanish-American War, the Philippine Insurrection, and even World War I did not require the Army to fight in large urban areas. Between the Civil War and World War II, the U.S. Army fought in several urban areas worldwide supporting U.S. commitments. These limited urban combat operations were small but essential parts of what are currently called urban stability operations. From 1900 to 1901, the Army provided public security for a sector of Peking, China of around 50,000 inhabitants. The Army conducted UO and, in the course of the operation, the 9th U.S. Infantry suffered 20-percent casualties while fighting in Tientsin. Punitive expeditions to places such as Siberia, Cuba, Philippines, Central America, and Mexico put the Army in various urban situations that required using military power, notably, the occupation and security of Vera Cruz, Mexico in 1914. In the context of these smaller-scale contingencies, UO became a staple of U.S. Army employment.

1-15. World War II forced the Army to grapple with the issues of large-scale urban combat almost immediately. In his 1941 defense of the Philippines, General MacArthur examined how to defend Manila.

Manila represented a large, modern, friendly urban area, which was the capital city of a close U.S. ally. Defending the urban area posed numerous challenges. Ultimately General MacArthur determined that he could best conduct its defense outside the city by defeating the enemy forces in combat on the invasion beaches or shortly after they landed. When Japanese forces defeated MacArthur's Philippine Army in a series of engagements, MacArthur had to decide how best to protect the friendly populace of Manila. He had two choices: abandoning the city or waging a costly defense that would likely result in the city's destruction, thousands of noncombatant casualties, and no operational advantage. He had little choice but to declare Manila an open city and move his forces to Bataan to wage an operational defense in the vain hope that a counteroffensive could relieve his isolated force. On 2 January 1942, Japanese forces entered Manila unopposed.

1-16. Had General MacArthur decided to defend Manila, his forces would have found scant doctrine in the Army regarding how to fight in an urban area. Doctrine for urban operations did not appear until early 1944, when faced with the possibility of fighting through the larger urban areas of Western Europe. At this time the U.S. Army published FM 31-50. This manual had the first formal discussion of how the Army viewed urban combat. It was based on the Army's limited experiences in the Mediterranean theater and the study of German and Soviet experiences on the Eastern front.

1-17. FM 31-50 emphasized a deliberate pace, individual and small unit initiative, the liberal use of direct and indirect firepower, and decentralized command and execution. It focused on the urban area (as opposed to the environment); however, it did include policies towards the noncombatants. The manual was also focused at the regimental combat team level. Complementing the doctrine of FM 31-50 was the 1944 operations manual, FM 100-5. This latter manual rightly emphasized the importance of combined arms actions and the need for extensive reconnaissance of prepared and defended cities. The Army successfully implemented this doctrine in several major instances of urban combat, most notably the capture of the first German city, Aachen, and hundreds of small-scale urban assaults on cities, towns, and villages across France, the Benelux, and Germany. Army forces also successfully employed this urban combat doctrine during the liberation of Manila in 1945.

1-18. The legacy of this era of Army operations was an effective tactical solution to urban offensive combat: isolate the urban area, seize a foothold, and expand the foothold block by block until occupying the entire urban area and destroying the enemy. The doctrine's emphasis on firepower kept friendly casualties to a minimum. Unfortunately, when enemy forces stoutly defended the urban area, the emphasis on firepower resulted in the area's virtual destruction and high casualties among noncombatants.

1-19. The doctrinal approach honed in World War II remained the accepted Army approach to urban combat to the century's end. The last successful implementation occurred during the Korean War and the final liberation of Seoul. The Vietnam conflict did not offer the Army opportunities or the requirement to practice urban combat or test and refine doctrine on a large scale. The largest urban battle, Hue, was a chaotic tactical battle that validated most of the historical lessons of urban combat without generating any new doctrinal insights for large-scale urban warfare.

1-20. From the mid-1950s through the 1990s, the Army conducted UO in the U.S. in support of civil authorities during civil unrest and anti-Vietnam protests. Some operations involved numerous active and Reserve Component forces engaged in restoring public order. The Detroit riots of 1967 and the Los Angeles riots of 1992 required the commitments of active and National Guard units. In 1968, the Army deployed over 35,000 troops to Washington D.C., Chicago, and Baltimore following the death of Dr. Martin Luther King, Jr.

1-21. In the 1970s and 1980s, Army doctrine predominantly focused on urban areas and successfully fighting a conventional ground war against Soviet and Warsaw Pact forces in Central Europe. FM 90-10 (1979) described how to conduct urban operations against Soviet forces in Germany. Its concepts were never tested other than in simulation, and its approach to urban combat was not substantially different from that practiced by the Army since World War II. Despite previous doctrine's admonition to avoid cities, the Army has had to fight in them in diverse circumstances.

1-22. In the 1990s and early 21st century, commanders and planners began to more carefully consider the impact of urban environments on the overall conduct of campaigns and major operations. Maneuver warfare characterized Operation Desert Storm in 1991; however, urban environments became prominent during subsequent operations in Somalia, the Balkans, and again in Iraq. (During Operation Iraqi Freedom in 2003, for example, Baghdad was considered as a strategic center of gravity for planning.) This was a result of adversaries recognizing the asymmetric advantages to be gained by fighting superior U.S. forces in an urban environment as well as a general worldwide trend of increasing urbanization. The evolving nature of Army operations in the urban environment was recognized in the Army doctrine of full spectrum operations. No longer could combat operations be considered exclusively as a separate phase of the major operation. Instead, full spectrum operations—combat and stability or civil support—would be conducted simultaneously. It also became clear to Army commanders that while one type of operation can often dominate for a period, simultaneous, full-spectrum urban operations were now the norm and an in-depth understanding of complex urban societies essential to overall success.

MODERN ARMY URBAN OPERATIONS

1-23. Modern urban operations will be full spectrum operations executed jointly and will often contain a multinational and interagency component. They will span the entire range of possible applications of military power. At higher echelons, these separations are often viewed as levels or scales of intensity. For the tactical units conducting urban operations, these divisions appear indistinct, as the intensity is often high despite where the operation falls within the range of military operations.

JOINT, INTERAGENCY, AND MULTINATIONAL

1-24. The urban environment is too multifaceted for a single-service, single-agency, or single-dimensional solution. Generating desired effects and avoiding unintended negative consequences in this complex environment requires careful integration of joint (and often multinational) forces and interagency capabilities throughout all phases of the operation. Effective interagency collaboration will help plan effects, supporting actions, and measures of effectiveness to ensure that military actions complement diplomatic, economic, and informational activities.

1-25. Joint urban operations (JUOs) in which Army forces are a major component will be primarily land operations. These operations may take place within the context of a joint campaign conducted by a joint force land component commander or a joint task force (JTF) commander. Or they may be an Army operation under an ARFOR commander who himself operates for a JFC, depending on the organization of the theater's joint command structure. In the later case, the JFC will manage joint issues in the urban area.

1-26. The JFC conducting JUOs will focus on effectively organizing his forces for UO and tasking them in accordance with their service capabilities. His guide for the conduct of the JUO will be the joint operational tasks described in JP 3-0. JP 3 06 will provide the JTF commander specific guidance regarding the conduct of joint operational tasks in the urban environment. Army commanders will execute tasks assigned by the JFC and advise him on using Army forces and capabilities. Army commanders will also ensure that Army UO are nested within the JFC's concept of operations. Also, the ARFOR commander will request support through the JFC from other service and functional commanders who have urban capabilities critical to the success of Army UO. See Appendix D for more information on joint capabilities in an urban environment.

1-27. The military conflict will not be an end to itself. It is inevitably a means of transition from a perceived unsatisfactory state of affairs to an improved end. Therefore, urban operations will require the careful orchestration of military and civilian capabilities to achieve success. An early identification of potential requirements will allow for the proper allocation of responsibilities between military and civilian agencies and a clear understanding of what is possible for each to achieve. Such understanding will preclude the establishment of false perceptions from all concerned, but particularly from the urban population and the public at large.

1-28. Interagency cooperation will include the Army as part of the Department of Defense, elements of the Department of State, and other various governmental and nongovernmental organizations. Urban operations demand the expertise and abilities of various organizations and a synergistic unity of effort if strategic goals aimed at bringing about a more satisfactory political, social, diplomatic, economic, and military situation are to be achieved. However, there will often be a gap in time before bringing many governmental capabilities to bear that the Army and other military forces—to include coalition partners—will have to fill. Routine interagency training, planning, and coordination will help to decrease that gap in time to its minimum. As it is likely that civilian organization leaders will be in charge during some phases of the overall operation, they should lead planning at those times. A careful survey of urban requirements and rapid communication to participating civilian organizations will also help speed civilian capabilities to where they are needed most.

Winning the Peace in Iraq:
The Requirement for Full Spectrum Operations

We found that if we concentrated solely on establishing a large security force and targeted counterinsurgent combat operations—and only after that was accomplished, worked toward establishing a sustainable infrastructure supported by a strong government developing a free-market system—we would have waited too long. The outcome of a sequential plan allowed insurgent leaders to gain a competitive advantage through solidifying the psychological and structural support of the populace.

Further, those who viewed the attainment of security solely as a function of military action alone were mistaken. A gun on every street corner, although visually appealing, provides only a short-term solution and does not equate to long-term security grounded in democratic process. Our observation was born not from idealism, but because it creates the essence of true security, protecting not only our soldiers, but Iraq, the region, and, consequently, our homeland.

Major General Peter Chiarelli
Commander, 1st Cavalry Division

FULL SPECTRUM OPERATIONS

1-29. Army forces will conduct full spectrum operations within urban areas. Army commanders conduct full spectrum operations abroad by executing offensive, defensive, and stability urban operations as part of an the joint, interagency, and multinational effort described above. Army forces within the United States and its territories conduct full spectrum operations by combining offensive, defensive, and civil support operations to support homeland security. The situation will mandate that one type of operation—offense, defense, stability, or civil support—dominates the urban operation. Depending on whether the operation is overseas or within the United States, however, commanders will often find themselves executing offensive and defensive operations and stability or civil support operations simultaneously. In fact, waiting until all combat operations are concluded before beginning stability or civil support operations often results in lost, sometimes irretrievable, opportunities. Unsurprisingly, the mission determines the dominant type of operation, with the other types of Army operations conducted to shape the AO for mission success. The dominant type of operation will vary between different urban areas even in the same campaign.

Offense

1-30. Against a large conventional enemy in a major urban area with a large civil population present, offensive operations will likely require a greater commitment of Army resources than in other environments. Urban offensive operations will also incur the greatest risks to Army forces and noncombatants. Within defensive and stability or civil support operations, forces may conduct tactical offensive UO, such as counterattacks to maintain the initiative or raids to eliminate elements disrupting the stability or civil support operation.

Defense

1-31. Defensive UO are generally conducted as a shaping operation within a larger major operation. These temporary operations often set conditions for successful offensive operations, stability operations, or civil support operations. Commanders often conduct defensive UO within other types of operations to protect essential facilities in the urban area, protect flanks against counterattack, prevent the breakout of isolated enemies, or protect valuable supply bases or vulnerable convoy routes. Army forces conducting defensive UO must creatively use the environment to enhance their combat power.

1-32. In UO, essential facilities will likely include urban ports and airfields required by sister services to support ground operations. Therefore, Army commanders will need to carefully plan integrated air and base defense operations with air and naval component commanders. Protection requirements increase dramatically as air and naval assets are most vulnerable when aircraft are on the ground and ships are docked at port. Without a carefully synchronized and integrated base defense plan, Army commanders risk sporadic or total loss of support by air and naval forces—forces essential to successful UO.

Stability

1-33. Stability operations in an urban environment require offensive and defensive operations, combined with other tasks unique to each stability operation. Army forces conduct urban stability operations for various reasons, including noncombatant evacuation operations, peace operations, or support to insurgencies (see Chapter 9). Urban stability operations will require an offensive capability to destroy any military capability that overtly threatens its objectives before that military threat can adversely affect the operation. Army forces employ defensive capabilities to safeguard themselves as well as secure critical places, populations, or infrastructure in the urban area. Commanders may also employ defensive capabilities to separate and protect one faction from another. Various stability operations will also require the distribution of food or aid and the protection or assistance of agencies conducting economic or humanitarian activities.

Civil Support

1-34. Army civil support operations in an urban environment aid other U.S. agencies and organizations in mitigating the consequences of natural and man-made disasters. In response to the disaster, civil support operations require the equipment, personnel, or organizational abilities of Army forces rather than the Army's combat capabilities. In a civil support mission, these capabilities often involve Army transportation, medical, quartermaster, or engineer forces. Unless conducted in conjunction with homeland defense, urban civil support operations will seldom require combat; however, commanders must determine if force protection threats exist that could hamper Army civil support operations. During homeland security operations, defensive and offensive capabilities will be required to defeat hostile armed forces.

PREPARING FOR FUTURE URBAN OPERATIONS

1-35. To operate successfully in a complex urban environment requires a thorough understanding of the urban environment and rigorous, realistic UO training. Training should be conducted by the complete combined arms team and cover full spectrum operations to include appropriate tactics, techniques, and procedures (see FM 3-06.11 and TC 90-1). It should also replicate—

- The psychological impact of intense, close combat against a well-trained, relentless, and adaptive enemy.
- The effects of noncombatants—including governmental and nongovernmental organizations and agencies—in close proximity to Army forces. This necessitates—
  - An in-depth understanding of culture and its effects on perceptions.
  - An understanding of civil administration and governance.
  - The ability to mediate and negotiate with civilians including the ability to effectively communicate through an interpreter.
  - The development and use of flexible, effective, and understandable rules of engagement.
- A complex intelligence environment requiring lower-echelon units to collect and forward essential information to higher echelons for rapid synthesis into timely and useable intelligence for all levels of command. Understanding the multifaceted urban environment necessitates a bottom-fed approach to developing intelligence (instead of a top-fed approach more common and efficient for open terrain and conventional threats). It also emphasizes the need for intelligence reach and a truly collaborative approach to the development and sharing of intelligence.
- The communications challenges imposed by the environment as well as the need to transmit large volumes of information and data.
- The medical and logistic problems associated with operations in an urban area including constant threat interdiction against lines of communications and sustainment bases.

Developing Joint, Interagency, and Multinational Operability

1-36. UO training extends from the individual Soldier to the joint level including the integration of conventional and special operations forces. Preparedness also includes enhancing interoperability in regards to urban multinational and interagency operations. Training, as well as campaign and contingency plan creation, should include significant and sustained participation by civilian agencies. Fruitful multinational and interagency relationships must be cultivated before the onset of operations, that is, before Soldiers and their coalition counterparts are making decisions—many with strategic implications and often when they are under fire. Joint, interagency, and multinational collaboration will help design effects, supporting actions, and measures of effectiveness necessary to ensure that military actions in urban environments complement the diplomatic, informational, and economic activities necessary to achieve strategic objectives.

Conducting Live, Virtual, and Constructive Training

1-37. Force preparedness mandates integrating the actual use of urban terrain, exercises at urban training sites, simulations, or any combination into tactical- and operational-level intra- and interservice training. This type of multi-faceted training will help commanders develop a better understanding of the complexity of the urban environment and enable them to execute missions across full spectrum operations. Careful use of these facilities will also allow Army forces to fully integrate urban operations within self-development, institutional, and operational training.

Ensuring Every Soldier is an Urban Warrior

1-38. In a complex urban environment, every Soldier—regardless of branch or military occupational specialty—must be committed and prepared to close with and kill or capture threat forces in an urban environment. Every Soldier must also be prepared to effectively interact with the urban area's noncombatant population and assist in his unit's intelligence collection efforts. Each urban operation will be unique and commanders—given an opportunity to surge and conduct mission-specific post-mobilization and in-theater training—will need to conduct an analysis of the tasks requiring emphasis or modification to fit the conditions of the operational environment. In UO, every soldier will likely be required to—

- Perform advanced rifle marksmanship to include advanced firing positions, short-range marksmanship, and night firing techniques (unassisted and with the use of optics).
- Operate their unit's crew-served weapons.
- Conduct urban reconnaissance and combat patrolling.
- Enter and clear buildings and rooms as part of an urban attack or cordon and search operation.
- Defend an urban area.
- Act as a member of a combat convoy (including specific drivers training).
- Control civil disturbances.
- Navigate in an urban area.

1-39. While not all-inclusive and necessarily urban-specific, other critical individual and collective UO tasks (often modified for the urban environment) might include—

- Conduct troop-leading procedures.
- React to contact, ambush, snipers, indirect fire, and improvised explosive devices.
- Establish an observation point, personnel or vehicle checkpoint, or roadblock.
- Secure a disabled vehicle or downed aircraft.
- Call for indirect fire and close air support.
- Create and employ explosive charges.
- Handle detainees and enemy prisoners of war.
- Treat and evacuate casualties.
- Accurately report information.
- Understand the society and culture specific to the area of operations.
- Use basic commands and phrases in the region's dominant language.
- Conduct tactical questioning.
- Interact with the media.
- Conduct thorough after-action reviews.

1-40. Commanders must understand that all Soldiers will require urban-specific equipment to conduct many urban-specific tasks. Critically, commanders cannot expect supporting forces to fight alongside and support ground maneuver units in an urban environment without the appropriate equipment and training.

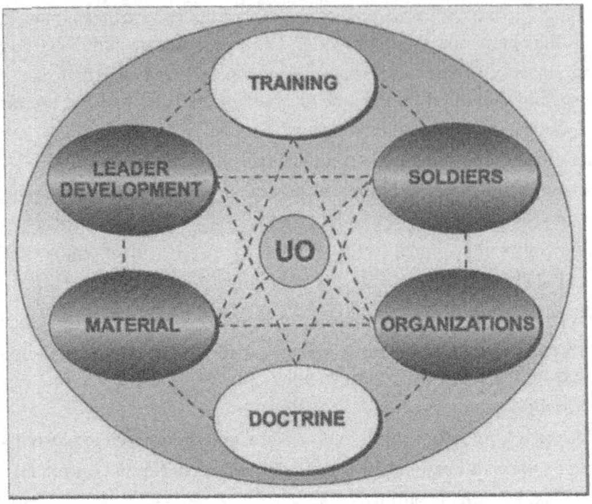

Figure 1-2. UO and DOTMLPF

Continual Adaptation and Innovation

1-41. Realistic UO training (as well as the conduct of real world operations) has the added benefit of identifying operational requirements and resultant changes necessary in our doctrine, organizations, training, materiel, leadership and education, personnel, and facilities (DOTMLPF) (see figure 1-2 and FM 100-11). While technology (materiel) and organizational changes are important, confident Soldiers (personnel) remain the decisive means for success. Technology and organizational changes will be a critical enabler to attain better understanding of the urban environment, enhance command and control on the noncontiguous battlefield, and achieve the agile, simultaneous, and precise effects required in UO. In the future, technology may lead to a radically new operational concept and approach to UO. Still, competent leaders and well-trained, disciplined, and adaptive Soldiers will remain the decisive means for the Army to succeed in this complex and multidimensional urban environment.

Chapter 2
Understanding the Urban Environment

From a planning perspective, commanders view cities not just as a topographic feature but as dynamic entities that include hostile forces, local population, and infrastructure. Planning for urban operations requires careful IPB, with particular emphasis on the three-dimensional nature of the topography and the intricate social structure of the population.

FM 3-0

Of all the environments in which to conduct operations, the urban environment confronts Army commanders with a combination of difficulties rarely found elsewhere. Its distinct characteristics result from an intricate topography and high population density. The topography's complexity stems from the man-made features and supporting infrastructure superimposed on the natural terrain. Hundreds, thousands, or millions of civilians may be near or intermingled with soldiers—friendly and enemy. This second factor, and the human dimension it represents, is potentially the most important and perplexing for commanders and their staffs to understand and evaluate. To this end, this chapter provides information essential to understanding the urban environment and conducting an effective intelligence preparation of the battlefield (see FM 2-01.3, FMI 2-91.4, and Appendix B).

Although urban areas possess general similarities, each environment is distinct and will react to and affect the presence and operations of Army forces differently. A tactical technique effective in one area may not be effective in another area due to physical differences, such as street patterns or the type of building construction. An Army policy popular with one urban group may cause resentment and hostility in another due to diverse cultural differences. All difficulties potentially exist, and they increase the complexity for Army forces operating in urban areas. These difficulties range from conventional military forces to disease and starvation (see Chapter 3) to a pervasive media—often acutely present in intricate combinations. Thus, commanders at all levels must make extraordinary efforts to assess and understand their particular urban environment to plan, prepare for, and execute effective urban operations (UO).

A COMPLEX ENVIRONMENT

2-1. Urban areas vary depending on their history, the cultures of their inhabitants, their economic development, the local climate, available building materials, and many other factors. This variety exists not only among urban areas but also within any particular area. The ever-changing mix of natural and man-made features in urban areas present commanders with some of the most difficult terrain in which to conduct military operations.

2-2. Although urban areas possess similar characteristics, no two are identical. The sprawl of Los Angeles, for example, bears little physical resemblance to New Delhi. Societal characteristics most significantly affect each area's uniqueness and complexity. While complex, information about the terrain, its potential effects on operations, and how it changes over time may be determined with some degree of certainty. However, the human dimension is much more difficult to understand and assess, particularly its effects on military operations. Like any environment, the side that can best understand and exploit the positive and mitigate the negative effects of the urban environment has the best chance of success.

2-3. Whether a large metropolis or a small village, each urban environment has identifiable components that constantly change and interact. This "system of systems" consists of the terrain, the society, and the infrastructure that links the two (see figure 2-1).

2-4. These systems are not separate and distinct categories, but rather overlapping and interdependent systems, acting dynamically with each other. Thoroughly analyzing these elements, along with the other factors of mission, enemy, weather, troops and support available, and time—

- Contributes to the accuracy of the commanders' situational understanding.
- Potentially lessens the number and cost of close combat engagements.
- Allows commanders to develop courses of action that apply appropriate resources against decisive points.

2-5. In stability operations and civil support operations, this understanding allows commanders to engage and dominate the decisive points critical to maintaining peace or restoring normalcy to the urban environment. Although each system is categorized into subordinate components or subsystems, commanders must often "step back" and visualize each system, the complex urban environment, and their area of operations (AO) as a unified whole. This "systems thinking" aids commanders in uncovering key relationships and intersections that can help reveal centers of gravity (COGs) and decisive points.

2-6. To comprehend the urban environment and its components to the fullest extent possible, commanders must carefully integrate and employ tactical reconnaissance forces, special operations forces (SOF)—to include psychological operations (PSYOP) and civil affairs units—and a myriad of other human intelligence (HUMINT) assets and regional, language, and cultural experts. The societal aspects and integrating infrastructure will challenge commanders' assessment and understanding. These aspects will also require greater dependence on nonmilitary and nongovernmental organizations (NGOs) and host-nation agencies for their information, knowledge, and expertise. This last consideration requires commanders to develop effective techniques and procedures for coordinating, interacting, and, to the greatest extent possible, synchronizing activities with these agencies.

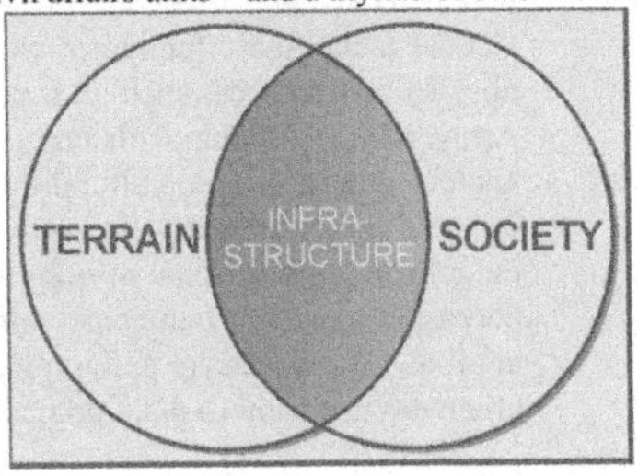

Figure 2-1. Keys to understanding the urban environment

URBAN TERRAIN

2-7. Although complex and difficult to penetrate with many intelligence, surveillance, and reconnaissance (ISR) assets, the terrain is the most recognizable aspect of an urban area. Truly understanding it, however, requires comprehending its multidimensional nature. The terrain consists of natural and man-made features, with man-made features dominating; an analysis considers both. Buildings, streets, and other infrastructure have varied patterns, forms, and sizes. The infinite ways in which these factors can intertwine make it difficult to describe a "typical" urban area. However, these various factors provide a framework for understanding the complex terrain in an urban area. Furthermore, man-made features significantly affect military systems and Soldiers, and thus tactics and operations. General effects on urban operations are discussed in this chapter. Specific effects on warfighting functions (see Chapters 4 and 10) and the spectrum of operations (see Chapters 7, 8, and 9) are interwoven throughout the manual.

MULTIDIMENSIONAL BATTLEFIELD

2-8. Urban areas present an extraordinary blend of horizontal, vertical, interior, exterior, and subterranean forms superimposed on the natural relief, drainage, and vegetation. An urban area may appear dwarfed on a map by the surrounding countryside. In fact, the size and extent of the urban area of operations is many times that of a similarly sized portion of undeveloped natural terrain. A multi-storied building may take up the same surface area as a small field, but each story or floor contains approximately an equal area as the ground upon which it sits. In effect, a ten-story building can have eleven times more defensible area than "bare" ground—ten floors and the roof. It is the sheer volume and density created by this urban geometry that makes UO resource intensive in time, manpower, and materiel.

2-9. Like natural disasters, UO can radically alter the physical characteristics of the urban terrain in ways not experienced in other environments. They may cause (either intentionally or not) uncontrollable fires or the loss of electricity. A power outage can cause flooding (especially in subsurface areas) by shutting down pumping stations. Entire buildings may be destroyed, eliminating reference points, leaving large piles of rubble, altering fields of fire, and making movement and transportation extremely difficult. Additionally, buildings and other urban structures, damaged but not destroyed, can become (or remain) effective obstacles and possible booby traps. Even without enemy exploitation, their weakened construction and unstable structure increase the risk of injury to Soldiers and civilians moving within them. (Engineer expertise will often be needed to determine whether the buildings can support occupation by Army forces or civilians.) Yet, even the total collapse of a building may not eliminate its defenders. Of additional concern, the likely presence of toxic industrial material (TIM) can create additional obstacles and health hazards.

2-10. Commanders in other environments normally address the depth, breadth, and height of their AO in terms of two areas: airspace and surface. In an urban environment, they broaden their scope to include supersurface and subsurface areas (see figure 2-2) that voluminously extend the commanders area of operations. Although spatially separated, each area may be used as an avenue of approach or mobility corridor, line of communications (LOC), or engagement area.

2-11. Supersurface and subsurface areas magnify the complexity of the urban physical environment. Commanders must consider activities that occur outside buildings and subterranean areas (the external space) as well as the activities happening unseen in buildings and subterranean systems (the internal space). This internal volume further challenges command, control, and intelligence collection activities and increases the combat power required to conduct UO. Commanders must develop methods and techniques to help themselves, their staffs, and their subordinate commanders and staffs to represent, visualize, and reference these multiple dimensions. Increasing the difficulty, such dimensions can change rapidly simply due to continued urban growth or, as described earlier, the effects of nature and UO themselves.

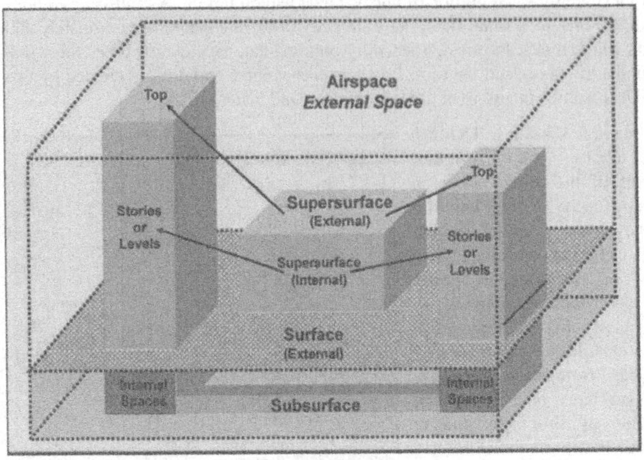

Figure 2-2. The multidimensional urban battlefield

Airspace

2-12. As in all other environments, aircraft and aerial munitions use the airspace as rapid avenues of approach in urbanized areas. Forces can use aviation assets for observation and reconnaissance, aerial attack, or high-speed insertion and extraction of Soldiers, supplies, and equipment. Some surface obstacles in an urban area, such as rubble, do not affect flight (though they may prevent the take-off and landing of aircraft). Buildings of varying height and the increased density of towers, signs, power lines, and other urban constructions, however, create obstacles to flight and the trajectory of many munitions (masking). Similarly, these obstacles can restrict a pilot's line of sight as well as physically limit low-altitude maneuverability in the urban airspace. Excellent cover and concealment afforded enemy gunners in an urban area increases aviation vulnerability to small arms and man-portable air defense systems (MANPADS), particularly when supporting ground forces. The potential for a high volume of air traffic (military and civilian) over and within urban airspace (including fixed-wing, rotary-wing, and unmanned aircraft systems) may become another significant hazard and necessitate increased airspace command and control measures.

Surface

2-13. Surface areas apply to exterior ground-level areas, such as parking lots, airfields, highways, streets, sidewalks, fields, and parks. They often provide primary avenues of approach and the means for rapid advance. However, buildings and other structures often canalize forces moving along them. As such, obstacles on urban surface areas usually have more effect than those in open terrain since bypass often requires entering and transiting buildings or radical changes to selected routes. Where urban areas border the ocean or sea, large lakes, and major rivers, the surface of these bodies of water may provide key friendly and threat avenues of approach or essential LOCs—a significant consideration for Army commanders. As such, amphibious, river-crossing, and river operations may be integral parts of the overall urban operation (see FM 3-05.212 and FM 3-90.12).

2-14. Larger open areas—such as stadiums, sports fields, school playgrounds, and parking lots—are often critical areas during urban operations. They can provide locations for displaced civilians, interrogation centers, and prisoner of war holding facilities. These areas also can afford suitable aircraft landing and pickup zones and artillery firing locations. They can provide logistic support areas and aerial resupply possibilities because they are often centrally located. Finally, large open areas (and immense or unusually shaped structures) within urban areas are often easier to see—especially from the air—and can serve as excellent target reference points from which to shift or control fires.

Supersurface

2-15. These areas include the internal floors or levels (intrasurface areas) and external roofs or tops of buildings, stadiums, towers, or other vertical structures. They can provide cover and concealment; limit or enhance observation and fields of fire; and restrict, canalize, or block movement. However, forces can move within and between supersurface areas creating additional, though normally secondary, avenues of approach. Rooftops may offer ideal locations for landing helicopters for small-scale air assaults and aerial resupply. First, however, engineers must analyze buildings for their structural integrity and obstacles. Such obstacles include electrical wires, antennas, and enemy-emplaced mines (although personnel may be inserted by jumping, rappelling, or fast roping from a hovering helicopter and extracted by hoist mechanisms). Some rooftops are designed as helipads. Roofs and other supersurface areas may also provide excellent locations for snipers; lightweight, handheld antitank weapons; MANPADS; and communications retransmission sites. They enable top-down attacks against the weakest points of armored vehicles and unsuspecting aircraft. Overall, elevated firing positions reduce the value of any cover in surrounding open areas and permit engagement at close range with less risk of immediate close assault. This area (and the subsurface area) requires commanders to think, plan, and execute ground operations vertically as well as horizontally. In this latter regard, UO share strong similarities with mountain operations (see FM 3-97.6).

Subsurface

2-16. Subsurface areas are below the surface level. They may serve as secondary and, in fewer instances, primary avenues of approach at lower tactical levels. When thoroughly reconnoitered and controlled, they offer excellent covered and concealed LOCs for moving supplies and evacuating casualties. They may also provide sites for caching and stockpiling supplies. Subsurface areas include subterranean areas such as subways, mines, tunnels, sewers, drainage systems, cellars, civil defense shelters, and other various underground utility systems. In older cities, they may include ancient hand-dug tunnels and catacombs. Both attacker and defender can use subsurface areas to gain surprise and maneuver against the rear and flanks of a threat and to conduct ambushes. However, these areas are often the most restrictive and easiest to defend or block. Their effectiveness depends on superior knowledge of their existence and overall design. Army commanders may also need to consider potential avenues of approach afforded by the subsurface areas of rivers and major bodies of water that border urban areas. This particularly applies when operating as part of a joint task force (JTF) task organized with SOF or when opposing a threat with similar capabilities.

MAJOR URBAN PATTERNS

2-17. Four major urban patterns (satellite, network, linear, and segment) can influence UO (see figure 2-3). Central to two of the patterns (satellite and network) is the hub or dominant urban area or pattern around which outlying urban areas or patterns radiate. (A segmented urban area, because it tends to be a larger urban area, can often be a hub.) In offensive and defensive operations, the hub serves as a pivot or strong point; as such, it can become a major obstacle to an attacker. If the attacker chooses to bypass the urban area (hub) located along his axis of advance without first isolating the area, he may expose his flank or LOC to attack from the hub as well as dependent urban areas or subordinate satellite patterns. Because the focus of stability and civil support operations is normally on people, commanders should understand the value and influence of the hub to the economic, political, or cultural well being of the surrounding area. Generally the larger the hub, the greater influence it has on satellite urban areas and surrounding rural areas. Commanders must remember that urban areas are not islands; all are connected to the surrounding rural (and other urban) areas through fluid and permeable boundaries and LOCs.

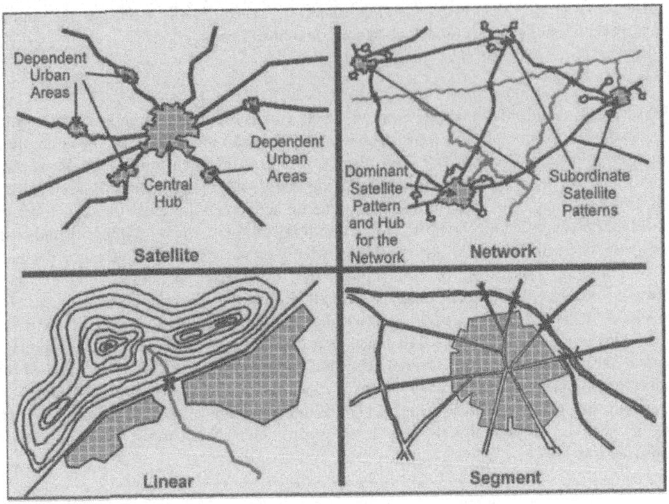

Figure 2-3. Major urban patterns

Satellite Pattern

2-18. This common pattern consists of a central hub surrounded by smaller, dependent urban areas. LOCs tend to converge on the hub. Outlying areas often support the principal urban area at the hub with means of reinforcement, resupply, and evacuation. In some instances, they may serve as mutually supporting battle positions. Commanders should consider the effects of the outlying urban areas on operations within the hub, and, conversely, the effects of operations within the hub on outlying urban areas. Information operations (IO), for example, targeted primarily at key leaders and other civilians located within the hub of a satellite pattern may subsequently influence civilians in outlying urban areas and achieve necessary effects without having to commit specific resources to these outlying areas.

Network Pattern

2-19. The network pattern represents the interlocking of the primary hubs of subordinate satellite patterns. Its elements are more self-sufficient and less supportive of each other, although a dominant hub may exist. Major LOCs in a network extend more than in a satellite pattern and take more of a rectangular rather than a convergent form. Its natural terrain may vary more than in a single satellite array. Operations in one area may or may not easily influence, or be influenced by, other urban areas in the pattern.

Linear Pattern

2-20. Potentially a subelement of the previous two patterns, the linear pattern may form one ray of the satellite pattern or be found along connecting links between the hubs of a network. Most frequently, this pattern results from the stringing of minor urban areas along a confined natural terrain corridor, such as an elongated valley, a body of water, or a man-made communications route. In offensive and defensive operations, this latter form of the linear pattern facilitates developing a series of strong defensive positions in depth, effectively blocking or delaying an attacking force moving along the canalized terrain.

Segment Pattern

2-21. When dominant natural terrain, such as a river or man-made features (canals, major highways, or railways), divides an urban area, it creates a segmented pattern. This pattern often makes it easier for commanders to assign areas of operations to subordinate commanders. However, this pattern may fragment operations and increase risk to an operation requiring mutual support between subordinate units. Still, the segmented urban areas may allow commanders to isolate threats more easily in these areas and focus operations within segments that contain their decisive points. Although an integral part of the whole (the urban area), each segment may develop distinct social, economic, cultural, and political characteristics. This social segmenting may benefit commanders faced with limited assets to influence or control the urban populace. After thoroughly analyzing the society, they may be able to focus IO and populace and resources control measures against only specific segments that affect decisive operations. Commanders may need only to isolate other segments or may need to just monitor for any significant changes in the attitudes, beliefs, or actions of the civilians located there.

LESSER STREET PATTERNS

2-22. Lesser patterns in the urban area result from the layout of the streets, roads, highways, and other thoroughfares. They evolve from influences of natural terrain, the original designer's personal prejudices, and the changing needs of the inhabitants. Street patterns (and widths) influence all warfighting functions; however, they greatly affect movement and maneuver, command and control, and sustainment. (In some portions of older Middle Eastern urban areas, the labyrinths of streets were designed only to allow two loaded donkeys to pass each other; tanks are too wide.) Urban areas can display any of three basic patterns and their combinations: radial, grid, and irregular (see figure 2-4).

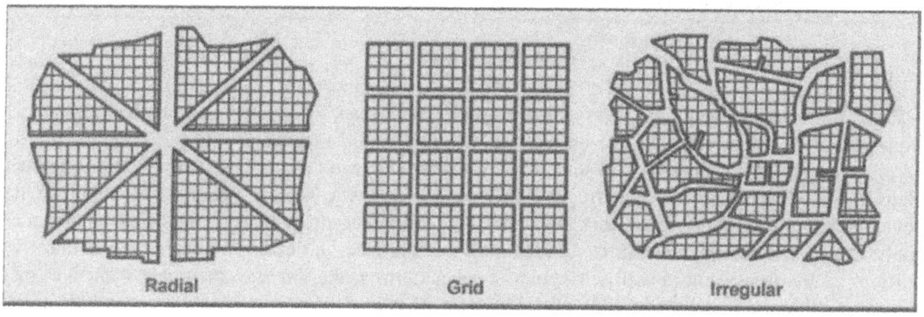

Figure 2-4. Basic internal street patterns

Radial

2-23. Societies of highly concentrated religious or secular power often construct urban areas with a radial design: all primary thoroughfares radiating out from the center of power. Urban areas with this design may signal an important historical aspect in the overall analysis of the urban society. Terrain permitting, these streets may extend outward in a complete circle or may form a semicircle or arc when a focal point abuts a natural barrier, such as a coastline or mountain. To increase mobility and traffic flow, societies often add concentric loops or rings to larger radial patterns. Unless commanders carefully plan boundaries, routes, and axes of advance, their subordinate units' movement or maneuver may be inadvertently funneled toward the center of urban areas with this pattern resulting in congestion, loss of momentum, and an increased potential for ambush or fratricide.

Grid

2-24. The most adaptable and universal form for urban areas is the grid or rectangular pattern: lines of streets at right angles to one another forming blocks similar to the pattern of a chessboard. A grid pattern can fill in and eventually take over an original radial pattern. Grid patterns often appear to ease the assignment of boundaries for subordinates units. However, commanders also consider how the natural terrain influences operations and the establishment of graphic control measures. They also consider the influence of the buildings and other structures lining these streets, such as their height and construction, before assigning boundaries and developing other control measures. Commanders should also consider the following when developing urban graphic control measures:

- Describing boundaries, phase lines, checkpoints, and other graphic control measures by easily recognizable features is as important in urban areas as elsewhere. While easily identifiable urban structures such as unusually tall or oddly-shaped buildings, cemeteries, stadiums, or prominent rail or highway interchanges can be useful references, available natural features are a better descriptor than man-made features that may be altered or unrecognizable. As an aid to air-to-ground coordination, commanders should select features that can be identified by both ground and air forces. Those that help in controlling ground forces may not be easily visible from the air and vice versa.
- Commanders should also consider whether a boundary along an easily-recognizable terrain feature, such as a river, will also be easy to identify by the threat who may seek to "find the seam" and exploit the likely control and coordination difficulties associated with boundaries, especially between higher-level units. This often requires commanders to carefully position a control measure away from the key feature to provide a designated subordinate force with the terrain and space necessary to control the feature. On the other hand, commanders working closely with local authorities during stability and civil support operations may not need to thoroughly understand the physical effect of street patterns on the assignment of boundaries as they might for combat UO. Instead, commanders may choose to assign boundaries overlaid on existing geopolitical boundaries used by local agencies to increase interoperability and aid in unity of effort.

Irregular

2-25. In most urban areas, regardless of the original intent, plan, or vision, existing street patterns emerge from successive plans overlaid one on another. Some are well planned to fit with previous plans while others a haphazard response to explosive urban growth. The result may mix patterns. Urban engineers and planners may specifically design irregular patterns for aesthetic reasons (as in many suburban housing developments) or to conform to marked terrain relief. Irregular street patterns may alert commanders and analysts that the underlying natural terrain may exert greater influence over operations than in other portions of the urban area. Finally, irregular street patterns make the movement and maneuver of forces less predictable. However, a labyrinth of irregular or "twisting" street patterns may increase the possibility of fratricide particularly for units that are trained or accustomed only to grid patterns.

An Urban Model

2-26. Throughout the world, urban areas have similar form and function. In form, urban areas contain like characteristics, readily divisible into distinct sections or areas. Functionally, they tend to be the centers of population, finance, politics, transportation, industry, and culture. While urban areas may be modeled by several different means, figure 2-5 illustrates the general forms and internal functions. Some forms and functions may overlap. For example, high-rise buildings are located in core areas as well as in outlying areas and may be used for residential purposes. With the rapid urbanization associated with developing nations, the areas displayed in this urban model often manifest themselves less clearly there than in developed nations.

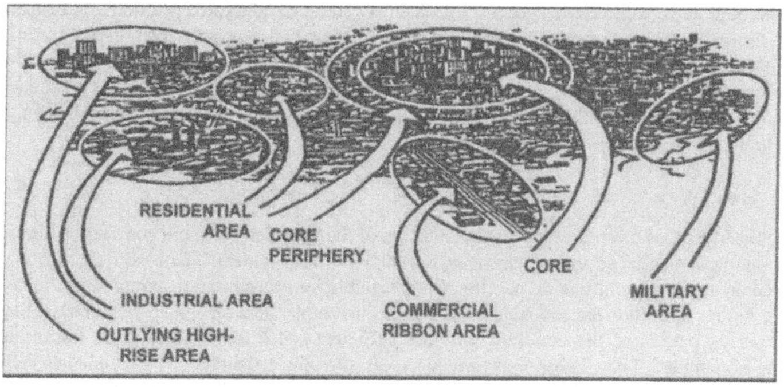

Figure 2-5. Urban functional zones

2-27. This analysis helps to determine, in general terms, potential advantages and disadvantages each portion of the urban area may have toward accomplishing the urban operation. However, construction materials and methods can vary drastically. Commanders and their staff will often need to identify specific building types and construction and understand weapons effects on them. If a commander desires precise effects, the chosen munitions or weapons system must be sufficiently accurate, capable of penetrating the target structure (without exiting the other side), and achieve effects within. Often noncombatants, critical infrastructure, or protected targets are in the vicinity. Commanders may need to determine if the surrounding walls or structures will sufficiently absorb or negate the blast or thermal effects of the weapon. Regardless, understanding the structure and composition of buildings and other structures in the urban AO may be necessary to allow commanders to determine the best means to accomplish the mission.

Core

2-28. The core is the heart of the urban area, the downtown or central business district. Relatively small and compact, it contains a large percentage of the urban area's shops, offices, and public institutions. Often, it houses the headquarters for commercial and financial activities and contains important cultural, historical, and governmental buildings. These activities prefer the core because of its accessibility. As the focal point of the transportation network, residents find the core the easiest part of the urban area to reach. It normally has the densest concentration of multistory buildings and subterranean features (underground parking garages, underground shopping centers, and basements).

2-29. High-rise buildings, varying greatly in height (possibly 50 stories above ground and four stories below ground), make up the cores of many of today's urban areas. Buildings routinely abut one another, with little or no setback from the sidewalks. Building height and density (except in outlying high-rise areas) often decreases from the core to the edge of the residential areas, while the amount of open areas frequently increases. Modern urban planning allows for more open spaces between buildings than found in the cores of older urban areas. Most core areas have undergone constant redevelopment resulting in various types of construction. Commonly, brick buildings abound in the oldest part of the core; framed, heavy-clad structures in the next oldest part; and a concentration of framed, light-clad buildings in the newest part. The outer edge of the core, the core periphery, has ordinarily undergone less change than the core resulting in buildings of uniform height (commonly two to three stories in towns and five to ten stories in larger urban areas).

2-30. Generally, offensive operations focused in core areas (even when effectively isolated) will require greater resources—particularly manpower, time, and information—than in many other parts of the urban area. Mounted maneuver often proves more difficult in core areas because of fewer open areas, buildings closer to the streets, and an increased density of civilian vehicles. Razed buildings in central core areas (especially high-rise buildings) become greater obstacles to mobility as they can collapse on and easily block thoroughfares. Rubble piles provide excellent covered and concealed positions for dismounted threat forces. Consequently, commanders often use more dismounted forces as part of their combined arms operations. Conversely, the core may be critical to urban defensive operations, particularly older areas of heavier construction that afford greater protection. Despite potential difficulties, the core area may be key to accomplishing many stability or civil support missions since it houses much of the human activity that occurs in the urban area.

Industrial Area

2-31. Industrial areas often develop on the outskirts of the urban areas where commercial transportation is easiest (along airfields and major sea, river, rail, and highway routes). The road networks in and around industrial areas are generally more developed and suitable for transportation assets. These areas will likely displace farther from the core and residential areas as urban planners recognize the potential threat of TIM. The dispersed pattern of the buildings provides sufficient space for large cargoes, trucks, and materiel handling equipment. These areas may provide ideal sites for sustainment bases and maintenance sites. While older heavier-clad structures may be found, new construction generally consists of low, large, flat-roofed factory and warehouse buildings with large parking areas and work yards. These structures generally have steel frame and lightweight exterior walls. Multistory structures usually have reinforced concrete floors and ceilings.

Toxic Industrial Chemical	Industrial/Commercial Uses
Ammonia	Commercial Refrigerant, Fertilizer and Food Production, Petroleum, Explosives, Other Chemicals
Arsine	Semiconductor Industry
Boron Trichloride	Organic Catalyst, Soldering Magnesium
Boron Trifluoride	Chemical Catalyst, Aluminum Refining
Carbon Disulfide	Industrial Solvent, Dry Cleaning, Agriculture, Petroleum, Electroplating
Chlorine	Potable Water, Disinfectants, Metal Treatment, Plastics & Rubber
Diborane	Plastics and Rubber
Ethylene Oxide	Industrial Alcohols, Fumigant, Industrial Sterilant
Fluorine	Uranium Processing, Rocket Fuel
Formaldehyde	Plastics, Fertilizers, Preservative/Corrosion Inhibitor, Fungicide and Germicide, Pesticide, Pharmaceuticals
Fuming Nitric Acid	Fertilizers, Explosives, Metal Processing, Pesticides, Rocket Fuel
Hydrogen Bromide	Chemical Industry, Pharmaceuticals
Hydrogen Chloride	Fabrics, Semiconductors
Hydrogen Cyanide	Pesticides, Other Chemicals, Pharmaceuticals, Electroplating
Hydrogen Fluoride	Glass Production, Chemical Catalyst
Hydrogen Sulfide	Metallurgy, Agricultural Disinfectant
Phosgene	Dyes, Pharmaceuticals, Herbicides & Insecticides
Phosphorus Trichloride	Metallurgy, Pesticides and Germicides, Gasoline Additive
Sulfur Dioxide	Paper, Food Processing, Ice Production, Disinfectant, Leather Processing
Sulfuric Acid	Fertilizers, Petroleum, Iron and Steel Production, Battery Electrolyte
Tungsten	Electronics, Other Chemicals

Figure 2-6. Some toxic industrial chemicals and their industrial or commercial uses

2-32. Toxic industrial chemicals and other TIM may be transported through an urban area (by rail, barge, truck, or pipeline) or found stored throughout. However, larger concentrations will exist in industrial areas, and their presence should concern Army forces operating near them. Some TIM are heavier than air and tend to settle in low-lying and subsurface areas.

2-33. Each year, over 70,000 different chemicals are produced, processed, or consumed globally. An estimated 25,000 commercial facilities around the world produce, process, or store chemicals that have a legitimate industrial use yet are also classified as chemical warfare agents. Many other chemicals (not classified as weapons) may still be sufficiently hazardous to pose a considerable threat to Army forces and civilians in urban areas as choking agents or asphyxiates, flammables or incendiaries, water contaminants, low-grade blister or nerve agents, or debilitating irritants. These chemicals can be released either accidentally or deliberately. For example, on 2 December 1984, nearly 40 tons of methylisocyanate used to produce pesticides leaked from a storage tank at Bhopal, India, killing thousands and injuring hundreds of thousands. Figure 2-6 contains a small sampling of other toxic industrial chemicals along with their industrial or commercial usage that commanders may encounter in an urban area. The most common chemicals that may pose an immediate risk to Army forces are highly toxic irritant gases such as ammonia, chlorine, hydrogen chloride, and sulfur dioxide.

2-34. Standard chemical defense equipment may not protect against (and chemical detection devices may fail to detect) many toxic industrial chemicals. Therefore, the risk to Soldiers operating near the chemicals may increase. Commanders must vigilantly identify these potential hazards, carefully consider them as part of their overall vulnerability analysis, factor the analysis into their risk assessment, and execute necessary contamination avoidance measures. (Local urban fire fighters may be a critical source of information for determining the likely locations of TIM.) Any assessment includes the chance that toxic industrial chemicals may be deliberately released by a threat to gain advantage or accidentally released by friendly actions (see FM 3-11.9, FM 3-11.14, FM 3-11.21, and the U.S. Department of Transportation's current version of the Emergency Response Guidebook).

Outlying High-Rise Area

2-35. High-rise areas consist of multistoried apartments, commercial offices, and businesses separated by large open areas, such as parking lots, parks, and individual one-story buildings. High-rise buildings are generally of framed, light-clad construction with thin walls of brick, lightweight concrete, or glass. The automobile, mass transit systems, and improved road networks encourage these areas to grow and function further from the urban core.

2-36. Similar to the urban core, units given the mission to clear these areas, or even portions therein, will need more resources—most notably personnel and time—to accomplish their mission. Commanders should consider courses of action that isolate these entire areas, multiple sections within these areas, or even individual buildings before assigning tasks. Without careful consideration and analysis, some tasks in these areas could unintentionally—but rapidly—drain a unit's resources or unhinge other portions of the major operation. When defending, commanders who can integrate these areas in the defense will present the attacker with similar resource problems and may be appropriate in a defense to delay. However, defending commanders must ensure that the defense is arranged so that this portion cannot be easily isolated and bypassed. Defensive positions in structures may require extensive reinforcement due to light-clad construction.

Residential Area

2-37. Residential areas can be found dispersed throughout the urban area; however, large suburban areas (or sprawl) normally form on the outskirts. Residential areas often consist of row houses or single-family dwellings set in a grid or ringed pattern in a planned development project. Yards, gardens, trees, and fences may separate the buildings in a residential area. In some areas of the world, residential areas may be located in high walled compounds with houses built right up to the edge of the street. Modern residential construction is more often of light-clad, framed wood construction, or brick, however; residential homes formed by poured or pre-cast concrete can be found throughout many parts of the world. The combined population of surrounding suburban areas often far outnumbers that of the urban area proper. Specific suburbs tend toward homogeneity based on ethnicity, religion, economics, or some other social aspect. Commanders must locate and analyze these areas to determine their impact on operations—often the most critical importance is the people located there (see the subsequent discussion in this chapter on the urban society).

2-38. In offensive and defensive operations, commanders should determine whether operations pose an unacceptable physical risk to civilians. If so, they may have to relocate civilians to a safer area, perhaps another residential area. If not, commanders may implement a "stay-put" policy for that area and attempt to isolate the effects of the operation from them. During civil support and stability operations, residential locations may be the initial focal point for operations since most of the permanent population is located there.

2-39. This area also contains an urban phenomenon known as shantytowns. These areas are commonly on unoccupied, low-value land in and around many urban areas in underdeveloped countries. Shantytowns may contain over 50-percent of the total urban population. They usually lack streets and public utilities. The lean-to structures tend to be irregularly laid out, connected by walking paths, and made of any scrap material available: lumber, brick, sheet metal, cardboard, cloth, or vegetation. The random arrangement of structures, the absence of formal street naming and numbering, and often the lack of easily identifiable buildings and terrain create challenges. These challenges include navigation, coordination, and the transmission of accurate information and intelligence. Depending on the operation, the temporary nature of the structures can also mean that mobility can be either more or less restricted than other sections of the urban area. A military force may easily knock down and traverse structures without affecting mobility at all. However, their destruction may cause unacceptable civilian casualties, in which case mobility becomes more restrictive as the narrow paths often do not accommodate vehicular traffic. Similarly, the makeshift materials inhibit weapons effects less than many other parts of the urban area built more solidly. A tank round, for example, may go much farther and injure many more noncombatants than in an area where the primary building material is stone. Regardless, commanders should consider the effects of their operations in this area, to include vehicles and weapons, as the weak structures increase the risk of fratricide, civilian casualties, and large, rapidly spreading fires.

Commercial Ribbon Area

2-40. Commercial ribbon areas are rows of stores, shops, and restaurants built along both sides of major streets that run through and between urban areas. These same types of areas often develop along the roads that connect one urban area to another (strip areas). The buildings uniformly stand two to three stories tall (about one story taller than the dwellings on the streets behind them).

Military Area

2-41. Fortifications and military installations may be found in or near urban areas throughout the world. Historically, they may have been the "seed" responsible for initiating the growth of the present-day urban area. Many countries possess long coastlines and borders with potentially hostile neighbors. To meet their defensive needs, they developed coastal and border defense works that include extensive subsurface facilities, many contiguous to urban areas. North Korea, for example, has built numerous hardened artillery, missile, and command and control facilities along both its coasts and the demilitarized zone which separates them from South Korea. While some fortifications may be inactive (particularly ancient fortifications), they can be rapidly activated and modified by threats (or Army forces) to accomplish their original role. Even if not activated, they may still serve as choke points and major obstacles to movement and maneuver.

2-42. Permanent-type fortifications can be made of earth, wood, rock, brick, concrete, steel-reinforced concrete, or any combination of the above. Some contemporary constructions have been built subsurface and employ heavy armor, major caliber weapons, internal communications, service facilities, and chemical, biological, radiological, and nuclear overpressure systems. Because they have been built specifically for military purposes, commanders and planners should carefully consider the effects of these military constructions on the conduct of UO.

URBAN SOCIETY

2-43. Although intricate, understanding the urban terrain is relatively straightforward in comparison to comprehending the multifaceted nature of urban society. UO often require Army forces to operate in close proximity to a high density of civilians. Even evacuated areas can have a stay-behind population in the tens of thousands. This population's presence, attitudes, actions, communications with the media, and needs may affect the conduct of operations. Homogeneity decreases drastically as the size of the urban area increases. Commanders must take into account the characteristics of a population whose beliefs and interests vary based on factors (see figure 2-7). Analysis and understanding of these societal factors is critical to a successful IO campaign and, thus, the entire operation.

2-44. Civilian populations continually influence, to varying degrees, operations conducted in an urban area. Thoroughly understanding these societal aspects and avoiding "mirror-imaging"—overlaying one's own values and thought processes on top of the person or group one is trying to understand—will help to accurately anticipate civilian actions and response. For example, while bribery and nepotism are not accepted norms in U.S. society, they may be customary practices in the foreign urban environment in which Army forces are operating. If commanders attempt to understand incidences of bribery and nepotism using their own societal norms and values as a basis, they are likely to misinterpret the significance of the events and implement an inappropriate course of action. Critically, however, the development of an in-depth understanding of the urban society does not mean that commanders and their Soldiers should adopt the culture of the foreign urban society. While understanding and respect of the urban area's culture is crucial to successful UO, Soldiers function best in dealing with the urban society when they act in accordance with American culture and values.

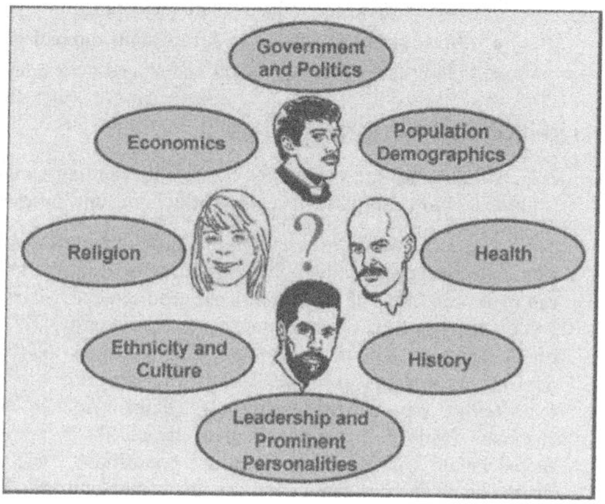

Figure 2-7. Key aspects of the urban society

POTENTIAL CENTER OF GRAVITY

2-45. A COG during an urban operation, particularly in stability and civil support operations, may be the civilian inhabitants themselves—specifically their behavior. However, supportive behavior is generally an advantage in any type of operation. Correspondingly, neutral behavior toward friendly forces is an advantage over hostile behavior. (However, a neutral population normally will not readily provide information to Army forces—often critical to understanding the urban environment and threat. In this light, neutrality can be seen as a disadvantage.) To influence or control their behavior, commanders must first understand the society's complex nature and character. Second, they must understand and accept that every military action (or inaction) may influence the relationship between the urban population and Army forces, and, by extension, mission success or failure. Lastly, they must understand that Army forces may play only a supporting (but essential) role as part of an integrated and synchronized multiagency effort focusing all aspects of national power—diplomatic, informational, military, and economic. With this awareness, commanders can take one or more actions:

- Coordinate and plan operations.
- Implement effective civil-military programs.

- Take the immediate action necessary to maintain support of a friendly populace, neutralize or gain the support of hostile or neutral factions, or do any combination of these activities to achieve precise effects and accomplish the mission.

Without this understanding, commanders increase the risk that their actions, particularly concerning the urban population, may not have the intended and necessary effects.

2-46. Although the factor of civil considerations takes on added significance in UO, it is just one that commanders evaluate. Sometimes, as in most urban counterinsurgency operations, it may be the most important factor to consider as a COG. At other times it may be the least important as to be almost negligible. Its importance is not constant; it changes over time (like all factors). At the beginning of the operation, civil considerations may not be essential to mission accomplishment, but as the operation progresses this factor's importance to success may increase. In other circumstances, the opposite may be true. Overall, commanders should consider three objectives regarding the civilians of the urban area:

- Minimize their interference with urban operations. In offensive and defensive operations this means moving them away from combat operations or establishing measures to shield them from its effects. In all operations, it often requires centralizing them in one or more locations—which may mean keeping them where they are.
- Maximize their support of Army, joint, and multinational forces and government agencies.
- Observe the necessary legal, moral, and humanitarian obligations.

General Population Size

2-47. Urban areas are commonly classified according to the general size of their population instead of landmass. Figure 2-8 lists one method of defining and categorizing urban areas by various population sizes.

2-48. These categories are useful to establish commonality and standardize terms that shape ideas, discussion, and concepts. Smaller populations usually suggest homogeneity among the inhabitants. Homogeneity can make consensus or compromise easier to achieve because fewer opposing viewpoints exist. Given this homogeneity, effects of change are more certain and often easier to determine. However, homogenous does not mean identical. If major social divisions do exist (either physical or ideological), commanders can more easily determine those divisions and their fundamental causes with smaller populations. Treating an urban population as a completely homogenous entity can lead to false assumptions, cultural misunderstandings, and poor situational understanding.

Category	Population
Village	3,000 or less.
Town	Over 3,000 to 100,000.
City	Over 100,000 to 1 million.
Metropolis	Over 1 million to 10 million.
Megalopolis	Over 10 million.

Figure 2-8. Urban areas by population size

2-49. As urban areas expand, the urban patterns begin to blur and the social complexity increases. For example, as satellite patterns continue to grow, the LOCs between a central hub and outlying urban areas may develop and begin to assume a linear urban pattern. Simultaneously, a hub and outlying urban areas may continue to expand until they merge into a single, large metropolis. On a larger scale, a network pattern can grow and unite as a single megalopolis. This growth physically unites smaller urban areas but cannot force conformity of needs and beliefs. It also increases the physical and social complexity of an urban area.

Group Size, Location, and Composition

2-50. Understanding how specific elements of the urban society affect operations (and vice versa), begins with analyzing their size, location, and composition (see figure 2-9). Because commanders must minimize civilian casualties, size and location (without regard to composition) are important initial demographic considerations. After determining the presence and numbers of civilians relative to decisive points, commanders can then decide whether civilian proximity and density represent a significant risk to their mission—refugees clogging LOCs, for example. If civilians are the primary focus of the operation, as in many stability and civil support operations, this same analysis may help to determine decisive points. In this analysis, commanders should consider that urban areas, on many levels, are in constant motion. The densities of circulating people and other traffic often vary according to cultural events or the time of day, such as religious holidays or sporting events and rush hours or market times. Therefore in planning urban operations, commanders must consider the timing, rhythms, or patterns of the population and their vehicular movements in the urban area. Identifying and understanding trends and patterns of activity (and disruptions to them) may provide critical information to commanders.

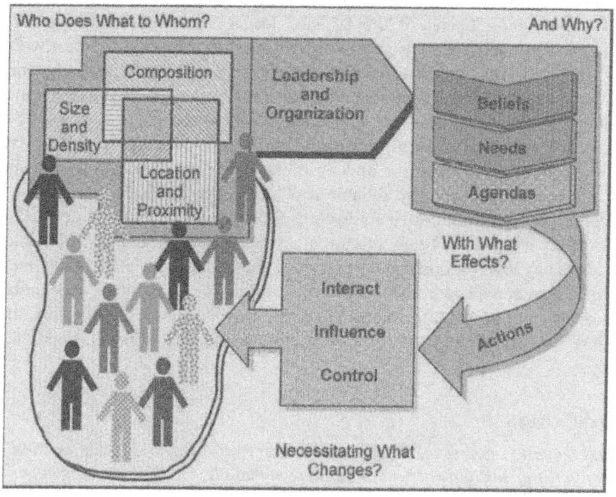

Figure 2-9. Simplified analysis of urban society

2-51. Commanders normally determine the composition of, or the identifiable groups or organizations within, the civilian urban population. Groups may be categorized by race, religion, national origin, tribe, clan, economic or social class, party affiliation, education level, union memberships, age, gender, occupation, or any other significant social demographic. Physical and ideological overlaps (and divisions) often exist between groups. Overlaps may provide early focus for analysis and suggest ways to affect more than one group simultaneously. In some cases, groups may have radically different ideologies but are (or can be) united by a single characteristic. Commanders must understand the intricacies of "who does what to whom." Such understanding furthers identifying the urban society's sources of power, influence (both formal and informal), and decisive points that hold the keys to controlling or protecting this potential COG. (See also the discussion of competing power structures in Chapter 3.) Commanders should have expert, detailed, and current knowledge and information to avoid developing simple formulas of social interaction that may actively mislead and add to a flawed course of action.

Leadership and Organization

2-52. Commanders must also understand how authority and responsibility is held or shared within and between each of the identified groups; they must understand leadership and the social hierarchy. For groups to exert meaningful influence, leadership provides vision, direction, and organized coherence. This leadership can be a function of personality as well as organization. Some groups depend on a charismatic leader to provide cohesion; although in some cultures, the spokesman is not the leader. Others de-emphasize individual leadership and provide redundancy and replacement in decision making. Others combine elements of both these types of leadership and organization. Based solely on personality, a leader may centralize power or, while still being in ultimate control, decentralize decision making and execution to subordinates. In contrast, a single person may head a group while a ruling council actually makes and executes policy. Groups centered on one leader (which may or may not be the officially designated leader) can often produce decisions and initiate actions rapidly but are vulnerable to disruptions if key personalities are removed or co-opted. Groups with shared or redundant leadership take longer to make decisions yet are more resistant to change and outside influence.

2-53. In UO, particularly stability operations, Army commanders and leaders at all levels will devote considerable effort to identifying and cultivating relationships with civilian leaders in their AO. This civilian leadership will include political, religious, tribal or clan, ethnic, and economic leaders. Commanders should consider that their attention toward and discussion with identified leaders may increase (or, some instance, decrease) the targeted leaders' prestige and power. While this may be intentional, commanders must often ensure that the leaders that they chose to deal with are the legitimate and accepted in the eyes of the urban population. Otherwise, they may further imbalance an already weak power structure and exacerbate the situation. In unique circumstances, commanders may need to identify and interact with the leadership of criminal organizations. These civilian leaders will be the means to affect change and allow the urban populace to understand and accept the purpose behind Army operations. As importantly, they will be the conduit for understanding the urban society's sentiments, perceptions, and reactions to Army operations. As such, reliable and trustworthy linguists that can not only interpret the language but can serve as cultural advisors to Army forces will be in great demand. Keeping the local population objectively informed not only of current Army operations but also of the intent and desired end state will often be a key task for commanders. A poor communications effort and a lack of understanding of intent may alienate a commander and cause significant problems in future negotiations. It is imperative that communications be clear and effective and that all Army leaders are aware of its implications. See FM 4-05.401 and FM 3-07 for the effective use of interpreters and the conduct of productive meetings and negotiations.

Interests and Actions

Me and Somalia against the world, me and my clan against Somalia, me and my family against the clan, me and my brother against my family, me against my brother.

Somali Proverb

2-54. Identifying and analyzing groups also helps commanders focus on specific segments of the urban society to determine their beliefs, needs, and agendas. It also helps commanders determine how those interests motivate groups to future action (or inaction)—previous patterns of activity are critical in this regard. This analysis seeks to determine why groups (and their leaders) act as they do. Commanders should consider political, economic, cultural, and religious factors in this analysis. These factors affect all groups to some extent and often provide the basis for their beliefs, needs (actual or perceived), and subsequent behavior. Size and location considerations also apply to each group to help determine to what extent its beliefs or ideologies, needs, and actions may impact the urban operation. However, size and proximity may not accurately indicate actual or potential capabilities. Individuals, small groups, and groups located some distance from the actual conduct of the urban operation may be able to influence large portions of the population. These individuals or groups may have a capability disproportionate to their size and proximity—especially against objectives that are not terrain oriented (as in the case of many stability operations).

Interaction, Influence, or Control

2-55. Commanders must cultivate an understanding of a group's—

- Size, location (and proximity to operations), and composition (to include leadership and organization).
- Interests.
- Capabilities.
- Potential actions (intent) and their effects—if any—on operations.

2-56. Then they can develop or modify courses of action as appropriate. Certain courses of action may be needed to improve the interaction between Army forces and civilians (and between other agencies) to accomplish common goals. Others may be needed to influence favorable support, stabilize neutral groups, or neutralize hostile groups. Still others may require more forceful means to control and protect—never to punish—civilians. Courses of action may include—

- Establishing buffer zones and restricted areas.
- Setting up checkpoints and roadblocks with other travel restrictions on people and goods.
- Screening civilians.
- Conducting negotiations (directly or as a mediator).
- Providing or protecting rations, water, and other critical resources.
- Restoring or improving specific, key infrastructure.
- Enforcing curfews.
- Inspecting facilities.
- Directing amnesty programs.
- Conducting internment and resettlement operations.
- Maintaining a "stay-put" policy.

2-57. Commanders must remember that many measures will require significant resources that may initially be beyond the capabilities of the Army force to impose and enforce. (Where possible, commanders should attempt to use host-nation security forces and local urban law enforcement to control activities.) Other elements of the environment, terrain and infrastructure, may fragment efforts and make it difficult to impose controls throughout the area. A careful assessment and understanding of the urban society's interests (beliefs, needs, and agendas) is essential before implementing any populace and resources control measures. Otherwise inappropriate controls, particularly if they are improperly perceived as punishment, may only aggravate the situation. Finally, an appropriate course of action may require no specific action towards the urban society. In most cases, training and discipline, grounded in cultural understanding and sensitivity, will help mitigate many potential adverse effects resulting from military-civilian interaction. (Commanders should review FM 3-19.15 and FM 3-19.40 for additional civilian control measures and considerations.)

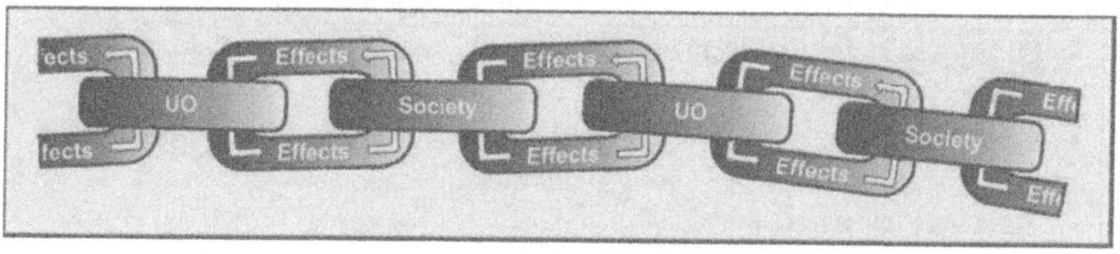

Figure 2-10. UO society chain of effects

A Chain of Effects

2-58. Since urban society is so dynamic and the relationship between various elements so complex, commanders must continually assess how their operations will affect the society and vice versa (see figure 2-10). Specifically, they assess how effectively their measures improve interaction with, influence of, and control over civilians. There is always a difference between intended and actual effects of a specific course of action. Nowhere is this more prominent than dealing with the urban society. This chain of effects frustrates understanding during UO. Therefore, commanders must continuously monitor these effects to make decisions and modifications while planning, preparing, executing, sustaining, and transitioning UO. (Commanders may conduct informal and formal polling activities to assess and understand changing civilian perceptions.) Initially certain aspects of the society, such as religion, may not affect the operation. However, if the threat successfully shapes the perceptions of the urban populace that Army forces are biased against their religion or religious beliefs (or at least critical segments are affected by propaganda), this element may become extremely important. In this instance, the urban commander may need to adjust his IO (to include PSYOP), public affairs (PA) activities, and CMO to counter this propaganda while diverting other combat power to control the populace. Overall, commanders must understand and account for second, third, and higher order effects of their actions and decisions.

Understanding the Effects of Unit and Soldier Actions in Iraq

During OPERATION IRAQI FREEDOM, an armor task force commander described his methodology as that of plotting and measuring everything. After a positive or negative event, he would have his staff evaluate all actions they had conducted before, during, and after the event. This would allow him to correlate activities with outcomes and develop tactics, techniques, and procedures for future success. While not all actions easily correlated to events, this methodological approach helped establish a base line of comparative success for his task force.

Some units assessed the Iraqi culture and developed practical guidelines or rules of interaction for their soldiers to follow as they conducted their missions. These guidelines encompassed a list of actions that their Soldiers should and should not do when dealing with the Iraqi populace. The guiding principle behind these actions was not only to "win the hearts and minds" of the Iraqi people but, more accurately, to establish legitimacy and credibility through appropriate Soldier actions that reflected a combined understanding of the mission and the Iraqi culture.

DO:

- Separate the men from the women and children during cordon and searches.
- Talk to the oldest or most senior male when civilian cooperation is needed.
- Bring something (for example, water, food, or replacement locks) to give residents to compensate for damages when the search ends up being a "dry hole."
- Learn Arabic phrases; say hello and greet people on the street.
- Use female Soldiers to search and talk to Iraqi females.
- Cordon and knock whenever possible instead of kicking in doors needlessly.
- Understand that in Iraq, women do not have the same social status as men.

DO NOT:

- Force a detainee's head to the ground or use your feet to hold someone to the ground.
- Stare at or touch Iraqi women.
- Show obvious disrespect to suspects in front of family members.
- Manhandle the Koran or other religious objects.
- Commit to anything; instead of agreeing to (or denying) an Iraqi's request, it is good practice to say "In'sha Allah" (If God wills it).

URBAN INFRASTRUCTURE

2-59. Urban infrastructures are those systems that support urban inhabitants and their economy. They form the essential link between the physical terrain and the urban society. During urban stability and civil support operations, restoration or repair of urban infrastructure will often be decisive to mission accomplishment. During urban combat operations, destroying, controlling, or protecting vital parts of the infrastructure may be a necessary shaping operation that can isolate a threat from potential sources of support. A threat force operating in an urban area may rely on the area's water, electricity, and sources of bulk fuel to support his forces. This is true particularly when his bases or facilities are physically located in or near the area. Isolating this threat from these sources may require him to generate his own electricity and transport his own water and fuel from outside the urban area. To transport supplies, the threat may rely on roads, airfields, sea or river lanes, and rail lines. Controlling these critical transportation nodes may prevent the threat from resupplying his forces. The control of key radio, television, and newspaper facilities may isolate him from the urban populace (another potential source of support).

INTERDEPENDENCE

2-60. Commanders must understand that destroying or disrupting any portion of the urban infrastructure can have a cascading effect (either intentional or unintentional) on the other elements of the infrastructure. Yet, they may be able to gain an operational advantage while minimizing unwanted and unintended effects. Commanders can control, seize, or secure an essential facility or structure by using precision munitions, electronic disruption of communications, or SOF and conventional ground forces. To gain this advantage, commanders will rely more on the expertise of Army engineer and civil affairs units; local urban engineers, planners, and public works employees; and others with infrastructure-specific expertise. After understanding the technical aspects of the area's systems and subsystems, commanders can then develop the best course of action.

A SYSTEM OF SYSTEMS

2-61. Hundreds of systems may exist. Each system has a critical role in the smooth functioning of the urban area. Simple or complex, all systems fit into six broad categories (see figure 2-11). Commanders should analyze key facilities in each category and determine their role and importance throughout all phases of the urban operation. As there is much overlap between infrastructure systems, this analysis considers each system individually and in relation to others to determine an appropriate course of action toward it.

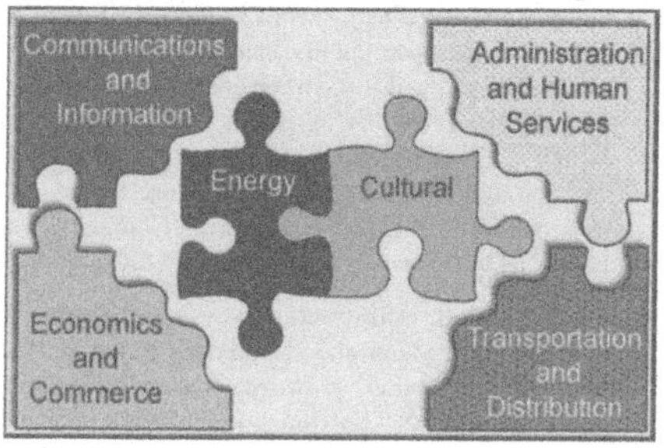

Figure 2-11. Urban infrastructure

A COMBINATION OF STRUCTURES AND PEOPLE

2-62. As depicted earlier in figure 2-1, each element of the infrastructure consists of both a physical (terrain) and human component. For example, the physical component of the electrical segment of the energy infrastructure consists of power stations, substations, a distribution network of lines and wires, and necessary vehicles and repair supplies and equipment. The human component of this same segment consists of the supervisors, engineers, linemen, electricians, and others who operate the system. Commanders must understand and recognize the physical and human components in their assessments.

POTENTIAL IMPACT ON FUTURE OPERATIONS

2-63. Destroying or incapacitating of any of these elements may impact future operations and inhabitants of the urban area. Destroying urban infrastructure during initial phases of an operation may require commanders to assume responsibility for repair, maintenance and clean up, and operation of those same facilities later. Although exceptions will exist, commanders cannot destroy or significantly damage the infrastructure of a foreign urban center during operations and expect the population to remain friendly to U.S. or allied forces. On the other hand, early repair or restoration of critical or essential infrastructure may improve civil-military relations, speed transition back to competent civilian authorities, and, overall, aid in successful mission accomplishment. Still, support from the urban society (albeit of increased importance in UO) is only one factor that commanders weigh while developing appropriate courses of action.

RESOURCE INTENSIVE

2-64. Requirements to protect, restore, or maintain critical infrastructure may divert substantial amounts of resources and manpower needed elsewhere and place additional constraints on subordinate commanders. Civilian infrastructure is often more difficult to secure and defend than military infrastructure. The potentially large and sprawling nature of many systems (such as water, power, transportation, communications, and government), make their protection a challenge. Yet, the infrastructure of an urban area may provide commanders with essential logistics and support. Therefore, the initial expenditure of time and other resources may be necessary to support concurrent or future operations. Legal considerations, however, may affect using the infrastructure and acquiring the urban area's goods and services. Commanders, their staffs, and subordinates (often down to the individual soldier) must know their limits concerning Army authority to commandeer civilian supplies or equipment to facilitate mission accomplishment (see the legal support discussion in Chapter 9). In stability and civil support operations, the safeguard or restoration of critical urban infrastructure for military or civilian use may be a decisive point in the overall operation.

2-65. Keys to understanding the magnitude of the resources and manpower required to restore the infrastructure are an initial infrastructure assessment and, as soon as practical afterward, a detailed infrastructure survey. An initial assessment provides the commander immediate feedback concerning the status of basic services needed to meet the urgent needs of the urban population. The systems assessed are based on mission, enemy, terrain and weather, troops and support available, time available, civil considerations, and the commander's vision of the overall end state. The infrastructure assessment, while typically performed by engineers, may be accomplished by, or in conjunction with, others with sufficient expertise to provide the type and quality of information required. These others may include civil affairs, medical, and chemical personnel. Those tasked with this assessment should routinely consult other Army and coalition forces and governmental and nongovernmental agencies currently operating in the urban area as well as the urban civilian leadership for their informed input.

2-66. While an infrastructure assessment functions to support the resolution of immediate challenges to urban reconstruction and restoration, it also provides the initial basis for determining the conditions for successful transition. However, commanders and planners must continually expand and refine their understanding. As a necessary follow-on, commanders initiate a detailed infrastructure survey. This survey is normally conducted by U.S. Army Corps of Engineers personnel assigned to forward engineer support teams. As with the assessment, the commander should incorporate other technical specialty personnel in the survey team to enhance the quality and accuracy of the product (see FM 3-34.250).

COMMUNICATIONS AND INFORMATION

2-67. This system is comprised of the facilities and the formal and informal means to transmit information and data from place to place. Understanding communication and information infrastructure of an urban area is important because it ultimately controls the flow of information to the population and the enemy. It includes—

- Telecommunications, such as telephone (to include wireless), telegraph, radio, television, and computer systems.
- Police, fire, and rescue communications systems.
- Public address, loudspeaker, and emergency alert systems.
- The postal system.
- Newspapers, magazines, billboards and posters, banners, graffiti, and other forms of print media.
- The informal human interaction that conveys information such as messengers, open-air speeches and protests, and everyday conversations.
- Other inventive informal means such as burning tires and honking horns.

2-68. Perhaps more than any other element of the infrastructure, communications and information link all the other elements in an interdependent "system of systems." It is a critical enabler that helps coordinate, organize, and manage urban activities and influence and control the urban society. Army commanders are acutely aware of the impact that a loss or degradation in communications has on their own operations. The urban environment experiences similar impacts to communication failures; however, urban governments and administrations are generally less prepared to deal with a collapsed communications and information infrastructure than are trained Army forces.

2-69. Militarily, a functioning urban communications and information system can serve as an alternate for both friendly and threat forces and can be easily secured with civilian, off-the-shelf technologies. Threats may make use of commercial systems intertwined with legitimate civilian users, making it unpalatable to prevent use of these assets. Forces can also use these systems to influence public opinion, gain intelligence information, support deception efforts, or otherwise support IO.

Increasing Impact of Computers

2-70. In many urban areas, computers link other elements of the urban infrastructure. They link functions and systems in the urban area and connect the area to other parts of the world. This latter aspect creates important implications for commanders of a major operation. Operations involving this cybernetic function may produce undesirable effects on a greater scale than initially intended. For example, commanders may be able to close or obstruct an urban area's banking system; however, this system may impact the international monetary exchange with unwanted or even unknown effects. The authority to conduct these types of IO will often be retained at the strategic level.

Pervasive Media

2-71. The media is central to the communications and information infrastructure and a critical operational concern. Compared to other operational environments (jungles, deserts, mountains, and cold weather areas), the media has more access to urban operations. This is due largely to airports, sea and river ports, and major road networks; ready access to power sources and telecommunications facilities; as well as access to existing local media structures. Hence, media presence may be pervasive and IO even more critical to success in UO than operations in many other environments.

A Complex Relationship

2-72. A complex relationship exists among information, the public, and policy formulation. Although the degree and manner in which public opinion shapes government policy are difficult to accurately determine, negative visual images of military operations presented by the media can change political objectives and, subsequently, military objectives. As important, media reporting can influence civilian activity in an urban AO to either the advantage or disadvantage of the commander.

Induce Cooperation Through Credibility

2-73. Commanders do not control the media; however, they monitor the flow of information that the news media receives and subsequently reports. Consequently, commanders should plan and execute PA operations that will induce cooperation between the media and Army forces. Successful relations between urban Army forces and the news media evolve from regular interaction based on credibility and trust. More information is usually better than less, except when the release of such information may jeopardize security and the success of the operations and threaten the safety of Soldiers. However, commanders cannot simply withhold information to protect the command from embarrassment. They consider media interests as part of the normal planning process and work to ensure that information presented to the news media is accurate, timely, and consistent with operations security. Since the media will likely arrive in the urban area before the conduct of operations, early deployment of PA assets may be critical. Commanders should synchronize PA activities with CMO and PSYOP. Such action eliminates duplicated effort and ensures a unity of purpose consistent with the IO concept of support (see Chapter 4).

2-74. Failure to provide sufficient information can hamper a commander's ability to conduct the mission. Commanders cannot refuse to deal with particular news media because they consistently report a negative image of Army forces and operations. Poor relationships with any media can result in inaccurate and even biased reporting. Such reporting can cause a public reaction that influences the ability to achieve operational objectives. During the Russian 1994-95 battle against Chechen separatists in Grozny, for example, the Russian military refused to communicate with reporters. The media reported primarily from the perspective of the Chechen rebels. This encouraged both local and international support for the rebels. It also allowed the Chechens, who lacked sophisticated information systems, to use the media to broadcast operational guidance to their forces. (During their second Chechnya campaign of 1999-2000, Russia learned this lesson well and the Russian view of the war dominated domestic public opinion.) On the other hand, successfully engaging the media can serve as a force multiplier. The Army's open and responsive interaction with the media during peacekeeping operations in Bosnian urban areas helped to explain the challenges and successes of Army forces in the Balkans to the public. This helped maintain domestic, international, and local political support for NATO operations and, with a successful command information program, helped maintain Soldiers' morale.

Transportation and Distribution

2-75. This element of the infrastructure consists of—

- Networked highways and railways to include bridges, subways and tunnels, underpasses and overpasses, ferries, and fords.
- Ports, harbors, and inland waterways.
- Airports, seaplane stations, and heliports.
- Mass transit.
- Cableways and tramways.
- Transport companies and delivery services that facilitate the movement of supplies, equipment, and people.

Similar to communications and information, this facet provides the physical link to all other elements of the infrastructure.

2-76. Army forces deploying into a theater of operations depend on ports and airfields; seizure and protection of these critical transportation nodes may impact the projection of combat power. Once in theater, transportation and distribution systems in the urban area can contribute greatly to the movement of forces, maneuver, and logistic operations throughout the entire AO. Control of decisive points in this infrastructure may be important to the military operation and to the normal functioning of the urban area (and surrounding rural areas). Supplies traveling through the transportation and distribution system may be military-specific supplies (such as ammunition and repair parts) and supplies for both the military and urban population (such as food, medicine, oil, and gas). The system may also support the movement of military forces and the urban area's population (for which it was designed). Therefore, commanders of a major operation may have to develop innovative methods that limit the transit of threat supplies and reinforcements while facilitating the movement of their own resources and those of civilians. This last consideration attempts to minimize hardship and promote normalcy in the urban area and will increase in significance as the need for legitimacy increases.

2-77. Most urban areas (particularly in developing countries) have two forms of transportation and distribution systems that exist simultaneously: a formal system and an informal or paratransit system. Large organizations, bureaucracy, imported technology, scheduled services, and fixed fares or rates characterize formal systems. Low barriers to entry; family and individual entrepreneur organizations; adapted technology; flexible routes, destinations, and times of service; and negotiated prices characterize the informal system. The informal system is more decentralized and covers a much greater portion of the urban area than the formal system. The informal transportation and distribution system often includes a waterborne element, is more likely to function through turbulence and conflict, and can extend hundreds of kilometers beyond the urban area. Accordingly, commanders should understand both systems to establish effective movement control.

ENERGY

2-78. The energy system provides the power to run the urban area. It consists of the industries and facilities that produce, store, and distribute electricity, coal, oil, wood, and natural gas. This area also encompasses alternate energy sources, such as nuclear, solar, hydroelectric, and geothermal power. Energy is needed for industrial production and is therefore vital to economics and commerce. Among many other things, this system also provides the fuels to heat, cool, and light homes and hospitals, cook and preserve food, power communications, and run the transportation necessary to move people and their supplies throughout the urban area. Loss of an important energy source such as electricity or gasoline, especially for those accustomed to having it, will become an immense area of discontent that the commander of a major urban operation will need to quickly address. Therefore some threats, particularly terrorists and insurgents, may actively target this element of the urban area's infrastructure to erode support for civilian authorities and Army forces.

2-79. Sources of energy may be tens or hundreds of miles away from the urban area itself. Therefore, commanders may exert control without applying combat power directly to the urban area itself by controlling or destroying the source (power generation or refinement plant) or the method of distribution (pipelines or power lines). With electrical energy that cannot be stored in any sizable amount, the latter may be the best means as most major urban areas receive this energy from more than one source in a network of power grids. However, control may be as simple as securing a power station or plant and turning off switches or removing a vital component that could later be restored. On the other hand, lengthy pipelines and power lines may compound security and protection of this element of the infrastructure.

2-80. The number of nations that have invested in nuclear power and nuclear research is increasing. With this increase, the potential for Army forces to operate in urban areas that include (or are near) these facilities also increases. Damage to one of these facilities and potential radiation hazards will present special challenges to commanders of a major operation. To safeguard friendly forces and civilians, commanders will need to employ a blend of peacetime and tactical nuclear contamination avoidance principles (see FM 3-11.14).

ECONOMICS AND COMMERCE

2-81. This system encompasses—

- Business and financial centers to include stores, shops, restaurants, hotels, marketplaces, banks, trading centers, and business offices.
- Recreational facilities such as amusement parks, golf courses, and stadiums.
- Outlying industrial, mineral, and agricultural features to include strip malls, farms, food processing and storage centers, manufacturing plants, mines, and mills.

2-82. An essential aspect of this area during operations may be the political sensitivity of U.S. or allied industries investing and operating in a foreign country, particularly during stability operations. An enemy or a disgruntled civilian population may attack or disrupt commercial activities as a political statement against the United States or our allies. Food production facilities also may assist commanders in Army food services and may be an essential concern during relief operations. During long-term stability operations, visible, material, and tangible economic progress consisting of the creation or restoration (and protection) of businesses, agriculture, and overall jobs will often be critical to—

- Generating or maintaining the urban population's support to Army forces and operations.
- Reducing support to threat forces and operations to include eliminating civilians as a potential manpower pool for insurgent or terrorist organizations and activities.
- Lowering other hostile civilian activities such as protests and riots.
- Transitioning the urban area back to legitimate civilian responsibility and control.

2-83. This element of the infrastructure also consists of the production and storage of toxic industrial chemicals used in agriculture (insecticides, herbicides, and fertilizers), manufacturing, cleaning, and research (to include biological agents). Fertilizer plants may be of specific concern as they contribute to providing a key material in terrorist and insurgent bomb-making activities. A thorough analysis of this element of the infrastructure may also be essential to understanding how urban insurgencies are funded and supported. This helps commanders to understand the true organization of the insurgency as well as to suggest methods to isolate insurgents from their economic or financial support. In their overall assessment of this area of the infrastructure, commanders should also consider the activities and influence of criminal organizations or elements.

Administration and Human Services

2-84. This wide-ranging system covers urban administrative organizations and service functions concerned with an urban area's public governance, health, safety, and welfare. Together, it encompasses—

- Governmental services that include embassies and diplomatic organizations.
- Activities that manage vital records, such as birth certificates and deeds.
- The judicial system.
- Hospitals and other medical services and facilities.
- Public housing and shelter.
- Water supply systems.
- Waste and hazardous material storage and processing facilities.
- Emergency and first-responder services such as police, fire, and rescue.
- Prisons.
- Welfare and social service systems.

Cultural

2-85. This system encompasses many organizations and structures that provide the urban populace with its social identity and reflect its culture. (This infrastructure system overlaps with many recreational facilities included under the economics and commerce infrastructure. For example, an urban society may radically follow soccer matches and teams. Hence, soccer stadiums relate to the society's cultural infrastructure.) Some of these facilities, particularly religious structures, will be protected targets and others may require security and law enforcement protection from looting and pilferage. However, commanders will need to quickly educate, inform, and continually remind the urban populace (and the media) that cultural infrastructure may lose its protected status when used by threats for military purposes. Cultural infrastructure may include—

- Religious organizations, places of worship, and shrines.
- Schools and universities.
- Museums and archeological sites.
- Historic monuments.
- Libraries.
- Theaters.

RESTORING AND PROTECTING ESSENTIAL SERVICES

2-86. Losing the support of essential elements of the infrastructure will have an immediate, destabilizing, and life-threatening impact on the inhabitants of the urban area. In stability and civil support operations, numerous parts of the administrative and human services and energy infrastructure often rise to critical importance before all other elements. Again, however, complete restoration of these essential services is often a lengthy, resource-intensive civil-military operation. (Following the end of major combat operations of OPERATION IRAQI FREEDOM, most units developed one or more of their logical lines of operations oriented along the restoration or improvement of urban infrastructure and essential services. An acronym used by many units to focus and track critical activities within this line of operation was sewer, water, electricity, and trash (SWET). Later, other units modified this acronym to SWEAT and then SWEAT-MS to include concerns for restoring academics [or schools], revitalizing medical facilities, and establishing security [police and host-nation security forces].) Of critical importance will be a simultaneous IO campaign that includes efforts to help ensure that the urban population develops realistic expectations about Army abilities to restore their essential urban infrastructure.

Understanding the Urban Environment: Paris – 1944

The summer of 1944 confronted German General Dietrich von Choltitz with a dilemma. As military commander of greater Paris, he was to eliminate French Resistance internal to the city while defending against approaching Allied units, missions for which he had insufficient forces. (Notably, General Eisenhower, the commander of the Allied forces, wanted to bypass Paris to sustain the offensive. Seizing the French capital would task his forces with the support of tens of thousands of civilians. Eisenhower was nevertheless ordered to capture the city. A political decision resulted in civilian assistance taking precedence over combat operations.)

Choltitz's situation was further complicated by Hitler's demand that he destroy the city, an action the general saw as needlessly destructive (and infeasible given his scant resources). Choltitz's seniors directed the preparation, and later the destruction, of Paris's 45 Seine River bridges. They were the only remaining crossing points over that waterway given Allied bombing of others outside the French capital. Premature destruction would trap German forces defending to their north, a second-order effect that Choltitz used to justify his disobedience of orders demanding the bridges' demolition.

The German general also recognized that some mission-critical elements were part of Paris's social rather than physical infrastructure: the leadership of the various resistance groups and the relationships between them. Choltitz understood that he lacked resources to defeat the many separate factions; he therefore chose the unorthodox (asymmetric) approach of accepting an intermediary's offer of a truce with these groups. Such an agreement provided some measure of the stability needed while Choltitz awaited promised reinforcements. Further, he realized that the resistance factions were by no means united in their goals. Communist elements sought a much different end than those looking toward a de Gaulle-led postwar government. A truce thus set the French Communists (who sought an uprising so as to legitimize their claims to power) against others trying to buy time until Allied forces arrived, forces that included Free French units supportive of de Gaulle.

Although his defense of the capital failed, Choltitz succeeded in harboring his available resources, reducing the effectiveness of the resistance organizations fighting his soldiers, and maintaining withdrawal routes for units north of the Seine. The German commander's analysis in support of these efforts was effective in part because of his insightful (1) identification of critical points that included elements of terrain, citizenry, and infrastructure; (2) understanding of the relationships between these parts; and (3) use of an asymmetric approach to address his lack of sufficient force to otherwise handle the urban densities that challenged him.

Chapter 3

Understanding the Urban Threat

... [T]he United States could be forced to intervene in unexpected crises against opponents with a wide range of capabilities. Moreover, these interventions may take place in distant regions where urban environments, other complex terrain, and varied climatic conditions present major operational challenges.

Quadrennial Defense Review Report, 30 September 2001

As the strategic environment has become less stable, more uncertain, and more dangerous, Army forces must be trained and ready to address persistent and evolving urban threats. These threats range from regional conventional military forces, paramilitary forces, guerrillas, and insurgents to terrorists, criminal groups, and angry crowds. These threats can hide in plain sight and become indistinguishable from the noncombatant urban population that may help shield, protect, and sustain them. Although uncertain about events, Army forces can be clear about trends. Increasingly, the Army will face threats that severely differ in doctrine, organization, and equipment, are skilled at developing and adapting techniques to counter Army tactics, techniques, and procedures (TTP), and can fully interact with the three other components of the urban battlefield—terrain, society, and infrastructure. In urban operations, commanders must broaden their concept of the threat to include natural disasters, hunger and starvation, and rampant disease. Further, commanders must plan to contend with many passive urban threats, such as psychological illnesses and toxic industrial materials (TIM). These threats may be found in isolation, but most likely commanders will encounter them in various combinations. Moreover, each new threat will pose a different combination and likely have new capabilities that previous opponents lacked.

ASYMMETRICAL AND ADAPTIVE

3-1. An emphasis on asymmetry to offset U.S. military capabilities has emerged as a significant trend among potential threats and become an integral part of threat objectives and tactics (discussed below). Though, this is not to imply that all future urban threats will fight and operate asymmetrically. Some may be asymmetrical; others may not. Future threats may be a symmetric conventional military; others may be paramilitary, insurgent, or even a nonmilitary threat such as a criminal organization. Asymmetry results when one opponent has dissimilar capabilities—values, organization, training, or equipment—that the other cannot immediately counter. It is not a new concept. It naturally evolves from a sound mission, enemy, terrain and weather, troops and support available, time available, civil considerations (METT TC) analysis by an intelligent, freethinking, and adaptive threat. These asymmetric approaches will include the most advanced, commercially-available technology innovatively applied and mixed with crude, simple, and unsophisticated weapons and TTP. The success of these approaches often depends upon the predictability of the forces on which they are employed. Thus, Army commanders must consider whether their operations and activities lend themselves to pattern analysis by threat forces and, if so, they may need to vary their battle rhythm and take other steps necessary to become less predictable.

WEAPONS OF MASS DESTRUCTION

3-2. A chief asymmetric means of engaging the national power of the United States is to employ weapons of mass destruction (WMD) against the United States or its allies. These weapons can be used against military forces and include high-yield explosives as well as nuclear, biological, and chemical weapons. Operations in urban areas may require concentrating forces and may create a lucrative target for a threat that possesses fewer numbers and less equipment.

3-3. A threat's WMD use will adversely affect the Army's abilities to conduct urban operations (UO) to various degrees. For example, the intervening structures and the effects of urban microclimates complicate the ability to detect and identify radiological, chemical, or biological attacks from a standoff distance. Also, the individual Soldier's ability to recognize his leaders, understand oral and visual commands, and operate increasingly sophisticated equipment is difficult when wearing protective clothing and equipment—particularly if his training proficiency is low. Despite the increased challenges and complexity, Army forces have the best (or better) training and equipment necessary to respond to such an attack compared with most armies around the world or with any civilian organization.

3-4. Although initial casualties could be high, the public can accept military casualties before those of civilians. Therefore, threats may gain an initial tactical advantage but would achieve less asymmetric benefit by directly attacking Army forces. Instead, they may attempt to achieve a strategic advantage by employing WMD against U.S. or allied civilian populations. In doing so, threats hope to use political sensitivity to high civilian casualties to reduce popular support for the United States or its allies. The chance of these attacks occurring in an urban area increases because—

- The area facilitates weapon effects and camouflages delivery means.
- The dense civilian population ensures a high casualty rate.
- The attack (or even the threat of attack) often will receive more publicity and public attention.
- The urban infrastructure is especially vulnerable to WMD, (particularly the systems of the economics and commerce infrastructure located in large urban areas) and may have far-reaching national and global effects.

THREAT OBJECTIVES

3-5. The threat will seek to achieve several key objectives when opposing Army forces operating in an urban environment (see figure 3-1). These objectives focus more on how a threat might fight in an urban area rather than specifically who the threat might be or in what region of the world the conflict might occur. These objectives are more easily achieved in an urban environment due to—

- The high costs in time, material, and manpower involved in UO.
- The limiting effects of urban areas on many technological advantages.
- The proximity of airfields and ports to urban areas.
- The potential moral dilemmas created by exposing numerous civilians to harm or injury.
- The ability of the media to observe and report the threat's version of events.

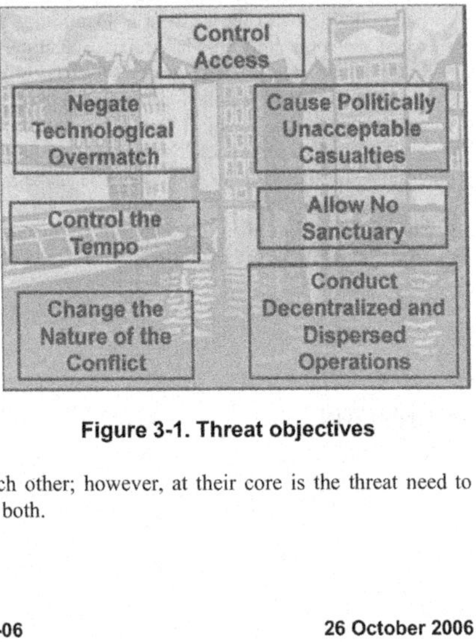

Figure 3-1. Threat objectives

3-6. These objectives complement and overlap each other; however, at their core is the threat need to defeat an enemy of superior numbers, technology, or both.

Control Access

... Iraq made no direct effort to impede the buildup in the theater. Planners should consider what might have happened if Iraq had attempted a strategy based on denying access to the region. Planners might also wonder what the outcome would have been if Iraq had attacked US forces in Kuwait before they were ready for the running start. The point is that the conditions in [US Central Command] in 2003 are unlikely to be replicated elsewhere.

On Point: The United States Army in Operation IRAQI FREEDOM

3-7. The Army may not be located where future conflicts are fought. Thus, the Army maintains the ability to rapidly project and sustain combat power over long distances and time spans. This capability demands that Army forces quickly gain and maintain control of seaports or aerial ports of embarkation or debarkation, particularly where the density of U.S. basing and en route infrastructure is low. Commanders gain control of these ports by unopposed (assisted or unassisted) or forcible entry operations. In either case, these phased-entry operations may present potential vulnerabilities, particularly—

- Unsuitable composition of initial or early entry forces lacking necessary combat power for immediate decisive operations.
- Initial command and control difficulties and an immature situational understanding.
- Weak initial port or base defenses without the necessary integration of naval and air component forces.
- Lack of developed logistic support.

3-8. Consequently, threats may attack during initial force projection operations to oppose, delay, disrupt, or otherwise manipulate the build-up of essential combat power into a theater of operations. These attacks may occur anywhere deploying Army forces are located, at overseas bases, at home stations, and even in military communities. Increasingly, deployment facilities such as airfields and ports exist as integral components of urban areas. Threats will invariably use the complex and concealing nature of these urban areas, coupled with the vulnerabilities, to create favorable conditions for their attacks.

Negate Technological Overmatch

3-9. Threats will always strive to force engagements at a time and place most advantageous to them. They may locate military forces and vital military capabilities in urban areas to achieve sanctuary from the effects of Army capabilities and make Army forces and systems more vulnerable to less-sophisticated weapons.

3-10. The clutter of the physical structures, electromagnetic radiation, and population diminishes Army capabilities. This clutter makes it difficult for Army forces to acquire and effectively engage targets at long ranges. In urban areas, the terrain often allows a threat to operate in closer proximity to friendly forces. Therefore, the threat may "hug" friendly forces to avoid the effects of high-firepower standoff weapon systems and degrade their ability to gain or maintain a thorough common operational picture. Additionally, this threat tactic attempts to inhibit friendly commanders from employing some weapon systems and munitions for fear of fratricide. Threats will also seek to hide and operate in culturally sensitive areas and sites, such as places of worship and schools, to establish temporary safe havens and refuge from Army forces operating within the urban environment.

Control the Tempo

3-11. Threats will try to achieve a decisive advantage by setting and controlling the tempo. To prevent the Army's entry into theater, threats may try to create a high operational tempo to take advantage of the inherent weaknesses in power projection operations outlined earlier. As other efforts deny entry, threats may seize the initiative, achieve surprise, and exploit the tempo differential by attacking with heavy conventional forces potentially possessing greater firepower and more rapid ground mobility than the Army's initial-entry forces.

3-12. If they cannot deny entry or end the conflict quickly, threats may use any preparations made in the initial high-tempo period to prolong the event, aiming to degrade U.S. or allied commitment. (Some threats may conduct operationally planning in terms of decades rather than years.) The complex nature of the urban environment slows operations conducted in, around, and over these areas. Threats may maximize this characteristic by fighting (conventionally or unconventionally) for key urban complexes and infrastructure, forcing friendly forces to operate within these areas. If Army operations focus on one or more urban areas, the overall campaign slows. However, even when UO make up only one component of a much larger campaign, they may consume valuable resources needed for other operations and delay the entire campaign. (See also the discussion of Tempo in Chapter 7.)

Tempo

The battle for Aachen, Germany, in the fall of 1944, developed during the U.S. First Army's offensive to breach the Westwall fortifications. Aachen, the ancient capital of Charlemagne, had symbolic political and psychological significance for the Germans and Americans. Furthermore, it was the first city on German soil to face an assault by the Allies. Consequently, the symbolic importance of this first major battle in Germany ensured bitter resistance against American attackers. The Germans surrendered only after the city was destroyed. Expected to take a few days, instead, the battle took weeks. Although the Army had achieved a clear tactical victory, the German defense of Aachen cost the First Army valuable time and resources, and delayed the planned attack to the Rhine River.

Change the Nature of the Conflict

3-13. Threats may attempt to change the fundamental nature of the urban conflict to exploit ambiguous or tenuous political-military objectives. Many nations must gain and maintain domestic popular support to use their armies for political objectives. The threat may attempt to change the nature of the conflict by modifying its strategy and tactics, the environment, or any combination, ultimately hoping to reduce friendly popular support. For example, introducing an urban terrorist threat to U.S. civilians or Soldiers not directly engaged in operations changes the nature of the conflict. This type of threat may not have been an initial consideration, and this change may reverse public support for the operation. Another example, growing U.S. coalition combat power may cause the threat to switch from open maneuver warfare to urban insurgency operations to avoid decisive combat with superior forces and achieve a stalemate. Originally expecting a quick solution or victory, the political leadership may now envision a longer deployment with less chance of lasting success.

Cause Politically Unacceptable Casualties

3-14. Threat forces may gain an advantage against superior friendly forces by capitalizing on a perceived weakness of many Western nations: the inability to endure continuous losses or casualties for other than vital national interests or losses for which they are psychologically unprepared. A secondary U.S. interest may equate to national survival on the part of a threat. Therefore, the threat (particularly with fanatical leadership) may willingly sacrifice excessive amounts of money, equipment, and people (soldiers and civilians) to achieve victory. Threats may attempt to weaken U.S. resolve and national will to sustain the deployment or conflict by inflicting highly visible, embarrassing, and if possible, large losses on Army forces, even at the cost of similar losses to themselves. Many threat forces will use UO to inflict mass casualties and destroy premier Army weapon and information systems. The physical characteristics of the urban environment support these ambush techniques. Light infantry or insurgents with readily obtainable, hand-held antiarmor weapons can effectively attack armored vehicles and helicopters, no matter how sophisticated, in an urban area.

ALLOW NO SANCTUARY

3-15. Threats will attempt to deny Army forces safe haven anytime and anywhere. Terrorism and insurgency may be some of the tactics used to deny sanctuary to Army forces. They will attack Army forces anywhere, particularly while operating in urban areas where the fear from being attacked from any quarter is often greater. Threats may be or employ state-sponsored or independent terrorists, well equipped and motivated to accomplish their assigned missions.

3-16. Military buildings, facilities, and installations in urban areas are particularly vulnerable to high-yield explosive munitions as well as other clever means to create large explosions. The close-in nature of urban areas, large populations, and high volume of vehicle traffic provide a good environment for target reconnaissance, explosives positioning (conventional or high-yield), and cover for an attack. These attacks will likely be preceded by extensive, careful reconnaissance, necessitating a solid friendly counterterrorism, counterinsurgency, and counterintelligence effort.

CONDUCT DECENTRALIZED AND DISPERSED OPERATIONS

3-17. To a certain extent, decentralized and dispersed operations are an integral part of all threat objectives. However, this concept warrants separate emphasis as an objective since threat forces will likely place great significance on it on future urban battlefields. Both dispersed and decentralized approaches seek to reduce threat vulnerabilities to air power and precision-guided munitions (PGM) while increasing their agility, flexibility, and overall maneuverability in an urban environment.

3-18. Urban terrain tends to fragment and separate forces that operate in it. Threat forces recognize this characteristic, accept it, and make it work to their advantage. They conduct operations from dispersed urban locations to reduce their vulnerability to friendly decisive operations and massed firepower. Although separated, threat forces will attempt to retain the ability to assemble and mass quickly so to strike as opportunities present themselves. Once threat forces complete the operation, they will return to separate locations to avoid potential counterattack. The fluidity and seemingly disjointed appearance of these UO threats will challenge friendly efforts to conduct templating and pattern analysis. Ambushes (air and ground) will be used to deny friendly ground and air reconnaissance of their dispersed locations.

3-19. Dispersed operations normally depend on good command and control to achieve synchronization and massed effects. Threat forces also understand the debilitating effects of the urban terrain on communications and the execution of operations. When they cannot mass their forces or effects, they will depend on decentralized operations to achieve their objectives. They will operate autonomously, guided only by a higher authority's purpose and intent. These operations make them even less vulnerable to massed attacks and PGM as smaller threat forces do not present an objective or target that will allow friendly decisive operations. Again, pattern analysis and templating, although essential, will be extremely difficult. This objective may prolongs the conflict but is central to achieving the other threat objectives.

3-20. When the threat employs decentralized operations, is effectively isolated, or both, Army and coalition forces are much more likely to be able to share lessons learned and emergent TTP than a dispersed threat. Threat isolation, either imposed by Army forces or as a result of the threat's desire to remain concealed from Army intelligence, surveillance, and reconnaissance (ISR) capabilities, reduces the threat's ability to freely and easily exchange information and quickly adapt. As learning organizations, this allows Army forces to develop dominance over a threat's weapons and tactics for longer, albeit temporary, periods of time. This ability to rapidly train and learn is an advantage the Army and coalition forces must maintain while denying the same to future threats.

URBAN THREAT TACTICS

3-21. Urban areas provide a potentially casualty-producing and stress-inducing environment ideally suited for using specific urban tactics. Moreover, urban areas can provide threats with an unmatched degree of cover and concealment from friendly ISR and firepower systems. Understanding how these tactics will be used often requires a concurrent understanding of the threat's culture, customs, language, and philosophical ideals. While active urban threats may vary widely, many techniques will be common to all.

Figure 3-2 outlines a set of tactics available to potential threats opposing mission accomplishment in urban areas. Army forces may use many of these tactics, except those that violate the law, ethics, and morals, to defeat urban threats. Moreover, using asymmetric means is not the sole domain of the threat. Army commanders can also leverage capabilities, create conditions, and plan operations to develop their own asymmetric advantages to accomplish the mission.

USE THE POPULATION TO ADVANTAGE

3-22. Many urban areas may be too large to evacuate completely (if at all). Even if desirable, a military force may have no place to safeguard and secure the inhabitants. Therefore, future UO may see large segments of the populace remain. Offensive and defensive operations may be constrained not only by the terrain, but also by the presence of many civilians. Army forces involved in urban stability and civil support operations will certainly conduct missions in and amongst the residents. These residents may restrict operations and, when gathered in large numbers, may (even without initial hostile intent) present a critical force protection issue for the commander.

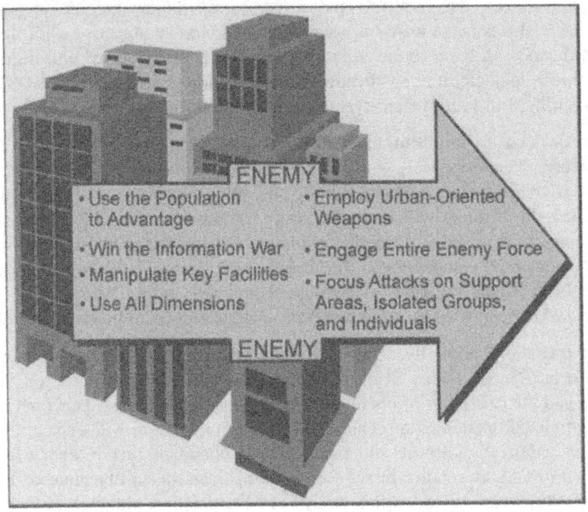

Figure 3-2. Urban threat tactics

Use as Key Terrain and Concealment

3-23. From the threat standpoint, the populace is similar to key terrain: the side that manages it best has an advantage. Threat forces may gain this advantage by using civilians as camouflage, concealment, and a means of deception. Guerrilla and terrorist elements may look no different from any other member of the community. Many foreign conventional and paramilitary troops—in addition to terrorists and insurgents—often have a "civilian" look. Western military forces originally adopted the clean-shaven and close-cut hair standards to combat disease and infection, but future opponents may not adhere to those standards. Instead, they may adopt grooming standards, civilian-looking clothing, and other "nonmilitary" characteristics to make themselves indistinguishable from the civilians. To avoid defeat or capture, threats will often throw down their weapons and attempt to blend into the urban population. Therefore, commanders must adopt effective populace and resource control measures to limit, identify, and track threat movements through the urban area of operation.

Identifying Threats from Noncombatants

During Russia's 1994-95 conflict with Chechnya, Russian forces had difficulty identifying Chechen guerrilla forces from Grozny's noncombatant population. Because their dress and ethnic appearance was identical to that of the urban populace, Chechen soldiers could freely walk around the city, suddenly disappear, and then abruptly reappear firing their weapons from basements, windows, or dark alleyways. To distinguish fighters from peaceful city dwellers, Russian forces began looking at men's shoulders to see if they were bruised (from firing weapons) and their forearms to see if there was burned hair or flesh (from the extraction of cartridges). They closely examined their clothing and smelled for gunpowder residue. To identify a Chechen artilleryman, Russian soldiers checked for glossy spots left by artillery and mortar rounds on the bends and cuffs of sleeves. They also turned pockets inside out to check for a shiny, silvery-leaden hue indicating the former presence of small arms ammunition. Russian forces also recognized a grenade launcher operator or mortar man from fibers and crumpled pieces of gun cotton on their clothing. U.S. Army commanders may need to develop similar, imaginative techniques and rapidly integrate technological means to identify and track threat forces such as—

- Military working dogs, ion spray, and electronic vapor tracers to locate explosives (or their former presence).
- Biometrics—measurable biological characteristics—such as fingerprints, voiceprints, facial scans, and retina scans.
- Electronic and chemical tagging devices.
- Monitoring hospitals for injuries that indicate participation in combat activities. For example, improper use of rocket-propelled grenade launchers may result in significant burns to the lower part of the body due to the back blast.

Gain Cover, Protection, and Increased Mobility

3-24. Threat forces may attempt to gain cover by using the urban inhabitants as human shields. With this increase in protection, they simultaneously increase their mobility. They recognize the Army's focus on developing and applying rules of engagement (ROE). They will take advantage of the restraining effects of international law and the Army ethical values to enhance their mobility in proximity to friendly positions. Knowing the Army's reluctance to cause noncombatant casualties and collateral damage, threats may operate in areas containing civilians and essential facilities to restrict the Army's use of massed or nonprecision firepower. For example, they may use school buses and ambulances to transport fighters and equipment about the urban battlefield. They may also employ "rent-a-crowds"—civilians paid or incited to demonstrate against military forces—armed only with sticks, stones, and Molotov cocktails (a potential asymmetric challenge). Threats can also hide military equipment within sensitive infrastructure—for example, caching weapons in places of worship, schools, or medical facilities. As Army and coalition forces potentially respond by targeting and firing upon these locations, threats will make use of these reactions for their possible propaganda value.

Make Moral Responsibilities a Weakness

3-25. Depending on their successes, threats may use these tactics and skillful information operations that attack national will and coalition sensitivities in an attempt to force the Army to establish more restrictive ROE. For example, a teenaged fighter coerced or enticed to fire a rifle or rocket-propelled grenade launcher can be quickly stripped of his weapon once dead and displayed at the local hospital as a child victim of coalition forces. Threat forces may also take advantage of the Army's moral responsibilities. By herding refugees into friendly controlled areas, threat forces try to make the civilians a burden on the Army's logistic and security resources. Similarly, some threats may attack to kill NGOs and force the remainder of them out of the area to achieve the same burden effect. At the same time, these threats hope to cause the population to lose confidence in the Army's (and potentially the government's) ability to protect and provide them essential services.

3-26. Threat forces, on the other hand, may not abide by international agreements, such as the Geneva conventions. Threats may not take prisoners unless they can be ransomed or made part of a local prisoner exchange. They may even execute friendly prisoners in front of the media to show their "strength" and, more importantly, to cause friendly forces to overreact and lose their legitimacy. Threat forces can then use such an overreaction to unite others with their cause. Pretending to be civilian noncombatants, they may also feign injury or pretend to need assistance (such as a disabled taxi) with the primary motive of luring Army forces into a deliberate ambush. While the threat may gain a short-term advantage by using ethical responsibilities as a weapon, the Army commander's strict adherence to moral principles will, in the long run, contribute greatly to threat defeat—particularly during long-term urban stability operations.

Acquire Intelligence and Logistic Support

3-27. Indigenous threat forces can normally use the local population for intelligence and logistic support far more effectively than can an alien army. Threat forces may manipulate local hires serving among U.S. Soldiers, such as those contracted by the Army for base operation purposes or translator duties. In addition, refugees moving through friendly controlled sectors may provide the threat with information on friendly dispositions, readiness, and intent. (Women and children may be specifically used for this purpose since U.S. Soldiers' cultural bias can create an incorrect perception that these noncombatants pose no threat to security.) Even friendly residents may become unwitting or unwilling informants, providing an enemy or a hostile with vital information on friendly activities, dispositions, and capabilities. However, a threat employing particularly cruel, abusive, or repressive measures may easily turn certain groups in the urban area against them, even when they share a common history, culture, and ethnicity with the civilians. This is more likely in those areas with higher population densities.

3-28. Threat forces may also seek to use some nongovernmental organizations (NGOs). They may try to obtain relief supplies either through the organizations' legitimate relief operations or as a target for theft. Some organizations may even be fronts for weapons, food, ammunition, money, and fighters. For example, during Russia's second conflict in Chechnya (1999-2000), documents purportedly found in Grozny by the Russians listed nations such as Sudan, Nigeria, Niger, and Ivory Coast as sending fighters to Chechnya under the guise of the International Islamic Relief Organization. (Chechen fighters also disguised themselves as Red Cross workers.) This deception increases the need for strict security and force protection measures, close coordination with NGOs operating in urban areas, and closer monitoring of suspect organizations' activities by civil affairs personnel.

WIN THE INFORMATION WAR

3-29. Threat forces will try to win the information war as much as they will directly oppose Army UO. Threat urban campaigns need not be tactical military successes. They need only to weaken legitimacy and make the opposition's campaign appear unpalatable to domestic, U.S., and world support. As a critical part of a threat's overall information operations, they will use the ever-present media to tell their story. Portable video cameras, commercial radios, and cellular telephones, available and easily concealed, will be as important to many threat actors as weapons and ammunition. Internet access, already firmly established in many urban areas, provides the means to easily disseminate threat propaganda, misinformation, and disinformation through web sites and electronic mail. The internet also allows threats to conduct their own distance learning and education. Threats will make good use of the instantaneity and global reach of television and radio broadcasts, particularly local and international media, to obtain information of immediate tactical value and broadcast their own operational and tactical guidance. Hackers, covered and concealed in the interior spaces of the urban area, may gain access to U.S. sites to manipulate and obtain information to the threat's advantage.

Information and the Media

The media coverage of the urban battle for Hue, South Vietnam, although only one of hundreds of different attacks of the Tet Offensive, affected the will of both the American people and their political leadership. On January 31, 1968, two North Vietnamese Army (NVA)/Vietcong (VC) regiments and two sapper battalions, moving rapidly and with the element of surprise, attacked and seized part of the walled city (Citadel) of Hue. It was the third largest city in South Vietnam, the former capital of a united Vietnam, the capital of Thua Thien province, and a spiritual and cultural center. Initially intending to hold the city for seven days, the NVA/VC retained portions of the city for approximately three weeks against determined U.S. and South Vietnamese attempts to retake it.

Hue marked a revolution in the coverage of war by modern media. It was the first time Americans could sit at home and watch an ongoing battle on the evening news. One of the most intense and savage battles of the Vietnam conflict, it was televised every evening for almost a month. Although the battle for Hue was a tactical victory for the United States, the North Vietnamese clearly achieved strategic success by searing the American consciousness with the high costs of urban warfare. Had U.S. leaders made winning the information war a central part of the overall campaign plan—for example, exposing the American people to the NVA's brutality by publicizing the civilian executions in Hue—civilian support for the war may have been bolstered and a different outcome achieved. See Chapter 7 for a more detailed account of the battle for Hue.

Manipulate Key Facilities

3-30. Threat forces will attempt to identify and quickly seize control of critical components of the urban area to help shape the operational environment. Urban telephone exchanges can provide threats with simple and reliable communications that can be easily secured with off-the-shelf technologies. Existing public address and loudspeaker systems can be used to incite riots or openly command and control dispersed threat forces. (Local language may provide threats with a measure of communications security.) Sewage treatment plants and flood control machinery can be used to implement WMD strategies or to make sections of the urban area uninhabitable. Media stations significantly improve the information operations abilities of the controlling force. Power generation and transmission sites provide means to control significant aspects of civilian society over a large area. Insurgents will attempt to destroy key facilities and critical infrastructure in order to shape the populace's perception that the existing government is incapable of providing for their safety and meeting their basic needs.

Use All Dimensions

3-31. Threats will think and operate throughout the depth, breadth, and height (including supersurface and subsurface areas) of the urban environment. However, threats will often choose to fight from within interior spaces rather than out in the streets and other open areas. Conventional lateral boundaries will often not apply as threat forces control some stories of the same building while friendly forces control others. The use of all three dimensions of an urban area makes identification, reporting, and targeting of enemy locations more difficult for forces accustomed and trained to acquire and engage targets in primarily two dimensions. Because of these target identification and acquisition difficulties, threats will likely use decoys in the urban area to cause erroneous assessments of their combat capabilities, strength, and disposition. They will also attempt to use these decoys to absorb expensive and limited precision-guided munitions as well as cause misallocation of other critical Army resources.

3-32. Roofs and other supersurface areas provide urban threats with excellent observation points and battle positions above the maximum elevation of many weapons. Shots from upper floors strike armored vehicles in vulnerable points. Basements and other subsurface areas also provide firing points below many weapons' minimum depressions and strike at another weakness in most armor. Sewers and subways may provide covered and concealed access throughout the area of operations.

EMPLOY URBAN-ORIENTED WEAPONS

3-33. Whether purpose-built or adapted, many weapons are more useful in an urban environment while others may have significant disadvantages. Urban threat weapons are much like the nature of urbanization and the urban environment: inventive and varied. Many threats will integrate widely available off-the-shelf technologies into their weapon systems and armed forces. However, sniper rifles and small, man-portable, fire-and-forget weapons and demolitions and other improvised explosive devices (IEDs), to include suicide and car bombs, will likely dominate the urban environment. Figure 3-3 lists examples of threat weapons favored in UO.

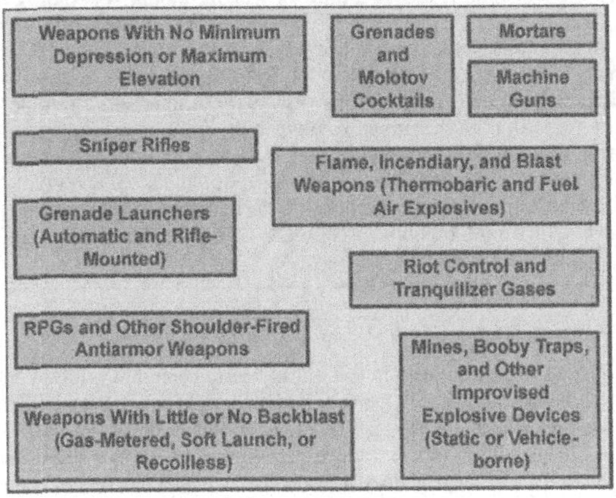

Figure 3-3. Favored threat weapons

ENGAGE ENTIRE ENEMY FORCE

3-34. Threats may attempt to keep all or significant portions of Army forces engaged in continuous operations to increase their susceptibility to stress induced illnesses. UO, by their nature, produce an inordinate number of combat-stress casualties. Continuous operations exacerbate this problem. Threat forces that employ this tactic will often maintain a large reserve or achieve respite by hiding among the civilian population to minimize the psychological impacts on their own forces.

3-35. To accomplish this key tactic, threat UO will likely involve decentralized maneuver, precision fires, and simultaneous operations involving unconventional and special purpose forces and guerilla, insurgent, and terrorist tactics. Threats may also attempt to bait Army forces by ambush or other means with a primary focus of attacking response or reaction forces. Overall, threat forces will take advantage of any exposed weakness in weapons, equipment, and TTP and they will cultivate the flexibility to engage in battles as favorable opportunities present themselves.

FOCUS ATTACKS ON SUPPORT AREAS, ISOLATED GROUPS, AND INDIVIDUALS

3-36. To supplement the previous tactic, threat forces will seek to target support areas, small groups, leaders and their headquarters, and unaware Soldiers. Their focus on resupply and convoy operations, casualty evacuation, and other sustainment activities, coupled with the compartmented terrain, navigational challenges, and multiple three-dimensional avenues of approach often makes these locations and Soldiers more susceptible to surprise raids and ambushes. Attacks on these areas and groups are conducted to erode the Army's ability to sustain UO, to inflict maximum casualties, and to induce psychological stress. These attacks can be mitigated by careful, regular evaluation of choke points and other restrictive terrain, regular awareness training for units and individuals operating in or transiting through potential incident-prone areas, well-planned and protected convoys, and thorough after-action analysis of incidents.

NEGATIVE EFFECTS OF URBANIZATION

3-37. Many urban areas are the engines for increased industrialism and economic growth as an expanding population provides the labor for manufacturing and service needs. However, rapid and inadequately planned growth can result in undesirable consequences. Uncontrolled urbanization may result in an infrastructure and economic base unable to support the growing population. A large transient, ill-housed, and idle population in a close geographic space may produce strife. Classes, cultures, ethnic groups, and races that might otherwise peacefully coexist can clash under the stress of survival. Uncontrolled urban growth has resulted in the negative effects listed in figure 3-4. In many urban stability and civil support operations, these may be the primary "threats" to mission accomplishment.

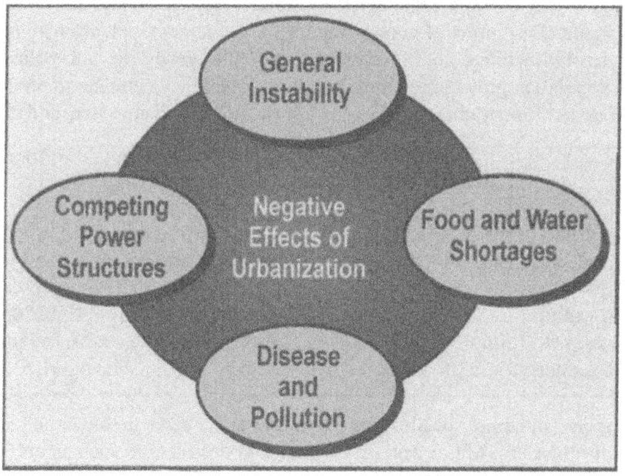

Figure 3-4. Negative effects of urbanization

3-38. Not all urban areas prevail as inherently unstable or hotbeds for unrest. Urban growth due to migration may remove sources of conflict, or it may provide the catalyst for violence. Commanders should recognize the possible effects of uncontrolled urbanization. During their intelligence preparation of the battlefield (IPB), they should determine if these conditions exist. Throughout mission analysis and the development of courses of action, commanders should consider the impact (if any) on their operations. At the same time, they must recognize that UO may create similar problems that may affect the current operation as well as the overall campaign.

General Instability

3-39. Urbanization can enhance stability by generating industrialization and economic growth resulting in more jobs, a higher overall standard of living, and an educated, relatively satisfied populace. However, the population dynamics associated with urbanization can also have an opposite, destabilizing effect. Radical population growth may create overcrowding and generate or aggravate resource and quality of life issues. Intense and destructive competition for employment, housing, and social status may develop in this climate of economic deprivation. The inability of some governments to handle these problems—

- Makes their urban areas potential sources of unrest.
- Increases the likelihood of the Army's involvement in stability operations.
- Complicates all operations conducted in such an urban environment.

Weak civil administrations—perhaps weakened by the effects of previous UO—have difficulty controlling their society, safeguarding their military armaments, and preventing their urban areas from serving as sanctuaries to terrorists, insurgents, and criminal organizations.

3-40. Urbanization in developing countries warrants more concern. Intense migration and growth, coupled with the forced closeness of people once separated by the rural countryside, may stress already struggling institutions, hasten conflict, and lead to overall instability. Commanders must understand that UO, depending on the operation, may either cause massive population movement out of or into urban areas.

3-41. Urban areas with a large youth population may also help to generate conditions for instability. Rural-to-urban migrants tend to be relatively young. In 1999, Cairo, for example, had more than 40 percent of its population younger than 15 years. Young urban populations generate enormous demands for social resources, primarily education and jobs. Even a strong urban economy may fold under the economic expectations of a tremendous influx of young migrants. Disorder and violence may result as hostiles (many nonstate actors) easily mobilize and manipulate the idle young to act politically and criminally. Urbanization and population growth are more dangerous when they combine to produce a cohort of young urban dwellers separated from traditional social controls, such as village elders and clan leaders.

Cultural and Religious Instability

The 1992 bombing of the Babri Masid Mosque in Ayodya, India, enflamed an already intense cultural and religious rivalry between Hindus and Muslims and led to rioting throughout many Indian urban areas. Of the 1,500 who died in conflicts and riots, almost 95 percent died in urban areas. The violence struck Ahmedabab and Bombay most seriously, with related acts of murder, gang rapes, and arson occurring months after the destruction of the mosque.

3-42. Ethnic, religious, and other social issues may become the vents for anger and frustration produced by the high tension of urban life. Major acts of violence and destruction, such as occurred in 1992 in India, can directly threaten a nation's security. Army forces may have to conduct large-scale, stability operations to promote peace and protect national interests. In these cases, all levels of command will be particularly concerned with maintaining impartiality and perceived legitimacy.

Food and Water Shortages

3-43. Rapid urbanization, primarily in developing nations, may lead to severe food shortages that could influence Army forces (or lead to their use). Such shortages may cause instability, massive migration, revolts, or increased support of armed opposition groups. Armed factions may target NGOs that supply aid as a means of furthering dissatisfaction among the populace. In effect, food may become a weapon. Deployed troops may need to provide or support humanitarian food aid networks to keep the humanitarian situation from escalating.

3-44. However, planners must be wary of the unintended consequences of well-intentioned humanitarian assistance operations. For example, providing free, safe food may alleviate starvation, but could also undercut the local agricultural system by reducing demand in the market. If the food is distributed through urban centers, urbanization could increase, further reducing the food supply and adding to the existing strains on the infrastructure. Unless they are trying to use them as a means to separate combatants from noncombatants, commanders should normally use centralized feeding centers as a last resort. Instead, Army forces should bring the food closest to the population to encourage civilians to stay in their homes. If safe areas or camps are created, they should be designed for use over as short a time as is feasible. The general rule should be to return the urban population to their homes as soon as possible. Army forces conducting domestic support or foreign humanitarian assistance operations that cannot maintain the safe food supplies may find the frustrations and hostility of the local population focused on them.

3-45. Water shortages (and quality) are becoming a serious problem in many regions. Sewage, industrial waste, other forms of pollution and deliberate contamination pose threats to existing water supplies. Commanders operating in an urban environment need to know the water supply origins and its treatment, purification, distribution, and vulnerabilities. Before beginning operations, commanders must know if they are providing water for the noncombatants as well as their own forces. Across the spectrum of operations, controlling and protecting a limited water supply is, or may become, an essential operational consideration during UO.

Food and Water Shortages

Countries as varied as Indonesia and Algeria exported their food surpluses only two generations ago but now import up to two-thirds of their basic staples. This cycle has resulted in many countries, which once exported agricultural products, facing the growing cost of imports to feed their urban populations. Estimates predict that in the 2010 timeframe, at least 65 countries (including 30 of Africa's 51 countries) may depend completely on food imports. For some countries, the outlook is even worse. The Democratic Republic of Congo (Zaire), once a net food exporter, faces severe malnutrition and utter starvation for a large portion of its population.

Other estimates predict that by 2025, 2.7 to 3.5 billion people will live in water-deficient countries. Even now in urban areas such as Jakarta, Indonesia, for example, urban authorities cannot provide the necessary water to support its growing population. Less than half of Jakarta's population is supplied by the urban area's water infrastructure; the remainder of the area's residents rely on water from wells and other groundwater supplies that lie only a few feet below the surface. These alternate sources need to be sufficiently boiled or residents risk increased disease. In the northern part of Jakarta, land is actually sinking as the urban area's groundwater supply is overused by its residents. Houses on low-lying areas have to be rebuilt every few years to keep them above sea level. Meanwhile, saltwater from the Java Sea is seeping into the land polluting the remaining groundwater and further exacerbating the situation.

Disease and Pollution

3-46. Urban areas frequently spawn epidemics; therefore, widespread disease may pose a significant threat to Army forces that operate there. In many developing nations, rapid urbanization has occurred without a corresponding upgrade, expansion, or even development of adequate sewage and water systems. Some urban areas have only one toilet for every 750 people or more. In these areas, hundreds of thousands live much as they would in poor villages, yet so confined as to ensure high transmission rates for airborne, waterborne, sexually transmitted, and contact-transmitted microbes.

3-47. In urban areas lacking adequate trash and waste management infrastructure, insect-spread diseases proliferate. Mosquitoes that breed in polluted water, open water tanks, and irrigated urban gardens carry malaria and dengue fever—the leading causes of sickness and death from infectious disease in Latin America and Africa. The problem compounds with growing numbers of bacteria resistant to various antibiotics, a shortage of trained medical personnel, inadequate or insufficient medical facilities and supplies, and unclean agricultural and food-processing practices.

3-48. Pollution also creates critical health problems in developing areas and a potential health risk for intervening Army forces. For example, urban areas in China have recorded five to ten times the levels of sulfur dioxide found in the air of urban areas in the developed world. In parts of Poland, toxic waste has so polluted the land and water that ten percent of the babies have birth defects. Pollution may cause immediate health problems but more often, the insidious effects appear months or years after exposure. As discussed earlier, UO may contribute, either intentionally or unintentionally, to an increase in pollution. Destruction of industrial complexes that use, produce, and store hazardous material may produce toxic gas and smoke pollutants and create significant health concerns to exposed Soldiers and civilians. Soldiers forced to fight within the confines of subsurface areas such as sewers and drainage systems further increases their exposure to contaminants.

3-49. Commanders should initiate force health protection planning early, including analysis of the medical threat and other critical medical information requirements during the IPB process. A medical surveillance system should monitor the daily status of Army personnel throughout the operation. In preparation, all personnel must receive a predeployment medical examination. This exam establishes an accurate baseline health status of the force and ensures that Army forces do not introduce new diseases to an urban area, possibly exacerbating the situation. Conversely, Soldiers not immune to native viruses or possessing a weakened immune system due to continuous operations and the stress associated with UO may put Army forces at a significant disadvantage. An outbreak of plague during an operation would have an effect similar to a chemical or biological attack. The closer that Army forces operate to civilians (the humanitarian assistance operations conducted in Port-au-Prince, Haiti, and Mogadishu, Somalia, for example), the more probable that these situations may occur. (See Chapter 9.)

COMPETING POWER STRUCTURES

3-50. Many groups can exist that become strong enough to rival the power of the governing officials and eventually turn the urban area into a system of divergent and competing power structures. These groups can consist of insurgent forces, a merchant class or an economic elite, criminal organizations, or some other significant source of power such as religious organizations (see Chapter 10), clans, or tribes (figure 3-5). In the absence of a legitimate authority, armed factions headed by "warlords" may vie to fill the power void. Sometimes these groups or organizations, normally at odds with each other, may form alliances to achieve specific goals. Commanders must recognize, identify, and understand these alternate urban power bases and, if necessary, develop engagement strategies to neutralize or harness them to accomplish the Army mission. Four of these groups, not previously covered or covered in a subsequent chapter, are discussed in greater detail below.

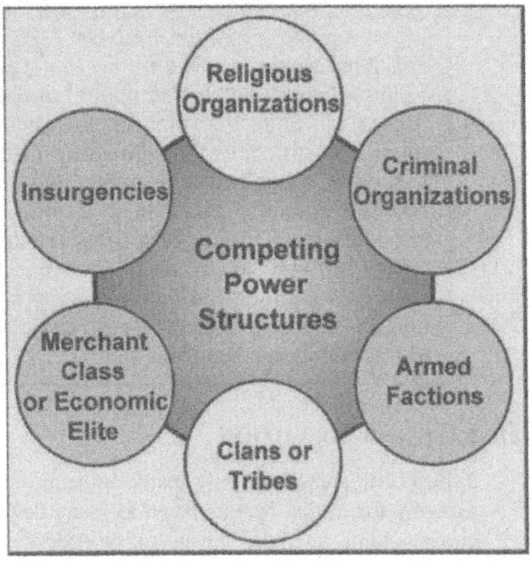

Figure 3-5. Competing power structures

Urban Insurgencies

3-51. As urban migration increases in the developing world, rural guerrillas appear to follow. This transition of insurgencies from rural to urban areas occurs because urban areas offer a rich field of targets for insurgent attacks. People immediately notice any disruption of urban infrastructure, thus having great propaganda value. A concentrated urban population whose allegiance and support are critical to the success of an insurgency is often more susceptible to propaganda, political organization, and terrorism. Insurgents can easily arrange mass demonstrations using available communications facilities, both overt and covert. Travel is effortless and large urban populations provide cover and concealment. On the whole, urban areas, particularly those suffering the effects of general instability outlined earlier, may provide a fertile environment for guerrillas to apply formerly rural insurgent strategies. However, even with a primarily rural-based insurgency, operations in urban areas will often be essential to the insurgents' strategy as they offer them distinct opportunities to disrupt, discredit, and demoralize the government (see FM 3-05.20 and FMI 3-07.22).

[illegible]

Merchant Class or Economic Elite

3-52. Urban areas normally possess a merchant class or economic elite as part of their social structure. In some urban areas, they may carry more power than the local or central state government. They may isolate themselves physically and socially from the sprawling poor yet wield enormous power over the country's political and economic activities. The degree of economic separation between the merchant class and the poor may be small but still socially or politically significant.

3-53. In a vastly impoverished area where the economy of the urban area is severely disrupted, the merchant class will often continue to operate and function and, as a result, achieve a measure of influence. To continue to operate under acute economic turmoil, they may form alliances in criminal organizations and secure loyalties within the government. Outside resources introduced into a crisis area (such as food, water, fuel, and pharmaceuticals) take on increased value, may replace currency as the medium for exchange, and often become the means to amass and hold wealth. One primary way to obtain wealth is to steal it.

3-54. In some turbulent situations that lead to the need for stability operations, commanders may harness the power of the merchant class as a force for peace and stability instead of one that uses crime to achieve economic goals. For example, in a relief situation, instead of competing with the merchant class by distributing food directly to the needy and possibly creating an environment of looting and black marketeering, it may be possible to monetize food. Food assistance from donor governments could be sold to merchants at an attractive price so they have a reliable source of supply. This could, in turn, create a healthy economic system and separate merchants from criminals and gangs.

Criminal Organizations

3-55. Organized criminal groups have grown common in urban areas; have also become an important part of the urban social structure (gangs for example); and can exert considerable influence on governments, people, and military forces conducting UO. These organizations can threaten the successful completion of urban operations both directly and indirectly. Criminals and criminal organizations may directly target Army forces, stealing supplies and extorting money or contracts. Criminals may also violently confront Army forces during mission execution. During many UO, particularly during or following combat, civil disturbances, or large natural disasters, looting (organized or unorganized) may become of critical concern.

For a portion of the population, crime may be the only reliable source of income and resources to support their families. Therefore, UO will often require a combined law enforcement and military response.

3-56. Some large criminal organizations relying on international connections often have better resources and equipment than their insurgent counterparts. (A prime example is drug traffickers.) Their large financial resources, long-reaching connections, and ruthlessness provide them the means to corrupt or intimidate local officials and government institutions. Their tactics parallel those of insurgents. They have developed an intuitive cultural understanding of slum neighborhoods and the ability to lure civilians into criminal activities. They have also mastered the management of mobs. They recruit teenagers and young adults in their efforts against rivals and authorities, just as insurgents muster armies from the youth of rural villages. In many developing nations, there exists an alliance between insurgents and organized criminal groups. In these alliances, the insurgents defend the criminals and the criminals fund the insurgents.

3-57. Excessive criminal activity can undermine the commander's efforts to establish a sense of security among the remainder of the urban populace. Commanders will rely on the ability of their military police (MP) units to assess and understand the potential influence that crime and criminal organizations may have on UO. MPs—in coordination with local police—will need to quickly identify law enforcement gaps to include dysfunctional police organizations and stations and inadequate prison systems. Urban police records may be an important intelligence database to assist in this assessment and may be particularly useful in counterterrorism and counterinsurgency operations. Success in ensuring that law and order is maintained or reestablishing law and order will directly affect the population's perceptions and support of Army operations. Also, MP intelligence operations will contribute immeasurably to the overall IPB effort and aid in understanding of the complexities of the urban society (see FM 3-19.50).

Crime and Criminal Organizations

Crime and poverty plague urban areas such as Rio de Janeiro, Brazil's second largest urban area, and would affect military operations conducted in their limits. Rio has some of the nation's highest negative urban indicators: the largest number of slum dwellers (1 million), the highest murder rate (1 of 700 residents per year), and the highest kidnapping rate (4 per week). In 1989, the homicide rate of the urban area was three times higher than New York City's, and the rate of urban violence has continued to rise. Therefore, law enforcement management may be a critical issue for Army forces operating in urban environments similar to that of Rio de Janeiro.

However, criminal elements or organizations may not always work against Army commanders. They can be co-opted or influenced to serve friendly objectives. For example, during World War II the U.S. Navy worked covertly with the Mafia in New York City to secure the New York harbor from German U-boats believed to be torpedoing ships there. The Mafia controlled most dock activities in New York harbor and was perfectly positioned to monitor other subversive waterfront activity. This capability provided needed information to the Navy for its counterintelligence and security tasks. New York civil authorities, therefore, agreed to permit a Navy-Mafia alliance to operate at the port for the greater good of the country. Although the Mafia was not the preferred ally of the Navy, it had the capability to protect U.S. ships and the interest (patriotism) to help in the war effort. In those circumstances, the temporary alliance worked (see also the civilian threat discussion in Appendix B).

Armed Factions

3-58. A characteristic of some stability operations has been the deterioration or complete collapse of political authority in the country or urban area in crisis. In some cases, "warlords" or the leaders of various armed factions have attempted to fill the power vacuum (see Appendix C for an example). These individuals often have no particular claim to legitimacy. Their power issues from their weapons, not necessarily from their political skills, human services provided, or popular consent (although they have some popular support to remain in their relative position of authority). Others, however, may already have a substantial base of support (for example, a religious base) and then seek to achieve their goals through violent means. In dealing with the leaders of these armed factions during stability operations, it may appear that there are three general courses of action. The first is to aggressively locate and target the leaders and their forces militarily. The second is to ignore them completely, and the third is to meet and work with them diplomatically. Depending on METT-TC, commanders may reduce some of the risks involved in these first three courses of action by combining the last two to create a fourth approach. Nevertheless, the technique chosen must clearly support political and military objectives and lead to the desired end state.

3-59. In some cases, the leaders of armed factions pose a direct threat to Army forces, the urban populace, and the overall accomplishment of the mission. Therefore, it may become necessary to target them and their forces for capture or destruction. Yet, commanders should generally avoid framing the success of a major operation on the death or capture of one or more individuals. Even with increased ISR capabilities, commanders may find an individual manhunt difficult or impossible to achieve. Instead, it may be better to seek to isolate warlords and their forces militarily, while simultaneously seeking to marginalize their influence through a combination of political, economic, and informational means. This does not necessarily mean that commanders must abandon the search if the death or capture the faction leader will contribute to achieving the mission. However, commanders must maintain the capability to strike with speed and precision if the threat leader's location is discovered in the clutter of the complex urban environment.

3-60. In some situations, commanders may attempt to marginalize warlords by ignoring them. However, refusal to acknowledge warlords may increase the threat to Army forces and NGOs. Their militias may attack Army forces to achieve recognition or simply due to misunderstanding or inherent friction between armed forces. On the other hand, dealing with them may provide legitimacy to the exclusion of other elements of the urban population such as professional groups (for example, doctors or teachers), religious leaders, and traditional clan or tribal chiefs—which may have a greater claim to legitimacy and better form the foundation for a reconstituted urban society.

3-61. A compromise between the last two options may offer the best chance for success. In this fourth course of action, commanders can generally recognize these warlords and work with them diplomatically to decrease the risk of armed conflict and civilian casualties; however, this recognition can be kept at staff levels to avoid bestowing any perceived legitimacy on them. Instead, commanders themselves may visibly meet the other elements of society that have a more legitimate claim to political, social, or economic leadership. Inevitably, commanders may need to meet with warlords. In those circumstances, clan or tribal elders, and others who represent traditional authority, should attend the meetings. Moreover, commanders should ask for and give deliberate consideration to their opinions above those of the leaders of the armed factions.

This page intentionally left blank.

Chapter 4

Understanding the Urban Environment's Effects on Warfighting Functions and Tactics

War is, above all things, an art, employing science in all its branches as its servant, but depending first and chiefly upon the skill of the artisan. It has its own rules, but not one of them is rigid and invariable. As new implements are devised, new methods result in its mechanical execution; but over and above all its mechanical appliances, it rests upon the complex factors of human nature, which cannot be reduced to formulas and rules.

Captain Francis V. Greene, 1883

Commanders of major operations should understand the potential effects that the urban environment may have on warfighting functions. They should also understand the possible effects that the urban environment may have on lower-level tactics to properly plan, prepare, and execute major operations that may include UO. Otherwise, commanders may ask their subordinates to achieve effects, accomplish objectives, or adhere to a timetable that is unsupportable due to the constraints imposed by the urban environment. Commanders and their staffs must do more than simply understand the impossible; rather, they must apply the art and science of warfighting to the urban environment and determine what it will take to make it possible.

WARFIGHTING FUNCTIONS

4-1. Understanding the potential effects of the urban environment on warfighting functions permits the urban commander to better visualize his operational environment. With this appreciation, he can conduct a more thorough assessment and thereby determine the most efficient and effective means of employing Army forces. The staff should be intimately familiar with effects in their area of expertise and use that knowledge to understand the problem and develop creative and innovative solutions to achieve their commander's intent.

WARFIGHTING FUNCTIONS

- Intelligence
- Movement and maneuver
- Fire support
- Protection
- Sustainment
- Command and control

INTELLIGENCE

4-2. The intelligence function facilitates understanding of the threat and the environment. The urban environment affects this critical function in many ways. Impacts of the environment on the intelligence function include degraded reconnaissance capability; more difficult IPB process; and an increased importance of credible HUMINT (including the contribution of local civilian liaisons), and an established intelligence reach capability. The Army forces' response to these effects can result in timely, accurate, and actionable intelligence that permits the effective application of other warfighting functions to the mission within the urban environment (see FM 2-0).

Degraded Reconnaissance and Surveillance Capability

4-3. The physical environment creates a major challenge to the intelligence function. The man-made construction in the urban areas provides nearly complete cover and concealment for threats. Although improving many sensor capabilities cannot penetrate the subsurface facilities and much of the space within supersurface areas. The mass of buildings can also defuse electronic signatures. Tall buildings shield movement within urban canyons from aerial observation except from directly overhead. Urban threats may be less technology dependent and may thwart some signals intelligence efforts simply by turning off their radios and using messengers. Threat forces will likely use elements of the civilian telecommunications infrastructure for C2. These systems may include traditional landline phones, cellular telephones, and computer-to-computer or Internet data communications. Most urban telecommunications systems use buried fiber or cables or employ modern digital signaling technology. Such systems are difficult to intercept and exploit at the tactical level.

4-4. From the above, it is evident that these characteristics make it more difficult for the intelligence function to use electronic means to determine threat dispositions and, in offensive and defensive UO, identify decisive points leading to centers of gravity. While the environment limits some typical collection methods, all enemy electronic and human activity creates some form of observable signature and exposes the enemy to potential collection. Seeking ways to take advantage of these vulnerabilities will provide the commander an information advantage over his opponent.

Challenging IPB Process

4-5. The sheer complexity of the environment also challenges the intelligence function. The intelligence function applies the IPB process to the urban environment in accordance with Army doctrine (see Appendix B). With more data points for the IPB process to identify, evaluate, and monitor, this application becomes more demanding. The human and societal aspects of the environment and the physical complexity primarily cause this difference. Relationships between aspects of the environment, built on an immense infrastructure of formal and informal systems connecting the population to the urban area, are usually less familiar to analysts. Thus, the urban environment often requires more specifically-focused intelligence resources to plan, prepare for, execute, and assess operations than in other environments.

4-6. Compounding the challenges is the relative incongruity of all urban environments. No two urban areas are alike physically, in population, or in infrastructure. Thus, experience in one urban area with a particular population and pattern of infrastructure does not readily transfer to another urban area. Any experience in UO is valuable and normally serves as a starting point for analysis, but the intelligence function cannot assume (and treat as fact) that patterns of behavior and the relationships in one urban area mirror another urban area. The opposite is as likely to hold true. The intelligence function will have to study each urban area individually to determine how it works and understand its complex relationships.

4-7. Each characteristic of the urban environment—terrain, society, and infrastructure—is dynamic and can change radically in response to UO or external influences. Civilian populations pose a special challenge to commanders conducting UO. Civilians react to, interact with, and influence to varying degrees Army forces. Commanders must know and account for the potential influence these populations may have on their operations. Intelligence analysts must revisit or continuously monitor the critical points looking for changes, relationships, and patterns.

4-8. The actions of Army forces will affect, positively or negatively, their relationship with the urban population and, hence, mission success. NGOs may deliberately or inadvertently influence civilians. The intelligence function can monitor and predict the reactions of the civil population. However, accurate predictive analysis of a large population requires specific training and extensive cultural and regional expertise.

Increased Importance of Human Intelligence

> ***HUMINT*** is the collection of foreign information—by a trained HUMINT collector—from people and multimedia to identify elements, intentions, composition, strength, dispositions, tactics, equipment, personnel, and capabilities. It uses human sources as a tool, and a variety of collection methods, both passively and actively, to collect information. FM 2-0

4-9. The intelligence function adjusts to the degradation of its technical intelligence gathering systems by increasing emphasis on HUMINT in UO. HUMINT operations may be the primary and most productive intelligence source in UO. In urban offensive and defensive operations, HUMINT gathers information from refugees, immigrants and former citizens (especially previous civil administrators), civilian contractors, and military personnel who have operated in the area. Credible intelligence of this type can help meet requirements, provide more detail, and alleviate some of the need to physically penetrate the urban area with reconnaissance forces. In many urban operations where HUMINT is the primary source of intelligence, acting on single-source reporting is a constant pitfall. Yet, situations may arise where commanders must weigh the consequences of inaction against any potential negative consequences resulting from acting on uncorroborated, single-source information. (See also the Human Capabilities discussion under the Urban ISR portion of Chapter 5.)

4-10. In urban stability operations, HUMINT identifies threats and monitors the intentions and attitudes of the population. A chief source of information contributing to the development of accurate HUMINT, particularly at the tactical level, is reconnaissance forces—especially small-unit dismounted patrols. Urban reconnaissance forces and patrols should be thoroughly and routinely debriefed by unit intelligence personnel to obtain information that aids in developing a clearer picture of the threat and the urban environment. Reliable and trustworthy HUMINT is particularly important in foreign internal defense, counterterrorism, and support to counterdrug operations. Leaders must organize intelligence resources appropriately, and learn and apply valuable techniques, such as pattern and link analysis (see FM 34-3). Additionally, Soldiers, as part of reconnaissance and patrolling training, should be taught to handle captured documents, weapons, material, and equipment as legal evidence much like military and civilian police. Proper "evidence" handling is often a critical intelligence concern in counterterrorism and counterinsurgency operations.

Developing Local Liaisons

4-11. Whenever Soldiers encounter the urban populace, the resulting interaction may become an important source of information the commander can use to answer questions about the threat and the urban environment. While military intelligence units are the primary collectors and processors of HUMINT, commanders are not likely to have enough trained HUMINT Soldiers to satisfy their requirements—particularly in a larger urban environment and during longer-term stability operations. Therefore, commanders may need to cultivate and establish local civilian associations to provide relevant information for decision making and to support the overall HUMINT effort.

4-12. Urban liaisons can be developed through positive civil-military interaction with the urban populace. Critical information may be acquired through interface with the urban leadership (both formal and informal), administration officials, business owners, host-nation support workers, inhabitants along a unit's patrol route, pedestrians at a checkpoint, civilian detainees, or any other human source willing to volunteer information to Army forces or who respond positively to tactical questioning. (Noncombatants are never coerced to provide information.) Commanders may also direct unit leaders to conduct liaisons with specific local leaders and key members of the community to obtain command directed information. Critical information may also come from other U.S. and coalition forces and intelligence organizations operating near or within the commander's AO. To this end, commanders should ensure that collectors operating in an urban area coordinate and deconflict activities and, if possible, outbrief subordinate, geographically-responsible commanders with any relevant information that may affect their current operations. Any relevant information obtained incident to civilian liaison activities should be routinely provided to intelligence staffs not only to gain assistance in verifying the credibility of the information but to share the information with all affected echelons and units.

4-13. NGOs operating in urban areas can also be especially beneficial resources for credible and relevant information about the urban environment. (However, they are generally not a good source for information about the threat since providing such information can violate their neutrality thereby making it difficult for them to achieve their humanitarian aid objectives.) During the 1999 fighting in Kosovo, for example, the Red Cross provided the most accurate figures regarding the number of Kosovar refugees, helping U.S. and other coalition forces to estimate the appropriate level of support required to handle their needs. In addition to a developed understanding of the current needs of the local urban populace, NGOs may also have—

- A network of influential associations.
- Historical archives.
- Extensive understanding of the urban infrastructure.
- Key knowledge of political and economic influences.
- A keen awareness of significant changes in the urban environment.
- Insight into the current security situation.
- Up-to-date web sites and maps.

4-14. While productive civilian associations may become long term, they should not be confused with HUMINT source operations. Only trained HUMINT personnel can recruit and task sources to seek out threat information. Information obtained from these societal connections is normally incidental to other civil-military relationships. For example, as part of infrastructure repair in an urban stability operation, a commander may be instrumental in obtaining a generator for a local hospital. Within the context of this relationship, the commander may develop a rapport with one or more of the hospital's administrators or health practitioners. These civilians may be inclined to provide valuable information about the threat and the urban environment—often on a continuing basis. In any civil-military relationship, however, commanders ensure that the information provided is not tied to promises of assistance or that such assistance is in any way perceived as a means to purchase civilian loyalty.

4-15. Commanders also understand that repeated interaction with any one individual may put that individual and his family in danger from threat forces. Before this potential danger becomes a reality, they should refer their civilian connections to trained HUMINT personnel for protection and continued exploitation. In addition to civilian protection considerations, commanders may also deem it necessary to turn their civilian associations over to trained HUMINT collectors anytime during the relationship if they consider the information that the contact is providing (or may provide) is credible, relevant, and—

- Provides essential threat information on a repetitive basis.
- Helps answer higher-level CCIR.
- Affects operations in another AO.
- Requires interrogation or monetary compensation to obtain.

However, turning a liaison over to trained HUMINT teams does not necessarily preclude maintaining a continued, albeit a more guarded, relationship with the individual.

4-16. In developing these civilian liaisons essential to understanding the urban environment, commanders must avoid the distinct possibility of conducting unofficial source operations by non-HUMINT Soldiers. While prohibited by regulatory guidance (see Defense Intelligence Agency Manual 58-11 for requirements and restrictions for source operations), such actions also run the additional risks of—

- Obtaining unevaluated information that cannot be crosschecked with and verified by other sources of information.
- Creating inequities that result from illegally rewarding contracts which can undermine HUMINT Soldiers who are constrained by intelligence contingency fund regulations.
- Disrupting ongoing HUMINT operations when different sources are seen to be treated differently by non-HUMINT Soldiers vice HUMINT Soldiers.
- Providing non-HUMINT and HUMINT Soldiers with the same information potentially leading to a false confirmation of information.

- Increasing the likelihood that untrained Soldiers may fall victim to a threat deception and misinformation.

Established Intelligence Reach

4-17. Understanding the complex urban environment, particularly the infrastructure and the society, will require more sources of information beyond a unit's organic intelligence capabilities. Therefore, commanders will have to make extensive use of intelligence reach to access information and conduct collaboration and information sharing with other units, organizations, and individual subject matter experts. Before deployment (and throughout the operation), units should establish a comprehensive directory of intelligence reach resources. These resources may include national, joint, Army, foreign, commercial, and university research programs. (Prior to deployment for OPERATION IRAQI FREEDOM, some units established contacts within the local community outside their bases such as police, fire department, and government officials that expanded their reach once in theater—particularly for information regarding civilian infrastructure and urban administration.) Once deployed, intelligence reach includes effective information sharing and collaboration among adjacent units, sister services, coalition partners, and other governmental and nongovernmental agencies operating in the area. Effective information sharing and collaboration requires common network analysis software and databases to be used among all Army forces and, if possible, other governmental agencies.

Movement and Maneuver

4-18. Army maneuver forces—infantry, armor, cavalry, and attack aviation—move to achieve a position of advantage. Entire urban areas or specific threat forces located within may be isolated from affecting other operations and then bypassed. However, when the situation requires entering the urban area to accomplish the mission—such as when bypassed urban threat forces interdict critical lines of communications (LOCs), the environment will significantly affect the Army's ability to move and maneuver (see figure 4-1). These negative effects include canalization, compartmentalization, and increased vulnerability. However, tactics and techniques equip Army forces to overcome these challenges and maneuver successfully. One tactic, effective combined arms task organization, includes an increased dismounted maneuver capability combined with armor and combat engineers, continuous operations, and technological enhancements.

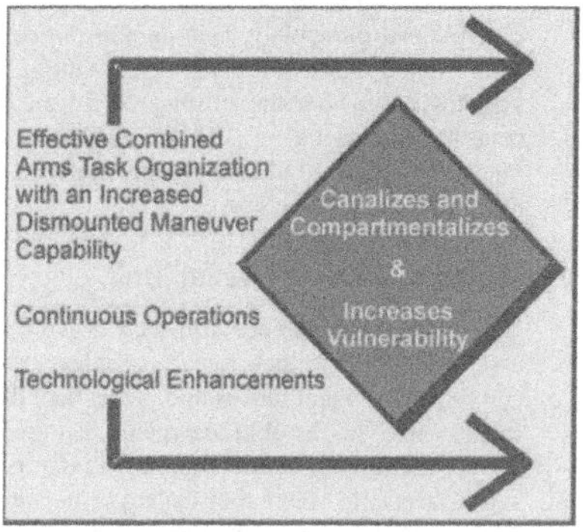

Figure 4-1. Urban maneuver challenges and means to overcome them

4-19. In all UO, mobility operations may allow civilian traffic and commerce to resume, letting the urban area return to some semblance of normalcy (often a critical objective). In stability operations, mobility often focuses on keeping lines of communications open and reducing the threat of explosive hazards to Soldiers and civilians. In civil support operations, mobility may focus on removing storm debris or reducing obstacles caused by destroyed property.

Canalization and Compartmentalization

4-20. The urban terrain will often canalize and compartment forces—and their fires—moving and maneuvering through it. Buildings pose obstacles to both mounted and dismounted movement, forcing units to be canalized along streets. The buildings also block movement between streets, thus compartmenting units. Fires are canalized into open and unmasked areas where vision is unobstructed, producing concentrated fire zones and areas, especially at road intersections and in front of defended positions. Hence, changing directions, repositioning committed forces, reinforcing forces in contact, bypassing threats, and maneuvering to the threat flank become extremely difficult. Units often breach obstacles to help solve this problem. Using helicopters to quickly move forces, both forward into contact and to rear areas as part of repositioning, also permits Army forces to overcome some terrain constraints.

4-21. The canalized and compartmented effects can slow movement and maneuver considerably. However, slowed movement also results from the physically demanding soldier tasks required in an urban environment. Soldiers operate dismounted across rubble and hard surfaces. Operating in three dimensions, they constantly move up the supersurface areas of building interiors and down into basements, cellars, and other subsurface areas. They breach many obstacles and use upper-body strength, ropes, and ladders to scale heights. The inability to see into the next room, floor, or building magnifies stress. The resulting fatigue slows the overall rate of Army force movement and maneuver.

Increased Vulnerability

4-22. The urban environment increases the vulnerability of Army forces executing movement and maneuver in offensive, defensive, and stability or civil support operations. Both the physical terrain and the urban population provide threat cover and concealment. Air movement and maneuver is vulnerable for many of the same reasons. In offensive or defensive operations, enemy forces can remain undetected in buildings and in position to ambush Army forces. Consistent with METT-TC, Army forces should clear buildings along maneuver routes prior to mounted movement along those axes. Failure to clear routes (and effectively mark cleared portions) may expose mounted movement to ambush at close range. Movement back across streets and obstacles may be difficult particularly if the element of surprise was essential in the initial crossing or breach. The same buildings also provide cover and concealment to enemy air defense capabilities—particularly man-portable air defense systems that can be fired from multiple positions hidden amongst the clutter of fires, lights, smoke and dust, and easily concealed and transported in civilian vehicles throughout the urban area. In all operations, but especially stability operations, civilians can conceal threat elements. The threat can then initiate offensive operations against Army forces from close range and where ROE will hamper applying combat power. Thus, maneuver through a dense population can be a high-risk operation.

Combined Arms Task Organization

4-23. Effective combined arms task organization ensures that forces are task organized with infantry—the essential building block for all organizations conducting UO. Infantry protects mounted elements as the combined arms unit moves and maneuvers through the urban area. (In some urban situations, mechanized infantry may not be able to provide dismounted support beyond support to its own vehicles—tanks may require the support of additional light infantry). The infantry destroys the enemy in buildings, bunkers, and subsurface areas where they cannot be defeated by mounted forces and prevents infiltration of threat forces back into hard-won urban terrain. Field artillery aids in dismounted and mounted (to include air) maneuver by suppressing known and suspected enemy positions with precision fires. Attack aviation make best use of their standoff capabilities (see later fire support discussion) and aircraft speed to conduct running and diving fires; in UO, hovering fire is generally avoided. Armored elements protect Soldiers from small arms fire and destroy or suppress enemy positions with precise, direct fire. Carefully protected artillery may also be used in this direct fire role. Armored forces and attack helicopters also can facilitate maneuver through shock action that can have a psychological effect, particularly against less well-trained threats and, in discrete instances, hostile crowds. (Although, commanders consider that the "intimidation" value of any method erodes quickly with its repetitive use.)

4-24. Combined arms also ensure that combat engineers support dismounted maneuver by assisting in covered and concealed maneuver through buildings and off exposed streets. In addition to combat engineers, explosive ordnance disposal teams, military police, chemical personnel, and other with essential expertise to conduct mobility missions significantly reduces mobility and maneuver challenges (see FM 3-34.2). Urban buildings are often obstacles to movement and mobility. Combat engineers, trained and equipped for UO, can turn these obstacles into an advantage by breaching them with "mouse holes" made by explosives, sledgehammers, bulldozers or armored vehicles, or high-strength (diamond or carbide-tipped) cutting devices. These breaches permit dismounted movement through buildings under both cover and concealment.

4-25. Combat engineers must also be trained and equipped to facilitate mounted mobility in the urban environment. Generally, buildings restrict mounted movement to the compartmented and canalized streets. Threats can block streets with roadblocks ranging from sophisticated log and concrete cribs reinforced with antitank and antipersonnel mines to the expedient use of cars, buses, and trucks to create obstacles. Combat engineers must be capable of breaching these obstacles to maintain the coherence of the combined arms team (mounted and dismounted). Combat engineers should be forward, often task organized down to platoon level, and have the expertise and equipment to rapidly reduce point obstacles. It even may be necessary that every armored vehicle (or section of two vehicles) be task organized with an associated engineer squad and vehicle. Because of the increased density and hardness of many urban building and construction materials, heavy engineer equipment (such as the D9 bulldozer) will be in great demand to accomplish mobility (and countermobility, and survivability) functions in an urban environment. However, commanders will need to consider increased protection requirements and the availability of equipment transport to move these slower-moving engineer assets around the urban battlefield.

4-26. A major difference of UO combined arms is in proportion and organization. Although based an accurate METT-TC assessment, UO will often require an increased proportion of dismounted infantry and engineer capabilities; armor may not be required in the same high numbers. As significant, genuine combined arms urban operations are required at lower tactical levels where small, well trained and led units will dominate. Commonly, company level will require true combined arms capability and may include combat engineers, military intelligence, reconnaissance, and artillery. Combined arms teams can then form at platoon and squad levels. Because of this, larger units will need more CA, military intelligence, and combat engineers than habitually attached for combat in more open or less restrictive terrain.

4-27. In determining the appropriate task organization, commanders consider a suitable span of control for subordinate commanders. They also consider the potential of dissipating a unit's combat power, capabilities, and synergy by breaking a unit up into smaller units in an attempt to ensure subordinate maneuver units have a complete combined arms capability. For example, an additional engineer battalion may be task organized to a brigade combat team. In turn, the brigade combat team may task organize this battalion into engineer companies under the control of their subordinate maneuver battalions. If this type of organization continues, maneuver companies may end up with an engineer platoon and, in the end; maneuver platoons each having an engineer squad. Ultimately, a combined arms capability may have been established at lower tactical levels but the parent maneuver unit (in this example, the brigade combat team) may have lost the ability to conduct larger engineer operations without having to re-task organize and potentially disrupt current operations and established relationships. As a guide, urban commanders may consider task organizing to create combined arms organizations at lower tactical levels when operations are predominately offensive or defensive, and bringing those assets back under their own control when the operation transitions to predominately stability or civil support operations.

Continuous Operations and Technology Enhancements

4-28. Two other means to improve Army forces' ability to move and maneuver in urban terrain is through continuous operations and the leveraging of technology, such as the Army's night operations capability. Historically, urban battles have been fought primarily during daylight because of technological limitations and fatigue. By utilizing night vision technologies, accurate situational understanding, a common operational picture (COP), training, and rotated units, Army forces can defeat threats who use the same soldiers in day and night operations and who are less well-equipped and adept at night operations. Night operations are also a means of mitigating the air defense threat against air maneuver. Continuous operations through night maneuver with fresh forces are challenging, but it can overcome many advantages that a stationary force has against maneuver in the urban environment. However, commanders should also consider that streetlights, fires, background illumination (as well as dark building interiors without ambient light), the increased heat absorption of many urban structures, and the skillful use of searchlights by threat forces may limit the effectiveness of night vision devices and make thermal imagery identification difficult.

Countermobility

4-29. Countermobility capabilities in urban terrain are also an essential consideration in all UO. In defensive operations, commanders use countermobility capability to control where the enemy moves in the urban area. Repositioning defensive forces in the urban area can be difficult and obstacles are essential to limiting the enemy's maneuver options. During offensive operations, countermobility protects exposed flanks and air assaulting forces from counterattack. In stability and civil support operations, countermobility operations may take the form of constructing barriers to assist in populace and resources control at critical urban locations.

FIRE SUPPORT

4-30. The fire support function includes the collective and coordinated use of several means to attack targets in the urban area (see Appendix D for joint capabilities). These means include target acquisition data, indirect fire weapons, rotary- and fixed-wing aircraft, offensive information operations (IO), and other lethal and nonlethal means. The urban environment affects these components of the fires function and their employment.

Target Acquisition

4-31. Target acquisition in an urban environment faces several challenges. First, forces have difficulty penetrating the urban environment's increased cover and concealment using sensors and reconnaissance. Acquiring targeting information and tracking targets throughout the depth of the urban area may prove challenging. Moving personnel or vehicular targets are normally easiest to acquire. However, the cover and concealment provided by urban terrain gives moving targets short exposure times requiring firing systems to act rapidly on targeting data. Targeting of opposing indirect fire units by acquisition radar may work more effectively in urban terrain because of the necessary high angles of indirect fire. The urban environment presents similar difficulties for battle damage assessment.

4-32. Targeting challenges are met by innovatively integrating reconnaissance capabilities. These capabilities include SOF, cavalry, unmanned aircraft systems, and aerial observers as well as the standard reconnaissance assets. More artillery systems may need to be used to ensure the responsiveness (rather than the weight) of fires. Positioning numerous artillery systems reduces the dead space (as discussed below) and permits units to establish more direct sensor-to-shooter links.

The Targeting Process

4-33. Heightened concerns for collateral damage will require that commanders pay particular attention to their targeting process. This process ensures that all available combat power, both lethal and nonlethal (including offensive IO), is effectively integrated and synchronized to accomplish the mission. Commanders ensure that techniques and procedures are in place, rehearsed, and understood by all members of their staffs. Additionally, the C2 system must be responsive and agile; otherwise, an elusive and adaptable threat will likely disappear before units can employ the appropriate weapon systems. In an urban area, even 10-digit grid coordinates may not be sufficient to accurately identify targets as buildings may be connected to each other—often throughout the entire block. Target locations, in addition to grid coordinates, may need to routinely include the street address, number of stories, shape, color, or any other distinguishing characteristic essential for ground and air forces to achieve targeting precision. A common urban reference system with graphics, reference points, and other control measures adequate for both ground and air forces may also help facilitate identification of targets and facilitate the rapid clearance of fires.

4-34. Greater concerns exist for the safety and health (environmental matters) of the urban populace and the protection of critical infrastructure and cultural structures. Hence, CA and staff judge advocates (see Chapters 5 and 10 respectively) will play a greater role for the expert advice they can provide regarding these elements of the urban environment. Nonetheless, all members of the staff ensure that operations minimize collateral damage. That responsibility does not end with identifying potential collateral damage; the goal, as always, is successful mission accomplishment. Again, staffs are guided by the commander's intent and work to develop courses of action that incorporate collateral damage concerns (short- and long-term) yet accomplish the mission. This requires a keen understanding of the legal issues and both friendly and enemy weapon systems' effects in an urban environment.

Urban Effects on Fire Support Systems

4-35. Both the physical and human components of the urban area affect how units use fire support weapon systems. The physical aspects of the urban environment, such as the heights and concentration of buildings, may cause significant masking and dead space (see figure 4-2). Intervening buildings that stand three or more stories tall hinder close indirect fire support. Tall buildings can potentially mask several blocks of area along the gun-target line of artillery. (For low-angle artillery fire, dead space is about five times the height of the building behind which the target sits.) The potential for collateral damage to adjacent buildings may also prevent engagement with artillery. Such damage might cause noncombatant and friendly troop casualties and unintentional (and unwanted) rubbling. Commanders can offset these effects by carefully placing artillery positions, repositioning artillery as targets change, and using mortars. Mortars have a steep angle of fall and short minimum ranges as a high-angle alternative to field artillery fire. (In comparison to artillery, dead space for mortar fire is only about one-half the height of the building.) For fixed-winged aircraft, precision munitions and weapons with low explosive yields and near-vertical impact angles resulting in bomb burial can significantly reduce collateral damage. Collateral damage concerns may also cause commanders to—

- Maintain approval authority for some sensitive or protected targets (churches or mosques, for example) at higher echelons of command.
- Restrict attacks to certain times of day.
- Give warning prior to an attack so that noncombatants can evacuate the area.
- Incorporate indigenous forces into the operation.
- Abort an attack unless the required level of precision effects can be achieved.
- Prepare specific branches and sequels to the IO plan to inform the populace why the collateral damage was justified. These plans may include filming or otherwise documenting the operation to thwart threat propaganda and claims of excessive collateral damage.
- Develop and rehearse detailed staff battle drills that address clearance of fires in an urban environment.

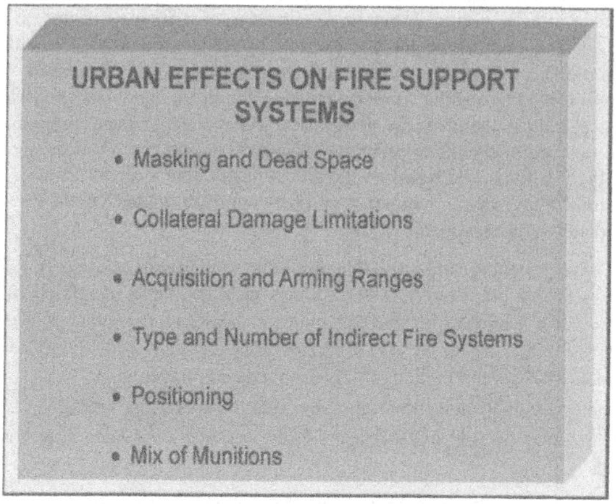

Figure 4-2. Urban effects on fire support systems

4-36. Vertical structures interrupt line of sight (LOS) and create corridors of visibility along street axes. The result is thereby shortened acquisition and arming ranges for supporting fires from attack helicopters and subsequently affected engagement techniques and delivery options. Pilots maintain LOS long enough to acquire targets, achieve weapons delivery solutions, and fly to those parameters. For example, tube-launched, optically tracked, wire-guided heavy antitank missile systems require 65 meters to arm. Similarly, the Hellfire missile requires at least 500 meters to reliably arm and stabilize on the intended target. Thus, attack helicopters firing from longer ranges actually improve the probability of a hit. Poor weather and heavy smoke and dust rising from urban fires and explosions may hinder target identification, laser designation, and guidance for rotary- and fixed-winged aircraft. Poor air-to-ground communications may also hinder effective use of airpower. The close proximity of friendly units and noncombatants requires units to agree on, thoroughly disseminate, and rehearse clear techniques and procedures for marking target and friendly locations. The ability for ground units to "talk-on" aircraft using a common reference system described earlier helps expedite aerial target acquisition (and helps mitigate potential fratricide). FM 3-06.1 details other aviation TTP in an urban environment.

4-37. The urban environment also affects the type and number of indirect fire weapon systems employed. Commanders may prefer high-angle fire because of its ability to fire in close proximity to friendly occupied buildings. Tactically, commanders may consider reinforcing units in UO with mortar platoons from reserve units. This will increase the number of systems available to support maneuver units. Unguided Multiple Launch Rocket Systems (MLRSs) may be of limited use in urban areas due to their exceptional destructive capabilities and the potential for collateral damage. However, commanders may use unguided MLRSs to effectively isolate the urban area from outside influence. Commanders may also employ field artillery systems as independent sections, particularly self-propelled systems, in the direct-fire role; decreasing volume and increasing precision of artillery fire helps minimize collateral damage. While discretely applying the effects of high-explosive and concrete-piercing munitions, these self-propelled systems take advantage of the mobility and limited protection of their armored vehicles.

4-38. The urban area may also affect the positioning of artillery. Sufficient space may not exist to place battery or platoon positions with the proper unmasked gun line. This may mandate moving and positioning artillery in sections while still massing fires on specific targets. Commanders must protect artillery systems, particularly when organized into small sections. Threats to artillery include raids and snipers. Therefore, maneuver and firing units will have to place increased emphasis on securing their positions and other appropriate force protection measures.

4-39. The mix of munitions used by indirect fire systems will change somewhat in urban areas. Units will likely request more precision-guided munitions (PGM) for artillery systems to target small enemy positions, such as snipers or machine guns, while limiting collateral damage. Currently, only conventional tube artillery, not mortars, has this capability. However, large expanses of polished, flat reflective surfaces common in urban areas may degrade laser designation for these munitions (as well as attack helicopter PGM). The vertical nature amplifies the geometrical constraints of many precision munitions. Remote designators need to be close enough to accurately designate but far enough away not to be acquired by the PGM during its flight path. PGMs based on the global positioning system (for instance, guided MLRS or the Air Force's joint direct attack munitions) or other optically guided PGMs may be more effective if urban terrain hinders laser designation.

4-40. The urban environment greatly affects the use of nonprecision munitions. Building height may cause variable time fuses to arm prematurely. Tall buildings may also mask the effects of illumination rounds. Units may choose not to use dual-purpose conventional munitions if (similar considerations apply to Air Force cluster bombs)—

- The enemy has several building floors for overhead protection.
- Dismounted friendly units need rapid access to the area being fired on.
- Large numbers of civilians will operate in the target areas soon after combat operations have ceased.

4-41. Depending on the building construction, commanders may prohibit or limit illumination, smoke, and other munitions because of fire hazards. In particular instances, they may specifically use them for that effect. Structure fires in an urban area are difficult to control and may affect friendly units. Conventional high-explosive munitions may work best against concrete, steel, stone, and other reinforced structures. When not used in the direct-fire role, a greater mass of indirect fire is often required to achieve desired effects. Commanders balance firepower and collateral damage since the rubbling caused by massive indirect fires may adversely affect a unit's ability to maneuver and provide a threat with additional cover and concealment.

4-42. Nonlethal weapons, munitions, and devices can help commanders maintain the desired balance of force protection, mission accomplishment, and safety of noncombatants by expanding the number of options available when deadly force may be problematic. As additional nonlethal capabilities are developed, they are routinely considered for their applicability to UO. In determining their use and employment, commanders, in addition to any previous experience at using these weapons, munitions, and devices, consider—

- **Risk**. The use of nonlethal weapons in situations where lethal force is more appropriate may drastically increase the risk to Army forces.
- **Threat Perspective**. A threat may interpret the use of nonlethal weapons as a reluctance to use force and embolden him to adopt courses of action that he would not otherwise use.
- **Legal Concerns**. Laws or international agreements may restrict or prohibit their use (see Chapter 10).
- **Environmental Concerns**. Environmental interests may also limit their use.
- **Public Opinion**. The apparent suffering caused by nonlethal weapons, especially when there are no combat casualties with which to contrast it, may arouse adverse public opinion.

PROTECTION

4-43. The protection function includes those tasks and systems that preserve the force so that commanders can apply maximum combat power. Preserving the force includes enhancing survivability and properly planned and executed air and missile defense as well as defensive IO (see IO discussion in Chapter 5) and chemical, biological, radiological, nuclear, and high-yield explosive (CBRNE) counterproliferation and consequence management activities (see CBRNE discussion in Chapter 9).

Survivability

4-44. Survivability in the urban environment is a significant force multiplier. Properly positioned Army forces can take advantage of the increased survivability afforded by the physical terrain. Even a limited engineer effort can significantly enhance the combat power of small Army forces. In stability operations, properly planned and constructed survivability positions can enable small groups of Soldiers to withstand the assaults of large mobs, sniping, and indirect fire. Well-protected support bases are often critically essential to minimizing casualties during long-term stability operations and can become a key engineer task.

4-45. While executing major combat operations or campaigns, in particular defensive operations, well planned and resourced engineer efforts can enhance the survivability characteristics of the urban area. These efforts, though still requiring significant time and materiel, can establish defensive strong points more quickly and with greater protection than can be done in more open terrain. Skillfully integrating the strong point into the urban defense greatly increases the overall effectiveness of the defense disproportionately to the number of forces actually occupying the strong point (see Chapter 8).

4-46. Commanders increase survivability by ensuring that all Soldiers have necessary protective equipment and are trained and disciplined in their use. In addition to standard equipment such as helmets, gloves, boots, and chemical protective overgarments, commanders should ensure, as necessary, availability of other protective equipment and material such as—

- Body armor.
- Goggles or ballistic eye protection.
- Knee and elbow protectors.
- Riot control equipment such as batons, face masks, and shields.

- Barrier material such as preformed concrete barriers, wire, sandbags, and fencing material.
- Up-armored or hardened vehicles.
- Fire extinguishers and other fire-fighting equipment.
- Immunizations.

4-47. The Army's urban survivability operations can become complex if the Army is tasked to support survivability operations for civilians. Such operations can range from constructing civil defense shelters or evacuating the population to assisting the population in preparing for or reacting to the use of weapons of mass destruction. However, Army forces are not organized or equipped to support a major urban area's requirements as well as its own mission needs. Normally, Army forces can render this type of support only as a focused mission using a unique, specially equipped task organization.

Air and Missile Defense

4-48. The air and missile defense protects the force from air surveillance and air and missile attack. This system uses—

- The careful massing of air and missile defense combat power at points critical to the urban operation.
- The proper mix of air defense weapon and sensor systems.
- Matched (or greater) mobility to the supported force.
- The integration of the air defense plan into the overall urban operation.
- The integration of Army systems with those of joint and multinational forces.

4-49. Properly planned and executed air and missile defense prevents air threats from interdicting friendly forces and frees the commander to synchronize maneuver and other elements of firepower. Even in a major combat operation or campaign, the enemy will likely have limited air and missile capabilities and so seek to achieve the greatest payoff for the use of these systems. Attacking Army forces and facilities promises the greatest likelihood of achieving results, making urban areas the most likely targets for air and missile attack.

Rotary- and Fixed-Winged Aircraft

4-50. Enemy rotary-wing aircraft can be used in various roles to include air assault, fire support, and combat service support. Some threats may use unmanned aircraft systems to obtain intelligence and target acquisition data on friendly forces. Increased air mobility limitations and targeting difficulties may cause enemy fixed-wing aircraft to target key logistics, C2 nodes, and troop concentrations outside the urban area, simultaneously attacking key infrastructure both in and out of the urban area.

Increased Missile Threat

4-51. The intermediate range missile capability of potential threats has increased to be the most likely air threat to an urban area. Urban areas, particularly friendly or allied, make the most attractive targets because of the sometimes-limited accuracy of these systems. By firing missiles at an urban area, a threat seeks three possible objectives:

- Inflict casualties and materiel damage on military forces.
- Inflict casualties and materiel damage on the urban population.
- Undermine the confidence or trust of the civil population (particularly if allied) in the ability of Army forces to protect them.

4-52. If facing a missile threat, commanders conducting UO work closely with civil authorities (as well as joint and multinational forces) to integrate the Army warning system with civil defense mechanisms. Similarly, Army forces may support urban agencies reacting to a missile attack with medical and medical evacuation support, survivor recovery and assistance in damaged areas, and crowd control augmentation of local police forces. Before such an attack, Army engineers might assist and advise urban officials on how to construct shelters.

Increased Security of Assets

4-53. When defending against an air or missile threat in a neutral or hostile urban environment, air defense assets are concerned with security. Separating air defense locations from high population and traffic centers, as well as augmenting these positions with defending forces, can prevent or defeat threat efforts to neutralize them. Additionally, increased density of UO means increased concentration of all friendly and enemy systems engaged in air and counter-air operations. This density may increase friend and foe identification challenges, air space management challenges, and the overall risk in the conduct of air operations. Finally, limited air defense assets, difficulties in providing mutual support between systems, potential mobility limitations, and other effects of the urban environment increase the need for (and effectiveness of) a combined arms approach to air defense (see FM 44-8).

SUSTAINMENT

4-54. The sustainment function incorporates support activities and technical service specialties, to include maximizing available urban infrastructure and contracted logistics support. It provides the physical means with which forces operate. Properly conducted, the sustainment function ensures freedom of action, extends operational reach, and prolongs endurance. Commanders conducting sustainment to support full spectrum operations must understand the diverse logistic requirements of units conducting UO. They must also understand how the environment (to include the population) can impact sustainment support. These requirements range from minimal to extensive, requiring Army forces to potentially provide or coordinate all life support essentials to a large urban population.

4-55. Commanders and staffs consider and plan for Army sustainment operations that are based in a major urban area. These operations are located in major urban areas to exploit air- and seaports, maintenance and storage facilities, transportation networks, host-nation contracting opportunities, and labor support. These operations are also UO. The commander gains additional factors to consider from basing the sustainment operation in an urban environment. See Chapter 10 for a detailed discussion of urban sustainment considerations.

COMMAND AND CONTROL

> *Fighting in a city is much more involved than fighting in the field. Here the "big chiefs" have practically no influence on the officers and squad leaders commanding the units and subunits.*
>
> Soviet General Vasili Chuikov
> during the 1942-43 Battle for Stalingrad

4-56. The command and control function is the related tasks and systems that support the commander in exercising authority and direction. The urban environment influences both the commander and his C2 system (which includes INFOSYS). The leader's ability to physically see the battlefield, his interaction with the human component of the environment, and his intellectual flexibility in the face of change all impact the mission. The C2 system faces difficulties placed on the tactical Internet and system hardware by the urban environment, by the increased volume of information, and by requirements to support the dynamic decision making necessary to execute successful UO.

Unity of Command

4-57. Although severely challenged, the principle of unity of command remains essential to UO. However, the number of tasks and the size of the urban area often require that Army forces operate noncontiguously. Noncontiguous operations stress the C2 system and challenge the commander's ability to unify the actions of his subordinates, apply the full force of his combat power, and achieve success. To apply this crucial principle in an urban environment requires centralized planning, mission orders, and highly decentralized execution. The method of C2 that best supports UO is mission command (see FM 6-0). Mission command permits subordinates to be innovative and operate independently according to clear orders and intent as well as clearly articulated ROE. These orders and ROE guide subordinates to make the right decision when facing—

- A determined, resolute, and adaptive threat.
- A complex, multidimensional battlefield.
- Intermittent or complete loss of communications.
- Numerous potentially hostile civilians close to military operations.
- The constant critique of the media and military pundits.

4-58. Decentralized execution allows commanders to focus on the overall situation—a situation that requires constant assessment and coordination with other forces and agencies—instead of the numerous details of lower-level tactical situations. Fundamentally, this concept of C2 requires commanders who can accept risk and trust in the initiative, judgment, and tactical and technical competence of their subordinate leaders. Many times, it requires commanders to exercise a degree of patience as subordinate commanders and leaders apply mental agility to novel situations.

Political and Media Impact

4-59. Commanders of a major operation consider how the need to maintain a heightened awareness of the political situation may affect their exercise of C2. A magnified political awareness and media sensitivity may create a desire to micromanage and rely solely on detailed command. Reliance on this method may create tactical leaders afraid to act decisively and with speed and determination—waiting instead for expected guidance from a higher-level commander. Threats may capitalize on this hesitation by conducting operations faster than Army forces can react. Mission orders that express the overarching political objectives and the impact of inappropriate actions, combined with training and trust, will decrease the need for detailed command. Leaders must reduce a complex political concept to its simplest form, particularly at the small-unit level. Even a basic understanding will help curtail potentially damaging political actions and enable subordinates to make the often instantaneous decisions required in UO—decisions that support military and political objectives.

Commander's Visualization

> *I heard small-arms fire and RPG explosions and felt shrapnel hit the vehicle.... Land navigation at this time was impossible; every time I tried to look out, I was thrown in a different direction.... At this time, I was totally disoriented and had not realized we were on our own.*
>
> Captain Mark Hollis
> "Platoon Under Fire"

4-60. Leaders at all levels need to see the battlefield to lead Soldiers, make effective decisions, and give direction. Sensors and other surveillance and reconnaissance assets alone cannot provide all the information regarding the urban environment that commanders will need. The focus of lead elements narrows rapidly once in contact with a hostile force limiting their assessment to the local area. Therefore, tactical commanders will not be able to observe operations from long, stand-off ranges. Their personal observation remains as critical in urban areas as elsewhere and helps to preclude commanders from demanding their subordinates accomplish a task or advance at a rate inconsistent with the immediate situation. In urban offensive and defensive operations, seeing the battlefield requires that commanders move themselves and their command posts forward to positions that may be more exposed to risk. Thus, commanders modify their C2 system capabilities to make them smaller, reduce their signature, and increase their mobility. Because of the greater threat to C2, security efforts may be more intense.

4-61. In stability operations, commanders often intervene personally to reassure the urban population and community and faction leaders about the intentions of Army forces. In these type operations, threats may attack leaders to gain the greatest payoff with the least expenditure of resources. Commanders carefully evaluate risk and potential benefits of such exposure. These risks however, cannot stop them from seeing the battlefield, personally intervening in situations as appropriate, and leading their Soldiers.

4-62. Commander's visualization also requires having an accurate understanding of friendly and enemy locations, detailed maps, other appropriate intelligence products, and INFOSYS that accurately depict the urban environment and help establish a COP. The reliability of these items is as important to planning major operations as it is to tactical-level operations. The commander of the major operation ensure that subordinate tactical-level commanders have the necessary products to achieve accurate situational understanding and dominate the urban environment as subordinate commands often lack the personnel or assets to develop these products. Frequently, satellite or aerial imagery is requested to compensate for the drastic changes that can occur due to UO, natural disasters, and outdated or imprecise maps. (Even maps developed and maintained by urban area's administrative activities may not be up-to-date. Extensive and continually expanding shantytowns, for example, may not be mapped at all. Maps may have even been purposefully distorted or critical detail intentionally omitted. The systems used to transliterate some languages such as Arabic and Chinese to Anglicized alphabets often result in the same location being spelled several—and frequently considerably different—ways. Maps may also assign names to features that are completely different than those used by locals to refer to them.)

4-63. Other critical intelligence products needed in the COP include overlays or gridded reference graphics. (Whenever possible, gridded reference graphics should conform to standard military grid reference system formats to reduce the probability of error when entering target coordinates into targeting systems that use global positioning systems.) These products should be developed and distributed to all participants prior to the UO. Overall, their focus should be on ease of reference and usefulness for all forces—ground and air (see Appendix B). Overlays and graphics can also portray important societal information or urban infrastructure, such as—

- Religious, ethnic, racial, or other significant and identifiable social divisions.
- Locations of police, fire, and emergency medical services and their boundaries or zones of coverage.
- Protected structures such as places of worship, hospitals, or other historical and culturally significant buildings or locations.
- Underground subway, tunnel, sewer, or water systems.
- Bridges, elevated roadways, and rail lines.
- Electrical generation (to include nuclear) and gas storage and production facilities and their distribution lines.
- Water and sewage treatment facilities.
- Telephone exchanges and television and radio stations.
- Toxic industrial material locations.

Mental Flexibility

4-64. Commanders conducting UO must remain mentally flexible. Situations can change rapidly because of the complexity of the human dimension. Typical of the change is a stability operation that suddenly requires the use of force. Commanders must be capable of quickly adjusting their mental focus from a noncombat to combat situation. Equally important is dealing with populations during combat operations. Consequently, commanders must also be capable of rapidly adjusting plans and orders for sudden stability or civil support tasks that emerge during or soon after a combat mission. In developing their vision, commanders must consider the second- and third-order effects of UO.

Information Systems

4-65. The urban environment also challenges INFOSYS that support the commander, especially communications. Urban structures, materials, densities, and configurations (such as urban canyons) and power constraints associated with man-portable radios significantly degrade frequency modulation (FM) communications. This causes problems at brigade-level and below where commanders rely heavily on constant FM radio contact with subordinates. Tactical communication problems might also cause an inability to maintain a COP, to give orders and guidance, to request support, or to coordinate and synchronize elements of the combined arms team. Communication problems in urban areas can prevent the achievement of information superiority and contribute directly to mission failure. In UO, allocating critical or high-value communication assets will be significant and essential to the main effort.

4-66. In an urban environment, units and staffs properly prepare for and mitigate the communication problems in urban areas (see figure 4-3). Adequate communications, in most cases, are ensured by—

- Training in and use of retransmission and relay sites and equipment, which may include unmanned aircraft systems (UAS).
- Airborne command posts, satellite communications, high-frequency radios, and other redundant communication platforms and systems.
- Careful positioning of commanders, command posts, and antennas to take advantage of urban terrain characteristics.
- Detailed communications analysis for movement from one AO to another due to the likely density of units operating in the urban environment.

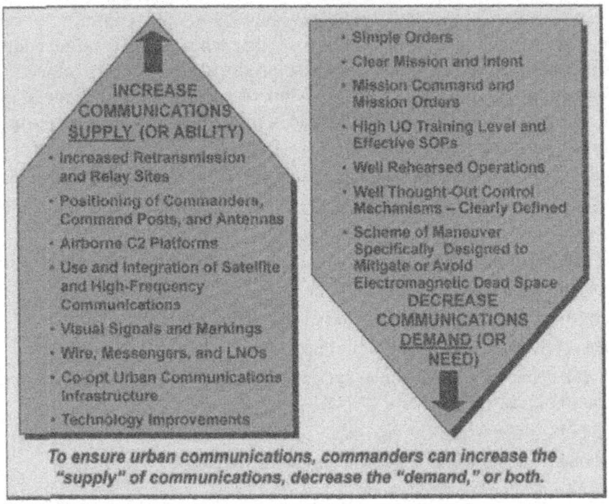

Figure 4-3. Methods to overcome urban communications challenges

4-67. Standing operating procedures (SOPs) for visual markings (both day and night) may assist in command and control. These SOPs indicate unit locations and other essential information. They coordinate with units across common boundaries. Given adequate consideration to limitations on multinational capabilities, these SOPs may assist in command and control and preclude fratricide incidents resulting from loss of FM communications. However, visual signals, including pyrotechnics, are less effective in buildings and enclosed spaces.

4-68. In defensive, stability, or civil support operations, positions often do not change as frequently as in offensive operations. Urban commanders then rely more on military wire (properly camouflaged amongst the civilian communications infrastructure), commercial communications, and messengers. Even in combat, some if not all of the urban area's organic communications structure remains intact for Army use. For example, every building may have one or more telephone distribution boxes that can control hundreds of individual telephone lines. Setting up wire communications using these points is relatively simple but, like all wire communications, is susceptible to wire-tapping. Cellular telephones can usually work well in urban areas; however, locating and destroying the repeater stations or other land-based elements of the cellular telephone system (or the effects of natural disasters) can easily disable them. (Cell phones may be a critical and only means to rapidly communicate with key civilian organizations and important community leaders.) Consequently, the C2 system may use these alternatives to FM communications but with proper operations and physical security procedures in place.

Example of Simple Communications Innovation: Israel's Six-Day War – 1967

In the 1967 battle for Jerusalem, the Israeli Defense Force placed flags on top of cleared buildings so that aircraft providing close air support could monitor the Israeli forward line of troops. They also used a spotlight during the night to mark specific buildings as close air support targets.

4-69. Units will likely use multiple means to communicate throughout the urban area. Hence, commanders emphasize proper operations security procedures (OPSEC) despite the level of security provided by the communications system. This emphasis helps to lessen the probability that Soldiers will inadvertently compromise essential information as they switch from one mode of communications to another (for example, from secure FM radio to unsecured cellular telephones or from classified to unclassified Internet domains).

4-70. Command posts above brigade-level ensure that they can communicate in an urban area without significant disruption. In stability and civil support operations, immediate and reliable communications between tactical and strategic levels may be necessary. Higher commanders anticipate that although the urban area does not significantly challenge their INFOSYS, the area may severely challenge systems at the lower tactical levels. For this reason, information flow from lower to higher may take longer. If the situation is not acceptable, the higher headquarters takes steps to mitigate it, such as increasing the number of liaison officers operating with units engaged in decisive operations. In some instances, the scheme of maneuver may be specifically designed to account for communications interference, propagation characteristics, and electromagnetic dead space. (However, this will require more time, resources, and a detailed urban communications IPB.)

4-71. Finally, urban areas can overload the INFOSYS with information. UO across the spectrum of operations can generate large volumes of information when crises threaten. This sheer volume can easily overwhelm UO commanders and command posts, and the information conduit connecting the two. Training prepares command posts to handle this volume of information and to filter the critical from the merely informative. Staffs must create products (visual or textual) that help their commanders understand the urban environment, not just present them information to know.

KEY TACTICAL CONSIDERATIONS

4-72. Commanders and planners of major UO must thoroughly understand the tactical urban battle. Especially, they must understand the effects of the environment on men, equipment, and systems. The complexity of urban environment changes and often compresses many tactical factors typically considered in the planning process. Figure 4-4 shows some of these compressed factors, however, commanders and their staffs should carefully review FM 3-06.11 for doctrine to support tactical urban operations.

Figure 4-4. Compressed tactical factors

TIME

4-73. The time available to think and act is compressed in urban combat operations. The tactical engagements that comprise battles and major UO are often quick and decisive; therefore, higher-level commanders require the ability to conduct battle command on the move so that their decision making remains correspondingly fast. The impact of decisions (or lack of) and the outcome of battle can occur in mere minutes. Often the amount of information and the number of decisions can overwhelm the overall ability of INFOSYS to respond. Commanders have little time to influence tactical actions with resources kept in reserve. Reserves and fire support assets are close to the point of decision so that they can respond in time to make a difference. The terrain causes C2 challenges that further inhibit commanders from responding quickly to changes in the situation. Small unit leaders receive training that emphasizes understanding the commander's intent so that they can recognize tactical opportunities and can act quickly to take advantage of them.

DISTANCES AND DENSITY

4-74. Distances in UO are compressed to correspond to the density of threat forces and noncombatants. In open terrain, squads, platoons, and companies may be able to control or influence thousands of meters of space. In UO, large buildings can absorb the efforts of several companies or battalions. Crowds of thousands can assemble in areas of a few hundred meters requiring correspondingly large forces for control. Maximum engagement ranges, as influenced by the urban terrain, are usually closer. Units may require field artillery for direct fire at targets ranging fewer than a hundred meters. Commanders and staffs understand the telescoping nature of the battlefield, the density of threat forces, and the density of noncombatants. In addition to the actual conduct of urban tactical operations, these factors will directly affect training, planning, force deployment, and strength.

4-75. Time-distance considerations are especially important throughout planning cycles. Though distances may be short, the physical nature of the environment can drastically change the planning factors for unit movements. The advance of a battalion may be measured in hundreds of meters per day. Thus, all time and distance calculations that relate to sequencing of forces, synchronizing combat power and other capacities, and making decisions require reevaluation based on the urban conditions.

4-76. Although it may appear counterintuitive, airspace above the urban area may be also be severely compressed as multiple fixed-wing, rotary-wing, and indirect fires (including multinational assets) compete for the same space. Increasing this density is the use of airspace by civilian aircraft and the proliferation of tactical UAS. Due to the potential for a high volume of air traffic, commanders and planers pay close attention to the integration and deconfliction of airspace over urban areas. Commanders should consider specific techniques and procedures for UAS which may include—

- Defining airspace control areas.
- Regulating flight times.
- Creating altitude restrictions.
- Including UAS on the air tasking order.

COMBAT POWER

4-77. The urban terrain can also compress combat power. This terrain increases the utility and effects of some weapons and systems, increasing overall combat power. One system that dramatically demonstrates this effect is the sniper. In open terrain, snipers slightly influence operations. In UO, snipers—well concealed, positioned, and protected—can take on significance disproportionate to their combat capability in other situations (see FM 23-10).

4-78. The density of ground combat power in a given size area is also increased because of the effect of the terrain on ranges. The complex terrain precludes standoff engagement from extended ranges by dispersed forces. Commanders often position weapon systems closer together and at shorter ranges to mass effects on the same target. Thus, commanders may position armored vehicles, which typically position themselves Hundreds of meters from friendly troops and other vehicles, within a few meters of each other to provide mutual support. Targets, which in open terrain are engaged at thousands of meters, are engaged in tens of meters on the urban battlefield.

4-79. The dense clutter of the urban environment also affects target acquisition. Systems, such as radar optimized for open terrain, will not be able to acquire targets as effectively. Decreased acquisition capability equates to diminished combat power. It may also require increasing the density of acquisition systems to compensate for reduced capability.

4-80. Finally, the density of combat power may also increase the vulnerability of Army forces. Many Army systems are protected from enemy systems at longer ranges. The number of enemy systems that can threaten Army forces at a short range increases dramatically. Lack of dispersal will make it more likely that multiple Army systems can be targeted by a single enemy threat.

LEVELS OF WAR

4-81. The levels of war are also compressed in the urban area. The tactical actions of individuals and small units can directly influence operational and even national and strategic objectives. Conversely, the decisions of the President can directly affect the conduct of tactical operations. UO have short cause and effect links between the tactical, operational, and strategic levels of operations. Because of the close media scrutiny of UO, the President can sometimes observe the actions of platoons in real time. For example, the media may film a platoon applying nonlethal force for crowd control. The President can view that film on the nightly news before the platoon even disengages from the action, much less reports formally through the various levels of command. If appropriate, the President can decide and direct the strategic and operational commanders to adjust ROE before the platoon has reported. Therefore, commanders at all levels must understand the urban environment's potential compressive effects on the levels of war. A major impact of these effects can be a lower tolerance for tactical errors and a greater need for detailed planning and precision in execution and weapons' effects (lethal and nonlethal).

DECISION MAKING

4-82. The nature of the urban environment compresses the time available to make decisions and increases the number of decisions to make. This is particularly true at the lower tactical levels. Units observing an urban AO face more potential unknowns than in other situations. A large structure presents many more potential firing positions that are observed than simpler terrain. Movement in one of those windows forces the soldier or unit to quickly make a decision regarding the nature of the target—deciding whether it is a threat or a noncombatant. Incorporating combatant and noncombatant discriminatory considerations into all live-fire training will improve Soldiers' ability to make these critical judgment decisions. Overall, lack of understanding regarding the urban environment requires commanders to rely more on analytic decision making, while greater understanding facilitates adaptability and allows commanders to make rapid and intuitive decisions while still accounting for second, third, and higher-order effects (see Figure 4-5 and FM 6-0). Realistically, however, all decisions made during urban operations will likely require a combination of analytic and intuitive decision-making abilities.

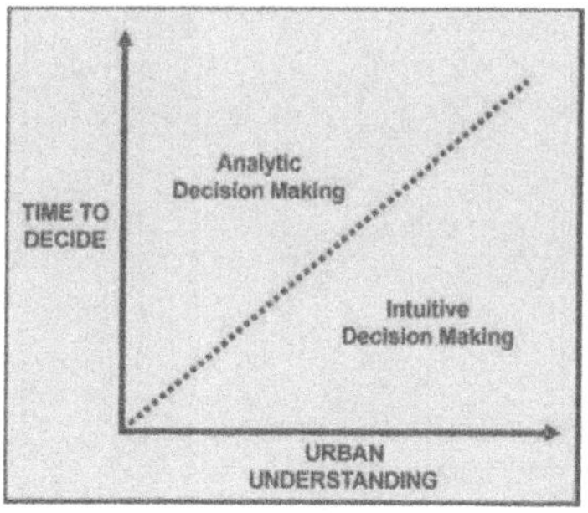

Figure 4-5. Urban understanding and decision making

This page intentionally left blank.

Chapter 5

Contemplating Urban Operations

We based all our further calculations on the most unfavorable assumptions: the inevitability of heavy and prolonged fighting in the streets of Berlin, the possibility of German counter-attacks from outside the ring of encirclement from the west and south-west, restoration of the enemy's defence to the west of Berlin and the consequent need to continue the offensive.

General of the Army, S. M. Shtemenko
describing operational level planning for taking Berlin
The Soviet General Staff at War

In any potential situation and in any area, Army commanders will likely need to assess and understand the relevance and impact of one or more urban areas on their operations. They will also need to determine whether full spectrum urban operations (UO) will be essential to mission accomplishment. UO may be the commander's sole focus or only one of several tasks nested in an even larger operation. Although UO potentially can be conducted as a single battle, engagement, or strike, they will more often be conducted as a major operation requiring joint resources. Such actions result from the increasing sizes of urban areas. Army commanders of a major urban operation must then ensure that UO clearly support the operational objectives of the joint force commander (JFC), requesting and appropriately integrating critical joint resources. Whether the urban operation is the major operation itself or one of many tasks in a larger operation, Army commanders must understand and thoroughly shape the conditions so subordinate tactical commanders can engage and dominate in this complex environment.

A ***major operation*** is a series of tactical actions (battles, engagements, strikes) conducted by various combat forces of a single or several services, coordinated in time and place, to accomplish operational, and sometimes strategic objectives in an operational area.

DETERMINING THE NECESSITY OF URBAN OPERATIONS

5-1. Early in planning, commanders of a major operation must address the necessity and feasibility of conducting operations in urban areas located throughout their areas of operations (AOs). Chapter 1 discussed strategic and operational considerations that compel forces to operate in urban areas. These reasons include the location and intent of the threat force; critical infrastructure or capabilities that are operationally or strategically valuable; the geographic location of an urban area; and the area's political, economic, or cultural significance. Additionally, humanitarian concerns may require the control of an urban area or necessitate operations within it. Several considerations exist, however, that may make UO unnecessary, unwarranted, or prohibitively inefficient. When determining whether to operate in an urban environment, commanders must consider the operational (and accidental) risks and balance them with mission benefits. The factors shown in Figure 5-1 highlight some measures to evaluate the risks associated with UO.

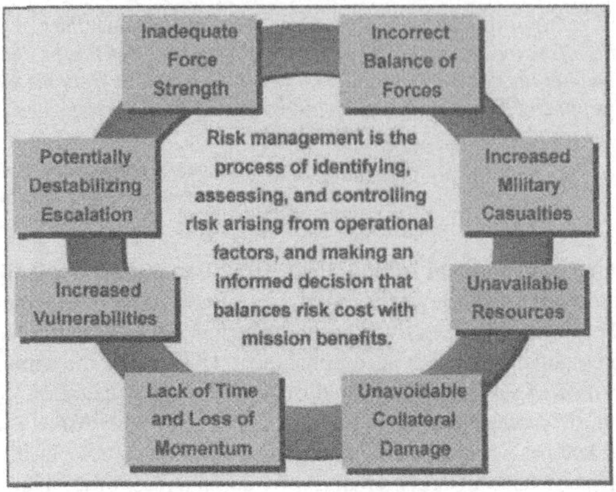

Figure 5-1. Risk management and the associated risks with urban operations

INADEQUATE FORCE STRENGTH

5-2. When facing prospective UO, commanders must consider if they have the necessary troops available to conduct the operation properly and within acceptable risk. Under normal circumstances, large urban areas require many forces merely to establish control. The New York City police department has over forty thousand officers and hundreds of support staff simply to conduct peacetime law enforcement. Major UO, particularly those that are opposed, will often require a significant number of forces. Offensive missions, for example, can require three to five times greater troop density than for similar missions in open terrain. If commanders lack sufficient force to conduct effective operations, they may postpone or consider not initiating those operations until they have the necessary strength. Commanders should also add to their analysis the requirements for troop strength elsewhere in the AO. Additionally, commanders must consider the number (and type) of forces required to make the transition from urban offensive and defensive to stability or civil support operations when determining overall force requirements—not just the number of forces required to realize objectives for major combat operations.

INCORRECT BALANCE OF FORCES

5-3. Along with force strength, commanders must consider the type and balance of forces available. This consideration includes an assessment of their level of training in urban operations. Generally, UO put a premium on well-trained, dismounted infantry units. The operational or tactical necessity to clear threat forces from the dense urban environment, hold hard-won terrain, and interact with the urban population greatly increase dismounted requirements. Therefore, Army forces conducting UO are often force tailored to include a larger infantry component. In addition, special operations forces (SOF) are invaluable in UO. SOF include psychological operations (PSYOP) and civil affairs (CA) forces. They should always be considered as part of the task organization.

5-4. UO must include combined arms to ensure tactical success in combat. Although masses of heavy forces may not be required, successful UO require the complementary effects of all Army forces. Even if an urban operation is unlikely to involve offensive and defensive operations, armor, combat engineers, and field artillery may be essential to mobility and force protection. In urban stability and civil support operations, successful mission accomplishment requires more robust CA organizations. They are also valuable in urban offensive and defensive operations. While commanders may have sufficient combat and combat support forces, they may lack enough sustainment forces to provide the support necessary to maintain the tempo. Again, commanders without balanced types of forces, to include their proficiency in operating in urban environments, should consider alternatives to UO or delaying UO until proper force types are trained and available in sufficient numbers.

Increased Military Casualties

5-5. Casualties in UO are more likely than in operations in other environments. In urban offense and defense, friendly and threat forces often engage at close range with little space to maneuver. The urban terrain provides numerous advantages to the urban defender; higher casualties occur among troops on the offensive, where frontal assaults may be the only tactical option. Conversely, defenders with limited ability to withdraw can also suffer high casualties when isolated and attacked. Casualties can be more difficult to prevent in urban stability operations because of the dense complex terrain, the close proximity of the urban population, and the possible difficulty in distinguishing friend from foe. The potential for high casualties and the subsequent need for casualty evacuation under difficult circumstances make the positioning and availability of adequate medical resources another important consideration. Additionally, high-intensity urban combat and the potential for increased stress casualties may require additional units to allow for adequate unit rotations so that Soldiers receive the rest they require.

5-6. Though casualties occur in all operations, commanders should recognize the likelihood of more casualties during large-scale or high-intensity UO. During the battle for Hue in 1968, for example, many company-size units suffered more than 60 percent casualties in only a few days of offensive operations. Commanders conducting urban stability operations must know the casualty risk and how it relates to national and strategic objectives. While a lower risk normally exists in stability operations than in offensive and defensive operations, just one casualty may adversely impact the success of the stability mission. A realistic understanding of the risk and the nature of casualties resulting from UO critically affect the decision-making process. If commanders assess the casualty risk as high, they must ensure that their higher headquarters understands their assessment and that the objectives sought within the urban area are commensurate with the anticipated risk.

Unavailable Resources

5-7. Offensive and defensive operations in an urban environment put a premium on certain types of munitions and equipment. Forces may want to use vast amounts of precision munitions in the urban environment. At the tactical level, they will likely use more munitions than during operations in other environments. These munitions include—

- Grenades (fragmentation, concussion, stun, riot control, and smoke).
- Mortar ammunition (due to its rate of fire, responsiveness, and high-angle fire characteristic).
- Explosives.
- Small arms.

5-8. Soldiers need access to special equipment necessary to execute small-unit tactics effectively. In urban stability and civil support operations, this equipment may include antiriot gear, such as batons, protective clothing, and other nonlethal crowd-control devices. In urban offensive and defensive operations, special equipment can include sniper rifles, scaling ladders, knee and elbow pads, and breaching equipment such as door busters, bolt cutters, and sledgehammers. Soldiers can conduct UO with standard clothing and military equipment. However, failure to equip them with the right types and quantities of munitions and special equipment will make mission success more difficult and costly. When commanders consider whether to conduct UO, they must evaluate the ability of logistics to provide the necessary resources (see Chapter 10). Considerations must include the ability to supply all Soldiers regardless of branch or military occupational specialty with urban-specific equipment.

Unavoidable Collateral Damage

5-9. UO require an expanded view of risk assessment. When considering risk to Army, joint, and multinational forces, commanders must also analyze the risk to the area's population and infrastructure. This comprehensive analysis includes the second- and third-order effects of significant civil casualties and infrastructure damage. Collateral damage can influence world and domestic opinion of military operations and thus directly affect ongoing operations. It also influences the postconflict physical environment and attitudes of the population. Negative impressions of the civilian population caused by collateral damage can take generations to overcome. Destroying an urban area to save it is not a viable course of action for Army commanders. The density of civilian populations in urban areas and the multidimensional nature of the environment make it more likely that even accurate attacks with precision weapons will injure noncombatants. Unavoidable collateral damage of sufficient magnitude may justify avoiding UO, which, though it may be tactically successful, may run counter to national and strategic objectives.

LACK OF TIME AND LOSS OF MOMENTUM

5-10. Commanders conducting major operations must analyze the time required to conduct UO successfully. UO can be time consuming and can require larger quantities of resources. The density of the environment, the need for additional time to conduct a thorough reconnaissance, the additional stress and physical exertion imposed on Army forces operating in urban areas, and the potential requirements to care for the needs of the urban population consume time and slow momentum. Commanders cannot permit UO conducted as a shaping operation to divert resources from the decisive operation. Nor can they allow UO to interrupt critical time lines, unnecessarily slow tempo, or delay the overall operation. Threat forces may conduct UO with the primary purpose of causing these effects. Commanders must recognize that time generally works against political and military objectives and, hence, they must develop plans and operations to avoid or minimize UO that might unacceptably delay or disrupt the larger, decisive operation. Once commanders achieve major combat objectives, however, they will likely need to shift resources and focus on the urban areas that they previously isolated and bypassed.

INCREASED VULNERABILITIES

5-11. Commanders must weigh the potential for increased vulnerabilities when executing UO. The density of the environment makes protection (safety, field discipline, force protection, and especially fratricide avoidance) much more difficult. Forces operating in a large urban area increase their risk of isolation and defeat in detail. Joint capabilities, such as air power, may work less effectively to support a close urban battle than in some other environments. Thus, responding to unexpected situations or augmenting disadvantageous force ratios when applying joint capabilities is significantly more difficult. Although organized, trained, and equipped for success in any environment, the Army vulnerability to weapons of mass destruction (WMD) increases when forces concentrate to conduct UO. Commanders may consider not committing forces or limiting the size of a force committed to an urban area because of increased vulnerability to (and likelihood of) attack by WMD.

5-12. Fratricide avoidance is a matter of concern for commanders in all operations. The complex urban terrain and density of participating forces coupled with typical battlefield effects—smoke, dust, burning fires—and weather effects—fog, snow, rain, and clouds—immensely increase the potential for urban fratricide. Additionally, safety requirements make it difficult to fully replicate training conditions that allow leaders to become more aware of the conditions that contribute to fratricide. The effects of fratricide can be devastating to UO and spread deeply within the Army force. Critical effects include—

- Needless loss of combat power.
- Decreased confidence in leadership, weapons, and equipment. These lead to a loss in initiative and aggressiveness, failure to use supporting combat systems, and hesitation to conduct limited visibility operations.
- Disrupted operations and decreased tempo.
- General degradation of cohesion and morale.

5-13. Therefore, commanders should increase fratricide awareness and emphasis on prevention measures during UO. Causes can be procedural, technical, or a combination of the two and include—

- Combat identification failures due to poor situational understanding, lack of communication, failure to effectively coordinate, and short engagement ranges coupled with the need for quick reaction.

- Location errors involving either the target or enemy forces due to poor situational understanding.
- Inappropriate command and control and fire support coordinating measures; a failure to receive, understand, or adhere to these measures.
- Imprecise weapons and munitions effects such as, an antitank round that penetrates several walls before exploding near friendly forces.

Exacerbating these difficulties will be the likelihood of Army forces conducting operations with (or within proximity of) SOF, coalition forces, and indigenous security forces including local police. (During OPERATION IRAQI FREEDOM, the head of the Australian Defence Forces SOF in Iraq made it a point to carefully coordinate with U.S. Army forces—down to the company level—whenever his personnel would be in an American AO. He felt the danger of fratricide was a much greater risk than the loss of operational security.)

Potentially Destabilizing Escalation

5-14. In the urban environment, Army forces cannot avoid close contact with enemy forces and civilians that may potentially become hostiles. In urban stability and civil support operations, commanders should consider the chance of this contact escalating into confrontation and violence, which may become destabilizing. This consideration may delay, limit, or altogether preclude UO using Army forces.

Consider Alternatives and Risk Reduction Measures

5-15. Since UO are often high risk, commanders should consider courses of action that provide alternatives. When the objective of an urban operation is a facility, commanders should consider replicating that facility outside of the urban area. For example, a critical requirement for an airfield to sustain operations may lead commanders to consider UO to seize or secure one located in an urban area. However, if adequate resources exist (especially time and adequate general engineering support), Army forces may build an airfield outside of the urban area and eliminate the need to conduct the urban operation. Similarly, logistics over-the-shore operations may be an alternative to seizing a port facility. In some situations, the objective of UO may be to protect a political organization such as a government. Relocating the government, its institutions, and its personnel to a safer area may be possible. Commanders can also design an operation to avoid an urban area. For example, if an urban area dominates a particular avenue of approach, use a different avenue of approach. Using a different avenue of approach differs from isolating and bypassing because the entire operation specifically makes the urban area irrelevant.

5-16. If commanders execute UO, they must assess potential hazards, and then they develop controls to either eliminate or reduce the risks to Army forces. The first means to offset risk is always to ensure a thorough understanding of the urban environment and its effects on operations by all members of the force. Other measures to mitigate risk may include—

- Detailed planning to include thorough intelligence preparation of the battlefield and the development of appropriate branches and sequels.
- Integrated, accurate, and timely intelligence, surveillance, and reconnaissance (ISR).
- Clear missions and intent, which includes a well-articulated end state that looks beyond the cessation of combat operations.
- Sufficient reserves and rotation of forces.
- Vigilant physical security precautions to include increased use of barriers and other defenses, particularly when urban areas are used as support areas.
- Operative communications and other information systems (INFOSYS).
- Effective populace and resources control measures.
- Comprehensive and flexible rules of engagement (ROE) continuously reviewed to ensure they remain adequate for the situation.
- Sufficient control measures (which often include a common urban reference system) and standard marking and identification techniques that adequately consider limited visibility concerns for both air and ground forces. Measures should allow commanders to satisfactorily

control UO and minimize fratricide without unreasonably restricting subordinate commanders' ability to accomplish assigned missions. Commanders must ensure that all subordinate units thoroughly disseminate any approved nonstandard reference systems.

- Proper targeting procedures (including effective fire support coordinating measures and a streamlined legal review of targets), positive identification of targets, and controlled clearance of fires. The goal is achievement of precise (yet rapid) effects with both lethal and nonlethal means. In close air support, positive air-to-ground communications are essential to coordinate and authenticate markings.
- Well-synchronized information operations (IO) that begin before introducing Army forces into the urban environment and well through transition. Commanders should emphasize vigilant operations security (OPSEC) particularly when operating closely with the media, nongovernmental organizations (NGOs), and elements of the civilian population.
- Active and effective integrating, synchronizing, and coordinating among all forces, agencies, and organizations involved in the operation. Commanders should allow adequate planning and rehearsal time for subordinates.
- Responsive, sustainable, and flexible urban sustainment.
- Forces well trained in joint, interagency, multinational, and combined arms UO.
- The creation of adaptable, learning organizations. This requires thorough after-action analyses conducted during actual operations as well as after training exercises. In addition to official Army sites (such as the Center for Army Lessons Learned and the Battle Command Knowledge System), commanders must create a unit-level system to allow hard-won, lessons learned and tactics developed to be immediately passed on to other units and soldiers—even in the midst of an operation. This system may be technology-based, procedural, or both. (During OPERATION IRAQI FREEDOM, the 1st Cavalry Division developed an effective web-based knowledge network—called CAVNET—that allowed them to actively capture and share lessons learned among subordinate units.)

INTEGRATION INTO LAND OPERATIONS

5-17. The commander of the major operation, after determining that urban operations are required, will then integrate the urban operation into his overall operation. He does this by articulating his intent and concept for the urban operation to his subordinates. The commander of the major operation must also set the conditions for successful tactical urban operations by his subordinates. He should define ROE, focus ISR efforts, task organize his capabilities, ensure information superiority, design the operational framework, and initiate and sustain effective coordination with other agencies and organizations (see FM 6-0).

CONCEPT OF THE OPERATION

5-18. The commander's concept of the operation should address all operationally important urban areas in his AO. It should also articulate his vision of the urban operation through directions to his staff and subordinates. Subordinate commanders should address urban areas that the higher commander does not specifically address. The commander's concept should discuss each urban area in terms of task and purpose (see FM 5-0). The commander should also describes his vision of the situation's end state in terms of—

- The threat.
- The urban environment (terrain, society, and infrastructure).
- Friendly forces.
- The conditions necessary to transition control of urban areas within his AO to another agency or back to legitimate civilian control.

Rules of Engagement

5-19. National- or joint-level command authorities may develop urban-specific ROE. If not, Army commanders, as part of their assessment, must determine if urban-specific ROE are required for their situation and provide supplemental ROE. However, commanders must forward any conflicts or incongruities to their higher headquarters for immediate resolution.

5-20. Developing effective ROE relies on thoroughly understanding the national and strategic environment and objectives. It also relies on understanding how to conduct urban operations at the tactical level including weapons effects. For example, broad ROE may result in significant collateral damage and civilian casualties. Even in a major operation or campaign, significant collateral damage caused during UO can make postcombat operations difficult. Such damage may even change national and international public opinion or threaten the achievement of national and strategic objectives. In contrast, restrictive ROE can hamper tactical operations causing mission failure, higher friendly casualties, or both. ROE are often part of essential elements of friendly information (EEFI), protected to reduce the potential for threat exploitation. Even in a limited urban operation, ROE will frequently need to change as circumstances warrant. Therefore, commanders should plan ROE "branches" for anticipated changes in the operational environment.

5-21. In urban operations, ROE are flexible, detailed, and understandable. They should preclude the indiscriminate use of deadly force while allowing soldiers latitude to finish the mission and defend themselves. ROE should recognize that the urban area is not homogenous and may vary according to the key elements of the threat and environment: terrain, society, and infrastructure. To be effective, ROE are consistent throughout the force (an increased challenge in multinational urban operations), and soldiers are thoroughly trained and familiar with them.

Enemy Considerations

5-22. The nature of an urban enemy affects ROE as well. Commanders must consider the type of enemy weapon systems, the degree of defensive preparation, the ability to target enemy vulnerabilities with precision systems, and the ability to distinguish combatant from noncombatant.

Terrain Considerations

5-23. ROE may vary according to the terrain or physical attributes of an urban area. Physical factors may drive the ROE to preclude certain types of munitions. For example, if the construction of a portion of the area is sensitive to fire, then ROE may preclude using incendiary munitions in that area. The ROE may lift this prohibition when units move into areas of masonry construction. Toxic industrial chemicals or radiological contaminants in an industrial area may also affect ROE.

Societal Considerations

5-24. The societal or human dimension of the urban environment will often affect ROE the most. Commanders must base the ROE development on a thorough understanding of the civilian population and threat. They evaluate the loyalty of the population, its dynamic involvement in activities that affects the operation, and its size and physical location. A population that is present and supports Army forces will likely elicit more restrictive ROE than a hostile population actively supporting forces opposing the Army forces. A neutral population, not actively involved in behavior affecting Army forces, supports consideration of more restrictive ROE. In all cases, ROE conforms to the law of war. However, ROE may be much more restrictive than the law of war requires.

5-25. The location of the population also affects ROE. The evacuation or consolidation of noncombatants into controlled, safe areas may result in less restrictive ROE. A U.S. or allied population that remains in the urban area conducting routine business in and amongst Army forces during noncombat UO will normally require the most stringent ROE.

Infrastructure Considerations

5-26. Commanders must consider the urban infrastructure when developing ROE. Urban infrastructure vital to current or future Army operations may dictate that commanders adjust ROE to ensure that critical elements of the infrastructure remain intact during the conduct of operations. If Army forces conduct an urban operation to capture port facilities, for example, the ROE should address damage to the key facilities that are the objective of the operation.

RESOURCE ALLOCATION

5-27. Commanders of a major operation must ensure that subordinate tactical commanders have the resources necessary to conduct UO effectively. They assign appropriate forces to subordinate commanders tasked to conduct UO; support them with Army forces at the operational level; and request and coordinate their support by joint resources.

Task Organization

5-28. Task organizing subordinate units for urban operations will depend largely on the nature of the operation. Some units, however, are always part of the task organization to ensure the success of UO. Infantry, CA, aviation, military police, PSYOP, military intelligence, and engineers are units required for all urban operations across full spectrum operations. Other type forces—such as armor, artillery, and chemical—have essential roles in specific types of urban operations but may be less applicable across the spectrum of operations. Commanders and staffs of a major operation must understand their mission, the particular urban environment in which they operate, and the general effects of the environment across the warfighting functions to allocate the appropriate forces to their tactical commanders. See Chapter 4 for details.

Operational-Level Support

5-29. Commanders of a major operation should also support the tactical commander with forces remaining under their direct control. These forces can include Army SOF, such as CA, PSYOP, and Special Forces, ground and air cavalry, aviation, logistics, engineers, and communications support. These forces may not be under operational control of the supported command, but their efforts are synchronized and coordinated.

Coordinating and Requesting Joint Support

5-30. Commanders of a major operation will often provide forces to the JFC as well as receive assets. They will also coordinate for and integrate joint assets to support the tactical battle. These assets will usually include air support, such as close air support, tactical airlift, and aerial reconnaissance and surveillance. Intelligence support comes in the form of reach to strategic and national intelligence capabilities and to space-based systems. This reach to space assets provides reliable, robust long-range communications, environmental monitoring, ISR, positioning and navigation, and warning of enemy missile launch. Special operations capabilities can assist the tactical mission with special reconnaissance and direct action (including special operations aviation) against high-payoff targets. Joint resources also provide the Army forces augmentation by Marine ground forces. In coastal areas, Navy forces and Coast Guard elements assist Army forces with security, sealift, and fire support. Additionally, the potential agility, reach, and power projection provided by Navy forces and platforms, combined with their self-sustaining capability, can play a vital role in supporting UO. This is especially true when UO are conducted in the littorals, require forcible entry operations, and when host nation infrastructure is severely taxed or destroyed. Commanders of a major operation will coordinate with the JFC regarding available joint resources and their allocation. They then ensure that their efforts coordinate with and complement those of tactical Army forces in the urban area. Appendix D discusses the potential contribution of joint capabilities to Army UO.

URBAN ISR

5-31. Commanders at all levels require accurate and timely information to conduct assessments for successful urban operations. This is critical to planning and execution. Senior commanders have a large role in coordinating the urban ISR effort. National strategic sources (as well as open sources) provide most of the information that commanders and staffs require on the characteristics of the human dimension, the physical properties of the terrain, and the infrastructure. The general characteristics of these aspects of the urban environment do not change drastically over time, with one exception. Military operations or natural disasters can change physical characteristics radically. Analysts can obtain crucial information through diligent research of intelligence databases and open sources. However, the disposition and composition of the urban threat is time sensitive and not likely to be discovered through this type of investigation. Due to the effects of the urban environment, deceptive efforts may influence the threat more easily. The urban population is dynamic and critical aspects of this dimension must be updated or confirmed as a prelude to urban operations. Surveillance and reconnaissance (including geospatial assets) provide relevant information regarding threat dispositions, composition and the state of the population, and the specifics of the urban terrain. Successful urban operations depend on the successful conduct of urban reconnaissance. (See also the earlier discussion of effects on the intelligence and command and control warfighting functions in Chapter 4. While counterintelligence (CI) is an essential ISR component, it will be covered later in this chapter under IO.)

Challenges

5-32. The most significant challenge to urban ISR is physical. The physical organization and complexity of the urban terrain, both man-made and natural, challenges national strategic, operational, and tactical ISR capabilities. Commanders should understand the challenges when planning and allocating time and resources to their ISR efforts. They acknowledge that subordinate commanders will face similar challenges. Therefore, commanders consider subordinate capabilities, limitations, and needs when planning, requesting, allocating, and prioritizing ISR assets and capabilities.

Imagery Capabilities

5-33. A significant national, strategic, and operational ISR capability is imagery. Even at the tactical level, units employ unmanned aircraft systems (UAS) and vehicle and dismounted electro-optics, commercial cameras (still and video), and co-opt existing civilian surveillance systems to gain situational understanding. However, the structures of the urban area significantly degrade the information that imagery acquires and may make it susceptible to physical deception measures.

5-34. Imagery, however, is an excellent source regarding the arrangement and nature of many other physical aspects. It can be used to create updated mapping products and provide significant detail of major portions of the infrastructure. Imagery can also reveal what may be happening in structures through detailed study of patterns and other exterior indicators. Yet, the bulk of a skillful threat's forces, well positioned and concealed inside or underneath structures in the urban area, are largely immune from rapid detection by overhead imaging systems. The volume of civilian movement in an urban area will itself provide a degree of camouflage and increase the difficulty of employing pattern analysis. The success in 1999 of the Yugoslavian army concealing heavy forces when confronting NATO indicates the limits of these assets to penetrate an urban area. This leaves a requirement to supplement overhead imagery in the urban environment with a reliable vehicle-mounted or handheld imaging system. For example, thermal sights on a vehicle patrolling an urban street during limited visibility may note the hot engine of a vehicle on the side of the road, possibly indicating suspicious activity. Simply providing patrols with digital cameras or video cameras can also help provide needed information and fill in information gaps.

5-35. The requirement for high altitude and low oblique imagery in more open terrain has been supplemented by a growing need for detailed, close-up, and nearly vertical overhead shots of specific portions of the urban area. This forces collection managers and system operators to move expensive and scarce imagery collection systems closer to the target or to employ flight profiles that place them at greater risk from threat air defense systems. While gun-camera images from aircraft can provide a stand-off reconnaissance platform, UAS imagery may be one of the fastest, least risky methods by which commanders can conduct reconnaissance to update and verify current maps of the urban area and determine routes, obstacles such as damaged and destroyed buildings, and intact and destroyed bridges and other significant infrastructure. Increased UAS proliferation will require greater attention to airspace command and control procedures as well as the movement and protection of ground control stations.

5-36. As the media will likely be present in the urban environment, commanders may consider devoting intelligence resources (to include linguists) to monitoring the various local, national and international sources, particularly the medium of television. The media coverage of threat forces may provide important intelligence and relevant information of immediate tactical value. Commanders, however, should always be cognizant that threats may recognize that Army forces will be using this as a source of intelligence and, feeling no compulsion to be truthful, may use it for their own deception means.

Electronic Capabilities

5-37. The physical attributes of the urban area also diminish the effectiveness of electronic ISR capabilities. Buildings and other structures significantly disrupt radio communications in an urban area. Buildings not only make tactical radio communications difficult for the user, they also make them difficult to locate, intercept, and jam. The range and clarity of frequency modulation (FM) signals significantly diminish when antennas are located inside buildings or when buildings block line of sight between the source and receiving station. To mitigate this effect on signals intelligence means, detection capabilities often move closer to the transmission source. Thus, the threat's vulnerability to compromise by means of his FM and other wireless communications in an urban environment is much less than in many other environments.

5-38. Without losing tactical surprise and increasing risk, units cannot effectively use many electronic detection and surveillance capabilities until urban combat is imminent or perhaps already begun. Ground surveillance radars will have limited uses in urban areas because of the lack of wide-open spaces in which they best operate, but can be used along the periphery of urban areas to detect infiltration and exfiltration of threat forces. However, remote systems such as the remotely monitored battlefield sensor system and the platoon early warning device may have great utility in monitoring many of the numerous avenues of approach that cannot be kept under human surveillance due to manpower constraints. Among many urban applications, remote sensors and early warning devices can be used to monitor subsurface passageways, the entrances and exits to buildings, fire escapes, the perimeters of sustainment bases, and covert foot trails between urban and rural areas.

Human Capabilities

5-39. The limits on imagery and electronic ISR capabilities place a premium on human-based visual reconnaissance. Commanders have three types of human reconnaissance assets to augment electronic reconnaissance resources: special reconnaissance, conventional combat reconnaissance, and human intelligence (HUMINT) gathered by military intelligence from individuals and multimedia. The urban environment poses several challenges to these capabilities.

5-40. The urban area challenges special reconnaissance in several ways. First is the access to the urban area. Although avenues of approach may be numerous, concealed avenues of approach into a defended urban area may be limited and thoroughly covered. Air access is also more difficult because aircraft are detected more easily, airspace is smaller, drop and landing zones are limited or not secure, and more air defense systems probably exist. Still, special reconnaissance efforts to penetrate the urban area can be successful using unconventional techniques including high-altitude low-opening parachutes or underwater penetration.

5-41. Special reconnaissance then faces a second challenge: moving in and identifying targets in the urban area. Stealth movement in an occupied urban area is exceptionally difficult. Repositioning to new or alternate positions is also dangerous. The soldiers' ability to conceal themselves among the civil population can mitigate some of these challenges but includes inherent risks of a different nature. Also difficult is establishing observation positions that provide a field of view of several targets.

5-42. Lastly, special reconnaissance may face navigational and reporting challenges. Special reconnaissance's ability to locate themselves and communicate critical locations and routes are challenged by—

- Differences in language and numbering systems. Road signs, if they exist at all, may be removed or deliberately altered to confuse.
- Irregular street patterns.
- Outdated maps.
- Intervening structures that impede communications and global positioning systems.
- Changes to the landscape due to the effects of UO or natural disasters.
- Featureless shantytowns.

5-43. Conventional reconnaissance faces many of the same challenges as special reconnaissance. Conventional reconnaissance also may lack the advantage of surprise and the special equipment and training that provides special reconnaissance stealth capability. Conventional reconnaissance is not likely to operate undetected by the civilian population. Given the constraints discussed above on other sources, conventional reconnaissance units will likely begin their mission with much less information than they would have on threat dispositions in a less complex environment. Commanders may choose to have their reconnaissance elements fight for information in the urban area. While this high-risk option is more favorable under fluid conditions, it can be used at any time. It requires careful planning, rehearsal, and formulation of information requirements.

5-44. As previously indicated, human intelligence will be one of the most valuable sources for information regarding the situation inside an urban area. HUMINT may take advantage of the proximity and large numbers of potential informants to gather information about threat activities and capabilities. It is especially valuable because it can address all elements of the environment. HUMINT sources can describe political and religious nuances that may be significant to commanders. Such information is useful for insights regarding the human dimension but extremely difficult to obtain from other means. This intelligence also can describe the infrastructure uncovering essential details of how the infrastructure functions. Obtaining good HUMINT requires skilled interrogators and linguists—often in large numbers. Commanders must know and account for some of the possible shortcomings of HUMINT:

- It is susceptible to the influence of the threat; the threat can threaten and influence the source as well as local interpreters. Commander may need to consider that people may not want to talk through a local interpreter because they are providing information they might not want others to know; anonymity may be essential. On the other hand, a local interpreter may provide community access. Yet, those trusted by the locals may not be truthful with Army forces. Conversely, those truthful with Army forces may not be trusted by the indigenous population particularly if these interpreters come from another country, region, or demographic group. In such cases, commanders may need more than one interpreter.
- It is limited by the accuracy of the source and interpreter's perceptions. A local interpreter may have difficulty understanding and translating military topics or the intricacies of urban infrastructure. Some local translators may want to please both participants, so they may shade the conversation, telling each side what they want to hear. Finally, accuracy may be affected by the collector's own perceptions if not firmly grounded in an understanding of social, cultural, and other local conditions.
- Multiple echelons collecting from a given area may inadvertently collect from the same sources and thereby give the false impression of independent sources confirming the same data. To help mitigate this possibility, commanders should ensure that all HUMINT sources are properly recorded on intelligence source registers.
- It may not be timely. The process of identifying and cultivating a source (particularly in an environment where most civilians support threat forces), gathering information, analyzing the information, and providing the intelligence to commanders can be extremely time consuming.
- Some informants may come from unscrupulous or sordid elements of the urban society and may also have their own agenda. They may attempt to use protection afforded them by their relationship with Army forces to conduct activities (even atrocities) that will compromise

political and military objectives. In some cases, a local interpreter's own agenda may influence his translation. (For example, a local schoolteacher used as a translator may report to the collector that the source believes that repairing schools in their urban area is critical, when actually the source may have said that potable water is most needed.)

Conducting Urban ISR

5-45. To be successful, ISR efforts (national to tactical level) are exceptionally comprehensive and synchronized. Moreover, success necessitates the integration of all ISR sources into operational and tactical planning. This requires that ISR assets deploy and execute early, diversify, properly focus, and integrate under a comprehensive ISR plan. Successful ISR also requires flexibility to adapt to the operational and tactical needs of the commander (see Figure 5-2). Commanders must ensure that the appropriate intelligence architecture (including robust links to joint and multinational elements) is established and tested prior to execution of ISR effort.

Early Deployment

5-46. One of the first requirements for effective urban ISR is the early deployment and employment of assets. The complex urban terrain presents a significant challenge. Commanders should consider that ISR assets will normally take longer to gather data amid the complexity.

5-47. Limited national, strategic, and operational imagery intelligence (IMINT) and signals intelligence (SIGINT) capabilities should be considered and quickly requested. If they are approved, they are tasked and deployed or repositioned to begin urban ISR operations. This takes time. Fortunately, spacing the ISR effort over time permits the analysis of the information or data as it is received. Such time also permits subsequently refining the ISR effort before all assets are committed.

5-48. SOF or conventional units will require significantly more time to execute reconnaissance missions and maintain an acceptable survivability rate. Urban reconnaissance operations require additional time for stealthy insertion into the urban area. IMINT and SIGINT capabilities are used to identify possible locations of high-value targets and corresponding observation positions; this helps minimize time-consuming and high-risk repositioning in the urban area. Again, reconnaissance units may require extensive time to observe from observation positions for indicators of threat activity and disposition and identify patterns.

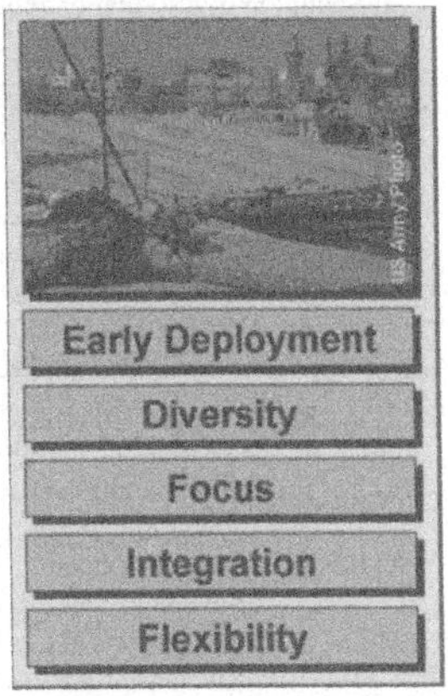

Figure 5-2. Urban ISR considerations

5-49. As conventional combat forces prepare to commit to the urban area, conventional reconnaissance precedes their actions. Conventional reconnaissance will often be a slow and methodical effort. Such forces may need time to reconnoiter the interior of structures for snipers and other small threat teams. They may also need time to deploy and destroy snipers and small delaying elements and to breach harassing obstacles. If necessary, they may need time to mass the combat power necessary to fight through enemy security and continue the reconnaissance.

Diversity

5-50. No single ISR capability can solve the riddle of the urban defense. In order to gain an accurate common operational picture of the complex urban terrain necessary to focus combat power on decisive points, commanders must employ diverse ISR capabilities. These capabilities will each contribute pieces of relevant information to permit identifying operational objectives and leveraging tactical combat power to achieve those objectives quickly. Higher-level commanders should understand that tactical reconnaissance capabilities alone often cannot provide all the tactical information required for success at lower echelons.

5-51. Using diverse capabilities challenges the threat's ability to conduct deception and otherwise defeat the friendly ISR effort. A threat that focuses on minimizing his vulnerability to overhead imagery (for example, satellites) may increase his reliance on communications and thus his vulnerability to SIGINT. At the same time, he may decrease his ability to detect the actions of ground reconnaissance units. Conversely, a threat that actively campaigns to detect ground reconnaissance may make himself more vulnerable to SIGINT and IMINT. Moreover, as Army forces enter urban areas, threat activity and susceptibility to SIGNINT often increases.

5-52. Diverse capabilities also facilitate the tactical ISR effort. Tactical reconnaissance units often consist of small dismounted teams and small combined arms teams with a dismounted element and an armor-protected mounted element. Combat engineers with their technical expertise and breaching capability are essential to the combined arms reconnaissance effort. The teams' movements are synchronized and coordinated with other assets, such as UAS and air cavalry reconnaissance. These teams use several movement techniques including infiltration, with the primary objective of conducting zone reconnaissance along key axes that support brigade and battalion actions against decisive points. To accomplish this mission, reconnaissance reconnoiters the proposed routes and alternate approaches. This supports deception and contingency planning. (Reconnaissance of alternate routes and objectives also applies to aerial reconnaissance; helicopters and UAS are not invisible. They can alert a threat to impending operations.) Infiltration of dismounted reconnaissance is made easier when a threat focuses on combined arms reconnaissance teams. Aerial reconnaissance, such as air cavalry and UAS, provides early warning of threat elements to ground reconnaissance, identifies obstacles and ambush sites, and helps select the routes for ground reconnaissance. Air elements may also reduce the mobility of counterreconnaissance forces.

Focus

5-53. Another key to successful ISR is the ability to focus the assets on commander's critical information requirements (CCIR). This focus begins with mission analysis and the commander's initial planning guidance. It is incrementally refined throughout planning and execution as each ISR effort provides information and permits more specific focus in subsequent efforts. The size and complexity of the urban environment require that the ISR effort center strictly on decisive points or centers of gravity (COGs). Therefore, the overall ISR effort will have two major focuses. The first is to uncover and confirm information on the decisive points and COG. The second is to determine the approaches (physical, psychological, or both) leading to the decisive points. The first focus will likely drive ISR in support of major operations. The second focus will likely provide the impetus for tactical ISR efforts. For example, special operations reconnaissance might focus on a major command center that controls the entire urban area and that is one of the higher-echelon command's CCIR. Tactical reconnaissance might focus on the nature of the defense along a particular avenue of approach to the objective.

Integration

5-54. Another important aspect of urban ISR is integration. All reconnaissance capabilities provide both distinctive information as well as information that confirms and adds to that coming from other sources. Essential to urban ISR is the link between all of these sources, either directly or through an integrating headquarters.

5-55. ISR operations must be vertically and horizontally linked. Vertical links ensure that ISR operations among the various levels of command are complementary and that the information flow between these levels is rapid. Horizontal links ensure that forces operating in close proximity (particularly adjacent units), where areas of interest overlap, can rapidly share results of their individual ISR efforts. Together, this helps ensure that all Army forces share a common operational picture and permits the greatest flexibility and survivability of ISR resources.

5-56. ISR operations also are integrated into the planning system, especially the targeting process. As part of targeting, positioned reconnaissance and surveillance elements may become the trigger and terminal control for applying precision fires when appropriate and after considering the risks of compromise of the position or platform. ISR operations and plans must also be synchronized with the air tasking order, as many surveillance platforms are coordinated on the same timeline.

Flexibility

5-57. The urban ISR effort must be more flexible than in other operations. This flexibility permits the ISR effort to meet unforeseen circumstances and to deal with the challenges of the urban environment. As indicated previously, the urban environment is particularly difficult to penetrate. The practical effects of this characteristic are that—

- The initial ISR effort may not be as successful as in other operations.
- More intelligence requirements may be discovered later while executing ISR operations than otherwise.
- The threat may be more successful in active counterreconnaissance because of the concealment advantages of the urban environment (hiding in structures as well as among the urban population).

Therefore, tactical and operational commanders should consider requesting greater than usual ISR support from higher headquarters. Higher headquarters should be proactive in augmenting units conducting urban operations with additional ISR assets. Additionally, ISR assets remaining under the control of the higher headquarters must respond more quickly to the CCIR of supported commanders. Sequencing reconnaissance missions over time provides flexibility by creating uncommitted reconnaissance assets.

5-58. Time sequencing of ISR assets is essential to flexibility. It makes ISR assets more survivable and allows the intelligence cycle to mature the CCIR. It also creates a ready ISR capability to augment committed forces in critical areas if required or diverts them around centers of threat resistance. If not required, the original ISR tasks can be executed as envisioned in planning. Cueing allows a high-value ISR asset to be able to respond to multiple targets based on an ongoing assessment of the overall reconnaissance effort and the changing CCIR. Redundancy permits the effort to overcome line of sight restrictions, the destruction of an ISR asset, and the ability to combine ISR resources to create combat power if required. Maximizing the ISR effort requires applying all available ISR assets to support the urban operation. Additionally, assets—such as air defense artillery and field artillery radars and engineer reconnaissance teams—are integrated into the ISR effort. In urban operations, units will also commit infantry and armor elements (plus their organic reconnaissance elements) into the tactical reconnaissance effort. These units increase the dismount capability and the ability of reconnaissance elements to fight for information and fight through security zones.

Information Operations

5-59. Information operations are an integral part of all Army operations and a critical component in creating and maintaining information superiority. The information environment is the sum of individuals, organizations, or systems that collect, process, and disseminate information; it also includes the information itself. In UO, the information environment is extremely dense due to the proliferation of INFOSYS and widespread access to those systems. In urban operations, commanders must consider how the urban environment, particularly the human component, uniquely relates to executing IO. They must also anticipate the threat's information operation campaign and preclude successful operations on its part. Overall in regard to the urban society, commanders must convince the urban inhabitants of the inevitable success of UO, the legitimacy of Army actions, and the beneficial effects that Army success will eventually generate.

5-60. Because urban operations are likely to be joint, interagency, and multinational, commanders must guard against various forms of information fratricide as a multitude of actors plan and conduct IO in the urban area. For example, the lead governmental agency's key IO objective may be to establish the legitimacy of newly formed civilian authorities. In an attempt to develop close relationships with the civilian populace, Army commanders may continue to work closely with traditional, informal leaders to the exclusion of the new authority. These actions, while they are often conducted out of practical and immediate necessity, may run counter to the lead agency's goal. Overall, Army commanders must nest their IO campaign objectives and themes within those of the lead agency, aggressively coordinate with other governmental agencies and coalition partners, and synchronize activities down to the tactical level to prevent working at odds and avoid information fratricide. However, the process must not be so centralized and rigid that subordinate units lose the flexibility necessary to develop their own products and themes essential to addressing their own AO and the particular situation that confronts them.

5-61. IO are executed using core and supporting elements and related activities (see Figure 5-3 and FM 3-13). The elements of IO are employed in either an offensive or defensive role. Many aspects of IO are not affected differently in an urban environment from any other environment. The following sections outline some IO considerations unique to urban operations.

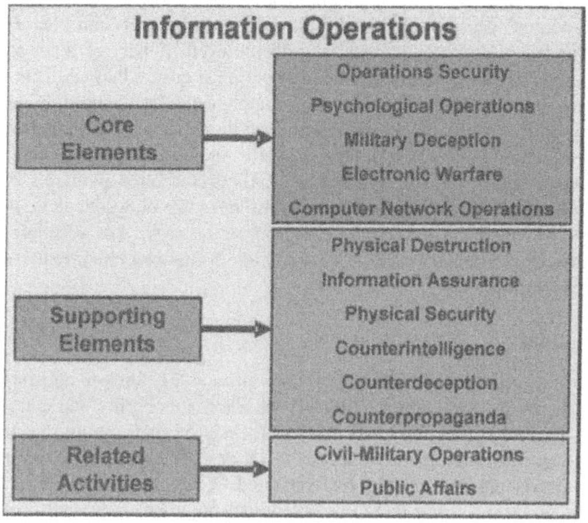

Figure 5-3. IO elements and related activities

Operations Security

5-62. In the urban environment, Army forces can leverage existing urban infrastructure, including the communications and information infrastructure, to enhance Army operations. The danger in integrating these systems is violating OPSEC. Commands ensure that Army forces use only approved systems and proper safeguards exist. Commands also supervise subordinate units for inadvertent breaches of OPSEC policies when using existing urban systems. Finally, established OPSEC procedures must be reasonable and practical. Overly complicated and stringent procedures increase the likelihood that unintentional OPSEC violations may occur as Army forces seek to rapidly accomplish the mission.

5-63. Of particular concern are computers (see also the discussion of Computer Network Operations below). During longer-term UO, commanders will constantly be upgrading and improving the quality of life for their soldiers through the establishment of well-resourced forward operating and sustainment bases. This will likely include the provision of unclassified access to the Internet and e-mail allowing Soldiers to communicate with friends and families around the world. E-mail addresses, facsimile numbers, cell phone numbers, photographs, and other sensitive (but unclassified) information can be valuable sources of threat information. Soldiers must be periodically trained and reminded of everyday procedures to combat potential threat exploitation of these unclassified sources (such as not opening e-mails from unknown sources as they may harbor Trojan programs and viruses).

5-64. To compound concerns, commanders will also establish protected systems to disseminate tactical lessons learned to individual soldiers. This creates the potential that Soldiers can inadvertently release classified information about ongoing operations to threat intelligence systems. In addition to OPSEC training and persistent awareness, commanders must establish procedures to shutdown unclassified Internet and e-mail access immediately, both before and after critical events, to mitigate potential OPSEC violations. (Commanders may also need to randomly shutdown Internet and e-mail access as part of their overall OPSEC measures to preclude such information blackouts serving as a potential warning of impending operations.) Technical surveillance countermeasures—the identification of technical collection activities conducted by threat intelligence entities—will be an important CI service to identify this and other technical OPSEC concerns.

5-65. The close proximity of Army operations to a civil population, particularly in stability and civil support operations, makes Army activities themselves an additional OPSEC concern. Hostile civilians or other threats integrated into the urban population may have more chances to observe Army activities closely. Such observations can provide insight into tactics, techniques, and procedures (TTP) and expose operational vulnerabilities. However, threats may coerce even friendly civilians to provide a threat with EEFI. Therefore, commanders in an urban environment must ensure that civilians cannot observe critical TTP. Any observable patterns and TTP vary and are supplemented with deception efforts. Physical security is increasingly important in urban areas to control civilians' access. Although many urban operations require close coordination with NGOs, commanders should screen information provided to these organizations to protect EEFI. Release of EEFI to NGOs is controlled and done with full recognition and understanding of potential consequences—the benefits must far outweigh the risks involved. (Even the best-intentioned NGO might inadvertently compromise security. For example, by revealing that a commander has closed a particular route the next day, the NGO may unwittingly reveal the objective of a pending offensive operation.)

Psychological Operations

5-66. PSYOP aim to influence the behavior and attitude of foreign audiences, both military and noncombatant, in the urban environment. PSYOP are a force multiplier and contribute in many ways to mission success (see FM 3-05.30). Effective PSYOP focuses on transmitting selected messages to specific individuals and groups in order to influence their actions. Their ability to influence the attitudes and disposition of the urban population cannot be overstated. While the complexity of the societal component of the urban environment can make PSYOP challenging, the urban society also offers many options and resources. Potentially, PSYOP (with other political and economic actions) may help limit or preclude the use of military force in urban areas. In some circumstances, UO may be relevant to the major operation only in terms of their psychological effect.

5-67. The positive influence created by PSYOP is often useful in developing an effective HUMINT capability particularly in an urban area where many civilians actively or passively support the threat. Persuading and influencing a few to support friendly forces may pay great dividends. These few supporters may allow Army forces to penetrate the urban area and obtain essential information. Such information can apply to threat capabilities, threat intentions, and even the urban environment itself.

5-68. PSYOP, combined with other elements of offensive IO, aid in isolation of a threat—a critical shaping action for any urban operation. For example, commanders may use PSYOP to inform civilians about new food distribution points located away from urban combat operations. This action supports the UO fundamental of separating combatants from noncombatants and helps to further isolate the threat (both physically and psychologically) from the civilian populace. Aside from projecting a positive image of friendly forces over threat forces, PSYOP can also isolate the threat by identifying and exploiting ethnic, cultural, religious, and economic differences between the elements of the civilian populace and threat forces as well as the differences among supportive and unsupportive civilian factions. The complexity of the urban environment enables quick changes in opinion or attitude. Commanders must continually evaluate the results of PSYOP for mission relevance.

Military Deception

5-69. Urban operations present numerous challenges to tactical commanders; however, higher-level commanders may help to mitigate some challenges through effective military deception Commanders can use military deception efforts designed to mislead threat decision makers as to friendly force disposition, capabilities, vulnerabilities, and intentions. Military deception actions may allow commanders to achieve tactical surprise or improve relative combat power at a selected location. For example, allowing the threat to observe certain activities on a selected avenue of approach may cause the threat to shift his forces (and effort) to the area perceived to be threatened. (This movement may also aid in determining the overall disposition of threat forces and intentions.) Repositioned forces or effort to activities or locations that are not decisive to the achievement of friendly objectives, combined with other IO designed to overwhelm threat information and intelligence systems, may create the force and tempo differential necessary to achieve success. Commanders should tailor urban deception plans to the specific urban area, paying close attention to the societal characteristics of the target population.

Electronic Warfare

5-70. Electronic warfare (EW) includes all actions that use electromagnetic or directed energy weapons to control the electromagnetic spectrum or to attack a threat. Conducting EW in urban areas seeks to achieve much the same results as in other environments. A major consideration in urban areas is collateral effects on portions of the urban infrastructure that rely on the electromagnetic spectrum for service. Thus, precision is a major factor in planning for EW operations. For example, EW attacking a threat's television broadcasts avoids affecting the television broadcasts of neutral or friendly television. Likewise, EW attacking military communications in a large urban area avoids adversely affecting the area's police and other emergency service communications. Urban offensive and defensive operations will have the least restrictions on EW operations while urban stability may have significant constraints on using EW capabilities.

Computer Network Operations

5-71. Computer network operations (CNO) include computer network attack (CNA), computer network defense (CND), and computer network exploitation (CNE). CNO are not applicable to units at corps and below. Echelons above corps (EAC) units will conduct CNA and CNE. If tactical units require either of these network supports, they will request it of EAC units.

Computer Network Defense

5-72. In urban operations, CND will require extreme measures to protect and defend the computers and networks from disruption, denial, degradation, or destruction. The nature of the urban environment and configuration of computer networks provides the threat with many opportunities to interdict local area networks (LANs) unless monitored by military forces. LANs controlled by military forces are normally more secure than the civilian infrastructure. Commanders should prepare for opportunities by the threat to insert misinformation.

Computer Network Attack

5-73. Considerations regarding the execution of CNA in urban operations are similar to those of EW: CNAs that do not discriminate can disrupt vital civilian systems. However, possible adverse effects on the civilian infrastructure can be much larger—potentially on a global scale. In the short term, CNAs may serve to enhance immediate combat operations but have a debilitating effect on the efficiency of follow-on urban stability operations. Because of these far-reaching effects, tactical units do not execute CNA. CNA is requested of EAC units. EAC units will receive all requests from lower echelons, carefully consider second- and third-order effects of CNA, and work to ensure its precise application.

Computer Network Exploitation

5-74. CNE consists of enabling operations and intelligence collection to gather data from target or adversary automated INFOSYS or networks. Tactical units do not have the capability for CNE. CNE contributes to intelligence collection at EAC. In UO, CNE will be centrally controlled.

Physical Destruction

5-75. Physical destruction includes those actions—including direct and indirect fires from air, land, sea, space, and Special Forces—taken with, to augment, or supplement IO actions. Like many other IO elements, major concerns with employing physical destruction in UO are precision and follow-on effects. Thus, commanders using physical destruction to support IO must adhere to the same constraints as all other urban fires.

Information Assurance

5-76. Information assurance in UO takes on an added dimension. As with other operations, availability of information means timely, reliable access to data and services by authorized users. In UO, the timeliness of information may be restricted because structures block the transmission waves. The need for retransmission facilities may overwhelm the signal community. The reliability can be questioned because of the blockage between units and communications nodes. Unauthorized users may intercept the communications and input misinformation or disinformation. Commanders must protect the integrity of all information from unauthorized changes, including destruction. INFOSYS with integrity operate correctly, consistently, and accurately. The authentication of information may be accomplished by sophisticated electronic means. However, it is more likely that communications-electronics operating instructions authentication tables will authenticate the information. Commanders should consider the confidential nature of all information in UO and establish procedures to protect the information from unauthorized disclosure. Of additional concern, the density of the infrastructure in urban areas may inhibit receipt by the intended individual or unit. Therefore, commanders must develop alternative means to determine if the message was received.

Physical Security

5-77. Physical security consists of those actions and measures to safeguard personnel; to prevent unauthorized access to equipment, installations, material, and documents; and to safeguard them against espionage, sabotage, damage, and theft. The joint, interagency, and multinational nature of urban operations increases the need for physical security measures, especially with organizations that do not emphasize physical and operational security within their own organizations. The potential for locating sustainment bases in or near urban areas and the use of indigenous personnel on these facilities to accomplish a multitude of tasks mandates that commanders carefully consider the effectiveness of their physical security measures. In some stability operations where Army forces must recruit and train security or police forces, and then conduct joint operations and patrols with them, commanders must examine how best to establish close working relations while maintaining essential physical (and operations) security.

Counterintelligence

5-78. The urban environment, particularly in stability operations, is ideal for espionage, other intelligence activities, sabotage, or assassination. Threats can approach, conduct reconnaissance, and escape under the concealment offered by the urban environment. The dense environment and electromagnetic clutter serves to hide technological collection means. To counter threat intelligence collection activities, the Army employs CI forces. They provide analysis of the threat's human, imagery, signals, and measurement and signature intelligence capabilities to develop countermeasures against them (see FM 34-60).

5-79. In conjunction with military police, engineers, medical personnel, and other as required, CI help create urban threat and vulnerability assessments necessary in developing comprehensive force protection measures and evaluating their effectiveness. CI personnel can also supplement the screening of linguists and other critical host-nation personnel employed by Army forces. As Soldiers operate in close proximity to the urban society, threats will attempt to use three, time-tested tools to influence them: money, sex, and drugs. Periodic CI refresher training remains crucial to decrease the effectiveness of these threat tactics.

Counterdeception

5-80. In UO, threat forces can easily accomplish deception operations. The force that controls the area above and below ground will have freedom of movement. Deception aimed at friendly commanders will cause them to deploy combat power at the wrong place and the wrong time. Counterdeception by friendly commanders will identify and exploit threat attempts to mislead friendly forces. Counterdeception is difficult. Cultures of certain rhetoric and actions are more predisposed to deception than others. Knowing a threat's previous deception methods is important. Dismissing tactical indicators because they conflict with preconceptions may allow a hostile deception operation that plays on the preconception to succeed.

Counterpropaganda

Four hostile newspapers are more to be feared than a thousand bayonets.

Napoleon Bonaparte

5-81. Because propaganda is aimed at both combatants and noncombatants, UO are especially concerned with its use. Propaganda can rapidly and dramatically affect the attitudes of the urban population and will probably occur (or increase significantly) after urban operations have begun. Thus, it can create situations in the human dimension of the environment quite different from those discovered in the pre-operations assessment. Counterpropaganda is, therefore, essential to urban operations. To negate, deflect, or destroy the threat's propaganda capability, counterpropaganda requires—

- Monitoring and evaluating the effectiveness the threat's propaganda efforts.
- Determining innovative methods using all Army force capabilities, especially PSYOP and PA units.

5-82. In particular, humanitarian assistance, when coupled with an effective information operations campaign, can be a powerful counterpropaganda tool to sway public support. For example, media coverage of Army forces directly providing or coordinating NGO humanitarian efforts to feed, shelter, and provide medical care to the urban population can help counter threat efforts to discredit Army forces as an evil occupying force. This type of coverage can be used to turn opinion against the threat and could also prove an effective tool in gaining favorable international support.

Public Affairs

5-83. A closely related activity to IO is public affairs (PA). PA influences urban operations by transmitting information through the media to internal (urban Army forces as well as the urban civilian populace) and external audiences. At higher levels of command, PA can help maintain popular national support for the urban operation by clarifying the links between strategic goals and operational objectives. At both the operational and tactical levels, it links Army units, the urban inhabitants, the U.S. and international public, and the media. The media will often focus on the negative actions that are ongoing in an AO; hence, the urban commander's challenge will be to publicize his unit's good news stories. Commanders should not wait for media inquiries and interview requests. Instead, they should aggressively plan public affairs to get the Army's story out to the media and, ultimately, the public.

5-84. PA can help determine potential media issues that may influence planned UO. It can also aid commanders in understanding the impact of UO on the environment (particularly its citizens) and other agencies and organizations operating in the urban area. PA also helps to counter rumors, uncertainty, fear, loneliness, confusion, and other factors that cause stress (to both soldiers and civilians) and undermine effective UO. If the populace does not understand the mission or the Army forces' capabilities, false expectations may be created that Army forces may not be able to meet. PA can help prepare the American public for the possibility of high casualty rates. Overall, PA supports urban commanders in their goals to achieve information superiority and preserve public support.

5-85. The density of information sources and reporters in UO ensures that all Army activities will be subject to media and public scrutiny. Many reporters will congregate in cities for their own comfort and take advantage of established communications networks. Urban areas are densely populated and, together with Army forces and NGOs operating there, will present the greatest number of human-interest stories.

The local urban or host-nation media, however, will often have their own agendas developed over a longer period of time. This local media may also have a greater influence over the urban population than the international media. The indigenous media may not follow international norms. Commanders are responsible for understanding the media (particularly the local media), its role, and its potential for negative or positive influence.

5-86. Commanders cannot allow themselves to be intimidated by the media. Media sources are going to get and report their story one way or another. Therefore, commanders should support open and independent reporting and grant access to their units as early and as far forward as the situation permits. Successful units often go beyond simply accommodating the media to actually integrating them into their units. In the future, embedding of media may become the standard for reporting Army operations. Other considerations for improving urban media relations include:

- Encouraging the media to report on all ongoing aspects of the unit's urban operations—not exclusively on combat operations. This may create a unit responsibility to protect journalists that participate in covering suggested activities.
- Hosting media luncheons or other events for key journalists. This provides commanders the ability to get the whole story out to the press from their unit's perspective. To gain media interest and unit credibility, this will often require key leadership participation.
- Identifying Soldiers that speak the local and international journalists' native language. Even if journalists speak English as a second language, they may overlook or misunderstand important points due to language subtleties. Additionally, people tend to be friendlier to those that speak their own language.
- Identifying selected Soldiers at lower tactical levels to act as PA representatives for their unit as an additional duty. However, commanders must provide them with appropriate training and resources to accomplish specific PA tasks.
- Writing articles for local newspapers that emphasize the unit's good news. Depending on the situation, subordinate units may be tasked to write stories about the good accomplishments in their AO. Commanders, with appropriate OPSEC considerations, may hire locals to assist in this effort. However, commanders must take extra precautions to ensure that these efforts are not perceived as covert propaganda or as undermining free and independent news reporting.

5-87. PA does not distort, direct, or manipulate information. Its effectiveness stems directly from establishing and maintaining credibility with the urban population and media. PA protects the overall integrity of the Army as an institution by ensuring that the Army is recognized as the most reliable source for official information among all other competing sources. Commanders must synchronize PA with the core and supporting elements of IO (particularly PSYOP and counterpropaganda) to ensure that all Army sources send only one message. Urban commanders must plan for the media and integrate PA into their decision-making and (through IO) targeting processes. Similar to lethal targeting, commanders should consider target audiences and coverage areas for the various new agencies and mediums. They also pay attention to print media circulation statistics and television and radio broadcast times. For example, an urban populace may tend to read morning newspapers vice late editions and watch or listen to nightly news instead of morning or daytime programming. Additionally, low literacy rates among the urban population may indicate that television and radio are better mediums to use to influence the urban population than print media such as newspapers, magazines, leaflets, and posters.

5-88. The PA principles listed in Figure 5-4 and addressed in FM 46-1 summarize PA. They serve as useful guides toward planning and executing PA operations regardless of the environment. However, the principles of "practice security at the source," and "truth is paramount" particularly apply to the urban environment. The compartmented nature of most UO impede commanders' and their PA officers' ability to be at all places where the media will likely be. Friendly inaccuracies, intentional or not, will help fuel the threat information campaign. Therefore, all soldiers are trained, provided with clear and understandable PA guidance, and prepared to accurately communicate to the civilian media. The keys at all levels are understanding, prepared acceptance, and truthfulness tempered with an essential concern for OPSEC.

- Truth is Paramount
- If News is Out, It is Out
- Public Affairs Must be Deployed Early
- Not All News is Good News
- Practice Security at the Source
- Media are Not the Enemy
- Telling Our Story is Good for the Army
- Soldiers and Families Come First

Figure 5-4. Public affairs principles

Civil-Military Operations

5-89. Civil-military operations (CMO) are critical and will be an essential, implied task of any urban operation. They are included here as a closely related activity of IO. Commanders use CMO to establish, maintain, influence, or exploit relations to achieve operational objectives. These relations are among military forces, civilian authorities and organizations (both governmental and nongovernmental), and the civilian population. CMO range from support to combat operations to assisting in establishing political, economic, and social stability. In some instances, commanders may need to be prepared to assume temporary responsibility for the functions and capacities of the urban government. Like PA, effective CMO is based on establishing and maintaining credibility and trust with the urban populace and civilian organizations operating in the urban environment. Because of the potential importance of CMO, commanders may consider placing an assistant commander, executive officer, or some other senior leader to monitor, guide, liaise with other units and organizations, and, overall, ensure that the appropriate command emphasis is placed on these activities.

Civil Affairs

5-90. Civil affairs (CA) units and teams will be critical during UO. While any military force can conduct CMO, CA units are specifically organized, trained, and equipped to conduct activities in support of CMO. They have experience in planning and conducting CMO, a regional focus (which includes enhanced cultural awareness and language training), civilian technical expertise that parallel common urban government and administration functions, and experience in negotiation and mediation. Such experience ensures relevant support to commanders conducting urban operations. (As it is likely that there will not be enough CA forces to cover all requirements, commanders should survey their Soldiers to determine the whether any possess civilian expertise that may be beneficial to CMO.) CA units organize their skills into six functional specialty areas (see Figure 5-5). Commanders should use these skills, often unfamiliar to most military personnel, to—

- Develop their situational understanding of the urban environment (particularly the infrastructure and society).
- Plan CMO to support UO.
- Achieve many of the fundamentals of UO.

Note: To ensure optimal integration, commanders must establish clear command and support relationships between CA and maneuver forces.

Figure 5-5. Civil affairs functional specialties

5-91. In addition to providing essential information for understanding the urban environment, CA personnel and activities help shape the battlefield, engage a civil problem, consolidate gains, and transition to a legitimate civil authority. Specifically, CA teams and CMO help urban commanders—

- Minimize civilian interference with UO and the impact of urban operations on the populace and infrastructure. CA personnel can help establish and run a civil-military operations center (CMOC), discussed later in this chapter, to coordinate UO with civilian agencies (both governmental and nongovernmental), other services, and multinational partners.
- Provide advice and assistance to restore or rehabilitate portions of the infrastructure.
- Plan, supervise, and execute necessary populace and resources control measures (in close coordination with military police units) until no longer required or the urban operation is completed.
- When requested or when military necessity or legitimate directives require, establish all or portions of the urban civil administration.
- Determine available supplies and services in the urban area and if necessary assist in negotiating their acquisition. CA also help commanders assess the capability, dependability, and willingness of urban sources to provide and sustain identified needs as well as to calculate the impact of using them on other aspects of the urban environment.
- In conjunction with the SJA, fulfill the Army's responsibilities toward the urban population under international, host-nation, and U.S. law.
- Plan and conduct the transition of control for the urban area or operation to another military or civilian governmental or nongovernmental organization or agency.

5-92. As part of CMO, stability or civil support operations will be critical as commanders begin to restore essential services to meet the emergency needs of a potentially devastated urban population. Besides repairing or restoring vital services and other critical infrastructure, these and other projects may also stimulate the urban area's economy and provide work for the unemployed. (Employed workers are less likely to be co-opted by terrorists, insurgents, and criminal gangs.) As such, they can be a critical stabilizing tool for commanders as they attempt to transition control and responsibility to other agencies or to legitimate civilian control. These projects may include—

- Providing clean water.
- Restoring waste management functions (less-extensive sewer repairs and trash collection and disposal activities).
- Improving medical care.
- Rehabilitating schools.
- Assisting in the establishment of local governments.
- Improving security and emergency services (police, fire, and rescue).
- Providing electrical generators for hospitals, clinics, and factories.
- Restoring irrigation to surrounding farmland.
- Repairing important cultural infrastructure.

5-93. As part of the initial planning process, CA units conduct an *area assessment*, which can provide commanders with essential information about the environment (see FM 41-10). Commanders should integrate this initial assessment into the overall urban-focused IPB process. To help analyze civil considerations in any environment, commanders and staffs can consider many characteristics such as areas, structures, capabilities, organizations, people, and events (ASCOPE). These characteristics easily align with an urban area's three main characteristics of terrain, society, and infrastructure; and, like them, they are overlapped and interdependent (see FM 6-0). Of these, the category of events is a critical addition to urban characteristics previously categorized and discussed. Significant spikes (and lulls) to threat and civilian activity can be expected to occur in conjunction with important religious, historic, and political events (for example elections). Therefore, commanders should determine, track, and pay particular attention to these events when planning operations and managing transitions (see also the Key Event Chart discussion in Appendix B, paragraph B-43).

5-94. Similar to PA operations, CMO are related to information operations. The nature of CMO and the need for CA personnel to develop and maintain close relationships with the urban population put CA personnel in a favorable position to collect information. CA personnel work daily with civilians, their equipment, and their records that may be prime sources of information. If used correctly, CA personnel can complement the intelligence collection process necessary to understand the dynamic societal component of the urban environment and detect significant changes. However, CA personnel are not, and cannot appear as, intelligence agents; otherwise, it will undermine their ability to interact with the civilian community. Examples of information available to CA units include government documents, libraries, and archives; files of newspapers and periodicals; industrial and commercial records; and technical equipment, blueprints, and plans.

5-95. Overall, CA expertise is critical to a commander's understanding of the complexities of the infrastructure and societal components of the urban area. These components (together with the terrain or physical component of the urban area) interconnect. CA forces help identify and understand the relationships and interactions between these urban components. From this understanding, commanders can anticipate how specific military actions affect the urban environment and the subsequent reactions. CA personnel help commanders predict and consider the second- and third-order effects and reactions as well as the long-term consequences. Understanding these long-term consequences helps in achieving stability and ensuring a smoother and speedier transition of the urban area back to civilian responsibility and control. Oppositely, an unconsidered short-term solution may, in the end, turn out to be the solution requiring the greatest amount of time. Poor solutions may exacerbate the situation. Others may contribute little to conflict resolution and waste precious resources (particularly time) until a more reasoned approach is formulated—one that should have been the first solution.

Civil-Military Reconstruction Efforts

5-96. Reconstruction efforts and activities, an integral part of CMO, should complement NGO efforts and are conducted in concert with larger urban area and nationwide projects often headed by the U.S. Army Corps of Engineers (USACE) and the U.S. Agency for International Development (USAID). For example, extensive, large-scale sewer projects or major repairs to the urban area's electrical infrastructure would be left to the USACE and USAID. When planning and conducting urban reconstruction, commanders should—

- **Keep expectations realistic.** With so much to do, it is easy for Army forces to become overcommitted. Commanders must focus on the achievable, build off successes, and create constructive forward process.
- **Seek opportunities.** While commanders think long-term, they should look for catalytic investment of resources. Minimal resources provided at the right time can generate significant activities resulting in increased momentum toward the desired end state.
- **Align resources with local needs.** In analyzing the merit of a project, commanders must check to determine if it is supported and desired by the urban populace. Local participation in the process reinforces ownership and acceptance and eventual transition of responsibility.
- **Balance speed and quality.** Commanders must balance the achievement of rapid response with quality construction and repairs.
- **Ensure efforts do not undermine locals.** Army reconstruction efforts should not undermine the growth and legitimacy of local institutions and authorities.
- **Integrate in the targeting process.** Commanders should routinely include approval of civil-military reconstruction products as part of the information operation's contribution to the overall targeting process.

INTEGRATION OF CONVENTIONAL AND SPECIAL OPERATIONS FORCES

5-97. One important Army and joint resource that commanders of a major operation can use to influence urban operations is SOF. Several types of these forces exist (including the CA forces discussed above), each with unique and complementary capabilities. They can be extremely valuable in UO for their ability to execute discrete missions with a higher degree of precision than conventional forces, to provide information, and to enhance cultural understanding. However, the challenges of using SOF include command and control; integration; coordination with conventional forces that will normally command, control, and conduct the bulk of UO tasks; and the imprudent inclination to use SOF forces for conventional purposes. The density and complexity of UO make close coordination and synchronization of conventional forces and SOF essential to mission success. The nature of the environment dictates that both forces will work in close proximity to each other; the separation in space and time between SOF and conventional forces will often be much less in urban areas than in other environments. Overall, the nature of the environment demands a synergistic combination of capabilities to achieve effects on the threat and mission success.

5-98. Successfully integrating SOF occurs with proper integration into, or coordination with, the command structure of the force conducting the UO. SOF within a theater (less CA and tactical PSYOP) ordinarily falls under joint command and control. Therefore, the commander of the major operation responsible for an urban area, if he is not a JFC, will have to coordinate through the JFC to integrate SOF capabilities into the UO. Examples of critical coordination elements include:

- Boundaries.
- No-fire areas. (If assigned, SOF liaison officers (LNOs) must be routinely included in targeting meetings.)
- Coordination points.
- The exchange of key intelligence and information.
- Requirements to support personnel recovery contingencies.

Conventional and Special Forces Integration

On 20 April 2003 during OPERATION IRAQI FREEDOM (OIF), an Army Special Forces (SF) operational detachment alpha (ODA) was patrolling through villages and secondary roads east-southeast of An Najaf. A local Iraqi approached the team and volunteered the location of a senior Ba'ath Party official in the nearby town of Ghamas. Unfortunately, he did not provide an exact location but rather an approximate location and the name of the family dwelling. The ODA coordinated with their company commander and nearby elements of the 101st Airborne Division to conduct a raid on the residence at dawn the next day. Joint planning, preparation, and rehearsals among the units occurred throughout the night.

At 0430, the raid convoy departed its base en route to Ghamas. The small task force consisted of an assault team, a security team, a command and control element, and three blocking forces composed of Soldiers from a scout platoon and antiarmor company of the 101st. At 0515, the convoy hit the release point outside Ghamas. The 101st vehicles moved to their blocking positions at three bridges surrounding the town. The remainder of the task force moved into the town. An interpreter quickly found a local guide who knew where the house was located. He took the SF team to a walled two-story dwelling with a courtyard. The security team isolated the objective and provided overwatch while the assault team forcibly seized the house and apprehended three adult males. Among them was Abd Hamden, the target of the raid and a senior Ba'ath Party official from Baghdad.

After a tactical interrogation, one of the men provided the location of a Fedayeen major nearby. The SF and conventional force raided this house minutes later but only found the major's relatives. As the Soldiers left the town with their captives, they were cheered by the locals throughout the town. Although only one small example, the extensive coordination between SOF and conventional forces during urban operations was a tremendous success during OIF.

SYNCHRONIZATION OF ACTIVITIES WITH OTHER AGENCIES

5-99. The population density of the urban environment, its economic and political importance, and its life-supporting infrastructure attracts many types of organizations. These organizations include—

- Other U.S. governmental agencies.
- International governmental organizations.
- Allied and neutral national governments.
- Allied and coalition forces.
- Local governmental agencies and politicians.
- NGOs.

5-100. Even in a major operation or campaign, many organizations operate in the area as long as possible before combat or as soon as possible after combat. Therefore, coordination with these organizations sharing the urban AO will be essential to achieve synchronization; however, effective synchronization is challenging, time consuming, and manpower intensive. The staffs of larger headquarters (divisions or higher) normally have the breadth of resources and experience to best conduct the necessary synchronization. They can effectively use or manage the organizations interested in the urban area and mitigate their potential adverse effects on UO. By taking on as much of the synchronization requirements as possible, the operational headquarters permits its tactical subordinates to remain focused on accomplishing their tactical missions. The higher headquarters should assume as much of the burden of synchronization as possible. However, the density of the urban environment will often require that smaller tactical units coordinate and synchronize their activities with other agencies and the local civilian leadership (formal and informal) simply because of their physical presence in the units' AOs. In urban stability and civil support operations, mission accomplishment will require effective civil-military coordination and synchronization activities and measures at all levels as either a specified or implied task.

CIVIL-MILITARY OPERATIONS CENTERS

5-101. To coordinate activities among the varied agencies and organizations operating in an urban area and the local population, urban commanders can establish a CMOC. The CMOC synchronizes Army activities and resources with the efforts and resources of all others involved (see FM 41-10). This can be particularly important in stability and civil support operations where combat operations are not the dominant characteristic of the major operation. CMOCs can be established at all levels of command. Hence, more than one CMOC may exist in an AO, particularly large urban areas. CMOCs may be organized in various ways and include representatives from as many agencies as required to facilitate the flow of information among all concerned parties. Commanders must still ensure that force protection and OPSEC requirements are not compromised. Effective CMOCs can serve as clearinghouses for the receipt and validation of all civilian requests for support, can aid in prioritizing efforts and eliminating redundancy, can decrease the potential for inappropriate displays of wealth by one or more of the participating organizations, and, most importantly, can reduce wasting the urban commander's scarce resources.

LIAISON OFFICERS

5-102. LNOs—sufficiently experienced and adequately trained in liaison duties and functions—are necessary to deal with the other agencies that have interests in the urban area. Army LNOs work with the lead agency or other organizations (including local civilian agencies such as police) that the commander has identified as critical to mission success. Together they work to rapidly establish unity of effort and maintain coordination, often before a CMOC is established. The additional coordination afforded by the physical presence of LNOs within these organizations may be required even after the CMOC is fully functional. When commanders lack enough LNOs to meet requirements, they should prioritize and often assign a single LNO to several organizations. That LNO will then share his time and presence to those organizations based on the situation and his commander's guidance.

COMMANDER'S PERSONAL INVOLVEMENT

5-103. Overall, establishing a close relationship with other agencies and the urban civilian population will often be a major, positive factor in successful mission accomplishment, particularly in urban stability operations. Despite internally established command and control relationships, commanders that develop a direct and personal relationship with the leaders and staff of other agencies can often avoid conflict, win support, foster trust, and help eliminate the "us versus them" mentality that can frustrate cooperation among Army forces and civilian organizations.

Chapter 6

Foundations for Urban Operations

Utilities such as electricity and water are as much weapons of war as rifles, artillery pieces or fighter aircraft. . . . In the case of Manila, where there was a noncombatant, civilian population of one million in place, it was the attacker's aim to capture the utilities which the defender planned to destroy.

The Battle for Manila

Commanders conducting major urban operations (UO) use their ability to visualize how doctrine and military capabilities are applied within the context of the urban environment. An operational framework is the basic foundation for this visualization. In turn, this visualization forms the basis of operational design and decision making. To accurately visualize, describe, and direct the conduct of UO, commanders and their staffs must understand the basic fundamentals applicable to most UO.

URBAN OPERATIONAL FRAMEWORK

6-1. Army leaders who have an urban area in their area of operations (AO) or are assigned missions in an urban area follow an urban operational framework. They identify the portion of the urban area essential to mission success, shape the area, precisely mass the effects of combat power to rapidly dominate the area, protect and strengthen initial gains without losing momentum, and then transition control of the area to another agency. This framework divides into five essential components: understand, shape, engage, consolidate, and transition. These five components provide a means for conceptualizing the application of Army combat power and capabilities in the urban environment.

6-2. The urban operational framework assists commanders in visualizing urban operations. This framework is simply an aid to the commander. It is not sequential, nor is it a planner's tool for phasing an operation. Commanders should combine the urban operational framework with—

- The principles of war.
- The tenets of Army operations.
- The components of operational design.
- Considerations for stability operations and civil support operations.
- Sustainment characteristics.
- Running estimates.
- Commander's critical information requirements (CCIR).
- Each commander's experience.

The framework contributes to the visualizing, describing, and directing aspects of leadership that make commanders the catalysts of the operations process (see Figure 6-1). In the same manner, the urban operational framework contributes to the overall operations process (see FM 3-0).

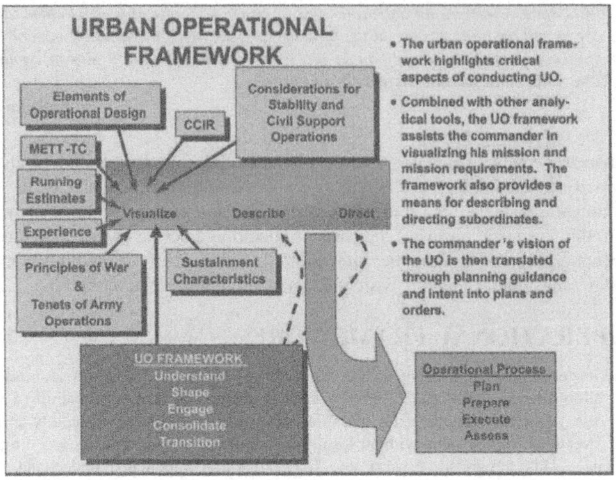

Figure 6-1. The urban operational framework and battle command

UNDERSTAND

6-3. Understanding requires continuous assessment—throughout planning, preparation, and execution—of the current situation and progress of an operation, and the evaluation of it against measures of effectiveness to make decisions and adjustments. Commanders use visualization, staff officers use running estimates, and all use the intelligence preparation of the battlefield (IPB) process to assess and understand the urban environment. Commanders and staffs begin to understand by observing and then collecting information about the situation. They observe and learn about the urban environment (terrain, society, and infrastructure), and other factors of METT-TC—mission, enemy, weather, troops and support available, and time available. They use intelligence, surveillance, and reconnaissance means; information systems (INFOSYS); and reports from other headquarters, services, organizations, and agencies. Then they orient themselves to the situation and achieve situational understanding based on a common operational picture (COP) and continuously updated CCIR. Largely, the ability to rapidly and accurately achieve a holistic understanding of the urban environment contributes to the commanders' abilities to seize, retain, and exploit the initiative during UO.

Disproportionately Critical

6-4. The Army operations process requires continuous assessment and understanding; it precedes and guides every activity. In UO, however, in-depth understanding is disproportionately critical for several reasons. First, each urban environment is unique. Other environments can be studied and their characteristics quantified in a general manner with accuracy. This is fundamentally not true of different urban areas. The characteristics and experience in one urban area often have limited value and application to an urban area elsewhere. This characteristic sets UO apart from operations in other environments.

Extremely Dynamic

6-5. The urban environment is also extremely dynamic. Either deliberate destruction or collateral damage can quickly alter physical aspects of the urban environment. The human aspect is even more dynamic and potentially volatile. A friendly civil population, for example, can become hostile almost instantaneously.

These dynamics (combined with initial difficulty of understanding and describing this unique environment) make it difficult for commanders and staffs to initially develop and maintain a COP and establish situational understanding. Furthermore, public reaction to media coverage of the urban operation and political changes influence national security strategy and objectives. Such changes can affect the basic nature of an operation, especially after it has commenced. Anticipating these potential effects and developing appropriate branches and sequels based on an accurate understanding often determines how quickly commanders can achieve the desired end state.

Risk Assessment

6-6. As in any environment, UO pose both tactical and accident risks. However, the level of uncertainty, ambiguity, and friction can often be higher than that of many other environments. Such challenges increase the probability and severity of a potential loss due to the presence of the enemy, a hostile civilian group, or some other hazardous condition within the urban environment (see Determining the Necessity of Urban Operations in Chapter 5). Therefore, commanders must—

- Identify, assess, and understand hazards that may be encountered in executing their missions.
- Develop and implement clear and practical control measures to eliminate unnecessary risk.
- Continuously supervise and assess to ensure measures are properly understood, executed, and remain appropriate as the situation changes.

6-7. Risk decisions are commanders' business. However, staffs, subordinate leaders, and even individual Soldiers must also understand the risk management process and continuously look for hazards at their level or within their area of expertise. Any risks identified (with recommended risk reduction measures) must be quickly elevated to the appropriate level within the chain of command (see FM 3-100.12 and FM 5-19).

Complex and Resource Intensive

6-8. The urban environment is the most complex of all the environments in which the Army conducts operations. It is often comprised of a diverse civil population and complex, ill-defined physical components. A sophisticated structure of functional, social, cultural, economic, political, and informational institutions unites it. Thus, the analysis to understand the environment is also complex and time and resource intensive. The nuances of the urban environment can take years to uncover. Hence, constant analysis of the environment requires greater command attention and resources. Accurately understanding the environment is a prerequisite to shaping it, and both understanding and shaping activities are crucial to effectively engage the elements of the urban environment critical to success.

SHAPE

6-9. Shaping operations, part of all Army operations, are essential to successful UO. They set the conditions for decisive operations at the tactical level in the urban area. Rapid action, minimum friendly casualties, and acceptable collateral damage distinguish this success when the AO is properly shaped. Failure to adequately shape the urban AO creates unacceptable risk. The commander of a major urban operation has several resources with which to begin shaping the AO. Important capabilities include—

- Fires.
- Information operations.
- Special operations capabilities.
- The maneuver of major subordinate units.

6-10. Critical urban shaping operations may include actions taken to achieve or prevent isolation, understand the environment, maintain freedom of action, protect the force, develop cooperative relationships with the urban population, and train Army forces for sustained UO.

Isolation

6-11. Isolation of an urban environment is often the most critical component of shaping operations. Commanders who's AO includes operationally significant urban areas often conduct many shaping operations to isolate, or prevent isolation of, those areas from other parts of the AO. Likewise, commanders operating in the urban area focus on isolating decisive points and objectives in the urban area or averting isolation of points that are critical to maintaining their own freedom of action. Isolation is usually the key shaping action that affects UO. It applies across full spectrum operations. Most successful UO have effectively isolated the urban area. Failure to do so often contributed to difficult or failed UO. In fact, the relationship between successful isolation and successful UO is so great that the threat often opposes external isolation actions more strongly than operations executed in the urban area (or critical areas within). In some situations, the success of isolation efforts has been decisive. This occurs when the isolation of the urban area compels a defending enemy to withdraw or to surrender before beginning or completing decisive operations. In UO that are opposed, Army forces attempt to isolate the threat three ways: physically, electronically, and, as a resultant combination of these first two, psychologically (see Figure 6-2).

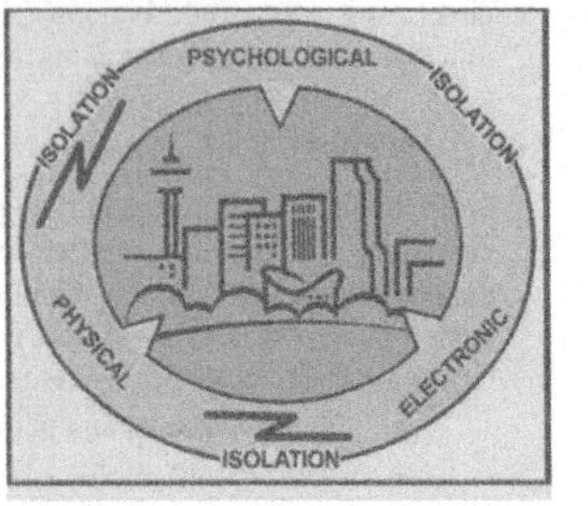

Figure 6-2. Urban isolation

Physical Isolation

6-12. In offensive UO, physical isolation keeps the threat from receiving information, supplies, and reinforcement while preventing him from withdrawing or breaking out. Conversely, a defending Army force attempts to avoid its own physical isolation. Simultaneously, this force conducts operations to isolate the threat outside, as they enter, or at selected locations in the urban area. Physical isolation can occur at all levels. In many situations, the commander of a major combat operation may attempt to isolate the entire urban area and all enemy forces defending or attacking it. At the tactical level, forces isolate and attack individual decisive points often using a cordon technique. In stability operations, physical isolation may be more subtly focused on isolating less obvious decisive points, such as a hostile civilian group's individual leaders. In many operations, isolation may be temporary and synchronized to facilitate a decisive operation elsewhere. To effectively isolate an urban area, air, space, and sea forces are often necessary additions to the capabilities of ground forces.

Electronic Isolation

6-13. Electronic isolation is achieved through offensive information operations (IO). Electronic warfare (particularly two of its components: electronic warfare support and electronic attack) and computer network attack are critical to electronic isolation (see FM 3-13 and the Information Operations discussion in Chapter 5). At the operational level, offensive IO aims to quickly and effectively control the information flow into and out of an urban area. This isolation separates the threat's command and control (C2) system in the urban area from its operational and strategic leadership outside the urban area. Offensive IO also focuses on preventing the threat from communicating with civilians through television, radio, telephone, and computer systems. At the tactical level, IO aim to isolate the threat's combat capability from its C2 and leadership within the urban area, thus preventing unity of effort within the urban area. Defensive IO can prevent isolation of friendly forces defending in an urban area.

Psychological Isolation

6-14. Psychological isolation is a function of physical actions, electronic warfare, and other forms of IO, especially military deception and psychological operations. Psychological isolation denies the threat political and military allies. It separates the enemy or hostile civilian group from the friendly population, nongovernmental organizations (NGOs) operating in the urban area, and from political leaders who may consider supporting Army forces. Psychological isolation destroys the morale of individual enemy soldiers or hostile civilians. It creates a feeling of isolation and hopelessness in the mind of the threat. It undermines the threat leadership's ability to think, plan, decide, and act. As such, it undermines the confidence of the threat in their leadership. On the other hand, IO, as well as the disciplined conduct of Army personnel, can help to forge legitimacy for Army operations. In stability operations, psychologically isolating the threat can result in the friendly urban population and NGOs positively supporting Army operations.

Other Shaping Actions

6-15. Other shaping actions can include the proper sequencing and deployment of forces, reconnaissance operations, and force protection. These actions contribute equally to the success of any urban operation. Commanders must understand how the urban environment affects their ability to accomplish these shaping actions. However, civil-military operations (CMO), another closely related activity of IO, are important to shaping the urban operational environment for decisive operations. The specific civil-military task can vary greatly and may include affecting a cooperative relationship with the civil political system, protecting portions of the civil population or infrastructure, or establishing refugee camps or safe areas for noncombatants. This is most true in stability and civil support operations. Successful CMO also can contribute to the psychological isolation of the threat. (See Civil-Military Operations in Chapter 5 for a more detailed discussion.)

Training and Education

6-16. Finally, Army commanders must consider that critical shaping actions often occur prior to the urban operation in the form of professional education, and home-station and in-theater training. Commanders can enhance training through joint, interagency, multinational, and combined arms exercises and effective rehearsals.

Engage

6-17. In UO, Army forces engage by appropriately applying their full range of capabilities against decisive points leading to centers of gravity. Successful engagement takes advantage of the Army force's superior training, leadership, and, within the constraints of the environment, equipment and technology. Successful engagement also requires the establishment of necessary levels of control and influence over all or portions of the urban environment until responsibilities can be transferred to other legitimate military or civilian control. Engagement may range from the overwhelming, yet precise, application of combat power to defeat a threat, to large-scale humanitarian assistance operations, to unobtrusive advice and assistance to the urban security forces. It may require brief engagement in a relatively small portion of an urban area or with small segments of the population to long-term engagement throughout multiple, large urban areas and with vast sectors of the population. The commander's ability to engage in a specific urban operation depends, as always, on the situation and the assigned mission.

Offense: Attack Decisive Points

6-18. In offensive UO, forces successfully engage by striking at the enemy's center of gravity using multiple offensive actions from unexpected directions and throughout all dimensions. Army forces aim to engage identifiable decisive points. Successful efforts against decisive points lead to effects on the center of gravity. The center of gravity will differ in each offensive situation. It may be an individual enemy leader, the enemy's combat power, the enemy's communications capability, or a physical structure of cultural, political, or economic significance.

Defense: Deny Vital Functions and Critical Infrastructure

6-19. In defensive UO, engage translates into denying the enemy control of the vital functions and critical infrastructure of the urban area. Forces achieve this by leveraging the defensive advantages of the urban terrain, defending essential areas in depth, using economy of force in nonessential areas, controlling the enemy direction of attack with natural and man-made obstacles, and retaining the initiative through counterattacks.

Stability and Civil Support: Apply Innovation and Imagination

6-20. The ability to engage in urban stability operations hinges on the type of stability operation commanders execute. In a noncombatant evacuation operation, forces limit engagement to finite geographic areas and times. In contrast, a peace operation may require engagement of a large urban area for an extended time. In this operation, engage is defined as using the array of Army capabilities to create specific conditions among the belligerents. Thus, the techniques used for engagement in stability operations vary according to the situation and as situations mature during long-term operations.

6-21. In urban civil support operations, engagement is accomplished by innovative and subtle application of Army capabilities. Since Army forces usually support other agencies that lead the operation, effective engagement results from carefully and discretely applying Army capabilities to the tasks assigned by the lead agency. In a humanitarian relief situation, Army forces may be tasked to transport supplies in the urban area. Successful engagement in this activity then becomes the goal of Army forces and may be achieved by providing, managing, and protecting transportation assets and clearing and maintaining sufficient lines of communications (LOCs).

CONSOLIDATE

6-22. Army forces consolidate to protect and strengthen initial gains and ensure retention of the initiative. Consolidation includes actions taken to eliminate or neutralize isolated or bypassed threat forces (including the processing of enemy prisoners and civilian detainees) to increase security and protect LOCs. It includes the rapid repositioning and, as necessary, reorganization of forces and intelligence, reconnaissance, and surveillance (ISR) assets. Consolidation may also include activities in support of the civilian population such as relocating displaced civilians, reestablishing law and order, humanitarian assistance and relief operations, and restoration of key urban infrastructure. A significant consolidation effort may be the reduction or elimination numerous explosive hazards resulting from previous urban combat operations. During consolidation, commanders often coordinate with NGOs to establish unity of effort in the execution of stability or civil support operations. Consolidation begins with the initiation of the urban operation and follows each critical activity.

TRANSITION

6-23. When planning UO, commanders ensure that they plan, prepare for, and manage transitions. Transitions are movements from one phase of an operation to another or a change in responsibility for all or portions of the urban environment from one unit, organization, or agency to another. They involve significant changes in the type of operation, concept of the operation, mission, situation, task organization, forces, resource allocation and support arrangements, or C2. Transitions occur in all operations, but in UO they occur with greater frequency and intensity, are more complex, and often involve agencies other than U.S. military organizations. For example, a successful attack may transition to a defend mission that includes not only defense tasks but also stability tasks. All operations will likely include a transition of responsibility for some aspect of the urban environment to (or back to) a legitimate civilian authority. Unless planned and executed effectively, transitions can reduce the tempo of the urban operation, slow its momentum, and cede the initiative to the threat.

Mental and Physical Preparation

6-24. Transitions occur as conditions warrant. They can be carefully planned and controlled, or they can be quick and dramatic, such as the swift transformation of a stability operation into offense or defense—and back again. Units prepare mentally and physically to address rapid transitions. Accordingly, plans include branches and sequels that address anticipated or possible transition points. When the dominant type of operation changes from an offense to stability or civil support, the types of units originally conducting the UO may no longer be appropriate. A large mobile reserve may permit increased flexibility to react to unplanned transition requirements. Operations or units in one part of an urban area may transition before operations or units in a different part of the same urban area. This will require commanders to exercise careful command and control and execute various types of operations and associated tasks simultaneously.

Transition to Legitimate Civilian Authorities or Agencies

6-25. Combat operations are not an end to themselves. They are inevitably a means of transition, an activity for moving from an unsatisfactory state of affairs to an improved end (as determined by national objectives). In UO, a distinct aspect of transition is the requirement to quickly and efficiently transition the major portions of Army responsibilities to civil agencies. Some tasks to which units will transition are not traditional combat tasks but rather stability tasks more closely associated with CMO. In stability or civil support operations this is often a near-term critical mission objective. In these operations, commanders aim to alleviate the circumstances requiring Army forces and ensure that other civilian agencies assume the functions provided by Army forces. Following combat operations, civilian agencies are assisted and encouraged to quickly resume specific support activities—such as providing sanitary services, food services, law enforcement, and health services—because of their high demand on Army resources.

6-26. Transitioning responsibility for the urban environment (or selected portions) to legitimate civilian agencies or authorities requires commanders to make an accurate assessment and develop an in-depth understanding of the civilian organization's ability to accept and handle the responsibility. Commanders consider not only resources available but whether the organization has the appropriate leadership and technical expertise as well as a vision in consonance with U.S. objectives. Commanders should avoid forcing a transition only to alleviate the burden on Army forces. On the other hand, they may need to encourage acceptance of responsibility while continuing to provide "behind-the-scenes" advice, assistance, and resources. Transitioning responsibility prematurely may alleviate the commander's immediate resource concerns but may, in the longer term, negatively impact the achievement of strategic objectives.

Clearly Visualize and Describe the End State

6-27. Army UO conclude when Army forces depart and have no further mission requirements in the urban area. At the outset, commanders visualize and describe the intended end state of a unit's execution of UO. Commanders then clarify and update this visualization as the political or strategic situation is refined or changes. Importantly, commanders must create a command climate and procedures that allow their staffs and subordinate commanders the ability to ask questions that will enable all to share in an accurate perception of the commander's visualization. This enables subordinate units to identify likely transitions and ensures that current operational planning takes into account second- and third-order effects. As long as an active Army AO contains an urban area, some type of urban operation will exist. After urban combat successfully ends, combat forces may move on. Support forces conducting sustaining operations may then occupy the area and continue to conduct a different form of UO.

Applying the Urban Operational Framework: Panama – 1989

The U.S. conducted OPERATION JUST CAUSE in December 1989 to remove the illegal ruler of Panama, Manuel Noriega, and to restore that country to a democracy. It also conducted the operation to ensure the safety of a substantial number of U.S. personnel as well as the security of U.S. interests in Panama. The major focus of JUST CAUSE was in Panama City, the country's capital. Most operations occurred in this large urban area, one of the numerous smaller urban areas, or the urban-like military bases. These bases increased the AO and were directly linked to operations in the capital city. This successful operation illustrates how commanders can apply the urban operational framework to visualize, describe, and direct the critical aspects of urban operations.

Understand

The synchronization achieved during the operation may have obscured the challenges faced in the initial assessment process in Panama. However, it was not as simple as it may have seemed. Using the framework of the urban environment, U.S. forces required details of the physical characteristics of the environment, the infrastructure, and the human dimension including the capabilities of the Panamanian military.

Because Army forces had a long history in Panama, commanders clearly understood the physical challenges and layout of critical urban areas (see Figure 6-3), particularly Panama City. They also understood how the infrastructure in each urban area functioned and which parts was a key element to success. Examples of key portions of the infrastructure included the Madden Dam, which controlled the water flow through the Panama Canal, and the Cerro Azul television tower, which was the main Panamanian broadcast tower.

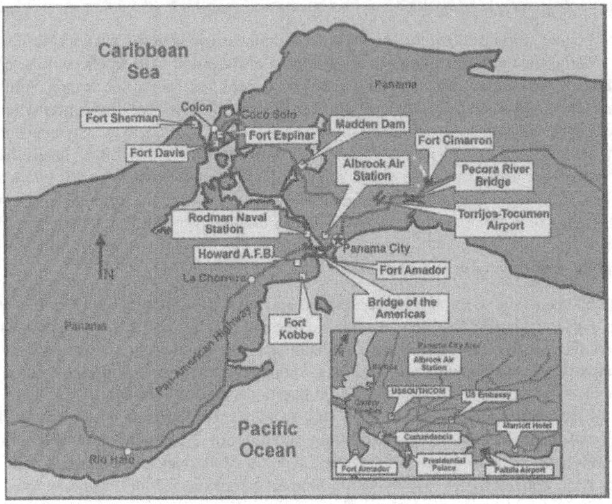

Figure 6-3. Panama

Collecting information and developing intelligence on the human elements of the urban environment was critical to operational success and a challenge. Because the Panamanian Defense Force (PDF) had traditionally been an ally of the United States, Army forces did not have a systemic database that adequately depicted their order of battle and their true capabilities. Additionally, much of the situation in Panama was colored in political terms making it more difficult for traditional military sources to evaluate the status of PDF forces. For example, Army planners needed to know if PDF military units (when faced with a formidable U.S. force) would fight at all for Noriega and if they did fight, how hard and long would they fight. The answers depended largely on their political loyalty to Noriega and on the individual loyalty of the unit officers to the Panamanian president. Thus, Army commanders needed to understand the military characteristics of PDF units and their political affiliations and tendencies.

Because transition from combat to noncombat tasks would be critical to achieving all objectives, particularly the restoration of democracy, Army forces also needed an accurate assessment of the political opposition to Noriega—including that opposition's capabilities and vulnerabilities. Again, Army forces were required to make assessments outside those needed solely for combat operations. Ultimately, assessing the political opposition's vulnerabilities led to assigning Army units to protect them throughout the operation so that they could serve as a foundation for a new democratic government.

Finally, the commander's assessment included an evaluation (often subjective) of the attitudes and disposition of the Panamanian people. Human intelligence (HUMINT) was the primary source of information on the population. Army forces had good access to the population because of their close proximity and historical ties to Panama. Many Soldiers were married to Panamanians, and the Army had total access to local media and to prominent individuals.

National imagery and special operations forces (SOF) also contributed to the ability of Army forces to assess the urban environments of JUST CAUSE. All units executing operations had detailed satellite photos of objective areas. Additionally, key objectives were placed under SOF surveillance well in advance. This surveillance revealed unit readiness, vulnerabilities, detailed disposition, and other patterns critical to mission success. The combination of the two capabilities allowed units to plan and achieve the synchronization necessary for such a complex urban operation.

Shape

During OPERATION JUST CAUSE, commanders conducted numerous shaping operations to establish the conditions for the decisive operations. Many operations were designed to control information, such as an assault on the Azul television tower identified during the assessment of the infrastructure. Planners designed many shaping operations to isolate various garrisons and PDF units. An example of tactical isolation was the plan for the Pacora River Bridge to prevent reinforcements from reaching the garrison at Torrijos-Tocumen Airport.

Operational isolation was achieved through the Ranger Regiment's and 82nd Airborne Division's assault on targets at Rio Hato in the west and Fort Cimarron in the east. These actions in conjunction with the securing of Maddam Dam had the primary objective of isolating Panama City. They were also the largest of the major actions occurring during OPERATION JUST CAUSE. The airborne assault was also the largest airborne operation conducted by U.S. forces since World War II. This large-scale shaping operation demonstrates that shaping operations are critical to mission success and can be more resource intensive than the actual operations that achieve domination.

Engage

U.S. Army forces achieved successful engagement in OPERATION JUST CAUSE by establishing unchallenged military control over Panama City and eliminating Noriega's capability to challenge that control. Toward this end, the operation attacked two decisive points. The first was the assault on the PDF headquarters located in Panama City: the Comandancia. The second was the operation undertaken to locate and seize Noriega himself.

Three battalions of task force (TF) Bayonet (5-87th Infantry, 1-508 Infantry [Airborne], and 4-6th Infantry [Mechanized]) executed the attack on the Comandancia and Fort Amador. They were also tasked to protect the American Embassy in downtown Panama City. To execute these missions, they moved from various staging areas located throughout the city to their assigned objectives using air assault, mounted, and dismounted approaches. The ground movement through the city proved to be the most difficult and hazardous part of the mission due to the vulnerability of the troops in their armored personnel carriers and trucks. The dismounted movement was slower than the mounted movement but allowed the Soldiers greater cover and concealment.

The strongest opposition to TF Bayonet occurred at the Comandancia. Elements of three PDF companies and two public order companies held out for three hours. The troops moving to Comandancia were subject to a large volume of sniper fire, and in the assault, unidentified indirect fire caused significant casualties among the mechanized forces. TF Bayonet forces, supported by airborne armored reconnaissance vehicles and Hellfire missiles from Apache helicopters, captured the

Comandancia. Commanders noted in particular the precision of the supporting fires from attack helicopters. The assault by fire from supporting AC 130 gunships severely damaged much of the reinforced Comandancia building.

Simultaneously, SOF attacked several targets where Noriega might be located. These initial attacks were unsuccessful. However, many subsequent actions neutralized Noriega's influence and eventually resulted in his apprehension on 3 January 1990. These actions included the well-organized and relentless manhunt conducted by SOF units, the isolation of Panama City itself, population control efforts, sophisticated IO, and cooperation with other U.S. agencies.

Consolidate and Transition

OPERATION JUST CAUSE demonstrated the vital need for a thought-out plan that adequately addresses consolidation and the transition from combat to noncombat before commanders initiate operations. Normally in complex UO, commanders cannot leave the details of transition until after the operation has begun without unacceptable risk to overall mission accomplishment. The follow-on stability operation, OPERATION BLIND LOGIC (later renamed OPERATION PROMOTE LIBERTY) began 24 hours after the initial assault and thus both operations were occurring simultaneously. This simultaneity of different types of operations is typical in major operations conducted in a large urban area. The stability operation involved more time than the combat operation and continued well after the close of OPERATION JUST CAUSE and after most of the major combat units had redeployed. It involved significant resources without the same level of risk to U.S. forces as the combat operations.

Civil affairs (CA) were a dominant part of the transition from combat to stability operations. The 96th Civil Affairs battalion was central to this operation. CA forces established a civil police force, emergency food distribution, property protection, production and distribution of newspapers, cleanup of the city, and building support for a new civil government. Most tasks were coordinated through Army CA forces and executed by other Army forces under the supervision of CA.

IO were also a major aspect of affecting a stable transition and successful postcombat operations. These operations built support for the U.S. operation among the population. They emphasized that the U.S. conflict was with Noriega and not the Panamanian people and that the U.S. forces would depart as soon as a new Panamanian government could take over.

Other U.S. agencies played critical roles in stability operations in Panama. The U.S. Drug Enforcement Agency and Justice Department were important to the negotiations that led to Noriega's capture. The U.S. State Department helped to negotiate for Noriega and develop military policies and plans during stability operations. The American Embassy advised commanders regarding the large diplomatic community that existed in Panama City.

FUNDAMENTALS OF URBAN OPERATIONS

6-28. UO often differ from one operation to the next. However, some fundamentals apply to UO regardless of the mission, geographical location, or level of command. Some of these fundamentals are not exclusive to urban environments. Yet, they are particularly relevant to an environment dominated by man-made structures and a dense noncombatant population (see figure 6-4). Appendix A provides an historical example of how many of these fundamentals applied to an actual conflict situation. Vitally, these fundamentals help to ensure that every action taken by a commander operating in an urban environment contributes to the desired end-state of the major operation.

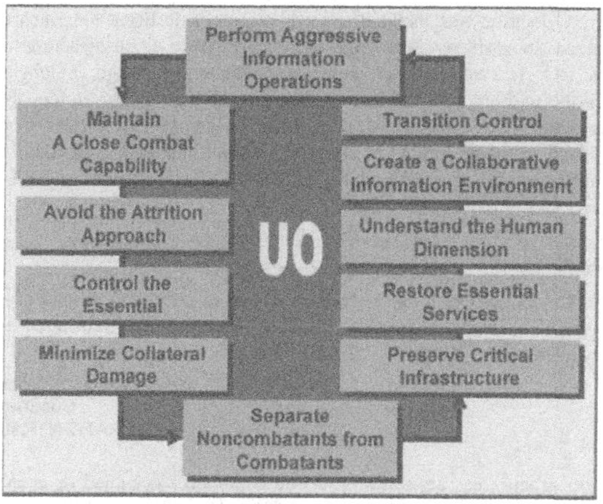

Figure 6-4. Fundamentals of urban operations

PERFORM AGGRESSIVE INFORMATION OPERATIONS

6-29. Information operations aimed at influencing non-Army sources of information are critical in UO across the spectrum of operations. Because of the density of noncombatants and information sources, the media, the public, allies, coalition partners, neutral nations, and strategic leadership will likely scrutinize how Army forces participate in UO. The proliferation of cell phones, video cameras, Internet capability, and media outlets ensure close observation of the activities of Army forces. With information sources rapidly expanding, public information of Army operations will often be available faster than the internal military INFOSYS can process it. Army forces should aggressively integrate IO into every facet and at all levels of the operation to manage perceptions and mitigate unintended consequences. Under media scrutiny, the actions of one Soldier may have significant strategic implications. IO aim to make the information accurate; placed in the proper context of the Army's mission; and available to all interested parties: the public, the media, and other agencies. Ensuring that the urban population understands how they and their urban area fit within the commander's vision, intent, and end state will be a critical task.

MAINTAIN A CLOSE COMBAT CAPABILITY

6-30. Close combat is required in all offensive and defensive UO. This core capability must also be present and visible in urban stability operations and may be required in urban civil support operations. Close combat in any urban operation is resource intensive, requires properly trained and equipped forces, and has the potential for high casualties. However, the ability to decisively close with and destroy enemy forces as a combined arms team remains essential. Potential threats respect strength and, instead, seek to attack ill-prepared or complacent Army forces—not necessarily forces that are better protected or armored, but those less likely or capable of inflicting casualties against them. In stability and civil support operations, a lack of respect and fear of Army forces can hinder recovery as much as the ill-advised use of forces. Hence, all Soldiers, regardless of branch or military occupational specialty, must be properly equipped and trained to fight in an urban environment. This ability allows Army forces to deter aggression, compel compliance, morally and physically dominate a threat and destroy his means to resist, and terminate urban conflicts on the Army commander's terms.

Avoid the Attrition Approach

6-31. Previous Army doctrine was inclined towards a systematic linear approach to urban combat. This approach emphasized standoff weapons and firepower. Army force structure does not support this approach towards UO. It can result in significant collateral damage, a lengthy operation, and an inconsistency with the political situation and strategic objectives. Enemy forces that defend urban areas want Army forces to adopt this approach because of the likely costs in resources. Commanders should only consider this approach to urban combat as an exception and justified by unique circumstances. Instead, commanders should seek to achieve precise, intended effects against multiple decisive points that overwhelm a threat's ability to react effectively.

Control the Essential

I'm talking about attacking those things from which the regime draws power but being very careful about it so that we don't get large bodies of young Americans caught up in house-to-house Berlin, World War II-type scenario.

Lt. Gen. William C. Wallace
Commander, V Corps
OPERATION IRAQI FREEDOM

6-32. Many modern urban areas are too large to be completely occupied or even effectively controlled without an enormous force. Therefore, Army forces focus their efforts on controlling only the essentials to mission accomplishment. At a minimum, this requires control of key terrain. Key terrain is terrain whose possession or control provides a marked advantage to one side or another. In the urban environment, commanders determine key terrain based on its functional, political, economic, or social significance. A power station or a place of worship may be key terrain.

6-33. All principles of war can apply to UO. The principle of mass and the principle of economy of force (in addition to the principle of unity of command discussed in the previous chapter) are particularly important in guiding UO and providing mission focus. Army forces mass combat power only to control those requirements essential for mission success. This permits conservation of combat power. It also implies economy of force and associated risk in those areas where Army forces choose not to exercise control.

Minimize Collateral Damage

6-34. Forces should integrate precision fires, IO, and nonlethal tactical systems consistent with mission accomplishment while decreasing the potential for collateral damage. Commanders develop unique rules of engagement (ROE) for each urban operation and provide necessary firepower constraints. IO and nonlethal systems may compensate for some restrictions, especially in stability or civil support operations. Commanders continually assess the short- and long-term effects of operations and firepower on the population, infrastructure, subsequent missions, and national and strategic objectives. They must also consider what, if any, provisions should be made to amend or address potential collateral damage. Overall, commanders must balance restraint and precision with speed and overwhelming combat power. By avoiding unnecessary harm to all elements of the urban environment, commanders retain the moral high ground and help sustain legitimacy for their operations. Minimization of collateral damage allows civilians to continue to provide for their own needs or the rapid return of the urban area to civilian self-sufficiency.

Separate Noncombatants from Combatants

6-35. Promptly separating noncombatants from combatants (psychologically and physically) may make the operation more efficient and diminish some of the threat's potential asymmetrical advantages. This separation also may reduce restrictions on the use of firepower, enhance force protection, and strip the threat from its popular support base. This important task becomes more difficult when the threat is an unconventional force that can mix with civilians.

6-36. In more recent operations, threats have sought to integrate their military capabilities as closely as possible into the civilian population and infrastructure. In these conditions, commanders must increase their efforts to discriminate between the two. Soldiers managing violence in this setting require the highest level of individual and organizational discipline and judgment. Soldiers will require the mental agility to separate their aggression toward threats from the noncombatant civilian population. The training, effort, and command emphasis in this area is as important as fully successful results. Such efforts strongly impact national and international perceptions of the operation.

Preserve Critical Infrastructure

6-37. Commanders should analyze the urban area to identify critical infrastructure. They should attempt to preserve and protect the critical elements for postcombat sustainment operations, stability or civil support operations, or the overall health and well-being of the indigenous population. Urban areas remain in the AO after combat operations have ceased. Postcombat UO are unavoidable. Different from simply avoiding collateral damage, Army forces may have to initiate actions to prevent an enemy or a hostile civilian group from removing or destroying critical infrastructure and assets. This may include cultural infrastructure such as religious and historical places. In some cases, preserving the infrastructure and the urban society's sources of economic and cultural wealth may be the assigned objective of the urban operation.

Restore Essential Services

6-38. Army forces plan to restore essential services that may fail to function before or during an operation. Essential services include power, food, water, sewage, medical care, and security and law enforcement. When planning for and conducting Army UO, units can use nonlethal and less destructive munitions and capabilities to keep potentially vital infrastructure intact. Initially, Army forces may be the only force able to restore or provide essential services and commanders must plan accordingly. Failure to do so can result in serious health problems for the civilians, which can affect the health of Army forces and negatively impact overall mission success. Army forces seek to transfer responsibility for providing essential services to other agencies, NGOs, or the local government as quickly and effectively as possible. Despite the potential causes for the failure or destruction of essential services, commanders also seek to ensure that civilians continually perceive Army restoration activities as assistance rather than an Army requirement. Otherwise, the civilian populace may be slow to accept or resume responsibility for their urban area.

Understand the Human Dimension

> *Do not try to do too much with your own hands. Better the Arabs do it tolerably than that you do it perfectly. It is their war, and you are to help them, not to win it for them. Actually, also, under the very odd conditions of Arabia, your practical work will not be as good as, perhaps, you think it is. It may take them longer and it may not be as good as you think, but if it is theirs, it will be better.*
>
> T. E. Lawrence
> "Twenty-Seven Articles"

6-39. Commanders carefully consider and manage the perceptions, allegiance, and morale of the civilians. Their assessment of the environment needs to accurately identify the attitudes of the people toward Army forces. Operational guidance to subordinates—including ROE, protection, sustainment operations, and fraternization—is based on this assessment. Commanders expect and consider the demographic variance in the attitudes of an urban population. They cannot inadvertently apply Western cultural norms to a non-Western urban population. Commanders can only make reliable assessments based on a thorough understanding and appreciation of the local society and their culture. Developing this ability also requires the additional ability to effectively share information among all echelons of command.

6-40. Sound policies, proper discipline, adequate consideration for local culture, and rapid engagement of local urban leaders will positively affect the attitudes of the population toward Army forces. Additionally, well-conceived and executed IO will enhance the position of Army forces relative to the urban population. Even during high-intensity urban combat, heightened awareness of and sensitivity toward the civilians can lead to a better postcombat situation than if civil considerations were unobserved or diminished in importance. An improved postcombat situation enhances transition. As the environment of conflict becomes more complex, the human dimension (and associated moral aspects) takes on greater importance and may have the greatest potential for affecting the successful outcome of UO. Therefore, the human aspect and development of cultural acuity creates a discrete planning factor.

Create a Collaborative Information Environment

6-41. The complexity of the urban environment, particularly the human dimension, requires rapid information sharing—providing and receiving—from the national level to the tactical level, among Army headquarters at each echelon, with sister services and coalition partners, and with participating governmental and (at appropriate times) nongovernmental agencies. The analysis of urban information into the relevant intelligence necessary to refine and deepen a commander's understanding of the urban environment and its infrastructure of systems also demands collaboration among the various information sources and consumers. Therefore, commanders seek to establish streamlined procedures, develop commonality among databases, and make best use of existing information systems to disseminate and receive the necessary intelligence and relevant information for subordinates and partner organizations and agencies to exercise effective leadership and decision making and establish synergistic unity of effort in this multifaceted environment.

Transition Control

6-42. Because UO are resource intensive, commanders should plan to end them as quickly as possible, yet consistently with successful mission accomplishment. (Depending on the factors of METT-TC, successful transition may take a few days or many years to achieve.) The end state of all UO transfers control of the urban area to another agency or returns it to legitimate civilian control and responsibility. Rapid transition releases Army resources for use elsewhere and improves the civilian morale and disposition toward Army forces. This requires the successful completion of the Army force mission and a thorough transition plan. The transition plan may include returning control of the urban area to another agency a portion at a time as conditions permit. A successful transition plan should consider early alignment of military capabilities with existing urban governmental and administrative organizations, agencies, structures, and districts. Transition planning must be conducted before the onset of operations and continually adjusted as the situation develops.

Chapter 7

Urban Offensive Operations

... Capture Suez City "provided it does not become a Stalingrad situation."

Order to the Adan Armored Division
prior to its 1973 attack on Suez City
On the Banks of the Suez

Offensive urban operations (UO) are one of the most challenging operations that military forces can undertake. Campaigns and wars have sometimes hinged on their success or failure. Costly in resources, even when successful, they are not lightly entered into. Once engaged, they are executed rapidly and decisively. For reasons already discussed, threat forces defending in UO may gain advantages from the environment while Army force capabilities may diminish. Despite the challenges, Army forces conduct successful urban offensive operations by combining the Army's existing offensive doctrine with a thorough understanding of the environment.

PURPOSE OF URBAN OFFENSIVE OPERATIONS

7-1. Like all offensive operations, urban offensive operations are designed to impose the will of commanders on the threat. The urban offense often aims to destroy, defeat, or neutralize a threat force. However, the purpose may be to achieve some effect relating to the population or infrastructure of the urban area. Army forces may conduct offensive operations to secure a port or a communications center, to eliminate a threat to a friendly government or the urban population, or to deny the threat use of urban infrastructure. No matter the purpose, commanders must use a combined arms approach for successful urban offensive operations.

OFFENSIVE CHARACTERISTICS APPLIED TO URBAN OPERATIONS

7-2. All offensive operations, including those in urban areas, contain the characteristics of surprise, concentration, tempo, and audacity (see FM 3-0). Commanders consider and incorporate these characteristics in their offensive UO plans.

SURPRISE

7-3. Army forces can achieve offensive surprise at two levels: operational and tactical. In urban offensive operations, operational surprise can be decisive. The goal is to attack the urban area before the threat expects it, from a direction he doesn't expect, or in a manner he doesn't expect. In UO, this requires an attack against urban areas that the threat believes will provide sanctuary from the technological advantages of Army forces. Usually, urban areas that meet this criterion are not easily accessible. Army forces launch an attack against these urban complexes in different ways: through a vertical assault using airborne or air assault forces, through an amphibious assault, or through a penetration followed by a rapid and deep advance. All three attacks aim to achieve surprise and to deny the threat time to prepare and establish a defense. Surprise in a major urban operation prevents a threat from falling back to occupy prepared positions in and around an urban area.

7-4. At lower tactical levels, forces achieve surprise by attacking asymmetrically. An asymmetric method attacks the threat so he cannot respond effectively. This may be achieved by using special operations forces

(SOF) against a threat prepared for a conventional attack, by attacking decisively with heavy forces when the threat expects an effort by light forces or SOF, or by leveraging Army forces' extensive information operations (IO) capability. Offensive IO—primarily using IO elements of deception, electronic warfare, and operations security (OPSEC)—can help achieve surprise at all levels (see Chapter 4). Attacking at night surprises the threat and maximizes the Army forces' training, command and control (C2), and technological advantages. Attacking from unexpected or multiple directions achieves surprise by leveraging Army information systems (INFOSYS) and superior synchronization of combat power and capabilities.

CONCENTRATION

7-5. In UO, the attacking force creates a major advantage by concentrating the effects of combat power at the point and time of its choosing. The area and its compartmented effects naturally disperse and dissipate combat capability. Often commanders must position considerable amounts of stay-behind forces to protect hard-fought gains. The environment also hinders repositioning forces rapidly. Such effects can work equally against defending and attacking forces. However, in a well-prepared defense, the defender often has the advantage of interior lines. The defender can reinforce or reposition forces more quickly using covered and concealed routes (such as, sewers, tunnels, or prepared holes made in walls). Successful UO need synchronized air and ground maneuver with overwhelming effects from fires at decisive points on the urban battlefield. To achieve proper synchronization and precise effects, commanders should consider the unique time and distance relationships set by the environment.

TEMPO

7-6. Tempo is the rate of military action; it differs between UO and operations in more open terrain. Often, the primary purpose of the threat's urban defense is to take advantage of that differential to disrupt the rapidity and overall tempo of the Army's major operation. Additionally, the complexity and the potential for increased risk in an urban environment may invoke a cautious and methodical response on the part of commanders and their staffs thereby exacerbating tempo differences. While careful and meticulous preparation and planning are required in UO, commanders conducting major operations that include urban areas must strive to maintain an active tempo in offensive operations through synchronization of combat power and anticipation of threat reactions. A high tempo allows Army forces to achieve surprise and quickly gain positions of advantage. Controlling operational tempo and not allowing the different tempo of UO to adversely affect other operations is a challenge that commanders of major operations must overcome.

Operational Context of Urban Operations: Brittany Ports – 1944

The plan for the invasion of Normandy, France, in June 1944 was meticulously developed. The plan not only addressed the invasion itself, but also contained detailed planning for the campaign to follow. A major concern of the detailed campaign planning was logistics. To address this critical concern, and specifically the problem of ports to supply the allied armies once ashore, the pre-invasion planning called for the major ports of the French province of Brittany—Brest, Lorient, and Saint Nazaire—to be objectives of General Patton's Third Army, once it was activated.

Early August 1944, almost two months after the successful Normandy invasion, the operational situation significantly differed from that envisioned by the D-Day planners. General Montgomery's Twenty-first Army Group was still fighting in the Bocage of Normandy. In contrast, General Bradley's Twelfth Army Group had just achieved a major breakthrough at Saint Lo, secured the Cotentin Peninsula, and reached the city of Avranches. Here was a decision point. Bradley and Eisenhower had to decide whether to adhere to the original plan and turn west with Patton's forces to secure the peninsula or to take advantage of the breakout at Saint Lo and turn east to exploit the disruption of the German defenses.

Ultimately they reached a compromise. General Middleton's VIII Corps was tasked to secure the peninsula, and the bulk of Patton's Army, three Army corps, was turned northeast to exploit the operational collapse of the main German defenses. (See figure 7-1.)

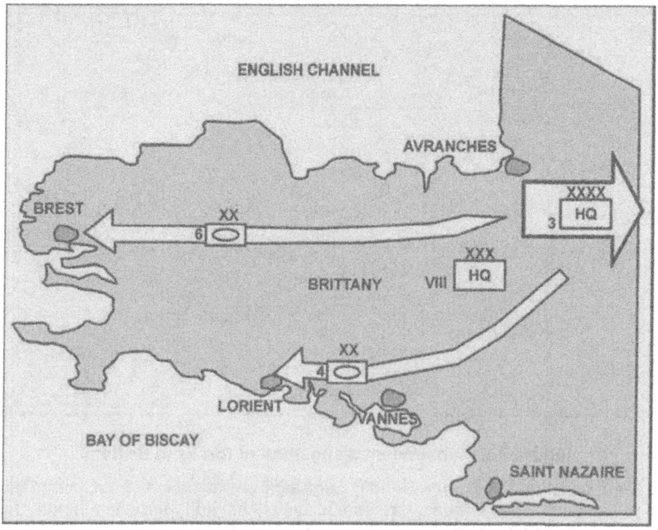

Figure 7-1. Initial attack in Brittany

Middleton's corps sprinted into the peninsula with the 4th and 6th Armored Divisions leading the way. However, poor communications, disagreements between commands, and contradictory orders caused the corps to hesitate before pushing the two divisions to continue to exploit toward the ports. The result: the 6th Armored Division missed an opportunity to seize Brest against light resistance by one day. The 4th Armored Division, after capturing the smaller port of Vannes, was also frustrated on the approaches to Lorient. The American reaction to the inability to rapidly seize the ports demonstrated an understanding of changing circumstances. The 6th Armored Division turned the attack at Brest to the 8th Infantry Division and then relieved the 4th Armored Division at Lorient. The 4th Armored was moved to rejoin the rest of Third Army exploiting to the east and north. Ultimately Brest fell to VIII Corps on 19 September after a 43-day siege by three infantry divisions. The victory yielded 36,000 German prisoners of war (POWs). However, the German defense and demolitions of the port left the port without an impact on the logistic situation of the allies. Brest cost the U.S. Army almost 10,000 casualties and the commitment of significant supplies. The experience convinced commanders to surround and bypass the other major Brittany ports. Lorient and Saint Nazaire remained under German control, deep in allied territory, until the war ended ten months later (see figure 7-2).

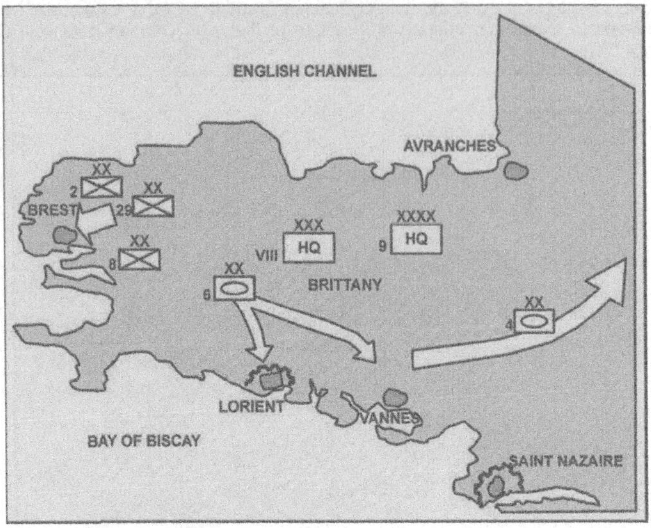

Figure 7-2. Subsequent disposition of forces in Brittany

The operational lessons of the Brittany campaign are numerous. First, commanders are responsible to continually assess assumptions and decisions made during planning based on the changing circumstances of the battlefield. This includes the planning decision to conduct urban offensive operations. When the allies arrived at the Brittany Peninsula, the focus of the operational maneuver was no longer securing logistics facilities but exploiting the breakthrough at Saint Lo and the disintegrating the German defense. The bulk of Third Army then was turned to the north and east rather than west into the peninsula.

The Brest experience also demonstrates that the costs of urban offensive operations are continually assessed against the operational value of the objective. This lesson was applied to the cities of Lorient and Saint Nazaire. The cities were never seized from the Germans because their logistic value failed to warrant the required resources to take them. German retention of the ports had no major adverse effect on the overall campaign.

Another lesson is that commanders cannot allow urban operations to disrupt the tempo of other offensive operations. One German goal of defending the ports was to disrupt the rapid tempo of the U.S. exploitation. They failed to achieve this goal because General Bradley continued the exploitation with the bulk of Third Army and executed the original plan with only a single corps.

Finally, commanders cannot allow emotion to color their decision to conduct or continue UO. The failure of 6th Armored Division to seize Brest rapidly caused some commanders to believe that Brest had to be captured because the prestige of the Army was committed to the battle. Costs of the continuing combat operations to seize Brest were significant. These resources might have been better committed elsewhere in the theater.

7-7. Tactical tempo is also important in urban combat. Because of the complex terrain, defending forces can rapidly occupy and defend from a position of strength. Once Army forces initiate tactical offensive operations, they cannot allow the threat to set the tempo of the operation. Instead, attacking forces seek to maintain a high tempo of operations. However, the tactical tempo of urban operations differs from operations in other terrain. Not necessarily slow, it requires a careful balance of preparation, speed, and security. In terms of unit fatigue, resource consumption, and contact with the threat, the tempo of most urban offensive operations may be rated as very high. On the other hand, in distances traveled and time consumed to achieve objectives, the tempo of many urban offensive operations might be rated as slow. The urban battlefield's density concentrates activity and consumes resources in a relatively small area. The lack of terrain seized or secured is not to be construed to mean a low tempo in the battle. Although the tempo may seem excruciatingly slow at higher levels of command and exceeding fast at lower, tactical levels, in reality, the natural tempo of urban operations is not faster or slower than other types of operations, merely different. Creating and operating at a tempo faster than an opponent can maintain, however, can favor forces which are better led, trained, prepared, and resourced.

7-8. A high tactical tempo in urban offensive operations challenges logisticians to provide for the increased consumption of munitions and degrades soldiers' physical capabilities. Commanders must anticipate these challenges and develop the means and abilities to overcome them. In the past, these challenges forced commanders to conduct urban offensives cyclically. They used night and other periods of limited visibility to resupply, rest, and refit forces. The environment influenced the tempo of their operations. This type of "battle rhythm" resulted in the forces spending each new day attacking a rested threat that was in a well-prepared position.

7-9. Army forces must maintain the tempo. Offensive operations continue even during darkness. Moreover, Army forces increase the tempo of operations at night to leverage the limited visibility capabilities, increased situational understanding, training, and INFOSYS that give an advantage to Army forces in all environments. To overcome the physical impact of the environment on soldiers, commanders can retain a large reserve to rotate, continuing offensive operations at night. The force that fights in daylight becomes the reserve, rests, and conducts sustaining operations while another force fights at night. Army forces can then maintain the tempo of operations and leverage technological advantages in urban offensive combat.

7-10. Tempo in UO does not necessarily mean speed. Offensive operations balance speed, security, and adequate firepower. Commanders must plan for the complex tactical environment and the requirements to secure flanks and airspace as the operation progresses. Mission orders allow subordinate units to make the most of tactical advantages and fleeting opportunities.

AUDACITY

7-11. Audacity is a simple plan of action, boldly executed. Superb execution and calculated risk exemplify it. In an urban attack, a thorough understanding of the physical terrain can mitigate risk. The terrain's complexity can be studied to reveal advantages to the attacker. Audacity can also be embodied in an operation by inventively integrating and coordinating the direct action tasks of SOF throughout the operation. Combining SOF actions with conventional attacks can asymmetrically unhinge a threat's defensive plan. Well trained Soldiers confident in their ability to execute urban offensive operations foster audacity.

URBAN OFFENSIVE OPERATIONS AND BATTLEFIELD ORGANIZATION

7-12. Urban offensive operations, like all operations, are arranged using the overall battlefield organization of decisive, shaping, and sustaining operations. Each operation is essential to the success of an urban offensive, and usually two or more of these operations occur simultaneously. Decisive operations are attacks that conclusively determine the outcome of UO. These attacks strike at a series of decisive points and directly lead to neutralizing the threat's center of gravity. Shaping operations in urban offensive operations create the conditions for decisive operations. In UO, much of the shaping effort focuses on isolation, which is critical in both major operations and tactical battles and engagements. Sustaining operations in urban offensive operations ensure freedom of action. They occur throughout the area of operations (AO) and for the duration of the operation.

DECISIVE OPERATIONS

7-13. A tactical commander fights decisive urban combat, whereas commanders conducting a larger major operation influence urban combat by setting the conditions for tactical success. Higher commanders may directly influence urban offensive operations by operational maneuver, by coordinating joint fires, by closely coordinating conventional forces, or with SOF.

7-14. Tactical urban offensive operations quickly devolve into small-unit tactics of squads, platoons, and companies seizing their objectives. The compartmented effect of the terrain and the obstacles to command and control of small units, especially once they enter close combat inside buildings or underground, often restricts the higher commander's ability to influence operations. Commanders can influence the actions of subordinates by clearly identifying the decisive points leading to the center of gravity; using mission orders (as discussed in Chapter 5); developing effective task organizations; synchronizing their decisive, shaping, and sustaining operations; and managing transitions.

7-15. Like all operations, successful decisive operations in UO depend on identifying the decisive points so the forces can destroy or neutralize the threat's center of gravity. Seizing a key structure or system that makes the threat's defense untenable; interdicting a key resupply route that effectively isolates the threat force from his primary source of support; or isolating the threat so that his force can no longer influence friendly activity may be more effective than his outright destruction.

7-16. Commanders must select the right subordinate force for the mission and balance it with appropriate attachments. Higher commanders do not direct how to organize the small tactical combined arms teams, but they must ensure that subordinates have the proper balance of forces from which to form these teams. Successful urban offensive operations require small tactical combined arms teams. Urban offensive operations require abundant infantry as the base of this force. However, successful urban combat requires a combined arms approach adjusted for the conditions of the environment. In urban offensive operations, it will normally not be a question of whether include armor and mechanized elements but, rather, how best to accomplish this task organization. Precision-capable artillery systems generally support urban operations better than unguided rocket artillery.

7-17. Divisions entering urban combat may require additional resources. These resources include military intelligence support in the form of linguists, human intelligence (HUMINT) specialists, and unmanned aircraft systems (UAS). Engineering assets will be at a premium; the task organization of a task force executing the decisive operation may require a one-to-one ratio of engineer units to combat units. Corps and higher engineering support may be necessary to meet these requirements and to repair vital and specialized infrastructure. A tailored and dedicated support battalion or group may need to assist in providing anticipated support to a displaced and stressed civil population. Finally, divisional CA units may require augmentation to deal with NGOs and civilian government issues.

7-18. Successfully conducting decisive operations in the urban environment requires properly synchronizing the application of all available combat power. Army forces have a major advantage in the command and control of operations. Commanders can use this advantage to attack numerous decisive points simultaneously or in rapid succession. They also use it to attack each individual decisive point from as many directions and with as many different complementary capabilities as possible. Commanders must completely understand urban environmental effects on warfighting functions to envision and execute the bold and imaginative operations required. Significantly, these operations require that C2 systems account for the mitigating effects of the environment as execution occurs.

7-19. Properly synchronized actions considerably enhance the relative value of the combat power applied at the decisive points. They present to the threat more requirements than he has resources with which to respond. Synchronized IO and multiple maneuver actions paralyze the threat's decision-making capacity with information overload combined with attacks on his C2 systems. Additionally, well-synchronized actions limit the time the threat has to make decisions and forces him into bad decisions. In the urban environment, these effects are enhanced because C2 systems are already strained, poor decisions are harder to retrieve, and units that do not react effectively are more easily isolated and destroyed.

Shaping Operations

7-20. Shaping operations that support the urban attack separate into those focused on isolating the threat and all others. Army forces isolate the threat to ensure successful urban offensive operations. Depending on the threat reaction to isolation efforts and the nature of the threat center of gravity, this task may become decisive. Other shaping operations include those common to all offensive operations and others unique to urban operations. Unique urban shaping operations may include securing a foothold in a well-fortified defensive sector, securing key infrastructure, or protecting noncombatants. Because of the nature of UO, shaping operations may consume a much larger proportion of the force than during other operations and may take place both inside and outside the urban area (see Applying the Urban Operational Framework: Panama in Chapter 5). By successfully isolating a threat force, the force needed to conduct the decisive operation may be relatively small.

Sustaining Operations

7-21. Commanders conducting urban offensive operations must ensure security of the sustaining operation and bases; in many situations, protecting lines of communications (LOCs) and sustaining operations may be the greatest vulnerability of the attacking force. Those supporting an urban offensive are tailored to the urban environment and are well forward. Ideally, the supporting forces closely follow the combat forces and move within or just outside the urban area as soon as they secure an area. Operating in the urban area during offensive operations allows the sustaining operation to take advantage of the defensive attributes of the environment for security purposes.

7-22. Counterattacks against sustaining operations may take the form of special operations activities aimed at LOCs leading to or within the urban area. Choke points—such as bridges, tunnels, and mountain passes—are vulnerable to these attacks and may require combat forces to protect them. Threat forces attack the LOC to blunt the Army's combat power advantage in the urban area.

7-23. Attacks against the LOC into the urban area may also attempt to isolate the attacking Army forces from its sustainment base. Isolated forces in an urban area are greatly disadvantaged. Commanders must plan and aggressively execute strong measures to protect their LOC, even if it requires reduced combat power to execute their offensive operation.

7-24. Sustaining operations anticipate the volume and unique logistics requirements of urban operations. Specialized individual equipment—such as grappling hooks, ladders, and pads—is identified and provided to troops in quantity before they are needed. Forces stockpile and distribute their attacking units' special munitions requirements including small arms, explosives, and grenades of all types, precision artillery munitions, and mortar ammunition. Forces also supply transport to move the resources rapidly forward, both to and through the urban environment. Sustaining operations cannot rely on "operational pauses" to execute their tasks. Commanders must plan to continuously supply resources and capabilities to the most forward combatants as offensive operations advance.

7-25. Sustaining operations also anticipate the growth of sustainment requirements as Army forces secure and take responsibility for large portions of the urban area. The success of Army urban offensive operations will often uncover the civil population in former threat occupied areas. It may attract the civil population from sections of the urban area where the Army is not operating to areas occupied by Army forces. Rural populations may migrate to the urban area as the result of successful Army offensive operations.

7-26. Army forces may be required to take initial responsibility to provide for the urban population. This consideration is integrated into logistics planning and organization from the start of the planning process. To be successful and efficient in such a situation, sustainment planning includes Army civil affairs (CA) specialists and local government representatives. It also integrates and consults with the international community and nongovernmental organizations (NGOs) that might augment or supplement Army sustainment capabilities.

FORMS AND TYPES OF URBAN OFFENSE

7-27. Traditional forms of offensive maneuver include envelopment, turning movement, infiltration, penetration, and frontal attack. Traditional types of offensive operations are movement to contact, attack, exploitation, and pursuit. These traditional forms and types listed apply to urban combat. Some have greater application to an urban environment than others do. Moreover, success will belong to commanders who imaginatively combine and sequence these forms and types throughout the depth, breadth, and height of the urban battlefield. This is true at the lowest tactical level and in major operations.

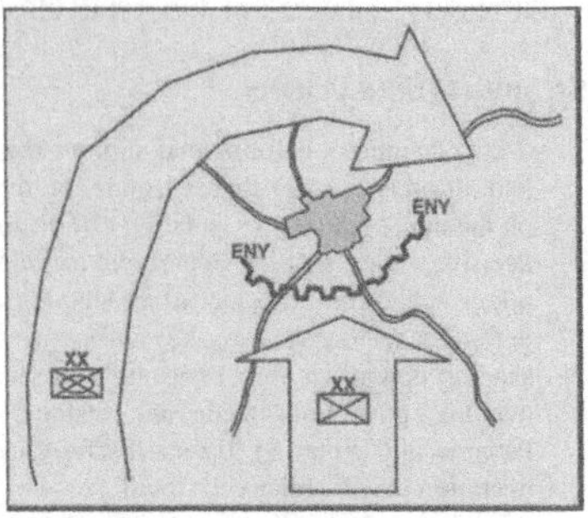

Figure 7-3. Envelopment isolates an urban area

FORMS OF OFFENSIVE MANEUVER

Envelopment

7-28. The envelopment is the ideal maneuver for isolating threat elements in the urban area or isolating the area itself. A deep envelopment effectively isolates the defending forces and sets the conditions for attacking the urban area from the flank or rear. Yet, enveloping an objective or threat force in the urban area is often harder since achieving speed of maneuver in the environment is so difficult (see figure 7-3). Vertical envelopment, however, works effectively if Army fires can effectively suppress or neutralize the threat air defense.

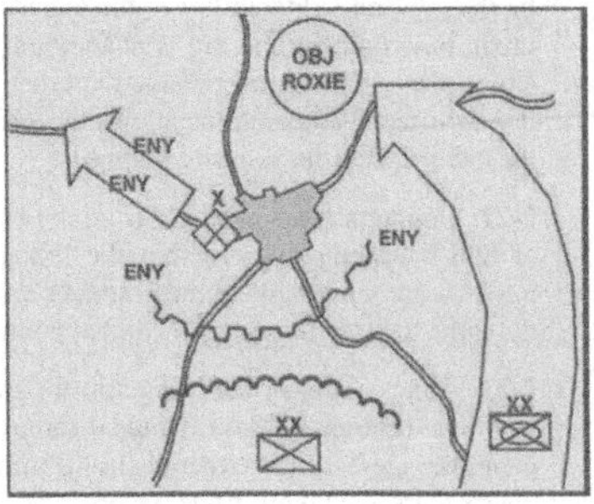

Figure 7-4. Turning movement

Turning Movement

7-29. Turning movements can also be extremely effective in major operations (see figure 7-4). By controlling key LOCs into the urban area, Army forces can force the threat to abandon the urban area entirely. These movements may also force the threat to fight in the open to regain control of LOCs.

Infiltration

7-30. Infiltration secures key objectives in the urban area while avoiding unnecessary combat with threat defensive forces on conditions favorable to them (see figure 7-5). This technique seeks to avoid the threat's defense using stealthy, clandestine movement through all dimensions of the urban area to occupy positions of

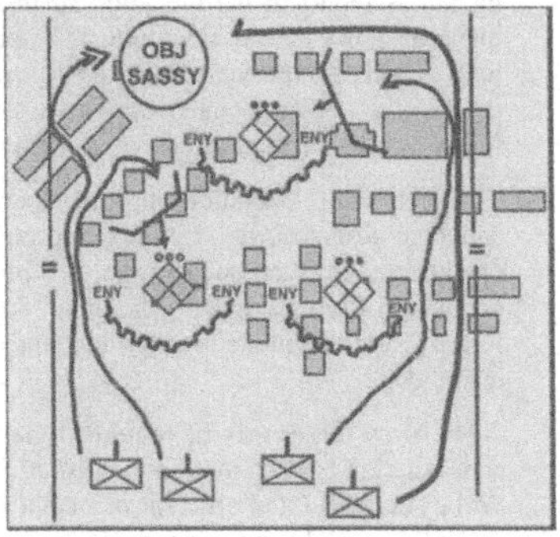

F

igure 7-5. Infiltration

advantage in the threat's rear (or elsewhere).It depends on the careful selection of objectives that threaten the integrity of the threat's defense and a superior common operational picture (COP). Well-planned and resourced deception operations may potentially play a critical role in masking the movement of infiltrating forces. The difficulty of infiltration attacks increases with the size and number of units involved. It is also more difficult when Army forces face a hostile civilian population. Under such circumstances, infiltration by conventional forces may be impossible. Armored forces are generally inappropriate for infiltration operations. However, they may infiltrate large urban areas if the threat is not established in strength and had insufficient time to prepare defenses.

Penetration

7-31. Penetration is often the most useful form of attack against a prepared and comprehensive urban defense (see figure 7-6). It focuses on successfully attacking a decisive point or on segmenting or fragmenting the defense—thereby weakening it and allowing for piecemeal destruction. (The decisive point may be a relatively weak or undefended area that allows Army forces to establish a foothold for attacks on the remainder of the urban area.) Ideally in urban combat, multiple penetrations in all dimensions are focused at the same decisive point or on several decisive points simultaneously. In urban combat, the flanks of a penetration attack are secure, and resources are positioned to exploit the penetration once achieved. Although always a combined arms team, rapid penetrations are enhanced by the potential speed, firepower, and shock action of armored and mechanized forces.

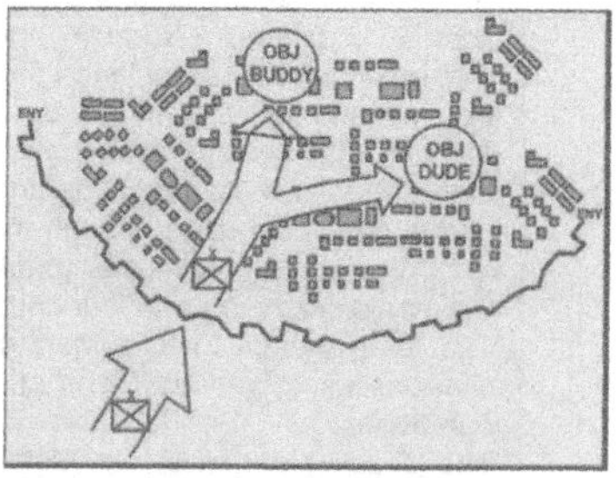

Figure 7-6. Penetration

7-32. Importantly, commanders must consider required actions and resources that must be applied following a successful (or unsuccessful) penetration. A penetration may result in the rapid collapse and defeat of the threat defense and complete capitulation. On the other hand, success may cause threat forces to withdraw but leave significant stay-behind forces (or disperse into the urban population as an insurgent-type force) which may necessitate methodical room-to-room clearance operations by significant dismounted forces. Additionally, securing portions (or all) of an urban area requires occupation by Army forces to prevent re-infiltration of threat forces thereby further increasing manpower requirements. Based on the factors of the situation, commanders may conclude that over time methodical clearance operations conducted from the outset, while frontloading time and resource requirements, will be less costly than a penetration followed by systematic clearing operations.

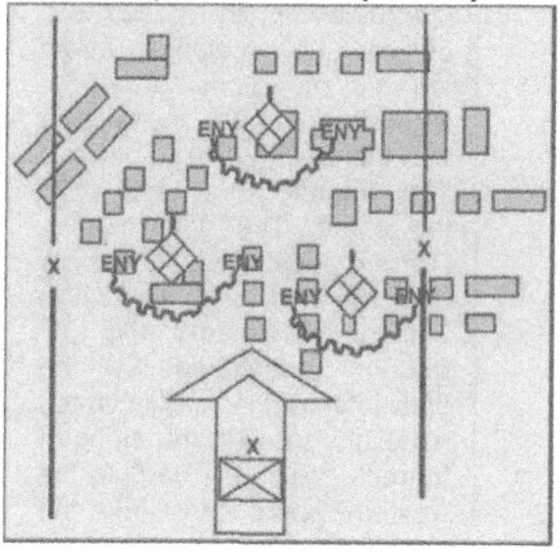

Figure 7-7. Frontal attack

Frontal Attack

7-33. For the commander of a major operation, the frontal attack is generally the least favorable form of maneuver against an urban area unless the threat is at an obvious disadvantage in organization; numbers, training, weapons capabilities, and overall combat power (see figure 7-7). Frontal attacks require many resources to execute properly, risk dispersing combat power into nonessential portions of the area, and risk exposing more of the force than necessary to threat fires. In urban offensive combat, forces most effectively use the frontal attack at the lowest tactical level once they set conditions to ensure that they have achieved overwhelming combat power. Then the force of the frontal attack overwhelms the threat with speed and coordinated and synchronized combat power at the point of attack. The assigned frontage for units conducting an attack on an urban area depends upon the size and type of the buildings and the anticipated threat disposition. Generally, a company attacks on a one- to two-block front and a battalion on a two- to four-block front (based on city blocks averaging 175 meters in width).

Forms of Attack in the Urban Offense: Metz – 1944

In November 1944, the U.S. Third Army launched its final effort to take the French city of Metz from the defending Germans. This was the Army's third attempt. The first attempt had been a surprise, mounted attack. This was followed by a series of piecemeal infantry assaults on the surrounding fortresses. Finally, a deliberate effort was made to take the city in a coordinated effort by XX Corps.

The initial failures stemmed from a shortage of resources on the U.S. side, to include fuel and units—especially infantry. This added to the ad-hoc nature of the first two efforts. The third effort, though more deliberate, was still constrained by resources. XX Corps could only muster three nearly full-strength infantry divisions to attack the German's defending with four under-strength divisions.

The third attempt to take Metz demonstrates how a corps operates with multiple divisions using various forms of attack to achieve its objective in urban offensive operations.

The opening phase of the Metz battle had attacks by the 90th and 5th Infantry Divisions to envelop the city from the north and south (see figure 7-8). This isolated the city and ensured the garrison could not escape nor be reinforced. The garrison was under orders from Hitler not to conduct a breakout and to resist to the last man; thus, German forces strongly counterattacked both wings of the envelopment to prevent isolation.

The second phase of the operation was the penetration of the city defenses from multiple directions. The 5th Infantry Division penetrated into the city with the 10th and 11th Infantry Regiments from the south. The 95th Infantry Division penetrated into the city from the north with the 377th Infantry Regiment and TF Bacon. Simultaneously, the 95th Infantry Division infiltrated battalions through the string of fortress positions guarding the western approaches into the city, isolated, and bypassed these positions with its other two regiments.

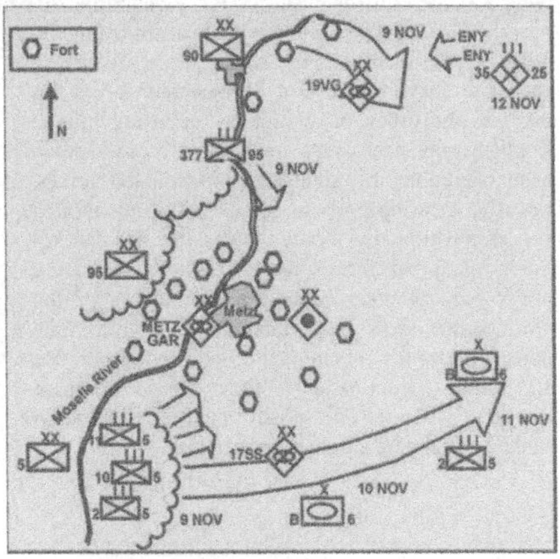

Figure 7-8. Metz envelopment

The final reduction of the defense was a series of battalion frontal attacks, which took place against the last remaining strongholds within the city. Even in these final engagements, however, the infantry battalions isolated, bypassed, and then attacked from the flanks and rear whenever possible (see figure 7-9).

The city was declared secured on 19 November. However, at that point more than a half-dozen of the fortresses had yet to be reduced. The 95th Infantry Division, after a four-day rest, quickly moved forward to rejoin the still rapidly advancing corps forward elements. Elements of the 5th Infantry Division remained in siege posture around the remaining strong points, the last of which surrendered on 19 December 1944 when it ran out of food. U.S. forces made no efforts to attempt to assault these bypassed fortresses although extensive psychological operations (PSYOP) were used.

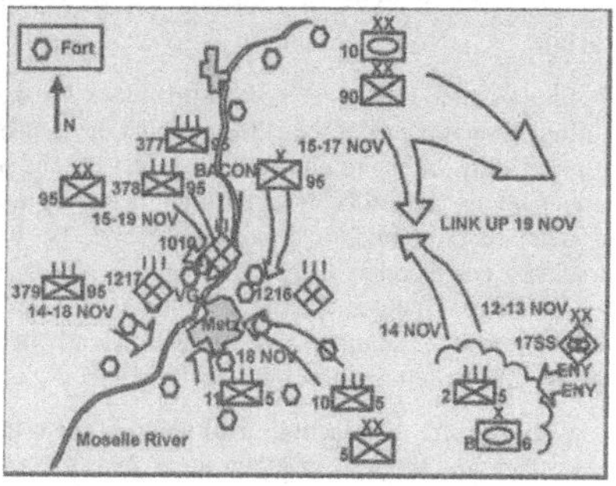

Figure 7-9. Metz final assault

TYPES OF OFFENSIVE OPERATIONS

Movement to Contact

7-34. In an urban area where the threat situation is vague, Army forces will often conduct a movement to contact to establish or regain threat contact and develop the situation. A movement to contact in an urban area occurs as both sides try to establish their influence or control over a contested urban area. The situation determines whether the movement to contact or its specific technique, the search and attack, is appropriate. A conventional force-oriented movement to contact will likely take place when friendly and threat conventional forces attempt to establish control simultaneously. Initially, neither side is defensive. The friendly force aims to quickly locate and fix the threat while establishing control of the urban area and its key infrastructure. The search and attack technique works well when a smaller threat has established a noncontiguous defense in an urban area. This operation is characterized by the friendly point defense of key infrastructure, robust reconnaissance, and rapidly concentrated combat power to fix and defeat or destroy threat resistance once located.

7-35. A meeting engagement often results from the movement to contact. It occurs when a moving force that is partially deployed for battle collides with and engages a threat at an unexpected time and place. In a meeting engagement in an urban area, the unit that reacts most quickly and decisively will likely win. Rapid and accurate decision making depends heavily on understanding the nature of the urban area and its impact on operations. Thus, in a meeting engagement, commanders should quickly assess the impact and role of all components of the urban environment (terrain, infrastructure, and society) on the operation. To this end, responsive reconnaissance is important. This permits accurate decision making regarding where to attack, where to defend, and how to allocate resources. Situational understanding enhanced by robust digital INFOSYS that provide an accurate COP also facilitates the rapid reaction of Army units and a synchronized response. This reaction and response allow Army forces to seize the initiative and dominate the threat.

Attack

7-36. The attack is the most common and likely offensive operation that Army forces will conduct in an urban environment. Commanders conducting major operations and commanders of large tactical units usually execute deliberate attacks. In the urban environment, units larger than battalion-size rarely conduct hasty attacks. Hasty attacks are common below company level as units use their initiative to take advantage of tactical opportunities. However, larger units will conduct hasty attacks when threat defenses are disrupted or unprepared, to take advantage of an unexpected situation, and to prevent the threat from establishing or re-establishing a coherent defense.

Exploitation

7-37. Exploitation follows a successful attack to disrupt the threat in depth. Commanders of major operations should consider focusing exploitation attacks on urban areas. A threat defeated in an attack will attempt to rally units, reinforce with reserves, and reorganize his defense. With its information and communications capability, transportation network, and defensive attributes, the urban area is the natural focal point to reestablish a disrupted defense. By establishing urban centers as the objectives of the exploitation, commanders deny the threat the sanctuary needed to reorganize and reestablish his defense. The exploitation focuses on the urban area as well as on the remnants of the threat. A successful exploitation to seize an urban area works efficiently because the attack preempts the defense and denies the threat the full advantages of urban terrain.

7-38. Commanders conducting exploitation must acknowledge the vulnerability of their forces to counterattack and ambush in urban areas. An urban area provides ideal cover and concealment to hide threat reserves, reinforcements, or reorganized forces. Constrictions of routes into and through the urban area make exploitation forces a potentially dense target and limit maneuver options. Robust and well-coordinated reconnaissance, tactical dispersal, and use of advance guard security forces protect against this threat (see Defensive Combat Power: Suez City vignette in Chapter 8).

Pursuit

7-39. The pursuit is designed to destroy threat forces attempting to escape. It focuses on the threat and not on urban areas. When conducting a pursuit, Army forces move through undefended urban areas and, if possible, bypass those in which threat forces successfully take refuge. The threat will likely attempt to use urban areas to disrupt the pursuit and permit the threat main body to escape. Commanders can prevent escape by denying the threat the time to establish forces in urban areas that cannot be bypassed. The agility of Army aviation forces for attack, reconnaissance, and transportation is essential to execute a successful pursuit around and through urban areas.

URBAN OFFENSIVE CONSIDERATIONS

7-40. The urban operational framework (understand, shape, engage, consolidate, and transition) provides a structure for developing considerations unique to urban offensive operations. The considerations vary depending on the situation and scale of the operation. Some considerations applicable to major operations that include an urban area will also be considerations at the tactical level focused in the urban area. However, no set rules exist. All urban operations are unique. Issues addressed at the operational level in one situation may be addressed in a new situation only at the tactical level. Under the right circumstances, a consideration may become an operational issue, a tactical issue, or a combination of the two. The following identifies some planning and execution issues that commanders conducting major operations should address.

UNDERSTAND

7-41. The first requirement, and a continuing requirement throughout the conduct of urban operations, is the assessment and understanding of the situation. Commanders should base this understanding on detailed information regarding the particular urban area. Since the threat will likely dominate or control most of the urban area during the planning phase of offensive operations, achieving an accurate understanding of the urban environment will be difficult. A comprehensive intelligence, surveillance, and reconnaissance (ISR) effort in support of a rigorous intelligence preparation of the battlefield (IPB) process overcomes this obstacle.

Integrated Intelligence, Surveillance, and Reconnaissance

7-42. The commander of a major operation that includes an urban area, unlike his subordinate commanders, can target reconnaissance deep into his AO and area of interest. This begins the application of ISR resources against the urban area that may lead to decisive ground operations. This ISR effort and the understanding it supports continue as long as the urban area remains in the AO. Commanders of major operations initially direct ISR assets on those information requirements that support determining whether or not to conduct urban offensive operations. Once decided, ISR resources shift to support the planning and execution of the operation in the urban area.

7-43. The first resources that a senior commander can use are national and strategic sensors. He requests them through the appropriate joint force commander. The commander aggressively pursues full use of these systems to begin building an initial database for analyzing the significant aspects of the terrain; key infrastructure considerations; the status and disposition of the population; and the size, type, disposition, and intentions of threat forces in the area.

7-44. Simultaneously, multiple intelligence sources contribute to the database. The sources collect, process, store, display, and disseminate the relevant information on large urban areas through open and classified resources. These information sources include—

- Historical research.
- Travel brochures that include cultural information and recent maps.
- Classified debriefings of diplomats, businesses, DOD personnel, and allies.
- Military maps and special geospatial products of the urban area.
- Previous intelligence assessments of the country, government, and population.
- Reachback to appropriate economic, political, cultural, and infrastructure subject matter experts not in the commanders AO.

The gathering and analysis of HUMINT plays a critical part of this process and assists commanders in understanding ethnic, cultural, religious, economic, political, and other societal and infrastructural facets of the environment.

7-45. As the intelligence and the national reconnaissance and surveillance efforts progress, commanders will insert, if available and feasible, Special Forces reconnaissance assets into the urban environment. These elements will seek to confirm or deny the information received from imagery intelligence (IMINT), signals intelligence (SIGINT), and HUMINT sources. Among many factors, using SOF depends on their availability, the particular urban area, the area's ethnic composition, and the relationship between the urban population and the threat. Other joint operational reconnaissance and surveillance assets that higher-echelon commanders may have available might include the Joint Surveillance Target Attack Radar System, Guard Rail targeting aircraft, UAS, and space-based systems.

7-46. The commander's staff will use all sources of information—IMINT and SIGINT sensors, HUMINT, historical research, and reconnaissance—to refine his ability to understand the urban environment. Digitally linking subordinate commanders with information sources helps to develop a COP essential to their situational understanding of the urban environment. The IPB process guides this assessment. As operations progress, additional reconnaissance and surveillance assets may become available. These may include UAS, long-range reconnaissance and surveillance units, counterfire radar, and air and ground cavalry. As these assets are employed, they are linked into the net of sources sharing information and further refine a common situational understanding of the environment.

Focused Assessment Efforts

7-47. In urban offensive operations, the tactical commander's assessment focuses on defeating the threat in the urban area within the constraints of the environment. Toward this end, identifying and assessing decisive points to attack is a commander's priority assessment task. Some unique aspects of the urban environment also require the focus of the commander's assessment efforts. These include the character of the urban defense, collateral damage considerations, and the effects of the environment on friendly and threat courses of action.

Character of the Urban Defense

7-48. To be both efficient and effective, Army urban offensive operations focus on what is decisive. Decisive points for an urban attack depend primarily on the mission within the urban area. They can vary widely in composition and size. Since commanders only focus on the essential, they may determine the decisive point to be a single building or a limited sector of an urban area. It could be an entire system within the urban infrastructure such as communications and information, or a limited subsystem of the transportation and distribution infrastructure such as a single airfield. Sometimes what is decisive in the urban area is the threat military capability, but even this large an objective, when carefully analyzed, may not require destruction of all threat forces or control of the entire urban area. Decisive points relate directly to the threat's center of gravity and to mission success. Some decisive points related to the urban threat's center of gravity may be physically located outside the urban area.

7-49. To gain specifics on threat dispositions within the urban area requires reconnaissance capability to see into the depths of the area and the intelligence capability to determine the threat's likely defensive course of action. With this information, commanders can determine decisive points and apply Army combat power discretely against them. Effective urban offensive operations require detailed situational understanding of an area of interest that extends well beyond the perimeter of the urban area.

7-50. Commanders see throughout the depth of the urban area using several actions (see figure 7-10). First, they evaluate sensor data and imagery. This guides targeting of special reconnaissance. Simultaneously, HUMINT is conducted using any persons who might know the urban area and threat. This includes civilians (allies, aides, neutrals, obstacles, and hostiles) and POWs. Finally, tactical conventional reconnaissance assets including reconnaissance forces, aviation, artillery radar, signals intelligence, and UAS are directed at the urban area. All these sources and data are linked through digital INFOSYS to provide commanders and their subordinates with improved situational understanding and a COP.

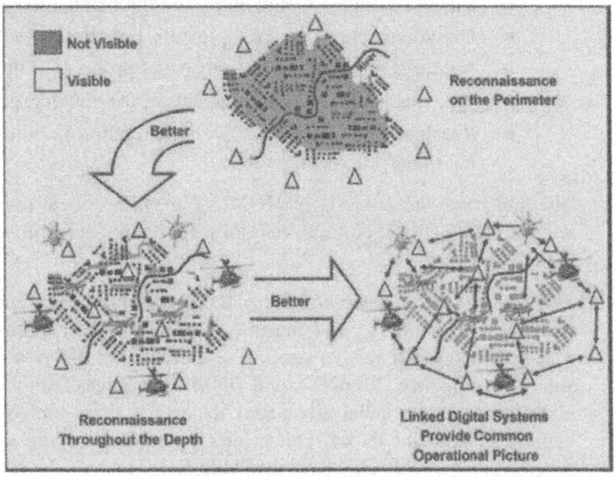

Figure 7-10. Required urban reconnaissance capabilities

Collateral Damage Considerations

7-51. Commanders must also assess the collateral damage risks that their operations may include. This assessment helps to initially determine the viability of a course of action. However, commanders should reassess their courses of action at frequent intervals in urban offensive operations based on known information to determine if the original evaluations remain valid. This reassessment minimizes potential collateral damage from a change in mission or a change in a course of action. Many aspects of the environment can change during mission execution.

The Environment's Effects on Courses of Action

7-52. The urban environment's unique aspects can significantly impact the course of action chosen by Army forces and the threat. Commanders must assess these effects in planning, but they must also verify and monitor these effects as forces execute offensive missions. In particular, commanders will want to confirm the civilian population's locations, beliefs, and actions and to monitor any changes. They will need to validate terrain considerations and monitor the effects of any changes due to rubble and other damage to structures. In urban terrain, dead space, cover, and concealment can only be identified physically and may change considerably as operations are executed.

SHAPE

7-53. Commanders of major operations have a primary contribution to urban operations: the planning and conduct of effective shaping operations that set the conditions for subordinate tactical success. In urban operations, isolation will be a critical condition. Effective isolation will require persistent, continuous surveillance and reconnaissance, innovative use of fires and maneuver (including effective force allocation decisions), and well-established sensor-to-shooter links. These efforts—combined and synchronized with SOF direct actions, IO that minimize noncombatant influences, and necessary shaping attacks (particularly the seizure of a foothold)—establish the conditions necessary for the subsequent offensive domination of the area.

Isolation is Essential

7-54. One key to success in the history of urban operations has been the effective isolation of the threat force (see figure 7-11). This applies today and equally well to major urban offensive operations as it does to smaller-unit attacks. This isolation not only denies access to the urban area from outside but also contains threat forces within. In a modern metropolis or megalopolis, this can appear a daunting task. Operational isolation requires dominating all physical and electronic contact between the threat in the urban area and supporting threat forces outside the urban area. This does not necessarily require physically encircling the urban area, but it does require that Army forces be able to exert control over the area's entire perimeter, as well as decisive points within. For a sprawling urban area, successful isolation may require the commitment of a large amount of resources.

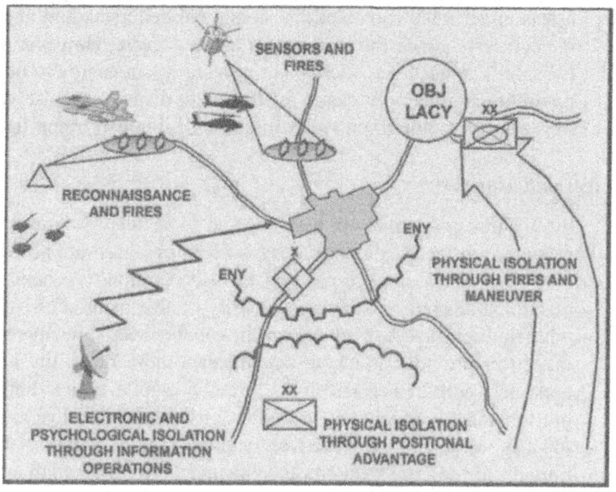

Figure 7-11. Shaping through isolation

7-55. Successful isolation of the urban area depends as much on the nature of the threat as it does on any other factor. A conventional threat in a large urban area may be much easier to isolate than an insurgent threat in a much smaller urban area. The forces needed in the former situation may be less than those needed in the latter. The more the characteristics of the threat are conventional in nature generally the easier it will be to isolate him using standard combat methods and equipment. Isolating a more unconventional force requires many of the same techniques as used against conventional forces. However, it also requires a much greater ability to simultaneously conduct offensive IO, to integrate CA units and civil-military operations (CMO), and to work with allies, nongovernmental organizations, and local authorities. Fundamentally, isolating a less conventional threat puts increased emphasis on separating combatants from noncombatants.

Offensive Isolation Objectives

7-56. Isolation seeks to achieve two primary objectives with respect to defeating a threat's urban defense:

- Weaken the overall coherence of his defense.
- Limit or manipulate his maneuver options.

Isolating the threat in the urban area from external support, as well as isolating him from sources of support within the urban area, weakens his overall defense. The defense is weakened through a combination of attrition (the threat cannot replace his losses) and the diversion of his combat power from the defense to operations to counter the isolation effort. Isolation can also prevent the threat from shifting his forces to reinforce decisive points in the urban area or to conduct counterattacks.

7-57. Commanders may choose not to isolate the urban area completely—or at least make it appear so to threat forces. Instead, they may afford the threat an apparent means of escape, create the conditions for its use through effective fire and maneuver against the defenders, and then destroy the threat through various ambush methods. While friendly forces may be able to move undetected to appropriate ambush sites, it is more likely that this technique will necessitate rapidly mobile air and ground forces moving along carefully chosen routes through the urban area. Commanders must consider maintaining the ability to complete the isolation of the urban area to prevent reinforcement and escape of urban threat forces—particularly if the ambush attack does not achieve desired effects.

Persistent Surveillance

7-58. Persistent surveillance of the urban area is essential to all types of actions used to isolate an urban area and as complete as resources will allow. Surveillance of the urban area relies on either reconnaissance forces or sensors continuously observing or monitoring urban avenues of approach. This network of ISR assets updates the commander's situational understanding and provides the means to quickly identify and, if necessary, attack threat elements as they move. However, particularly with sensors, commanders must consider that each detection is not necessarily an enemy to be attacked. Noncombatant activity clutters the environment making it easier for threats to disguise themselves and increases the burden (and the number of resources required) on Army forces to distinguish friend from foe.

Fires and Maneuver

7-59. Fires and maneuver may be used to achieve isolation, either singly or in combination. (As always, effective obstacles, monitored by sensors or observation, are integral to any isolation technique.) First, attacking forces can pre-position themselves along avenues of approach to deny entry and exit through positional advantage. Relying primarily on this method of isolation, particularly around a large urban area with multiple avenues of approach, can be overly resource intensive. Instead, the pairing of fires and maneuver provides attacking commanders more flexibility and allows them to isolate several avenues of approach with fewer resources. Highly mobile attack helicopters, operating outside threat-controlled portions of the urban area, are ideal for this purpose. Inside threat-controlled areas, it is more difficult to identify, eliminate, or effectively suppress the air defense threat. The threats may have numerous man-portable air defense weapons and enhanced effects of small arms used for air defense. Therefore, the risk to using this equipment may outweigh the potential benefits. However, mobile ground units—such as an air assault (subject to the same air defense threat considerations as attack aviation), armored, or mechanized forces—can also rapidly move to attack and destroy a threat moving in or out of an urban area. Potential disadvantages of the combined, fires and maneuver, option are that—

- Critical assets, on standby and dedicated to isolation efforts, may be unavailable for other missions.
- The attacking force may not locate the threat in time to complete its mission (an inherent risk to any attack).

7-60. Another alternative relies on indirect or joint fires alone to destroy the threat force. Its disadvantage is that fires alone often cannot completely destroy or stop a determined force from moving into or out of an urban area. Although targets and avenues of approach will require continual surveillance, it is usually a less resource-intensive option than those that include maneuver. It also does not normally require fires assets to remain on standby to accomplish the mission. However, fires must be able to reliably and quickly respond. For Army field artillery units and naval gunfire, the units must also be in range, which requires careful positioning. A skilled threat can avoid interdiction fires by using the geometry of the area to identify gaps due to obstructing terrain or the firing unit's range limitations. It can also use concealment and weather to avoid observation. However, effective sensor-to-shooter links throughout the urban battlefield will reduce the threat's ability to hide (see figure 7-12). A resolute threat may risk significant losses to fires to prevent isolation or may attempt to use noncombatants as a shield. Ultimately, commanders should use innovative combinations of all techniques discussed. Some units will physically block key avenues of approach. Surveillance will monitor less important routes and avenues. Artillery fires, joint fires, and maneuver units will then respond to the results of surveillance depending on the circumstances.

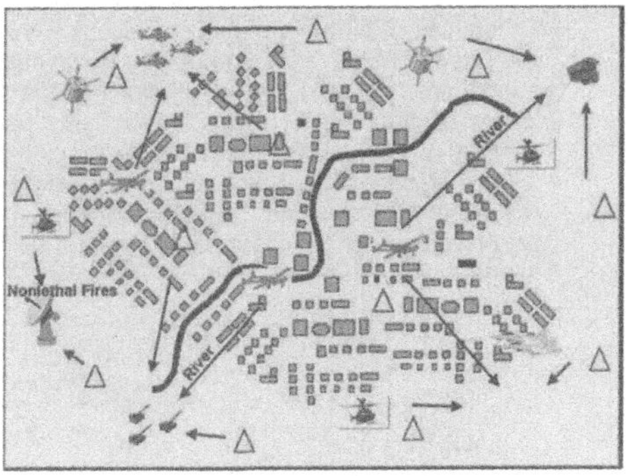

Figure 7-12. Critical sensor-to-shooter links

Threat Reactions

7-61. The reaction of the threat to the effects of isolation will depend on his mission, morale, force structure, and overall campaign plan. The threat may recognize isolation actions early and withdraw from the urban area before isolation is completed instead of risking destruction. On the other hand, the threat, based on a different or flawed assessment (perhaps a perception shaped by the Army force commander), may choose to—

- Continue to defend (or hide) and conduct local ambushes and counterattacks.
- Attack to break into the urban area or infiltrate forces and supplies in.
- Attack to break out of the urban area or infiltrate forces out.
- Execute any combination of the above (see figure 7-13).

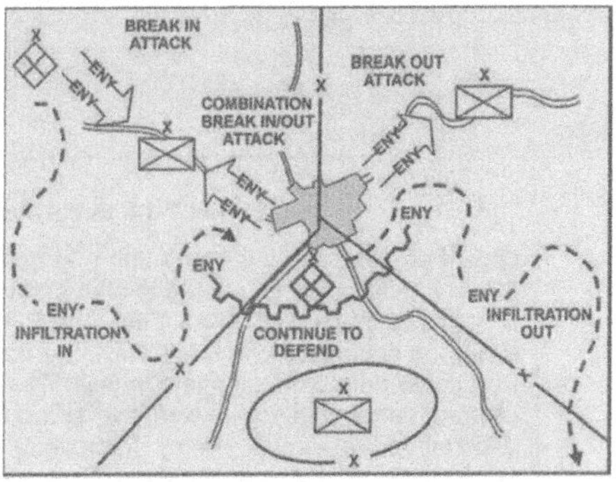

Figure 7-13. Reactions to isolation

7-62. Attacking commanders must consider how the threat leadership's subsequent actions will affect the continuance of overall offensive operations. They deliberate many considerations, to include—

- The allocation of more forces to the shaping operations to isolate the urban area.
- The allocation of more combat power to achieve rapid penetration and seizure of objectives to take advantage of developing threat dispositions in the urban area.

Isolating an Urban Area: Hue, Vietnam – 1968

On 31 January 1968, the 4th and 6th North Vietnamese Army (NVA) regiments and attached NVA and Vietcong (VC) sapper battalions attempted to seize control of Hue from the South Vietnamese Army's (ARVN) 1st Division as part of the North Vietnamese Tet Offensive. The attack, which was launched with complete surprise, successfully established temporary control over most of the city and facilitated the occupation of strong defensive positions within the city's ancient fortress known as the Citadel. However, it did not lead to the capture of the headquarters of the ARVN 1st Division located in the Citadel nor the U.S. military assistance command—Vietnam compound located in the southeast part of the city.

Both ARVN and U.S. forces swiftly reacted to the NVA attack. Within 24 hours, ARVN infantry and cavalry units counterattacked to recapture the Citadel. South of the Perfume River, the 1st Marine Division began attacking to clear the southern portion of the city (see figure 7-14).

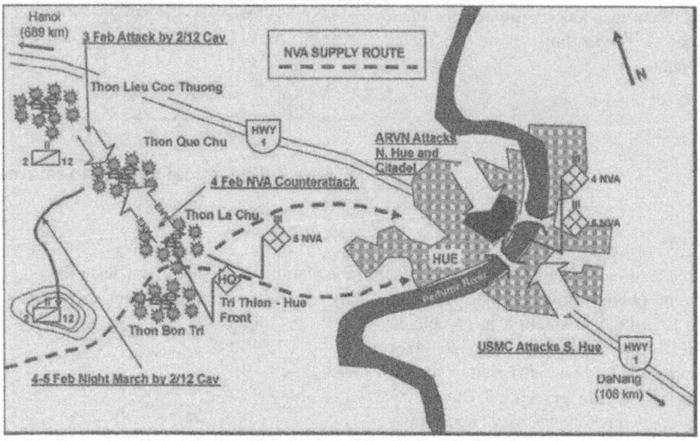

Figure 7-14. Initial attack to isolate Hue

On 2 February, the first Army unit was committed to Hue. The 2-12th Cavalry (an infantry battalion) was ordered to attack southeast along Highway 1 and clear enemy interdicting that route. On 3 February, 2-12th Cavalry began their attack moving along the south side of the highway. The plan was to clear a series of small hamlets in succession en route to the city itself. The first hamlet, Thon Lieu Coc Thuong, was cleared easily. Moving toward the second hamlet, Thon Que Chu, the battalion encountered a strong enemy force that was well entrenched. Air and artillery supported the battalion as it attacked and captured the hamlet against strong resistance. The battalion dug in for the night and prepared to resume the attack against the third hamlet, Thon La Chu, on 4 February.

Unknown to the 2-12th Cavalry, Thon La Chu was the headquarters of the Tri Thien-Hue Front, and it was defended by the NVA's 5th Infantry Regiment. Also important, the hamlet sat astride the NVA's primary supply route to the regiments fighting in Hue. The NVA was determined that 2-12th Cavalry's attack would not succeed; the NVA was determined to destroy the threat to its LOCs.

As the 2-12th Cavalry prepared to resume the attack on 4 February, the NVA fiercely counterattacked with all three battalions of the 5th NVA Regiment. It soon became apparent that 2-12th Cavalry could not continue to attack. As the day continued and the NVA pressure steadily increased, the battalion began to lose its position in Thon Que Chu. To avoid destruction, the battalion broke contact.

As darkness fell on 4 February, 2-12th Cavalry finally broke contact with the NVA. However, instead of retreating north back to its start line, the battalion moved west and then south to good defensive high ground deeper in the NVA AO. Eleven hours later the battalion was set in its new defensive position.

At dawn on 5 February, 2-12th Cavalry was established 5 kilometers west of Hue. The battalion soon observed enemy forces and supplies moving toward Hue. From its high ground position, the battalion directed artillery and air strikes against the NVA forces. By its bold move to bypass the 5th NVA Regiment, the battalion now held excellent position to direct fires on the primary NVA supply line into Hue. These fires were the first step toward isolating the NVA in Hue.

The fires controlled by the 2-12th Cavalry shut down the NVA LOCs into Hue during the daytime. However, under the cover of darkness supplies and reinforcements were still entering the city (see figure 7-15). It became apparent that the isolation of the NVA in Hue would require the capture of Thong La Chu. The problem facing American forces was concentrating combat power against the NVA. All U.S. units at this time were actively engaged against the numerous NVA attacks that constituted the NVA's 31 January Tet Offensive.

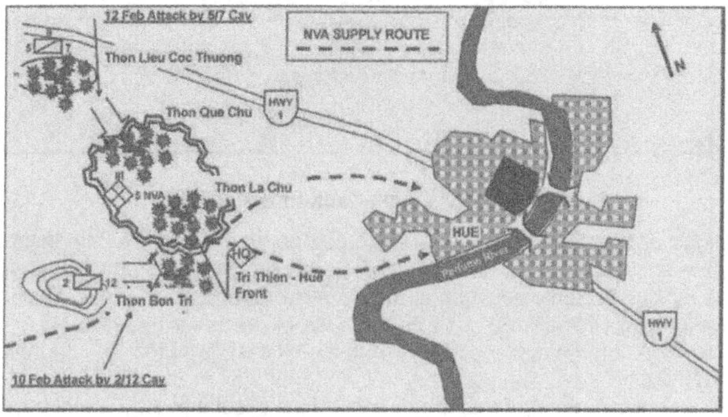

Figure 7-15. Subsequent attack to isolate Hue

The first additional American unit was not available until 12 February when the 5 7th Cavalry attacked Thong Que Chu much like the 2-12th Cavalry had attacked previously. The 5-7th Cavalry had even less success against the alert 5th NVA. The 5-7th Cavalry was forced to occupy defensive positions in Thon Lieu Coc Thuong and await the build up of combat power before it could continue to attack. In the interval, 2-12th Cavalry had moved off the high ground and captured the hamlet of Thon Bon Tri, south of the 5th NVA Regiment.

On 21 February, the 1st Cavalry Division had moved enough resources to the area to launch an effective attack to isolate Hue (see figure 7-16). In addition to the 5-7th and 2-12th Cavalry, the 1-7th Cavalry arrived in the AO and the 2-501st Airborne Infantry of the 101st Airborne Division was attached. On 21 February, after a combined artillery, air, and naval gunfire bombardment, the four battalions attacked the Thon La Chu stronghold. Elements of the 5th NVA Regiment were either destroyed in place or fled northeast. The next day resistance in Hue was noticeably lighter. U.S. Marine and ARVN units began the last phase of fighting to recapture the Imperial Palace. On 26 February, the North Vietnamese flag was removed from the Citadel and the ARVN I Corps declared the city secured.

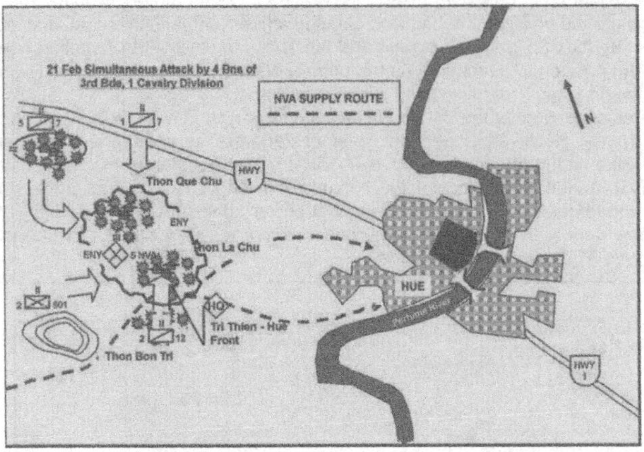

Figure 7-16. Final attack to isolate Hue

The actions of the 1st Cavalry Division forces northwest of Hue demonstrated the importance and the difficulty of isolating an enemy fighting in an urban area. Isolating Hue was difficult not only because of the dispersion and surprise with which the Tet Offensive caught U.S. forces, but also because of the tenacity of the NVA. At least one-third of the combat power of the NVA in the Hue AO was focused on maintaining access to the city.

Although only one aspect of this urban battle, Hue's isolation had an immediate and important, if not decisive, impact on operations. It not only resulted in restriction and then elimination of supplies and reinforcements, but it also immediately impacted the conduct of the defending NVA forces. Isolation caused an immediate drop in NVA morale and changed the nature of the defense. Once the enemy was isolated from external support and retreat, the objective of the NVA in the city changed from defending to avoiding destruction and attempting to infiltrate out of the city.

Civilian Reactions

7-63. Commanders must also consider the potential effects on and reactions and perceptions of the population living in the urban areas that they choose to isolate and bypass—either as a direct effect or as a response of the threat force being isolated. Isolation to reduce the threat's ability to sustain itself will likely have similar (and worse) effects on the civilian population remaining in the isolated area. (If food and water are in short supply, threat forces may take from noncombatants to satisfy their needs, leaving civilians to starve.) Isolation may also create a collapse of civilian authority within an urban area as it becomes apparent that the military arm of their government is suffering defeat. Due to their isolation, elements of the population may completely usurp the governmental and administrative functions of the former regime and establish their own local control, or the population may lapse into lawlessness. Returning later, Army commanders may find that these self-governing residents are proud of their accomplishments and, in some instances, less willing to allow Army forces to assume control since they may be perceived as having done nothing to earn that privilege. Alternatively, as witnessed in some urban areas during OPERATION IRAQI FREEDOM in 2003, a power vacuum may lead to intra-urban conflicts among rival factions coupled with general public disorder, looting, and destruction of the infrastructure (see also the discussion of Competing Power Structures in Chapter 3).

Direct Action by Special Operations Forces

7-64. Although SOF in urban offensive operations will likely conduct essential reconnaissance, they also have a direct action capability to shape the offensive operation (see figure 7-17). Special Forces and Rangers can use direct action capabilities to attack targets to help isolate the urban area or to directly support decisive actions subsequently or simultaneously executed by conventional forces. Successful attacks against urban infrastructure, such as transportation or communications centers, further the area's physical and electronic isolation. Direct action against command centers, logistics bases, and air defense assets can contribute to the success of conventional attacks by destroying or disrupting key threat capabilities. Direct action can also secure key targets such as airports, power stations, and television stations necessary for subsequent operations. Direct action by Special Forces and Rangers in these operations can help achieve precision and reduce potential damage to the target or noncombatant casualties.

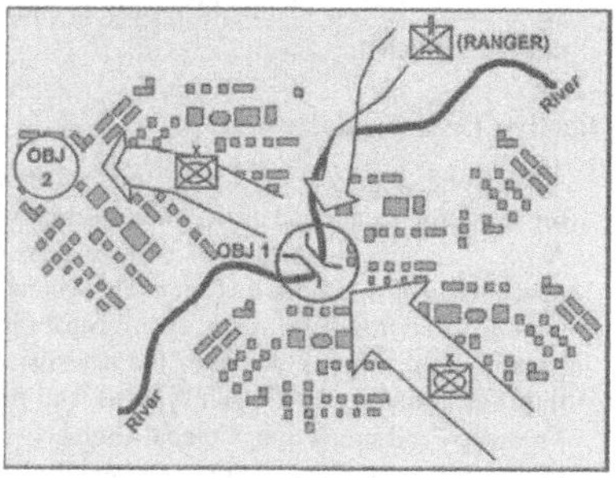

Figure 7-17. Coordination of SOF and conventional capabilities

Information Operations

7-65. Regardless of how Army forces physically isolate the urban area, they combine physical isolation with offensive IO to electronically isolate the threat and undermine his morale. Electronic isolation will cut off communications between forces in the urban area from their higher command to deny both from knowing the other's status. IO combined with isolation may persuade the threat's higher command or leadership that its forces located in the urban area are defeated. Thus, the command or leadership's intentions to break through to the besieged threat forces may be affected. PSYOP can undermine the morale of the threat in the area and reinforce electronic isolation and perceptions of abandonment. IO can be used to reduce any loyalty the civil population may have to the threat. IO can also ensure that civilians have the information that minimizes their exposure to combat and, as a result, overall noncombatant casualties. In addition, IO aim to deceive the threat regarding the time and place of Army force operations and intentions.

Detailed Leader Reconnaissance

7-66. Army commanders must clearly see the urban environment to understand the challenges facing their brigades, battalions, companies, platoons, and squads. Urban terrain can be deceptive until viewed from the soldier's perspective. Commanders are responsible to intimately know the conditions to allocate resources effectively to subordinate units. Often, particularly at battalion level and above, commanders will not be able to command and control dispersed forces from positions forward, but be forced by the terrain to rely on semifixed command posts. Detailed leader reconnaissance of the AO by commanders, their staff, and their subordinates before the mission can compensate for this challenge. This reconnaissance will give commanders a personal feel for the challenges of the terrain and will facilitate more accurate planning and better decision making during operations.

Mission Orders

7-67. Before contact, commanders can mitigate some terrain challenges to effective C2 using mission orders. Subordinates have mission orders to take advantage of opportunities before C2 systems can adversely impact the environment. To see the battle and provide effective and timely direction, tactical leaders will follow closely behind units as they assault buildings, floors, and rooms. Thus, only the most mobile INFOSYS can accompany tactical leaders into combat, and they will suffer the degrading effects of the environment. Mission orders permit rapid and decisive execution without commanders intervening at battalion level and above. Higher-level commanders can facilitate mission orders through their subordinates by articulating their desired end state, clearly stating their intent, and building flexibility into the overall plan.

Effective Task Organization

7-68. Commanders can shape urban offensive operations through effective and innovative task organization. Combined arms, often starting with an infantry base, are essential to success and may be the Army's asymmetric means of defeating an urban threat. Urban attacks will quickly break down into noncontiguous firefights between small units. To achieve the tactical agility for mission success in this nonlinear environment, many Army capabilities are task organized down to the company, platoon, and squad levels. Infantry provides the decisive capability to enter buildings and other structures to ensure threat destruction. Tanks, gun systems, and fighting vehicles provide additional speed and mobility, direct firepower, and protection. Combat engineers provide specialized breaching and reconnaissance capability. Field artillery provides the indirect (and if necessary, direct) firepower. Such mobility and firepower create the conditions necessary for the dismounted infantry to close with and destroy a covered threat in an urban defense. When a threat skillfully uses the urban area to limit ground maneuver, vertical envelopment or aerial attack using precision-guided munitions from Army aviation may circumvent his defenses and achieve necessary effects. Generally, ground systems used within the urban area will not be able to operate independently from dismounted infantry. The infantry will be required to protect armor and mechanized systems from close antiarmor weapons, particularly when those weapons are in well-prepared positions located throughout the urban area but especially on rooftops and in basements.

7-69. In urban offensive operations, direct fire support can be critical. Armor vehicle munitions types do not always achieve decisive effects against some urban structures. In some cases, field artillery high explosive munitions work better than armor for direct fire support of infantry. Large caliber (105 or 155mm) high explosives directly fired at a structure often produce a more severe shock effect than tank and fighting vehicle cannon and machine guns produce. Artillery is also able to achieve higher elevation than armor and engage threats located at greater heights.

7-70. However, commanders must view artillery as not just a weapon but a weapon system. As such, artillery should normally be placed under tactical control (TACON) of maneuver commanders, such as a platoon of three guns TACON to a company or a battery to a battalion, not just one gun to a company or other maneuver unit. Self-propelled artillery has some of the mobility characteristics of armor; however, it provides minimal ballistic protection from fragmentation for the crew. Although these systems seem formidable, they provide less crew protection than a Bradley fighting vehicle, for example, and contain large amounts of onboard ammunition and propellant. They are susceptible to catastrophic destruction by heavy automatic weapons, light cannon, and antitank fire. Therefore, infantry units must carefully secure and protect these systems (even more so than armored vehicles) when employed in urban offensive operations, particularly when forward in the direct fire role.

Creative Task Organization: Using Artillery in the Direct Fire Role

Task organizing artillery to permit its use in a direct fire role demonstrates the innovative task organization required for urban operations. The following provides three historical examples of task organizing and using field artillery for a direct fire role.

In 1944, U.S. Army units of the 1st Infantry Division were assigned to attack and seize the German city of Aachen. The city's internal defense included bunkers designed to serve as air raid shelters. These positions, buildings of stone, seemed impervious to direct fire tank weapons, demolitions, and small arms. To reduce the positions, the 1st Infantry Division relied on the artillery's direct fire.

Field artillery used this way had physical and psychological effects on the defenders. The 26th Regimental Combat Team's history of the battle describes the German reaction to the artillery pieces:

The chief shock to the defenders, Colonel Wilck (Aachen defense commander) said, came from the self-propelled 155s and tanks. The colonel spoke with considerable consternation of the 155mm self-propelled rifles. A shell from one of the guns, he said, pierced three houses completely before exploding and wrecking a fourth.

The 26th Infantry Regiment also described how the artillery, one piece attached to each assaulting infantry battalion, helped the infantry to penetrate buildings.

With solid blocks of buildings comprising most of the city, there wasn't any easy way to get at the Germans in the buildings. The eight-inch gun solved the problem. Beginning on the eastern outskirts the gun would plow a round into the side of the built up block of buildings at about ground level. One shell would usually open an entrance into the first tier of floors, i.e. the first building. Then several more shells were fired through the first hole. Thus a tunnel would be rapidly made all the way to the next cross street. Soldiers could then rush the newly formed entrance, clear the upper floors with hand grenades and rifles and then move on to the next building to repeat the process. When a block or square, was thus completely cleared of Germans—soldiers, skulkers, or even snipers—the next square was treated in the same way, working forward square by square, right and left, thereby avoiding nearly all exposure in the streets.

In 1982, Israeli forces invaded southern Lebanon to destroy base camps of the Palestine Liberation Organization (PLO). This operation involved significant fighting in urban areas including major operations in Beirut. Artillery, firing in a direct fire role, played a major part of the tactical solution. Artillery was particularly effective in the 33-day siege of Beirut. During this siege, Israeli forces used artillery in its traditional role as well as in the direct fire role.

The Israeli army was committed to a policy of disproportionate response during the Beirut siege. When fired on with small arms, crew-served weapons, tanks, or indirect artillery, the Israeli forces responded with intense, high-caliber direct and indirect fire from tanks and artillery positioned around the city. Many firing positions were on heights to the south and southwest that dominated much of the city. These positions had almost unrestricted fields of view. Israeli artillery fired from these positions directly into high-rise buildings concealing PLO gunners and snipers. The artillery, using direct fire, destroyed entire floors, collapsed floors on top of each other, and completely removed some upper floors. Such a response, as in Aachen in 1944, had as much a psychological impact as it did a physical impact on the PLO defenders.

In the early hours of 20 December 1989, the United States launched OPERATION JUST CAUSE. One of this operation's objectives was removing the Panamanian dictator, Manuel Noriega. U.S. forces carefully planned using all fires before the operation to minimize casualties and collateral damage. Part of this detailed fire planning called for applying artillery in a direct fire role.

The Panamanian Defense Force's (PDF) 5th Rifle Company based at Fort Amador was one of the key objectives of U.S. forces at the start of hostilities. This unit was high priority because it was the closest PDF unit to Noriega's headquarters, the Comandancia. U.S. forces expected the 500-man company to react first to

> OPERATION JUST CAUSE by reinforcing the defense of the Comandancia. It also posed a threat to U.S. military dependents housed at Fort Amador.
>
> To quickly neutralize this force, the United States assembled a three-company force composed of A, B, and headquarters elements of 1-508th Infantry (Airborne), supported by 105mm towed howitzers of 320th Field Artillery and M113 armored personnel carriers. The howitzers and the personnel carriers were covertly pre-positioned at the fort. At approximately 0100, helicopters transported the two airborne rifle companies into position. The howitzers then suppressed any personnel in the PDF-controlled buildings on Fort Amador while demonstrating the firepower of the U.S. task force. They used direct fire into the PDF barracks. The impact of the 105mm high explosives and .50-caliber fire from the M113s convinced the PDF infantry to give up after token resistance. Following the direct fire, U.S. infantry assaulted and cleared the dozen PDF buildings, finding that most occupants had fled or surrendered. For more details of OPERATION JUST CAUSE, see Applying the Urban Operational Framework: Panama in Chapter 5.
>
> ---
>
> The three examples cited indicate the importance of the situation-dependent and innovative task organization of artillery and its use in the direct fire role. Using artillery helps overcome some challenges of offensive operations in the urban environment, and it has an important psychological effect on a defending threat. Such task organization takes a traditional tool of a higher-level tactical commander and uses it to directly influence the company-level battle. This philosophy of task organization can be applied to other types of forces—not just artillery. PSYOP teams, interpreters, CA specialists, armor, engineers, and reconnaissance teams may require task organization different from traditional organization. The compartmented urban environment drives the requirements for these assets lower in the tactical scheme than in open operations. Consequently, commanders must understand and account for more of these assets for UO than for operations in less restrictive environments.

7-71. Army aviation may also be inventively task organized. It can support urban operations with lift, attack, and reconnaissance capabilities. Tactical commanders down to company may use all these capabilities to positively influence ground close combat. Army attack and reconnaissance aircraft can provide flank security for attacking ground forces. Attack aircraft may also provide direct fire support to individual platoons or squads. Lift may move entire battalions as part of brigade operations, or it may move single squads to a position of advantage (such as a roof) as part of a small unit assault. Army aviation can assist with C2 by providing airborne retransmission capability, airborne command posts, and the confirmed status and position of friendly forces. However, Army aviation is a limited and high-value asset; commanders should review its use in innovative task organizations. It is particularly vulnerable to urban air defense threats unless used over terrain secured by ground forces. From these positions, aircraft can use enhanced sensors to conduct reconnaissance and use precision weapons with standoff capability.

Shaping Attacks

7-72. In a large urban area, the defending threat cannot be strong everywhere. Shaping operations can also take the form of attacks against vulnerable positions to force the threat to maneuver and redeploy his forces in the urban area. This prevents him from merely defending from prepared positions. Forcing the threat to move negates many of the defensive advantages of urban terrain, confirms his dispositions, exposes vulnerable flanks, and permits target acquisition and engagement with precision standoff fires.

7-73. A critical shaping operation in urban offensive operations is usually an initial attack to seize a foothold. Once Army forces establish this foothold, they accrue some of the defensive advantages of urban terrain. From this protected location, Army forces continue offensive operations and have a position of advantage against neighboring threat defensive positions.

ENGAGE

7-74. Commanders may employ several methods to decisively engage elements of the urban area during offensive operations. These include—

- Rapid maneuver.
- Appropriate use of SOF.
- Precise application of fires and effects.
- Proper balance of speed and security.

7-75. None is unique to UO. Their effective execution, however, allows Army commanders to dominate in this challenging environment by effectively using resources with the least amount of collateral damage. Overall, decisive engagement results from urban offensive operations when forces achieve the objective of the assigned mission and establish preeminent control over the necessary terrain, population, and infrastructure. Largely, the Army commander's ability to engage is based on superior situational understanding and the correct application of unit strengths to the challenges found in the urban environment.

Rapid and Bold Maneuver

7-76. Commanders of major operations may have or create the opportunity to seize an urban area with rapid and bold maneuver. Such maneuver requires striking while the area remains relatively undefended—essentially preempting an effective defense. This opportunity occurs when the urban area is well to the rear of defending threat forces or before the onset of hostilities. Under such conditions, an attack requires striking deep behind threat forces or striking quickly with little time for the threat to make deliberate preparations. Attacks under such conditions may entail significant risk; however, the potential benefit of audacious offensive operations may be well worth possible losses. Three potential ways to accomplish such attacks (and their combinations) are:

- Airborne or air assault.
- Amphibious assault.
- Rapid penetration followed by an exceptionally aggressive exploitation, for example, a heavy force using shock, armor protection, and mobility.

7-77. Commanders should analyze all potential urban operations to seek an opportunity or advantage to apply rapid and bold operational maneuver to the task. Using operational maneuver to avoid urban combat against an established threat defense potentially marks a significant operational achievement and can have decisive strategic consequences. Just influencing the threat's morale can positively affect all future operations. However, commanders must evaluate the challenges of such a course of action. These challenges may include—

- Sustaining the operation.
- Avoiding isolation and piecemeal destruction.
- Successfully conducting shaping attacks.
- Achieving the necessary tactical, operational, and strategic surprise.

7-78. Commanders can also build on the shaping effects of isolating the urban area internally and externally by attacking urban decisive points from multiple directions. They can attack multiple decisive points either simultaneously or in a systematic, synchronized manner. This complicates the threat's situational understanding of the urban environment, further impedes his decision making, and allows Army commanders to dictate the tempo.

Bold Operational Maneuver to Seize an Urban Area: Inchon and Seoul, Korea – 1950

In August 1950, UN forces in Korea were desperately attempting to stave off defeat by establishing a viable defense in southern Korea. This defense, known as the Pusan Perimeter, was the focus of the world's attention. The commander of U.S. Forces Far East, General Douglas MacArthur, was not focused on the U.S. Eighth Army in Pusan. Instead, he focused on how to dramatically and decisively transition to the offense—the recapture of the South Korean capital of Seoul.

Seoul was more than just the South Korean political capital, although that was an important consideration. Seoul was the key to the Korean transportation system, the hub of the national communications system, and the center of the country's economy and culture. It was also strategically placed astride maneuver corridors to the north, south, and northeast. It had been Korea's capital since 1394 when King Taejo Yi Sung-Gye established it as the center of the Chosun dynasty. It had been captured by the surprise communist invasion of the south in June 1950, and the key to success in Korea was its recapture. The question was how.

The obvious military solution to the Korean situation in August 1950 was to build military strength within the Pusan Perimeter and then, when sufficiently strong, to breakout and counterattack north and recapture Seoul. The disadvantage of this course of action was that the counterattack would be through hundreds of miles of mountainous terrain, through several other major cities, and across numerous rivers. The North Korean People's Army (NKPA) would be able to fall back from one mountain defense and one river line to another and would have weeks or months to prepare the defenses of Seoul.

MacArthur's solution was to delay a conventional counterattack, boldly maneuver by sea deep into the flanks of the NKPA, land on the Korean west coast at Inchon, and rapidly seize Seoul before the enemy could react (see figure 7-18). This course of action had numerous disadvantages: achieving surprise; assembling a landing force trained in amphibious operations; few good landing sites; supplying a large force once landed; and needing to simultaneously continue to defend and prepare a counterattack with the Eighth Army from Pusan. MacArthur understood the challenges of the operation but believed that the potential value compensated for the risk.

The potential value of a rapid assault on Seoul through a deep amphibious operation was immense. The operation would trap the bulk of the NKPA in South Korea and facilitate its destruction. It would also capture Seoul before the NKPA could prepare a defense. The quick recapture would immeasurably impact the morale of UN forces and South Korea. MacArthur felt the payoff of success justified the risks and challenges of the operation.

On 15 September, elements of the 1st Marine Division landed in Inchon taking the NKPA completely by surprise. On D+1 they were joined by elements of the 7th Infantry Division, both divisions under the control of U.S. X Corps. Naval gunfire and Marine close air support flying from Navy aircraft carriers supported the landings. Seven days later, the 5th Marine Regiment had battled 25 miles to the outskirts of Seoul and began attacking the city from the north. Three days later, on 25 September, the 1st Marine Division and elements of the 7th Infantry Division had defeated the 10,000 defenders of the NKPA's 18th Rifle Division.

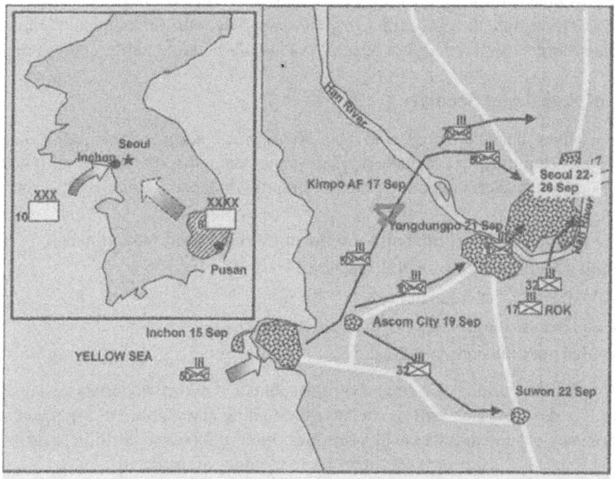

Figure 7-18. Inchon-Seoul campaign, September 1950

> Most of MacArthur's predictions for the Inchon-Seoul campaign were accurate. Superb staff work, excellent logistics operations, and unmatched training permitted the UN forces to assemble a trained landing force, land it over one of the most difficult shores in the history of military amphibious operations, and keep it supported. More importantly, the operation achieved total strategic and operational surprise. Seoul was recaptured quickly and, although resistance was fierce, the NKPA could not react fast enough to influence the outcome. In conjunction with the Eighth Army counterattack from Pusan, 75 percent of the NKPA was destroyed. The Inchon-Seoul campaign was an important turning point in the war, and had not the Chinese People's Army intervened two months later, it would have been the decisive campaign of the war.

Appropriate Use of Special Operations Forces

7-79. Sometimes Army forces can decisively engage (instead of simply shape) the urban area using the direct action capability of SOF. When the threat fails to develop a comprehensive defense and does not possess large, capable conventional forces, then Army forces can achieve operational surprise. Commanders, by synchronizing conventional and SOF effects, may actively control offensive operations to dominate the area. Importantly, however, conventional ground forces must be available to quickly assume the mission because SOF units acting as the primary striking force have limited logistic capability to sustain long-term operations. OPERATION JUST CAUSE offers several examples of this type of synchronization (see Applying the Urban Operational Framework: Panama in Chapter 6).

Precise Application of Fires and Effects

7-80. Precisely applied fires and the massed effects of combat power characterize successful urban attacks. The fires can be direct fire from combined heavy or light ground teams; direct or indirect fires from supporting Army aviation standing off from the target and any possible air defense threat; precision indirect fires from conventional tube artillery; or direct and indirect fires from supporting joint assets including Air Force and naval assets. All efforts strive to reduce collateral damage around the point of attack, consistent with mission success. Forces use fires to deny the threat the ability to maneuver in the urban area and to destroy the threat when he attempts to maneuver. When the threat exposes himself by moving, the environment no longer protects him, and fires can effectively engage him. Overall, precise fires and effects demonstrate the power of Army forces and help the urban population understand that only legitimate military targets are the focus of attacks, potentially building public support of Army UO.

Proper Balance of Speed and Security

7-81. Attacking units balance speed and security. Forces secure flanks as units advance, control dominating terrain (buildings), evacuate civilians, and keep the integrity and synchronization of the combined arms team. Obstacles are anticipated and rapidly breached. Commanders can choose avenues of approach to—

- Provide cover and concealment for following aviation and support units.
- Permit travel by all classes of vehicles.
- Easily defend from counterattack.
- Avoid nonessential centers of threat resistance.
- Avoid population concentrations.

7-82. Army aviation is a critical resource to protect flanks. Another important resource is engineers who seal off surface and subsurface entries and avenues along the flanks of the attack. Finally, as in all offensive operations, ground and air cavalry are ideal mobile forces to perform security in an economy of force role along flanks allowing decisive forces more freedom of maneuver.

CONSOLIDATE

7-83. Commanders, at all levels, must consolidate to strengthen their position during urban offensive operations without loss of momentum. They take the steps necessary to make permanent any temporary battlefield successes while maintaining relentless pressure on threat forces. Consolidation provides security and protection, facilitates reorganization, and allows forces to prepare for counterattack.

Repositioning of Forces

7-84. Following seizure of the objective, commanders normally consolidate by adjusting and repositioning forces. While UO are likely to be noncontiguous, commanders reposition ISR assets (including observation posts and reconnaissance patrols) to maintain contact with the threat, to establish contact with nearby friendly units and ensure that there are no exploitable gaps or seams, and to help maintain freedom of action. Physical occupation of the terrain (as well as continued reconnaissance) will provide commanders with a fuller understanding of the urban environment. With this enhanced understanding, commanders may need to adjust boundaries and other control measures to better adapt to the effects of urban terrain features such as canals, subway tunnels, raised roadways, and tall buildings. As necessary, commanders may also need to reposition communications assets and C2 facilities to enable subsequent operations.

7-85. Immediately after the conduct of successful urban operations, units must also be alert to the potential for rapid and violent counterattacks. Defenders may launch a quick counterattack to regain terrain before the offensive forces have consolidated and fully assumed the defensive advantages of the urban terrain. Delaying a counterattack in UO, even for a few minutes, permits the environment's advantages to shift to the successful attacker. Thus, attacking units anticipate this reaction and reposition forces such that they are prepared to defeat it.

7-86. As necessary, commanders should reposition armor and artillery (and other fire support assets) to account for the changing situation and battlefield geometry. Commanders should also consider consolidating and repositioning armored and artillery forces in positions—either inside or outside the urban area—to add significant combat power to a hasty defense, to defeat threat counterattacks, or to allow for quick resumption of the attack. If integrated into a hasty defense inside the urban area, these forces will require continued infantry protection. As a mobile counterattack force positioned inside the urban area, armored forces will also require careful selection of attack positions and counterattack routes. Damage to buildings and infrastructure may limit maneuver and the use of direct and indirect weapon systems.

Repositioning forces outside the urban area may contribute to strengthening or reestablishing the isolation of the urban area and will make better use of range and standoff capability, allow them to take full advantage of their speed and mobility, and decrease their need for additional protective support.

Eliminating Pockets of Resistance

7-87. Focusing on controlling the essential and attacking decisive points means that attacking units will have bypassed some elements of the threat's defense. Within the bounds of the initial plan and current situation, commanders should consider whether it is necessary to devote the time and resources to immediately clear and eliminate remaining threat forces and pockets of resistance with all or parts of the attacking force or leave the task to follow-on forces. In their deliberations, commanders should determine if remaining threat forces are capable of consolidating and mounting effective counterattacks before follow-on units can engage the remaining threat and take advantage of their dispersion and disunity. As another part of their deliberations, commanders should also consider if remaining threat forces will likely (and immediately) endanger—

- The urban inhabitants.
- Lines of communications.
- Critical resources within the urban area necessary to support the inhabitants or subsequent operations.

Reorganization

7-88. Reorganization includes all measures taken by the commander to maintain combat effectiveness or return to a specified level of combat capability. As necessary, these actions include—

- Redistributing or cross-leveling supplies and equipment until resupply can be accomplished.
- Replacing key personnel and combining units or crews to form mission-capable ones.
- Integrating replacement soldiers and systems into the unit and matching operational systems with available crews.
- Recovering, treating, and evacuating casualties, enemy prisoners of war, and damaged equipment.
- Conducting training and disseminating critical lessons learned.
- Other actions to reestablish unit cohesion.

TRANSITION

7-89. Effective transitioning allows commanders to continue full spectrum operations in the urban area and elsewhere in the AO without unnecessary delays. Commanders can accomplish this by thorough planning—including appropriate branches and sequels (revised as the situation changes)—that gives adequate consideration to post-offensive organizational, training, psychological, and civilian requirements. If properly prepared, commanders can anticipate rather than react to potential mission changes.

Early and Concurrent Transition Planning

7-90. Commanders can ensure smooth transitions of urban offensive operations by planning for post-offensive operations early. Based on the mission envisioned, they determine which subordinates and what type of force structure to use. Post-offensive missions, like all urban operations, encompass full spectrum operations. At the successful conclusion of offensive operations, Army forces transition to some type of stability or civil support operation. Commanders may leave the subordinate unit in place to execute the new mission, may reorganize the subordinate unit for the mission, or relieve the unit that just completed offensive operations with a new unit.

Changes to Task Organization

7-91. Commanders should consider the organization of forces following offensive operations. Hostile civilians may require significant combat forces or military police forces to maintain stability. On the other hand, friendly civilians may require a minimum of military police or combat forces, but significant logistic support. Commanders must carefully consider the urban situation before deciding how to use combat forces that recently participated in a high-intensity offensive operation.

Training and Psychological Considerations

7-92. Many Army combat tasks may not support follow-on stability or civil support operations without considerable modification (if at all). Often, noncombat skills—not normally part of a unit's mission essential task list such as negotiating or mediating skills—will be required. However, the greatest modification required applies to each soldier's mental outlook. Forces that transition directly from combat to stability operations may not be psychologically prepared for a rapid and drastic change of mission. Commanders cannot expect troops who have just completed high-intensity offensive operations to rapidly adjust and exercise the sensitivity and judgment required in most stability operations. This especially applies if the population is hostile to Army forces. If possible, combat forces assisting in stability operations, particularly in hostile civilian situations, should not have had recent experience in high-intensity urban operations and they should have trained for the mission.

Return to Civil Agencies

7-93. Commanders of major operations also have the critical role of transitioning aspects of the urban offensive operation to civil, allied, NGOs, and other agencies as appropriate. Transition planning is detailed and aims to quickly return as much civil control of the area as is feasible after the attack. Beyond local civil control, outside civil agencies and NGOs assume tasks as completely and as rapidly as possible. These organizations are consulted and integrated into the planning process as early as possible. Commanders should begin planning for transition simultaneously with planning for offensive operations. They consider the feasibility of relinquishing control of urban areas to civil government, law enforcement, or NGOs even before completing offensive operations. During the conduct of urban operations, these transition operations are closely synchronized with the execution of the attack.

Transition to a New Mission

7-94. In urban offensive operations, like other offensive missions, the change in mission after a successful urban attack may be to a hasty defense or a continuation of offensive operations outside the area. However, in urban offensive operations the mission will just as likely rapidly change to a civil support or stability mission. This is particularly true if the unit has had special training and is task organized for urban operations. Transition to stability or civil support operations is often accompanied by a transition in roles from supported to supporting.

7-95. Even more challenging than transition at the end of the mission is transition during the conduct of the mission. Soldiers may have a difficultly transitioning from stability or civil support to offense and defense, and back again multiple times during an urban offensive operation. Soldiers may apply the tactics, techniques, and procedures of urban offensive operations directly to the stability or civil support missions with potentially disastrous results. Commanders need to segregate missions in time and space, and, if sufficient forces exist, by unit. To this end, commanders should consider permanently designating specific units to conduct civil-military and humanitarian support tasks and avoid rapid mission changes that rotate units (particularly at company level and below) between violent and nonviolent tasks. However, commanders may not have that luxury and may need to rely heavily on preparatory training (including the inculcation of Army values) and strong unit leadership to mitigate potential difficulties.

Chapter 8
Urban Defensive Operations

Generally, a modern city magnifies the power of the defender and robs the attacker of his advantages in firepower and mobility. A city can ingest an invading army, paralyze it for weeks on end, and grind it down to a state of ineffectiveness.

"Military Operations on Urbanized Terrain:
The 2d Battalion, 26th Infantry, at Aachen, October 1944"

The skillful defense of an urban area can decisively affect a campaign. The urban area offers many advantages to defending forces. An adroit defender can use the advantages of the urban environment to negate combat power disparities, blunt the tempo of an attack, attrit threat forces, and sap the morale of attacking troops. The defender gains an opportunity to concentrate resources, reconstitute attrited units, and transition to the offense. A successful defense of an urban area can also deny the threat vital resources. Defense in the urban environment is an essential Army capability and can significantly affect the outcome of entire campaigns and the achievement of national objectives.

PURPOSE OF URBAN DEFENSIVE OPERATIONS

8-1. Army forces defend urban areas for various reasons: defeating a threat attack, buying time, economizing forces, protecting an ally's political institutions and economic infrastructure, protecting an urban population, shaping conditions for decisive offensive operations, and shaping conditions for executing stability or civil support operations. During force projection operations, urban areas may be used as initial lodgment areas that Army commanders may need to defend at the outset until they build sufficient combat power. Usually two or more of these purposes apply to the urban defense. Urban defensive operations provide commanders great opportunities to turn the environment's characteristics to the advantage of Army forces. Urban areas are ideal for defensive operations and greatly enhance the combat power of defending units.

DEFENSIVE CHARACTERISTICS APPLIED TO URBAN OPERATIONS

8-2. There are five general characteristics of the successful defense: preparation, security, disruption, massing effects, and flexibility. All apply to the successful urban defense and to the higher commander supporting a subordinate defending in the urban area.

PREPARATION

8-3. The urban area suits the defense since the area's physical characteristics naturally enhance the combat power of defending units. These characteristics include protection, obstacles, and concealment. Urban terrain provides superb defensive positions with minimum preparation. With deliberate preparation, urban defensive positions can rapidly become strong points.

8-4. One primary characteristic of urban terrain that enhances the defense is protection. With little or no advance preparation, buildings, subsurface structures, and walls protect Soldiers from direct and indirect fire, interdict indirect fire, limit observation, and limit engagement ranges (requiring Soldiers to be skilled at combat in close quarters and quick-fire techniques). Nearly all buildings provide some ballistic protection from direct and indirect fire. Mason and stone buildings with basements and cellars can protect Soldiers from most fires except the largest caliber or tonnage bomb. Minimal additional preparation turns them into formidable, defensive strong points.

8-5. Buildings in urban areas, because of their height and close proximity, also can protect Soldiers by masking them from indirect fire. The height of a building may interdict the flight path of an artillery round, rocket, missile, or bomb at a point short of the intended target. Masking protects static defending forces and protects forces moving along routes bordered with tall buildings that form urban "canyons". These protected routes can be used for sustainment, counterattacks, and maneuver.

8-6. Structurally significant buildings in an urban area can create major obstacles to maneuver. These obstacles immediately canalize maneuver into existing streets and routes without any preparation by the defense. These obstacles then become kill zones for well-positioned and sited defensive forces. Minimal obstacle construction as point obstacles blocking streets and routes can further restrict the maneuver options of the attacking force. Rubble from structures collapsing into streets after fires (intentional or unintentional) can also block routes.

8-7. Buildings also conceal the location, disposition, and intent of the defense. They limit visual observation to the external perimeter of the urban area. They degrade radar and electronic position identifiers and decrease the utility of overhead imagery. The physical aspect of the urban environment greatly enhances the defense by degrading the opposition's intelligence, surveillance, and reconnaissance (ISR) capabilities. Buildings can conceal static defensive positions and the maneuver of defensive forces in the urban area. Although the environment constrains defensive mobility in much the same manner as offensive mobility, the defender has the time and opportunity to conduct careful reconnaissance and select and prepare routes. This gives the defender the ability to move reserves, maneuver counterattack forces, and plan sustainment without observation. Careful preparation provides the defender a mobility advantage over attacking forces.

SECURITY

8-8. The urban area can be an advantage or a disadvantage to the security of defending forces. This largely depends on the nature of the human dimension of the environment. If the population is evacuated or allied with Army forces, then the environment may assist in the security of defending Army forces. However, if the population is present and hostile, then the environment may make security difficult.

8-9. The physical aspects of the urban environment, uninfluenced by the human dimension, may assist in the security of defending Army forces. The combat power of small security forces manning observation posts is greatly enhanced. Forces can more easily restrict and monitor avenues of approach for threat reconnaissance. Defending forces positioned mostly in structures are difficult to locate.

8-10. The physical aspects of the environment may also present some security challenges, primarily with observation. The compartmented terrain limits the field of observation from any one point. The defense may require more security forces to adequately observe the mounted and dismounted avenues to prevent infiltration. Threat forces that successfully infiltrate will be more difficult to locate. These forces will gain numerous hide positions for small reconnaissance units in complex terrain and the effect the terrain has in masking electronic signatures.

8-11. Friendly civilians in the urban area can help identify threat forces attempting to conduct reconnaissance. Civilian activity will also help to mask defense preparations. However, a hostile element of the population may pass intelligence information to the threat. They may assist threat reconnaissance to infiltrate the urban area or provide guides, manpower, or resource support for threat forces. Commanders should take measures to ensure strict control of hostile populations. If resources permit, commanders may consider removing potentially hostile civilians from the area.

DISRUPTION

8-12. The urban environment's attributes assist defending Army forces to disrupt the attacker. It does this through compartmentalization, inhibiting command and control (C2), and facilitating counterattacks.

8-13. The physical aspects of the urban area force the attacking threat into compartmented urban canyons that make mutual support between attacking threat columns difficult. Shifting resources from one portion of the threat attack to another is also difficult. Physically, the urban area disrupts tactical communications making synchronization of combat power difficult.

8-14. The urban terrain hinders the mobility capabilities of the defense. However, careful planning, preparation, and rehearsals can facilitate more rapid movement of larger forces. Defending forces can assemble counterattacks undetected, move them along covered and concealed routes, and achieve surprise at the point of the counterattack. Attacking forces, using the compartmented terrain, often leave forward elements in position to be isolated or expose long and vulnerable flanks to friendly counterattack and interdiction.

MASSING EFFECTS

8-15. The urban environment allows defenders to better protect their centers of gravity and decisive points. The restrictive terrain reduces the attacker's maneuver options. Defenders can position forces in protected and mutually supportive positions oriented on deadly engagement areas. Relatively few well-positioned defenders can generate significant combat power. Without the positional advantage and the corresponding protective effects of the terrain, attacking forces often mass numbers to achieve the necessary combat power.

8-16. Knowing the complex terrain permits defending forces to plan engagement areas that maximize the effects of their combat power. Defending forces can remove fences, walls, rooftops, and even entire buildings to facilitate fields of fire and unmask indirect fire flight paths. Forces carefully choose firing positions for indirect fire systems so that flight paths travel between buildings into engagement areas. By leveraging this knowledge of the terrain, numerically inferior defenders can synchronize devastating fires on offensive forces that are forced by terrain and reinforcing obstacles to mass in confined spaces where fires can have the greatest effect.

FLEXIBILITY

8-17. Defensive flexibility results from detailed planning and mental agility. Commanders develop defensive flexibility by ensuring that plans adequately address branches and sequels that include alternate and subsequent positions and emphasize counterattack options. The urban area facilitates defensive flexibility because the urban terrain can be quickly adapted for defensive operations with little or no preparation. The effect is similar to having multiple, prepared positions on nearly every possible approach. The urban area can also permit rapid, covered movement on interior lines. This permits swift movement to and occupation of strong defensive positions with little or no preparation. The defense also has more flexibility since defenders often know and better understand the urban terrain's effects on operations. Normally, defenders will not get lost as easily, will know complex lines of sight and masking effects, and will best understand the ballistic characteristics of individual structures.

8-18. Mental agility allows commanders to see that the best urban defense may actually be to defend outside of the area. Such a defense mitigates the danger to the urban population and potentially reduces collateral damage. It takes advantage of Army long-range engagement capabilities and denies the threat the opportunity to "hug" Army forces or noncombatants as protection from fires. This defense may be appropriate when Army forces have enough resources to defend more open terrain, when time permits deploying extensive obstacles and constructing protected positions, and when natural terrain such as river obstacles aids the defense.

URBAN DEFENSIVE OPERATIONS AND BATTLEFIELD ORGANIZATION

8-19. Urban defensive operations are organized within the overall battlefield organization of sustaining, shaping, and decisive operations. The success of urban defense depends on each operation, but commanders must synchronize these simultaneous operations as one action. Sustaining operations in defensive urban operations (UO) ensure freedom of action. Critically, urban sustaining operations ensure security of the lines of communications and establish effective movement control. Shaping operations in defensive UO create the conditions for decisive operations. Shaping operations vary greatly depending on the type of defense. For example, in a mobile defense the shaping operation may be the fixing force. In contrast, in an area defense the fixed defense may be the decisive operation. In the urban defense, decisive operations focus on accomplishing the commander's mission. The decisive operation may not defeat the threat's main effort, and it may not prevent threat occupation of large portions of the urban area if those tasks are not essential to mission accomplishment. For example, if the defense's objective is to protect a critical communications node, then, depending on the commander's overall intent, threat actions to secure an airfield elsewhere may not be important.

TYPES OF URBAN DEFENSE

8-20. Commanders can view urban area defensive operations two ways: as conducting a major defensive operation with an urban area in their area of operations (AO) and as defending entirely in an urban area.

AREA DEFENSE

8-21. At the operational level, an area defense may include both urban areas and open maneuver areas. The most common defense in an urban area and the most suitable for the characteristics of this distinct environment is the area defense. As a defensive operation, the area defense concentrates on denying threat forces access to designated terrain for a specific time rather than destroying the threat outright. Although an area defense in an urban area does not directly seek to destroy or defeat attacking threat forces, as an objective it does aim to force culmination of the threat's attack. The urban area defense often works effectively to exhaust threat resources and shape conditions for a transition to offensive operations. The urban area may also be used as a strong point to force threat movement in a different direction or to fix threat forces as part of a large, mobile defense taking place in the AO outside the urban area.

MOBILE DEFENSE

8-22. A mobile defense can operate in an urban area but only under specific conditions. It focuses on destroying or defeating the threat through a decisive attack by a striking force. It requires the defender to have greater mobility than the attacker. To shape a mobility advantage, the urban defender effectively uses the terrain and correctly task organizes his forces' mobility. The principles of applying the mobile defense in the urban area remain the same: a small fixing force stops the threat and limits any ability to maneuver while a striking force quickly maneuvers and counterattacks to destroy him.

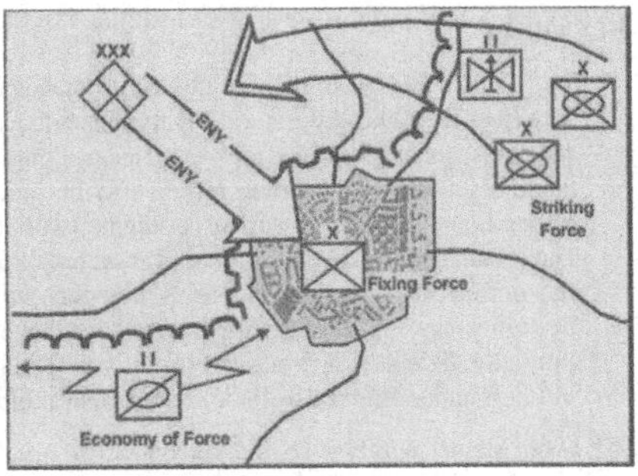

Figure 8-1. An urban area incorporated into a larger mobile defense

8-23. One key to executing a mobile defense in the urban area is to entice a threat force into the depths of the urban area where it begins to lose mobility options. A well-placed fixing force augmented with man-made obstacles and taking advantage of the naturally constrictive terrain can stop a much larger force. If the attacking force is largely mounted and armored, its mobility in the urban area may be reduced to less than that of dismounted infantry. In addition, if the attacking force's movement into the urban area is mounted and rapid, the commander's situational understanding also diminishes. Then the striking force, consisting of dismounted infantry forces, can execute the counterattack with surprise from multiple directions and dimensions (subsurface, surface, supersurface, and airspace). Man-portable antiarmor weapons—firing from flanks and top down and supported by precision indirect fires from both organic and joint systems—can rapidly destroy the threat.

8-24. From the perspective of commanders of the major operation, the urban environment can help defending forces achieve a mobility advantage over an attacker in a broader sense. Defending commanders can attempt to shape the battlefield so that the attacker commits significant resources into an urban area, where his maneuver capabilities are reduced (see figure 8-1). A disproportionately small defending force, which relies on the defensive combat power advantages of the urban environment, reduces and fixes the attacker's maneuver capabilities. Other defending forces mass outside the urban area then strike the threat with a combined mobility and firepower advantage.

Urban Defense in a Major Operation: Stalingrad – 1942 to 1943

The German and Soviet battle for Stalingrad in late 1942 and into early 1943 illustrates how a tactical urban area defense integrates into a larger mobile defense. Stalingrad was a battle fought on a huge scale: it involved army groups on both sides and thousands of square kilometers. Though the city was relatively small, it remained the focus of both German and Soviet offensive and defensive operations during the six-month battle.

In the summer of 1942, the Germans launched a strategic offensive in southern Russia. Its goal was the valuable oil fields of the Caucasus. German forces turning south into the Caucasus exposed a vulnerable flank to Soviet forces positioned between the Don and Volga Rivers. For the German Caucasus operation to succeed, it had to destroy Soviet forces between the Don and Volga, establish a good defensive line, and capture Stalingrad. This city would anchor the German defense and interdict the critical flow of supplies from the Caspian Sea via the Volga River into central Russia. Stalingrad, by virtue of its name, also had important political and cultural value to the Germans and Soviets.

The opening phases of the German offensive were successful; German forces—the 6th Army and 4th Panzer Army—entered the outskirts of Stalingrad in late August 1942 (see figure 8-2).

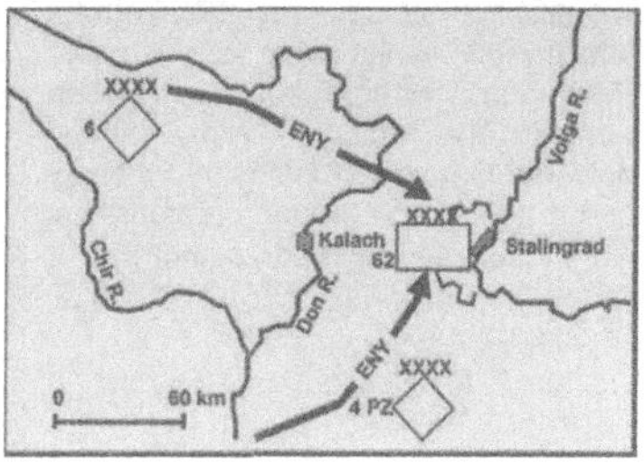

Figure 8-2. German attacks to seize Stalingrad

After a month of intense fighting, the Germans possessed nearly 90 percent of the city. At this point, the 6th Army commanded all German forces in and around Stalingrad. The Soviet 62nd Army's defense was reduced to a front only a few hundred meters deep and a couple of kilometers long on the banks of the Volga. The Soviet defenses hinged on fortress-like concrete industrial buildings and the fanatical bravery and tenacity of soldiers and civilians fighting in the city's remains (see figure 8-3 on page 8-6).

Beginning in mid-September, the Soviet command began looking at how to convert the defense of Stalingrad into an operational opportunity. During October and November, the 62nd Army held on to its toehold in Stalingrad. While maintaining the defense of the 62nd Army, the Soviets secretly began to build up strength on both flanks of the German 6th Army. The Germans increased their vulnerability by securing the German 6th Army's flanks with less capable Romanian, Hungarian, and Italian armies. Also, the 6th Army moved powerful German divisions into the city and rotated with German divisions that were exhausted by urban combat.

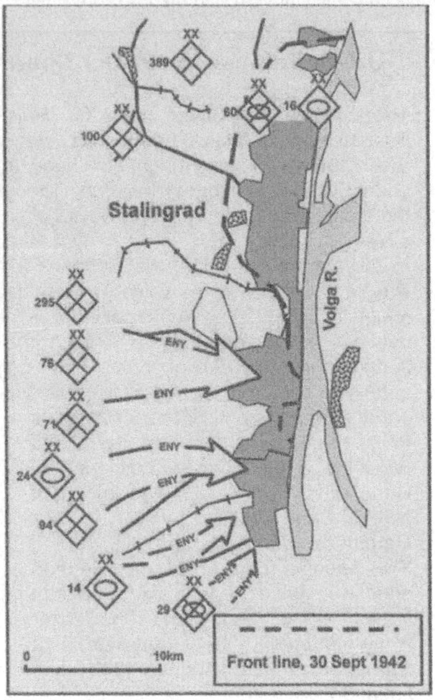

Figure 8-3. German attacks to seize Stalingrad, September 1942

On 19 November, the Soviets launched Operation Uranus that attacked two Romanian armies with seven Soviet armies. Simultaneously, the 8th Russian Army attacked to aid the 62nd Army in further fixing the German 6th Army. Within five days, the Soviet armies of the Don Front, Southwest Front, and Stalingrad Front met near the city of Kalach and sealed the fate of the German 6th Army's 300,000 troops in Stalingrad (see figure 8-4).

On the third day of the Soviet offensive, when encirclement seemed inevitable but not yet complete, the 6th Army commander asked permission to withdraw from the trap. The German high command denied permission believing that the Army could be supplied by air and then a renewed offensive could break through to the city. On 12 December, the German LVII Panzer Corps launched an offensive north to break through to Stalingrad. This offensive made progress until another Soviet offensive on 16 December forced its cancellation. This ended any hope of recovering Stalingrad and the 6th Army. On 31 January 1943, the 6th Army surrendered after sustaining losses of almost two-thirds of its strength. The Soviets took over 100,000 prisoners.

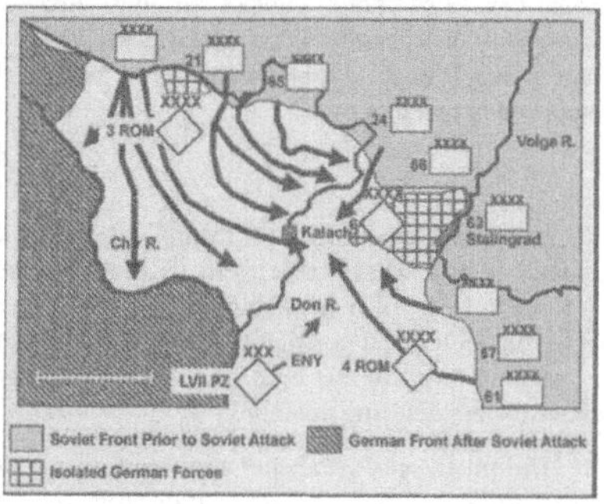

Figure 8-4. Soviet attacks trap German 6th Army

Many lessons emerge from the successful defense of Stalingrad. Tactically, the defense showed how using the terrain of a modern industrial city wisely could increase the combat power of an inferior, defending force and reduce the maneuver options of a mobile, modern attacking force. Another element in the Soviet's tactical success was the Germans' inability to isolate the defenders. The Germans never threatened the Soviet supply bases east of the Volga and, despite German air superiority, the Soviets continuously supplied and reinforced the 62nd Army across the Volga River. Also, Soviet artillery west of the river was able to fire in support of Soviet forces and was never threatened with ground attack.

At the operational level, the Soviets demonstrated a keen understanding of using an urban area within the context of a mobile defense. The 62nd Army's stubborn area defense of Stalingrad drew the bulk of the German combat power into the urban area where they were fixed by a smaller and quantitatively inferior defending force. This allowed the Soviets to build combat power outside the urban area. The Soviets set the conditions for a mobile defense by positioning powerful Soviet armor forces in open terrain outside the urban area against quantitatively inferior German allied forces. In OPERATION URANUS, the mobile defense's strike force destroyed the enemy outside the urban area and trapped the greater part of the best enemy formations inside the urban area. The trapped units were then subjected to dwindling resources and extensive psychological operations, further isolated into pockets, and defeated in detail.

RETROGRADE

8-25. A retrograde involves organized movement away from the threat. Retrograde operations include withdrawals, delays, and retirements. These defensive operations often occur in an urban environment. The urban environment enhances the defending force's ability to conduct retrograde operations successfully (see figure 8-5).

8-26. The cover and concealment afforded by the urban environment facilitates withdrawals where friendly forces attempt to break contact with the threat and move away. The environment also restricts threat reconnaissance, which is less able to detect friendly forces moving out of position, and presents excellent opportunities for deception actions. Finally, a small security force's ability to remain concealed until contact in the urban environment significantly slows threat attempts to regain contact once Army forces have broken contact and begun to move.

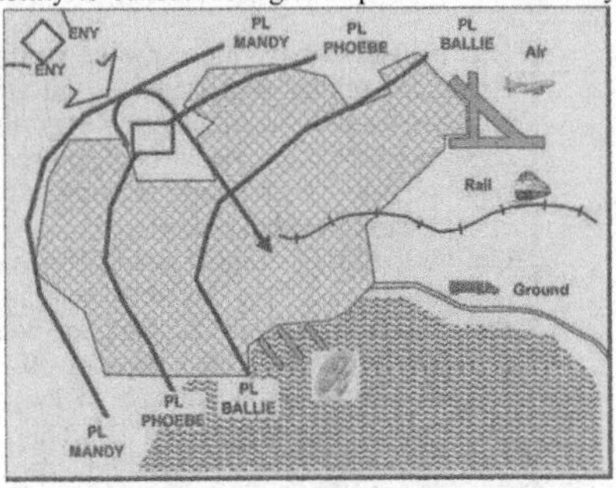

Figure 8-5. Retrograde through an urban area

8-27. The urban environment's natural cover and concealment, as well as the compartmented effects, facilitates delays. Delays can effectively draw the threat into the urban area for subsequent counterattack or as an integral part of a withdrawal under threat pressure. Delaying units can quickly displace from one covered and concealed position to another; the repositioning options are vast. Compartmented effects force the attacking threat to move on well-defined and easily interdicted routes and limit the threat's ability to flank or bypass delaying positions.

8-28. The urban area's transportation and distribution network facilitates retiring forces that are not in contact. Properly used, the urban transportation system can quickly move large forces and associated resources, using port facilities, airfields, railheads, and well-developed road networks.

URBAN DEFENSIVE CONSIDERATIONS

What is the position about London? I have a very clear view that we should fight every inch of it, and that it would devour quite a large invading army.

Winston Churchill
War in the Streets

8-29. The urban operational framework—understand, shape, engage, consolidate, and transition—provides structure to developing considerations for defensive operations. The considerations can vary depending on the level of war at which the operation is conducted, the type of defense, and the situation. Most issues discussed may, in the right circumstances, apply to both commanders conducting major UO and commanders at lower tactical levels of command.

UNDERSTAND

8-30. The commander defending in the urban area must assess many factors. His mission statement and guidance from higher commanders help him focus his assessment. If the mission is to deny a threat access to port facilities in an urban area, the commander's assessment will be focused much differently than if the mission is to deny the threat control over the entire urban area. The mission, enemy, terrain and weather, troops and support available, time available, civil considerations (METT TC) structure guides the commander's assessment. Of these, the impacts of the threat and environment—to include the terrain, weather, and civil considerations—are significant to the commander's understanding of urban defensive operations.

The Threat

8-31. In the urban defense, a key element is the commander's understanding of the threat. One of his primary concerns is to determine the attacker's general scheme, methodology, or concept. Overall, the attacker may take one of two approaches. The most obvious would be a direct approach aimed at seizing the objectives in the area by a frontal attack. A more sophisticated approach would be indirect and begin by isolating Army forces defending the urban area. Innumerable combinations of these two extremes exist, but the threat's intentions toward the urban area will favor one approach over another. The defending Army commander (whose AO includes but is not limited to the urban area) conducts defensive planning, particularly his allocation of forces, based on this initial assessment of threat intentions. This assessment determines whether the commander's primary concern is preventing isolation by defeating threat efforts outside the area or defeating a threat attacking the urban area directly. For the higher commander, this assessment determines how he allocates forces in and outside the urban area. For the commander in the urban area, this assessment clarifies threats to sustainment operations and helps shape how he arrays his forces.

The Environment's Defensive Characteristics

8-32. A second key assessment is the defensive qualities of the urban environment. This understanding, as in any defensive scenario, is based on mission requirements and on a systemic analysis of the terrain in terms of observation and fields of fire, avenues of approach, key terrain, obstacles, and cover and concealment (OAKOC). It is also based on potential chemical, biological, radiological, nuclear and fire hazards that may be present in the urban area. This understanding accounts for the unique characteristics of urban terrain, population, and infrastructure as discussed in Chapter 2.

8-33. Generally, units occupy less terrain in urban areas than more open areas. For example, an infantry company, which might occupy 1,500 to 2,000 meters in open terrain, is usually restricted to a frontage of 300 to 800 meters in urban areas. The density of building in the urban area, building sizes and heights, construction materials, rubble, and street patterns will dictate the actual frontage of units; however, for initial planning purposes, figure 8-6 provides approximate frontages and depths for units defending in an urban area.

UNIT	Frontage (Blocks*)	Depth (Blocks*)
Battalion	4 – 8	3 – 6
Company	3 – 4	2 – 3
Platoon	1 – 2	1
*Average block is 175 meters		

Figure 8-6. Approximate defensive frontages and depths

SHAPE

8-34. Commanders of a major operation shape the urban battle according to the type of defense they are attempting to conduct. If conducting an area defense or retrograde, they use shaping actions like those for any defensive action. Important shaping actions that apply to all defensive UO include—

- Preventing or defeating isolation.
- Separating attacking forces from supporting resources.
- Creating a mobility advantage.
- Applying economy of force measures.
- Effectively managing the urban population.
- Planning counterattacks.

Preventing or Defeating Isolation

8-35. Failure to prevent isolation of the urban area can rapidly lead to the failure of the entire urban defense. Its importance cannot be overstated. In planning the defense, commanders must anticipate that the threat will attempt to isolate the urban area. Defensive planning addresses in detail defeating threat attacks aimed at isolation of the urban area. Commanders may defeat this effort by allocating sufficient defending forces outside the urban area to prevent its isolation. Defensive information operations (IO) based on deception can also be used to mislead the threat regarding the defensive array in and outside the urban area. Such information can convince the threat that a direct attack against the urban area is the most favorable approach.

8-36. If the threat has successfully isolated the urban area, commanders of a major operation have several courses of actions. Three options are an exfiltration, a breakout attack by forces defending the urban area, or an attack by forces outside the urban area to relieve the siege. A fourth option combines the last two: counterattacks from both inside and outside the urban area to rupture the isolation (see breakout operations in FM 3-90). Time is critical to the success of either operation. Commanders should plan for both contingencies to ensure rapid execution if necessary. Delay permits threat forces surrounding the urban area to prepare defenses, reorganize their attacking force, retain the initiative, and continue offensive operations. The passage of time also reduces the resources of defending forces and their ability to breakout. Therefore, commanders and staff of a major operation must vigilantly avoid isolation when Army forces are defending urban areas in their AO.

Separating Attacking Forces from Supporting Resources

8-37. Commanders of the major operation primarily use fires and IO for separating in space and time threat forces attacking the urban area from echelons and resources in support. The purpose of this shaping action is the same as for any conventional area defense. It aims to allow the defending forces to defeat the threat piecemeal as they arrive in the urban area without support and already disrupted by deep fires and IO against information systems. This separation and disruption of the threat also sets the conditions for a mobile defense if commanders choose to execute that type of defense. These operations also prevent the threat commander from synchronizing and massing his combat power at the decisive point in the close battle.

8-38. If the urban area is part of a major mobile defense operation, the urban defense becomes the fixing force. Commanders can shape the defense to encourage the threat to attack into the urban area. They can lure the threat using a combination of techniques depending on the situation. They may make the urban area appear only lightly defended while other alternative courses of action appear strongly defended by friendly forces. Placing the bulk of the defending forces in concealed positions well within the urban area and positioning security forces on the periphery of the urban area portray a weak defense. In other situations, the opposite is true. If the urban area is an important objective to the threat, friendly forces can make the urban area appear heavily defended, thus ensuring that he commits sufficient combat power to the urban area to overwhelm the defense. Both cases have the same objective: to cause a major commitment of threat forces in the urban area. Once this commitment is made, the mobile defense striking force attacks and defeats the threat outside the urban area. This isolates the threat in the urban area and facilitates its destruction.

8-39. In the urban tactical battle, many shaping actions mirror those in all defensive operations. The size and complexity of the urban area prevent defending forces from being strong everywhere; shaping operations designed to engage the threat on terms advantageous to the defense have particular importance. Shaping actions include reconnaissance and security operations, passages of lines, and movement of reserve forces prior to their commitment. In addition, shaping operations critical to urban defense include mobility and countermobility operations, offensive IO, economy of force operations, and population management operations.

Creating a Mobility Advantage

8-40. In urban terrain, countermobility operations can greatly influence bringing the threat into the engagement areas of defending forces. Countermobility operations—based on understanding the urban transportation system, design, and construction characteristics—can be unusually effective (see Chapter 2). Demolitions can have important implications for creating impassable obstacles in urban canyons as well as for clearing fields of fire where necessary. Careful engineer planning can make the already constrictive terrain virtually impassable to mounted forces where appropriate, thus denying the threat combined arms capabilities. Countermobility operations in urban terrain drastically increase the defense's ability to shape the attacker's approach and to increase the combat power ratios in favor of the defense. As with all aspects of UO, countermobility considers collateral damage and the second- and third-order effects of obstacle construction.

8-41. Well-conceived mobility operations in urban terrain can provide defending forces mobility superiority over attacking forces. This is achieved by carefully selecting routes between primary, alternate, and subsequent positions, and for moving reserves and counterattack forces. These routes are reconnoitered, cleared, and marked before the operation. They maximize the cover and concealment characteristics of the terrain. Using demolitions, lanes, and innovative obstacles denies the defense of these same routes.

Applying Economy of Force Measures

8-42. Economy of force is extremely important to effective tactical urban defense. A megalopolis is too large and too easily accessible for defending forces to be strong everywhere. Economy of force enables the defending force to mass effects at decisive points. Forces used in an economy of force role execute security missions and take advantage of obstacles, mobility, and firepower to portray greater combat power than they actually possess. They prevent the threat from determining the actual disposition and strength of the friendly defense. If, contrary to expectations, they are strongly attacked, their mobility—stemming from a mounted maneuver capability, planning, and an intimate knowledge of the terrain—allows them to delay until reserves can meet the threat. Security forces in an economy of force role take position in parts of the urban area where the threat is less likely to attack.

Defensive Combat Power: Suez – 1973

At the end of October, the Israeli Army was in the midst of effective counterattack against the Egyptian Army. The Israelis had success attacking west across the Suez Canal. Their armored divisions were attempting to achieve several objectives, to include destroying Egyptian air defense sites and completing the encirclement of the Egyptian 3rd Army, which was trapped on the canal's east side.

To completely encircle the Egyptian 3rd Army, the Israelis had to seize all possible crossing sites to it from the canal's west bank and the Red Sea. Also, as international negotiations towards a cease-fire progressed, the Israeli government wanted to capture as much Egyptian territory as possible to improve their negotiating position after hostilities.

Consequently, the Israeli Adan Armored Division was tasked to seize the Egyptian Red Sea port of Suez on the morning of 24 October. A cease-fire was to begin at 0700, and the Israeli intent was to be decisively engaged in the city by that time and then consolidate their position as part of the cease-fire compliance.

The Adan Division plan to seize Suez was a two-part operation. Each of the division's armored brigades would have a role. The 460th Brigade would attack west of the city and complete the city's encirclement. Simultaneously, the 217th Brigade would attack in columns of battalions through the city to seize three key intersections in the city. This was in accordance with standard Israeli armored doctrine for fighting in an urban area. The 217th Brigade would seize its objectives through speed, firepower, and shock action. Once the objectives were seized, infantry and armored teams would continue attacking from the secured objectives to mop up and destroy pockets of resistance. The Israeli commanders expected to demoralize the defending Egyptians—two infantry battalions and one antitank company—by this rapid attack. The armored division commander was specifically advised by his commander to avoid a "Stalingrad" situation.

The attack got off to an ominous beginning as mist greatly inhibited a scheduled aerial bombardment in support of the attack. The 217th Brigade began its attack without infantry support and was quickly stopped by antitank missiles and antitank fire. Infantry was quickly integrated into the brigade and the attack resumed.

At the first objective, the Israelis encountered their first problems. A withering barrage of small arms, antitank missiles, and antitank fire hit the lead tank battalion, including direct fire from SU-23 anti-aircraft guns. Virtually all the officers and tank commanders in the tank battalion were killed or wounded, and several tanks were destroyed. Disabled vehicles blocked portions of the road, and vehicles that turned on to secondary roads were ambushed and destroyed. The battalion, however, successfully fought its way through the first brigade objective and on to the final brigade objective.

Hastily attached paratroop infantry in company strength were next in column following the tanks. They were traveling in buses and trucks. As the lead tank battalion took fire, the paratroopers dismounted, and attempted to secure adjacent buildings. The tank battalion's action of fighting through the objective caused the paratroopers to mount up and also attempt to move through the objective. Because of their soft skinned vehicles the paratroopers were unable to remain mounted and again dismounted, assaulted, and secured several buildings that they could defend. Once inside the buildings, the paratroopers found they were cut off, pinned down, and unable to evacuate their considerable casualties, which included the battalion commander. The paratroopers were on the initial brigade objective but were unable to maneuver and were taking casualties.

A second paratroop company also dismounted and quickly became stalled in house-to-house fighting. The brigade reconnaissance company in M113 personnel carriers brought up the rear of the brigade column and lost several vehicles and was also unable to advance.

By 1100 the Israeli attack culminated. Elements of the 217th Brigade were on all three of the brigade's objectives in the city. However, the armored battalion, which had achieved the deepest penetration, was without infantry support and under severe antitank fire. Both paratroop companies were isolated and pinned down. In addition, an attempt to link up with the paratroopers had failed. At the same time, the civilian population of the city began to react. They erected impromptu barriers, ambushed isolated Israeli troops, and carried supplies and information to Egyptian forces.

The Israeli division commander ordered the brigade to break contact and fight its way out of the city. The armored battalion was able to fight its way out in daylight. The paratroop companies were forced to wait until darkness and then infiltrated out of the city carrying their wounded with them. Israeli casualties totaled 88 killed and hundreds wounded in addition to 28 combat vehicles destroyed. Egyptian casualties were unknown but not believed to be significant.

The fight for Suez effectively demonstrates numerous urban defensive techniques. It also vividly demonstrates the significant effect on defensive combat power of the urban environment.

The Egyptian defense demonstrates how the compartmented urban terrain restricts the mobility and the massing of firepower of armored forces. Trapped in column on the road, the Israelis were unable to mass fire on particular targets nor effectively synchronize and coordinate their fires. The short-range engagement, also a characteristic of urban combat, reduced the Israeli armor protection and eliminated the Israeli armor's ability to keep out of small arms range. Thus, hand-held antiarmor weapons were more effective in an urban area. Additionally, Egyptian small arms and sniper fire critically affected Israeli C2 by successfully targeting leaders.

The Egyptian defenders effectively isolated the mounted Israelis by defending and planning engagement areas in depth. The Egyptians synchronized so that they engaged the entire Israeli force simultaneously. This forced the Israelis to fight in multiple directions. It also separated the Israeli infantry from the armor and prevented the formation of combined arms teams necessary for effective urban offensive operations.

Suez also demonstrated how civilians come to the advantage of the defense. After the battle was joined, the population—by threatening isolated pockets of Israelis and building barricades—helped prevent the Israelis from reorganizing while in contact and hindered the Israelis breaking contact. The population was also a valuable source of intelligence for the Egyptians and precluded covert Israeli movement in daytime.

Suez shows the ability of a well-placed defense in depth to fix a superior force in an urban area. Despite the Israeli commander's caution to avoid a "Stalingrad," the Israeli division, brigade, and battalion commanders were quickly trapped and unable to easily break contact. Even a successful defense on the perimeter of the city would not have been nearly as effective, as the Israelis would have easily broken contact once the strength of the defense was recognized.

Another key to the success of the Egyptian defense was the Israelis' inadequate reconnaissance. While the Israelis knew the approximate size of the defending forces, they had no idea of the Egyptian dispositions. In this case, time prevented adequate reconnaissance. Key to a successful defense is adequate security to obscure defense dispositions, which permits surprise and shock effect.

The Suez defense was a decisive defeat of elite Israeli forces by regular infantry units inferior in training, morale, and numbers. Total disaster was averted only because of the professionalism of the Israeli armored forces and paratroopers that permitted them to continue to fight and eventually exfiltrate the urban trap. The Israeli forces thus escaped total destruction. Suez strongly demonstrates how the enhancing effects of the urban environment on defensive combat power are significant enough to permit inferior regular forces to defeat elite formations. Since

> the 1973 Suez battle, U.S. forces in Mogadishu, Somalia, and Russian forces in Grozny, Chechnya have faced similar urban defensive ambushes.

Effectively Managing the Urban Population

8-43. Another way to shape the urban defensive battle is population management. In most cases, defending force commanders are in the urban area before combat. This time gives them the chance to manage civilians. Consequently, they can better manage and protect the population (a legal requirement) and gain more freedom of action for their forces.

8-44. Managing the civilians during the defense is a function of the size, disposition, and needs of the population and the resources available to the commander. Requesting higher support or coordinating with nongovernmental organizations and the local civil leadership for support may make up shortages of resources. Resources devoted to population management are carefully weighed against availability, military mission requirements, and possible collateral damage affecting tactical, operational, or strategic success. It may prove impractical to evacuate an urban area's population; still, commanders should attempt to create and move most civilians to protected areas. Moving the population allows defending forces to more liberally apply fires, emplace obstacles, and relieve combat units and support units of requirements to continue life support for civilians while executing combat operations. Overall, effective civil-military operations can turn a friendly (or a neutral) population into an effective force multiplier providing support to every warfighting function.

Planning Counterattacks

8-45. Counterattacks are also an important tool in shaping the battlefield for defensive success. Counterattacks as a shaping tool have two applications: retaining the initiative and separating forces. However, opportunity for effective counterattacks will be brief and, therefore, timing will be critical. If conducted too soon, the counterattack may expend resources required later; if conducted too late, it may not be effective. Commanders should understand the effect of the urban environment on time-distance relationships; otherwise, the timing of the attack may be upset and the operation desynchronized. Additionally, commanders should develop plans beyond the counterattack to exploit potential success.

ENGAGE

8-46. Engaging the urban area in a defensive operation requires decisively defeating the threat's attacks. Defensive forces use the terrain to their advantage, employ precision supporting fires, and use direct fire from protected positions aligned against carefully selected avenues of approaches and kill zones. The combat power of the defense augmented by shaping actions and the characteristics of urban terrain force culmination of the threat attack. Like urban offensive operations, effective engagement in urban defensive operations typically results from successful actions at the tactical level of war. These actions include—

- Performing aggressive ISR.
- Creating depth.
- Executing an effective obstacle plan.
- Conducting coordinated counterattacks.

Performing Aggressive ISR

8-47. ISR efforts of the defender are focused initially on identifying relevant information about the location and nature of the threat's main effort. Once identified, the defender's ISR focus shifts to assessing the rate at which the threat attack moves to its culminating point. Indicators of culmination may be physical fatigue of Soldiers, a breakdown in C2 capability, difficulty providing logistics support, or the increasing time required to reorganize small units to attack. When that culmination is achieved, friendly forces counterattack before the threat has a chance to transition to a hasty defense.

Creating Depth

8-48. Depth in the defense is the key to forcing the threat to culminate. The urban defense cannot allow itself to be penetrated nor permit forward elements to be destroyed. The defense is designed with the greatest depth possible. Defending forces weaken the threat to the fullest extent possible by attack from each position but not permit themselves to be destroyed by fires or close assault. Instead, as threat combat power builds up against individual positions, the use of mission orders permits subordinate leaders to disengage on their own initiative and move on preplanned routes to subsequent positions. Positions are designed to be mutually supporting—withdrawing from one position to a subsequent one while supporting positions cover by fires. The attacker is constantly forced to deploy and reorganize without being able to achieve decisive effects against the defender.

Executing an Effective Obstacle Plan

8-49. Obstacles in the urban defense are designed to break up the threat's combined arms capability. Separating dismounted forces from mounted forces disrupts the cohesion of the attacker and reduces his combat power. It also exposes his individual elements to the effects of asymmetric counterattack. The leading threat dismounted force can be effectively counterattacked by a friendly combined arms element while the threat armored force remains vulnerable to antiarmor attack by dismounted forces.

Conducting Coordinated Counterattacks

8-50. The counterattack is one of the key actions of the urban defense. However, commanders do not counterattack unless there is a reasonable chance of success. As the attacker moves into the depth of the urban area, his forces become fatigued, attrited, and increasingly disorganized. He likely also creates an increasingly long and exposed flank. At all levels, forces defending in urban terrain look for opportunities to counterattack. As the offensive force reaches the culmination point where it can no longer continue to attack with the available forces, the defensive commander executes a planned and coordinated counterattack. The counterattack aims to regain the initiative and to make the threat fight in multiple directions. Infiltration using superior knowledge of the terrain (including supersurface and subsurface capabilities) and the skillful use of stay-behind forces permits attacking the threat throughout the depth of his formations. Small-scale counterattacks focus on C2 and sustainment capabilities. These counterattacks can set the conditions for a deliberate attack leading to the ultimate destruction of the attacking threat force.

CONSOLIDATE

8-51. Consolidation is as important to urban defensive operations as to offensive operations. Many of the same consolidation considerations for the urban offense apply equally well to the defense. Commanders should reinforce or reposition maneuver forces and fire support assets on the urban battlefield based on weaknesses uncovered during rehearsals and opportunities discovered during actual execution. While maximizing the many advantages of the urban defense, the commander of the urban defense aggressively seeks ways of attriting and weakening enemy forces before they enter the urban area and close combat is initiated. Commanders should combine the static and mobile elements of the defense to strengthen their positions in relation to the enemy while seeking every opportunity to transition to urban offensive operations. As in urban offensive operations, commanders should conduct any necessary reorganization actions that they were unable to accomplish during execution.

TRANSITION

8-52. Transitions in urban defensive operations occur at all levels. As with offensive operations, commanders of major operations should address which units are assigned to continue to operate in the area after defensive operations have ceased. In defensive UO, this task is not as challenging as an occupation mission during urban offensive operations. The psychology of troops defending an urban area differs from those attacking into it. Defending forces become accustomed to the environment, having experience in the environment before combat. In terms of training, it is easier for follow-on missions to be assigned to a unit that has successfully defended the urban area. This course of action takes advantage of the defending unit's experience in the area and its relationships with other agencies—agencies that were operating alongside the units before and possibly during the defense. In defensive operations, regardless of the civilians' attitudes, policies regarding that population are established before the successful defense, and the command likely has experience executing operations with civil authorities and other agencies. Thus, these relationships are neither new nor as significant an issue as in offensive operations. Therefore, commanders must be prepared to execute various stability or civil support operations or use a successful defense to springboard into more decisive offensive operations elsewhere in the commanders' AO.

Transition Emphasis to Stability or Civil Support Operations

8-53. At the end of a successful urban defense, operational commanders should generally expect civil authority, control, and jurisdiction to increase. Additionally, the civil population may be anxious to return. Defensive combat will require virtually complete military control of the urban area; however, after the successful defense, a rapid transition will occur from military control to civil or joint military and civil control afterward. Although the full spectrum of Army operations will often be conducted simultaneously in UO, commanders will transition emphasis to stability or civil support operations. Important transition tasks will include demilitarizing munitions, clearing obstacles, and searching for isolated threat pockets of resistance. Conclusion of the defensive operations will also require transition to joint civil-military tasks, such as evaluating structures for safety, restoring essential services, and possibly creating joint law enforcement. Commanders of major operations, using a civil-military operations center and G9, should anticipate these requirements and begin early preparations to ensure a smooth, successful transition.

Transition to Offensive Operations

8-54. Units that have successfully defended the urban area may then transition to offensive operations. A rapid transition to offensive operations will require identification, preparation, and training of units designated to assume missions as the defending units leave the urban area. This preparation emphasizes continuity of policies and relationships already established. A relief in place occurs. The new occupying units provide not only a continuity of policy, but also a continuity of attitude toward the urban area, its population, and its institutions.

Chapter 9

Urban Stability and Civil Support Operations

The fundamental shared aims between stability and civil support operations and any operation conducted in an urban environment are the increased significance and influence of the civil population and nonmilitary organizations. Often, no military victory is to be achieved. The center of gravity for these operations normally cannot be attacked through military means alone; the Army (and the military in general) is often but one tool supporting a larger, civil-focused effort. When conducted, military defeat of an armed adversary is only a phase in a larger campaign, the overall objectives of which will be to bring about a more satisfactory political, social, diplomatic, economic, and military situation than existed prior to the conflict. Without a tightly coordinated civil-military effort, overall success will be difficult or impossible to achieve. Commanders who can understand and cope with the complexities of stability or civil support operations gain insights that directly apply to executing any urban operation. Urban stability or civil support operations may complement urban offensive and defensive operations, or may dominate the overall operation. Army forces may need to conduct offensive and defensive operations to defend themselves or destroy urban threats seeking to prevent the decisive stability or civil support mission. During hostilities, urban stability operations may keep armed conflict from spreading, encourage coalition and alliance partners, and secure the civilian population's support in unstable urban areas (and the surrounding rural areas under their influence). Following hostilities, urban stability operations may provide a secure environment for civil authorities to rebuild. Urban civil support operations can range from transporting, feeding, and sheltering the population made homeless as a result of man-made or natural disasters to conducting civil search and rescue and providing urgent medical care.

PURPOSE OF URBAN STABILITY AND CIVIL SUPPORT OPERATIONS

9-1. Army forces conduct stability and civil support operations to deter war, resolve conflict, promote peace, strengthen democratic processes, retain U.S. influence or access abroad, assist U.S. civil authorities, and support moral and legal imperatives. Stability operations promote and sustain regional and global stability. Civil support operations meet the urgent needs of designated U.S. civilians until civil authorities can accomplish these tasks without military assistance. Nearly every urban operation will involve some type or form of stability operation or civil support operation combined and sequenced with offensive and defensive operations.

CHARACTERISTICS OF URBAN STABILITY AND CIVIL SUPPORT OPERATIONS

9-2. Worldwide urbanization, migration trends from rural to urban areas, and more centralized populations in urban areas increase the chance that Army forces will conduct stability or civil support operations in or near urban areas. Simply put, many people live in urban areas, and their welfare will be the primary reason for conducting these operations. Urban areas that serve as economic and government centers (the ideal location for U.S. and allied embassies) are often the focal point for many threat activities. Therefore, Army forces may need to conduct stability operations in these foreign cities to counter those threats. Additionally, urban areas may contain the resources and infrastructure to support both types of operations, regardless of whether the overall focus is in urban or rural areas. Repairing or restoring the infrastructure may be a critical task in accomplishing either type of operation. Supported governmental and nongovernmental agencies are not as logistically self-sufficient as the Army. As such, these agencies will often need to center their operations in and around urban areas to use the area's infrastructure to support themselves and their objectives. These agencies may require military protection to accomplish their missions. Figure 9-1 lists some defining characteristics of these wide-ranging operations.

- Long or Short Duration
- Joint and Interagency
- Unilateral or Multinational
- Increased Civil-Military and Legal Considerations
- Greater Potential for Ambiguity
- Increased Constraints Necessitating More Restrictive ROE
- Amplified Need for Cultural and Political Sensitivity

Figure 9-1. Characteristics of stability and reconstruction and support operations

9-3. Stability and civil support operations are diverse, varied in duration, and unilateral or multinational. Stability operations are conducted overseas as part of a campaign or major operation, while civil support operations are conducted within the United States as part of homeland security. Like all urban operations (UO), they are usually joint. Unlike urban offensive and defensive operations, they are more often interagency operations and require more restrictive rules of engagement (ROE). The multiplicity of actors involved usually increases the scope and scale of required coordination and communication. In urban stability or civil support operations, adverse conditions arising from natural or man-made disasters or other endemic conditions—such as human suffering, disease, violations of human rights, or privation—will significantly modify the urban environment. Unresolved political issues and tenuous agreements, difficulties discriminating combatants from noncombatants or between parties of a dispute, and the absence of basic law and order all serve to complicate an already complex and uncertain environment. Civil-military and legal considerations take on added significance in all urban operations (see Civil-Military Operations and Legal Support in Chapters 5 and 10 respectively), but even more so in urban stability and civil support operations. Finally, recognizing, defining, and achieving the desired end state is often more difficult than in offensive and defensive operations.

9-4. Overall, commanders of major operations involving urban stability and civil support operations should not expect clear guidance. They must learn, adapt, and live with ambiguity. They cannot expect to operate in a political vacuum (even commanders at the tactical level); rather, they should expect to work alongside both governmental and nongovernmental leaders and organizations. Commanders conducting UO should not expect an easily identifiable enemy located across a clearly demarcated line. In fact in many peace operations, they and their Soldiers must resist the need to have an enemy—difficult at best when one side or another (or both) may be sniping at them. Commanders should also expect changing and additional missions and tasks, without being allowed to use every means at their disposal to carry out those missions. Many tasks required may be ones for which their units have never, or rarely, trained. Finally, commanders should expect to show restraint with a keen sensitivity to political considerations and to alien cultures, either or both of which they might find confusing or even repugnant.

URBAN STABILITY AND CIVIL SUPPORT OPERATIONS AND BATTLEFIELD ORGANIZATION

9-5. Each urban stability operation or civil support operation will be distinct from one another. These operations differ even more when applied to a specific urban area. Due to the complexity of the environment, commanders must carefully arrange their forces and operations according to purpose, time, and space to accomplish the mission. In most UO the terrain, the dense population (military and civilian), and the participating organizations will further complicate this arrangement.

9-6. The support and assistance that Army forces will provide during these operations is only temporary (although it may be of long duration) and this facet requires particular emphasis. Commanders should plan and execute both of these types of operations with that essential consideration always in mind. Eventually, the government and administration—either foreign or domestic—must secure and support their population by themselves. Therefore, commanders must envision and set the conditions that allow for the transition of control and responsibility to shift to legitimate civilian authorities. While commanders may only provide assistance and support based on specific and well planned civilian requests, more often commanders will need to determine requirements in collaboration with competent civilian authorities and agencies or, in some cases, with little or no initial civilian assistance at all. Ultimately, transition planning should occur as an integral part of the overall operational planning and include collaboration with appropriate civilian agencies and organization as early as possible to ensure a seamless transition to civilian control without major setbacks and loss of forward momentum.

STABILITY OPERATIONS

9-7. Stability operations establish, sustain, and exploit security and control over foreign areas, populations, and resources. Urban areas will be decisive to accomplishing many types of stability operations because urban areas are the centers of population, culture, economy, and government. Much of the support provided by Army forces will aim to assist local, regional, or national governments to restore essential services and infrastructure and reestablish civil order and authority. The location of civilian authorities in urban areas will, by necessity, be a dominating factor in accomplishing the mission. As importantly, many stability operations—enforcing peace in Bosnia for example—will require interacting with, influencing, controlling, or protecting all or parts of the civilian population. Assessing, understanding, and gaining the support of civilians in key economic, cultural, or political urban areas may influence surrounding regions (smaller urban areas and the rural countryside) and may be decisive to achieving overall stability objectives.

9-8. Stability operations may involve both coercive and cooperative actions. They are conducted in situations in which legitimate civil authority is unable to provide the necessary security and control for the urban population as a result of—

- Deliberate operations to influence a regime change.
- Offensive or defensive operations or natural disasters resulting in ineffective civil authorities, conditions beyond the capabilities of the foreign urban government, or both.

CIVIL SUPPORT OPERATIONS

9-9. Civil support operations are an integral part of homeland security which aim to protect U.S. citizens and infrastructure from conventional and unconventional threats and mitigate the effects of man-made and natural disasters. (The other part to homeland security is homeland defense.) Army forces conduct civil support operations by providing Army resources, expertise, and capabilities in support of a lead agency—which may be to Department of Defense—to address the consequences of man-made or natural accidents beyond the capabilities of local or regional U.S. civilian authorities. Such a situation could result from a chemical, biological, radiological, nuclear, and high-yield explosive (CBRNE) incident, hurricane, flood, or civil disturbance in a domestic urban area.

CBRNE Incidents

9-10. CBRNE incidents can be disastrous and are of particular concern for both stability and civil support UO. In urban areas, the potential for catastrophic loss of life and property is enormous. The Army categorizes CBRNE incidents separately from other natural and man-made disasters because it has specific expertise with these weapons. CBRNE incidents usually result from a military or terrorist threat (adding a law enforcement dimension to the disaster). Other urban CBRNE considerations include the following:

- Subways and other subsurface areas offer ideal areas for limited chemical or biological attacks. Nuclear attack (and high-yield explosives) can produce tragic results due to the effects of collapsing structures, flying debris, and fires. Dispersion patterns are affected by the urban terrain and are more difficult to predict and monitor. Large-scale incidents may produce hundreds of thousands of casualties, but even a limited attack may require evacuating and screening large numbers of civilians. Requirements for medical support, basic life support, and, if necessary, decontamination may quickly overwhelm the Army force's capabilities even with augmentation.
- Similar to natural disasters, panic and disorder may accompany the CBRNE event. Fleeing civilians may clog elements of the transportation and distribution infrastructure. Physical destruction may also affect other components of the infrastructure of critical and immediate concern, such as energy and administration and human services (water, sanitation, medical, fire fighting, and law enforcement). Because all elements of the infrastructure may be affected, the overall recovery time may be lengthened and the effects broadened to include much of the surrounding area. The effects of a single urban CBRNE event potentially could be felt nationally or globally.

Decisive Operations

9-11. In urban stability operations, decisive operations may take many years and include multiple actions before achieving the desired end state. This particularly applies to the strategic and operational levels. Oppositely, decisive operations involved in an urban civil support operation for mitigating or reducing disease, hunger, privation, and the effects of disasters normally achieve faster results. However, any operation that attacks the underlying cause and seeks to prevent or relieve such conditions will usually take longer. In urban areas, establishing law and order to protect critical infrastructure and the inhabitants from lawlessness and violence is often critical and also the decisive operation.

Shaping Operations

9-12. Shaping operations establish and maintain the conditions for executing decisive operations. In urban stability or civil support operations, shaping operations always include information operations (IO) that influence perceptions and maintain legitimacy. Often, various participants, and their potentially divergent missions and methods, are involved. Army commanders must coordinate their planning and efforts (early and continuously) to ensure that their decisive, shaping, or sustaining operations are not working against other agencies' efforts and operations—agencies that may have the lead role in the operation. Thus, a critical shaping operation may be to establish the coordination to help develop a common purpose and direction among agencies, particularly those that may experience continuous personnel turnover during the conduct of a lengthy operation. In some instances and with some organizations and agencies, particularly nongovernmental organizations (NGOs), genuine unity of effort may not be achievable; however, recognizing the differences in aims and goals will allow Army commanders to conduct operations with less friction. Commanders should actively request and include NGOs and appropriate governmental agencies in mission readiness exercises or any other training for stability or civil support operations.

SUSTAINING OPERATIONS

9-13. Sustaining operations enable decisive and shaping operations and include logistics, rear area and base security, movement control, terrain management, and infrastructure development. Sustainment bases, especially those located in urban areas, become an attractive target for hostile civilians; therefore, commanders must actively and aggressively protect these bases as well as lines of communications (see Chapter 10).

CONSIDERATIONS FOR URBAN STABILITY AND CIVIL SUPPORT OPERATIONS

In wars of intervention the essentials are to secure a general who is both a statesman and a soldier; to have clear stipulations with the allies as to the part to be taken by each in the principal operations; finally, to agree upon an objective point which shall be in harmony with the common interests.

Lieutenant General Antoine-Henri,
Baron de Jomini

9-14. The urban operational framework (understand, shape, engage, consolidate, and transition) provides a structure for developing considerations unique to urban stability and civil support operations. Many considerations presented in urban offensive and defensive operations apply to urban stability and civil support operations, particularly those that address how to understand the urban and overall operational environment. Because the situations in which stability and civil support operations normally occur share strong similarities with any urban environment, many of these considerations are closely linked to the urban fundamentals presented in Chapter 5. Taken together, commanders will often find them useful in conducting UO throughout full spectrum operations. Appendix C has a historical example of how to apply the urban operational framework to a stability operation.

UNDERSTAND

9-15. In urban stability or civil support operations, commanders must carefully assess and understand the political dimension of the operational environment, as well as their role and the media's part in managing information. These operations are inherently tied to the exercise of diplomatic power. All operations in urban areas are often the focus of the media and thus gain considerable public and political attention. Therefore, military objectives in urban stability or civil support operations are more directly linked with political objectives. The relationship between the levels of war—strategic, operational, and tactical—is often closer than in urban offensive and defensive operations. Military objectives must be carefully nested within political objectives. Commanders must ensure that the ways and means to accomplish their objectives, to include security and force protection measures, will hold up to media scrutiny and are appropriate for the situation and environment. All levels of command must understand the link between political and military objectives, to include a basic understanding at the Soldier level. One uncoordinated, undisciplined, or inappropriate action, even at the lowest level, could negate months or years of previous, disciplined effort. Commanders must balance security and force protection measures with mission accomplishment. Ineffective measures can put Soldiers at too great a risk and jeopardize the mission. Conversely, overly stringent measures may make it difficult for forces to interact with the population closely—essential in many of these operations. Finally, commanders will need a thorough assessment of the governmental and nongovernmental organizations and agencies that will be operating in or near urban areas that fall within their area of operations.

Political and Military Objectives

9-16. Commanders must translate political objectives into military objectives that are clear and achievable (clear tasks and purposes) and can lead to the desired end state. (In most stability operations, defeat of a threat is not the ultimate desired end state.) Political objectives may be vague making it difficult for commanders to conduct their mission analysis. This applies to tactical- and even operational-level commanders, unskilled at higher level, strategic political-military assessments. Each type of stability operation or civil support operation is distinct, often unfamiliar to the executing unit, and unique to the specific situation. These factors often make it difficult to confidently determine the specific tasks that will lead to mission success. Therefore, commanders must also establish measures of effectiveness that aid in understanding and evaluating progress and help gauge mission accomplishment.

> ***Note***: Commanders should consult the US Agency for International Development's (USAID) Field Operations Guide for Disaster Assessment and Response when conducting their assessments and developing measures of effectiveness for many urban relief operations.

9-17. These measures should be quantifiable (in some circumstances, a qualitative assessment may be most appropriate) and link cause with effect. They help determine the changes required and are essential to the assessment cycle required for urban stability or civil support operations. In a humanitarian relief operation to aid the starving, commanders could determine that the decisive effort is delivering safe food to the urban area. To judge success or effectiveness, they could determine that the appropriate measure is the number of food trucks dispatched daily to each distribution site; the more trucks, the more effective the efforts. However, this measure must correlate with the overarching measure of effectiveness: decline in the mortality rate. If no significant decrease in deaths due to starvation occurs, they may need to reassess and modify the tasks or measures of effectiveness. A better measure may be to track the amount of food consumed by those in need instead of simply counting the number of trucks dispatched. Areas around which measures of effectiveness can be formed for many stability or civil support operations (including the example above) and which will help return most societies to some degree of normalcy and self-sufficiency include:

- Restoring law and order.
- Decreasing morbidity and mortality rates.
- Securing safe food and water.
- Restoring critical infrastructure.
- Resettling the population.
- Reestablishing economic activity.

Commanders can often develop measures of effectiveness to address these areas in terms of establishing or restoring security and providing logistics. (Figure 9-2 provides example measures of effectiveness from the strategic to tactical levels for a possible stability operation.)

9-18. Political objectives are fluid and modified in response to new domestic and international events or circumstances. Thus, assessment is continuous, and commanders must adjust their own objectives and subsequent missions accordingly. In urban stability or civil support operations, commanders often develop military objectives that support or align with the objectives of another agency that has overall responsibility for the urban operation. In this supporting role, commanders may receive numerous requests for Soldier and materiel assistance from the supported agency and other supporting agencies operating in the urban area (to include elements of the urban population). With such unclear lines of authority and areas of responsibility, they must ensure that the tasks, missions, or requested Army resources fall clearly in the intended scope and purpose of the Army's participation in the operation. They must not develop or execute missions based on inadequate or false assumptions, misinterpreted intent, or well-meaning but erroneously interpreted laws or regulations by any organization, to include even the lead agency. When missions appear outside their scope, commanders must quickly relay their assessment to their higher headquarters for immediate resolution. The commander's goal is not to limit or slow military participation but to contribute as intended and in consonance with political objectives and the law.

Security and Force Protection Measures

9-19. Commanders must plan for and continually assess the security of their forces operating in an urban area as well as constantly review protection measures. Establishing a robust intelligence—particularly human intelligence (HUMINT)—network that can determine the intentions and capabilities of the threat and the urban populace is the basis for establishing force protection for Army forces operating in the urban environment. However, many such operations, particularly stability operations, require extra time to forge a lasting change. Over time, and particularly in peacetime when objectives center on helping others and avoiding violence, even the complex urban environment may seem benign. Without continued, aggressive command emphasis, Soldiers may become lulled into complacency. Also, during periods of transition or transfer of authority from one unit or organization to another, departing Soldiers' focus often shifts to redeployment activities and away from force protection concerns. It is usually then that Army forces are most vulnerable to terrorist and insurgent tactics, such as bombings, kidnappings, ambushes, raids, and other forms of urban violence.

Measures of Effectiveness: The Need to Measure Progress		
Strategic Level Criteria	**Operational Level Criteria**	**Tactical Level Criteria**
Accountable to the American people for defining and measuring progress towards defeating terrorism and meeting national security goals.	Accountable to the strategic level for measuring operational success and providing linkage to strategic goals.	Responsible to the operational level for measuring tactical success and providing linkage to operational goals.
Examples: • Prevention of the insurgency from receiving aid or resources from other international groups. • A functioning national government. • Amount of international support and aid to reconstruction. • Number of nations contributing manpower to coalition forces.	Examples: • Host-nation security forces trained and equipped. • Denial of the merging of insurgent forces with terrorist groups. • Amount and distribution of— • Electricity. • Liquid propane gas. • Gasoline. • Functioning provincial governments.	Examples: • Number of insurgent forces in the AO neutralized. • Reduced attacks on coalition forces in the AO. • Reduced civilian-on-civilian violence in the AO. • Host-nation security force recruitment goals met. • Host-nation security force training goals met. • Number of reliable human intelligence walk-ins. • Amount of unexploded explosive ordnance cleared. • Functioning neighborhood and district advisory councils.

Figure 9-2. Example measures of effectiveness

9-20. Although force protection will not ensure successful urban stability or civil support operations, improper assessment and inadequate force protection measures can cause the operation to fail. In either operation, keeping a neutral attitude toward all elements of the urban population, while maintaining the appropriate defensive posture, enhances security. One threat principle discussed in Chapter 3 was that threats would seek to cause politically unacceptable casualties. An improper threat assessment and a lapse in security at the tactical level could result in casualties. That result could affect strategy by influencing domestic popular support and subsequently national leadership decisions and policy.

9-21. Emphasizing security and force protection measures does not mean isolating Soldiers from contact with the urban population. On the contrary, commanders should balance survivability with mobility according to the factors of METT-TC—mission, enemy, terrain and weather, troops and support available, time available, civil considerations. Survivability measures—such as sandbagging, hardening, or fortifying buildings and installations, particularly where large numbers of Soldiers are fed and billeted—will be necessary and require considerable command attention. On the other hand, mobility operations are essential in preserving freedom of action and denying a threat the opportunity to observe, plan, and attack urban forces. Continual Army presence in the urban area may provide the urban population a sense of security and allow Soldiers to develop a detailed knowledge of the "patterns of life" in their assigned area of operation (AO). Armed with this knowledge, they can detect the absence of the normal or the presence of the abnormal that might indicate a potential threat. Overall, mission degradation and increased risk to the force can result if force protection measures prevent Army forces from conducting prudent missions and establishing an active and capable presence.

Assessment of Security and Force Protection: Belfast, Northern Ireland

Since 1969, Belfast has significantly affected the British military campaign for stabilizing the area. British operations in Belfast illustrate the difficulty of balancing the security and protection of forces with maintaining the stabilizing presence necessary to uphold law and order, minimize violence, and control the urban population.

British successes in protecting Belfast's infrastructure and government facilities from terrorist attacks compelled various terrorist cells, especially the Irish Republican Army and the Provisional Irish Republican Army, to attack more military targets.

At the time, British soldiers and bases presented relatively unprotected targets to these factions, and attacks against them solidified their legitimacy as an "army." In response, British commanders implemented extreme security and force protection measures—from ballistic protection vests and helmets to fortress-like operational bases and large unit patrols. These protection measures successfully decreased the violence against British soldiers in Belfast. However, they also decreased the soldiers' interaction with the population and their ability to stabilize the city. The large patrols protected the soldiers, but also inhibited effective saturation of neighborhoods. These patrols, coupled with fortress-like bases and bulky protective clothing, created an "us-versus-them" mentality among civilians and soldiers.

As force protection increased and stabilizing effects decreased, the terrorists were provided more targets of opportunity among the civilians and infrastructure. British commanders reassessed the situation, identified this "see-saw" effect, and adapted to strike a better balance between force protection and effective presence patrols. For example, British forces switched to four-man patrols to enable greater mobility and wore berets instead of helmets to appear less aggressive.

While the above historical example illustrates a critical aspect of stability and civil support operations, it subtly reveals another important lesson for Army commanders. Many of our future coalition partners and assisting governmental and nongovernmental organizations will bring key insights to UO (particularly dealing with urban societies) based on their historical and operational experience. Army commanders and their staffs should seek out and remain open to others' unique understandings that are based on hard-earned and well-analyzed lessons learned.

Participating Organizations and Agencies

9-22. Across the spectrum of urban operations, but more so in these operations, numerous NGOs may be involved in relieving adverse humanitarian conditions. Dense populations and infrastructure make an urban area a likely headquarters location for them. (In 1994 during OPERATION UPHOLD DEMOCRACY, for example, over 400 civilian agencies and relief organizations were operating in Haiti.) Therefore, commanders should assess all significant NGOs and governmental agencies operating (or likely to operate) in or near the urban area to include their—

- Functions, purposes, or agendas.
- Known headquarters and operating locations.
- Leadership or senior points of contact (including telephone numbers).
- Communications capabilities.
- Potential as a source for critical information.
- Financial abilities and constraints.
- Logistic resources: transportation, energy and fuel, food and water, clothing and shelter, and emergency medical and health care services.
- Law enforcement, fire fighting, and search and rescue capabilities.
- Refugee services.
- Engineering and construction capabilities.
- Other unique capabilities or expertise.
- Previous military, multinational, and interagency coordination experience and training.
- Rapport with the urban population.
- Relationship with the media.
- Biases or prejudices (especially towards participating U.S. or coalition forces, other civilian organizations, or elements of the urban society).

Commanders can then seek to determine the resources and capabilities that these organizations may bring and the possible problem areas to include resources or assistance they will likely need or request from Army forces. These organizations will be critical to meeting the population's immediate needs and minimizing the effects of collateral damage or disaster. However, commanders should consider whether a close relationship with any of these organizations will compromise the organization's appearance of neutrality (particularly threat perceptions) and adversely affect their ability to assist the population.

SHAPE

9-23. Commanders conduct many activities to shape the conditions for successful decisive operations. In urban stability and civil support operations, two activities rise to the forefront of importance: aggressive IO and security operations.

Vigorous Information Operations

9-24. IO, particularly psychological operations (PSYOP) and the related activities of civil affairs (CA) and public affairs, are essential to shape the urban environment for the successful conduct of stability or civil support operations. Vigorous IO can influence the perceptions, decisions, and will of the threat, the urban population, and other groups in support of the commander's mission. IO objectives are translated to IO tasks that are then executed to create the commander's desired effects in shaping the battlefield. These operations can isolate an urban threat from his sources of support; neutralize hostile urban populations or gain the support of neutral populations; and mitigate the effects of threat IO, misinformation, rumors, confusion, and apprehension. Developing effective measures of effectiveness is essential to a good IO campaign strategy. One of the most valuable methods for obtaining data for use in this process is face-to-face encounters with targeted audiences by unit patrols and HUMINT, PSYOP, and CA teams. A valuable technique may be to conduct periodic, unbiased surveys or opinion polls of the civilian population to determine changes in their perceptions and attitudes.

Security Operations

Protecting Civilians and Critical Infrastructure

9-25. Security for NGOs and civilians may also be an important shaping operation, particularly for stability and civil support operations. Commanders may need to provide security to civil agencies and NGOs located near or operating in the urban area so that these agencies can focus their relief efforts directly to the emergency. Commanders may also need to protect the urban population and critical infrastructure to maintain law and order if the urban area's security or police forces are nonexistent or incapacitated, or the urban area's security situation has undergone drastic change (such as the result of a natural disaster) and security or police forces require additional augmentation. (See also the discussion of the legal aspects of Civilians Accompanying the Force in Chapter 10.)

Preserving Resources

9-26. Just as forces may be at risk during urban stability or civil support operations, so may their resources. In urban areas of great need, supplies and equipment are extremely valuable. Criminal elements, insurgent forces, and people in need may try to steal weapons, ammunition, food, construction material, medical supplies, and fuel. Protecting these resources may become a critical shaping operation. Otherwise, Army forces and supporting agencies may lack the resources to accomplish their primary objectives or overall mission.

Prioritize Resources and Efforts

9-27. During UO, commanders will always face limited resources with which to shape the battlefield, conduct their decisive operations, and accomplish their objectives. They must continually prioritize, allocate, and apply those resources to achieve the desired end state. To this end, they may develop an order of merit list for proposed projects and constantly update it over time. To some degree, the local urban population will usually need to be protected and sustained by Army forces. Hence, commanders must tailor their objectives and shape their operations to achieve the greatest good for the largest number. Commanders first apply the urban fundamental of preserving critical infrastructure to reduce the disruption to the residents' health and welfare. Second, they apply the urban fundamental of restoring essential services, which includes prioritizing their efforts to provide vital services for the greatest number of inhabitants possible. In operations that include efforts to alleviate human suffering, the criticism for any participating organization is likely to be there is not enough being done or Army forces are not being responsive enough. Therefore, commanders must develop clear measures of effectiveness not only to determine necessary improvements to operational plans but also to demonstrate their Soldiers' hard work and sacrifice and U.S. commitment to the operation.

Improve the Urban Economy

9-28. When conducting reconstruction and infrastructure repair, commanders should consider using such activities to simultaneously improve the urban economy. Hiring local civilians and organizations to do reconstruction work helps satisfies urban job requirements, may inspire critical elements of the urban society to assume responsibility for the success or failure of urban restoration efforts, and may potentially reduce threat influence by diminishing their civilian sources of aid. Hiring indigenous personnel for short-term projects does not replace the need for long-term economic planning and the development of stable jobs. The overall reconstruction effort must be guided by the commander's vision of the end state. However, the commander's guidance and intent should be broad and expansive enough to allow responsive decision making by subordinate commanders based on their analysis of the urban population's needs in their assigned AO.

ENGAGE

If there is any lesson to be derived from the work of the regular troops in San Francisco, it is that nothing can take the place of training and discipline, and that self-control and patience are as important as courage.

Brigadier General Frederick Funston
commenting on the Army's assistance following
the 1906 San Francisco earthquake and fire

9-29. The focus of the Army is warfighting. Therefore, when Army commanders conduct many urban stability and civil support operations, they must adjust their concept of success. Commanders will most often find themselves in a supporting role and less often responsible for conducting decisive operations. They must accept this supporting function and capitalize on the professional values they have instilled in each Soldier, particularly the sense of duty to do what needs to be done despite difficulty, danger, and personal hardship. Commanders must also put accomplishing the overall mission ahead of individual desires to take the lead—desire often fulfilled by being the supported rather than supporting commander. In many stability and civil support operations, success may be described as settlement and compromise rather than victory. Yet, the Army's professionalism and values—combined with inherent adaptability, aggressive coordination, perseverance, and reasonable restraint—will allow Army forces to engage purposefully and dominate during complex urban stability or civil support operations.

Adaptability

9-30. Adaptability is particularly critical to urban stability or civil support operations because these operations relentlessly present complex challenges to commanders for which no prescribed solutions exist. Commanders often lack the experience and training that provide the basis for creating the unique solutions required for these operations. Since the primary purpose for the Army is to fight and win the nation's wars, the challenge then is to adapt urban warfighting skills to the unique stability or civil support situation.

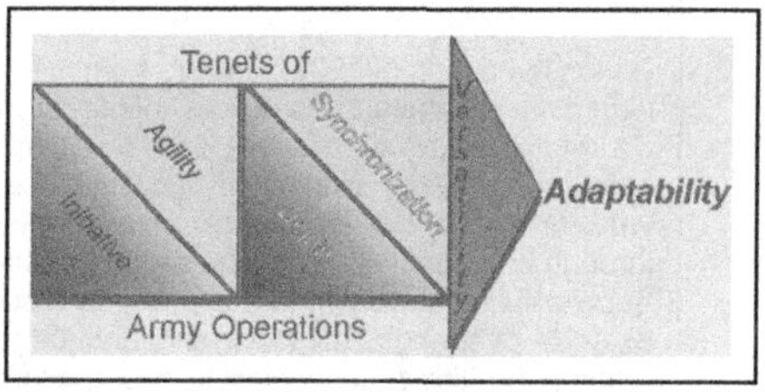

Figure 9-3. Adaptability

9-31. Doctrine (joint and Army) provides an inherent cohesion among the leaders of the Army and other services. Still, Army commanders conducting urban stability or civil support operations will often work with and support other agencies that have dissimilar purposes, methods, and professional languages. Army commanders must then capitalize on three of the five doctrinal tenets of Army operations: initiative, agility, and versatility (see figure 9-3 and FM 3-0). Commanders must bend as each situation and the urban environment demand without losing their orientation. They must thoroughly embrace the mission command philosophy of command and control addressed in FM 6-0 and Chapter 5 to encourage and allow subordinates to exercise creative and critical thinking required for planning and executing these UO. Commanders must also be alert to and recognize "good ideas," and effective tactics, techniques, and procedures, regardless of their source—other services, coalition partners, governmental and nongovernmental organizations, and even the threat—and adapt them for their own purposes in UO.

9-32. Adaptability also springs from detailed planning that carefully considers and realistically accounts for the extent of stability and civil support operations. Although no plan can account for every contingency and completely eliminate the unexpected, good plans—which include detailed civil considerations—provide platforms from which to adjust course more readily. Adequate planning allows commanders not only to react more effectively, but also to be forward-thinking and take actions that favorably steer the course of events.

Aggressive Coordination and Synchronization

9-33. In urban stability and civil support operations, the increased number of participants (both military and nonmilitary) and divergent missions and methods create a significant coordination and synchronization challenge. Significant potential for duplicated effort and working at cross-purposes exists. The success of UO often depends on establishing a successful working relationship with all groups operating in the urban area. The absence of unity of command among civil and military organizations does not prevent commanders from influencing other participants not under his direct command through persuasion, good leadership, and innovative ideas.

9-34. Commanders may consider establishing, as necessary, separate organizations for combat operations and for stability operations or civil support operations to increase coordination and enhance local, NGO, and international support. Further, aligning the unit or subordinate units with NGOs may contribute to establishing popular legitimacy for the operation and place greater pressure on threat forces. In some instances, commanders may consider organizing part of their staff around government, administrative, and infrastructure functions that mirror the urban area in which their forces are operating. Development of a mirror urban area organization may give greater legitimacy to the urban government or administration and ease transition of responsibility once the end state is achieved. Commanders must be mindful, however, that local groups seen as allying themselves with Army or coalition authorities will likely experience pressure to demonstrate their independence as established dates for redeployment or other critical events approach. In some instances that demonstration of independence may be violent.

Civil Support and Coordination with Civilian Authorities: Los Angeles – 1992

During the spring of 1992, Soldiers from the 40th Infantry Division, California National Guard were among the forces deployed to Los Angeles County to assist the California Highway Patrol, Los Angeles County Sheriffs, and civilian law enforcement. They worked to quell the riots that were sparked by the "not guilty" verdicts concerning four police officers who, following a lengthy high-speed chase through Los Angeles, were accused of brutally beating Rodney King.

Successful accomplishment of this civil support operation was attributed to the exercise of strong Army leadership and judgment at lower tactical levels, particularly among the unit's noncommissioned officers. An essential component of combat power, it was especially critical in executing noncontiguous and decentralized operations in the compartmented terrain of Los Angeles. As important, however, was the clear understanding that Army forces were to support civilian law enforcement—and not the other way around. The 40th Infantry Division aligned its area of operations with local law enforcement boundaries and relied heavily on police recommendations for the level at which Soldiers be armed (the need for magazines to be locked in weapons or rounds chambered).

One incident emphasized the need for coordination of command and control measures with civilian agencies even at the lowest tactical levels. To civilian law enforcement and Army forces, the command "Cover me" was interpreted the same: be prepared to shoot if necessary. However, when a police officer responding to a complaint of domestic abuse issued that command to an accompanying squad of Marines, they responded by immediately providing a supporting base of fire that narrowly missed some children at home. However, the Marines responded as they had been trained. This command meant something entirely different to them than for Army Soldiers and civilian law enforcement. Again, coordination at all levels is critical to the success of the operation (see also the vignette in Appendix B).

9-35. In the constraints imposed by METT-TC and operations security (OPSEC), commanders should seek to coordinate all tactical stability operations with other agencies and forces that share the urban environment. Importantly, they should seek to coordinate appropriate information and intelligence sharing with participating organizations. Commanders must strive to overcome difficulties, such as mutual suspicion, different values and motivations, and varying methods of organization and execution. Frequently, they must initiate cooperative efforts with participating civilian agencies and determine where their objectives and plans complement or conflict with those agencies. Commanders can then match Army force capabilities to the needs of the supported agencies. In situations leading to many urban civil support operations, confusion may initially make it difficult to ascertain specific priority requirements. Reconnaissance and liaison elements—heavily weighted with CA, engineers, and health support personnel—may need to be deployed immediately to determine what type of support Army forces provide. Overall, aggressive coordination will foster trust and make unity of effort possible in urban stability or civil support operations where unity of command is difficult or impossible to achieve.

Perseverance

9-36. The society is a major factor responsible for increasing the overall duration of urban operations. This particularly applies to urban stability operations where success often depends on changing people's fundamental beliefs and subsequent actions. Modifying behavior requires influence, sometimes with coercion or control, and perseverance. The urban population often must be convinced or persuaded to accept change. Necessary change may take as long as or longer than the evolution of the conflict. Decades of problems and their consequences cannot be immediately corrected. Frequently, the affected segments of the urban society must see that change is lasting and basic problems are being effectively addressed.

9-37. In most stability operations, success will not occur unless the host nation, not Army forces, ultimately prevails. The host urban administration must address the underlying problem or revise its policies toward the disaffected portions of the urban population. Otherwise, apparent successes will be short lived. The UO fundamental of understanding the human dimension is of paramount importance in applying this consideration. After all Army forces, particularly commanders and staff of major operations, understand the society's history and culture, they can begin to accurately identify the problem, understand root causes, quickly engage and assist key civilian leadership, and, overall, plan and execute successful Army UO.

Reasonable Restraint

9-38. Unlike offensive and defensive operations where commanders seek to apply overwhelming combat power at decisive points, restraint is more essential to success in urban stability and civil support operations. It involves employing combat power selectively, discriminately, and precisely (yet still at decisive points) in accordance with assigned missions and prescribed legal and policy limitations. Similar to the UO fundamentals of minimizing collateral damage and preserving critical infrastructure, restraint entails restrictions on using force. Commanders of major operations should issue or supplement ROE to guide the tactical application of combat power. Excessively or arbitrarily using force is never justified or tolerated by Army forces. Even unintentionally injuring or killing inhabitants and inadvertently destroying their property and infrastructure lessens legitimacy and the urban population's sympathy and support. Collateral damage may even cause some inhabitants to become hostile. In urban stability or civil support operations, even force against a violent opponent is minimized. Undue force often leads to commanders applying ever-increasing force to achieve the same results.

9-39. Although restraint is essential, Army forces, primarily during urban stability operations, must always be capable of decisive (albeit in some circumstances, limited) combat operations. This is in accordance with the UO fundamental of maintaining a close combat capability. This capability must be present and visible, yet displayed in a nonthreatening manner. A commander's intent normally includes demonstrating strength and resolve without provoking an unintended response. Army forces must be capable of moving quickly through the urban area and available on short notice. When necessary, Army forces must be prepared to apply combat power rapidly, forcefully, and decisively to prevent, end, or deter urban confrontations. Keeping this deterrent viable requires readiness, constant training, and rehearsals. It also requires active reconnaissance, superb OPSEC, a combined arms team, and timely and accurate intelligence.

Restraint: An Najaf, Iraq – 2003

On 31 March 2003, during OPERATION IRAQI FREEDOM, The 101st Airborne (Air Assault) Division transitioned from isolating An Najaf—among the holiest places for Shia Muslims in Iraq and the home of one of their leading holy men, the Grand Ayatollah Ali Hussein Sistani—to an attack to clear Iraqi fighters from the town. The decision to attack was made in response to increasing attacks against 101st forces situated near An Najaf and due to alarming reports that the Fedayeen were killing Iraqi family members to force adult males to fight coalition forces. Since An Najaf had an airfield, control of the town would also allow the 101st to obtain needed hardstands for their aircraft.

Ultimately, the division used two brigades to attack and clear the town. The 2nd Brigade Combat Team (BCT) attacked from the north and the 1st BCT—responsible for the initial penetration—attacked from the southwest. Using tanks, infantry fighting vehicles, and light infantry, the commander of the 1st BCT formed effective combined arms teams supported by artillery and air that successfully fought their way into the city.

On 3 April 2003, Soldiers of TF 2-327 Infantry turned a street corner to face a group of civilian men blocking their way and shouting in Arabic, "God is great." The crowd quickly grew into hundreds as they mistakenly thought the Soldiers were trying to capture Sistani and seize the Imam Ali Mosque. Someone in the crowd threw a rock at the Soldiers, which started a hailstorm of rocks; even the battalion commander was hit on the head, chest, and the corner of his sunglasses.

Based on information from a Free-Iraqi Fighter accompanying the battalion and his own observations during the attack, the commander of TF 2-327 believed that most of the people in An Najaf neither wanted to fight nor obstruct his unit's efforts. Consequently, he ordered his Soldiers to position themselves on one knee, smile, and point their weapons toward the ground. At this gesture, many of the Iraqis backed off and sat down, enabling the commander to identify the true troublemakers. He identified eight. In case these agitators were to produce weapons and start to shoot, the commander wanted to make sure that the remainder of the crowd would know from where the shooting would originate. Next, he ordered his Soldiers to withdraw to allow the tension to subside. With his own rifle pointed toward the ground, the battalion commander bowed to the crowd and led his Soldiers away. When tempers had calmed, the Grand Ayatollah Sistani issued a decree (fatwa) calling on the people of Najaf to welcome the American Soldiers.

The attack on An Najaf by the 101st provides an excellent example of well-led Soldiers capable of understanding cultural differences and possessing the discipline and mental flexibility to transition from the aggressive mindset required for high-intensity urban combat to the restraint essential for the initiation of stability operations. In his weekly radio address, President George W. Bush commented: "This gesture of respect helped defuse a dangerous situation and made our peaceful intentions clear."

CONSOLIDATE

9-40. As urban operations will often be full spectrum, many of the consolidation activities necessary to secure gains in urban offensive and defensive operations will be applicable to urban stability and civil support operations. However, the greatest obstacles to attaining strategic objectives will come after major urban combat operations. Therefore, emphasis will appropriately shift from actions to ensure the defeat of threat forces to those measures that address the needs of the urban population, manage their perceptions, and allow responsibility to shift from Army forces to legitimate indigenous civilian control (or the intermediate step to other military forces, governmental agencies, and organizations).

Continued Civilian and Infrastructure Protection

9-41. Following urban offensive or defensive operations, there will likely be a need to secure and protect the civilian population and much of the civilian infrastructure from the civilians themselves. After having minimized collateral damage and preserved critical infrastructure, commanders must implement measures to preclude looting and destruction of critical and essential infrastructure by the urban population and civilian-on-civilian violence. This may be as relatively simple as allowing the urban police force to return to work or may be as difficult as hiring, vetting, and training an indigenous police force. In the latter case commanders may need to determine—

- Number and operability of police stations.
- Responsibility for recruiting, hiring, training, and equipping the urban security or police force. The accountable unit, organization, or agency will need to consider a vetting process, suitable salaries and wages, and appropriate training standards.
- The appropriate responses toward those civilians who threaten, oppose, or harm the new police force.

9-42. Alternatively, civilian security firms (from inside or outside the urban area or country) can provide supplemental protection until indigenous police forces are fully functional. Further, the commander of the major urban operation may plan to manage expected instability primarily with Army forces. Often, this will require larger numbers of infantry, military police, and dismounted forces. Other populace and resource control measures such as curfews may also assist in civilian (including NGOs) and infrastructure protection. Previous shaping operations aimed at improving the local economy can also assist in this regard.

Resolute Legitimacy

9-43. Closely linked to the restraint described previously is legitimacy or the proper exercise of authority for reasonable purposes. Achieving or maintaining legitimacy during urban stability or civil support operations is essential in gaining and maintaining the support of the urban population. Commanders can ensure legitimacy by building consent among the population, projecting a credible force, and appropriately using that force. Perceptions play a key role in legitimacy, and skillful IO can shape perceptions. Commanders must send messages that are consistent with the actions of their forces. Generally, the urban population will accept violence for proper purposes if that force is used impartially. Perceptions that force is excessive or that certain groups are being favored over others can erode legitimacy and generate resentment, resistance, and, in some situations, violent acts of revenge.

9-44. In stability operations against an elusive insurgent threat, commanders may need to explain to urban residents why damage was necessary, apologize, or make near-instant restitution for some unsuccessful Army operations that may have been planned based on inaccurate or incomplete intelligence. (Soldiers may even make it a point to thank homeowners for allowing the search of their homes during cordon and search operations.) During urban operations, a single Soldier's misbehavior can significantly degrade a commander's ability to project an image of impartiality and legitimacy. Fortunately, disciplined Soldiers can contribute immeasurably to gaining and maintaining legitimacy, mitigating ill will, or otherwise winning the urban population's trust and confidence. In stability operations, the greater fight will often be to win the battle of perceptions and ideas instead of one to seize terrain and triumph over an enemy. Inconsistencies in message and behavior will provide threats with raw material for their propaganda and precipitate doubt in the minds of the urban populace who might otherwise support Army objectives.

Transition

9-45. Commanders of major operations are the focal point for synchronizing tactical stability operations or civil support operations with strategic diplomatic and political issues. They are also the critical links between national intelligence resources and the tactical commander. Because strategic, diplomatic, and political changes can quickly transition the type of urban operation, they must keep subordinate tactical commanders abreast of changes in intelligence, policy, and higher decisions. The potential to rapidly transition to urban combat operations emphasizes the need to maintain the capability to conduct close, urban combat. Failure to recognize changes and transition points may lead to UO that do not support the attainment of the overall objective and needlessly use resources, particularly Soldiers' lives. Therefore, Army forces on the ground in an urban stability operation must be more aware of the strategic environment than the threat and the civilian population, each of whom will have their own means of monitoring the national and international situation.

Legitimate and Capable Civilian Control

> *I met immediately with the [special operations forces liaison officer] to discuss how we would establish a new government that would be able to make instant progress. We decided to select a city council with the following departments: commerce, public works, social services, health, emergency services, education, public relations, and agriculture. A city council director would lead the council, and each department would also have a director in charge of its respective services. A critical step in selecting the city council was ensuring that there was a balance on the council between Altun Kupri's three major ethnic groups. An ethnic imbalance on the council or perceived favoritism would have decreased the legitimacy of the Coalition forces' efforts in the city and served as fuel for the fire of competition for power between the Turkomen and Kurds.*
>
> Captain Jeffrey B. Van Sickle
> "Stability Operations in Northern Iraq"

9-46. Commanders must maintain or enhance the credibility and legitimacy of the government and police of the urban area and, in the case of stability operations, of the host nation's military forces operating there. In accordance with the urban fundamental of transitioning control, urban commanders must conclude UO quickly and successfully, often to use assets elsewhere in their area of operations. This entails returning the control of the urban area back to civilian responsibility as soon as feasible. The host nation's military and the urban area's leadership and police are integrated into all aspects of the urban stability or civil support operations to maintain their legitimacy. They must be allowed (or influenced) to take the lead in developing and implementing solutions to their own problems. This requires commanders to transition from "leading from the front" to "leading from behind" in an advisory and assistance role. Effective transition to civilian control and responsibility requires commanders at all echelons who understand the basic operation of civil governments and the administration and management of key urban infrastructure.

9-47. If the host nation's leadership, military, and police are not up to the task, commanders can take steps to increase its capabilities through training, advice, and assistance by CA units or by other nongovernmental or governmental organizations and agencies. Sometimes, new leadership and a restructured police force may be required, particularly when corrupt and no longer trusted by the population. This candid assessment of the urban leadership's ability to govern, protect, and support itself is made early in the planning process. Only then can commanders ensure that resources and a well thought-out and coordinated plan (particularly with civilian organizations and agencies) are available for a speedy transition. IO will be paramount in these instances to ensure that the urban population sees the training and rebuilding process itself as legitimate. Throughout urban stability or civil support operations, commanders shape the conditions to successfully hand over all activities to urban civilian authorities.

Longer-Term Commitment

Maintain the Focus

9-48. Many stability operations often require perseverance and a longer-term U.S. commitment requiring operational endurance and the establishment of a sustainable battle rhythm. Lengthy operations will also require a transitional rotation of Army units into the area of operations to continue the mission. Considerations for these transitions are similar to a relief in place and battle handover (see FM 3-90) combined with considerations for deployment and redeployment. FM 41-10 contains comprehensive guidance on transition planning and coordination activities applicable to UO. In addition to any threat considerations, planning for urban transitions between units will often include emphasis and understanding of—

- Formal and informal civilian leadership and relationships.
- Government institutions and administrative functions.
- On-going reconstruction projects.
- The urban economy.
- Participating governmental and nongovernmental organizations and established relationships and cooperation activities, particularly information sharing.
- Significant key events affecting (or likely to affect) operations.
- Significant cultural lessons learned as a result of the outgoing unit's operations.

9-49. The commander of the major operation must ensure that the incoming unit also understands the political and strategic objectives behind the tasks that they will be required to accomplish. Otherwise, the new unit may begin to plan operations that are similar to those conducted by the previous unit without achieving the desired end state or accomplishing the mission.

Transition Trust

9-50. If units are successful in establishing and cultivating relationships with the urban populace and their leadership, and have built an effective level of trust, their anticipated departure from the AO as part of a unit rotation schedule may have detrimental effects on working relationships. As a result, local leaders may develop deep anxieties about the replacement force and their ability to "be as good" as the current command. Therefore, incumbent forces must take steps to identify successful and ongoing efforts and activities in which the new unit can continue to achieve success. In some instances, the current unit may delay a project or activity whose scheduled completion is close to the date of relief or the transfer of authority so as to allow the new unit immediate success. In other instances, projects or activities can be backward planned to ensure that the completion date is scheduled soon after the new unit arrives. The goal for the outgoing unit is to ensure that the new unit is readily accepted and that both units execute a seamless relief in place.

Modify Objectives to Match the Current Environment

9-51. While it is important that incoming commanders create a seamless transition with their outgoing counterparts, they must make full use of the opportunity to review the current political, strategic, and local urban environment to determine potential modifications to logical lines of operations and unit objectives. Otherwise, units may fall into the trap of executing a series of six-month to one-year rotations that do not significantly contribute to solidifying the conditions required to ultimately transition the urban environment back to legitimate civilian control. In operations that involve longer-term commitments, the potential exists for the same unit to conduct two (or more) rotations back into the same AO. Commanders must resist the desire to simply continue with previous lines of operations that, at the time, were deemed successful. The urban environment will change in response to Army actions and this change will necessitate subsequent operational adaptations.

Provide an End State Not an End Date

9-52. Though it may not be feasible, due to political constraints, commanders of major urban operations should fight the public desire to establish a fixed transition timetable. Instead, they should plan urban stability operations based on an attainable end state and not on a specified withdrawal or end date. Progress toward strategic (and operational) goals is susceptible to many changes and delay, particularly in coalitions which will likely be a part of most future urban operations. As such, an end date will only serve to unduly constrain commanders in the attainment of their urban operational objectives. Having set and announced a firm timetable, American and coalition publics will expect withdrawal to commence as scheduled and will not easily be convinced of the need to stay longer even if the situation changes drastically. In general, the public can understand and accept a logical and clearly articulated end state as long as commanders can demonstrate adequate progress based on well developed measures of effectiveness. As important, a timetable allows threats to adjust their plans to the anticipated withdrawal date. Instead of yielding to relentless diplomatic, informational, military, and economic pressure backed by firm political resolve, threats will likely exercise strategic and operational patience.

This page intentionally left blank.

Chapter 10

Urban Sustainment

Sustainment activities exist to enable the Army to initiate and sustain full spectrum operations. Inseparable from decisive and shaping operations, logistics is the central component of sustaining operations and provides the means for commanders to generate and maintain combat power and extend the operational reach of the force. In offensive and defensive operations, sustainment activities are not by themselves likely to be decisive or shaping, yet significantly contribute to those operations. In stability and civil support operations, when the critical objectives may be restoring the infrastructure and the welfare of civilians, sustainment activities will often be decisive. Their success will allow Army forces to dominate this complex environment. However, like all urban operations (UO), sustainment activities affect and are affected by the environment. The urban terrain, infrastructure, and existing resources, coupled with supportive civilians, may facilitate sustainment activities. In contrast, a poorly designed or damaged infrastructure and a hostile population may severely hamper sustainment activities. In the latter case, critical Army resources required elsewhere in the area of operations (AO) may be diverted to repair facilities and control and support the inhabitants of the urban areas.

URBAN SUSTAINMENT CHARACTERISTICS

10-1. Fundamental sustainment characteristics (see figure 10-1) guide prudent sustainment planning regardless of the environment. They provide commanders an excellent framework to analyze and help develop urban sustainment requirements, assess the impact of the environment on the provision of logistics, and gauge the effectiveness of urban sustainment activities.

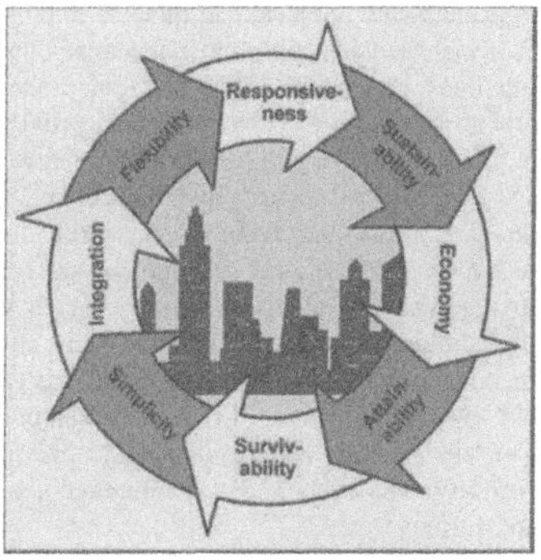

Figure 10-1. Sustainment characteristics

Responsiveness and Sustainability

10-2. UO require responsiveness and sustainability to establish and maintain the tempo necessary for success. Responsiveness—providing the right support in the right place at the right time—is the essential sustainment characteristic. It requires that sustainment commanders and planners accurately forecast urban operational requirements. Continuous urban operations will drain personnel, equipment, and supplies at rates vastly different than in other environments. Therefore, sustainability—the ability to maintain continuous support throughout all phases of the operation—will be a significant concern. Anticipation is critical to both responsiveness and sustainability. It requires that sustainment commanders and planners comprehend the potential effects that components of the urban environment (terrain, infrastructure, and society) may have on operations and sustainment, either benefiting or impeding UO. Effective urban operational and sustainment planning cannot be accomplished separately. Operational and sustainment planners, as well as sustainment operators, must be closely linked to aid in synchronizing and attaining responsiveness and sustainability. The key is often the right balance between efficiency and effectiveness.

Economy and Attainability

10-3. A thoughtful assessment and understanding of the urban environment can also help determine how specific urban areas can contribute to or frustrate the achievement of economy and attainability. Economy is providing the most efficient support at the least cost to accomplish the mission. Attainability means generating the minimum essential supplies and services necessary to begin operations. If available, obtaining support in the AO costs less than purchasing the supplies outside the area and then transporting them there. Critical resources may be available in urban areas to support the operation. However, relying on sources outside the established military sustainment system may create conflict with other sustainment characteristics. A strike by longshoremen, for example, may shut down port operations (at least temporarily) lowering responsiveness and sustainability.

Survivability

10-4. Survivability is being able to protect support functions from destruction or degradation. Commanders often choose to locate sustainment functions in an urban area because the buildings may better protect and conceal equipment, supplies, and people. Urban industrial areas and airports are frequently chosen as support areas because they offer this protection as well as sizeable warehouses, large parking areas, and materials handling equipment (MHE). Such areas facilitate the storage and movement of equipment and supplies. They may also provide readily available water, electricity, and other potentially useful urban resources and infrastructure. However, these areas may also contain toxic industrial materials (TIM) (see Chapter 2). These materials and chemicals in close proximity to support areas may unjustifiably increase the risk to survivability. Sustainment activities in any environment will always be targeted by threat forces. Furthermore, sustainment activities located in any type of confined urban area can offer lucrative targets for terrorists, insurgents, or even angry crowds and mobs. Therefore, no sustainment activity should be considered safe from attack during UO. (During OPERATION JUST CAUSE, Panamanian paramilitary forces and deserters even attacked marked ambulances). Although host-nation support may include assets to assist in defending sustainment units, bases, and lines of communications (LOCs), sustainment commanders must carefully consider if adequate protection measures can ensure survivability. The sustainment commander's greatest challenge to force protection may be complacency born of routine.

Base Security: Tan Son Nhut, Vietnam – 1968

Colonel Nam Truyen was the commander of the 9th Vietcong Division who planned and conducted the attack on the U.S. airbase at Tan Son Nhut during the 1968 Tet Offensive. He had previously entered the airbase during the 1967 Christmas cease-fire using forged identity papers to conduct his own personal and highly thorough reconnaissance.

> UO increase the likelihood that the indigenous population will be utilized as a workforce to supplement Army sustainment activities. Hence, strict operations security will be paramount. Commanders will need to carefully screen potential workers to determine their relative trustworthiness. Human intelligence (HUMINT) and counterintelligence (CI) assets will likely be required to assist in this effort. However, higher-level commanders should seek to take on as much of this burden as possible so that lower-level, tactical commanders can use their limited intelligence assets to accomplish other critical intelligence collection and analysis activities.
>
> All commanders with base protection responsibilities should implement random, changing force protection measures that focus on civilians—even those civilians who have, over time, earned a perceived measure of trust. (Civilians can be co-opted and coerced by threat forces.) As such, all civilian activity that seems out of the ordinary should be promptly investigated regardless of whether a specific force protection measure is in effect. Although leaders and Soldiers seek to establish mutual trust and respect between themselves and the local populace, commanders should normally err on the side of protecting their Soldiers.

SIMPLICITY

10-5. Simplicity is required in both planning and executing sustainment activities in this complex environment. Developing standard procedures among the Army, other services, and especially civilian governmental and nongovernmental agencies, liaison and open channels of communication, simple plans and orders, and extensive rehearsals contribute immeasurably to attaining this necessary characteristic.

INTEGRATION

10-6. The need for sustainment integration increases in UO due to its joint and multinational nature and greater numbers of other governmental and nongovernmental organizations (NGOs) and agencies operating in or near urban areas. More NGOs will likely exist because urban areas often contain most of a region's population. Most NGOs focus on people. Army forces and other military and nonmilitary groups must cooperate and coordinate their actions. Much of their coordination will revolve around sustainment activities. Cooperation and coordination will take advantage of each group's sustainment capabilities, help to avoid duplicated effort (contributing to economy), and create sustainment synergy. It will also help to curtail competition for the same urban resources and assist in developing a unified list of priorities. Such coordination will help ensure that other operations by one force or agency will not disrupt or destroy portions of the urban infrastructure critical to another's sustainment operations and the overall mission.

10-7. Success in UO requires combined arms and innovative task organizations. As such, UO will often result in the integration of heavy and light units necessitating a simultaneous integrated approach to sustainment. Gaining units must integrate forces into their maintenance systems and carefully consider potentially increased resource requirements. (For example, armor units consume greater amounts of Class III—bulk and petroleum—than a similar-sized light unit.) When integrating into a light unit, armored and mechanized units should bring as much of their critical parts and supplies as possible and immediately integrate into the gaining unit's logistical reporting system to ensure adequate resupply.

FLEXIBILITY

10-8. Lastly, commanders must develop flexibility. Although they and their staffs must thoroughly understand the urban environment essential to planning sustainment operations, they cannot possibly anticipate every eventuality. Urban commanders must possess the ability to exploit fleeting opportunities. Knowledge of the environment, particularly its infrastructure, can aid in developing innovative solutions to sustainment acquisition and distribution problems. Flexibility enables sustainment personnel to remain adaptive and responsive to the force commander's needs. Key to maintaining flexibility is the constant assessment of the sustainment situation and the readiness to modify or change procedures to adapt to the current conditions confronting Army forces operating within the urban environment.

10-9. Maneuver and sustainment commanders should consider and prioritize these characteristics as they visualize UO. Each characteristic does not affect every operation and urban area in the same way. The sustainment characteristics seldom exert equal influence, and their importance varies according to mission, enemy, terrain and weather, troops and support available, time available, civil considerations (METT-TC). Like the principles of war, commanders must not ignore the potential impact of sustainment characteristics and how their influence changes as the operation evolves (see FM 4-0).

LOGISTICS PREPARATION OF THE THEATER

> *[Sustainment] planning accounts for increased consumption, increased threats to lines of communications, and anticipated support to noncombatants...Urban operations place a premium on closely coordinated, combined arms teams and carefully protected [sustainment activities]. Urban operations are [logistic]-intensive, demanding large quantities of material and support for military forces and noncombatants displaced by operations.*
>
> FM 3-0

10-10. A thorough logistics preparation of the theater (LPT) is critical for an adaptable UO support plan. Sustainment planners conduct the LPT to understand the situation from a sustainment perspective and determine how best to support the maneuver commander's plan. However, decisions that impact the urban environment and the political situation require a combined operational and sustainment perspective. Sustainment planners must understand the urban environment, the fundamentals of UO, the proposed course of action, and the urban environment's effects on sustainment (as well as the other warfighting functions). Such knowledge allows the planners to develop a detailed estimate of support requirements. A thorough LPT helps commanders determine the most effective method of providing adequate, responsive support to meet support estimates while minimizing the sustainment footprint. Overall, it helps tie together UO requirements with acquisition and distribution. As with all operations, but particularly in a dynamic urban environment, this assessment process is continuous since requirements will change as the urban operation unfolds and matures.

SUPPORT TO INTELLIGENCE

10-11. The LPT resembles and runs parallel to the intelligence preparation of the battlefield (IPB). Products generated under IPB may be useful in the sustainment analysis. Conversely, the LPT may contribute to the IPB by identifying critical resources and infrastructure and assessing their potential to influence (positively or negatively) the operational plan. This information may warrant a course of action that includes offensive or defensive operations to seize, secure, or destroy those critical resources. In UO initially planned for other than sustainment reasons, the information may require altering the plan or imposing additional constraints to protect the identified resources. While these resources may or may not be critical to current operations, they are usually important to set or shape the conditions necessary for Army forces to consolidate and transition to subsequent missions or redeploy. This close relationship between IPB and LPT underscores the need to quickly and continuously involve sustainment personnel for their sustainment expertise and perspective in planning UO.

10-12. As sustainment units are among the most dispersed and omnipresent in any AO, sustainment Soldiers are potentially valuable intelligence collectors. Soldiers working in and around urban sustainment bases and executing frequent convoys through urban areas quickly become very familiar with the urban terrain and the routine associated with that terrain. As with all Soldiers, sustainment personnel must be incorporated into a comprehensive human intelligence collection system.

URBAN SUSTAINMENT INFORMATION

10-13. Figure 10-2 illustrates that a thorough analysis of the key components of urban areas in the commander's AO provides the data for an accurate LPT and subsequent UO sustainment plan (see Chapter 2 and Appendix B). Analyzing the urban terrain and infrastructure helps to determine—

- Geographic influences on consumption factors and on the provision of support (weather, climate, and topography).
- The availability of supplies, such as safe food, potable water, petroleum, electrical energy, barrier material, and compatible repair parts.
- The location of facilities, such as warehouses, cold-storage sites, water treatment facilities, manufacturing plants, hospitals, hotels for billeting, and waste treatment facilities.
- Transportation information, such as seaport and harbor facilities, airfields, rail and road networks, traffic flow patterns, choke points, and control problems.
- Locations and accessibility of maintenance facilities and equipment, and machine works for the possible fabrication of parts.
- The available general skills among the urban population, such as linguists, drivers, mechanics, MHE operators, longshoremen, and other vital trade skills.

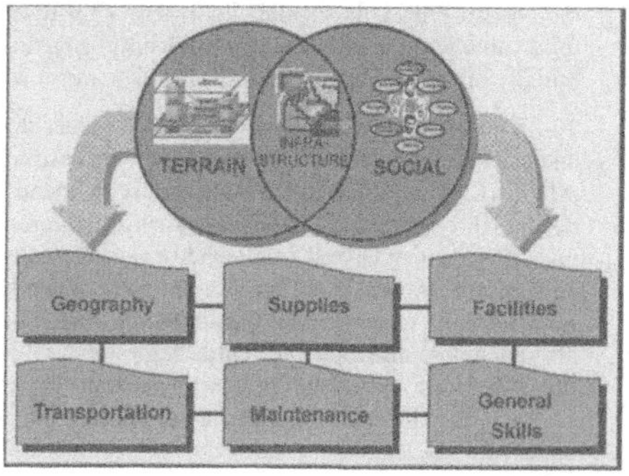

Figure 10-2. Urban environment and essential elements of sustainment information

Potential Restrictions

10-14. Commanders must be aware of restrictions that apply to the use of some non-U.S. resources. Security and requirements for U.S. national control dictate that only U.S. assets may perform certain services and functions. Therefore, some foreign urban-area capabilities, even if abundantly available, may not be used. These might include—

- Command and control of medical supply, service, maintenance, replacements, and communications.
- Triage of casualties for evacuation.
- Treatment of chemical, biological, radiological, nuclear, and nuclear (CBRN) casualties, as well as the decontamination of U.S. equipment, personnel, and remains.
- Identification and burial of deceased U.S. personnel.
- Veterinary subsistence inspection.
- Law and order operations over U.S. forces and U.S. military prisoner confinement operations, as well as accountability and security of enemy prisoners of war in U.S. custody.

Urban Societal Considerations

10-15. As in all aspects of UO, the urban society is a critical element of the LPT analysis. Sustainment planners must assess whether they can acquire and use urban resources and infrastructure without overly disrupting the urban society and their environment. If the resources are only sufficient for the urban area's inhabitants (and dependent populations in outlying areas), and the facilities cannot increase production to accommodate the needs of Army forces, then commanders may not be able to rely on those resources to support their operations. In fact, the opposite may be true. The effects of UO on the inhabitants, particularly during offensive and defensive operations, may place increased burdens on the Army's resources. For example, commanders may require civilians to evacuate all or parts of an urban area to safeguard them from the effects of planned combat operations. This precautionary measure may require Army forces to support civilian evacuees with temporary housing, food, water, clothing, and medical and sanitary facilities. It will also require carefully planned evacuation routes to ensure that the movement does not interfere with sustainment of the fight.

10-16. Sustainment planners must take into account the potential effects that the provision of supplies to the urban population may have on the operational environment (provided by Army forces or not). For example, an embargo on petroleum products was lifted after Army forces entered Haiti in the 1990s and a less repressive regime assumed governmental control. Few civilian vehicles had moved on the streets of the Port-au-Prince, the capital city, when fuel was scarce and the movement of mounted forces through the urban area was virtually unimpeded. This changed almost overnight once fuel supplies became available, initially making it very difficult for forces to move where required (unless transported by helicopter).

10-17. Sustainment planners must also consider the urban society's ability to restore their own facilities and provide for themselves (if necessary with assistance from Army forces). Throughout this analysis, civil affairs (CA) units can advise and assist in identifying and assessing urban supply systems, services, personnel, resources, and facilities. Critically, commanders must understand that purchasing local goods and services may have the unintended negative consequence of financially sustaining the most disruptive and violent factions in the area. Army forces must seek to purchase urban products and services that will not contribute to prolonging the conflict or crisis. In many stability or civil support operations, they must also attempt to distribute the contracts for goods and services purchased locally as fairly as possible among urban factions and ethnic groups to maintain impartiality and legitimacy. As part of their coordination efforts, commanders should attempt to achieve the cooperation of relief agencies and other NGOs in this endeavor.

10-18. Finally, sustainment planners must also identify potential threats and increased protection requirements that the urban society (criminals, gangs, and riotous mobs) may present, particularly when support units and activities are positioned in urban areas. The disposition or allegiance of the urban population is also important to consider. The infrastructure of an urban area may exhibit great potential to support the sustainment efforts of Army forces, but if the population is hostile or unreliable, use of the facilities and resources may be unfeasible.

Urban Support Areas and Bases

10-19. A major influence on the operation plan and its subsequent execution is often the proper identification and preparation of support areas and bases. The LPT helps commanders determine the need, advantages, and disadvantages of using urban areas in the AO as bases from which to provide support and conduct distribution operations. Ideally, these areas should support reception, staging, onward movement, and integration operations. They should also allow easy sea and air access, offer adequate protection and storage space, increase the throughput of supplies and equipment, and be accessible to multiple LOCs. Consequently, commanders often establish support areas and bases near seaports and airports that are part of a larger urban area. However, threats recognize the Army's need for ports and airfields and may devote substantial resources and combat power to defend them. Army commanders also face the challenge of integrating airfield and port defensive operations with air and naval component commanders to deny threats the ability to conduct stand-off attacks against aircraft and ships. Only an integrated, joint approach can ensure persistent air and sea support– essential not only to sustainment but to firepower as well—is not degraded by the threat. Therefore, planners may determine during the LPT that the risks of seizing or establishing urban lodgment areas may be too high (see Chapter 4). Instead, they may recommend building an airfield, conducting logistics over-the-shore operations, or constructing sustainment bases in more isolated—but more easily secured—locations.

Overall Assessment

10-20. As shown above, the LPT process and analysis help to determine if urban areas in the AO—

- Are suitable as areas and bases for support.
- Can contribute sufficient quantities of and are a dependable source for resources for the overall operation.
- May additionally drain the supported commander's resources.

10-21. The results of this process often serve as a basis for reviewing requirements for civilian contract support and host-nation support and for developing sustainment input into time-phased force and deployment data. This chapter focuses on the effects urban areas may have on accomplishing sustainment functions and related activities, particularly when sustainment units and activities are in urban areas.

ACHIEVING FORCE AGILITY

10-22. To maintain the responsiveness necessary to support full spectrum operations against a wide array of threats and operational environments, sustainment forces must remain agile, possessing the mental and physical agility to transition within or between types of operation with minimal augmentation, no break in contact, and no significant additional training. Responsiveness, flexibility, and economy are key characteristics that enable sustainment forces to support a dynamic combat force; they represent the capability of sustainment forces to provide the resources to initiate, sustain, and extend UO. Agile sustainment forces are characterized by modular force designs, the ability to tailor sustainment organizations for the supporting mission, and the ability to conduct split-based operations. These sustainment aspects are critical to achieving the agility necessary for Army forces to succeed in UO.

MODULAR DESIGN

10-23. Sustainment units are structured as modular organizations, or must possess the ability to modularize in order to conduct decentralized sustainment operations required in UO. At the company level, modularity enables each major company sub-element to assume a cross-section of the company's total capabilities, allowing commanders to employ individual modules to provide a support function, while the rest of the unit remains operational. Modularity enhances responsiveness while enabling support forces to project capabilities as far forward as possible into urban areas, yet with the minimum forward footprint necessary to sustain operations.

FORCE TAILORING

10-24. Sustainment force tailoring refers to the process of determining and deploying the right mix of capabilities to support the force or mission. Sustainment commanders may task organize specific capabilities into functional elements in order to maximize effectiveness and efficiency, and to minimize the sustainment footprint. During UO, the sustainment commander can tailor the support element required to accomplish a specific mission or task, thereby mitigating the risk associated with deploying a larger, more robust capability package forward into the urban area.

SPLIT-BASED OPERATIONS

10-25. Split-based operations refer to performing certain sustainment administrative and management functions outside the area of operations. During UO, Soldiers and civilians can perform personnel, materiel, and distribution management functions without deploying forward into the urban area if the information systems are adequate. This is essential to minimizing risk to support forces and the sustainment footprint in the urban area, and still ensures that all support requirements are met.

SUSTAINMENT FUNCTIONS

- Supply and Field Services
- Maintenance Support
- Transportation Support
- Force Health Protection
- Explosive Ordnance Disposal Support
- Human Resources Support
- Financial Management Support
- Religious Support
- Legal Support to Operations
- General Engineering Support

Figure 10-3. Sustainment functions

SUSTAINMENT FUNCTIONS

10-26. Sustainment consists of eleven functions that must be carefully planned, managed, and synchronized to provide responsive and efficient support to maneuver commanders (see figure 10-3).

Similar to the components of the urban environment, particularly its infrastructure, they overlap and are interdependent. The success of one function depends on the success of several others. Like urban infrastructure, they have two components: a physical component (supplies, equipment, and facilities) and a human component (the Soldiers and civilians who execute these functions). Like city mayors, commanders must plan, manage, and synchronize these functions to provide responsive and efficient sustainment to UO.

10-27. Commanders and planners should consider two essential aspects when addressing these sustainment functions. One aspect looks outward and one looks inward. The first aspect is how these functions can best support full-spectrum UO—the outward analysis. The second aspect is how the urban area affects the conduct of sustainment functions, particularly when those functions are located or performed in an urban area—the inward analysis.

SUPPLY AND FIELD SERVICES

10-28. The supply function involves acquiring, managing, receiving, storing, protecting, maintaining, salvaging, and distributing all classes of supply (except Class VIII) required to equip and sustain Army (and joint) forces. For UO, commanders may need to decide early in the planning cycle whether to stockpile resources forward or to rely on distribution-based sustainment management to satisfy requirements. Some specialized items identified below may not be available through the normal military supply system and may take sustainment personnel much longer to obtain or fabricate. Operation planners must quickly identify the special equipment and increased supply requirements for UO to give sustainment personnel the time necessary to acquire them.

10-29. Field services are those essential services required to enhance a Soldier's quality of life. They consist of clothing exchange, laundry and shower support, textile repair, mortuary affairs, preparation for aerial delivery, food services, billeting, and sanitation. The urban commander determines the need and priority of each service after careful METT-TC analysis. Some facilities such as shower, laundry, and cold storage may be available in the urban area through theater support contracting or provided by external support contracts such as the logistics civilian augmentation program (LOGCAP). Additionally, requirements to care for the urban population will increase requirements for field services immensely. In some circumstances, most notably urban stability or civil support operations, adequately protected field service units or activities will be critical and may be the only support provided.

Increased Urban Supply Requirements

Water

10-30. Water production and distribution will become more difficult for sustainment forces during UO. (Many parts of an urban area can become virtual deserts if public water supplies are cut off or contaminated due to deliberate military action or natural disasters.) Consequently, sustainment managers will be called upon to find more efficient and effective means of moving and distributing potable water, especially when bulk storage of water within the urban area may not be possible or feasible. In sustainment bases that do not have access to urban or other water sources, the transportation and storage of non-potable water may be as great a concern as potable water. Unless adequate storage is available for both potable and non-potable water, units may be forced to rely solely on potable water for all bulk requirements to mitigate health risk. Due to the compartmented nature of many urban operations, planners may need to ensure that additional containers for water (and fuel) are available to support dispersed stock at small-unit level, and that man-portable systems are available in sufficient quantities to ensure timely replenishment.

Class II

10-31. Clothing and individual equipment exposed to the urban environment (concrete, glass, and steel) experience wear and damage at accelerated rates. Units will also require items such as rope, grappling hooks, crowbars, ladders, chain saws, elbow- and kneepads, special vehicle and personnel body armor, fire-fighting equipment, packboards, carpentry tools, and other specialized items to conduct UO. Planners must recognize this and plan accordingly to ensure operations are not interrupted due to equipment shortfalls.

Class III (Bulk)

10-32. With the potential exception of aviation fuel, bulk Class III requirements for UO generally decrease at the maneuver unit level. However, increased fuel requirements for engineer and power-generation equipment attached to or operating with forward units may offset these decreases. Generally, fuel sources within the urban area (refinement facilities, gas stations, garages, and airfields) should not be utilized unless tested and approved for use by a fuel laboratory.

Class IV

10-33. Requirements for construction and barrier material (and the power tools to cut it) during UO will likely far exceed the capability of the sustainment force to provide. Commanders must prioritize distribution, defensive projects, and construction and repair efforts in order to maximize the efficient use of available Class IV. Every effort should be made to obtain supplemental Class IV material from the local economy, as long as that effort does not adversely affect operations or the restoration and improvement of critical urban infrastructure. Of note, specialized, prefabricated road barriers will also be in great demand to block roads, create checkpoints, reinforce defensive positions, and protect headquarters and sustainment activities.

Class V

10-34. Urban combat operations will often increase requirements for Class V. Ammunition consumption rates for urban environments have been as much as five to ten times greater than operations in other environments. Urban offensive and defensive operations will often increase overall requirements for certain types of ammunition (crew-served and small arms ammunition, mortars, antitank rounds, grenades, mines, and demolitions) while others may typically decrease (tank, artillery, attack helicopter). Since forces operating on urban terrain may not be capable of storing ammunition, an established system of distribution is vital to ensuring a constant flow of ammunition necessary to sustain continuous operations and retain the initiative.

Class IX

10-35. Requirements for repair parts generally decrease in UO, with the notable exception of those systems that experience increased ammunition consumption. The heavy use of crew-served and small arms, mortar, and antitank systems in the urban area causes wear and failure at significantly increased rates. Additionally, wheeled-vehicle tires and armored-vehicle track pads and road arms, road wheels, center guides, wedge bolts, and track blocks may wear rapidly on urban streets, rubble, and debris. Based on the tempo of operations, it may be more feasible to replace whole sets of tracks rather than have crews replace individual track pads. Hence, commanders may consider resourcing all wheeled vehicles with spare tires and creating containerized and rapidly transportable sets of track and track parts for speedy delivery to urban sustainment bases. On the whole, planners should consider the use of repair parts packages specifically configured for UO that can be transported by tailored, modular sustainment elements that directly support operations within the urban environment. During long-term stability operations, outgoing units may leave large portions of their prescribed load list in theater to ensure that relieving units have adequate supplies to maintain the tempo of operations.

Storage and Distribution

10-36. UO present challenges to storage and distribution that may not be experienced outside of the urban environment. Forces operating within these areas may not be static and, therefore, cannot store equipment and resources beyond what they can transport organically. Even forces conducting operations from secure locations within the urban area may not be able to store materiel, necessitating a continuous, coordinated distribution system to ensure forces are adequately resourced. During UO, sustainment forces may need to utilize unit or throughput distribution techniques to provide the level of support required to sustain the force.

10-37. Urban commanders must also consider the benefits of stockpiling resources forward against relying on the Army's distribution-based sustainment system. Stockpiling brings supplies close to urban forces and helps ensure supplies are immediately available to support potentially increased consumption rates. However, this method may burden the support structure that moves, handles, and protects large quantities of resources often on a repetitive basis. The Army's normal distribution-based sustainment system reduces this burden significantly; however, available transportation assets (sea, air, and ground) impact delivery response times. These transportation assets often combine military assets overlaid on the host nation's (and urban area's) transportation and distribution infrastructure. Although the civilian infrastructure may initially support the Army's distribution system, later effects of UO, such as destruction of equipment and facilities or loss of civilian workers, may degrade the system. Moreover, Army forces may have to share these assets with other military, multinational force, or civilian organizations participating in the urban operation, as well as with the indigenous civilian population at large. This shared system may put at risk the timely delivery of critical supplies to Army forces.

10-38. Sustainment planners must understand the urban environment and its effects on the proposed method of distribution. They must also understand how urban operations (to include sustainment UO) may affect the urban environment. Storing bulk fuel or ammunition in or near an urban area, for example, may increase the risk of fire and explosive hazard to civilians and Army forces. With this awareness, planners present the commander with an estimate that considers both risks and benefits. Depending on the particular area and other METT-TC factors, they may recommend one method of distribution then transition later to another or a combination of methods. Preconfigured resupply loads delivered as close as possible to where they are needed is vital to responsiveness. Only an agile distribution-based sustainment system will allow Army forces to be responsive and operationally effective across full spectrum operations in an urban environment.

Food Preparation

10-39. The daily feeding standard for Soldiers operating in urban areas remains the same: three quality meals per Soldier. UO creates an increased energy demand on Soldiers, requiring a caloric intake of about 5,000 calories each day. Since producing and delivering prepared meals to forward elements may be difficult or even impossible due to the urban environment, the innovative use of meal supplements can be critical to meeting the dietary requirements of the force. Energy drinks and bars, rich in electrolytes and carbohydrates, are an essential supplement to prepared meals and should be provided in quantities that support or exceed the requirements of the forces operating in the urban area.

10-40. Food (and water) may be available in the urban area; however, local sources must be tested, carefully monitored, and medically approved before consumption. Garbage disposal may be an important consideration in the urban area. Improper trash disposal may leave a signature trail (particularly during urban defensive operations) that may degrade sanitary conditions leading to increased disease and nonbattle injuries (DNBIs). Commanders must understand that food operations, if not properly positioned and secured, can become a focal point for the urban population. Strict policy regarding distribution and control of any Class I supplies (including waste products) will be enforced. Black marketers will be attracted to Army food service activities as well (even during offensive and defensive operations).

Water Purification

Vigilant Monitoring

10-41. Urban areas will often have a ready source of water to support the urban inhabitants and its infrastructure. However, this water may not be suitable for U.S. and allied forces (though the urban population may have developed immunity toward its microorganisms). The higher concentration of TIM in urban areas compounds this problem, as ground water is highly susceptible to chemical contamination, even supplies located miles away from the source of contamination. U.S. Soldiers should be trained and cautioned against using water from an urban area (to include ice and bottled water) until preventive medicine and veterinary personnel can determine its quality. When water quality is unknown, commanders ensure use of tactical water purification equipment that will upgrade it to the Army's water quality standards. Even if initial testing indicates that urban water sources are safe for Army forces, personnel must continuously monitor the water quality. However, Army water purification, storage, and transportation requirements for UO can be greatly reduced if the existing urban water supply can be integrated into sustainment operations. An early assessment of the feasibility of this course of action is critical to sustainment planning.

Potentially Greater Requirements

10-42. Offensive and defensive UO are often intense and can produce more casualties, including civilians. Consequently, medical facilities, already consumers of large volumes of water, may require even more water. Water purification, particularly in the urban areas of developing nations and during urban disaster relief operations, will be a critical and constant concern for Army forces.

Possible Key Terrain

10-43. Sites that can control the water of the urban area may be key terrain, providing not only a resource for Army forces, but also a means to control the threat, the civilians, or both. These sites may be the sources of the water—the river, lake, reservoir, or storage tanks—or the means that process and transport the water—pipelines, pumping stations, or treatment facilities. Many sites may be outside the urban area, as many large urban areas draw water from distant sources. The seizure of a pumping station or pipeline may make it possible for commanders to control water supplies without expending resources required to enter the urban area. To preserve critical infrastructure, commanders may increase security to protect these locations from contamination or destruction. Engineers also may need to restore, maintain, or operate existing water facilities damaged by the threat or disaster and to drill new wells and construct new water facilities.

Mortuary Affairs

10-44. Mortuary affairs provide the necessary care and disposition of deceased personnel. It supports the Army across the spectrum of operations. It may directly and suddenly impact (positively or negatively) the morale of the Soldiers and the American public and may influence relations with the civilian population in the AO. It can also affect the health of Soldiers and the urban populace. Overall, commanders must carefully plan evacuation routes and temporary collection and internment sites, adhering to local customs and traditions to lessen potentially negative consequences.

10-45. Units are responsible for recovering the remains of their own fatalities and evacuating them to the closest mortuary affairs collection point, usually located at the nearest support area. Urban governmental authorities, family, and representatives of appropriate nongovernmental organization are responsible (and should always be relied on to the greatest extent possible) for recovering and handling indigenous civilian remains. However, because of the sheer density of noncombatants in UO, commanders may also find themselves responsible for civilian remains. Deaths of civilians under Army control, such as urban evacuees and refugees at Army-operated sites, often obligate the Army to care for their remains including medical certification and records of death. High-intensity urban combat may result in civilian deaths, and health concerns will require Army forces to deal with civilian remains expeditiously. Commanders should consult local religious leaders, the Staff Judge Advocate (SJA), CA personnel, and chaplains to verify that they are abiding by law and customs. Overall, commanders must ensure that forces treat all deceased, including civilians, with dignity and respect (another important aspect of adhering to the urban fundamental of understanding the human dimension).

Aerial Delivery

10-46. Aerial delivery is the movement by fixed- or rotary-wing aircraft and delivery by the use of parachute or sling load of Soldiers, supplies, and equipment. As a vital link in the distribution system, it contributes to flexibility and provides the capability of supplying the force even when urban ground LOCs are disrupted. Forces use aerial delivery to deliver supplies and equipment when not possible or timely by other means. During urban stability or civil support operations, it is used extensively to move supplies to meet the urgent needs of a population in crisis. In all UO, delivery aircraft are highly vulnerable to small arms, rockets, and air defense systems. A threat may further decrease an already limited number of urban drop zones (aircraft may be able to avoid air defense systems, but ground forces may not be able to secure the drop zone and retrieve the cargo). Equipment and supplies transported by helicopter sling-load lessen the latter disadvantage. There are usually more available sites to deposit sling loads, such as rooftops (engineers may need to determine if they have sufficient structural integrity), parking lots, and athletic fields. However, load instability during flight may restrict a helicopter's airspeed and maneuver capabilities making it more vulnerable to small arms and man-portable air defense systems. For these reasons, aerial delivery of supplies in UO may be much less efficient than in many other environments.

MAINTENANCE SUPPORT

10-47. Maintenance entails keeping materiel in operating condition, returning it to service, or updating and upgrading its capability. It includes recovering and evacuating disabled equipment; replacing forward; performing preventive maintenance checks and services (PMCS); increasing battle damage assessment and repair (BDAR); and determining potential maintenance resources in the urban area.

Rapid Recovery Essential

10-48. Disabled vehicles produce obstacles to movement during UO, blocking narrow thoroughfares and creating vulnerable targets for threat forces. This makes rapid recovery operations essential to avoid clogging limited LOCs and mounted avenues with vehicle evacuation operations. Hence, maintenance personnel should be proficient in hastily securing unit maintenance collection sites near the damaged equipment, ensuring route security to and from the disabled vehicle, and methods of egress under fire. The increased potential for the task organization of armored units into smaller attachments during UO (often platoon-sized or smaller) will strain limited recovery assets and increase the need for sections to be well trained and resourced (to include adequate tow-bars and chains) for self-recovery operations. Units must maintain centralized and responsive control over these potentially critical recovery assets and position them as far forward as the situation allows.

Replace Forward

10-49. One of the guiding maintenance principles is to replace forward and fix rear. Maintenance activities, with a forward focus on system replacement, task and use the distribution and evacuation channels to push components and end items to the sustainment level for repair. However, the urban environment may make distribution and evacuation difficult or even impossible. Repairing equipment as far forward as possible will often be vital to maintaining the tempo necessary to succeed in UO. When recovery is required, equipment is moved only as far back as the point where repairs can be made. When selecting a maintenance site, commanders consider:

- Security and force protection measures.
- Sufficient area around equipment for lift or recovery vehicles.
- Availability of existing maintenance facilities or garages.

10-50. The unforgiving urban terrain will invariably increase damage to man-portable weapons and equipment, particularly electronic equipment sensitive to jarring. More frequent jarring can also translate into more frequent calibration requirements. Although by definition man-portable weapons and equipment are easier to transport, evacuating these systems may prove as difficult as evacuating vehicles and larger, heavier equipment. Therefore, field support maintenance support teams (MSTs) will frequently need to repair equipment at (or as near as possible to) the point where it was damaged. Equipment operators are responsible to properly diagnose the fault or damage. Such action ensures that the correct repair parts and maintenance personnel are sent forward to complete necessary repairs. In UO, particularly offensive and defensive operations, units may need to replace rather than repair equipment. Subsequently, sustainment personnel may need to plan for increased replacement of what might normally be repairable equipment, as well as plan for increased repair parts for man-portable items.

Add-On Protection Increases Wear

10-51. To increase protection against small arms, mines, rocket-propelled grenades, and lightweight antiarmor weapons, units may need to attach additional armor and improvised wire-mesh screens and cages to both wheeled and tracked vehicles operating in an urban environment. (Units can also sandbag vehicles to achieve a degree of increased protection.) These modifications, however, increase wear on brakes, springs, suspension, and tires, all of which are already vulnerable to the increased amount of debris present in the urban environment. Over time, this wear will begin to increase failure rates in major assemblies, such as engines, transmissions, and transfers. To offset the increase in wear and, subsequently repair parts requirements, commanders conducting UO should consider emphasizing these aspects during daily operator PMCS and increasing the frequency of scheduled services.

Increased Battle Damage Assessment and Repair

10-52. In UO, operators, crews, MSTs, and recovery teams execute BDAR far more than in other environments. BDAR quickly restores minimum essential combat capabilities for a specific mission (normally of short duration) or allows the equipment to self-recover by expediently fixing or bypassing components. Commanders may need to authorize supervised battlefield cannibalization and controlled exchange when units lack critical parts or cannot bring them forward. Sustainment commanders should consider the use of BDAR kits, specifically designed for the systems they support. The use of these kits, though providing only temporary repair, will extend the availability of combat systems until full repair capability is accessible.

Potential Urban Maintenance Resources

10-53. Although urban areas can complicate maintenance, they may contribute to this sustainment function. Analyzing the urban area in the commander's AO may reveal potential sources of parts, tools, equipment, and facilities, and the expertise necessary to fix equipment, repair components, and fabricate critical parts. Sites of potential value may include automotive repair shops, foundries, car dealerships, parts stores, junkyards, scrap-metal dealers, and machine shops. Urban areas may also serve as key sources for parts and facilities (and theater support contractor personnel) to repair automation and network communication equipment.

Host Nation Repair Operations: OPERATION IRAQI FREEDOM

During operations in Mosul, Iraq in the summer months of 2003, elements of the 2nd Brigade Combat Team (BCT), 101st Airborne Division (Air Assault) suffered from a decline in combat power that outpaced the ability of the distribution-based sustainment system to provide repair parts. As readiness across the BCT threatened to degrade operational capability, sustainment personnel from the brigade's forward support battalion (FSB)—drawing on lessons learned from the 1990-1991 Gulf War—scoured the local populace for repair parts, facilities, and expertise in an attempt to offset the losses suffered due to supply shortfalls.

Initially, local retailers provided limited repair parts and facilities to the FSB, but the effect was significant and readiness rates began to stabilize. As time passed, however, the continued wear on equipment operating in the urban environment, coupled with the harsh summer weather, began to take a toll on reparable components and major assemblies. With limited availability in theater, and order-ship times as long as 60 days, requisitions would not be filled before the effect on combat fleet readiness reached a critical point.

Recognizing a trend that threatened the success of the BCT's urban operations in Mosul, the FSB established a contract maintenance facility in the brigade support area, used its contingency contracting officer to hire a group of local mechanics certified through General Motors, and opened the first host nation repair facility in Iraq. The group repaired major assemblies and reparable components and systems, maintaining a failure rate consistent with major CONUS rebuild facilities. The effect was immediate: during the first four months of the program, a full two-thirds of the combat platforms repaired and returned to service operated with components or assemblies rebuilt in the host nation facility.

Over time, the program was expanded to support other divisional units in northern Iraq. This effort provided maneuver commanders with responsive maintenance support and demonstrated the effect of sustainment flexibility and sustainability in urban operations. Furthermore, this action met the division commander's intent of infusing much-needed funds into the local economy while strengthening American-Iraqi relations in the region.

TRANSPORTATION SUPPORT

10-54. Transportation supports the concept of the urban operation by moving and transferring units, Soldiers, equipment, and supplies. Transportation incorporates military, commercial, host-nation, and urban area capabilities to establish a flexible system that expands to meet the needs of the force. Transportation includes movement control, terminal operations, and mode operations.

10-55. Urban areas are often critical to transportation operations. These areas may serve as a lodgment or support area for entry of Army forces and sustainment supplies. The existing transportation and distribution infrastructure may be essential to reception, staging, and onward movement. Theater support contracts, as well as host-nation support agreements, may greatly increase the ability of Army forces to use the urban area's facilities, which may include docks, airfields, warehouses, and cargo handling equipment. Urban support may also include skilled urban workers, such as longshoremen, truck drivers, and MHE operators.

US Army Photo

Urban Terminals

10-56. In addition to serving as major seaports and aerial ports of debarkation, urban areas may provide additional terminals in the AO. Forces may use these terminals for further staging, loading, discharging, and transferring the handling between various inland transportation modes and carriers (motor, air, rail, and water). These urban terminals—with synchronized movement control—permit commanders to rapidly shift transportation modes and carriers. This increases flexibility and ensures the continued forward movement of equipment and supplies to support ongoing operations. Movement control, particularly in urban areas, relies heavily on support from military police in their maneuver and mobility support role. Without this support, urban LOCs may become congested, hinder movement and maneuver, and degrade force effectiveness (see FM 3-19.4). Urban commanders may need to establish multiple checkpoints, roadblocks, and traffic control points; restrict selected roads to military traffic; and reroute movement to unaffected road networks when civil support and refugee control operations compete for available routes. Again, military police operations are critical in this regard and will require continuous, close coordination with urban civilian police (if available).

Obstacles to Ground Transportation

10-57. Although urban areas can contribute to transportation operations, rubble and other damage can become obstacles to ground movement. Even in an undamaged urban environment, road and bridge weight restrictions, barriers and medians, steep embankments, large drainage systems, and other urban structures may limit transportation operations. Urban route maintenance, to include reinforcing bridges and constructing bypasses, may become a priority task for engineer units. Bypassed pockets of resistance and ambushes pose a constant threat along urban supply routes and routes in close proximity to urban areas. Urban convoy and resupply operations will often require increased security in the form of continuous route security operations, regular (daily, if necessary) mine and improvised explosive device (IED) clearance operations, numerous observation posts, attack helicopter support, a larger, more mobile tactical combat force, and adequate communications capabilities among convoy vehicles and with supporting forces and higher headquarters. While the level of security and force protection may vary, convoy operations are never considered an "administrative" operation (see FM 4-01.45). Security needs may increase manpower requirements for sustaining operations and potentially reduce resources from decisive operations. Moving critical supplies may require heavily armed convoys and armored vehicles. Drivers must be well trained in convoy operations, rehearsed, and alert. They must recognize and avoid potential mines and IEDs (stationary and vehicle-borne) and be able to react rapidly to ambushes. Additionally, commanders may need to institute deception measures such as randomly closing and opening supply routes to confuse threat pattern analysis. As before, aerial resupply can alleviate problems due to ground obstacles, but the air defense threat and proximity of threat forces may preclude their routine use.

Population Effects

10-58. The ability of Army forces to use vital urban transportation facilities depends largely on the civilians and the threat. The civilian population can affect the transportation system if they do not support the goals of Army operations. Urban transportation systems—such as ports, railroads, and rivers—require many specialists to operate. Without these specialists, the system's utility is degraded and may not function at all. In urban stability or civil support operations, Army forces will share the system with civilians and other agencies. Civilian authorities may refuse to allow Army forces to use any portions of an urban area's transportation system. Negotiating for access to that system under these circumstances then becomes a command priority.

Threat Effects

10-59. The threat can significantly affect urban transportation systems. Many are composed of smaller subsystems. Each subsystem is vulnerable to attack, which in turn often shuts down the whole system. A large canal system, for example, may have entrance and exit facilities, the canal itself, a means to pull the vessel along such as a locomotive engine, and the civilians that run each of these subsystems. Both an attacker and defender understand the components of the particular transportation system. If important to current or subsequent operations, defending forces then develop plans and allocate forces to protect these subsystems. Attacking forces, on the other hand, often avoid collateral damage to the system, while simultaneously preventing enemy destruction of the facilities.

Force Health Protection

10-60. Force health protection (FHP) consists of all services performed, provided, or arranged to promote, improve, conserve, or restore the mental or physical well-being of personnel in the Army and, as directed, for other services, agencies, and organizations. FHP conserves the force by—

- Preventing DNBIs and controlling combat stress.
- Clearing the urban area of casualties.
- Providing forward medical treatment and hospitalization and en route care during medical evacuation.
- Providing required veterinary, dental, and laboratory services.
- Ensuring that adequate Class VIII supplies, medical equipment, and blood are available.

10-61. FHP operations minimize the effects of wounds, injuries, disease, urban environmental hazards, and psychological stresses on unit effectiveness, readiness, and morale. Effective UO require acclimated Soldiers trained in specific urban tactics, techniques, and procedures. FHP helps maintain the health of urban forces, thereby conserving that trained manpower. Historically, urban combat operations have generated three- to six-times greater casualty rates than operations in any other type environment. (Greater enemy wounded and a need to provide care for them will also exacerbate this situation.) FHP operations that maintain the health and readiness of Army forces reduce the strain on the replacement and evacuation systems. Such care allows Soldiers to concentrate on the task at hand instead of the increased risks associated with UO.

10-62. As part of the overall LPT, commanders and medical planners must analyze and continuously assess the urban area. They must identify medical threats, required medical resources, and the quality and availability of medical facilities and resources (to include civilian medical personnel). This assessment prevents duplicated services and permits more effectively and efficiently organized medical resources.

10-63. FHP analysis also identifies available hospitals, clinics, medical treatment facilities, and medical supplies and equipment (including production facilities) in the urban area. It may also identify NGOs capable of providing medical services and supplies. (However, stringent federal regulations, standards of medical care, and a need for unavailable advanced technologies may limit their use by Army forces. But, this analysis will still be crucial in determining whether urban medical facilities can support civilian requirements.) FHP personnel keep abreast of the operational situation and its impact on FHP and the urban population. Peace operations, for example, may rapidly transition to high-intensity offensive and defensive operations requiring medical support able to handle potential mass casualty scenarios which may include civilians.

Care of Civilians

10-64. During combat operations, the military normally does not provide injured civilians with medical care, instead relying on civilian medical personnel to provide this service. However, based on METT-TC and requirements under Geneva Conventions, commanders may need to recover, evacuate, and treat numerous civilians (particularly in urban civil support or stability operations) until local civilian medical personnel and facilities can be reconstituted and supplied. In urban civil support operations involving weapons of mass destruction, the primary focus of Army support may be FHP.

10-65. If commanders provide civilian medical support, they adhere to the UO fundamental of transitioning control and transferring responsibilities for medical care of civilians to another agency or into civilian hands as soon as is practical. In UO, commanders consider and address the medical treatment of civilians (enemy and friendly) early in the planning process. Any initial assessment or survey teams therefore contain medical representatives. Of significant concern, commanders may need to ensure that adequate supplies and equipment are on hand to treat the urban population's children as combat health service support is normally focused on the treatment of adult Soldiers. Additionally, follow-on military care for civilians after treatment at a combat support hospital is often nonexistent or extremely limited to a case-by-case basis necessitating the quick restoration of civilian hospital capabilities. Finally, military health practitioners must also assess and consider cultural factors related to civilian medical treatment. (For example, male doctors may not be able to examine or treat indigenous women.) To alleviate confusion and preclude avoidable negative repercussions and unrealistic expectations, commanders may need to develop specific medical policies, directives, and standing operating procedures. These "medical rules of engagement" (ROE) ensure that subordinates know how much medical care they may provide to—

- The urban population.
- Other host-nation and third-country civilians.
- Coalition and host-nation forces.
- Contractor personnel.

10-66. Commanders should also consider that medical aid to the local urban population and other civilians can be a powerful influential tool. Properly planned and executed, its potential benefits for Army forces include enhanced security and intelligence collection created by increasing the willingness of civilians to interact with Soldiers. These effects cannot be left to happenstance. They should be a conscious part of the information operations campaign nested within the overall UO plan. Consideration must also be given, however, to mitigating the potential negative consequences involved in the inevitable reduction or curtailing of medical assistance by Army FHP personnel. As with any aspect of UO, providing care without considering the longer-term consequences could result in far greater disadvantages than benefits.

Disease and Nonbattle Injury

10-67. DNBI is a major medical threat during all operations and presents additional risk during UO. Urban pollution hazards and potential exposure to TIM may increase the risk to soldier health. Some urban areas, particularly those in developing countries, are already large sources of communicable diseases, such as tuberculosis, cholera, typhus, hepatitis, malaria, dengue, and acquired immune deficiency syndrome (AIDS). Other factors that increase the number of disease vectors and the potential for the transmission of infectious diseases might include—

- Physical damage or deterioration of urban infrastructure—such as electricity, water, and sewage services.
- Industries that use or produce hazardous materials.
- Increased habitats for rodents and other pests due to rubble, debris, and uncollected trash.
- Inability for the threat or local authorities to recover and dispose of human remains and animal carcasses.
- Greater threat of animal diseases transmittable to humans due to larger numbers of animals wounded or injured by the effects of UO.
- Greater susceptibility of urban civilians to infectious diseases—increasing the potential of person-to-person disease transmission—due to declining public health capabilities and nutrition levels.

10-68. The density of the environment extends these urban medical risks to Army forces. Commanders should establish a medical epidemiological surveillance system early to assess the health of the force and promptly identify unusual or local occurrences that may signal preventive medicine problems or the influence of biological or chemical agents. These potential hazards, particularly the release of TIM, will influence the type of medical supplies needed by medical personnel and will also necessitate critical planning and preparation for potential mass casualties (civilian as well as military).

10-69. Adhering to the UO fundamental of preserving critical infrastructure will likely prevent surges in DNBI. Preventive medicine personnel must identify the diseases and recommend control and preventive measures. In urban areas, they may also conduct civilian health screening, health education, and immunization programs. Medical screening of military personnel, particularly multinational forces, may be required to prevent introducing new diseases (especially drug-resistant strains) into an urban area. A new disease may tax the medical system and introduce a new medical problem into an area already in crisis. Lastly, field sanitation training (to include a general emphasis on hand washing; training in the use of barrier protection such as latex gloves when rendering care to any person; and animal, rodent, and pest control), equipment, and supplies are part of overall preventive medicine measures and considerations.

Combat Stress

10-70. Stress occurs in every operation and type of environment; while limited stress levels are beneficial, too much is harmful. Controlled combat stress can invoke positive, adaptive reactions such as loyalty, selflessness, and heroism. On the other hand, uncontrolled combat stress can result in negative, harmful behavior and combat stress casualties. Such behaviors and casualties—battle fatigue, misconduct stress behaviors, and post-traumatic stress disorder—can interfere with the unit's mission. Physical and mental factors leading to combat stress result from the environment and the deliberate enemy actions aimed at killing, wounding, or demoralizing Soldiers. However, many stressors are also generated from the Soldier's own leaders and mission demands.

Combat Stress: Chechnya – 1994 to 1996

Russia's 1994-1996 conflict with Chechnya, a republic in the southwestern part of the Russian Federation, produced an increased number of psychological trauma and combat stress casualties. One medical survey found 72 percent of the soldiers screened had some sort of psychological disorder symptoms. Of that, 46 percent exhibited asthenic depression (a weak, apathetic, or retarded motor state). The other 26 percent exhibited psychotic reactions such as high states of anxiety, excitement, or aggressiveness and a deterioration of moral values or interpersonal relations. The statistics showed more troops experienced combat stress disorders than during their 1980s war in Afghanistan. One primary difference was that in Chechnya, Russian forces conducted combat mostly in cities rather than in mountains, valleys, and other rural areas.

Combat always invokes fear in soldiers. However, poor training and planning, uncertainty in their cause, and urban populations that resented their presence exacerbated the psychological climate for the Russian forces in Chechnya. Acts of subversion and terrorism by Chechen guerrillas kept the Russians in a constant, high state of readiness and anxiety; the soldiers viewed every civilian—young or old, male or female—as a potential enemy. This psychological pressure was not simply a by-product, but an objective of information operations and a prime reason for taking the fight into the close confines of Chechnya's urban areas. Torture and mutilation of prisoners; immediate execution of captured pilots; imitative electromagnetic deception (Chechens mimicked Russian radio transmissions and directed Russian close air support against their own forces); and propaganda to convince civilians that Russia's actions had a religious bias against Muslims and Islam were conducted to exert intense, unremitting psychological pressure on Russian forces—with great success.

The characteristics of urban areas combined with Chechen insurgent activities and information operations, civilians that did not welcome foreigners, and an unpopular and poorly supported conflict with an open-ended mission reinforced the need for a national will to strengthen and support the urban fight. These characteristics also reinforce the need for clear objectives, proper leadership and training, and available medical assets able to properly diagnose and treat combat stress casualties.

10-71. During UO, compartmented urban combat leads to physical isolation while difficulties in transmitting radio signals lead to communications isolation; together, they create an overwhelming sense of loneliness. Snipers, mines, and IEDs combined with the closeness and high intensity of urban combat contribute to an unremitting fear of attack from any quarter that further increases stress casualties. Additionally, seeing and perhaps accidentally inflicting casualties on civilians (especially women and children) increase battle fatigue. If civilians are hostile or a threat uses the population as cover and concealment, then the potential for misconduct stress behaviors often increases. Urban areas may provide temptations for looting, alcohol and substance abuse, black marketeering, and harmful social interactions; these temptations may increase misconduct stress behaviors. Leaders can prevent or rapidly identify, successfully manage, and treat stress-related casualties (see FM 22-51 and FM 6-22.5) and prevent misconduct stress behaviors as well as potential violations of the law of war. To reduce the number of stress casualties, commanders must—

- Provide tough, realistic training in urban combat (and combat stress management).
- Plan for rotation of forces and adequate rest and relaxation opportunities.
- Establish effective ROE.
- Create unit cohesion.
- Ensure adequate mental health support is available.
- Provide strong, capable leadership. Leaders, forward on the battlefield, will determine whether the necessary discipline perseveres.

Evacuation

10-72. Transportation restrictions may preclude the medical evacuation (MEDEVAC) of urban casualties. MEDEVAC by air ambulance remains the preferred method; however, it may not be feasible. The threat from small arms fire, rocket-propelled grenades, and other shoulder-fired weapons may prohibit the employment of helicopters. Aeromedical evacuation can also be hampered by the lack of landing zones due to narrow streets, heavily damaged buildings, and barricaded or booby-trapped roofs. If the employment of air ambulances is not feasible, a ground evacuation operation will be necessary. In UO, ground ambulances should often be—

- Hardened to improve the safety of the crew and patients as well as to increase the survivability of the evacuation platform.
- Field-sited as close to the operation as is tactically and physically possible.
- Maneuverable in tight spaces and narrow streets.

10-73. Evacuation by ground ambulance may have their own set of obstacles to surmount such as rubble; debris; barricades; mines, booby-traps, and IEDs; damaged or destroyed vehicles, and flooded or snow-filled streets. If enough ground ambulances are unavailable due to increased casualties or the tactical situation does not allow, units will need to execute casualty evacuation (CASEVAC) operations.

10-74. In UO, CASEVAC may entail evacuation by litter and the expedient use of a vehicle of opportunity or other military vehicle not dedicated for medical evacuation. Based on the threat situation, CASEVAC may also be limited to armored vehicles, conducted only during limited visibility, or both. For evacuation by litter, units may require more litter bearers to move the injured to a point where they can be further evacuated by ground or air ambulance. Lengthy evacuation routes require more litter bearers, as multiple litter relay teams will be necessary to conserve energy and expedite evacuation. However, unless augmented or relieved of this responsibility by another unit, these litter bearers will come from the casualties' own unit thereby diminishing the unit's strength levels necessary to accomplish its primary mission. Instead, tactical commanders may need to wait until a lull in or a cessation of the battle occurs. Depending on projected casualty rates, higher-level commanders may augment units with additional personnel to perform evacuation or may assign maneuver units this mission. Evacuation personnel will require specific training in urban evacuation techniques (moving casualties from subsurface and supersurface levels to and along the surface level).

10-75. Overall, urban commanders at all echelons must develop detailed casualty evacuation plans that include the establishment of treatment facilities as far forward as the situation allows. Engineers are critical to clear routes for medical resupply and evacuation. Based on the threat situation, additional armored vehicles may need to accompany ambulances in a multi-vehicle convoy not only to and from the patient pickup point but also through to a higher level of care. Army forces may also need to plan for and acquire unique capabilities, equipment, and skills now typically seen in civilian urban search-and-rescue teams to clear debris and search for urban casualties. For example, specially trained dogs may play a vital role in locating victims among urban rubble.

Treatment

10-76. First aid training will have increased significance in UO. The compartmented nature of UO, transportation restrictions, communications difficulties, and the low density of trauma specialists (combat medics) may limit the urban casualty's initial treatment administered by nonmedical personnel or to self-treatment measures. To mitigate this risk, units identify and train combat lifesavers (CLS) to perform in the absence of trauma specialists. Since the likelihood of Army forces performing UO continues to increase, commanders strive to meet or exceed Army standards for the number of fully trained CLS required for their specific unit. CLS (and trauma specialist) training must be realistic, hands-on, and conducted at appropriate intervals to ensure CLS can perform lifesaving tasks to include starting intravenous lines. This increase in self, buddy, and CLS care, as well as longer delays in evacuation, may also increase requirements for additional first aid, CLS, and trauma specialist-carried supplies. (Ultimately, commanders may have to consider weighting the main effort with greater overall FHP support for units conducting UO than for units conducting operations in other environments.)

10-77. The increased potential for delayed evacuation during UO mandates that Army trauma specialists be skilled in prolonged casualty care. During the 3-4 October 1993 battle in Somalia, for example, seven trauma specialists managed 39 casualties for more than 14 hours before they could be evacuated. Evacuation delays significantly increase the potential for infection; therefore, trauma specialists should also be trained to administer antibiotics on the battlefield.

10-78. FHP personnel must be capable of recognizing and treating injuries due to IEDs and incendiary or thermobaric weapons (fuel-air explosives)—favored, urban-oriented threat weapons (see Chapter 3). Thermobaric weapons explode; create a cloud of volatile gases, liquids, or powders; and then ignite, creating an immense fireball consuming oxygen and creating enormous overpressure. When employed in an urban structure, the blast wave or overpressure of any explosive is greatly amplified. Injuries resulting from these weapons are massive burns, broken or crushed bones, concussions and traumatic brain injuries, missile injuries, and internal injuries. To counter this threat, medical personnel must be aware of the capabilities of these blast weapons, wounding patterns, and treatment methods.

10-79. The increased use of body armor during UO will help prevent penetrating chest and abdomen wounds. As such, FHP personnel should expect more groin, pelvis, head, and extremity injuries. Furthermore, when fighting soldiers that are known to use body armor, a threat (particularly snipers) can be expected to target the head and face more often than other anatomic areas resulting in more head injuries. Medical personnel must also anticipate and be prepared to treat an increase in hearing loss due to the firing of weapons (particularly recoilless) in enclosed spaces with little ventilation. Medical personnel (and engineers) should also be trained on how to properly remove rubble that has fallen on a Soldier or civilian so as not to increase the extent of his injuries. Lastly, FHP personnel will need to research and develop unique treatments for injuries due to an ever-increasing arsenal of nonlethal weapons.

Explosive Ordnance Disposal Support

10-80. Explosive ordnance disposal (EOD) support provides the capability to neutralize domestic or foreign conventional explosive hazards which include unexploded explosive ordnance (UXO), booby traps, IEDs, captured enemy ammunition (CEA), and bulk explosives. Such material and devices threaten military operations and military and civilian facilities, equipment, and personnel. Explosive hazards create a much greater risk during UO than operations in any other environment. Confined spaces, hard surfaces, and more personnel (both Soldiers and civilians) in the vicinity may magnify the detonating effects of explosives. Dense terrain makes all types of explosive hazards more difficult to locate. Moreover, terrorists select urban settings for their booby traps and IEDs to potentially kill and destroy more, thus gaining greater visibility for their message.

10-81. 10-81. EOD units detect, mark, identify, render safe, and dispose of explosive hazards. Importantly, EOD specialists will work closely with unit intelligence personnel to conduct explosive forensics—technical intelligence collection and exploitation—to help identify the makers of the devices as well as their ever-evolving methods of construction, placement, concealment, and detonation. Chemical reconnaissance with their array of chemical sensors will also work closely with EOD units to reduce the uncertainty of the chemical threat. This integrated forensic approach is a proactive method to take the advantage away from a terrorist or insurgent threat that seeks to employ these devices as a primary tactic.

10-82. Urban operations will increase EOD specialists' role as advisors and instructors on explosive hazards, protection measures, and disposal techniques. They will advise and train other Army forces, other services, multinational partners, and civilian authorities. EOD specialists will also work closely with public affairs and psychological operations personnel to increase public awareness of the horrific effects of these munitions and devices. This education process is designed not only to teach the urban population to identify and avoid explosive hazards, especially UXO, booby traps, and IEDs, but to gain civilian assistance in reporting their locations. (In this regard, rewards for information leading to the discovery of unexploded ordnance may be beneficial. However, the largest reward should normally be reserved for information leading to the arrest or capture of bomb makers.)

10-83. In an urban environment where there are large amounts of UXO, IEDs, and CEA, the sheer amount of explosives may overwhelm the capabilities of available EOD personnel. Because of their training in demolitions and explosives, commanders will often use combat engineers to augment the destruction and clearance effort. Engineers are normally given responsibility for UXO clearance and disposal while EOD teams use their technical skills for IEDs. Commanders may also consider embedding EOD expertise within engineer units to provide the knowledge necessary to determine the correct amount and type of explosive necessary to detonate UXO and to understand potential risks and effects of detonation of various munitions. For example, EOD specialists would better understand the second-order effects and risk reduction measures necessary to detonate a large missile containing numerous submunitions.

10-84. Unfortunately, there is no single device or technique that can eliminate every explosive hazard encountered. Clearance operations are dangerous and time-consuming. Based on METT-TC, commanders may consider the following options when dealing with these explosive hazards:

- Report, mark, and bypass the hazard.
- Employ tactical breaching procedures.
- Self-extract to alternate routes or positions.
- Restrict routes or positions from further use.
- Clear specific areas or positions, or conduct an extensive route clearance.
- Lastly, accept risk of casualties and continue with the assigned mission.

10-85. Effective tactics, techniques, and procedures (TTP) for addressing explosive hazards must be captured and shared with friendly forces and coalition partners. Commanders recognize, however, that adaptive threats will continually modify their own techniques to counter the unit's response. See FM 3-34.119, FM 20-32, FM 3-100.38, FM 4-30.51, and FM 4-30.16.

Human Resources Support

10-86. Human resources support (HRS) encompasses the following functions: manning the force, personnel support, and personnel services. These activities include personnel readiness; replacement operations; personnel accounting; personnel information management; casualty operations; essential personnel services; postal operations; morale, welfare, and recreation; and band support provided to Soldiers, their families, Department of Army civilians, and contractors.

10-87. Successful UO require HRS functions. Clearly, HRS focuses on caring for the needs of people: Soldiers and civilians who support them. Since a critical component of the urban environment is the population, these activities, when required, may also support them. For example, HRS elements, in conjunction with mortuary affairs, military police, and CA, may provide administrative support to assist in the accountability of displaced persons and civilian internees and the establishment of other populace and resources control measures. This support may include the provision of identification and registration cards, travel permits, and other documents, and administrative support to casualty operations. Human resources managers may also assist with community activities and coordination with the Red Cross. If necessary, a postal platoon may assist urban officials in training and reestablishing civilian postal operations. Overall, this type of assistance and support helps to care for the immediate needs of the urban population and, ultimately, speed transition back to legitimate civilian responsibility and control.

10-88. In offensive and defensive UO, HRS personnel will need to account for more casualties and more frequent reconstitution. In all UO, success relies on training individual replacements. Urban combat requires Soldiers skilled in specific urban combat TTP. They understand the societal aspects of the urban population and have training in crowd control to avoid escalating potentially volatile situations. Training these replacements prior to their actual deployment or while in the replacement system frees urban commanders from having to do so. It also helps reduce Soldier (and supporting civilian) isolation, anxiety, and fear associated with urban areas and operations.

FINANCIAL MANAGEMENT SUPPORT

10-89. Money and its management are often critical to the success of UO. As such, financial management support makes resources available when and where they are needed and assist the urban commander in maintaining fiscal responsibilities. Financial management operations are necessary to conduct the contracting necessary to restore and rebuild infrastructure and provide real-time information, accounting, and services. Financial management operations ensure that urban operational policies and procedures adhere to laws and regulations, develop command resource requirements, and leverage appropriate fund sources to meet them.

10-90. UO will likely include other U.S. military services, governmental agencies, and contractors; multinational forces; NGOs; and various elements of the host nation. So many actors complicate financial accounting, resource management, and cost capturing. Yet, these activities are vital to accomplishing UO and maintaining legal requirements. Army forces will receive support from and provide support to these participants. Financial managers provide crucial expertise to the urban commander on the obligation authority. They also furnish the documentation necessary to obtain and pay for local goods and services using contract or commercial vendor services. However, commanders (with financial management assistance) must assess the economic impact of UO on the economy. An appropriate analysis includes how well the urban economics and commerce infrastructure can support the deployed force and how an influx of U.S. currency may affect the overall economy either positively or negatively.

Labor, Reward, and Solatia Programs

10-91. Financial management elements can also provide joint pay support and non-U.S. pay support for host-nation employees and day laborers supporting UO, as well as for civilian internees. These units can support reward programs and solatia (financial compensation). Commanders can use reward programs to purchase weapons, radios, information, and other items from the urban inhabitants. However, they must use caution when paying for critical human intelligence. First, they must ensure that multiple collection agencies in the command avoid paying the same source for the same information and interpreting these repeated inputs as validation that information is accurate and reliable. Second, they must establish a price or pay scale so subordinate units (particularly in multinational UO) equally compensate their sources for information. Commanders may also need to make solatia to alleviate grief, suffering, and anxiety resulting from injuries and property or personal loss resulting from some aspect of UO. At other times, commanders make solatia to meet cultural expectations of the urban population. These payments are nominal in amount and made without an admission of liability by the Army. Significantly, they may help maintain the legitimacy of the operation and preserve the support of the urban populace.

Financial Support to Civil-Military Projects

> *...[I]n an endeavor like that in Iraq, money is ammunition. In fact, depending on the situation, money can be more important than real ammunition—and that has often been the case in Iraq since early April 2003 when Saddam's regime collapsed and the focus rapidly shifted to reconstruction, economic revival, and restoration of basic services. Once money is available, the challenge is to spend it effectively and quickly to rapidly achieve measurable results. This leads to a related observation that the money needs to*

be provided as soon as possible to the organizations that have the capability and capacity to spend it in such a manner.

Lieutenant General David H. Petraeus
Commander, 101st Airborne Division (Air Assault) and, later,
Commander, Multi-National Security Transition Command-Iraq
during OPERATION IRAQI FREEDOM

10-92. Other financial resources separate from operational funds may be appropriated to support civil-military operations (CMO) projects during stability operations. (During OPERATION IRAQI FREEDOM, these funds were referred to as commander's emergency response program [CERP] funds.) These funds are typically disbursed as bulk currency to tactical units, where the funds are allocated against lower-cost, high-impact projects that support the immediate needs of civil authorities and their citizens. As such, commanders must be prepared to receive, account, secure, and properly disburse these resources in accordance with established procedures. Commanders have an inherent responsibility for ensuring the proper use and disbursement of these funds and must execute appropriate judgment, fiscal responsibility, and supervisory oversight of any resources appropriated in support of CMO. While commanders establish appropriate guidelines to accomplish this fiscal responsibility, they must provide their subordinate commanders—who have the need, capacity, and capability to put these funds to use—reasonable flexibility in how they spend at least a portion of the money, so that they can address emerging needs within their AO.

Contracting Expertise

10-93. Due to the nature of UO, tactical-level units will be likely required to plan and initiate contracts not only for reconstruction projects but to supplement their own sustainment needs. Therefore, higher-level commands must be prepared to assist tactical-level units with contracting expertise. This expertise is required to overcome difficulties in delineating the scope of work, defining construction standards, and establishing viable delivery dates. For small projects and purchases, a standard contractual format may be sufficient. However, for large projects, commanders must provide their subordinates with knowledgeable contracting personnel, who can write contracts with clearly-understood specifications, create purchase orders that specify quality and delivery times, and help monitor work for quality control and assurance. Payment guidelines and standards among all services and members of a coalition are essential to eliminate competition and potential overpayment. Translators will be also essential to accurately translate requirements and standards in addition to trusted local agents to assist in selecting quality engineering and construction firms and workers. Overall, subordinate commanders need adequate contracting support and expertise to meet civilian needs as well as their expectations for timeliness and quality. Without it, commanders may risk losing critical opportunities to positively influence civilian perceptions. (See also Legal Support to Operations below.)

RELIGIOUS SUPPORT

10-94. Religious support entails providing and performing religious support operations for commanders to protect the free exercise of religion for Soldiers, family members, and authorized civilians. This includes personal delivery of rites, sacraments, ordinances, pastoral and spiritual care, and religious education. Such support also consists of:

- Advising the commander on matters of religion, morals, and morale as affected by religion.
- The impact of local religions on threat courses of action, the urban society, and, subsequently, the mission.
- The ethical impact of command decisions.

Moral and Ethical Climate

10-95. Chaplains can help commanders sense the moral and ethical climate in Army units and understand potential moral dilemmas associated with planned UO. The urban environment affects Soldiers' mental health and can increase combat stress casualties, especially misconduct stress behaviors. Chaplains are an important part of combat stress support. The chaplain's presence and faith can sustain Soldiers throughout periods of great trauma. Religious support contributes to the total well being of Soldiers and aids in their return to combat readiness. Chaplains help bolster Soldiers' moral and ethical behaviors through spiritual fitness training so that Soldiers may better cope with ambiguous moral and ethical situations. Chaplains can help identify ethical concerns before they become critical command problems. To this end, their observations help develop and modify ROE, which—due to ambiguity or overly restraining rules—may be part of the problem.

Influence of Local Religions

10-96. Chaplains advise commanders on matters of religion as it affects the Soldiers within their units. They also explain the influences of local religions on threat courses of action and the urban populace and, subsequently, their potential effects on Army forces and UO. Religion is a crucial aspect of assessing the societal component of the urban environment. Understanding the major tenets and concepts of the religions and the impact of faith on civilians' lives may help commanders understand what motivates the threat and the urban populace. (Historically, when religion was a critical factor in armed conflict, the level of lethality has been significantly higher.) To understand threat strategy and tactics, commanders understand their ideology, which, in many cases, involves strong religious beliefs. This understanding also helps commanders to appreciate the inhabitants' attitudes toward other races, religions, and cultures and to identify unacceptable kinds of social interaction (particularly between Soldiers and civilians). Failure to recognize and respect religious beliefs can rapidly erode the perceived legitimacy of the mission and increase the length of the operation. In addition to its influence on threat courses of action, a thorough analysis of the urban environment also includes the degree of influence religion and religious leaders have on the area's government and economy.

Legal Support to Operations

10-97. Legal support provides operational law support in all legal disciplines (including military justice, international law, civil law—composed of contract, fiscal, and environmental law, claims, and legal assistance). This support assists in command and control, sustainment, and HRS of UO. Legal considerations are important in any operation; they take on added significance during UO. They form the foundation for establishing ROE and are critical in the targeting process. (For example, with the increased likelihood that the threat will use the media in an attempt to categorize Army attacks as unlawful, legal support will be critical in determining the protected status of targets.) Adequate legal instruction and training may help subordinate commanders and their Soldiers better understand some aspects ROE and, with this understanding, help them make the rapid decisions often required in a complex urban environment.

10-98. Legal considerations also affect how units acquire goods and services from urban areas and provide support to other agencies and organizations operating in an area. The environment's complex nature requires commanders and their staffs to review and closely consider applicable legal constraints when developing and executing friendly courses of action. Most urban areas have a highly developed legal system nested within county, provincial, state, and national systems. The SJA support to commanders addresses this urban system in conjunction with other legal systems and their potential to affect, positively or negatively, the success of UO.

10-99. International, host-nation, and U.S. law and other regulatory guidelines may vary in their applicability by time and place; actions permissible in one situation may be prohibited in another. These exceptions and complexities increase requirements for SJA, often working with CA personnel, to identify and resolve technical legal issues. Therefore, the SJA must actively advise and participate in all aspects of UO from predeployment training and initial planning through transition and redeployment. FM 27-100 contains detailed legal guidelines affecting UO.

International and Host-Nation Law

10-100. International law consists primarily of agreements, treaties, and customary law to include the law known as the law of war (see FM 27-10). The law of war consists of four general principles applicable when conducting any operation but requiring particular attention during UO. Figure 10-4 lists and describes the four principles: military necessity, discrimination (or distinction), unnecessary suffering (or humanity), and proportionality.

10-101. International law may affect urban operational issues, such as the right of entry, base operations, use of urban infrastructure, and overflight and landing rights. Status-of-forces agreements (SOFAs) exist or can be negotiated to resolve legal issues, such as the status of Soldiers (and, as necessary, contractors) operating in foreign areas to include criminal and civil jurisdictions, taxation concerns, and claims for damages and injuries. Unless a SOFA or other convention exists, Soldiers operating in foreign urban areas have the same legal status as tourists; they are subject to the laws and judicial process of the host nation. (During armed conflict, however, Soldiers always have the rights afforded to them by the Geneva Conventions). Commanders are responsible for understanding the international and host-nation agreements and laws that influence foreign UO. If local law hinders the operation, commanders may be able to inform the local U.S. diplomatic mission and request that it negotiate a solution.

10-102. Commanders may encounter civilian resistance groups whose actions may range from providing enemies with sustainment support to actively fighting Army and coalition forces. Members of such resistance groups must be dealt with in accordance with applicable provisions of the law of war. Commanders should seek legal guidance concerning detention and disposition of persons participating in various acts harmful to friendly forces and detrimental to the mission.

U.S. Law

10-103. Commanders conducting UO must also comply with U.S. law whether it is in the form of a statute, executive order, regulation, or other directive from a federal branch or agency. U.S. law influences UO by governing the acquisition of supplies and services for Army forces, regulating the assistance that can be rendered to foreign nations, and controlling intelligence activities. The Posse Comitatus Act, for example, makes it a crime for Army forces to enforce civil law. Similarly, portions of the Foreign Assistance Act prohibit Soldiers from performing law enforcement activities in foreign urban areas. However, circumstances—expressly authorized by the Constitution, acts of Congress, and other exceptions to these statutes—exist that allow the Army to support civilian law enforcement. Although not nearly all-inclusive, the above demonstrates how U.S. law further complicates urban operations, particularly stability operations, and increases the need for proactive SJA advice and counsel in all facets of UO.

Military Necessity: The principle that justifies the use of measures not forbidden by international or domestic law necessary to rapidly achieve military objectives.

Discrimination: The principle of distinguishing between combatants (who may be attacked) and noncombatants (who may not be attacked).

Unnecessary Suffering: The principle that prohibits the use of weapons, projectiles, or other materials in a manner calculated to cause superfluous injury and unnecessary suffering.

Proportionality: The principle that injury to persons and damage to property incidental to military action, in the circumstances ruling at the time, must not be excessive in relation to the concrete and direct military advantage anticipated.

Figure 10-4. General principles of the law of war

Legal Aspects of Nonlethal Force

10-104. Nonlethal capabilities can augment the means of deadly force and extend urban firepower options. They enhance the commander's ability to apply force in proportion to the threat and to allow discrimination in its use. The range of nonlethal capabilities includes offensive information operations, smoke and obscurants, irritants (such as chemical riot control agents), nonpenetrating projectiles, high-pressure water devices, and directed-energy weapons. These continually expanding capabilities give commanders more options to confront situations that do not warrant deadly force but require Soldiers to employ overwhelming decisive power. However, nonlethal capabilities are subject to the same legal constraints as lethal force (in fact, some nonlethal capabilities can cause serious injury and death, particularly if not employed properly) and undergo the same legal review. Like lethal force, nonlethal capabilities show military necessity, distinguish between combatants and noncombatants, distinguish between military objectives and protected property, are used proportionally, and do not result in unnecessary suffering.

10-105. Of special note, commanders cannot employ chemical herbicides or riot control agents as a method of warfare without prior Presidential approval. However, the Secretary of Defense and regional combatant commanders have authority to use herbicides and riot control agents in limited circumstances during armed conflict and in peacetime operations. Considering the complexity of this issue, commanders of major operations must streamline the approval process so that subordinate commanders do not discount use of these agents simply because of the high-level approval authority. Subordinate commanders must, however, be able to anticipate and clearly justify their potential use in UO. Also, any approval does not mean that these agents must be employed.

Civilians Accompanying the Force

10-106. LOGCAP personnel and other civilian support contractors provide a variety of sustainment functions for the Army. These functions may range from providing unskilled labor, transportation support, and health care to technical support of sophisticated equipment and weapons systems. Commanders must ensure that civilians providing support in their AO are not placed in positions of jeopardy, but if they are, these civilians must understand the risks that they assume when they engage in activities that might be misconstrued as direct or active participation in hostilities.

10-107. This last caution indicates an important legal consideration for commanders. They must try to ensure that threat forces can distinguish civilian support contractors from combatants and thereby remove any justification for them to be targeted or attacked. (As Army forces seek to work closely with the members of governmental and nongovernmental agencies and organizations in urban stability operations, commanders may need to take steps necessary to mitigate the risk of attack caused by their involvement with Army forces.) In the event that threat forces capture civilian support contractors, this same concern helps to ensure that, as lawful civilians accompanying the force, they will be afforded prisoner of war status. As the role of contractor support to Army operations continues to expand, commanders must stay abreast of changes in the law (to include evolving interpretative distinctions) as it affects the use of contractors on the urban battlefield. Other important considerations for commanders with civilian contractors supporting their force are—

- **Minimize Exposure to Harm**. Civilian contractors should be assigned duties at locations that minimize their exposure to harm. Commanders will need to periodically review these locations since they are likely to change as the urban operation evolves and transitions.
- **Prohibit the Performance of Combat Functions**. Civilian support personnel should not be working in situations that involve combat operations where they might be conceived as combatants. As an example, they should not perform force protection for Army forces such as fortification construction and checkpoint security.
- **Prohibit the Use of Arms**. Civilians should not be armed, unless approved by the combatant or service component commander on a limited, by-exception basis for personal defense.
- **Provide Identification Cards**. Civilians should be provided, and carry on their person, a Geneva Convention identification card identifying them as civilians authorized to accompany military forces and entitling them to be treated, if captured, as prisoners of war.
- **Restrict the Wear of Military Uniforms**. Civilians should not normally wear distinctive U.S. military uniforms unless the combatant or service component commander authorizes them to do so. Regardless of their clothing, however, civilians will wear a symbol that establishes their civilian status.

General Engineering Support

10-108. General engineering support will be essential during UO. This support helps assess, construct, maintain, and restore essential LOCs and urban facilities to sustain Army forces, the urban population, or both. Using civilian resources and investing Army general engineering resources requires careful consideration by commanders and staff planners. Since all elements of the urban infrastructure interconnect, general engineering support touches each category to some degree. Figure 10-5 illustrates how urban-specific, general engineering tasks align primarily with the transportation and distribution, energy, and administration and human services components of the urban infrastructure. These engineering tasks are significant and readily apply to UO. The last two, providing fire fighting support and waste management, have not been previously addressed and require more specific consideration.

Construct, Maintain, or Restore

- Roads and Highways
- Over-the-Shore Facilities
- Ports
- Railroad Facilities
- Airports and Heliports
- Fixed Bridges
- Electric Power Facilities
- Petroleum Pipelines and Storage Facilities
- Water Facilities

and Provide

- Mobile Electric Power
- Construction Support
- Fire Fighting Support
- Waste Management

Figure 10-5. General engineering support tasks

The Use or Investment of Resources

10-109. During urban offensive and defensive operations, Army engineer units accomplish tasks to sustain or improve movement and mobility, protection, and sustainment of U.S. and allied forces. These units should maximize the existing urban facilities, host-nation support, civilian contractors, and joint engineer assets. However, commanders should consider how using urban facilities to support military forces may negatively affect the population. On the other hand, construction and repair may benefit both Army units and the urban inhabitants. Restoring the urban transportation network not only improves military LOCs, but may also allow needed commerce to resume. Repairing urban airfields or ports increases throughput capabilities for military supplies, facilitates medical evacuation operations to the support base, accelerates needed relief efforts, and allows international commerce to proceed. Commanders may first invest resources and conduct general engineering tasks to restore facilities for civilian use. Such actions stem future drains on operational resources or facilitate later transition of control back to civilian authorities. For example, repairing police stations, detention facilities, and marksmanship ranges may help urban governments reestablish law and order after completing urban offensive or defensive operations. During stability or civil support operations, the focus of general engineering will often be to support and assist the urban population rather than Army forces.

Fire Fighting Support

10-110. Fire protection and prevention, as well as fire fighting, takes on added significance during UO, particularly offensive and defensive operations. Most ordnance produces heat and flame. This, coupled with an abundance of combustible material (buildings, furniture, gasoline, oil, and propane), poses a serious risk to Soldiers, civilians, and the urban operation. Large shantytowns can exacerbate this problem. In highly combustible areas, commanders may even need to limit or preclude the use of small-arms tracer ammunition. Fire threats to urban areas can be categorized as—

- **Isolated Fires**. These are restricted to a single structure or a specific area within a structure.
- **Area Fires**. These fires consume two or more structures and may extend to encompass an entire block. Generally, streets will serve as firebreaks and help contain the fire within a single block.
- **Firestorms**. These are the most violent and dangerous fires capable of rapidly consuming large areas by creating windstorms and intense heat. Firestorms are often inextinguishable until they have consumed all available combustible materials.

- **Explosive Hazards**. These are present in areas containing fuels and chemicals as well as military explosive hazards.

10-111. When analyzing the administration and human services component of the infrastructure, commanders must determine the adequacy of existing civilian fire fighting support. A deteriorated or nonexistent infrastructure that cannot support the urban area will likely fail to handle the increased risk due to military operations. Commanders may need to provide fire fighting teams to support their own forces and civilians.

10-112. A military force task organized with multiple fire fighting teams (even with maximum use of available civilian fire-fighting assets) will only be able to fight some fires in the AO. Water distribution systems damaged during operations, chemical and other TIM, and hostile activities will further complicate and limit fire fighting capabilities. Commanders must develop priorities for equipment, facility, and infrastructure protection. All Soldiers need training in fire prevention and initial or immediate response fire fighting. Such training includes planning covered and concealed movement, withdrawal, and evacuation routes. Soldiers should be trained to identify and remove ignition and fuel sources and provided additional fire fighting material such as extinguishers, sand, and blankets (see FM 5-415 and the U.S. Department of Transportation's current version of the Emergency Response Guidebook).

Waste Management

10-113. Management of all forms of waste, particularly human, putrescible (such as food), and medical, may become a critical planning consideration for Army forces. This particularly applies if the urban waste management infrastructure was previously inadequate or damaged during UO or natural disasters, the Army force is operating in the urban area for an extended period of time, and a significant number of the urban population remains. Failure to adequately consider this aspect, possibly coupled with the decaying remains of humans and animals and an inadequate or tainted water supply (such as may be caused by flooding), may create unacceptable sanitary and hygiene conditions and subsequently increase DNBIs as well as civilian casualties.

Appendix A

Siege of Beirut: An Illustration of the Fundamentals of Urban Operations

The [Israeli Defense Forces] IDF had neither the strategy nor the experience nor the configuration of forces to fight and sustain a house-to-house campaign in Beirut.

Richard A. Gabriel
Operation Peace for Galilee: The Israeli-PLO War in Lebanon

OVERALL STRATEGIC SITUATION

A-1. In 1982, Israel launched OPERATION PEACE FOR GALILEE designed to destroy the Palestine Liberation Organization (PLO) presence in southern Lebanon. On 1 June, the IDF launched a massive assault across the border into southern Lebanon. The Israeli attack focused on the PLO, but the operations quickly involved major ground and air combat between Israel and Syrian forces.

A-2. In the first few weeks, Israeli forces quickly pushed back both the Syrians and the PLO. However, except for some PLO forces isolated in bypassed urban areas, such as Tyre and Sidon, most of the PLO fell back into Beirut (see Figure A-1). By 30 June, Israeli forces had reached the outskirts of southern Beirut, occupied East Beirut, isolated the city from Syria and the rest of Lebanon, and blockaded the sea approaches to the city. Even so, with most of the PLO intact inside and with significant military and political capability, the Israelis had yet to achieve the objective of OPERATION PEACE FOR GALILEE. The Israeli command had to make a decision. It had three choices: permit the PLO to operate in Beirut; execute a potentially costly assault of the PLO in the city; or lay siege to the city and use the siege to successfully achieve the objective. The Israelis opted for the latter.

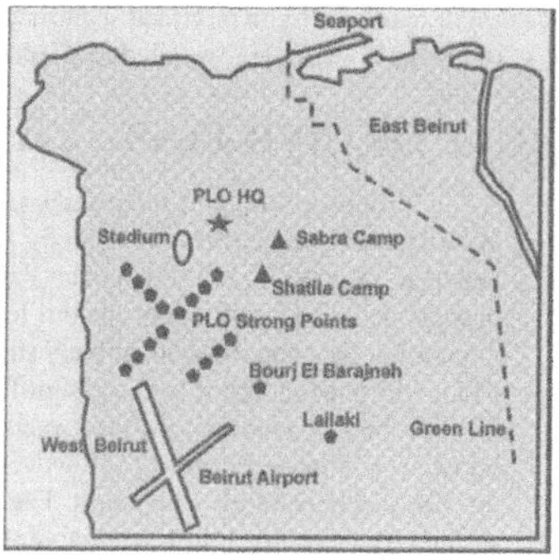

Figure A-1. The city of Beirut

Israeli Military Position

A-3. The Israelis had an excellent position around Beirut. They occupied high ground to the south and west, virtually dominating the entire city. Israeli naval forces controlled the seaward approaches to Beirut. The Israelis' position was also strong defensively, capable of defeating any attempt to break out of or into the city from northern Lebanon or Syria. The Israeli air force had total and complete air superiority. The Israelis controlled the water, fuel, and food sources of West Beirut. Although the PLO forces had stockpiles of food and supplies, the Israelis regulated the food, water, and generating power for the civil population.

A-4. Despite the superior positioning of Israeli forces, the IDF faced significant challenges to include the combat power of the PLO, Syria, and other threats in Beirut. Israeli doctrine and training did not emphasize urban operations. Additionally, Israel was constrained by its desire to limit collateral damage and friendly and noncombatant casualties. Organizationally, the Israeli army was not optimized to fight in urban terrain.

Armor and self-propelled artillery formations dominated the Israeli forces, and most Israeli infantry was mechanized. The Israeli forces had only a few elite formations of traditional dismounted infantry.

PLO MILITARY POSITION

A-5. Despite being surrounded and cut off from support, the PLO position in Beirut offered numerous advantages in addition to the characteristic advantages of urban defense. The PLO had long anticipated an Israeli invasion of southern Lebanon; it had had months to prepare bunkers, obstacles, and the defensive plan of Beirut and other urban areas. Approximately 14,000 Arab combatants in West Beirut readied to withstand the Israeli siege. This was done with the advice of Soviet, Syrian, and east European advisors. The preparation included stockpiling essential supplies in quantities sufficient to withstand a six-month siege. Also, the PLO fighters integrated into the civil populations of the urban areas. Often their families lived with them. The civil population itself was friendly and provided both information and concealment for PLO forces. PLO fighters were experienced in urban combat and knew the urban terrain intimately. PLO forces had been involved in urban fighting against Syrian conventional forces and Christian militias in Beirut several years prior to the Israeli invasion. Finally, the organization of the PLO—centered on small teams of fighters armed with machine-guns and antitank weapons, and trained in insurgent, hit-and-run tactics—was ideally suited to take maximum advantage of the urban environment.

ROLE OF CIVILIANS

A-6. Various ethnic and religious groups make up the civil population of southern Lebanon. However, West Beirut's population was heavily Palestinian and Lebanese. The civil population of West Beirut was between 350,000 and 500,000. The Palestinian population supported the PLO. The Lebanese population may be described as friendly neutral to the Israelis. Although unhappy under Palestinian dominance, this population was unwilling to actively support Israel. The civilian population was a logistic constraint on the PLO, which would have become significant had the siege lasted longer. The civilians in West Beirut were an even larger constraint on the Israelis. The presence of civilians significantly limited the ability of the Israelis to employ firepower. However, the Palestine refugee camps located in West Beirut were both civilian centers and military bases. The Israeli constraints on artillery and other systems against these parts of the city were much less restrictive than in other parts of West Beirut where the population was mostly Lebanese and where fewer key military targets existed.

A-7. The PLO knew of the Israeli aversion to causing civil casualties and purposely located key military centers, troop concentrations, and logistics and weapons systems in and amongst the population—particularly the refugee Palestinian population in the southern part of West Beirut. Tactically, they used the civilians to hide their forces and infiltrate Israeli positions.

A-8. The friendly Palestinian population provided intelligence to the PLO while the friendly Lebanese population provided intelligence for the IDF. Throughout the siege, the IDF maintained a policy of free passage out of Beirut for all civilians. This policy was strictly enforced and permitted no weapons to leave the city. Some estimates are that as many as 100,000 refugees took advantage of this policy.

INFORMATION OPERATIONS

A-9. The siege of Beirut involved using information operations (IO) to influence the media. PLO information operations were aimed at controlling the media and hence the international perception of the operation. This was done by carefully cultivating a select group of pro-PLO media years before hostilities even began. Once hostilities started, only these media sources were permitted to report from the besieged portions of the city, and they were only shown activities that portrayed the IDF negatively. The IDF did not vigorously counter the PLO plan. In fact, the IDF contributed to it by limiting media access to their activities. The PLO information operations had a successful impact. The international community was constantly pressuring the Israeli government to end hostilities. This put pressure on the IDF to conduct operations rapidly and to limit firepower and casualties.

CONDUCT OF THE URBAN OPERATIONS

A-10. The siege of Beirut began 1 July (see Figure A-2). By 4 July, Israeli forces occupied East Beirut, the Green Line separating East and West Beirut, and dominating positions south of the airport. IDF naval forces also controlled the sea west and north of Beirut. On 3 and 4 July, IDF artillery and naval fire began a regular campaign of firing on military targets throughout West Beirut. On 4 July, the IDF cut power and water to the city.

A-11. From 5 to 13 July, the Israeli fires continued to pound PLO targets in West Beirut. The PLO gave one significant response, firing on an Israeli position south of the city and causing several casualties. On 7 July, reacting to international pressure, the IDF returned power and water to West Beirut's civil population. On 11 July, the IDF launched its first attack, probing the southern portion of the airport with an armored task force (see Figure A-3). The PLO repulsed this attack and destroyed several IDF armored vehicles.

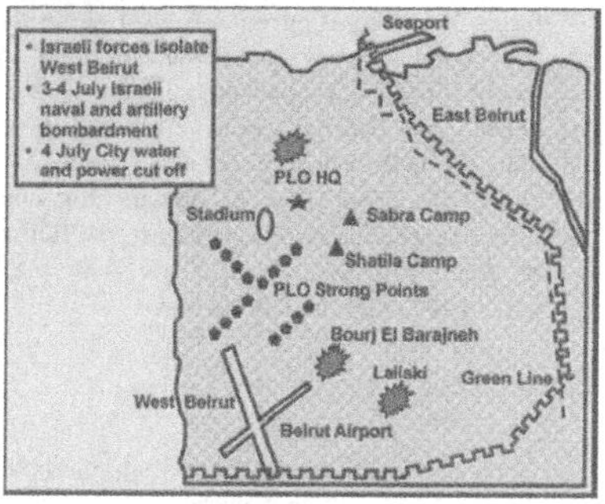

Figure A-2. Initial conduct of the urban operation

A-12. On 13 July, both sides entered into a cease-fire that lasted until 21 July. They began negotiations, mediated by international community, to end the siege. The PLO used this period to continue to fortify Beirut. The Israelis used the time to train their infantry and other arms in urban small unit tactics in Damour, a town the Israeli paratroopers had captured.

A-13. The cease-fire ended on 21 July as PLO forces launched three attacks on IDF rear areas. The Israelis responded with renewed and even more vigorous artillery, naval, and air bombardment of PLO positions in the city. The IDF attacks went on without respite until 30 July. On 28 July, the IDF renewed its ground attack in the south around the airport (see Figure A-4). This time IDF forces methodically advanced and captured a few hundred meters of ground establishing a toehold.

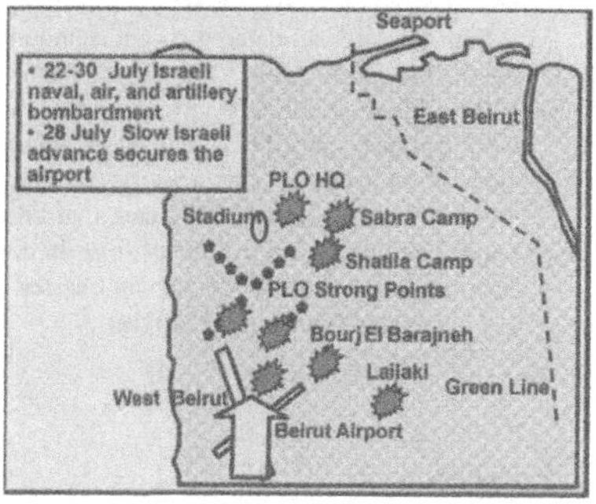

Figure A-3. Israeli probe of PLO defenses

A-14. The Israeli bombardment stopped on 31 July. However, on 1 August the IDF launched its first major ground attack, successfully seizing Beirut airport in the south (see Figure A-5). Israeli armored forces began massing on 2 August along the green line, simultaneously continuing the attack from the south to the outskirts of the Palestinian positions at Ouzai. On 3 August, the Israeli forces continued to reinforce both their southern attack forces and forces along the green line to prepare for continuing offensive operations. On 4 August, the IDF attacked at four different places. This was the much-anticipated major Israeli offensive.

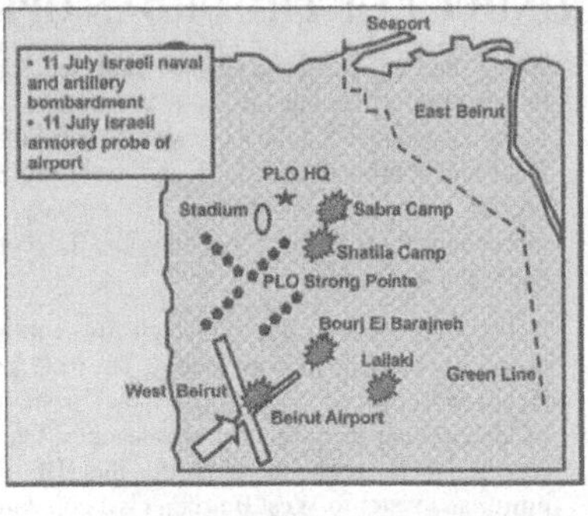

Figure A-4. Initial Israeli attack

A-15. The Israeli attack successfully disrupted the coherence of the PLO defense. The southern attack was the most successful: it pushed PLO forces back to their camps of Sabra and Shatila and threatened to overrun PLO headquarters. Along the green line the IDF attacked across three crossing points. All three attacks made modest gains against stiff resistance. For this day's offensive, the Israelis suffered 19 killed and 84 wounded, the highest single day total of the siege, bringing the total to 318 killed. Following the major attacks on 4 August, Israeli forces paused and, for four days, consolidated their gains and prepared to renew the offensive. Skirmishes and sniping continued, but without significant offensive action. On 9 August, the IDF renewed air and artillery attacks for four days. This activity culminated on 12 August with a massive aerial attack that killed over a hundred and wounded over 400—mostly civilians. A cease-fire started the next day and lasted until the PLO evacuated Beirut on 22 August.

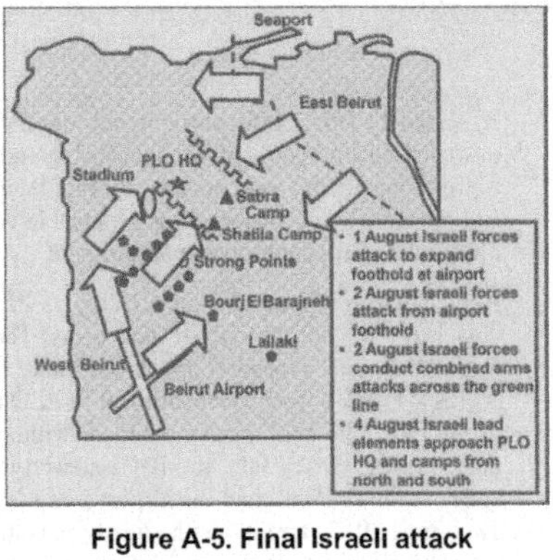

Figure A-5. Final Israeli attack

LESSONS

A-16. The Israeli siege of West Beirut was both a military and a political victory. However, the issue was in doubt until the last week of the siege. Military victory was never in question; the issue in doubt was whether the Israeli government could sustain military operations politically in the face of international and domestic opposition. On the other side, the PLO faced whether they could last militarily until a favorable political end could be negotiated. The answer was that the PLO's military situation became untenable before the Israeli political situation did.

A-17. This favorable military and political outcome stemmed from the careful balance of applying military force with political negotiation. The Israelis also balanced the type of tactics they employed against the domestic aversion to major friendly casualties and international concern with collateral damage.

Perform Aggressive Information Operations

A-18. The PLO devoted considerable resources and much planning on how to use IO to their best advantage. They chose to focus on media information sources as a means of influencing international and domestic opinion.

A-19. The PLO's carefully orchestrated misinformation and control of the media manipulated international sentiment. The major goal of this effort was to grossly exaggerate the claims of civilian casualties, damage, and number of refugees—and this was successfully accomplished. Actual casualties among the civilians were likely half of what the press reported during the battle. The failure of the IDF to present a believable and accurate account of operations to balance PLO efforts put tremendous pressure on the Israeli government to break off the siege. It was the PLO's primary hope for political victory.

A-20. In contrast to the overall weak performance in IO, the IDF excelled in psychological operations. IDF psychological operations attacked the morale of the PLO fighter and the Palestinian population. They were designed to wear down the will of the PLO to fight while convincing the PLO that the IDF would go to any extreme to win. Thus defeat was inevitable. The IDF used passive measures, such as leaflet drops and loudspeaker broadcasts. They used naval bombardment to emphasize the totality of the isolation of Beirut. To maintain high levels of stress, to deny sleep, and to emphasize their combat power, the IDF used constant naval, air, and artillery bombardment. They even employed sonic booms from low-flying aircraft to emphasize the IDF's dominance. These efforts helped to convince the PLO that the only alternative to negotiation on Israeli terms was complete destruction.

Maintain a Close Combat Capability

A-21. The ground combat during the siege of Beirut demonstrated that the lessons of tactical ground combat learned earlier in the twentieth century were still valid. Small combined arms teams built around infantry, but including armor and engineers, were the key to successful tactical combat. Artillery firing in direct fire support of infantry worked effectively as did the Vulcan air defense system. The Israeli tactical plan was sound. The Israelis attacked from multiple directions, segmented West Beirut into pieces, and then destroyed each individually. The plan's success strongly influenced the PLO willingness to negotiate. Tactical patience based on steady though slow progress toward decisive points limited both friendly and noncombatant casualties. In this case, the decisive points were PLO camps, strong points, and the PLO headquarters.

A-22. The Israeli willingness to execute close combat demonstrated throughout the siege, but especially in the attacks of 4 August, was decisive. Decisive ground combat was used sparingly, was successful and aimed at decisive points, and was timed carefully to impact on achieving the political objectives in negotiations. The PLO had hoped that their elaborate defensive preparations would have made Israeli assaults so costly as to convince the Israelis not to attack. That the Israelis could successfully attack the urban area convinced the PLO leadership that destruction of their forces was inevitable. For this reason they negotiated a cease-fire and a withdrawal on Israeli terms.

Avoid the Attrition Approach

A-23. The Israelis carefully focused their attacks on objects that were decisive and would have the greatest impact on the PLO: the known PLO headquarters and refugee centers. Other areas of West Beirut were essentially ignored. For example, the significant Syrian forces in West Beirut were not the focus of Israeli attention even though they had significant combat power. Selectively ignoring portions of the urban area allowed the Israelis to focus their combat power on the PLO and limit both friendly casualties and collateral damage.

Control the Essential and Preserve Critical Infrastructure

A-24. The Israeli siege assured Israeli control of the essential infrastructure of Beirut. The initial Israeli actions secured East Beirut and the city's water, power, and food supplies. The Israelis also dominated Beirut's international airport, closed all the sea access, and controlled all routes into and out of the city. They controlled and preserved all that was critical to operating the city and this put them in a commanding position when negotiating with the PLO.

Minimize Collateral Damage

A-25. The Israeli army took extraordinary steps to limit collateral damage, preserve critical infrastructure, and put in place stringent rules of engagement (ROE). They avoided randomly using grenades in house clearing, limited the use of massed artillery fires, and maximized the use of precision weapons. With this effort, the Israelis extensively used Maverick missiles because of their precise laser guidance and small warheads.

A-26. The strict ROE, however, conflicted with operational guidance that mandated that Israeli commanders minimize their own casualties and adhere to a rapid timetable. The nature of the environment made fighting slow. The concern for civilian casualties and damage to infrastructure declined as IDF casualties rose. They began to bring more field artillery to bear on Palestinian strong points and increasingly employed close air support. This tension underscores the delicate balance that Army commanders will face between minimizing collateral damage and protecting infrastructure while accomplishing the military objective with the least expenditure of resources—particularly soldiers. ROE is but one tool among many that a commander may employ to adhere to this UO fundamental.

Understand the Human Dimension

A-27. The Israelis had a noteworthy (although imperfect and at times flawed) ability to understand the human dimension during their operations against the PLO in Beirut. This was the result of two circumstances. First, the PLO was a threat with which the Israeli forces were familiar after literally decades of conflict. Second, through a close alliance and cooperation with Lebanese militia, the Israelis understood a great deal regarding the attitudes and disposition of the civil population both within and outside Beirut.

Separate Noncombatants from Combatants

A-28. Separating combatants from noncombatants was a difficult but important aspect of the Beirut operation. The Israelis made every effort to positively identify the military nature of all targets. They also operated a free passage system that permitted the passage of all civilians out of the city through Israeli lines. The need to impose cease-fires and open lanes for civilians to escape the fighting slowed IDF operations considerably. Additionally, Israeli assumptions that civilians in urban combat zones would abandon areas where fighting was taking place were incorrect. In many cases, civilians would try to stay in their homes, leaving only after the battle had begun. In contrast, the PLO tied their military operations closely to the civilian community to make targeting difficult. They also abstained from donning uniforms to make individual targeting difficult.

A-29. Earlier in OPERATION PEACE FOR GALILEE when the IDF attacked PLO forces located in Tyre, Israeli psychological operations convinced 30,000 Lebanese noncombatants to abandon their homes and move to beach locations outside the city. However, the IDF was subsequently unable to provide food, water, clothing, shelter, and sanitation for these displaced civilians. IDF commanders compounded the situation by interfering with the efforts by outside relief agencies to aid the displaced population (for fear that the PLO would somehow benefit). Predictably, many civilians tried to return to the city complicating IDF maneuver and targeting—that which the separation was designed to avoid. IDF commanders learned that, while separation is important, they must also adequately plan and prepare for the subsequent control, health, and welfare of the noncombatants they displace.

Restore Essential Services

A-30. Since essential services were under Israeli command, and had been since the beginning of the siege, the Israelis had the ability to easily restore these resources to West Beirut as soon as they adopted the cease-fire.

Transition Control

A-31. In the rear areas of the Israeli siege positions, the Israeli army immediately handed over civic and police responsibility to civil authorities. This policy of rapid transition to civil control within Israeli lines elevated the requirement for the Israeli army to act as an army of occupation. The Israeli army believed the efficient administration of local government and police and the resulting good will of the population more than compensated for the slightly increased force protection issues and the increased risk of PLO infiltration.

A-32. Upon the cease-fire agreement, Israeli forces withdrew to predetermined positions. International forces under United Nations (UN) control supervised the evacuation of the PLO and Syrian forces from Beirut. These actions were executed according to a meticulous plan developed by the Israeli negotiators and agreed to by the PLO. Israeli forces did not take over and occupy Beirut as a result of the 1982 siege (an occupation did occur later but as a result of changing situations).

SUMMARY

A-33. The Israeli siege of West Beirut demonstrates many of the most demanding challenges of urban combat. Apart from Israel's poor understanding of the strategic influence of the media on operations (a lesson that they learned and subsequently applied to future operations), the IDF's successful siege of Beirut emerged from their clear understanding of national strategic objectives and close coordination of diplomatic efforts with urban military operations. A key part of that synchronization of capabilities was the understanding that the efforts of the IDF would be enhanced if they left an escape option open to the PLO. This option was the PLO's supervised evacuation that occurred after the siege. Although the PLO was not physically destroyed, the evacuation without arms and to different host countries effectively shattered the PLO's military capability. Had Israel insisted on the physical destruction of the PLO in Beirut, the overall operation might have failed because that goal may not have been politically obtainable in view of the costs in casualties, collateral damage, and international opinion.

This page intentionally left blank.

Appendix B

An Urban Focus to the Intelligence Preparation of the Battlefield

Maneuvers that are possible and dispositions that are essential are indelibly written on the ground. Badly off, indeed, is the leader who is unable to read this writing. His lot must inevitably be one of blunder, defeat, and disaster.

Infantry in Battle

The complexity of the urban environment and increased number of variables (and their infinite combinations) increases the difficulty of providing timely, relevant, and effective intelligence support to urban operations (UO). Conducted effectively, however, the intelligence preparation of the battlefield (IPB) allows commanders to develop the situational understanding necessary to visualize, describe, and direct subordinates in successfully accomplishing the mission.

IPB is the systematic process of analyzing the threat and environment in a specific geographic area—the area of operations (AO) and its associated area of interest (see figure B-1). It provides the basis for intelligence support to current and future UO, drives the military decision-making process, and supports targeting and battle damage assessment. The procedure (as well as each of its four steps) is performed continuously throughout the planning, preparation, and execution of an urban operation.

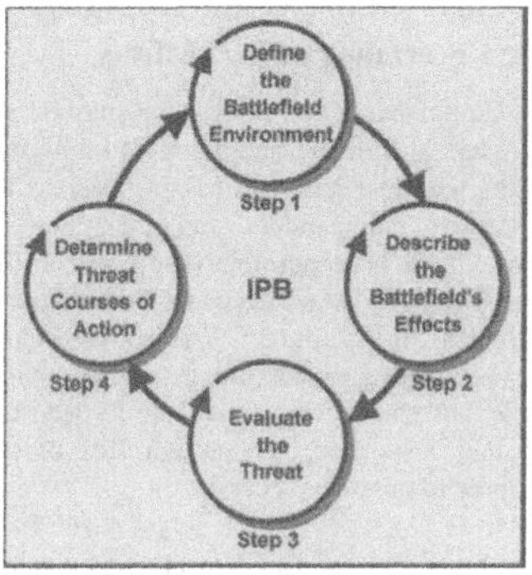

Figure B-1. The steps of IPB

UNAFFECTED PROCESS

B-1. The IPB process is useful at all echelons and remains constant regardless of the operation or environment. However, an urban focus to IPB stresses some aspects not normally emphasized for IPBs conducted for operations elsewhere. The complex urban mosaic is composed of the societal, cultural, or civil dimension of the urban environment; the overlapping and interdependent nature of the urban infrastructure; and the multidimensional terrain. This mosaic challenges the conduct of an urban-focused IPB. There is potential for full spectrum operations to be executed near-simultaneously as part of a single major operation occurring in one urban area with multiple transitions. Such multiplicity stresses the importance of a thorough, non-stop IPB cycle aggressively led by the commander and executed by the entire staff. Overall, the art of applying IPB to UO is in properly employing the steps to the specific operational environment. In UO, this translates to analyzing the significant characteristics of the environment, the role that its populace has in threat evaluation, and understanding how these affect the planning and execution of UO. FM 2-01.3 details how to conduct IPB; FM 34-3 has the processes and procedures for producing all-source intelligence. This appendix supplements the information found there; it does not replace it.

Increased Complexity

B-2. Uncovering intricate relationships takes time, careful analysis, and constant refinement to determine actual effects on friendly and threat courses of action (COAs). These relationships exist among—

- Urban population groups.
- The technical aspects of the infrastructure.
- The historical, cultural, political, or economic significance of the urban area in relation to surrounding urban and rural areas or the nation as a whole.
- The physical effects of the natural and man-made terrain.

B-3. A primary goal of any IPB is to accurately predict the threat's likely COA (step four—which may include political, social, religious, informational, economic, and military actions). Commanders can then develop their own COAs that maximize and apply combat power at decisive points. Understanding the decisive points in the urban operation allows commanders to select objectives that are clearly defined, decisive, and attainable.

Reducing Uncertainty and Its Effects

B-4. Commanders and their staffs may be unfamiliar with the intricacies of the urban environment and more adept at thinking and planning in other environments. Therefore, without detailed situational understanding, commanders may assign missions that their subordinate forces may not be able to achieve. As importantly, commanders and their staffs may miss critical opportunities because they appear overwhelming or impossible (and concede the initiative to the threat). They also may fail to anticipate potential threat COAs afforded by the distinctive urban environment. Commanders may fail to recognize that the least likely threat COA may be the one adopted precisely because it is least likely and, therefore, may be intended to maximize surprise. Misunderstanding the urban environment's effect on potential friendly and threat COAs may rapidly lead to mission failure and the unnecessary loss of Soldiers' lives and other resources. A thorough IPB of the urban environment can greatly reduce uncertainty and contribute to mission success.

Training, Experience, and Functional Area Expertise

B-5. Not all information about the urban environment is relevant to the situation and mission—hence the difficulty and the reason for IPB and intelligence analysis. Although it may appear daunting, institutional education, unit training, and experience at conducting intelligence support to UO will improve the ability to rapidly sort through all the potential information to separate the relevant from merely informative. (This applies to any new or difficult task.) The involvement and functional expertise of the entire staff will allow commanders to quickly identify the important elements of the environment affecting their operations. Fortunately, IPB as part of the entire intelligence process (plan, prepare, collect, process, and produce) is comprehensive enough to manage the seemingly overwhelming amounts of information coming from many sources. Accomplished properly, it allows commanders to recognize opportunities often without complete information.

B-6. As in any operational environment, tension exists between the desire to be methodical and the need to create the tempo necessary to seize, retain, and exploit the initiative necessary for decisive UO. Quickly defining the significant characteristics of the urban environment requiring in-depth evaluation (not only what we need to know but what is possible to know) allows rapid identification of intelligence gaps (what we know versus what we don't know). Such identification leads to information requirements and will drive the intelligence, surveillance, and reconnaissance (ISR) plan (how will we get the information we need). Commanders must carefully consider how to develop focused priority information requirements (PIR) to enable collectors to more easily weed relevant information from the plethora of information. Commanders can then make better decisions and implement them faster than urban threats can react.

Amplified Importance of Civil (Societal) Considerations

B-7. The Army focuses on warfighting. The experiences in urban operations gained at lower echelons often center on the tactics of urban offensive and defensive operations where the influences of terrain and enemy frequently dominate. At higher echelons, the terrain and enemy are still essential considerations, but the societal component of the urban environment is considered more closely and throughout the operational process. Moreover, the human or civil considerations gain importance in civil support or stability operations regardless of the echelon or level of command. In addition to the echelon and the type of operation, a similar relationship exists between the key elements of the urban environment and other situational factors. These factors can include where the operation lies within the range of operations or the level of war and the conventional or unconventional nature of the opposing threat. Figure B-2 graphically represents the varying significance of these elements to urban IPB. Overall, population effects are significant in how they impact the threat, Army forces, and overall accomplishment of strategic and operational goals.

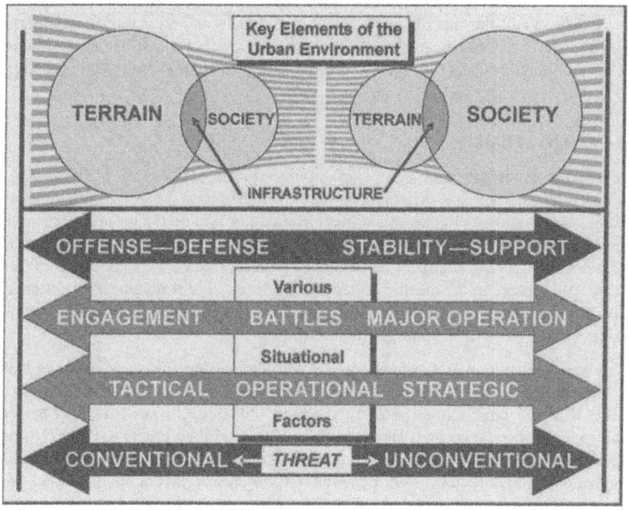

Figure B-2. Relativity of key urban environment elements

B-8. Describing the battlefield's effects—step two in IPB—ascribes meaning to the characteristics analyzed. It helps commanders understand how the environment enhances or degrades friendly and threat forces and capabilities. It also helps commanders understand how the environment supports the population. It also explains how changes in the "normal" urban environment (intentional or unintentional and because of threat or friendly activities) may affect the population. Included in this assessment are matters of perception. At each step of the IPB process, commanders must try to determine the urban society's perceptions of ongoing activities to ensure Army operations are viewed as intended. Throughout this process, commanders, staffs, and analysts cannot allow their biases—cultural, organizational, personal, or cognitive—to markedly influence or alter their assessment (see FM 34-3). This particularly applies when they analyze the societal aspect of the urban environment. With so many potential groups and varied interests in such a limited area, misperception is always a risk.

SIGNIFICANT CHARACTERISTICS

B-9. Urban intelligence analysis must include consideration of the urban environment's distinguishing attributes—man-made terrain, society, and infrastructure—as well as the underlying natural terrain (to include weather) and the threat. Because the urban environment is so complex, it is often useful to break it into categories. Then commanders can understand the intricacies of the environment that may affect their operations and assimilate this information into clear mental images. Commanders can then synthesize these images of the environment with the current status of friendly and threat forces and develop a desired end state. Then they can determine the most decisive sequence of activities that will move their forces from the current state to the end state. Identifying and understanding the environment's characteristics (from a friendly, threat, and noncombatant perspective) allows commanders to establish and maintain situational understanding. Then they can develop appropriate COAs and rules of engagement that will lead to decisive mission accomplishment.

B-10. Figures B-3, B-4, and B-5 are not intended to be all-encompassing lists of urban characteristics. Instead, they provide a starting point or outline useful for conducting an urban-focused IPB and analysis that can be modified to fit the specific operational environment and meet the commander's requirements. Commanders and staffs can compare the categories presented with those in the civil affairs area study and assessment format found in FM 41-10 and the IPB considerations for stability and civil support operations found in ST 2-91.1.

INTERCONNECTED SYSTEMS

B-11. Since the urban environment comprises an interconnected "system of systems," considerations among the key elements of the environment will overlap during an urban intelligence analysis. For example, boundaries, regions, or areas relate to a physical location on the ground. Hence, they have urban terrain implications. These boundaries, regions, or areas often stem from some historical, religious, political, administrative, or social aspect that could also be considered a characteristic of the urban society. Overlaps can also occur in a specific category, such as infrastructure. For instance, dams are a consideration for their potential effects on transportation and distribution (mobility), administration and human services (water supply), and energy (hydroelectric).

B-12. This overlap recognition is a critical concern for commanders and their staffs. In "taking apart" the urban environment and analyzing the pieces, commanders and staffs cannot lose perspective of how each piece interacts with any other and as part of the whole. Otherwise, their vision will be shortsighted, and they will fail to recognize the second- and third-order effects of their proposed COAs; the actual end state differing dramatically from the one envisioned by the commander. The increased density of combatants and noncombatants, infrastructure, and complex terrain means that a given action will likely have unintended consequences—positive or negative. Those consequences will be more widely felt and their impact will spread in less time than in other environments. These unintended results may have important strategic and operational consequences. The multiple ways these dynamic urban elements and characteristics combine make it necessary to approach each urban environment as a unique challenge for intelligence analysis.

URBAN TERRAIN AND WEATHER

Terrain

B-13. Earlier admonitions that civil considerations are more closely considered in UO do not necessarily mean that consideration for urban terrain is de-emphasized. In every urban operation, terrain and its effects on both threat and friendly forces must be assessed and understood. Then commanders can quickly choose and exploit the terrain (and weather conditions) that best supports their missions. Effective terrain analysis thoroughly assesses urban structures as well as the ground on which they stand (see figure B-3 and FM 5-33). An analysis of urban terrain first considers broader urban characteristics and effects and then progresses to a more detailed examination.

Natural Terrain

B-14. The natural terrain features beyond the urban area and beneath urban structures significantly influence unit operations. They dictate where buildings can be constructed, the slope and pattern of streets, and even the broad urban patterns that develop over longer periods of time, thereby influencing a unit's scheme of maneuver. The military aspects of terrain—observation and fields of fire, avenues of approach, key terrain, obstacles, and cover and concealment (OAKOC)—remain critical to the analysis of natural terrain in, under, and around urban areas. Fortunately, commanders and their staffs are normally accustomed to this type of analysis.

Forms and Functions

- Cores
- Industrial areas
  - Toxic industrial material production and storage facilities
  - Standard signs and markings for toxic chemicals
- Outlying high-rise areas
- Residential areas and shantytowns
- Commercial ribbon areas
- Forts and military bases

Broad Urban Patterns

- Types
  - Satellite
  - Network
  - Linear
  - Segment
- Dominant or central hub (if any)
- Area covered (square miles)

Street Patterns

- Basic Types
  - Radial
  - Grid
  - Irregular (planned and unplanned)
- Variations
  - Rayed
  - Radial-Ring
  - Contour-forming
  - Combined
- Widths

Construction and Placement

- Construction
  - Mass or framed
  - Light or heavy clad
  - Material (dirt, wood, stone, brick, cinder block, concrete, steel, and glass)
  - Density and thickness (roofs, floors, and interior and exterior walls)
  - Load bearing walls and columns
  - Height (floors)
  - Doors, windows, fire escapes, and other openings
  - Interior floor plan (including crawl spaces, elevators, and stairs)
- Placement
  - Random
  - Close-orderly block
  - Dispersed
- Ownership

Military Aspects of Terrain: OAKOC

- Observation and fields of fire
  - Smoke (fire), dust (explosions), and flying debris
  - Rubble
  - Engagement ranges (including minimum safe distances and backblast factors) and obliquity/angle (ricochets)
  - Elevation and depression considerations
  - Lasers and reflective concerns
- Avenues of approach (mobility corridors)
  - Airspace
  - Surface
  - Supersurface
  - Subsurface
- Key terrain
  - Landmarks
  - Buildings of significant cultural, social, political, historical, or economic significance
- Obstacles
  - Rubble and vehicles
  - Steep embankments
  - Medians
  - Inadequate bridges and overpasses (destroyed, weight-restricted, or narrow)
  - Tunnels and underpasses (destroyed or narrow)
  - Mines and roadside improvised explosive devices
  - Masking of fires
  - Burning buildings or other fire hazards
  - Rivers and lakes
- Cover and concealment
  - Building protection
  - Weapon penetration (single shot and multiple rounds) considerations
  - Rubble and vehicles

Figure B-3. Significant urban terrain characteristics

Analysis of an Urban Area's Underlying Terrain: Mitrovica, Kosovo – 1999

An urban area's underlying terrain provides many clues into its history, economy, society, and current situation. Mitrovica, Kosovo is an illustrative example. The Ibar River creates a natural line of communications through the middle of the city as well as an obstacle that bisects the urban area. This bisection naturally divides the two resident ethnic groups: Albanians and Serbs. The separation became significant at both the strategic and tactical levels during 1999 deployments to Kosovo. Army forces had to ensure that the Orthodox Church located south of the Ibar was accessible to Serbs residing in the north. North Atlantic Treaty Organization (NATO)

> peacekeepers built a footbridge across the river that allowed reliable, safe passage. The natural feature separating the two groups assisted NATO troops in maintaining stability in the region.

Man-Made Terrain

B-15. Building composition, frontages, placement, forms and functions, size, floor plans, and window and door locations affect maneuver, force positioning, and weapons deployment considerations. Angles, displacement, surface reflection, and antenna locations influence command and control. Structures also influence ISR operations. The increased density and volume created by man-made structures increases how much information commanders and their staffs collect and assess as well as the number of forces required. Building materials and construction will also influence force structures to include weapons and equipment required. Heavily constructed buildings combined with hot and cold extremes may affect target identification for thermal sights. Thick walls, for example, may make combat vehicle identification difficult by distorting hotspots. Additionally, the increased use of heaters and warming fires may clutter thermal sights with numerous hotspots. The ability to maneuver through the urban dimensions—airspace, supersurface, surface, and subsurface—and shoot through walls, ceilings, and floors also creates increased psychological stress. The physical characteristics of man-made terrain can also be analyzed using OAKOC.

Weather

B-16. Weather and its effects are often considered when examining the military aspects of terrain. Military aspects of weather include temperature (heat and cold), light conditions, precipitation (cloud cover, rain, snow, fog, and smog), and wind. Their military effects during UO are similar to any operational environment (see FM 34-81 and FM 34-81-1). Extremes of heat and cold affect weapon systems and the Soldiers that operate them. The extra luminescence provided by the ambient light of an urban area, unless controlled, may affect night vision capabilities and the ability of the Army to "own the night." Precipitation affects mobility and visibility. Smog inversion layers are common over cities. An inversion layer may trap smoke and chemicals in the air and increase ground and air temperatures to the detriment of Soldiers' health. (If the conditions are severe enough, it might require the use of protective masks.) Changes in temperature as a result of air inversions can also affect thermal sights during crossover periods of warm to cold and vice versa. Winds, which may increase as they are funneled through urban canyons, may—

- Increase other weather effects (for example, wind chill).
- Decrease visibility (blowing debris, sand, rain, and snow).
- Spread radiation, biological, and chemical hazards.
- Adversely affect low-altitude air mobility and airborne ISR platforms.

B-17. Commanders also analyze weather for its potential effect on civilians and civilian infrastructure as well as Soldiers and military equipment. Rain might create sewage overflow problems in urban areas with collapsed sewage infrastructure and in refugee camps, increasing disease and even creating panic. Rain and flooding may also make some subsurface areas impassable or extremely hazardous to civilian and military forces alike. Chemical agents and TIM may be washed into underground systems creating toxic hotspots. These effects become more pronounced as chemical agents or TIM are absorbed by brick and unsealed concrete sewers. Other weather effects on UO can include—

- Heavy snowfall in an urban area that may paralyze area transportation and distribution infrastructure, hindering the urban administration's ability to provide vital human services (police, fire fighting, medical, and rescue). Heavy rains and flooding may have similar effects especially on poorly designed and constructed roads or roads that have been damaged by tracked vehicles.
- Extreme hot and cold weather climates that increase the dependence (and military significance) of many elements of the infrastructure. For example, the energy infrastructure may be critical; without it, civilians may not be adequately cooled or heated or they may not be able to cook their food.

- In urban areas located in tropical regions, it can rain at the same time each day during the wet season. Threat forces may attack during these periods knowing aircraft will have difficulty responding. Bad weather also reduces the effectiveness of surveillance, direct and indirect fire, and logistic support.
- Inclement weather may preclude demonstrations or rallies by threats. Good weather may mean a maximum turnout of civilians for events such as festivals, sporting events, and other social, cultural, or religious activities.
- Severe weather may affect psychological, civil-military and humanitarian assistance operations. Heavy rains and severe dust storms may disrupt leaflet drops, construction projects, food and water distribution, and medical and veterinary assistance programs.

URBAN SOCIETY

B-18. This manual shows that societal considerations take on added importance in UO. Critical to operational success is knowing and understanding which groups live in an urban area, what relationships exist among them, and how each population group will respond to friendly and threat activities. Often determining any of this is very difficult. Cultural acuity is also essential in helping commanders and their staffs to view the urban area as the residents view it. The demographics presented depict what conditions exist, while the other categories help to explain the root causes or why conditions exist (see figure B-4 on page B-8). Other categories besides basic demographics that are important to gain this understanding include health, history, leadership, ethnicity and culture, religion, and government and politics.

From Cultural Awareness to Understanding and Competence

B-19. This societal understanding must go beyond a superficial awareness of gestures and taboos to developing a thorough understanding of the organizing principles that make the urban society hold together and function as a coherent entity—or not. In UO, particularly longer-term stability operations, commanders use this knowledge and insight to design and plan operations that work effectively within that society. For example in OPERATION IRAQI FREEDOM, Army units, as they shifted to long-term stability operations, were required to recruit, organize, equip, and train indigenous security forces in order to be able to transition security responsibility from U.S. forces to the burgeoning Iraqi government. Army commanders were successful—to a point. The first units that they built and trained attained a remarkable level of unity and tactical ability in a relatively short period of time. The newly formed units effectively fought alongside Army forces. However when prodded into conducting operations with minimal Army supervision, many Iraqi units lost unity and coherence. Army commanders initially failed to account for a basic Iraqi societal consideration: their society is patriarchal—each member or organization in Iraqi society always seeks a patron or sponsor to support them and provide for their needs. When Iraqi units trained and operated with Army forces, this relationship was satisfied. Operating on their own, however, Iraqi commanders and units lost this patronage without an equivalent Iraqi replacement. In hindsight, the establishment of an effective and more robust ministry of defense, either as a precursor or simultaneously with the establishment of tactical units, may have provided the Iraqi security forces the patronage that their culture required.

Resource the Societal Analysis

B-20. Aside from friendly and threat forces, the society is the only thinking component of the urban environment able to rapidly impact the urban operation. (Even people going about their daily routines can unwittingly hamper the mission.) Urban residents create conditions for restrictive rules of engagement, increase stress on Soldiers and logistic capabilities, and confuse threat identification (see Threat Considerations in this appendix). Demographic, health, safety, ethnic, and cultural concerns will be essential considerations in most UO. Other situational factors—the mission, enemy, and time available—dictate the balance between the level of detail and analysis to support the overall urban operation with the level of detail that commanders and their staffs can achieve. However, an analysis that fails to devote enough time and resources to understanding the urban society can find large elements of the population turned against the Army force. Detailed analysis of the urban society will help save time and ISR resources. While not necessarily a resource saver in the short term, a thorough understanding of the urban society and its culture ultimately contributes to faster achievement of strategic goals.

Population Demographics

- General population size
  - Village
  - Town
  - City
  - Metropolis
  - Megalopolis
- Group size based on race age, sex, political affiliation, economics, religion, tribe, clan, gang, criminal activities, or other significant grouping
  - Significant US or allied populations
  - Distribution, densities, and physical boundaries and overlaps
  - Majority, minority, and dominant groups
- Increasing or decreasing migration trends
  - Dislocated civilians
- Nongovernmental organizations (NGOs)
  - Local
  - National
  - International
- Languages (distribution, dialects, relationship to social structure)
- Educational levels and literacy rates
- Crime rates
- Birth and death rates
- Labor statistics and considerations
  - Skilled and unskilled
  - Imported and exported
  - Unemployment
  - Standard wages and per capita income
  - Workday and workweek norms

Health

- Diseases
- Nutritional deficiencies
- Local standards of care
- Pollution and environmental hazards (air, water, food, and soil)
- Health workers (types, numbers, and degree of skill)

History

- General and for a specific group
  - Internal or external
  - Recent conflicts
- Relationship with US, allies, and other participating multinational forces
- Applicable international treaties
- Status-of-forces agreements
- Antagonists/protagonists
- Heroes
- Events, facts, and dates considered important or celebrated
- Urban area's historical importance

Leadership and Prominent Personalities

- Identification, location, and prioritization of influential leaders (exploitation, evacuation, protection, etc.)
- Affiliation (ethnic, religion, military, government, industry, criminal, or entertainment)
- Education attained
- Organization and distribution of power
- Associations among different leaders and groups

Ethnicity and Culture

- Values, moral codes, taboos, and insults (verbal and non-verbal)
- Attitudes towards age, sex, and race (including same-sex interaction)
- Role of the clan, tribe, or family
- Biases between ethnic groups
- Privacy and individuality
- Recreation, entertainment, and humor
- Fatalism or self-determination
- Exchanges of gifts
- Displays of emotion
- Lines of authority
- Dating and marriage
- Greetings, leave-takings, and gestures
- Visiting practices
- Alcohol and drug use
- Important holidays, festivals, sporting, or entertainment events
- Eating and dietary practices
- Significance of animals and pets
- Urban-rural similarities and differences
- Driving habits
- Clothing

Religion

- Sects, divisions, and overlaps
- Religious biases and problems
- Relationship and influence on government, politics, economics, and education
- Impact on ethnic and cultural beliefs
- Key events or celebrations (daily, weekly, monthly, or annually)
- Funeral and burial practices

Government and Politics

- Present and past forms
- Organization and powers (executive, legislative, judicial, and administrative divisions)
- Scheduled elections and historical turnouts
- Degree of control over the population
  - Identification required
  - Border-crossing procedures
- Relationship with US or multinational governments, national governments, and criminal elements
- Political factions and boundaries
- Political traditions
- Grievances
- Censorship
- Nepotism and other clan, tribal, or social ties
- Civil defense and disaster preparedness (organization, plans, training, equipment, and resources)
  - Evacuation routes
- Legal system
  - System of laws
  - Applicable treaties
  - Courts and tribunals
  - Procedures
  - Records (birth and deeds)
- Property control
- Monetary system (formal and informal)
- Domestic and foreign trade
  - Taxation and tariffs
  - Customs requirements
  - Rationing and price controls
  - Economic performance and contribution to gross national product
- Economic aid
- Perception of relative deprivation
- Trade unions
- Competition with the black market and organized crime

Figure B-4. Significant urban societal characteristics

Transportation and Distribution

- Water
  - Shipyards and other port and harbor facilities
  - Inland waterways, canals, and locks
  - Offshore pipeline berths
  - Cargo storage and handling
  - Types and number of ships, boats, and ferries
  - Dams
- Streets and roads
  - Bridges and fords
  - Over- and underpasses
  - Raised embankments, tunnels, culverts, and other subterranean features (widths and clearances)
  - Parking areas (surface, subsurface, and supersurface)
  - Weight restrictions
  - Traffic light operations
  - Traffic patterns
  - Widths
  - Surface materials
- Rail
  - Lines
  - Terminals
  - Switchyards and junctions
  - Subways, bridges, elevated rail lines, and underpasses (clearances)
  - Track gauges
  - Types and number of rolling stock
  - Electrification
- Air
  - Airfields and runways (including capabilities)
  - Heliports and helipads (including rooftop)
  - Types and number of aircraft
  - Cargo storage and handling
- Trucking companies and delivery services
- Available material-handling equipment
- Rush hour and market time considerations
- Seasonal (weather) effects
- Rubble effects
- Impact of dislocated civilians and migration patterns
- Likely population congregation points
- Identifiable primary and alternate lines of communication

Economics and Commerce

- Industries
  - Types and locations
  - Important companies (including US or allied)
  - Military production facilities
- Sources of raw materials
- Use of toxic industrial materials and biological agents
  - Agriculture (insecticides, herbicides, and fertilizers)
  - Manufacturing
  - Cleaning
  - Research
- Food types, quantities, and sources
  - Requirements and availability
  - Storage and processing
  - Cleanliness standards
- Stores, shops, restaurants, hotels, and strip malls
- Recreation facilities
  - Outdoor and amusement parks
  - Stadiums and other sports facilities
- Machine shops
- Brick and lumber yards
- Banking and investment institutions

Administration and Human Services

- Police and fire protection
  - Headquarters, station, and key facilities locations
  - Organization and strengths
  - Equipment
  - Functions, authority, and jurisdictional boundaries
  - Contract guard services
- Welfare and public assistance
  - Monetary assistance
  - Orphanages
  - Elderly care facilities
- Water supply systems
  - Water sources and storage (lakes, reservoirs, cisterns, pools, and public baths)
  - Water treatment and quality
  - Pumping stations and other distribution methods (trucks, bottles)
  - Hydrant locations
- Snow removal capabilities
- Street light operations
- Health facilities
  - Hospitals
  - Emergency medical services
  - Mental institutions
  - Medical supplies and equipment
  - Research and pharmaceutical buildings
  - Blood banks
- Governmental buildings
  - Embassies
  - Capitol building
  - Legislative, judicial, and ministry buildings
  - Hall of records
- NGOs providing assistance
- Effects of military control measures on providing vital human services
- Waste and sanitation
  - Types (solid, sewage, and toxic)
  - Collection, processing, and disposal
  - Dumps or landfills
  - Drainage systems

Cultural

- Religious buildings and places of worship
- Shrines, monuments and other historical structures
- Schools
- Museums
- Theaters
- Libraries

Energy

- Types
  - Electric
  - Oil
  - Coal
  - Natural gas
  - Nuclear
  - Solar
  - Hydroelectric
  - Geothermal
- Facilities
  - Production and processing
  - Storage
- Distribution
  - Pipelines (above and below ground)
  - Power lines (overhead and underground)
  - Water, rail, and road
  - Potential hazards

Communication and Information

- Print media
  - Newspapers, periodicals, and pamphlets
  - Billboards and posters
  - Postal facilities
- Telephone facilities
  - Wire or wireless
  - Facsimile machines
- Telegraph facilities
- Radio facilities
- Police, fire, and rescue systems
- Security systems
- Television facilities
- Computers and the Internet
- Antennas, towers, relay stations, and lines (surface and subsurface)
- Integration of space-based capabilities
- Public forums and speech
- Low-technology media (car horns, drums, graffiti, and burning fires)
- Key media organizations and reporters
  - Local
  - International
  - US

Figure B-5. Significant urban infrastructure characteristics

URBAN INFRASTRUCTURE

B-21. Functional and analytical interdependence is readily apparent when examining urban infrastructures (see figure B-5 on page B-9). They are composed of physical structures or facilities and people. Hence, much of the analysis conducted for terrain and society can apply when assessing the urban infrastructure. For example, commanders, staffs, and analysts could not effectively assess the urban economic and commercial infrastructure without simultaneously considering labor. All aspects of the society relate and can be used to further analyze the urban work force since they are a sub-element of the urban society. Similarly, the OAKOC aspects used to evaluate terrain may also apply to the urban infrastructure, especially considerations of key terrain. Depending on the mission, elements of the urban infrastructure may become key terrain. While infrastructure analysis is not an engineer-specific task, Army engineers may provide the general and geospatial engineering expertise necessary to better understand the urban terrain and infrastructure.

THREAT CONSIDERATIONS

... [T]he adversaries of freedom ... send arms, agitators, aid, technicians and propaganda to every troubled area.... [S]ubversives and saboteurs and insurrectionists . ... [possess] the power to conscript talent and manpower for any purpose. and long experience in the techniques of violence and subversion....It is a contest of will and purpose as well as force and violence—a battle for minds and souls as well as lives and territory.

John F. Kennedy
Message to Congress, 5 May 1961

B-22. Chapter 3 outlines the instability and uncertainty of the strategic environment. Commanders, staffs, and analysts identify and analyze the threat's composition, strength, disposition, leadership, training, morale, weapons and capabilities, vulnerabilities, internal logistics and external support, doctrine (if any), strategy or modus operandi, and tactics. The threat can take a variety of forms:

- Conventional military forces.
- Paramilitary forces.
- Guerrillas and insurgents.
- Terrorists.
- Militia or special police organizations.

B-23. A general study of guerrilla and insurgent tactics, techniques, and procedures may prove beneficial to many types of operations regardless of the actual composition or type of threat forces. Insurgent strategies and tactics may work especially well in this complex environment and will likely be a part of any threat COA. Particularly, commanders should understand how a threat might restrict itself by the laws of land warfare and similar conventions, or exploit the use of these conventions to its own gain. Commanders can refer to FM 31-20-3 and FMI 3-07.22 for more information. For many of the above threats, no doctrinal templates may exist. Intelligence staffs will need to identify and track common threat patterns and tactics, techniques, and procedures as they develop and change during the course of the operation. Understanding culture will be critical to determining threat motives that drive decisions. Comman415081ders, staffs, and analysts must evaluate, update (or create), and manage threat databases early (and continuously) during the operational process.

URBAN ENVIRONMENTAL THREATS

B-24. While threats vary, they share a common characteristic: the capability and intent to conduct violence against Army forces to negatively influence mission accomplishment. These threats are often the most recognizable for forces trained for warfighting—these are often the enemy. In order for Army units to be able to conduct full spectrum urban operations, commanders broaden their concept of the threat when analyzing the urban environment's terrain, societal, and infrastructure characteristics. This analysis includes many environmental dangers (potentially affecting both sides of a conflict as well as noncombatants) such as—

- Natural disasters (earthquakes, tornados, hurricanes, volcanic eruptions, fires, floods, tidal waves or tsunamis, and heavy snows).
- Food shortages.
- Water shortages (both quantity and quality).
- Rampant disease.
- Pollution and toxic industrial materials.

B-25. A critical difference between the latter forms of threat and the former is the lack of intent to do harm. The latter may stand alone as threats, or these conditions may be created, initiated, or used by the enemy or a hostile civilian group as a weapon or tool. Threat analysis includes identifying and describing how each relevant characteristic of the area of operations can hinder mission accomplishment. This analysis, particularly during stability or civil support operations, will require extensive coordination and cooperation with urban civil authorities, law enforcement, and numerous governmental and nongovernmental organizations.

CIVILIANS

B-26. In a major combat operation or campaign where offensive and defensive operations are conducted against a conventional enemy, threat identification is more readily accomplished. However, adaptation may be necessary to even further broaden the threat concept to include specific elements of the urban society and, in some instances, nongovernmental organizations (NGOs) and other civilian agencies working in the urban area. In many stability operations, this modification can account for opposing armed forces that are not an enemy but are a threat to the mission. As discussed in Chapter 9, Army forces in many stability or civil support operations must avoid classifying or thinking of these threats as the enemy. Further, commanders ensure that Soldiers do not shift the animosity that they feel toward urban threats onto the civilian population. In OPERATION IRAQI FREEDOM, for example, some Soldiers began to refer to Iraqi civilians as "hajjis" in reference to the annual pilgrimage—the Hajj—that forms one of the tenets or pillars of Islamic faith. Those who have completed the pilgrimage are entitled to add the phrase al-Hajj or hajji (pilgrim) to their name. However, the Soldiers were not using the term in its proper context but rather to denigrate. In response, commanders prohibited the inappropriate use of this term as one of many measures to stress and develop cultural understanding and respect for the people that Army forces were there to assist.

Need for More Accurate Categories

B-27. Urban commanders recognize that threat evaluation is not solely a straightforward assessment of the capabilities of a known, armed enemy. Instead, they develop a system of categories that mitigates potential situational uncertainty and better indicates the level of threat (or utility) that civilian groups pose for Army forces conducting UO. Categories can range from a simple construct—threat, neutral, or friendly—to a more refined one (see figure B-6 on page B-12). In most UO, a more intricate approach is necessary to detect and monitor shifts in key relevant relationships. While this analysis is important in any urban operations, it is especially critical when Army forces are opposed by urban insurgent and guerrilla forces.

B-28. In any system, commanders consider that the classification of a group is relative to the perspective from which it is viewed. This is an especially important consideration in multinational and interagency UO. During operations in Somalia, for example, U.S. Army forces may have viewed a particular clan as a hostile element. The United Nations' Italian contingent, with their colonial background in the region, may have considered the same clan as neutral or even an ally.

Shifting and Overlapping Groups

B-29. Commanders recognize that no system of categorization will precisely classify any given group; no system can reflect the overall nature and complexity of the urban society. A single group may fit in a particular category. It may also have components in two or more categories simultaneously. Often, it can shift among categories during an operation. A given group may have individuals in it who have interests identical to or different from that group and these individual interests may change over time.

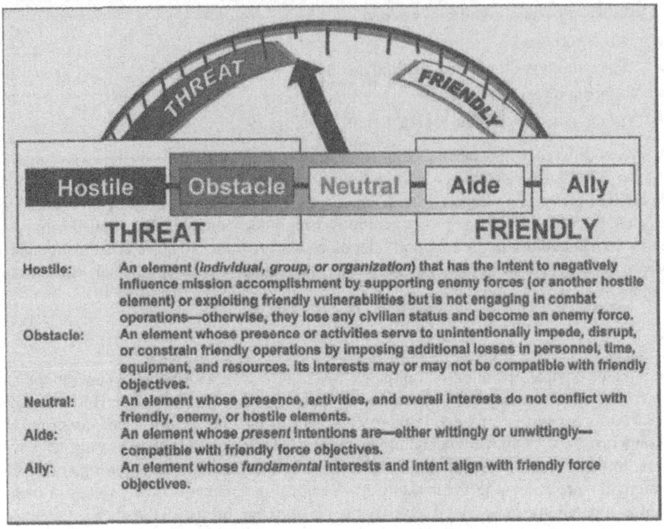

Figure B-6. A refined approach to civilian allegiance

B-30. A peace enforcement operation illustrates the varying nature of groups. An identified criminal group might be classified as an obstacle to the commander's mission because its illegal activities impede unit progress. Its compelling interest, however, is to make money rather than resist the actions of friendly forces. In the same operation, one of the armed belligerents may be intent on disrupting the peace process and would be, therefore, classified as a hostile. (Again, they are not an enemy unless the belligerents are engaged in planned combat operations against the peacekeeping force. Commanders should consider, however, that under international law, if any civilian takes a direct or active role in hostilities, that civilian can be subject to attack.) The belligerent force may finance the criminal organization to assist in further obstructing the peace mission. In this case, the criminal organization moves from being an obstacle to being a hostile.

B-31. This classification effort, therefore, is not a one-time undertaking; commanders must constantly review and update it. Groups or individuals can be influenced into assisting either the friendly or opposing force. People will also act opportunistically, shifting support and alliances as perceived advantages arise. Even seemingly passive and law abiding members of the urban society may conduct themselves in unexpected ways given the right conditions (mob violence, for example).

Shifting Civilian Interests and Intent

Among other applicable lessons (see also the vignette in Chapter 9), the Los Angeles riots of 1992 illustrate how urban population groups can shift their relative positions due to changing conditions in an urban AO. Several gangs exist in the Los Angeles area. Usually, these gangs are hostile to one another. During the riots, however, several rival gangs formed a "united front" against what was seen as a larger obstacle to their own interests: law enforcement. As a result, the hostile gangs became one another's aides during that time.

Similarities, Differences, Capabilities, and Vulnerabilities

B-32. Similarities and differences in interests and interdependencies between groups are often focal points for analysis (and the allocation of ISR assets). They may indicate how commanders may influence, coerce, or align civilian interests and intentions with mission objectives. Simultaneously, commanders should consider an analysis of the civilian element's (individual, group, or organization) capability to influence the accomplishment of friendly objectives. They should also consider civilians' vulnerabilities and dependencies. If a civilian group's fundamental interests align with friendly objectives and this group has the intent to assist friendly forces, it is clearly an ally. However, with limited or no capability, a specific group will not help move the commander any closer to achieving his desired end state and accomplishing the mission. Then the commander would normally limit the resources expended on this group to those necessary for maintaining their commitment to common goals and objectives.

B-33. In contrast, commanders may provide a group with resources to enhance or create the capability to assist in mission accomplishment. They may do this if they felt that the assistance gained (or reduction to threat support) exceeds the potential diminishment of their own force's capabilities from losing those same resources. Commanders would also consider the group's dependencies (such as food, infrastructure, and protection) and overall vulnerability to outside influence. If vulnerable to friendly influence or control (understanding urban societal considerations and matters of perception are critical in this regard), then forces are likely susceptible to enemy or hostile manipulation. Even if commanders can generate extra resources (and not significantly affect their own force's capabilities), they should still conduct this same cost-benefit analysis to determine which civilian group (if any) should receive resources.

Greatest Potential Nearer the Center

B-34. The most critical population sectors often are those nearest the middle of the spectrum shown in figure B-6, particularly if their capabilities (or potential) significantly degrade or enhance mission accomplishment. These are the obstacle, neutral, or aide categories. If their beliefs and interests can be adequately understood, commanders may have great chance to influence the population segment and significantly contribute to mission accomplishment. In the fight for information superiority, persuading a neutral population to become an aide may provide critical information required about the urban environment and threat forces operating within.

Political and Strategic Concerns

B-35. The aide category may be of significant political or strategic concern. An aide group may be invaluable for accomplishing intermediate objectives but become a vulnerability to accomplishing a larger stability operation. (Even an urban offensive or defensive operation is likely to transition to a stability operation.) For example, commanders may provide resources to a criminal organization to assist in defeating insurgent forces during urban combat operations. Once these forces have been subdued, the interests (monetary gain and defeat of the threat) linking friendly forces with this criminal element disappear. What may remain is a criminal organization with more power than a reconstituted or newly established law enforcement agency and a truly destabilizing force. This also illustrates the second- and third-order thinking that will be required of commanders and their staffs during UO.

URBAN INTELLIGENCE TOOLS AND PRODUCTS

B-36. Developing tools and products to assist with situational understanding is not the task of intelligence sections alone. Commanders and all elements of their staffs must develop and adapt products and tools suited to their particular requirements. Listed below are some of the analytic tools and products that may help meet those requirements in a complex environment (see figure B-7 on page B-14 and FMI 2-91.4). Standard tools and products include: modified combined obstacle overlays and doctrinal, situation, event, and decision support templates or matrices. In addition to these standards aids, staffs and analysts may develop or produce other innovative tools to assist commanders in their situational understanding of the complex urban environment. These tools may contain overlapping information as different types of information are compared to determine patterns among them. Staffs and analysts may also initiate requests for products (or information) from their higher headquarters or other agencies with the technical means or control over assets when the capability lies outside the Army force's means. The tools that may be developed or requested include—

- Imagery.
- Three-dimensional representations.
- Infrastructure blueprints.
- Hydrographic surveys.
- Psychological profiles.
- Matrices, diagrams, or charts.
- Various urban overlays.

B-37. There are many software applications available to the Army that can be used to conduct intelligence analysis and create many of the relevant products described above. These applications currently range from such programs as the Analyst Notebook and Crimelink which have link analysis, association matrix, and pattern analysis software tools to the Urban Tactical Planner developed by the Topographic Engineering Center as an operational planning tool. The focus of the following information, however, is the various types of tools that can be developed and used to understand the urban environment rather than the constantly evolving and changing software or hardware that may be used to create them.

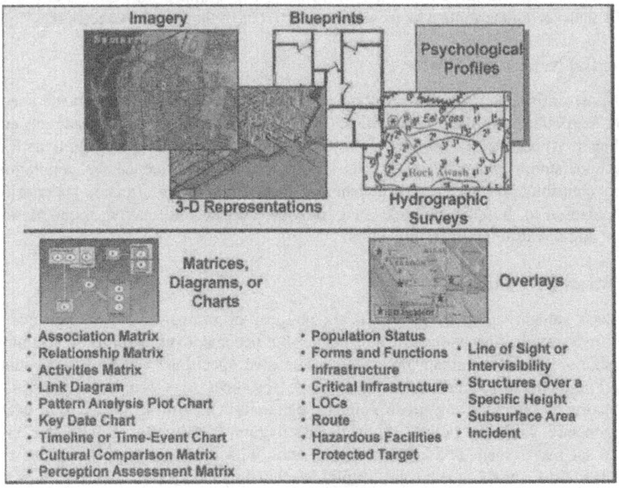

Figure B-7. Urban intelligence tools and products

IMAGERY

B-38. Recent satellite imagery or aerial photography will be required for most types of UO. Such images clarify vague and inaccurate maps and other graphic representations. Satellite assets provide responsive data input into the geographic information systems (GIS). (The National Geospatial Intelligence Agency [NGA] is one important source of imagery and map data.) GIS will often form the basis for creating the three-dimensional representations and the various overlays described below. Frequently updated (or continuous real-time) satellite or aerial imagery may be required for detailed pattern analysis and maintaining accurate situational understanding. For example, imagery taken during an area's rainy season may appear significantly altered during the summer months. Finally, digital, hand-held still and video cameras, particularly at the tactical level, will also be critical in developing and attaining situational understanding in a complex urban environment.

THREE-DIMENSIONAL REPRESENTATIONS

B-39. Often, physical or computer-generated (virtual) three-dimensional representations may be required to achieve situational understanding. These models or computer representations include specific sections of the urban area or specific buildings or structures. Such detail is particularly important for special operating forces and tactical-level units. These units require detail to achieve precision, increase the speed of the operation, and lessen friendly casualties and collateral damage.

INFRASTRUCTURE BLUEPRINTS

B-40. Urban police, fire, health, public utilities, city engineer, realty, and tourist agencies, and other urban organizations often maintain current blueprints and detailed maps. Such documents may prove useful to update or supplement military maps or to clarify the intricacies of a specific infrastructure. They may prove critical in operations that require detailed information to achieve the speed and precision required for success. Without such detail, analysts determine interior configurations based on a building's outward appearance. That task generally becomes more difficult as the building size increases.

HYDROGRAPHIC SURVEYS

B-41. Many urban areas are located along the world's littoral regions and major rivers. Therefore, commanders may need current hydrographic surveys to support amphibious, river crossing, riverine, and sustainment operations.

PSYCHOLOGICAL PROFILES

B-42. Psychological profiles analyze how key groups, leaders, or decision makers think or act—their attitudes, opinions, and views. They include an analysis of doctrine and strategy, culture, and historical patterns of behavior. The degree to which the attitudes, beliefs, and backgrounds of the military either reflect or conflict with the urban populace's (or civilian leadership's) core values is extremely important in this analysis. Psychological profiles help to assess the relative probability of a threat (or noncombatant group) adopting various COAs as well as evaluating a threat's vulnerability to deception. These profiles are derived from open-source intelligence as well as signals and human intelligence.

MATRICES, DIAGRAMS, AND CHARTS

B-43. Matrices, diagrams, and charts help to identify and understand key relationships among friendly and threat forces and other significant elements of the urban environment. While similar, each looks at information in a different way to uncover hidden patterns and connections useful in understanding the complex urban environment, particularly the threat and the urban society. These tools and products might include—

- **Association Matrix**. The association matrix helps identify the nature and relationship between individuals. Association matrices also help to identify those personalities and associations needing a more in-depth analysis necessary to establish the degree of relationship, contacts, or knowledge between individuals. Threat organizational structure and functions are uncovered as connections between personalities are made.

- **Relationship Matrix**. Relationship matrices are an extension of association matrices described above. They are used to examine the relationship between groups and organizations—threat forces, friendly forces, NGOs, the media, and the various elements of the urban population. Urban terrain and infrastructure may also be included as analytical factors to help expose relationships. These matrices graphically depict how human elements of the urban area interact with the physical elements to promote or degrade mission success. A significant relationship matrix may be a comparison of cultural perspectives—ideology, politics, religion, acceptable standards of living, and mores—between urban population groups and Army (and multinational) forces to help understand and accurately predict a civilian element's actions. The relationships discovered may suggest ways that commanders can shape the environment. For example, conflicting ideals or issues between two disparate threat forces may be emphasized to cause threats to focus their resources against each other rather than Army forces. Conversely, stressing common ideals identified between opposing noncombatant organizations may cause these groups to unite and work harmoniously toward a goal reflecting their commonalities.
- **Activities Matrix**. An activities matrix connects individuals to organizations, professions, events, activities, or addresses. Information from this matrix (combined with information from association and relationship matrices) also helps to link personalities, uncover the structure of an organization (threat or otherwise), and recognize differences, similarities, and dependencies for possible exploitation.
- **Link Diagram**. This tool graphically depicts many of the relationships and associations described above—people, groups, events, locations, or other factors deemed significant in any given situation. It helps commanders and analysts visualize how people and other factors are interrelated to determine key links.
- **Pattern Analysis Plot Chart**. This chart depicts the times and dates of a selected activity (such as ambushes, sniper and mortar attacks, bombings, and demonstrations) to search for patterns of activity for predictive purposes as well as to discern intent. This analysis can be conducted using a time-event wheel as well as by plotting events on maps using multiple historical overlays (analog or digital). (Compare with Incidents Overlay covered later in this chapter.) During pattern analysis, commanders and analysts consider not only what is occurring but also what is not occurring (or ceases to occur).
- **Key Event Chart**. In many urban operations, particularly stability operations, key holidays, historic events, and other significant cultural or political events can be extremely important for commanders to understand. More than just dates, these charts depict what can be expected to happen on each particular event. In Bosnia, for example, weddings are often held on Fridays and celebratory fire is a common occurrence. Understanding this cultural phenomenon could reduce collateral damage and accompanying civilian deaths.
- **Timeline or Time-Event Chart**. Timelines are a list of significant historical dates with relevant information and analysis that provide the commander with a record of past activities necessary to understand current operational conditions. A timeline may highlight a specific feature of the present situation such as population movements or political shifts or outline the general chronological record of the urban area, perhaps highlighting the activities of a certain population sector. Understanding past events also helps commanders predict future reactions to proposed COAs. Similarly to link diagrams, time-event charts display large amounts of information in a small space. These charts may help commanders discover larger-scale patterns of activity and relationships.
- **Cultural Comparison Matrix**. Commanders must avoid ethnocentricity—assuming that only one cultural perspective exists (often their own) and using that as the single lens with which to view the situation. To avoid this common obstacle to an accurate understanding of the urban society, commanders use this matrix to compare local ideology, attitudes, beliefs, and acceptable standards of living with associated U.S. and Army norms and values. This matrix not only highlights differences but also similarities.

- **Perception Assessment Matrix**. All of the above tools contribute to developing the perception assessment matrix; they allow commanders to see the urban environment from the perspective of its inhabitants. The perception assessment matrix is another tool that uses this refined perspective and the framework outlined in the matrix to predict the urban population's perception of a proposed COA. Although perceptions are not actions, they—more than reality—drive decision-making and resultant civilian activity. (As an example, legitimacy is a critical factor in developing many COAs that affect the civilian population, particularly during stability operations. This legitimacy must be viewed from the perspective of the inhabitants and not based on the Army commander's own culturally-shaped perceptions. Moreover, the legitimacy of an operation may need to be considered from other perspectives as well. These might include those of the American public, coalition partners and their publics, and those of neighboring states in the region.) This same matrix can later be used as a foundation to track and gauge the effectiveness of the chosen COA with favorable perception as one criterion of success. Commanders must be wary that perceptions are not fixed. They may change based on factors that the commander may not be able to control or influence. Therefore, commanders must continually monitor perceptions for deviation.

VARIOUS URBAN OVERLAYS

B-44. Staffs can produce various map overlays. These overlays depict physical locations of some aspect critical to the planning and conduct of the urban operation. Given adequate lead time, NGA can produce many overlays as an integrated map product (including satellite imagery). These overlays can include the—

- **Population Status Overlay**. This tool depicts the physical location of various groups identified by any significant social category such as religion, tribe, or language. During offensive and defensive operations, it may simply be where significant numbers of people are "huddled" or located throughout the battlefield. Population dispersal can vary significantly through the day, particularly at night, and must be considered as part of the overall analysis leading to the development of this tool.
- **Forms and Functions Overlay**. Based on the urban model, this overlay depicts the urban core or central business district, industrial areas, outlying high-rise areas, commercial ribbon areas, and residential areas, to include shantytowns.
- **Infrastructure Overlay**. This overlay is actually a series of overlays. It depicts identifiable subsystems in each form of urban infrastructure: communications and information, transportation and distribution, energy, economics and commerce, administration and human services, and cultural. Each subsystem can be broken down into more detail. Infrastructure data may be used to develop three other overlays—
  - **Critical Infrastructure Overlay**. This tool displays specific elements of the urban infrastructure that, if harmed, will adversely affect the living conditions of the urban society to the detriment of the mission. These elements may include power generation plants, water purification plants and pumping stations, and sewage treatment plants. This information could be coded as part of the overall infrastructure overlay.
  - **Lines of Communications (LOCs) Overlay**. The LOCs overlay highlights transportation systems and nodes, such as railways, road, trails, navigable waterways, airfields, and open areas for drop zones and landing zones. It also includes subsurface areas and routes such as sewage, drainage, and tunnels and considers movement between supersurface areas. The LOCs overlay and the route overlay (below) consider traffic conditions, times, and locations, to include potential points where significant portions of the urban population may congregate.
  - **Route Overlay**. This overlay emphasizes mobility information to assist commanders and planners in determining what forces and equipment can move along the urban area's mobility corridors. Pertinent data includes street names, patterns, and widths; bridge, underpass, and overpass locations; load capacities; potential sniper and ambush locations (which may be its own overlay); and key navigational landmarks such as major roads and highways, rivers and canals, cemeteries, bridges, stadiums, and churches. The structures over a specific height

overlay and subsurface overlay may assist in its development. As with the LOCs overlay, commanders, planners, and analysts think in all dimensions.

- **Line of Sight or Intervisibility Overlay**. This product creates a profile view (optical or electronic) of the terrain from the observer's location to other locations or targets. It can show trajectory or flight-line masking as well as obstructed or unobstructed signal pathways.
- **Structures over a Specific Height Overlay**. This level of detail may also be critical to communications, fires, and Army airspace command and control (air mobility corridors especially low-level flight profiles). Incorporated as part of this overlay, it may include floors or elevations above limitations for particular weapon systems at various distances from the structure.
- **Subsurface Area Overlay**. As an alternate to the building or structure height overlay, this product provides the locations of basements, underground parking garages, sewers, tunnels, subways, naturally-occurring subterranean formations, catacombs, and other subsurface areas. Similar to elevation "dead spaces," this overlay may show areas that exceed depression capabilities of weapon systems and potential threat ambush locations—again, affecting maneuver options.
- **Urban Logistic Resources Overlay**. This product identifies the locations of urban logistic resources that may contribute to mission accomplishment. It may contain specific warehouse sites, hospitals and medical supply locations, viable food stores, building material locations, fuel storage areas, car or truck lots, maintenance garages, and appliance warehouses. (NGO locations, taken from an NGO relationship matrix, may be an essential, overlapping element of this overlay.)
- **Hazardous Facilities Overlay**. This overlay identifies urban structures with known or suspected chemical, biological, or radiological features, such as nuclear power plants, fertilizer plants, oil refineries, pharmaceutical plants, and covert locations for producing weapons of mass destruction. These locations are critical to maneuver and fire planning.
- **Protected Target Overlay**. This overlay depicts terrain that should not be destroyed or attacked based on restrictions due to international, host-nation, or U.S. law and subsequent rules of engagement. These may include schools, hospitals, historical or other culturally significant monuments, and religious sites. This overlay may incorporate no-fire areas, such as special operations forces locations, critical infrastructure, logistic sources, and hazardous sites that must be protected as part of the commander's concept of the operation.
- **Incident Overlay**. Similar to the pattern analysis plot chart, this product depicts the location of different threat actions and types of tactics employed to uncover recurring routines, schemes, methods, tactics, or techniques and overall threat interests, objectives, or the desired end state.

B-45. The above IPB tools and products constitute a small sampling of what staffs and analysts can produce. They are limited only by their imaginations and mission needs (not all tools presented above may be relevant or necessary to every operation). Many products can be combined into a single product or each can generate further products of increasing level of detail. This is similar to transparent overlays positioned one atop another on a map. As discussed earlier, technology—software and hardware—will allow for more urban data to be combined, compared, analyzed, displayed, and shared—ultimately allowing scaleable products that can expand and contract to meet the intelligence requirements of any echelon of command down to the individual Soldier. The challenge will remain, however, to provide timely, accurate, complete, and relevant information in an understandable, usable, and share-capable form—share-capable among different levels of command as well as among multinational partners and various governmental and (as required) nongovernmental organizations—without overloading the commander.

Appendix C

Operations in Somalia: Applying the Urban Operational Framework to Stability Operations

It's impossible for an American mother to believe that a Somali mother would raise children to avenge the clan.

Major General Thomas M. Montgomery

GENERAL SITUATION

C-1. Following decades of political unrest and the fall of Somali dictator Siad Barre, a civil war broke out as 14 clans vied for power. The resulting nation composed of hostile social factions was held together by weak political alliances—none strong enough to unite and lead the country to national reconciliation. An ongoing drought led to famine and compounded the ethnic tensions and political instability. This volatile situation rapidly led to a phased U.S. involvement (see figure C-1). Army forces combined, sequenced, and proportionally emphasized the different types of operations to accomplish changing political objectives. Throughout all operations in Somalia, urban areas were critical to achieving mission success.

SOMALI OPERATIONS

INITIAL UN RESPONSE

C-2. The United Nations (UN) initially responded to requests for assistance from international relief organizations by sending supplies and other forms of humanitarian aid to Somalia. However, widespread looting, fighting between gangs, and other lawlessness prevented supplies from reaching the hungry and sick. Only 20 percent of the food entering the country reached the people who needed it. An estimated 25 percent of Somalia's 6 million people died of starvation or disease. In April 1992, the UN issued Security Council Resolution (UNSCR) 751 which authorized 50 unarmed observers, but the action had little effect. Under continuing pressure for additional measures to ensure the delivery of supplies and relief, the UN authorized 500 armed peacekeepers (furnished by Pakistan and transported by U.S. sea- and airlift) to protect humanitarian workers. The battalion's limited mission, designated UN Operations in Somalia I (UNOSOM I), encompassed safeguarding the unloading of ships and providing convoy security.

PROVIDE RELIEF (UNOSOM I)

C-3. In July 1992, the UN requested an increased airlift of supplies and the United States quickly responded. U.S. Central Command (USCENTCOM) activated joint task force (JTF) OPERATION PROVIDE RELIEF. Based on careful mission analysis, USCENTCOM limited the JTF's actions to—

- Deploying a humanitarian assistance survey team to assessing relief requirements.
- Providing an emergency airlift of supplies.
- Using Air Force cargo aircraft for daily relief sorties into Somalia.

C-4. USCENTCOM restricted the sorties to flying during daylight hours and to locations that would provide a permissive and safe environment. In mid-September 1992, the United States prudently expanded its role by stationing the amphibious ready group, USS Tarawa, offshore to provide support to the Pakistani security battalion and to provide security for U.S. airlift operations. The 11th Marine expeditionary unit (MEU) was on board the USS Tarawa to rapidly respond to any change in mission (see Appendix D for a description and the capabilities of a MEU).

Operation	Dates	UN Security Council Resolution	Relative Proportionality Between Types of Operations
PROVIDE RELIEF (UNOSOM I)	Aug 92 – Dec 92	UNSCR# 751 24 Apr 92	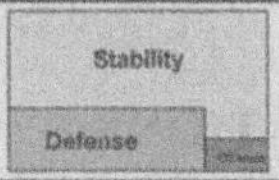
RESTORE HOPE (UNITAF)	Dec 92 – May 93	UNSCR# 794 3 Dec 92	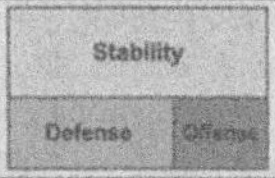
CONTINUED HOPE (UNOSOM II)	May 93 – Mar 94	UNSCR# 814 26 Mar 93	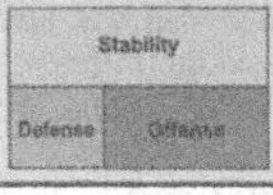

Figure C-1. Relative proportionality between types of operations

RESTORE HOPE (UNITAF)

C-5. By November 1992, the magnitude of the task, UN organizational deficiencies, and a continued lack of security precluded delivery of sufficient supplies to the needy. Notably, a ship laden with relief supplies was fired on in the harbor at Mogadishu, forcing its withdrawal before the supplies could be brought ashore, and a Pakistani peacekeeper was shot when his car was hijacked. Subsequently, the United States offered to provide forces and lead a UN-sponsored operation to reopen the flow of food to where it was needed most. In December 1992, the UN issued UNSCR 794, which authorized member states "to use all necessary means to establish a secure environment for humanitarian relief operations in Somalia" and demanded all "factions in Somalia immediately cease hostilities." To allay concerns of colonialism by a number of African countries, the UN Secretary-General was given oversight of the operation. The resolution also required soldiers to be withdrawn once order was restored; however, it provided no exit strategy. As clearly as possible, the USCENTCOM mission statement (below) for OPERATION RESTORE HOPE reflected the UN mandate:

> *When directed by the [President or the Secretary of Defense], [Commander, USCENTCOM] will conduct joint/combined military operations in Somalia to secure the major air and sea ports, key installations and food distribution points, to provide open and free passage of relief supplies, provide security for convoys and relief organization operations, and assist UN/[nongovernmental organizations] in providing humanitarian relief under UN auspices. Upon establishing a secure environment for uninterrupted relief operations, [Commander, USCENTCOM] terminates and transfers relief operations to UN peacekeeping forces.*

C-6. Mogadishu was the largest port in the country and the focal point of previous humanitarian relief activities of nongovernmental organizations (NGOs). It was also the headquarters of the coalition of 20 nations and over 30 active humanitarian relief organizations. As such, Mogadishu became the entry point for the operational buildup of the multinational force known as Unified Task Force (UNITAF) and the key logistic hub for all operations in Somalia. UNITAF immediately gained control over the flow of relief supplies into and through Mogadishu and stabilized the conflict among the clans. In less than a month, UNITAF forces expanded control over additional ports and interior airfields. They secured additional distribution sites in other key urban areas in the famine belt to include Baidoa, Baledogle, Gialalassi, Bardera, Belet Uen, Oddur, Marka, and the southern town of Kismayo (see figure C-2). With minimal force, the U.S.-led UNITAF established a secure environment that allowed relief to reach those in need, successfully fulfilling its limited—yet focused—mandate.

CONTINUED HOPE (UNOSOM II)

C-7. In March 1993, the UN issued UNSCR 814 establishing a permanent peacekeeping force, UNOSOM II. However, the orderly transition from UNITAF to UNOSOM II was repeatedly delayed until May 1993. (The UN Secretary-General urged the delay so that U.S. forces could effectively disarm bandits and rival clan factions in Somalia.) This resolution was significant in two critical aspects:

- It explicitly endorsed nation building with the specific objectives of rehabilitating the political institutions and economy of Somalia.
- It mandated the first ever UN-directed peace enforcement operation under the Chapter VII enforcement provisions of the Charter, including the requirement for UNOSOM II to disarm the Somali clans. The creation of a peaceful, secure environment included the northern region that had declared independence and had hereto been mostly ignored.

Figure C-2. Map of Somalia

C-8. These far-reaching objectives exceeded the limited mandate of UNITAF as well as those of any previous UN operation. Somali clan leaders rejected the shift from a peacekeeping operation to a peace enforcement operation. They perceived the UN as having lost its neutral position among rival factions. A more powerful clan leader, General Mohammed Farah Aideed (leader of the Habr Gidr clan), aggressively turned against the UN operation and began a radio and propaganda campaign. This campaign characterized UN soldiers as an occupation force trying to recolonize Somalia.

C-9. The mounting crisis erupted in June 1993. Aideed supporters killed 24 Pakistani soldiers and wounded 57 in an ambush while the soldiers were conducting a short-warning inspection of one of Aideed's weapons arsenals. UNSCR 837, passed the next day, called for immediately apprehending those responsible and quickly led to a manhunt for Aideed. The United States deployed 400 rangers and other special operations forces (SOF) personnel to aid in capturing Aideed, neutralizing his followers, and assisting the quick reaction force (QRF), composed of 10th Mountain Division units, in maintaining the peace around Mogadishu.

PHASED WITHDRAWAL

C-10. On 3 October 1993, elements of Task Force (TF) Ranger (a force of nearly 100 Rangers and SOF operators) executed a raid to capture some of Aideed's closest supporters. Although tactically successful, 2 helicopters were shot down, 75 soldiers were wounded, and 18 soldiers were killed accomplishing the mission. The U.S. deaths, as well as vivid scenes of mutilation to some of the soldiers, increased calls to Congress for withdrawing U.S. forces from Somalia. The President then ordered reinforcements to protect U.S. Forces, Somalia (USFORSOM) as they began a phased withdrawal with a 31 March deadline. The last contingent sailed from Mogadishu on 25 March, ending OPERATION CONTINUED HOPE and the overall U.S. mission in Somalia.

C-11. Although U.S. forces did not carry out the more ambitious UN goals of nation building, they executed their missions successfully, relieving untold suffering through humanitarian assistance with military skill and professionalism. Operations in Somalia occurred under unique circumstances, yet commanders may glean lessons applicable to future urban stability operations, as well as civil support operations. In any operations, commanders balance changing mission requirements and conditions.

UNDERSTAND

C-12. Although accomplished to varying degrees, U.S. forces failed to adequately assess and understand the urban environment, especially the society. Somali culture stresses the unity of the clan; alliances are made with other clans only when necessary to elicit some gain. Weapons, overt aggressiveness, and an unusual willingness to accept casualties are intrinsic parts of the Somali culture. Women and children are considered part of the clan's order of battle.

C-13. Early in the planning for OPERATION RESTORE HOPE, U.S. forces did recognize the limited transportation and distribution infrastructure in Mogadishu. The most notable was the limited or poor airport and harbor facilities and its impact on the ability of military forces and organizations to provide relief. Therefore, a naval construction battalion made major improvements in roads, warehouses, and other facilities that allowed more personnel, supplies, and equipment to join the relief effort faster.

UNDERSTANDING THE CLAN (THE HUMAN DIMENSION)

C-14. During OPERATION RESTORE HOPE, the UNITAF worked with the various clan leaders as the only recognized leadership remaining in the country. The UNITAF was under the leadership of LTG Robert B. Johnston and U.S. Ambassador to Somalia, Robert Oakley. In addition, UNITAF forces also tried to reestablish elements of the Somali National Police—one of the last respected institutions in the country that was not clan-based. This reinstated police force manned checkpoints throughout Mogadishu and provided crowd control at feeding centers. Largely because of this engagement strategy, the UNITAF succeeded in its missions of stabilizing the security situation and facilitating humanitarian relief. Before its termination, the UNITAF also worked with the 14 major Somali factions to agree to a plan for a transitional or transnational government.

C-15. The UN Special Representative of the Secretary-General, retired U.S. Navy Admiral Jonathon Howe, worked with the UNOSOM II commander, Turkish General Cevik Bir. During OPERATION CONTINUED HOPE, Howe and General Bir adopted a philosophy and operational strategy dissimilar to their UNITAF predecessors. Instead of engaging the clan leaders, Howe attempted to marginalize and isolate them. Howe initially attempted to ignore Aideed and other clan leaders in an attempt to decrease the warlord's power. Disregarding the long-established Somali cultural order, the UN felt that, in the interest of creating a representative, democratic Somali government, they would be better served by excluding the clan leadership. This decision ultimately set the stage for strategic failure.

THREAT STRATEGY AND TACTICS

C-16. During OPERATION RESTORE HOPE, U.S. forces also failed to properly analyze and understand their identified threat's intent and the impact that the urban environment would have on his strategy, operations, and tactics. The UN began to view eliminating Aideed's influence as a decisive point when creating an environment conducive to long-term conflict resolution. Aideed's objective, however, remained to consolidate control of the Somali nation under his leadership—his own brand of conflict resolution. He viewed the UN's operational center of gravity as the well-trained and technologically advanced American military forces, which he could not attack directly. He identified a potential American vulnerability—the inability to accept casualties for an operation not vital to national interests—since most Americans still viewed Somalia as a humanitarian effort. If he could convince the American public that the price for keeping troops in Somalia would be costly, or that their forces were hurting as many Somalis as they were helping, he believed they would withdraw their forces. If U.S. forces left, the powerless UN would leave soon after, allowing Aideed to consolidate Somalia under his leadership.

VULNERABILITY AND RISK ASSESSMENT

C-17. U.S. forces failed to assess, understand, and anticipate that Aideed would adopt this asymmetric approach and attack the American public's desire to remain involved in Somalia. By drawing U.S. forces into an urban fight on his home turf in Mogadishu, he could employ guerrilla insurgency tactics and use the urban area's noncombatants and its confining nature. Such tactics made it difficult for the U.S. forces to employ their technological superiority. If U.S. forces were unwilling to risk harming civilians, his forces could inflict heavy casualties on them, thereby degrading U.S. public support for operations in Somalia. If, on the other hand, the U.S. forces were willing to risk increased civilian casualties to protect themselves, those casualties would likely have the same effect.

C-18. However, an assessment and understanding of the Somali culture and society should have recognized the potential for Aideed's forces to use women and children as cover and concealment. Accordingly, the plan should have avoided entering the densely populated Bakara market district with such restrictive rules of engagement. As legitimacy is critical to stability operations, TF Ranger should have been prepared and authorized to employ nonlethal weapons, to include riot control gas, as an alternative to killing civilians or dying themselves.

C-19. U.S. forces also failed to assess and understand the critical vulnerability of their helicopters in an urban environment and the potential impact on their operations. TF Ranger underestimated the threat's ability to shoot down its helicopters even though they knew Somalis had attempted to use massed rocket-propelled grenade (RPG) fires during earlier raids. (Aideed brought in fundamentalist Islamic soldiers from Sudan, experienced in downing Russian helicopters in Afghanistan, to train his men in RPG firing techniques). In fact, the Somalis had succeeded in shooting down a UH-60 flying at rooftop level at night just one week prior to the battle. Instead, TF Ranger kept their most vulnerable helicopters, the MH-60 Blackhawks, loitering for forty minutes over the target area in an orbit that was well within Somali RPG range. The more maneuverable AH-6s and MH-6s could have provided the necessary fire support. Planning should have included a ready ground reaction force, properly task organized, for a downed helicopter and personnel recovery contingency.

C-20. Information operations considerations apply throughout the entire urban operational framework; however, operations security (OPSEC) is critical to both assessment and shaping. OPSEC requires continuous assessment throughout the urban operation particularly as it transitions among the spectrum of operations. As offensive operations grew during OPERATION CONTINUED HOPE, U.S. forces did little to protect essential elements of friendly information. Combined with the vulnerability of U.S. helicopters, Aideed's followers used U.S. forces' inattention to OPSEC measures to their advantage. The U.S. base in Mogadishu was open to public view and Somali contractors often moved about freely. Somalis had a clear view both day and night of the soldiers' billets. Whenever TF Ranger would prepare for a mission, the word rapidly spread through the city. On 3 October 1993, Aideed's followers immediately knew that aircraft had taken off and, based on their pattern analysis of TF Ranger's previous raids, RPG teams rushed to the rooftops along the flight paths of the task force's Blackhawks.

SHAPE

C-21. One of the most critical urban shaping operations is isolation. During OPERATION CONTINUED HOPE, U.S. forces largely discounted other essential elements of friendly information and did not establish significant public affairs and psychological operations (PSYOP) initiatives. In fact, Army forces lacked a public affairs organization altogether. Consequently, Aideed was not isolated from the support of the Somali people. This failure to shape the perceptions of the civilian populace coupled with the increased use of lethal force (discussed below) allowed Aideed to retain or create a sense of legitimacy and popular support. Ironically, the focus on Aideed helped create a popular figure among the warlords through which many noncompliant, hostile, and even neutral Somalis could rally behind.

C-22. During OPERATION RESTORE HOPE, Aideed conducted propaganda efforts through "Radio Aideed"—his own radio station. UNITAF countered these efforts with radio broadcasts. This technique proved so effective that Aideed called MG Anthony C. Zinni, UNITAF's director of operations, over to his house on several occasions to complain about UNITAF radio broadcasts. General Zinni responded, "If he didn't like what we said on the radio station, he ought to think about his radio station and we could mutually agree to lower the rhetoric." This approach worked.

ENGAGE

C-23. The complexity of urban operations requires unity of command to identify and effectively strike the center of gravity with overwhelming combat power or capabilities. Complex command and control relationships will only add to the complexity and inhibit a commander's ability to dominate and apply available combat power to accomplish assigned objectives. Stability operations as seen in Somalia required commanders to dominate within a supporting role and, throughout, required careful, measured restraint.

UNITY OF COMMAND (EFFORT)

C-24. During OPERATION RESTORE HOPE, UNITAF successfully met unity of command challenges through three innovations. First, they created a civil-military operations center (CMOC) to facilitate unity of effort between NGOs and military forces. Second, UNITAF divided the country into nine humanitarian relief sectors centered on critical urban areas that facilitated both relief distribution and military areas of responsibility. Third, to establish a reasonable span of control, nations that provided less than platoon-sized contingents were placed under the control of the Army, Marine Corps, and Air Force components.

C-25. On the other hand, during OPERATION CONTINUED HOPE, UNOSOM II command and control relationships made unity of command (effort) nearly impossible. The logistic components of USFORSOM were under UN operational control, while the QRF remained under USCENTCOM's combatant command—as was TF Ranger. However, the USCENTCOM commander was not in theater. He was not actively involved in planning TF Ranger's missions or in coordinating and integrating them with his other subordinate commands. It was left to TF Ranger to coordinate with the QRF as needed. Even in TF Ranger, there were dual chains of command between SOF operators and the Rangers. This underscores the need for close coordination and careful integration of SOF and conventional forces (see Chapter 4). It also emphasizes overall unity of command (or effort when command is not possible) among all forces operating in a single urban environment.

C-26. Following TF Ranger's 3 October mission, the command structure during OPERATION CONTINUED HOPE was further complicated with the new JTF-Somalia. This force was designed to protect U.S. forces during the withdrawal from Somalia. JTF-Somalia came under the operational control of USCENTCOM, but fell under the tactical control of USFORSOM. Neither the JTF nor USFORSOM controlled the naval forces that remained under USCENTCOM's operational control. However unity of effort (force protection and a rapid, orderly withdrawal) galvanized the command and fostered close coordination and cooperation among the semiautonomous units.

MEASURED RESTRAINT

C-27. During OPERATIONS PROVIDE RELIEF and RESTORE HOPE, U.S. forces dominated within their supporting roles. Their perseverance, adaptability, impartiality, and restraint allowed them to provide a stable, secure environment. Hence, relief organizations could provide the food and medical care necessary to reduce disease, malnourishment, and the overall mortality rate. However, during OPERATION CONTINUED HOPE, U.S. operations became increasingly aggressive under the UN mandate. However, peace enforcement also requires restraint and impartiality to successfully dominate and achieve political objectives. The increased use of force resulted in increased civilian casualties, which in turn reduced the Somalis' perception of U.S. legitimacy. As a result, most moderate Somalis began to side with Aideed and his supporters. Many Somalis, without effective information operations to shape their perceptions, felt that it was fine for foreign military forces to intervene in their country to feed the starving and even help establish a peaceful government, but not to purposefully target specific Somali leaders as criminals.

CONSOLIDATE AND TRANSITION

C-28. Across the spectrum of conflict, Army forces must be able to execute full spectrum operations not only sequentially but, as in the case of operations in Somalia, simultaneously. OPERATION PROVIDE RELIEF began primarily as a stability operation in the form of foreign humanitarian assistance. Later this operation progressed to include peacekeeping (another stability operation), defensive operations to protect

UN forces and relief supplies, and minimum offensive operations. As operations transitioned to OPERATION RESTORE HOPE, it became apparent that while foreign humanitarian assistance was still the principal operation, other operations were necessary. More forceful peacekeeping, show of force, and arms control (all forms of stability operations), and offensive and defensive operations grew increasingly necessary to establish a secure environment for uninterrupted relief operations. In the final phase of U.S. involvement during OPERATION CONTINUED HOPE, major changes to political objectives caused a major transition in the form of the stability operation to peace enforcement with an even greater increase in the use of force, offensively and defensively, to create a peaceful environment and conduct nation building. Finally, the TF Ranger raid demonstrated the need to maintain a robust, combined arms force capable of rapidly transitioning from stability operations (primarily peace operations) to combat operations.

SUMMARY

C-29. OPERATIONS PROVIDE RELIEF and RESTORE HOPE were unquestionably successes. Conversely, during OPERATION CONTINUED HOPE, the 3-4 October battle of Mogadishu (also known as the "Battle of the Black Sea") was a tactical success leading to an operational failure. TF Ranger succeeded in capturing 24 suspected Aideed supporters to include two of his key lieutenants. Arguably, given the appropriate response at the strategic level, it had the potential to be an operational success. After accompanying Ambassador Oakley to a meeting with Aideed soon after the battle, MG Zinni described Aideed as visibly shaken by the encounter. MG Zinni believed Aideed and his subordinate leadership were tired of the fighting and prepared to negotiate. Unfortunately, the U.S. strategic leadership failed to conduct the shaping actions necessary to inform and convince the American public (and its elected members of Congress) of the necessity of employing American forces to capture Aideed. The President was left with little recourse after the battle of Mogadishu but to avoid further military confrontation.

C-30. Despite this strategic failing, the operational commanders might have avoided the casualties, and any subsequent public and Congressional backlash, had they better communicated among themselves and worked with unity of effort. Recognizing the separate U.S. and UN chains of command, the UN Special Representative, along with the USCENTCOM, USFORSOM, and TF Ranger commanders, should have established the command and control architecture needed. This architecture would have integrated planning and execution for each urban operation conducted. These commanders failed to "operationalize" their plan. They did not properly link U.S. strategic objectives and concerns to the tactical plan. The TF Ranger mission was a direct operational attempt to obtain a strategic objective in a single tactical action. Yet, they failed to understand the lack of strategic groundwork, the threat's intent and capabilities, and the overall impact of the urban environment, to include the terrain and society, on the operation. Such an understanding may not have led to such a high-risk course of action but instead to one that de-emphasized military operations and emphasized a political solution that adequately considered the clans' influence.

This page intentionally left blank.

Appendix D

Joint and Multinational Urban Operations

[Joint force commanders] synchronize the actions of air, land, sea, space, and special operations forces to achieve strategic and operational objectives through integrated, joint campaigns and major operations. The goal is to increase the total effectiveness of the joint force, not necessarily to involve all forces or to involve all forces equally.

JP 3-0

As pointed out earlier, Army forces, brigade size and larger, will likely be required to conduct operations in, around, and over large urban areas in support of a joint force commander (JFC). The complexity of many urban environments, particularly those accessible from the sea, requires unique leveraging and integration of all the capabilities of U.S. military forces to successfully conduct the operation. This appendix discusses many of these capabilities; JP 3-06 details joint urban operations.

PURPOSE

D-1. In some situations, a major urban operation is required in an inland area where only Army forces are operating. Army commanders determine if the unique requirements of the urban environment require forming a joint task force (JTF) or, if not, request support by joint capabilities from the higher joint headquarters. Sometimes the nature of the operation is straightforward enough or the urban operation is on a small enough scale that conventional intraservice support relationships are sufficient to meet the mission requirements.

D-2. Most major urban operations (UO), however, require the close cooperation and application of joint service capabilities. A JTF may be designated to closely synchronize the efforts of all services and functions in an urban area designated as a joint operations area (JOA). If a large urban area falls in the context of an even larger ground force area of operations, a JTF dedicated to the urban operation may not be appropriate. These situations still require joint capabilities. In such cases, the responsible JFC designates support relations between major land units and joint functional commands. The major land units can consist of Army forces, Marine Corps forces, or joint forces land component command. The joint functional commands can consist of the joint special operations task force, joint psychological operations task force, or joint civil-military operations task force.

D-3. This appendix describes the roles of other services and joint combatant commands in UO. It provides an understanding that enables Army commanders to recommend when to form a JTF or to request support from the JFC. It also provides information so commanders can better coordinate their efforts with those of the JFC and the commanders of other services or components conducting UO. Lastly, this appendix describes some considerations when conducting UO with multinational forces.

SISTER SERVICE URBAN CAPABILITIES

D-4. Army forces conducting UO rely on other services and functional joint commands for specialized support in the urban environment. These capabilities are requested from and provided through the commanding JFC. Army forces request the assets and capabilities described in this annex through their higher headquarters to the joint command. The JFC determines if the assets will be made available, the appropriate command relationship, and the duration of the support. Army forces prepare to coordinate planning and execution with other services and to exchange liaison officers. These capabilities can greatly increase the Army's ability to understand, shape, engage, consolidate, and transition within the context of UO.

AIR FORCE

D-5. Air Force support is an important aspect of the Army force concept for urban operations. Air Force elements have a role to play in UO across the spectrum of operations.

D-6. Air Force intelligence, surveillance, and reconnaissance (ISR) systems contribute significantly to understanding the urban environment. These ISR systems include the E-8 Joint Surveillance Target Attack Radar System (JSTARS) (see figure D-1), U2 Dragon Lady, RC-135 Rivet Joint, or RQ-4 Global Hawk and RQ/MQ-1 Predator unmanned aerial systems. Air Force ISR systems can provide vital data to help assess threat intentions, threat dispositions, and an understanding of the civilian population. These systems also can downlink raw information in real-time to Army intelligence processing and display systems, such as the common ground station or division tactical exploitation system.

Figure D-1. USAF E-8 JSTARS platform

D-7. Air interdiction (AI) can be a vital component of shaping the urban operational environment. Often, AI of the avenues of approach into the urban area isolates the threat by diverting, disrupting, delaying, or destroying portions of the threat force before they can be used effectively against Army forces. AI is especially effective in major combat operations where restrictions on airpower are limited and the threat is more likely to be a conventionally equipped enemy. In 1991 during OPERATION DESERT STORM, AI helped prevent the Iraqi 5th Mechanized Division from reaching Khafji.

D-8. Precision-guided munitions (PGMs) delivered from combat aircraft can positively and directly affect the conduct of Army close combat actions in the urban area. Special munitions designed to penetrate reinforced concrete can provide unique support to land forces executing UO. Although laser-guided bombs are extremely accurate with an effective laser lock-on, environmental factors such as smoke, haze, and weather can adversely affect successful laser designation. If the launching aircraft can achieve a successful laser designation and lock-on, these weapons have devastating effects, penetrating deep into reinforced concrete before exploding with great force. If launched without a lock-on, or if the laser spot is lost, these weapons are unpredictable and can travel long distances before they impact. However, continual advancements in tactics and weapons systems, such as joint direct attack munition (JDAM) weapons, help decrease the adverse effects of these environmental factors. (For example, in 2003 during OPERATION IRAQI FREEDOM, the Air Force supported the Army's attack on An Najaf with JDAM-equipped F 16s and B 1s during a blinding sandstorm. With near zero visibility, JDAM weapons provided significant fire support, allowing Army forces to complete their mission and continue the attack north.) When PGMs are unavailable, general-purpose bombs from 500 to 2,000 pounds can also prove effective. However, there may be risk factors associated with their use such as collateral damage due to decreased accuracy, threat exposure to the delivering aircraft due to lower delivery altitudes, or reduced weapons effectiveness due to shallower penetration angles. (See FM 3-06.1 for a more detailed description of various aircraft weapons and capabilities.)

D-9. In addition to shaping and engaging the UO through firepower, commanders can use Air Force capabilities to improve and augment the urban transportation and distribution infrastructure. Air Force units can repair or improve airfields, revitalize civil aviation maintenance facilities, manage air-delivered cargo, and control civil and military air traffic. These latter capabilities particularly enhance urban stability and civil support operations. These capabilities may even be decisive. Air-delivered cargo and air traffic management, for example, were the decisive factors in U.S. forces' successful resistance of the Soviet blockade of Berlin in 1948. In OPERATION RESTORE HOPE, from December 1992 to May 1993, Air Force operations in Mogadishu were critical to airlifting and staging supplies and forces. The Air Force determined the ultimate success of the humanitarian assistance operation (see Appendix C).

D-10. In unique situations, such as the Berlin Blockade and OPERATION RESTORE HOPE, exercising Air Force urban capabilities may be the decisive action of the operation. Most often, however, Air Force capabilities will play a shaping role, sustaining role, or both in joint urban operations because of the inherent requirement to occupy terrain and interface with the population.

MARINE CORPS

D-11. The Marine Corps can assault across water obstacles into a defended urban environment. This capability is an invaluable tactical and operational tool. The mere threat of this capability can divert many enemy forces from other avenues of approach and obscure the true nature of an attack. The impact of the threat of amphibious assault was vividly demonstrated during OPERATION DESERT STORM where embarked Marine Forces diverted several Iraqi divisions to defensive positions along the coast and near Kuwait City.

D-12. The presence of Marine amphibious equipment, apart from Marine infantry, provides Army forces unique capabilities. In UO, the amphibious operation is often not an assault from the sea, but rather an assault river crossing. In 1950, the 7th U.S. Infantry Division used amphibious tractor support from the 1st Marine Division to conduct an assault river crossing of the Han River into downtown Seoul. See FM 3-31.1.

D-13. The worldwide deployment of Marine air-ground task forces (MAGTFs) enables a short notice response into any urban areas accessible from the sea. Typically, a deployed MAGTF is a Marine expeditionary unit (special operations capable) (MEU[SOC]). The MEU(SOC) can perform forcible entry operations, seize lodgments, and may execute these tasks anticipating reinforcement by Army or joint forces. They are also well positioned and equipped to rapidly reinforce Army forces already deployed in theater. The special-operations-capable training that these units accomplish before deploying includes urban warfare training and contributes to their value in UO. The MEU(SOC) is relatively small (its core unit is a Marine infantry battalion), is forward deployed, and has a wide spectrum of organic capabilities. It is an important asset in crisis stability and civil support operations.

D-14. The Marine expeditionary brigade is the MAGTF between a MEU and a Marine expeditionary force (the largest MAGTF). All MAGTFs include an aviation combat element, which is particularly skilled at providing accurate and timely close air support. The responsiveness and accuracy of Marine close air support aptly suits it to UO where the lethality of combat and the close range of engagements demand accurate and responsive fires.

D-15. Marine forces that conduct UO work well in littoral urban areas because of their unique relationship with naval forces and thus their capability to closely integrate land and sea operations. A supporting arms liaison team may be attached to Army forces at battalion level to provide ship-to-shore communications and coordination for naval gunfire support.

NAVY

D-16. Many major urban areas are accessible from the sea. Therefore, Army commanders should understand how sea power can influence and support UO. The Navy brings several major capabilities to UO. These include naval gunfire support, naval air support, and port and coastal security. In addition, the Navy can offer secure command post sites and meeting areas; provide limited emergency shelter and critical care for Soldiers and, as necessary, civilians; and store and transport critical supplies ashore providing flexibility in meeting support requirements. The ability to provide support from "over the horizon" not only enhances force protection but also the legitimacy of urban and host nation authorities during transition.

D-17. Naval surface fire support (which includes naval gunfire support) particularly applies to forcible entry operations in littoral urban areas. It provides an initial indirect fire support capability until Army forces land ashore. However, naval gunfire support lacks a precision munitions capability. In defensive operations, naval gunfire support can also add major fires, especially during retrograde operations through an urban area. Naval gunfire support of the 3rd Infantry Division was the primary fire support for the last two days of the X Corps retrograde operation out of the port of Hungnam, North Korea, in December 1950.

D-18. Destroyers and cruisers, which mount one and two 127mm MK45 lightweight gun systems respectively (see figure D-2), are capable of providing naval gunfire support. This gun system can provide a rate of fire of 20 rounds per minute per gun to a range of approximately 23 kilometers. Destroyers, cruisers, and attack submarines are capable of precision strikes with Tomahawk cruise missiles. Depending on the location of the urban area, naval air can provide responsive aviation support to UO alleviating the need for fixed-wing aviation bases.

Figure D-2. USN MK45 lightweight gun system

D-19. Due to its flat trajectory, terrain masking affects naval gunfire more than field artillery. Naval gunfire also results in large range probable errors (the dispersion pattern of the naval gun is roughly elliptical with the long axis in the direction of fire). Hence, coverage of targets such as roads and airfields is most effective when the gun-target line (GTL) coincides with the long axis of the target. Very close supporting fire can be delivered when the GTL is parallel to the front line of troops. Oppositely, a GTL perpendicular to the front trace can endanger friendly forces. Within the limits of hydrographic conditions, the ship can maneuver to achieve a better GTL, but ship movement also makes it difficult to adjust fire. Overall, naval and air threats, bad weather, and large range probable errors make naval gunfire difficult and can cause cancellation of supporting fires.

D-20. Water terminals located in urban areas are usually the debarkation points for the bulk of Army forces as well as a joint force theater logistics requirement. Army forces are responsible for water terminal operations while naval capabilities protect these strategic and operationally vital facilities.

D-21. Naval coastal warfare (NCW) is the responsibility of the JFC and is often exercised through the Navy component commander. He may assign a naval coastal warfare commander for an appropriate geographic area. NCW includes coastal sea control, port security, and harbor defense. While coastal sea control is conducted in the environment of the open seas, port security and harbor security include the urban environment. Port security is the safeguarding of vessels, harbors, ports, waterfront facilities, and cargo from internal threats. It includes destruction, loss, or injury from sabotage or other subversive acts; accidents; thefts; or other causes of similar nature. The Navy's role in protecting essential urban infrastructure is often key. This task is also important when executing stability operations if the threats against urban infrastructure will likely be unconventional.

D-22. Harbor defense protects harbor approaches, harbors, anchorages, and ports from external threats. Harbor defense focuses on the conventional defense of port infrastructure. It is a task appropriate in a major operation or campaign and often includes port security as a subtask. The JFC executes NCW (focused on harbor defense and port security) using combined Navy capabilities including surface warfare, aviation, and naval special operations. See also discussions in this appendix on transportation command and special operations command.

COAST GUARD

D-23. The Coast Guard, like the Navy, can significantly influence the conduct of UO when the urban area is accessible from the sea. The U.S. Coast Guard (USCG) is the federal authority for port security and harbor defense of domestic facilities. When directed by the President, the USCG can augment the Navy in operations overseas. Historically, the entire USCG was under Navy control during both World Wars I and II. USCG elements deployed overseas and operated under Navy control during the Vietnam War and during OPERATIONS DESERT SHIELD and DESERT STORM.

D-24. The USCG uses surface warfare systems and aerial reconnaissance systems to conduct its missions. Its air systems are unarmed, whereas its surface systems are armed for self-defense and law enforcement operations. The USCG is experienced and adept at supporting other agencies, local governments, and law enforcement. Its capabilities can best support Army UO in civil support operations or the civil support aspects of an operation. The USCG works effectively against an unconventional threat and threats with on-water capability.

URBAN FUNCTIONAL COMBATANT COMMAND CAPABILITIES

D-25. Three commands provide urban functional combatant command capabilities. Transportation Command manages the global transportation system. Strategic Command integrates plans for and employs space forces ad assets. Special Operations Command is responsible for providing special operations forces.

TRANSPORTATION COMMAND

D-26. U.S. Transportation Command (USTRANSCOM) provides strategic air, land, and sea transportation for the Department of Defense to deploy, employ, sustain, and redeploy U.S. military forces worldwide. USTRANSCOM provides global transportation management, using an integrated transportation system across the spectrum of operations through its transportation component commands (TCCs). The TCC consists of Air Mobility Command (AMC), Military Sealift Command (MSC), and Surface Deployment and Distribution Command (SDDC).

D-27. During urban operations, USTRANSCOM, through its TCC, can provide common-user terminal services in support of strategic transportation movements to a theater of operations. AMC provides common-user airlift, air refueling, and aeromedical evacuation services. It is the worldwide aerial port manager and, where designated, the operator of common-user aerial ports of embarkation and aerial ports of debarkation. MSC provides common-user sealift services between seaports of embarkation (SPOEs) and seaports of debarkation (SPODs). SDDC provides common-user ocean terminal service and, where designated, serves as the single port manager at SPOEs and SPODs. The urban transport system, in the form of railheads, ports, and airfields, is integral to many urban operations across the spectrum of Army operations and often the objective of Army UO. Army forces planning, preparing, and executing urban operations engage the supported geographic commander of a combatant command or his component commands in all aspects of conducting UO when the objective is transport related.

D-28. In offensive operations, commanders use transportation expertise to identify the urban transportation infrastructure (both the terrain and social [human] aspects) that is secured or that can affect current and future operations. This analysis also includes second- and third-order support systems. USTRANSCOM is consulted regarding the degree of acceptable damage that the system can sustain and still meet mission requirements. During execution, USTRANSCOM units may integrate into the operation so they can begin operating the transportation systems as early as possible. During the Inchon landing of September 1950, Army forces had begun rail operations on D+1 and port operations under way by D+3 of the forcible entry.

D-29. In defensive and stability or civil support operations, USTRANSCOM units safeguard and prevent disruption of the transport system by the conduct of defensive operations or stability tasks.

USTRANSCOM advises Army commanders of the impact of defensive tactics, techniques, and procedures (TTP) on USTRANSCOM operations as well as security requirements for USTRANSCOM facilities. The JFC provides guidance to deconflict any issues that may arise from these potentially divergent missions and tasks.

Strategic Command

D-30. The United States Strategic Command (USSTRATCOM) is the combatant command that integrates and synchronizes space operations, information operations, integrated missile defense, global command and control (C2); ISR, global strike, and strategic deterrence to provide. A unified source for greater understanding of specific threats around the world and the means to rapidly respond to those threats including weapons of mass destruction. USSTRATCOM operates assigned forces through its service component commands: U.S. Army Space Command/Army Forces Strategic Command; Air Force Space Command, and Naval Net Warfare Command.

D-31. USSTRATCOM provides critical space systems that offer global coverage and potential for real-time and near real-time support to urban operations. Space systems are unconstrained by political boundaries. Commanders can use space systems during peacetime or times of crisis to monitor an urban area before inserting friendly forces. Space systems enhance operations and assist commanders in overcoming some of the physical challenges of the urban environment. These enhancement operations include ISR; positioning and navigation; environmental monitoring; and communications. Space-based imagery and sensors are important ISR capabilities that contribute to situational understanding throughout the depth, breadth, and height of the urban area. ISR systems can provide route and target information for mission planning, locate presurveyed missile launch sites, detect camouflage, assess threat operations and movements, and warn of hostile acts and reconnaissance. Positioning and navigation systems also assist situational understanding through links to digital information systems (INFOSYS) while assisting tactical navigation in ambiguous terrain found in some urban areas. These enhancements can include precise location and position information for urban fires, ingress and egress routes, and rendezvous coordination. Environmental monitoring systems can provide weather and ionospheric information needed to assess weapon selection, air routes, ground and water trafficability, and communications. Communications systems provide secure, survivable links between elements of Army and joint forces to disseminate plans, orders, and warnings. These systems may form a critical link in the INFOSYS that transmit data to assessment centers and intelligence to key decision makers. However, some of the same environmental influences and degradation as ground-based systems may affect space-based systems. Commanders should refer to FM 3-14 for more information on space support to Army operations.

Special Operations Command

D-32. U.S. Special Operations Command (USSOCOM) exercises combatant command of all active and reserve special operations forces stationed in the United States. USSOCOM also provides trained and combat-ready special operations forces (SOF) to the geographical combatant commanders and, when directed by the President or the Secretary of Defense, command-designated special operations.

D-33. SOF provide commanders with capabilities critical to success in the urban environment. The density of this environment in both space and time requires the careful integration of SOF and conventional forces. Army forces conducting UO have a clear and unambiguous C2 relationship with the SOF in the urban area to ensure coordination, massing of effects, and unity of effort.

D-34. Each service has unique special operations (SO) capabilities. For example, SOF can identify and seize or destroy key terrain or infrastructure in denied areas; secure or capture key personnel; identify supportive civilians (aides and allies) who can facilitate future infiltrations and the cache of critical supplies; counter urban insurgencies; and conduct unconventional warfare in enemy-held urban areas. SOF can also emplace covert sensors and surveillance devices, provide clandestine intelligence collection, and provide target acquisition information or emplace beacons (for weapons guidance or navigation) in the highly restrictive terrain of the urban environment. In multinational UO, they can provide coalition support teams with trained, culturally aware, language proficient, military liaison personnel with organic communications connectivity. Army leaders understand the SO capabilities available to ensure that they request the right support for Army UO and to ensure unity of effort within the urban JOA.

Army SOF

D-35. The Army provides five types of SOF units to USSOCOM. Two of these, psychological operations (PSYOP) and civil affairs, are discussed in Chapter 4.

Ranger Forces

D-36. The U.S. Army 75th Ranger Regiment can rapidly deploy light infantry forces from company through regimental size. Rangers specialize in direct action (DA) SO missions and focus on airfield seizure and raids. Typically, Ranger units turn over their objectives to conventional units upon mission completion. The Ranger DA capability especially applies to UO because of the many critical infrastructures that often prompt Army forces to engage in UO. Ranger capabilities are ideal for seizing critical facilities to preempt their defense or destruction. The size and combat power of Ranger units permit execution of offensive and defensive operations against enemy conventional units for periods of limited duration; austere combat service support capabilities limit the Ranger regiment's ability to sustain combat action without extensive augmentation.

Special Forces

D-37. U.S. Army Special Forces Command (Airborne) trains and prepares Army Special Forces (SF) to deploy and execute operational requirements for geographical combatant commanders. SF units are small and capable of extended operations in remote and hostile locations. SF units execute seven basic missions: foreign internal defense (FID), unconventional warfare (UW), counterproliferation of weapons of mass destruction, special reconnaissance (SR), DA, counterterrorism, and information operations. Important collateral activities consist of coalition support, combat search and rescue, counterdrug activities, humanitarian demining activities, humanitarian assistance, security assistance, and special activities.

D-38. SR, DA, and coalition support are particularly important in UO. SF units may physically penetrate an urban area to conduct SR to determine threat strengths, dispositions, and intentions. In some situations, SF units, due to their language and cultural training, can gather accurate information regarding the disposition and attitudes of the population. SF special reconnaissance can also determine or verify the functional status of urban infrastructure as well as conduct target acquisition, area assessment, and poststrike reconnaissance.

D-39. SF teams can execute terminal guidance and with qualified joint tactical air controllers can control operations for Army and joint precision fires. Although these teams possess limited organic combat power for their short-duration, DA missions, they can bring significant effects to bear against high-payoff targets. Yet, SF units have no capability to conduct a sustained defense of such targets.

D-40. SF units can advise, train, and assist urban indigenous movements already in existence to conduct unconventional warfare and possibly accompany these groups into combat. The upsurge in urban insurgency and terrorism has caused worldwide concern since it is not confined to developing countries. Present day dissident groups are well aware of this situation. These groups realize that to be successful, they must center the insurgent activities on the major cities or political center of their countries.

Special Operations Aviation Forces

D-41. Army special operations aviation provided by the 160th Special Operations Aviation Regiment operates primarily to support SO missions. They execute insertion, extraction, and resupply missions to support SO. Similar to conventional aviation, this capability may be more vulnerable to concealed air defenses when operating over hostile or unsecured urban terrain than in many other environments.

Navy SOF

Sea-Air-Land Teams

D-42. Navy sea-air-land teams (SEALs) specialize in water approaches to targets. They operate in small, squad-size teams and have many of the same capabilities as Army Special Forces (see above). Navy SEALs do not typically have the cultural and language training of Army Special Forces. Their capability to insert from sea gives them a unique ability to penetrate into urban areas that are accessible from the sea.

Special Boat Units

D-43. Special boat units (SBUs) employ, operate, and maintain various surface combatant craft (see figure D-3) to conduct and support naval and joint special operations, riverine warfare, and coastal patrol and interdiction. The SBU can infiltrate and exfiltrate forces; provide small-caliber gunfire support; and conduct coastal patrol, surveillance, harassment, and interdiction of maritime lines of communications. These units are ideal in UO that include ports and rivers. They can assist in port security, conduct river patrols, and participate in harbor defense. They are well suited for preventing sea infiltration by unconventional threats.

Figure D-3. USN MK V special

Operations CraftAir Force SOF

D-44. The Air Force has three primary SOF elements: special operations air units, battlefield Airmen, and combat aviation advisory units.

Special Operations Wings and Groups

D-45. Air Force special operations aviation elements operate both fixed-wing and rotary-wing aircraft (which include variants of both C-130s and MH-53s). Air Force SOF conduct core missions which include specialized air mobility, precision aerospace fires, aerospace surface interface, personnel recovery operations, specialized refueling, combat aviation advisory, information operations, and PSYOP. As a result, Air Force SOF conduct infiltrations, exfiltrations, resupplies, close air support of SOF ground elements, electronic warfare, and aerial C2 support.

D-46. The AC-130 U and AC-130 H model aircraft (see figure D-4) are designed specifically to provide close air support to ground SOF. However, they will be an invaluable asset to any ground force operating in an urban environment. Hence, Army commanders seek opportunities to train with AC-130 gunships prior to the conduct of UO. AC-130s are armed with one 40mm autocannon and one 105mm howitzer. The AC-130 U is also armed with one 25mm autocannon. This aircraft has night capability and is extremely accurate. Its fires are responsive and can be decisive in close urban combat. Its cannon and howitzer are accurate enough to concentrate fire onto a single spot to create a rooftop breach that allows fire to be directed deeper into the building. The aircraft can provide excellent covert illumination with its infrared spotlight. The AC-130 is very vulnerable to air defense systems, which friendly forces must suppress or destroy to effectively use this system. (During OPERATION DESERT STORM in 1991, an AC-130 was shot down over Kuwait.)

Battlefield Airmen

D-47. Battlefield Airmen consist of special tactics teams (STTs) and special operations weather teams (SOWTs). Combat controllers and pararescue personnel constitute STTs. STTs are specially tailored to meet mission criteria and may vary from a small three-man team to a larger twenty-five-man element. STTs support the UO Army commanders by—

- Performing air-land-sea personnel recovery operations.
- Providing terminal attack control or guidance.
- Establishing and operating navigational aids and beacons.
- Providing liaison to ground commanders.
- Providing visual flight rules and limited instrument flight rules air traffic control.
- Providing positive control of the terminal objective area aviation environment during SOF operations.

Note: SOWTs are normally attached to Army SOF to provide weather observation and limited tactical forecasting.

Combat Aviation Advisory (CAA) Operations

D-48. CAA operations are part of an interagency process aimed toward freeing and protecting the host nation's population from subversion, lawlessness, and insurgency. Specifically, CAA operations are tailored to assess, train, advise, and assist foreign aviation forces in air operations employment and sustainability. CAA operations support geographical combatant commanders throughout the spectrum of conflict primarily by facilitating the availability, reliability, safety and interoperability, and integration of friendly and allied aviation forces supporting joint, interagency, and multinational forces and organizations. CAA operations provide assistance in the interrelated areas of FID, coalition support, UW, and humanitarian and disaster relief. CAA operations also include a liaison role in coalition support.

MULTINATIONAL CONSIDERATIONS

D-49. Army UO in foreign urban areas will often be joint and likely have a multinational component. When properly executed, integrating multinational forces into UO greatly enhances the operation's military (as well as political) effectiveness. Properly integrating multinational forces into UO requires a thorough understanding of both the urban environment and the nature of individual national forces. This understanding includes the political, cultural, and historical characteristics of the other national forces. Such understanding also includes the national force's doctrine and military capabilities, strengths, and weaknesses. In UO, these considerations are critical because these factors will alter the urban population's attitude toward multinational forces and the behavior of such forces as they interact with the urban population. Combining this understanding with effective C2 and an equal understanding of the urban environment results in effective multinational UO (see FM 100-8).

CULTURAL COMPATIBILITY

D-50. When evaluating and assigning UO tasks, Army commanders also consider the degree of interaction with the civilian population. A national force from a Muslim-majority country may work better with a civilian population that is also Muslim than a force having a different religion. In such a situation, the national force with the same religion as the urban population may be assigned tasks that require close relations with civilians. Army forces, in this situation, may be assigned tasks that are more remote from the population. National forces that have a national history of animosity to the civil population (or that sympathize with antagonists within the civilian population) are not used in tasks requiring diplomacy and close cooperation with the civilians or government.

Military Capabilities and Restrictions

D-51. Army forces are responsible to understand the military capabilities of national forces with which they work. Some national forces, as part of their normal capabilities, are adept at police functions that enable them to operate with little training in a law enforcement role. Other national forces specialize in small-unit, light infantry patrolling. These forces may be ideal in a stability operation. In contrast, a national force composed of conscripts and trained primarily in conventional warfare techniques may best work as a reinforcing force or may require extensive training before mission execution in an urban environment or a stability situation. Army commanders also consider the type of weapon systems with which participating multinational forces are equipped; they may be more or less effective in an urban environment. For example, some countries may still possess the Vulcan antiaircraft gun system (or similar weapon) that can be very useful in urban offensive and defensive operations.

D-52. Commanders in multinational UO must also understand nation-specific rules of engagement (ROE). Some nations will severely limit their participating military's use of lethal and nonlethal fires. On the other hand, other nations may not be as constrained as Army units. Ideally, commanders attempt to create ROE that are the same across the coalition but recognize that this may not be achievable. Hence, in an environment where minimization of collateral damage will be paramount, Army commanders will need to carefully consider national limitations on the employment of forces when planning and executing UO. Relatedly, commanders take into account each nation's ability to achieve precision effects with their available forces and systems.

Intelligence Sharing

D-53. Due to classification issues, sharing intelligence with coalition partners may be challenging; the United States may have close intelligence ties with some countries within the coalition and few or none with others. In many cases intelligence personnel from other countries have unique skills (and cultural perspectives) that can significantly contribute to the friendly intelligence effort. Reports from some coalition members may fill intelligence gaps for Army forces and the coalition as a whole. On the other hand, few countries have the sophisticated intelligence collection assets available to Army forces and information that Army units may provide could be critical to a coalition partner's force protection and overall mission success.

Information Management

D-54. National capabilities to collect, process, store, display, and disseminate information and establish a common operational picture vary considerably between coalition partners. As the capabilities of Army information systems increase, those of our future partners may not keep pace. Even when capabilities are similar, equipment may not be compatible. The complex nature of the urban environment and coalition language differences (and linguist shortages) will only serve to exacerbate incompatibilities and limitations. Therefore commanders must understand capabilities and limitations to ensure that orders, control measures, and information can be appropriately disseminated, displayed, and understood. Significantly, they must understand the coalition's ability to accurately locate and track friendly forces as well as identify friend from foe. Commanders must take the appropriate steps to ensure that essential intelligence and relevant information is provided to coalition partners in a form (and language) that they can accept and use. This will be critical to establishing the common operational picture necessary to facilitate rapid decision making and effective command and control across the coalition.

Logistic Requirements

D-55. Logistically, agreement among the multinational forces should include support consolidation whenever possible. Creating a multinational logistics office or coordination center works best when accomplished early. This office can coordinate local contracts as well as already agreed upon host-nation support. This coordination among participating nations will reduce the competition for local assets that could otherwise have detrimental effects on one or more participating nations. Additionally, many nations providing forces for a coalition effort may not be able to support them logistically and will rely almost entirely on the United States or another coalition army for logistic support. Finally, pooled resources such as transportation assets and maintenance personnel could potentially reduce deployment requirements for one or more coalition partners. As the coalition matures, the role of this office or center may be expanded to include command activities.

Source Notes

These are sources quoted or paraphrased in this publication. They are listed by page number. Quotations are identified by the first few word of the quote. Where a quote is embedded within a paragraph, the paragraph number is listed. Boldface indicates the title of historical vignettes.

Chapter 1–The Urban Outlook

1-1 "Today's security environment demands more...": FM 1, *The Army*, 14 June 2005: 1-19.

1-4 **Rome: A Microcosm of Urban Warfare**: R. Ernest Dupuy and Trevor N. Dupuy, *The Encyclopedia of Military History: From 3500 B.C. to the Present* (San Francisco: Harper & Row, 1977).

1-6 **The Three Block War**: as quoted in Russell W. Glenn, Steven Hartman, and Scott Gerwehr, *Urban Combat Service Support Operations: The Shoulders of Atlas* (Santa Monica, CA: RAND, 2003): 26.

1-8 **Winning the Peace in Iraq: The Requirement for Full Spectrum Operations**: Peter W. Chiarelli and Patrick R. Michaelis, "Winning the Peace: The Requirement for Full-Spectrum Operations," *Military Review* (July/August 2005): 4.

Chapter 2–Understanding the Urban Environment

2-1 "From a planning perspective, commanders view cities….": FM 3-0, *Operations*, 14 June 2001: 6-19.

2-16 "Me and Somalia Against the World": cited in Andrew S. Natsios, "Food Through Force: Humanitarian Intervention and U.S. Policy," *The Washington Quarterly* (winter 1994): 136.

2-18 **Understanding the Effects of Unit and Soldier Actions in Iraq**: *Initial Impressions Report: Operation Iraqi Freedom—Stability Operations-Support Operations,* Center for Army Lessons Learned (CALL) Newsletter No. 04-13 (Fort Leavenworth, KS: Center for Army Lessons Learned, U.S. Training and Doctrine Command [TRADOC], 2004): ii.

2-21 "Whoever coined the phrase… ": General Sir Rupert Smith, "Wars in Our Time–A Survey of Recent and Continuing Conflicts," *World Defence Systems*, volume 3:2 (London: Royal United Services Institute for Defence Studies, 2001): 121.

2-25 **Understanding the Urban Environment: Paris – 1944**: Russell W. Glenn, "Urban Combat is Complex," *Proceedings* (February 2002): 65. Russell W. Glenn, Steven Hartman, and Scott Gerwehr, *Urban Combat Service Support Operations: The Shoulders of Atlas* (Santa Monica, CA: RAND, 2003): 10-11.

Chapter 3–Understanding the Urban Threat

3-1 "…[T]he United States could be forced to intervene...": *Quadrennial Defense Review Report*, 30 September 2001: 6.

3-3 "…Iraq made no direct effort...": Gregory Fontenot, E. J. Degen, and David Tohn, *On Point: US Army in Operation IRAQI FREEDOM* (Fort Leavenworth, KS: Combat Studies Institute Press, 2004): 388.

3-4 **Tempo**: 26th Infantry Regimental Association, *Aachen. Military Operations in Urban Terrain* (Lititz, PA: 26th Infantry Regimental Association, 1999); Irving Werstein, *The Battle of Aachen* (New York: Thomas Y. Crowell Company, 1962); Charles Whiting, *Bloody Aachen* (New York: Stein and Day, 1976).

3-6 "Chechen fighters sometimes disguised...": Olga Oliker, *Russia's Chechen Wars 1994–2000: Lessons from Urban Combat* (Santa Monica, CA: RAND, 2001): 21.

3-7 **Identifying Threats from Noncombatants**: Olga Oliker, *Russia's Chechen Wars 1994–2000: Lessons from Urban Combat* (Santa Monica, CA: RAND, 2001).

3-9 **Information and the Media**: Eric Hammel, *Fire in the Streets: The Battle for Hue, Tet 1968* (New York: Dell Publishing, 1991).

3-12 **Cultural and Religious Instability**: Sid Heal, "Crowds, Mobs and Nonlethal Weapons," *Military Review* (March/April 2000): 45–50.

3-13 **Food and Water Shortages**: Leif R. Rosenberger, "The Strategic Importance of the World Food Supply," *Parameters* (spring 1997): 84–105.

3-15 **Insurgencies and the Urban Society**: Roger Trinquier, *Modern Warfare: A French View of Counterinsurgency* (New York: Praeger Publisher, 1964): 16-17.

3-16 **Crime and Criminal Organizations**: Eugene Linden, "The Exploding Cities of the Developing World," *Foreign Affairs* (January/February 1996): 52–65; Andrew S. Natsios, "Commander's Guidance: A Challenge of Complex Humanitarian Emergencies," *Parameters* (summer 1996): 50–66.

Chapter 4–Understanding the Urban Environment's Effects on Warfighting Functions and Tactics

4-1 "War is, above all things, an art...": Francis V. Greene, "The Important Improvements in the Art of War During the Past Twenty Years and Their Probable Effect on Future Military Operations," *Journal of the Military Service Institution of the United States* 4, no. 13 (1883): 41.

4-13 "Fighting in a city is much...": Vasili I. Chuikov, *The Battle for Stalingrad* (New York: Holt, Rinehart, and Winston, 1964): as cited by JP 3-06, *Doctrine for Urban Operations*, 16 September 2002: 1-8.

4-14 "I heard small-arms fire and...": Mark A. B. Hollis, "Platoon Under Fire," *Infantry* (January/April 1998): 29–30.

4-17 **Example of Simple Communications Innovation: Israel's Six-Day War – 1967**: Trevor N. Dupuy and Paul Martell, *Flawed Victory: The Arab-Israeli Conflict and the 1982 War in Lebanon* (Fairfax, VA: Hero Books, 1986); Richard A. Gabriel, *Operation Peace for Galilee: The Israeli-PLO War in Lebanon* (New York: Hill and Wang, 1984).

Chapter 5–Contemplating Urban Operations

5-1 "We based all our further calculations...": S. M. Shtemenko, *The Soviet General Staff at War, 1941-1945: Book One* (Moscow: Progress Publishers, 1981): 317–318.

5-19 "Four hostile newspapers...": Napoleon Bonaparte, Justin Wintle (ed.), *The Dictionary of War Quotations* (New York: The Free Press, 1989): 73.

5-24 **Conventional and Special Forces Integration**: Gregory Fontenot, E. J. Degen, and David Tohn, *On Point: US Army in Operation IRAQI FREEDOM* (Fort Leavenworth, KS: Combat Studies Institute Press, 2004): 403.

Chapter 6–Foundations for Urban Operations

6-1 "Utilities such as electricity…": Richard Connaughton, John Pimlott, and Duncan Anderson, *The Battle for Manila: The Most Devastating Untold Story of World War II* (Novata, CA: Presidio Press, 1995): 103.

6-7 **Applying the Urban Operational Framework: Panama – 1989**: Ronald H. Cole, *Operation Just Cause* (Washington DC: Joint History Office, 1995); Thomas Donnelly, Margaret Roth, and Caleb Baker, *Operation Just Cause: The Storming of Panama* (New York: Lexington Books, 1991); John T. Fishel, *The Fog of Peace: Planning and Executing the Restoration of Panama* (Carlisle, PA: Strategic Studies Institute, 1992); Malcolm McConnell, *Just Cause: The Real Story of America's High-Tech Invasion of Panama* (New York: St. Martin's Press, 1991); John Embry Parkerson, Jr., "United States Compliance with Humanitarian Law Respecting Civilians During Operation Just Cause," *Military Law Review* (1991): 31–140; Jennifer Morrison Taw, *Operation Just Cause: Lessons for Operations Other Than War* (Santa Monica, CA: RAND, 1996).

6-12 "I'm talking about attacking...": William S. Wallace quoted in Michael R. Gordon, "Baghdad Targets Picked if Hussein Holes Up There," *New York Times* (7 March 2003): A11.

6-13 "Do not try to..." T. E. Lawrence, "Twenty-Seven Articles," *The Arab Bulletin* (20 August 1917): as cited in David H. Petraeus, "Learning Counterinsurgency: Observations from Soldiering in Iraq," *Military Review* (January/February 2006): 4–5.

Chapter 7–Urban Offensive Operations

7-1 "…Capture Suez City...": Abraham Adan, *On the Banks of the Suez* (Novato, CA: Presidio Press, 1980): 409.

7-2 **Operational Context of Urban Operations: Brittany Ports – 1944**: Kent H. Butts, "The Strategic Importance of Water," *Parameters* (spring 1997): 65–83; Paul Carell, *Stalingrad: The Defeat of the German 6th Army* (Atglen, PA: Schieffer Publishing, 1993); R. Ernest Dupuy and Trevor N. Dupuy, *The Encyclopedia of Military History: From 3500 B.C. to the Present* (San Francisco: Harper & Row, 1977); Ralph Peters, "Our Soldiers, Their Cities," *Parameters* (spring 1996): 43–50; John J. Peterson, *Into the Cauldron* (Clinton, MD: Clavier House, 1973).

7-10 **Forms of Attack in the Urban Offense: Metz – 1944**: John Colby, *War From the Ground Up: The 90th Division in WWII* (Austin, TX: Nortex Press, 1991); Anthony Kemp, *The Unknown Battle: Metz, 1944* (New York: Stein and Day, 1981).

7-17 **Isolating an Urban Area: Hue, Vietnam – 1968**: James R. Arnold, *Tet Offensive 1968: Turning Point in Vietnam* (London: Osprey, 1990); Eric Hammel, *Fire in the Streets: The Battle for Hue, Tet 1968* (New York: Dell Publishing, 1991).

7-21 "Often what seems to be...": Russell W. Glenn, *An Attack on Duffer's Downtown* (Santa Monica, CA: RAND, 2001): 28.

7-22 **Creative Task Organization: Using Artillery in the Direct Fire Role**: 26th Infantry Regimental Association, *Aachen. Military Operations in Urban Terrain* (Lititz, PA: 26th Infantry Regimental Association, 1999); Ronald H. Cole, *Operation Just Cause* (Washington DC: Joint History Office, 1995); Thomas Donnelly, Margaret Roth, and Caleb Baker, *Operation Just Cause: The Storming of Panama* (New York: Lexington Books, 1991); David Eshel, *Mid-East Wars: The Lebanon War 1982* (Hod Hasharon, Israel: Eshel-Dramit, 1983); Richard A. Gabriel, *Operation Peace for Galilee: The Israeli-PLO War in Lebanon* (New York: Hill and Wang, 1984); Malcolm McConnell, *Just Cause: The Real Story of America's High-Tech Invasion of Panama* (New York: St. Martin's Press, 1991); Irving Werstein, *The Battle of Aachen* (New York: Thomas Y. Crowell Company, 1962); Charles Whiting, *Bloody Aachen* (New York: Stein and Day, 1976); Bruce Allen Watson, *Sieges: A Comparative Study* (Westport, CT: Praeger, 1993).

7-26 **Bold Operational Maneuver to Seize an Urban Area: Inchon and Seoul, Korea – 1950**: Robert Debs Heinl, Jr., *Victory at High Tide: The Inchon-Seoul Campaign* (Annapolis, MD: Nautical & Aviation Publishing Company of America, 1979); Douglas MacArthur, *Reminiscences* (New York: McGraw-Hill, 1964); Shelby L. Stanton, *Ten Corps in Korea, 1950* (Novato, CA: Presidio Press, 1989).

Chapter 8–Urban Defensive Operations

8-1 "Generally, a modern city magnifies...": Christopher R. Gabel, "Military Operations on Urbanized Terrain: The 2d Battalion, 26th Infantry, at Aachen, October 1944," *Urban Combat Operations: tactics, Techniques and Procedure.* Center for Army Lessons Learned (CALL) Newsletter No. 99-16 (Fort Leavenworth, KS: Center for Army Lessons Learned, U.S. Training and Doctrine Command [TRADOC], November 1999): 1-9.

8-5 **Urban Defense in a Major Operation: Stalingrad –1942 to 1943**: Anthony Beevor, *Stalingrad* (New York: Penguin Books, 1998); Vasili I. Chuikov, *The Battle for Stalingrad* (New York: Holt, Rinehart, and Winston, 1964); William Craig, *Enemy at the Gates: The Battle for Stalingrad* (New York: Reader's Digest Press, 1973).

8-8 "What is the position...": Michael Dewar, *War in the Streets: The Story of Urban Combat from Calais to Khafji* (New York: Sterling Publishers, 1992): 8.

8-10 **Defensive Combat Power: Suez – 1973**: Abraham Adan, *On the Banks of the Suez* (Novato, CA: Presidio Press, 1980).

Chapter 9–Urban Stability and Civil Support Operations

9-1 "The Rangers were bound...": Mark Bowden, *Black Hawk Down: A Story of Modern War* (New York: Atlantic Monthly Press, 1999): 18.

9-5 "In wars of intervention...": Antoine-Henri Jomini, "The Art of War," in *Roots of Strategy: Book 2* (Harrisburg, PA: Stackpole Books, 1987): 441.

9-8 **Assessment of Security and Force Protection: Belfast, Northern Ireland**: Michael Dewar, *British Army in Northern Ireland* (New York: Sterling Publishers, 1985); Desmond Hamill, *Pig in the Middle: The Army in Northern Ireland, 1969-1984* (London: Methuen, 1985).

9-11 "If there is any lesson...": Frederick Funston, "How the Army Worked to Save San Francisco," *Cosmopolitan Magazine* (July 1906). Available at http: //www.sfmuseum.org/1906/cosmo.html.

9-12 **Civil Support and Coordination with Civilian Authorities: Los Angeles – 1992**: James D. Delk, *Fires & Furies: The LA Riots* (Palm Springs, CA: Etc. Publications, 1995); William W. Mendel, *Combat in Cities: The LA Riots and Operation Rio* (Fort Leavenworth, KS: Foreign Military Studies Office, 1996).

9-14 **Restraint: An Najaf, Iraq – 2003**: Gregory Fontenot, E. J. Degen, and David Tohn, *On Point: US Army in Operation IRAQI FREEDOM* (Fort Leavenworth, KS: Combat Studies Institute Press, 2004).

9-16 "I met immediately with the...": Jeffrey B. Van Sickle, "Stability Operations in Northern Iraq: Task Force Altun Kupri," *Infantry* (January/February 2005): 26.

Chapter 10–Urban Sustainment

10-1 "Even supply is different…" Ralph Peters, "Our Soldiers, Their Cities," *Parameters* (spring 1996): 48.

10-2 **Base Security: Tan Son Nhut, Vietnam – 1968**: James R. Arnold, *Tet Offensive 1968: Turning Point in Vietnam* (London: Osprey, 1990); Keith William Nolan, *The Battle for Saigon: Tet 1968* (New York: Pocket Books, 1996).

10-4 "[Sustainment] planning accounts for...": FM 3-0, *Operations*, 14 June 2001: 6-19.

10-8 "Greater friendly force density would appear...": Russell W. Glenn, Steven Hartman, and Scott Gerwehr, *Urban Combat Service Support Operations: The Shoulders of Atlas* (Santa Monica, CA: RAND, 2003): 4.

10-13 **Host Nation Repair Operations: OPERATION IRAQI FREEDOM**: Steven Leonard, personal experience written specifically for FM 3-06.

10-18 **Combat Stress: Chechnya – 1994 to 1996**: Timothy L. Thomas and Charles P. O'Hara, "Combat Stress in Chechnya: 'The Equal Opportunity Disorder'," *Army Medical Department Journal* (January/March 2000): 46–53.

10-22 "…[I]n an endeavor like that in Iraq…": David H. Petraeus, "Learning Counterinsurgency: Observations from Soldiering in Iraq," *Military Review* (January/February 2006): 4–5.

Appendix A–Siege of Beirut: An Illustration of the Fundamentals of Urban Operations

A-1 **Siege of Beirut: An Illustration of the Fundamentals of Urban Operations**: Trevor N. Dupuy and Paul Martell, *Flawed Victory: The Arab-Israeli Conflict and the 1982 War in Lebanon* (Fairfax, VA: Hero Books, 1986); David Eshel, *Mid-East Wars: The Lebanon War 1982* (Hod Hasharon, Israel: Eshel-Dramit, 1983); Richard A. Gabriel, *Operation Peace for Galilee: The Israeli-PLO War in Lebanon* (New York: Hill and Wang, 1984).

A-1 "The [Israeli Defense Forces] IDF had...": Richard A. Gabriel, *Operation Peace for Galilee: The Israeli-PLO War in Lebanon* (New York: Hill and Wang, 1984): 128.

Appendix B–An Urban Focus to the Intelligence Preparation of the Battlefield

B-1 "Maneuvers that are possible...": *Infantry in Battle, 2d ed.* (Richmond, VA: Garret & Massie, 1939; Reprint, Fort Leavenworth, KS: U.S. Army Command and General Staff College, 1981): 69.

B-5 **Analysis of an Urban Area's Underlying Terrain: Mitrovica, Kosovo – 1999**: Jamison Jo Medby and Russell W. Glenn, *Street Smart: Intelligence Preparation of the Battlefield for Urban Operations* (Santa Monica, CA: RAND, 2002).

B-10 "…[T]he adversaries of freedom...": John F. Kennedy, Message to Congress, 25 May 1961. Available at <http://www.jfklibrary.org/Historical+Resources/Archives>.

B-12 **Shifting Civilian Interests and Intent**: James D. Delk, *Fires & Furies: The LA Riots* (Palm Springs, CA: Etc. Publications, 1995); Jamison Jo Medby and Russell W. Glenn, *Street Smart: Intelligence Preparation of the Battlefield for Urban Operations* (Santa Monica, CA: RAND, 2002).

Appendix C–Operations in Somalia: Applying the Urban Operational Framework to Stability Operations

C-1 **Operations in Somalia: Applying the Urban Operational Framework to Stability and Reconstruction Operations**: Kenneth Allard, *Somalia Operations: Lessons Learned* (Washington, DC: National Defense University Press, 1995).

C-1 "It's impossible for an American mother...": Thomas M. Montgomery, interview, "Ambush in Mogasdishu," *Frontline*, WGBH Boston, 29 September 1998, <http://www.pbs.org/wgbh/pages/frontline/programs/1998.html>.

C-2 C-5 …" to use all necessary…": *United Nations Security Council Resolution 794* (1992) (New York: United Nations, 1992), <http://daccessdds.un.org/doc/UNDOC/GEN/N92/772/11/PDF>.

C-2 C-5 "When directed by the…": CENTCOM mission statement as cited in Kenneth Allard, *Somalia Operations: Lessons Learned* (Washington, DC: National Defense University Press, 1995): 13-14.

Appendix D–Joint and Multinational Urban Operations

D-1 "[Joint force commanders] synchronize the actions…" JP 3-0, *Doctrine for Joint Operations*, 10 September 2001: II-4.

This page intentionally left blank.

Glossary

SECTION I – ACRONYMS AND ABBREVIATIONS

AI	air interdiction
AIDS	acquired immune deficiency syndrome
AMC	Air Mobility Command
AO	area of operations
ARVN	Army of the Republic of Viet Nam
ASCOPE	areas, structures, capabilities, organizations, people, and events
BCT	brigade combat team
BDAR	battle damage assessment and repair
BENELUX	Belgium, Netherlands, and Luxembourg
C2	command and control
CA	civil affairs
CAA	combat aviation advisory
CASEVAC	casualty evacuation
CBRN	chemical, biological, radiological, nuclear
CBRNE	chemical, biological, radiological, nuclear, high-yield explosive
CCIR	commander's critical information requirement
CI	counterintelligence
CEA	captured enemy ammunition
CERP	commander's emergency response program
CLS	combat lifesaver
CMO	civil-military operations
CMOC	civil-military operations center
CNA	computer network attack
CND	computer network defense
CNE	computer network exploitation
CNO	computer network operations
COA	course of action
COG	center of gravity
CONUS	continental United States
COP	common operational picture
DA	direct action
DNBI	disease and nonbattle injury
DOTMLPF	doctrine, organization, training, materiel, leadership and education, personnel, and facilities
EAC	echelons above corps
EEFI	essential elements of friendly information

EOD	explosive ordnance disposal
EW	electronic warfare
FHP	force health protection
FID	foreign internal defense
FM	field manual; frequency modulation
FMI	field manual, interim
FSB	forward support battalion
G-9	assistant chief of staff, civil affairs
GIS	geographic information system
GTL	gun-target line
HRS	human resources support
HUMINT	human intelligence
IDF	Israeli Defense Forces
IED	improvised explosive device
IMINT	imagery intelligence
INFOSYS	information system
IO	information operations
IPB	intelligence preparation of the battlefield
ISR	intelligence, surveillance, and reconnaissance
JDAM	Joint Direct Attack Munition
JFC	joint force commander
JOA	joint operations area
JP	joint publication
JTF	joint task force
JUO	joint urban operations
LAN	local area network
LNO	liaison officer
LOC	line of communications
LOGCAP	logistics civilian augmentation program
LOS	line of sight
LPT	logistics preparation of the theater
MAGTF	Marine air-ground task force
MANPADS	man-portable air defense system
MEDEVAC	medical evacuation
METT-TC	mission, enemy, terrain and weather, troops and support available, time available, civil considerations
MEU	Marine expeditionary unit
MEU(SOC)	Marine expeditionary unit (special operations capable)
MHE	materials handling equipment
MLRS	Multiple Launch Rocket System
MP	military police

MSC	Military Sealift Command
MST	maintenance support team
NATO	North Atlantic Treaty Organization
NCW	naval coastal warfare
NGA	National Geospatial-Intelligence Agency
NGO	nongovernmental organization
NKPA	North Korean People's Army
NVA	North Vietnamese Army
OAKOC	observation and fields of fire, avenues of approach, key terrain, obstacles, and cover and concealment
ODA	operational detachment alpha
OIF	Operation Iraqi Freedom
OPSEC	operations security
PA	public affairs
PDF	Panamanian Defense Force
PGM	precision-guided munitions
PLO	Palestinian Liberation Organization
PMCS	preventive maintenance checks and services
POW	prisoner of war
PSYOP	psychological operations
QRF	quick reaction force
ROE	rules of engagement
RPG	rocket-propelled grenade
SBU	special boat unit
SDDC	Surface Deployment and Distribution Command
SEAL	sea-air-land (team)
SF	special forces
SIGINT	signals intelligence
SJA	Staff Judge Advocate
SO	special operations
SOF	special operations forces
SOFA	status-of-forces agreement
SOP	standing operating procedure
SOWT	special operations weather team
SPOD	seaport of debarkation
SPE	seaport of embarkation
SR	special reconnaissance
STT	special tactics team
SWET	sewer, water, electricity, trash
SWEAT	sewer, water, electricity, academics trash
SWEAT-MS	sewer, water, electricity, academics trash-medical, security

TACON	tactical control
TC	training circular
TCC	transportation component command
TF	task force
TIM	toxic industrial material
TTP	tactics, techniques, and procedures
UAS	unmanned aircraft system
UN	United Nations
UNITAF	Unified Task Force
UNOSOM	United Nations Operations in Somalia
UNSCR	UN Security Council resolution
UO	urban operations
USACE	United States Army Corps of Engineers
USAID	United States Agency for International Development
USCENTCOM	United States Central Command
USFORSOM	United States Forces, Somalia
USCG	United States Coast Guard
USTRANSCOM	United States Transportation Command
UXO	unexploded ordnance
UW	unconventional warfare
VC	Viet Cong
WMD	weapons of mass destruction

SECTION II – Terms and Definitions

agility – The ability to move and adjust quickly and easily. (FM 3-0)

air interdiction – Air operations conducted to destroy, neutralize, or delay the enemy's military potential before it can be brought to bear effectively against friendly forces at such distance from friendly forces that detailed integration of each air mission with the fire and movement of friendly forces is not required. (JP 1-02)

area defense – A type of defensive operation that concentrates on denying enemy forces access to designated terrain for a specific time rather than destroying the enemy outright. (FM 3-0)

area of interest – That area of concern to the commander, including the area of influence, areas adjacent thereto, and extending into enemy territory to the objectives of current or planned operations. This area also includes areas occupied by enemy forces who could jeopardize the accomplishment of the mission. (JP 1-02)

area of operations – An operational area defined by the joint force commander for land and naval forces. Areas of operations do not typically encompass the entire operational area of the joint force commander, but should be large enough for component commanders to accomplish their missions and protect their forces. (JP 1-02)

ARFOR – The senior Army headquarters and all Army forces assigned or attached to a combatant command, subordinate joint force command, joint functional command, or multinational command. (FM 3-0)

assessment – (Army) The continuous monitoring of the current situation and progress of an operation.(FMI 5-0.1)

asymmetry – Dissimilarities in organization, equipment, doctrine, and values between other armed forces (formally organized or not) and US forces. Engagements are symmetric if forces, technologies, and weapons are similar; they are asymmetric if forces, technologies, and weapons are different, or if a resort to terrorism and rejection of more conventional rules of engagement are the norm. (FM 3-0)

attack – An offensive operation that destroys or defeats enemy forces, seizes and secures terrain, or both. (FM 3-0)

attainability - One of the eight characteristics of combat service support: generating the minimum essential supplies and services necessary to begin operations. (FM 4-0)

attrition - The reduction of the effectiveness of a force caused by loss of personnel and materiel. (JP 1-02)

avenue of approach – An air or ground route of an attacking force of a given size leading to its objective or to key terrain in its path. (JP 1-02)

axis of advance – (Army) The general area through which the bulk of a unit's combat power must move. (FM 3-90)

battle – A set of related engagements that lasts longer and involves larger forces than an engagement. (FM 3-0)

battlefield organization – The allocation of forces in the area of operations by purpose. It consists of three all-encompassing categories of operations: decisive, shaping, and sustaining. (FM 3-0)

battle rhythm – The sequencing of command and control activities within a headquarters and throughout the force to facilitate effective command and control. (FMI 5-0.1)

branch – (Army) A contingency plan or course of action (an option built into the basic plan or course of action) for changing the mission, disposition, orientation, or direction of movement of the force to aid success of the current operation, based on anticipated events, opportunities, or disruptions caused by enemy actions. Army forces prepare branches to exploit success and opportunities, or to counter disruptions caused by enemy actions. (FM 3-0)

breakout – An operation conducted by an encircled force to regain freedom of movement or contact with friendly units. It differs from other attacks only in that a simultaneous defense in other areas of the perimeter must be maintained. (FM 3-90)

buffer zone – A defined area controlled by a peace operations force from which disputing or belligerent forces have been excluded. A buffer zone is formed to create an area of separation between disputing or belligerent forces and reduce the risk of renewed conflict. Also called **area of separation** in some United Nations operations. (JP 1-02)

bypass – A tactical mission task in which the commander directs his unit to maneuver around an obstacle, position, or enemy force to maintain the momentum of the operation while deliberately avoiding combat with an enemy force. (FM 3-90)

campaign – A series of related military operations aimed at accomplishing a strategic or operational objective within a given time and space. (JP 1-02)

canalize – (Army) A tactical mission task in which the commander restricts enemy movement to a narrow zone by exploiting terrain coupled with the use of obstacles, fires, or friendly maneuver. (FM 3-90)

capability - The ability to execute a specified course of action (a capability may or may not be accompanied by an intention) (JP 1-02)

casualty evacuation – (Army) A term used by nonmedical units to refer to the movement of casualties aboard nonmedical vehicles or aircraft. (FM 8-10-6)

centers of gravity – Those characteristics, capabilities, or sources of power from which a military force derives its freedom of action, physical strength, or will to fight. (JP 1-02)

civil affairs – Designated Active and Reserve component forces and units organized, trained, and equipped specifically to conduct civil affairs activities and to support civil-military operations. (JP 1-02)

civil considerations – The influence of manmade infrastructure, civilian institutions, and attitudes and activities of the civilian leaders, populations, and organizations within an area of operations on the conduct of military operations. (FM 6-0)

civil disturbance – Group acts of violence and disorder prejudicial to public law and order. (JP 1-02)

civil-military operations – The activities of a commander that establish, maintain, influence, or exploit relations between military forces, governmental and nongovernmental civilian organizations and authorities, and the civilian populace in a friendly, neutral, or hostile operational area in order to facilitate military operations, to consolidate and achieve US objectives. Civil-military operations may include performance by military forces of activities and functions normally the responsibility of the local, regional, or national government. These activities may occur prior to, during, or subsequent to other military actions. They may also occur, if directed, in the absence of other military operations. Civil-military operations may be performed by designated civil affairs, by other military forces, or by a combination of civil affairs and other forces. (JP 1-02)

civil-military operations center – An ad hoc organization, normally established by the geographic combatant commander or subordinate joint force commander, to assist in the coordination of activities of engaged military forces, and other United States Government agencies, nongovernmental organizations, and regional and international organizations. There is no established structure, and its size and composition are situation dependent. (JP 1-02)

civil support operations – Domestic operations that address the consequences of man-made or natural accidents and incidents beyond the capabilities of civilian authorities. (FM 1)

clear – (Army) 1. A tactical mission task that requires the commander to remove all enemy forces and eliminate organized resistance within an assigned area. (FM 3-90) 2. The total elimination or neutralization of an obstacle that is usually performed by follow-on engineers and is not done under fire. (FM 3-34.2)

close air support – Air action by fixed- and rotary-wing aircraft against hostile targets that are in close proximity to friendly forces and that require detailed integration of each air mission with the fire and movement of those forces. (JP 1-02)

close combat – Combat carried out with direct fire weapons, supported by indirect fire, air-delivered fires, and nonlethal engagement means. Close combat defeats or destroys enemy forces or seizes and retains ground. (FM 3-0)

coalition – An ad hoc arrangement between two or more nations for common action. (JP 1-02)

collateral damage – Unintentional or incidental injury or damage to persons or objects that would not be lawful military targets in the circumstances ruling at the time. Such damage is not unlawful so long as it is not excessive in light of the overall military advantage anticipated from the attack. (JP 1-02)

combatant command – A unified or specified command with a broad continuing mission under a single commander established and so designated by the President, through the Secretary of Defense and with the advice and assistance of the Chairman of the Joint Chiefs of Staff. Combatant commands typically have geographic or functional responsibilities. (JP 1-02)

combat power – The total means of destructive and/or disruptive force which a military unit/formation can apply against the opponent at a given time. (JP 1-02)

combined arms – The synchronized or simultaneous application of several arms—such as infantry, armor, field artillery, engineers, air defense, and aviation—to achieve an effect on the enemy that is greater than if each arm were used against the enemy separately or in sequence. (FM 3-0)

combined arms team – (Army) Two or more arms mutually supporting one another, usually consisting of infantry, armor, cavalry, aviation, field artillery, air defense artillery, and engineers. (FM 3-90).

command and control system – (Army) The arrangement of personnel, information management, procedures, and equipment and facilities essential for the commander to conduct operations. (FM 6-0)

commander's critical information requirements – (Army) – Elements of information required by commanders that directly affect decisionmaking and dictate the successful execution of military operations. (FM 3-0)

commander's intent – (Army) A clear, concise statement of what the force must do and the conditions the force must meet to succeed with respect to the enemy, terrain, and civil considerations that represent the operation's desired end state. (FMI 5-0.1)

commander's visualization – The mental process of developing situational understanding, determining a desired end state, and envisioning how the force will achieve that end state. (FMI 5-0.1)

common operational picture – (Army) An operational picture tailored to the user's requirements, based on common data and information shared by more than one command. (FM 3-0)

computer network attack – Actions taken through the use of computer networks to disrupt, deny, degrade, or destroy information resident in computers or computer networks, or the computers and networks themselves. (JP 1-02)

computer network defense – Actions taken through computer networks to protect, monitor, analyze, detect and respond to unauthorized activity within Department of Defense information systems and computer networks. (JP 1-02)

concept of operations – (Army) How commanders see the actions of subordinate units fitting together to accomplish the mission. As a minimum, the description includes the scheme of maneuver and concept of fires. The concept of operations expands the commander's selected course of action and expresses how each element of the force will cooperate to accomplish the mission. (FM 3-0)

constraint – A restriction placed on the command by a higher command. A constraint dictates an action or inaction, thus restricting the freedom of action a subordinate commander has for planning. (FM 5-0)

control – (Army) 1. The regulation of forces and warfighting functions to accomplish the mission in accordance with the commander's intent. (FMI 5-0.1) 2. A tactical mission task that requires the commander to maintain physical influence over a specified area to prevent its use by an enemy. (FM 3-90)

controlled exchange – The removal of serviceable parts, components, or assemblies from unserviceable, economically reparable equipment and their immediate reuse in restoring a like item of equipment to a combat operable or serviceable condition. (FM 4-30.3)

control measure – A means of regulating forces or warfighting functions. (FMI 5-0.1)

counterattack – (Army) A form of attack by part or all of a defending force against an enemy attacking force with the general objective of denying the enemy his goal in attacking. See FM 3-0.

counterdeception – Efforts to negate, neutralize, diminish the effects of, or gain advantage from a foreign deception operation. Counterdeception does not include the intelligence function of identifying foreign deception operations. (JP 1-02)

counterdrug – Those active measures taken to detect, monitor, and counter the production, trafficking, and use of illegal drugs. (JP 1-02)

counterinsurgency – Those military, paramilitary, political, economic, psychological, and civic actions taken by a government to defeat insurgency. (JP 1-02)

counterintelligence – Information gathered and activities conducted to protect against espionage, other intelligence activities, sabotage, or assassinations conducted by or on behalf of foreign governments or elements thereof, foreign organizations, or foreign persons, or international terrorist activities. (JP 1-02)

countermobility operations – The construction of obstacles and emplacement of minefields to delay, disrupt, and destroy the enemy by reinforcement of the terrain. The primary purpose of countermobility operations is to slow or divert the enemy, to increase time for target acquisition, and to increase weapon effectiveness. (JP 1-02)

counterpropaganda – Programs of products and actions designed to nullify propaganda or mitigate its effects. (FM 3-05.30)

counterterrorism – Operations that include the offensive measures taken to prevent, deter, preempt, and respond to terrorism. (JP 1-02)

course of action – 1. Any sequence of activities that an individual or a unit may follow. 2. A possible plan open to an individual or a commander that would accomplish or is related to the accomplishment of a mission. 3. The scheme adopted to accomplish a job or mission. 4. A line of conduct in an engagement. 5. A product of the Joint Operation Planning and Execution System concept development phase. (JP 1-02)

cover – (Army) 1. Protection from the effects of fires. (FM 6-0) 2. A form of security operation whose primary task is to protect the main body by fighting to gain time while also observing and reporting information and preventing enemy ground observation of and direct fire against the main body. Unlike a screening or guard force, the covering force is a self-contained force capable of operating independently of the main body. (FM 3-90)

dead space – 1. An area within the maximum range of a weapon, radar, or observer, which cannot be covered by fire or observation from a particular position because of intervening obstacles, the nature of the ground, the characteristics of the trajectory, or the limitations of the pointing capabilities of the weapon. 2. An area or zone which is within range of a radio transmitter, but in which a signal is not received. 3. The volume of space above and around a gun or guided missile system into which it cannot fire because of mechanical or electronic limitations. (JP 1-02)

debarkation – The unloading of troops, equipment, or supplies from a ship or aircraft. (JP 1-02)

decision point – (Army) An event, area, or point in the battlespace where and when the friendly commander will make a critical decision. (FM 5-0)

decision support matrix – An aid used by the commander and staff to make battlefield decisions. This matrix is a staff product of the wargaming process that lists the decision point, location of the decision point, the criteria to be evaluated at the point of decision, the action or operations to occur at the decision point, and the unit or element that is to act and has responsibility to observe and report the information affecting the criteria for the decision. (FM 5-0)

decision support template – (Army) A staff product initially used in the wargaming process that graphically represents the decision points and projected situations, and indicates when, where, and under what conditions a decision is most likely to be required to initiate a specific activity or event. (FM 5-0)

decisive engagement – In land and naval warfare, an engagement in which a unit is considered fully committed and cannot maneuver or extricate itself. In the absence of outside assistance, the action must be fought to a conclusion and either won or lost with the forces at hand. (JP 1-02)

decisive operation – The operation that directly accomplishes the task assigned by the higher headquarters. Decisive operations conclusively determine the outcome of major operations, battles, and engagements. (FM 3-0)

decisive point – A geographic place, specific key event, critical system or function that allows commanders to gain a marked advantage over an enemy and greatly influence the outcome of an attack. (JP 1-02)

defeat – A tactical mission task that occurs when an enemy force has temporarily or permanently lost the physical means or the will to fight. The defeated force's commander is unwilling or unable to pursue his adopted course of action, thereby yielding to the friendly commander's will, and can no longer interfere to a significant degree with the actions of friendly forces. Defeat can result from the use of force or the threat of its use. (FM 3-90)

defeat in detail – Concentrating overwhelming combat power against separate parts of a force in sequence rather than defeating the entire force at once. (FM 3-90)

defense in depth – The siting of mutually supporting defense positions designed to absorb and progressively weaken attack, prevent initial observations of the whole position by the enemy, and to allow the commander to maneuver the reserve. (JP 1-02)

defensive information operations – The integration and coordination of policies and procedures, operations, personnel, and technology to protect and defend information and information systems. Defensive information operations ensure timely, accurate, and relevant information access while denying adversaries the opportunity to exploit friendly information and information systems for their own purposes. (FM 3-0)

defensive operations – Operations that defeat an enemy attack, buy time, economize forces, or develop conditions favorable for offensive operations. Defensive operations alone normally cannot achieve a decision. Their purpose is to create conditions for a counteroffensive that allows Army forces to regain the initiative. (FM 3-0)

delay – A form of retrograde in which a force under pressure trades space for time by slowing the enemy's momentum and inflicting maximum damage on the enemy without, in principle, becoming decisively engaged. (FM 3-90)

depth – The extension of operations in time, space, and resources. (FM 3-0)

destroy – 1. A tactical mission task that physically renders an enemy force combat-ineffective until it is reconstituted. 2. To damage a combat system so badly that it cannot perform any function or be restored to a usable condition without being entirely rebuilt. (FM 3-90)

direct action – Short-duration strikes and other small-scale offensive actions conducted as a special operation in hostile, denied, or other politically sensitive environments and which employ specialized military capabilities to seize, destroy, capture, exploit, recover, or damage designated targets. Direct action differs from conventional offensive actions in the level of physical and political risk, operational techniques, and the degree of discriminate and precise use of force to achieve specific objectives. (JP 1-02)

direct approach – The application of combat power directly against the enemy center of gravity or the enemy's principal strength. (FM 3-90)

direct fire – Fire delivered on a target using the target itself as a point of aim for either the weapon or the director. (JP 1-02)

disinformation – Information disseminated primarily by intelligence organizations or other covert agencies designed to distort information or deceive or influence United States decisionmakers, United States forces, coalition allies, key actors, or individuals via indirect or unconventional means. (FM 3-13)

displaced person – (Army) A civilian who is involuntarily outside the national boundaries of his or her country or as an internally displaced person is a civilian involuntarily outside his area or region within his country. (FM 2-0)

display – (Army) An information management activity: to represent relevant information in a usable, easily understood audio or visual form tailored to the needs of the user that conveys the common operational picture for decisionmaking and exercising command and control functions (FM 6-0)

disposition – 1. Distribution of the elements of a command within an area, usually the exact location of each unit headquarters and the deployment of forces subordinate to it. (JP 1-02)

disrupt – 1. A tactical mission task in which a commander integrates direct and indirect fires, terrain, and obstacles to upset an enemy's formation or tempo, interrupt his timetable, or cause his forces to commit prematurely or attack in piecemeal fashion. (FM 3-90) 2. An engineer obstacle effect that focuses fire planning and obstacle effort to cause the enemy to break up his formation and tempo, interrupt his timetable, commit breaching assets prematurely, and attack in a piecemeal effort. (FM 90-7)

distribution system – That complex of facilities, installations, methods, and procedures designed to receive, store, maintain, distribute, and control the flow of military materiel between the point of receipt into the military system and the point of issue to using activities and units. (JP 1-02)

doctrinal template – A model based on known or postulated adversary doctrine. Doctrinal templates illustrate the disposition and activity of adversary forces and assets conducting a particular operation unconstrained by the effects of the battlespace. They represent the application of adversary doctrine under ideal conditions. Ideally, doctrinal templates depict the threat's normal organization for combat, frontages, depths, boundaries and other control measures, assets available from other commands, objective depths, engagement areas, battle positions, and so forth. Doctrinal templates are usually scaled to allow ready use with geospatial products. (JP 1-02)

doctrine – Fundamental principles by which the military forces or elements thereof guide their actions in support of national objectives. It is authoritative but requires judgment in application. (JP 1-02)

economy – One of the eight characteristics of combat service support: providing the most efficient support at the least cost to accomplish the mission. (FM 4-0)

economy of force – One of the nine principles of war: allocate minimum essential combat power to secondary efforts. (FM 3-0)

effect – (Army) A result, outcome, or consequence of an action. (FMI 5-0.1)

electromagnetic deception – The deliberate radiation, re-radiation, alteration, suppression, absorption, denial, enhancement, or reflection of electromagnetic energy in a manner intended to convey misleading information to an enemy or enemy electromagnetic-dependent weapons, thereby degrading or neutralizing the enemy's combat capability. Among the types of electromagnetic deception are: a. **manipulative electromagnetic deception**—Actions to eliminate revealing, or convey misleading, electromagnetic telltale indicators that may be used by hostile forces; b. **simulative electromagnetic deception**—Actions to simulate friendly, notional, or actual capabilities to mislead hostile forces; c. **imitative electromagnetic deception**—The introduction of electromagnetic energy into enemy systems that imitates enemy emissions. (JP 1-02)

electromagnetic spectrum – The range of frequencies of electromagnetic radiation from zero to infinity. It is divided into 26 alphabetically designated bands. (JP 1-02)

electronic attack—That division of electronic warfare involving the use of electromagnetic energy, directed energy, or antiradiation weapons to attack personnel, facilities, or equipment with the intent of degrading, neutralizing, or destroying enemy combat capability and is considered a form of fires. Electronic attack includes: 1. actions taken to prevent or reduce an enemy's effective use of the electromagnetic spectrum, such as jamming and electromagnetic deception, and 2. employment of weapons that use either electromagnetic or directed energy as their primary destructive mechanism (lasers, radio frequency weapons, particle beams), or antiradiation weapons. (JP 1-02)

electronic warfare – Any military action involving the use of electromagnetic and directed energy to control the electromagnetic spectrum or to attack the enemy. The three major subdivisions within electronic warfare are: electronic attack, electronic protection, and electronic warfare support. (JP 1-02)

electronic warfare support—That division of electronic warfare involving actions tasked by, or under direct control of, an operational commander to search for, intercept, identify, and locate or localize sources of intentional and unintentional radiated electromagnetic energy for the purpose of immediate threat recognition, targeting, planning, and conduct of future operations. Thus, electronic warfare support provides information required for immediate decisions involving electronic warfare operations and other tactical actions such as threat avoidance, targeting, and homing. Electronic warfare support data can be used to produce signals intelligence, provide targeting for electronic or destructive attack, and produce measurement and signature intelligence. (JP 1-02)

embarkation – The process of putting personnel and/or vehicles and their associated stores and equipment into ships and/or aircraft. (JP 1-02)

encirclement – An operation where one force loses its freedom of maneuver because an opposing force is able to isolate it by controlling all ground lines of communications. (FM 3-0)

end state – (Army) At the operational and tactical levels, the conditions that, when achieved, accomplish the mission. At the operational level, these conditions attain the aims set for the campaign or major operation. (FM 3-0)

engagement – A tactical conflict, usually between opposing lower echelon maneuver forces. (JP 1-02)

envelopment – (Army) – A form of maneuver in which an attacking force seeks to avoid the principal enemy defenses by seizing objectives to the enemy rear to destroy the enemy in his current positions. At the tactical level, envelopments focus on seizing terrain, destroying specific enemy forces, and interdicting enemy withdrawal routes. (FM 3-0).

essential elements of friendly information – (Army) The critical aspects of a friendly operation that, if known by the enemy, would subsequently compromise, lead to failure, or limit success of the operation, and, therefore, must be protected from enemy detection. (FM 3-13)

essential task – A specified or implied task that must be executed to accomplish the mission. Essential tasks are always included in the unit's mission statement. (FM 5-0)

exfiltration – The removal of personnel or units from areas under enemy control by stealth, deception, surprise, or clandestine means. (JP 1-02)

exploitation – 1. Taking full advantage of success in military operations, following up initial gains, and making permanent the temporary effects already achieved. 2. An offensive operation that usually follows a successful attack and is designed to disorganize the enemy in depth. (JP 1-02)

explosive ordnance disposal – The detection, identification, on-site evaluation, rendering safe, recovery, and final disposal of unexploded explosive ordnance. It may also include explosive ordnance which has become hazardous by damage or deterioration. (JP 1-02)

firepower – 1. The amount of fire which may be delivered by a position, unit, or weapon system. 2. Ability to deliver fire. (JP 1-02)

fires – The effects of lethal or nonlethal weapons. (JP 1-02)

fire support coordinating measure – A measure employed by land or amphibious commanders to facilitate the rapid engagement of targets and simultaneously provide safeguards for friendly forces. (JP 1-02)

fix – (Army) 1. A tactical mission task where a commander prevents the enemy from moving any part of his force from a specific location for a specific period of time. 2. An engineer obstacle effect that focuses fire planning and obstacle effort to slow an attacker's movement within a specified area, normally an engagement area. (FM 3-90)

flexibility – One of the eight characteristics of combat service support: being able to adapt combat service support structures and procedures to changing situations, missions, and concepts of operations. (FM 4-0)

force projection – The ability to project the military element of national power from the continental United States (CONUS) or another theater in response to requirements for military operations. Force projection operations extend from mobilization and deployment of forces to redeployment to CONUS or home theater. (JP 1-02)

force protection – Actions taken to prevent or mitigate hostile actions against Department of Defense personnel (to include family members), resources, facilities, and critical information. These actions conserve the force's fighting potential so it can be applied at a decisive time and place and incorporates the coordinated and synchronized offensive and defensive measures to enable the effective employment of the joint force while degrading opportunities for the enemy. Force protection does not include actions to defeat the enemy or protect against accidents, weather, or disease. (JP 1-02)

force tailoring – The process of determining the right mix and sequence of units for a mission. (FM 3-0)

foreign humanitarian assistance – Programs conducted to relieve or reduce the results of natural or manmade disasters or other endemic conditions such as human pain, disease, hunger, or privation that might present a serious threat to life or that can result in great damage to or loss of property. Foreign humanitarian assistance (FHA) provided by US forces is limited in scope and duration. The foreign assistance provided is designed to supplement or complement the efforts of the host nation civil authorities or agencies that may have primary responsibility for providing FHA. FHA operations are those conducted outside the United States, its territories, and possessions. (JP 1-02)

foreign internal defense – Participation by civilian and military agencies of a government in any of the action programs taken by another government to free and protect its society from subversion, lawlessness, and insurgency. (JP 1-02)

fratricide – The unintentional killing or wounding of friendly personnel by friendly firepower. (FM 3-0)

frontal attack – (Army) A form of maneuver in which the attacking force seeks to destroy a weaker enemy force or fix a larger enemy force in place over a broad front. (FM 3-0)

full spectrum operations – The range of operations Army forces conduct in war and military operations other than war. (FM 3-0)

graphic control measure – A symbol used on maps and displays to regulate forces and warfighting functions. (FMI 5-0.1)

gun-target line – An imaginary straight line from gun to target. (JP 1-02)

hazardous material – Any substance which has a human health hazard associated with it. Special storage, use, handling, and shipment safety procedures and protocols must be followed to help protect against accidental exposure. Hazardous materials are specifically identified under federal law. (FM 3-100.4)

high angle fire – (Army) Fire delivered to clear an obstacle (such as a hill) that low-angle fire cannot, or fire delivered to attack targets on the reverse side of an obstacle (such as a hill) that cannot be attacked with low-angle or direct fire. (FM 6-30).

host nation – A nation that receives the forces and/or supplies of allied nations, coalition partners, and/or NATO organizations to be located on, to operate in, or to transit through its territory. (JP 1-02)

host-nation support – Civil and/or military assistance rendered by a nation to foreign forces within its territory during peacetime, crises or emergencies, or war based on agreements mutually concluded between nations. (JP 1-02)

human intelligence – A category of intelligence derived from information collected and provided by human sources. (JP 1-02)

imagery intelligence – Intelligence derived from the exploitation of collection by visual photography, infrared sensors, lasers, electro-optics, and radar sensors such as synthetic aperture radar wherein images of objects are reproduced optically or electronically on film, electronic display devices, or other media. (JP 1-02)

implied task – A task that must be performed to accomplish a specified task or mission, but is not stated in the higher headquarters order. (FM 5-0)

indirect fire – Fire delivered on a target that is not itself used as a point of aim for the weapons or the director. (JP 1-02)

infiltration – (Army) 1. A form of maneuver in which an attacking force conducts undetected movement through or into an area occupied by enemy forces to occupy a position of advantage in the enemy rear while exposing only small elements to enemy defensive fires. (FM 3-90)

information – (Army) 1. In the general sense, the meaning humans assign to data. 2. In the context of the cognitive hierarchy, data that have been processed to provide further meaning. (FM 6-0)

information fratricide – The results of employing information operations elements in a way that causes effects in the information environment that impede the conduct of friendly operations or adversely affect friendly forces. (FM 3-13)

information management – The provision of relevant information to the right person at the right time in a usable form to facilitate situational understanding and decisionmaking. It uses procedures and information systems to collect, process, store, display, and disseminate information. (FM 3-0)

information operations – (Army) The employment of the core capabilities of electronic warfare, computer network operations, psychological operations, military deception, and operations security, in concert with specified supporting and related capabilities, to affect and defend information and information systems and to influence decisionmaking. (FM 3-13)

information requirements – (Army) All information elements the commander and staff require to successfully conduct operations, that is, all elements necessary to address the factors of METT-TC. (FM 6-0)

information superiority – The operational advantage derived from the ability to collect, process, and disseminate an uninterrupted flow of information while exploiting or denying an adversary's ability to do the same. (JP 1-02)

information system – (Army) The equipment and facilities that collect, process, store, display, and disseminate information. This includes computers—hardware and software—and communications, as well as policies and procedures for their use. (FM 3-0)

insurgency – An organized movement aimed at the overthrow of a constituted government through the use of subversion and armed conflict. (JP 1-02)

integration – One of the eight characteristics of combat service support: the total inclusion of Army combat service support into the operations process (plan, prepare, execute, assess), as well as into other logistics comnponents of the unified force. (FM 4-0).

intelligence – 1. The product resulting from the collection, processing, integration, analysis, evaluation, and interpretation of available information concerning foreign countries or areas. 2. Information and knowledge about an adversary obtained through observation, investigation, analysis, or understanding. (JP 1-02)

intelligence preparation of the battlefield – The systematic, continuous process of analyzing the threat and environment in a specific geographic area. Intelligence preparation of the battlefield (IPB) is designed to support the staff estimate and military decisionmaking process. Most intelligence requirements are generated as a result of the IPB process and its interrelation with the decisionmaking process. (FM 34-130)

intelligence, surveillance, and reconnaissance – An activity that synchronizes and integrates the planning and operation of sensors, assets, and processing, exploitation, and dissemination systems in direct support of current and future operations. This is an integrated intelligence and operations function. (JP 1-02)

intelligence, surveillance, and reconnaissance plan – An integrated plan for collection of information from all available sources and analysis of that information to produce intelligence to meet requirements. Specifically, a logical plan for transforming priority intelligence requirements into orders or requests to reconnaissance and surveillance assets to collect pertinent information within a required time limit. (FM 34-3)

interagency – Activities or operations conducted by or through coordination with two or more agencies or an agency and one or more Services of the same nation. (FM 3-07)

interdict – A tactical mission task where the commander prevents, disrupts, or delays the enemy's use of an area or route. (FM 3-90)

interior lines – A force operates on interior lines when its operations diverge from a central point. (FM 3-0)

isolate – A tactical mission task that requires a unit to seal off—both physically and psychologically—an enemy from his sources of support, deny an enemy freedom of movement, and prevent an enemy unit from having contact with other enemy forces. (FM 3-90)

joint force – A general term applied to a force composed of significant elements, assigned or attached, of two or more Military Departments, operating under a single joint force commander. (JP 1-02)

joint force commander – A general term applied to a combatant commander, subunified commander, or joint task force commander authorized to exercise combatant command (command authority) or operational control over a joint force. (JP 1-02)

joint operations – A general term to describe military actions conducted by joint forces, or by Service forces in relationships (e.g., support, coordinating authority), which, of themselves, do not create joint forces. (JP 1-02)

joint task force – A joint force that is constituted and so designated by the Secretary of Defense, a combatant commander, a subunified commander, or an existing joint task force commander. (JP 1-02)

key terrain – Any locality, or area, the seizure or retention of which affords a marked advantage to either combatant. (JP 1-02)

landing zone – Any specified zone used for the landing of aircraft. JP 1-02)

law of war – That part of international law that regulates the conduct of armed hostilities. (JP 1-02)

liaison – That contact or intercommunication maintained between elements of military forces or other agencies to ensure mutual understanding and unity of purpose and action. (JP 1-02)

line of communications – A route, either land, water, and/or air, that connects an operating military force with a base of operations and along which supplies and military forces move. (JP 1-02)

lines of operations – (Army) A line that defines the orientation of the force in time and space, or purpose, in relation to an enemy or objective. (FMI 5-0.1).

logistics – The science of planning and carrying out the movement and maintenance of forces. In its most comprehensive sense, those aspects of military operations which deal with: a. design and development, acquisition, storage, movement, distribution, maintenance, evacuation, and disposition of materiel; b. movement, evacuation, and hospitalization of personnel; c. acquisition or construction, maintenance, operation, and disposition of facilities; and d. acquisition and furnishing of services. (JP 1-02)

logistics-over-the-shore operations – The loading and unloading of ships without the benefit of deep draft-capable, fixed port facilities; or as a means of moving forces close to tactical assembly areas dependent on threat force capabilities. (JP 1-02)

logistics preparation of the theater – Actions taken by combat service support personnel to optimize means—force structure, resources, and strategic lift—of supporting the joint force commander's plan. (FM 4-0)

major operation – A series of tactical actions (battles, engagements, strikes) conducted by various combat forces of a single or several Services, coordinated in time and place, to accomplish operational and, sometimes, strategic objectives in an operational area. These actions are conducted simultaneously or sequentially in accordance with a common plan and are controlled by a single commander. (JP 1-02)

maneuver – One of the nine principles of war: place the enemy in a disadvantageous position through the flexible application of combat power. (FM 3-0)

Marine air-ground task force – The Marine Corps principal organization for all missions across the range of military operations, composed of forces task-organized under a single commander capable of responding rapidly to a contingency anywhere in the world. The types of forces in the Marine air-ground task force (MAGTF) are functionally grouped into four core elements: a command element, an aviation combat element, a ground combat element, and a combat service support element. The four core elements are categories of forces, not formal commands. The basic structure of the MAGTF never varies, though the number, size, and type of Marine Corps units comprising each of its four elements will always be mission dependent. The flexibility of the organizational structure allows for one or more subordinate MAGTFs to be assigned. In a joint or multinational environment, other Service or multinational forces may be assigned or attached. (FM 1-02)

Marine expeditionary force – The largest Marine air-ground task force (MAGTF) and the Marine Corps principal warfighting organization, particularly for larger crises or contingencies. It is task-organized around a permanent command element and normally consists of one or more Marine divisions, Marine aircraft wings, and Marine force service support groups. The Marine expeditionary force is capable of missions across the range of military operations, including amphibious assault and sustained operations ashore in any environment. It can operate from a sea base, a land base, or both. In a joint or multinational environment, it may contain other Service or multinational forces assigned or attached to the MAGTF. (FM 1-02)

Marine expeditionary unit – A Marine air-ground task force (MAGTF) that is constructed around an infantry battalion reinforced, a helicopter squadron reinforced, and a task-organized combat service support element. It normally fulfills Marine Corps' forward sea-based deployment requirements. The Marine expeditionary unit provides an immediate reaction capability for crisis response and is capable of limited combat operations. In a joint or multinational environment, it may contain other Service or multinational forces assigned or attached to the MAGTF. (FM 1-02)

mass – One of the nine principles of war: the effects of combat power at the decisive place and time. (FM 3-0)

mass casualty – Any large number of casualties produced in a relatively short period of time, usually as the result of a single incident such as a military aircraft accident, hurricane, flood, earthquake, or armed attack that exceeds logistical support capabilities. (JP 1-02)

measure of effectiveness – (Army) A criterion used to assess changes in system behavior, capability, or operational environment that is tied to measuring the attainment of an end state, achievement of an objective, or creation of an effect. (FMI 5-0.1)

medical evacuation – The timely and efficient movement of the wounded, injured, or ill while providing en route medical care to and between medical treatment facilities. (FM 4-02)

meeting engagement – A combat action that occurs when a moving force engages an enemy at an unexpected time and place. (FM 3-0)

METT-TC – A memory aid used in two contexts: (1) In the context of information management, the major subject categories into which relevant information is grouped for military operations: mission, enemy, terrain and weather, troops and support available, time available, civil considerations. (2) In the context of tactics, the major factors considered during mission analysis. (FM 6-0)

military deception – Actions executed to deliberately mislead adversary military decisionmakers as to friendly military capabilities, intentions, and operations, thereby causing the adversary to take specific actions (or inactions) that will contribute to the accomplishment of the friendly mission. (JP 1-02)

military decisionmaking process – A process that integrates the activities of the commander, staff and subordinate commanders in developing and operation plan or order. It establishes procedures for analyzing a mission; developing, analyzing, and comparing courses of action; selecting the best course of action; and producing an operation plan or order (FMI 5-0.1).

misinformation – Incorrect information from any source that is released for unknown reasons or to solicit a response or interest from a nonpolitical or nonmilitary target. (FM 3-13)

mission – 1. The task, together with the purpose, that clearly indicates the action to be taken and the reason therefor. 2. In common usage, especially when applied to lower military units, a duty assigned to an individual or unit; a task. (JP 1-02)

mission command – The conduct of military operations through decentralized execution based upon mission orders for effective mission accomplishment. Successful mission command results from subordinate leaders at all echelons exercising disciplined initiative within the commander's intent to accomplish missions. It requires an environment of trust and mutual understanding. (FM 6-0)

mission orders – A technique for completing combat orders that allows subordinates maximum freedom of planning and action in accomplishing missions and leaves the "how" of mission accomplishment to the subordinate. (FM 6-0)

mobile defense – (Army) A type of defensive operation that concentrates on the destruction or defeat of the enemy through a decisive attack by a striking force. (FM 3-0)

mobility corridor – Areas where a force will be canalized due to terrain restrictions. They allow military forces to capitalize on the principles of mass and speed and are therefore relatively free of obstacles. (JP 1-02)

mobility operations – Obstacle reduction by maneuver and engineer units to reduce or negate the effects of existing or reinforcing obstacles. The objective is to maintain freedom of movement for maneuver units, weapon systems, and critical supplies. (FM 3-34)

modified combined obstacle overlay – A joint intelligence preparation of the battlespace product used to portray the effects of each battlespace dimension on military operations. It normally depicts militarily significant aspects of the battlespace environment, such as obstacles restricting military movement, key geography, and military objectives. (JP 1-02)

monitoring – (Army) Continuous observation of the current situation to identify opportunities for the force, threats to the force, gaps in information, and progress according to the plan or order. (FMI 5-0.1)

movement to contact – A form of the offensive designed to develop the situation and to establish or regain contact. (JP 1-02)

naval coastal warfare – Coastal sea control, harbor defense, and port security, executed both in coastal areas outside the United States in support of national policy and in the United States as part of this Nation's defense. (JP 1-02)

naval gunfire support – Fire provided by Navy surface gun systems in support of a unit or units tasked with achieving the commander's objectives. A subset of naval surface fire support. (JP 1-02)

naval surface fire support – Fire provided by Navy surface gun and missile systems in support of a unit or units. (JP 1-02)

neutral – (Army) 1. An individual, group of individuals, organization, or nation that is not hostile to or in any way supportive of only one belligerent force in a hostile environment. (FM 3-07)

no-fire area – A land area, designated by the appropriate commander, into which fires or their effects are prohibited. (JP 1-02)

noncombatant – 1. An individual in an area of combat operations who is not armed and is not participating in any activity in support of any of the factions or forces involved in combat. 2. An individual, such as chaplain or medical personnel, whose duties do not involve combat. (FM 3-07)

noncombatant evacuation operations – Operations directed by the Department of State, the Department of Defense, or other appropriate authority whereby noncombatants are evacuated from foreign countries when their lives are endangered by war, civil unrest, or natural disaster to safe havens or to the United States. (JP 1-02)

nongovernmental organization – A private, self-governing, not-for-profit organization dedicated to alleviating human suffering; and/or promoting education, health care, economic development, environmental protection, human rights, and conflict resolution; and/or encouraging the establishment of democratic institutions and civil society. (JP 1-02)

nonlethal fires – Any fires that do not directly seek the physical destruction of the intended target and are designed to impair, disrupt, or delay the performance of enemy operational forces, functions, and facilities. Psychological operations, electronic warfare (jamming), and other command and control countermeasures are all nonlethal fire options. (FM 6-20)

nonlethal weapons – Weapons that are explicitly designed and primarily employed so as to incapacitate personnel or materiel, while minimizing fatalities, permanent injury to personnel and undesired damage to property and the environment. a. Unlike conventional lethal weapons that destroy their targets through blast, penetration, and fragmentation, nonlethal weapons employ means other than gross physical destruction to prevent the target from functioning. b. Nonlethal weapons are intended to have one, or both, of the following characteristics: (1) They have relatively reversible effects on personnel or materiel. (2) They affect objects differently within their area of influence. (JP 1-02)

objective – (Army) 1. One of the nine principles of war: direct every military operation toward a clearly defined, decisive and attainable objective. (FM 3-0) 2. A location on the ground used to orient operations, phase operations, facilitate changes of direction, and provide for unity of effort. (FM 3-90)

offensive – One of the nine principles of war: seize, retain, and exploit the initiative. (FM 3-0)

offensive information operations – The integrated use of assigned and supporting capabilities and activities, mutually supported by intelligence, to affect enemy decisionmakers or to influence others to achieve or promote specific objectives. (FM 3-0)

offensive operations – Operations which aim at destroying or defeating an enemy. Their purpose is to impose US will on the enemy and achieve decisive victory. (FM 3-0)

operation – 1. A military action or the carrying out of a strategic, operational, tactical, service, training, or administrative military mission. 2. The process of carrying on combat, including movement, supply, attack, defense, and maneuvers needed to gain the objectives of any battle or campaign. (JP 1-02)

operational control – Command authority that may be exercised by commanders at any echelon at or below the level of combatant command. Operational control is inherent in combatant command (command authority) and may be delegated within the command. When forces are transferred between combatant commands, the command relationship the gaining commander will exercise (and the losing commander will relinquish) over these forces must be specified by the Secretary of Defense. Operational control is the authority to perform those functions of command over subordinate forces involving organizing and employing commands and forces, assigning tasks, designating objectives, and giving authoritative direction necessary to accomplish the mission. Operational control includes authoritative direction over all aspects of military operations and joint training necessary to accomplish the missions assigned to the command. Operational control should be exercised through the commanders of subordinate organizations. Normally this authority is exercised through subordinate joint force commanders and Service and/or functional component commanders. Operational control normally provides full authority to organize commands and forces and to employ those forces as the commander in operational control considers necessary to accomplish assigned missions; it does not, in and of itself, include authoritative direction for logistics or matters of administration, discipline, internal organization, or unit training. (JP 1-02)

operational framework – The arrangement of friendly forces and resources in time, space, and purpose with respect to each other and the enemy or situation. It consists of the area of operations, battlespace, and battlefield organization. (FM 3-0)

operations process – The major command and control activities performed during operations: planning, preparation, execution with continuous assessment. These activities occur continuously throughout an operation, overlapping and recurring as required. (FMI 5-0.1)

operations security – A process of identifying critical information and subsequently analyzing friendly actions attendant to military operations and other activities to: a. identify those actions that can be observed by adversary intelligence systems; b. determine indicators hostile intelligence systems might obtain that could be interpreted or pieced together to derive critical information in time to be useful to adversaries; and c. select and execute measures that eliminate or reduce to an acceptable level the vulnerabilities of friendly actions to adversary exploitation. (JP 1-02)

paramilitary forces – Forces or groups distinct from the regular armed forces of any country, but resembling them in organization, equipment, training, or mission. (JP 1-02)

patrol – A detachment of ground, sea, or air forces sent out for the purpose of gathering information or carrying out a destructive, harassing, mopping-up, or security mission. (JP 1-02)

peace enforcement – Application of military force, or the threat of its use, normally pursuant to international authorization, to compel compliance with resolutions or sanctions designed to maintain or restore peace and order. (JP 1-02)

peacekeeping – Military operations undertaken with the consent of all major parties to a dispute, designed to monitor and facilitate implementation of an agreement (ceasefire, truce, or other such agreement) and support diplomatic efforts to reach a long-term political settlement. (JP 1-02)

peace operations – A broad term that encompasses peacekeeping operations and peace enforcement operations conducted in support of diplomatic efforts to establish and maintain peace. (JP 1-02)

penetration – (Army) A form of maneuver in which an attacking force seeks to rupture enemy defenses on a narrow front to disrupt the defensive system. (FM 3-90).

perceptions – Mental images the commander wants the deception target to believe are real. (FM 3-13)

phase line – A line utilized for control and coordination of military operations, usually an easily identified feature in the operational area. (JP 1-02)

physical destruction – The application of combat power to destroy or degrade adversary forces, sources of information, command and control systems, and installations. It includes direct and indirect forces from ground, sea, and air forces. Also included are direct actions by special operations forces. (FM 3-13)

physical security – 1. That part of security concerned with physical measures designed to safeguard personnel; to prevent unauthorized access to equipment, installations, material, and documents; and to safeguard them against espionage, sabotage, damage, and theft. 2. In communications security, the component that results from all physical measures necessary to safeguard classified equipment, material, and documents from access thereto or observation thereof by unauthorized persons. (JP 1-02)

pickup zone – A geographic area used to pick up troops or equipment by helicopter. (FM 90-4)

plan – A design for a future or anticipated operation. (FM 5-0)

planning – The process by which commanders (and staffs, if available) translate the commander's visualization into a specific course of action for preparation and execution, focusing on the expected results. (FMI 5-0.1)

port of debarkation – (DOD) The geographic point at which cargo or personnel are discharged. This may be a seaport or aerial port of debarkation; for unit requirements, it may or may not coincide with the destination. (JP 1-02)

port of embarkation – (DOD) The geographic point in a routing scheme from which cargo and personnel depart. This may be a seaport or aerial port from which personnel and equipment flow to a port of debarkation; for unit and nonunit requirements, it may or may not coincide with the origin. (JP 1-02)

principles of war – Principles that provide general guidance for conducting war and military operations other than war at the strategic, operational, and tactical levels. The nine principles of war are: objective, offensive, mass, economy of force, maneuver, unity of command, security, surprise, and simplicity (FM 3-0)

propaganda – Any form of communication in support of national objectives designed to influence the opinions, emotions, attitudes, or behavior of any group in order to benefit the sponsor, either directly or indirectly. (JP 1-02)

psychological operations – Planned operations to convey selected information and indicators to foreign audiences to influence their emotions, motives, objective reasoning, and ultimately the behavior of foreign governments, organizations, groups, and individuals. The purpose of psychological operations is to induce or reinforce foreign attitudes and behavior favorable to the originator's objectives. (JP 1-02)

public affairs – Those public information, command information, and community relations activities directed toward both the external and internal publics with interest in the Department of Defense. (JP 1-02)

pursuit – An offensive operation designed to catch or cut off a hostile force attempting to escape, with the aim of destroying it. (JP 1-02)

push – In logistics, the delivery of a predetermined amount of supplies to a user on a scheduled basis without the user requesting them. (FM 4-0)

railhead – A point on a railway where loads are transferred between trains and other means of transport. (JP 1-02)

reachback – The process of obtaining products, services, and applications, or forces, or equipment, or material from organizations that are not forward deployed. (JP 1-02)

rear area – (Army) For any particular command, the area extending forward from its rear boundary to the rear of the area assigned to the next lower level of command. This area is provided primarily for the performance of support functions, and is where the majority of the echelon's sustaining functions occur. (FM 3-0)

refugee – A person who, by reason of real or imagined danger, has left his home country or country of nationality and is unwilling or unable to return. (JP 1-02)

relevant information – All information of importance to commanders and staffs in the exercise of command and control. (FM 3-0)

relief in place – (Army) An tactical enabling operation in which, by direction of higher authority, all or part of a unit is replaced in an area by the incoming unit. (FM 3-90)

reorganization – Action taken to shift internal resources within a degraded unit to increase its level of combat effectiveness. (FM 100-9)

reserve – Portion of a body of troops which is kept to the rear or withheld from action at the beginning of an engagement, in order to be available for a decisive movement. (JP 1-02)

responsiveness – One of the eight principles of combat service support: providing the right support at the right place at the right time. (FM 4-0)

retrograde – A type of defensive operation that involves organized movement away from the enemy. (FM 3-0)

riot control agent – (Army) A substance that produces temporary irritating or disabling physical effects that disappear within minutes of removal from exposure. There is no significant risk of permanent injury, and medical treatment is rarely required. (FM 3-11.11)

risk – (DOD) Probability and severity of loss linked to hazards. (JP 1-02)

risk management – The process of identifying, assessing, and controlling risk arising from operational factors, and making decisions that balance risk cost with mission benefits. (JP 1-02)

rules of engagement – Directives issued by competent military authority that delineate the circumstances and limitations under which United States forces will initiate and/or continue combat engagement with other forces encountered. (JP 1-02)

running estimate – A staff section's continuous assessment of current and future operations to determine if the current operation is proceeding according to the commander's intent and if future operations are supportable. (FMI 5-0.1)

search and attack – A technique of conducting a movement to contact that shares many of the characteristics of an area security mission. (FM 3-0)

secure – (Army) 1. A tactical mission task that involves preventing a unit, facility, or geographical location from being damaged or destroyed as a result of enemy action. (FM 3-90)

security – One of the nine principles of war: never permit the enemy to acquire an unexpected advantage (FM 3-0).

seize – (Army) A tactical mission task that involves taking possession of a designated area using overwhelming force. (FM 3-90)

sequel – (Army) An operation that follows the current operation. It is a future operation that anticipates possible outcomes – success, failure, or stalemate – of the current operation. (FM 3-0).

shaping operation – An operation at any echelon that creates and preserves conditions for the success of decisive operations. (FM 3-0)

show of force – An operation designed to demonstrate US resolve that involves increased visibility of US deployed forces in an attempt to defuse a specific situation that, if allowed to continue, may be detrimental to US interests or national objectives. (JP 1-02)

signals intelligence – 1. A category of intelligence comprising either individually or in combination all communications intelligence, electronic intelligence, and foreign instrumentation signals intelligence, however transmitted. 2. Intelligence derived from communications, electronics, and foreign instrumentation signals. (JP 1-02)

simplicity – One of the nine principles of war: prepare clear, uncomplicated plans, and clear, concise orders to ensure thorough understanding. (FM 3-0) [Note: "simplicity" is also one of the eight characteristics of combat service support.]

situational understanding – (Army) The product of applying analysis and judgment to the common operational picture to determine the relationship among the factors of METT-TC. (FM 3-0)

space operations – The employment of space system capabilities that provide the means to enhance command and control, facilitate the maneuver of forces, reduce the commander's uncertainty, and improve fire support, air defense, intelligence collection, and combat service support operations which will support strategic, operational, and tactical missions across the operational continuum in the near, mid, and far term. (FM 3-14)

special operations – Operations conducted in hostile, denied, or politically sensitive environments to achieve military, political, economic, and/or informational objectives employing military capabilities for which there is no broad conventional force requirement. These operations often require covert, clandestine, or low visibility capabilities. Special operations are applicable across the rnage of military operations. They can be conducted independently or in conjunction with operations of conventional forces or other government agencies and may include operations through, with, or by indigenous or surrogate forces. Special operations differ from conventional operations in degree of physical and political risk, operational techniques, mode of employment, independence from friendly support, and dependence on detailed operational intelligence and indigenous assets. (JP 1-02)

special operations forces – Those Active and Reserve Component forces of the Military Services designated by the Secretary of Defense and specifically organized, trained, and equipped to conduct and support special operations. (JP 1-02)

special reconnaissance – Reconnaissance and surveillance actions conducted by special operations forces in hostile, denied, or politically sensitive environments to collect or verify information of strategic or operational significance, employing military capabilities not normally found in conventional forces. These forces provide an additive capability for commanders and supplement other conventional reconnaissance and surveillance actions. (JP 1-02)

stability operations – Operations that promote and protect US national interests by influencing the threat, political, and information dimensions of the operational environment through a combination of peacetime developmental, cooperative activities and coercive actions in response to crisis. (FM 3-0)

status-of-forces agreement – An agreement that defines the legal position of a visiting military force deployed in the territory of a friendly state. Agreements delineating the status of visiting military forces may be bilateral or multilateral. Provisions pertaining to the status of visiting forces may be set forth in a separate agreement, or they may form a part of a more comprehensive agreement. These provisions describe how the authorities of a visiting force may control members of that force and the amenability of the force or its members to the local law or to the authority of local officials. To the extent that agreements delineate matters affecting the relations between a military force and civilian authorities and population, they may be considered as civil affairs agreements. (JP 1-02)

strategy – The art and science of developing and employing instruments of national power in a synchronized and integrated fashion to achieve theater, national and/or multinational objectives. (JP 1-02)

striking force – A committed force organized to conduct the decisive attack in a mobile defense. It normally comprises the maximum combat power available to the commander at the time of the attack. (FM 3-0)

strong point – (Army) A heavily fortified battle position tied to a natural or reinforcing obstacle to create an anchor for the defense or to deny the enemy decisive or key terrain. (FM 3-90)

surprise – One of the nine principles of war: strike the enemy at a time or place or in a manner for which he is unprepared. (FM 3-0)

survivability – (Army) 1. Concept which includes all aspects of protecting personnel, weapons, and supplies while simultaneously deceiving the enemy. Survivability tactics include building a good defense; employing frequent movement; using concealment, deception, and camouflage; and constructing fighting and protective positions for both individuals and equipment. Encompasses planning and locating position sites, designing adequate overhead cover, analyzing terrain conditions and construction materials, selecting excavation methods, and countering the effects of direct and indirect fire weapons. (FM 5-103) 2. One of the eight characteristics of combat service support: being able to shield support functions from destruction or degradation. (FM 4-0)

sustainability - One of the eight characteristics of combat service support: the ability to maintain continuous support, rear area and base security, movement control, terrain management, and infrastructure development. (FM 4-0)

sustaining operation – An operation at any echelon that enables shaping and decisive operations by providing combat service support, rear area and base security, movement control, terrain management, and infrastructure development. (FM 3-0)

synchronization – 1. The arrangement of military actions in time, space, and purpose to produce maximum relative combat power at a decisive place and time. See FM 3-0. 2. In the intelligence context, application of intelligence sources and methods in concert with the operation plan. (JP 1-02)

tactical combat force – A combat unit, with appropriate combat support and combat service support assets, that is assigned the mission of defeating Level III threats. (JP 1-02)

tactical control – Command authority over assigned or attached forces or commands, or military capability or forces made available for tasking, that is limited to the detailed direction and control of movements or maneuvers within the operational area necessary to accomplish missions or tasks assigned. Tactical control is inherent in operational control. Tactical control may be delegated to, and exercised at any level at or below the level of combatant command. When forces are transferred between combatant commands, the command relationship the gaining commander will exercise (and the losing commander will relinquish) over these forces must be specified by the Secretary of Defense. Tactical control provides sufficient authority for controlling and directing the application of force or tactical use of combat support assets within the assigned mission or task. (JP 1-02)

tactical level of war –The level of war at which battles and engagements are planned and executed to accomplish military objectives assigned to tactical units or task forces. Activities at this level focus on the ordered arrangement and maneuver of combat elements in relation to each other and to the enemy to achieve combat objectives. (JP 1-02)

target acquisition – The detection, identification, and location of a target in sufficient detail to permit the effective employment of weapons. (JP 1-02)

targeting – The process of selecting and prioritizing targets and matching the appropriate response to them, taking account of operational requirements and capabilities. (JP 1-02)

target reference point – An easily recognizable point on the ground (either natural or man-made) used to initiate, distribute, and control fires. Target reference points (TRPs) can also designate the center of an area where the commander plans to distribute or converge the fires of all his weapons rapidly. They are used by task force and below, and can further delineate sectors of fire within an engagement area. TRPs are designated using the standard target symbol and numbers issued by the fire support officer. Once designated, TRPs also constitute indirect fire targets. (FM 3-90)

task organization – A temporary grouping of forces designed to accomplish a particular mission. (FM 3-0)

task-organizing – (Army) The process of allocating available assets to subordinate commanders and establishing their command and support relationships. (FM 3-0)

tempo – The rate of military action. (FM 3-0)

terminal control – 1. The authority to direct the maneuver of aircraft which are delivering ordnance, pasengers, or cargo to a specific location or target. Terminal control is a type of air control. 2. Any electronic, mechanical, or visual control given to aircraft to facilitate target acquisition and resolution. (JP 1-02)

terrorism – The calculated use of unlawful violence or threat of unlawful violence to inculcate fear; intended to coerce or to intimidate governments or societies in the pursuit of goals that are generally political, religious, or ideological. (JP 1-02)

terrorist – An individual who uses violence, terror, and intimidation to achieve a result. (JP 1-02)

throughput distribution – The bypassing of one or more intermediate supply echelons in the supply system to avoid multiple handling. (FM 4-0)

time-phased force and deployment data – The Joint Operation Planning and Execution System database portion of an operation plan; it contains time-phased force data, non-unit-related cargo and personnel data, and movement data for the operation plan, including the following: a. In-place units; b. Units to be deployed to support the operation plan with a priority indicating the desired sequence for their arrival at the port of debarkation; c. Routing of forces to be deployed; d. Movement data associated with deploying forces; e. Estimates of non-unit related cargo and personnel movements to be conducted concurrently with the deployment of forces; and f. Estimate of transportation requirements that must be fulfilled by common-user lift resources as well as those requirements that can be fulfilled by assigned or attached transportation resources. (JP 1-02)

toxic industrial material – A generic term for toxic radioactive compounds in solid, liquid, aerosolized or gaseous form. These may be used, or stored for use, for industrial, commercial, medical, military, or domestic purposes. Toxic industrial materials may be chemical, biological, or radioactive and described as toxic industrial chemical, toxic industrial biological, or toxic industrial radiological. (FM 3-11.4)

track – 1. To display or record the successive positions of a moving object. 2. To keep a gun properly aimed, or to point continuously a target-locating instrument at a moving target. (JP 1-02)

troop leading procedures – A dynamic process used by small unit leaders to analyze a mission, develop a plan, and prepare for an operation. (FM 5-0)

turning movement – (Army) A form of maneuver in which the attacking force seeks to avoid the enemy's principal defensive positions by seizing objectives to the enemy rear and causing the enemy to move out of his current positions or divert major forces to meet the threat. (FM 3-0)

unconventional warfare – A broad spectrum of military and paramilitary operations, normally of long duration, predominantly conducted through, with, or by indigenous or surrogate forces who are organized, trained, equipped, supported, and directed in varying degrees by an external source. It includes, but is not limited to, guerrilla warfare, subversion, sabotage, intelligence activities, and unconventional assisted recovery. (JP 1-02)

unexploded explosive ordnance – Explosive ordnance which has been primed, fused, armed, or otherwise prepared for action, and which has been fired, dropped, launched, projected, or placed in such a manner as to constitute a hazard to operations, installations, personnel, or material, and remains unexploded either by malfunction or design or for any other cause. (JP 1-02)

unity of command – One of the nine principles of war: for every objective, ensure unity of effort under one responsible commander.

unity of effort – Coordination and cooperation among all military forces and other organizations toward a commonly recognized objective, even if the forces and nonmilitary organizations are not necessarily part of the same command structure. (FM 6-0)

unmanned aircraft system – A powered aerial vehicle that does not carry a human operator, uses aerodynamic forces to provide lift, can fly automomously or is remotely operated, can be expendable or recoverable, and can carry a lethal or nonlethal payload. Ballistic or semi-ballistic vehicles, cruise missiles, and artillery projectiles are not considered unmanned aircraft systems.

urban area – A topographical complex where manmade construction or high population density is the dominant feature. See also **urban environment**; **urban operations**. (FM 3-06)

urban environment – The physical urban area, as well as the complex and dynamic interaction among its key components—the terrain (natural and manmade), the population, and the supporting infrastructure—as an overlapping and interdependent system of systems. (FM 3-06)

urban operations – Offense, defense, stability, and support operations conducted in a topographical complex and adjacent natural terrain where manmade construction and high population density are the dominant features. (FM 3-0)

versatility – The ability of Army forces to meet the global, diverse mission requirements of full spectrum operations. (FM 3-0)

visualize – To create and think in mental images. (FM 6-0)

vulnerability – 1. The susceptibility of a nation or military force to any action by any means through which its war potential or combat effectiveness may be reduced or its will to fight diminished. 2. The characteristics of a system that cause it to suffer a definite degradation (incapability to perform the designated mission) as a result of having been subjected to a certain level of effects in an unnatural (manmade) hostile environment. 3. In information operations, a weakness in information system security design, procedures, implementation, or internal controls that could be exploited to gain unauthorized access to information or an information system. (JP 1-02)

warfighting function – A group of tasks and systems (people, organizations, information, and processes), united by a common purpose, that commanders use to accomplish missions and training objectives. The six warfighting functions are: a. **command and control** – the related tasks and systems that support the commander in exercising authority and direction; b. **fire support** – the related tasks and systems that provide collective and coordinated use of Army indirect fires, joint fires, and offensive information operations; c. **intelligence** – the related tasks and systems that facilitate understanding of the enemy, terrain, weather, and civil considerations; d. **movement and maneuver** – the related tasks and systems that move forces to achieve a position of advantage in relation to the enemy; e. **protection** – the related tasks and systems that preserve the force so the commander can apply maximum combat power; f. **sustainment** – the related tasks and systems that provide support and services to ensure freedom of action, extend operational reach, and prolon endurance. (FMI 5-0.1)

weapons of mass destruction – Weapons that are capable of a high order of destruction and/or of being used in such a manner as to destroy large numbers of people. Weapons of mass destruction can be high explosives or nuclear, biological, chemical, and radiological weapons, but exclude the means of transporting or propelling the weapon where such means is a separable and divisible part of the weapon. (JP 1-02)

This page intentionally left blank.

References

Military publications are listed by title. When a field manual has been published under a new number for the first time, the old number is provided in parenthesis after the new number.

REQUIRED PUBLICATIONS

These documents must be available to intended users of this publication.

JOINT PUBLICATIONS

Joint publications are available at <http://www.dtic.mil/doctrine/jel/>.

JP 1-02. *Department of Defense Dictionary of Military and Associated Terms*, 12 April 2001.

JP 3-06. *Doctrine for Joint Urban Operations*. 16 September 2002.

ARMY PUBLICATIONS

Army doctrinal publications are available at <http://www.apd.army.mil/>.

FM 1-02 (101-5-1). *Operational Terms and Graphics*. 21 September 2004.

FM 34-130. *Intelligence Preparation of the Battlefield*. 8 July 1994. FM 34-130 will be republished as FM 2-01.3

FM 3-0 (100-5). *Operations*. 14 June 2001.

FM 3-06.11 (90-10-1). *Combined Arms Operations in Urban Terrain*. 28 February 2002.

FM 3-07 (100-20). *Stability Operations and Support Operations*. 20 February 2003. When updated, FM 3-07 will be retitled *Stability Operations*.

FM 3-90. *Tactics*. 04 July 2001.

FM 5-19 (100-14). *Composite Risk Management*. 21 August 2006.

FM 7-0. *Training the Force*. 22 October 2002.

FM 7-1. *Battle Focused Training*. 15 September 2003.

FM 5-0 (101-5). *Army Planning and Orders Production*. 20 January 2005.

RELATED PUBLICATIONS

These sources contain relevant supplemental information.

JOINT PUBLICATIONS

Joint publications are available at <http://www.dtic.mil/doctrine/jel/>.

JP 3-0. *Doctrine for Joint Operations*. 10 September 2001.

JP 3-05. *Doctrine for Joint Special Operations*. 17 December 2003.

JP 3-08. *Interagency Coordination During Joint Operations* (2 Volumes). 9 October 1996.

JP 3-16. *Joint Doctrine for Multinational Operations*. 05 April 2000.

JP 4-0. *Doctrine for Logistic Support of Joint Operations*. 06 April 2000.

JP 5-00.2. *Joint Task Force Planning Guidance and Procedures*. 13 January 1999.

Joint Military Operations Historical Collection. 15 July 1997.

ARMY PUBLICATIONS

Army doctrinal publications are available at <http://www.apd.army.mil/>.

Army Regulations

FM 1. *The Army*. 14 June 2005.

FM 2-0 (34-1). *Intelligence*. 17 May 2004.

FM 2-22.3 (34-52). *Human Intelligence Collector Operations*. 6 September 2006.

FM 3-05 (100-25). *Army Special Operations Forces*. 20 September 2006.

FM 3-05.20 (31-20). *Special Forces Operations*. 1 April 2004.

FM 3-05.30 (33-1). *Psychological Operations*. 15 April 2005.

FM 3-05.105. *Multiservice Tactics, Techniques, and Procedures for Special Operations Forces in Nuclear, Biological, and Chemical Environments*. 28 September 2001.

FM 3-05.212 (TC 31-25). *Special Forces Waterborne Operations*. 31 August 2004.

FM 3-05.301 (33-1-1). *Psychological Operations Tactics, Techniques, and Procedures*. 31 December 2003.

FM 3-05.401 (41-10-1). *Civil Affairs Tactics, Techniques, and Procedures*. 23 September 2003.

FM 3-06 (90-10). *Urban Operations*. 01 June 2003.

FM 3-06.1. *Aviation Urban Operations: Multi-service Tactics, Techniques, and Procedures for Aviation Urban Operations*. 09 July 2005.

FM 3-07.2. *Antiterrorism*. (Future Publication).

FM 3-11.3 ( ). *Multiservice Tactics, Techniques, and Procedures for Chemical, Biological, Radiological, and Nuclear Contamination Avoidance*. 2 February 2006.

FM 3-11.9 (3-9). *Potential Military Chemical/Biological Agents and Compounds*. 10 January 2005.

FM 3-11.14 (3-14). *Multiservice Tactics, Techniques, and Procedures for Nuclear, Biological, and Chemical Vulnerability Assessment*. 28 November 2004.

FM 3-11.19 (3-19). *Multiservice Tactics, Techniques, and Procedures for Nuclear, Biological, and Chemical Reconnaissance*. 30 July 2004.

FM 3-11.21 (3-21). *Multiservice Tactics, Techniques, and Procedures for Nuclear, Biological, and Chemical Aspects of Consequence Management*. 12 December 2001.

FM 3-11.22. *Weapons of Mass Destruction Civil Support Team Tactics, Techniques, and Procedures*. 06 June 2003.

FM 3-13 (100-6). *Information Operations: Doctrine, Tactics, Techniques, and Procedures*. 28 November 2003.

FM 3-14 (100-18). *Space Support to Army Operations*. 18 May 2005.

FM 3-19.4 (19-4). *Military Police Leaders' Handbook*. 04 March 2002.

FM 3-19.15 (19-15). *Civil Disturbance Operations*. 18 April 2005.

FM 3-19.17. *Military Working Dogs*. 22 September 2005.

FM 3-19.40 (19-40). *Military Police Internment/Resettlement Operations*. 01 August 2001.

FM 3-19.50. *Police Intelligence Operations*. 21 July 2006.

FM 3-22.40 (90-40). *NLW: Tactical Employment of Nonlethal Weapons*. 15 January 2003.

FM 3-28. *Civil Support Operations*. (Future Publication).

FM 3-31. *Joint Force Land Component Commander Handbook (JFLCC)*. 13 December 2001.

FM 3-31.1 (90-31). *AMCI: Army and Marine Corps Integration in Joint Operations*. 21 November 2001.

FM 3-34 (5-100, 5-114). *Engineer Operations*. 02 January 2004.

FM 3-34.2. *Combined Arms Breaching Operations*. 31 August 2000. FM 3-34.2 will be republished as FM 3-90.11

FM 3-34.230 (5-105). *Topographic Operations*. 03 August 2000.

FM 3-61.1. *Public Affairs Tactics, Techniques and Procedures*. 01 October 2000.

FM 3-90.12 (90-13). *Combined Arms Gap Crossing Operations*. TBP.

FM 3-97.6 (90-6). *Mountain Operations*. 28 November 2000.

FM 3-100.12. *Risk Management: Multiservice Tactics, Techniques, and Procedures for Risk Management.* 15 February 2001.

FM 3-100.38. *UXO: Multi-service Tactics, Techniques and Procedures for Unexploded Ordnance Operations.* 18 August 2005.

FM 4-0 (100-10). *Combat Service Support.* 29 August 2003.

FM 4-01.45. *Tactical Convoy Operations: Multiservice Tactics, Techniques, and Procedures for Tactical Convoy Operations.* 24 March 2005.

FM 4-02.51 (8-51). *Combat and Operational Stress Control.* 6 June 2006.

FM 4-30.16. *EOD: Multiservice Procedures for Explosive Ordnance Disposal in a Joint Environment.* 27 October 2005.

FM 4-30.51. *Unexploded Ordnance (UXO) Procedures.* 13 July 2006.

FM 5-01.12 (90-41). *JTF LNO Integration: Multiservice Tactics, Techniques, and Procedures for Joint Task Force (JTF) Liaison Officer Integration.* 27 January 2003.

FM 5-33. *Terrain Analysis.* 8 September 1992. FM 5-33 will be republished as FM 3 34.330.

FM 5-103. *Survivability.* 10 June 1985. FM 5-103 will be republished as FM 5-34.112.

FM 5-104. *General Engineering.* 12 November 1986. FM 5-104 will be republished as FM 3-34.250.

FM 5-170. *Engineer Reconnaissance.* 05 May 1988. FM 5-170 will be republished as FM 3 34.212.

FM 5-415. *Fire-Fighting Operations.* 09 February 1999. FM 5-415 will be republished as FM 3-34.281.

FM 6-0. *Mission Command: Command and Control of Army Forces.* 11 August 2003.

FM 6-22.5 (22-9). *Combat Stress.* 23 June 2000.

FM 7-15. *The Army Universal Task List (Incl. changes 2).* 6 July 2006.

FM 20-32. *Mine/Countermine Operations.* 1 April 2005.

FM 22-51. *Leaders' Manual for Combat Stress Control.* 29 September 1994. FM 22-51 will be republished as FM 4-02.22.

FM 22-100. *Army Leadership: Be, Know, Do.* 31 August 1999. FM 22-100 will be republished as FM 6-22.

FM 23-10. *Sniper Training.* 17 August 1994. FM 23-10 will be republished as FM 3-22.10.

FM 27-10. *The Law of Land Warfare.* 18 July 1956. FM 27-10 will be republished as FM 1-04.10.

FM 27-100. *Legal Support to Operations.* 01 March 2000. FM 27-100 will be republished as FM 1-04.

FM 31-20-3. *Foreign Internal Defense Tactics, Techniques, and Procedures for Special Forces.* 20 September 1994. FM 31-20-3 will be republished as FM 3-05.202.

FM 31-50. *Combat in Fortified and Built-Up Areas.* March 1964.

FM 34-3. *Intelligence Analysis.* 15 March 1990. FM 34-3 will be republished as FM 2-33.4.

FM 34-60. *Counterintelligence.* 03 October 1995. FM 34-60 will be republished as FM 2-01.2.

FM 34-81. *Weather Support for Army Tactical Operations.* 31 August 1989. FM 34-81 will be republished as FM 2-33.2.

FM 34-81-1. *Battlefield Weather Effects.* 23 December 1992. FM 34-81-1 will be republished as FM 2-33.201.

FM 41-10. *Civil Affairs Operations.* 14 February 2000. FM 41-10 will be republished as FM 3-05.40.

FM 44-8. *Combined Arms for Air Defense.* 01 June 1999. FM 44-8 will be republished as FM 3-01.8.

FM 46-1. *Public Affairs Operations.* 30 May 1997. FM 46-1 will be republished as FM 3-61.

FM 90-8. *Counterguerrilla Operations.* 29 August 1986. FM 90-8 will be republished as FM 3-07.11.

FM 90-10. *Military Operations on Urbanized Terrain (MOUT) (How to Fight).* 15 August 1979.

FM 100-5. *Operations.* 1944.

FM 100-8. *The Army in Multinational Operations*. 24 November 1997. FM 100-8 will be republished as FM 3-16.

FM 100-11. *Force Integration*. 15 January 1998. FM 100-11 will be republished as FM 3-100.11.

FMI 2-91.4. *Intelligence Support to Operations in the Urban Environment*. 30 June 2005. FMI 2-91.4 will be republished as FM 2-91.4.

FMI 3-07.22. *Counterinsurgency Operations*. 01 October 2004. FMI 3-07.22 will be republished as FM 3-07.22.

FMI 3-34.119. *Improvised Explosive Device Defeat*. 21 September 2005.

FMI 4-30.50. *Modular Explosive Ordnance Disposal Operations*. 31 July 2006.

FMI 5-0.1. *The Operations Process*. 31 March 2006.

TC 3-15. *Nuclear Accident and Incident Response and Assistance (NAIRA)*. 27 December 1988.

TC 5-230. *Army Geospatial Guide for Commanders and Planners*. 28 November 2003.

TC 7-98-1. *Stability and Support Operations Training Support Package*. 05 June 1997.

TC 9-21-01. *Soldiers Improvised Explosive Device (IED) Awareness Guide: Iraq and Afghanistan Theaters of Operation*. 28 May 2004.

TC 90-1. *Training for Urban Operations*. 01 April 2002.

DEPARTMENT OF DEFENSE PUBLICATIONS

Defense Intelligence Agency Manual 58-11. *Defense Human Resources Intelligence Collection Management*. 20 August 1968.

Quadrennial Defense Review Report. 30 September 2001.

U.S. Army Training and Doctrine Command (TRADOC), Deputy Chief of Staff for Intelligence (DCSINT). *TRADOC DCSINT Handbook No. 1: A Military Guide to Terrorism in the Twenty-First Century* (Version2.0). Fort Leavenworth, KS: U.S. Government Printing Office, 2004.

NONMILITARY PUBLICATIONS

26th Infantry Regimental Association. *Aachen: Military Operations in Urban Terrain*. Lititz, PA: 26th Infantry Regimental Association, 1999.

Adan, Abraham. *On the Banks of the Suez*. Novato, CA: Presidio Press, 1980.

Akers, Frank H., Jr. and George B. Singleton. "Task Force Ranger: A Case Study Examining the Application of Advanced Technologies in Modern Urban Warfare." Oak Ridge, TN: National Security Program Office, November 2000.

Allard, Kenneth. *Somalia Operations: Lessons Learned*. Washington, DC: National Defense University Press, 1995.

Arnold, James R. *Tet Offensive 1968: Turning Point in Vietnam*. London: Osprey, 1990.

Bass, Bernard M. "Leading in the Army after Next." *Military Review* (March/April 1998): 46–58.

Beevor, Anthony. *Stalingrad*. New York: Penguin Books, 1998.

Bowden, Mark. *Black Hawk Down: A Story of Modern War*. New York: Atlantic Monthly Press, 1999.

Butts, Kent H. "The Strategic Importance of Water." *Parameters* (spring 1997): 65–83.

Carell, Paul. *Stalingrad: The Defeat of the German 6th Army*. Atglen, PA: Schieffer Publishing, 1993.

Celeski, Joseph D. "Joint Urban Operations: Special Forces in Urban Campaign Planning." *Special Warfare* (fall 2001): 2–13.

Chiarelli, Peter W. and Patrick R. Michaelis. "Winning the Peace: The Requirement for Full-Spectrum Operations." *Military Review* (July/August 2005): 4–17.

Chiarelli, Peter W., Patrick R. Michaelis, and Geoffrey A. Norman. "Armor in Urban Terrain: The Critical Enabler." *Armor* (March/April 2005): 7–12.

Chuikov, Vasili I. *The Battle for Stalingrad.* New York: Holt, Rinehart, and Winston, 1964.

Colby, John. *War From the Ground Up: The 90th Division in WWII.* Austin, TX: Nortex Press, 1991.

Cole, Ronald H. *Operation Just Cause.* Washington, DC: Joint History Office, 1995.

Connaughton, Richard, John Pimlott, and Duncan Anderson. *The Battle for Manila: The Most Devastating Untold Story of World War II.* Novata, CA: Presidio, 1995.

Craig, William. *Enemy at the Gates: The Battle for Stalingrad.* New York: Reader's Digest Press, 1973.

Delk, James D. *Fires & Furies: The LA Riots.* Palm Springs, CA: Etc. Publications, 1995.

Dewar, Michael. *British Army in Northern Ireland.* New York: Sterling Publishers, 1985.

______. *War in the Streets: The Story of Urban Combat from Calais to Khafji.* New York: Sterling Publishers, 1992.

Donnelly, Thomas, Margaret Roth, and Caleb Baker. *Operation Just Cause: The Storming of Panama.* New York: Lexington Books, 1991.

Dupuy, R. Ernest and Trevor N. Dupuy. *The Encyclopedia of Military History: From 3500 B.C. to the Present.* San Francisco: Harper & Row, 1977.

Dupuy, Trevor N. and Paul Martell. *Flawed Victory: The Arab-Israeli Conflict and the 1982 War in Lebanon.* Fairfax, VA: Hero Books, 1986.

Edwards, Sean J. A. *Freeing Mercury's Wings: Improving Tactical Communications in Cities.* Santa Monica, CA: RAND, 2001.

Eshel, David. *Mid-East Wars: The Lebanon War 1982.* Hod Hasharon, Israel: Eshel-Dramit, 1983.

Fishel, John T. *The Fog of Peace: Planning and Executing the Restoration of Panama.* Carlisle, PA: Strategic Studies Institute, 1992.

Fontenot, Gregory, E. J. Degen, and David Tohn. *On Point: US Army in Operation IRAQI FREEDOM.* Fort Leavenworth, KS: Combat Studies Institute Press, 2004.

Funston, Frederick. "How the Army Worked to Save San Francisco." *Cosmopolitan Magazine* (July 1906): <http://www.sfmuseum.org/1906/cosmo.html>.

Gabel, Christopher R. "Military Operations on Urbanized Terrain: The 2d Battalion, 26th Infantry, at Aachen, October 1944." *Urban Combat Operations: tactics, Techniques and Procedure.* Center for Army Lessons Learned (CALL) Newsletter No. 99-16. Fort Leavenworth, KS: Center for Army Lessons Learned, U.S. Training and Doctrine Command [TRADOC], November 1999: 1-9–1-13.

Gabriel, Richard A. *Operation Peace for Galilee: The Israeli-PLO War in Lebanon.* New York: Hill and Wang, 1984.

Glenn, Russell W. *An Attack on Duffer's Downtown.* Santa Monica, CA: RAND, 2001.

______. *Heavy Matter: Urban Operations' Density of Challenges.* Santa Monica, CA: RAND, 2000.

______. "Urban Combat is Complex." *Proceedings* (February 2002): 62–65.

______. "…We Band of Brothers": *The Call for Joint Urban Operations Doctrine.* Santa Monica, CA: RAND, 1999.

Glenn, Russell W., Steven Hartman, and Scott Gerwehr. *Urban Combat Service Support Operations: The Shoulders of Atlas.* Santa Monica, CA: RAND, 2003.

Gordon, Michael R. "Baghdad Targets Picked if Hussein Holes Up There." *New York Times* (7 March 2003): 1A2.

Grau, Lester W. "Something Old, Something New: Guerrillas, Terrorists, and Intelligence Analysis." *Military Review* (July/August 2004): 42–49.

Grau, Lester W. and Timothy Smith. *A 'Crushing' Victory: Fuel-Air Explosives and Grozny 2000.* Fort Leavenworth, KS: Foreign Military Studies Office, 2000.

Hamill, Desmond. *Pig in the Middle: The Army in Northern Ireland, 1969–1984.* London: Methuen, 1985.

Hammel, Eric. *Fire in the Streets: The Battle for Hue, Tet 1968*. New York: Dell Publishing, 1991.

Heal, Sid. "Crowds, Mobs and Nonlethal Weapons." *Military Review* (March/April 2000): 45–50.

Heinl, Robert Debs, Jr. *Victory at High Tide: The Inchon-Seoul Campaign*. Annapolis, MD: Nautical & Aviation Publishing Company of America, 1979.

Herzog, Cham. *The War of Atonement, October, 1973*. Boston: Little, Brown and Company, 1975.

Higham, Robin. *Bayonets in the Streets: The Use of Troops in Civil Disturbances*. Lawrence, KS: University Press of Kansas, 1969.

Hoar, Joseph P. "A CINC's Perspective." *JFQ Forum* (autumn 1993): 56–63.

Hollis, Mark A. B. "Platoon Under Fire." *Infantry* (January/April 1998): 27–34.

Infantry in Battle. 2d ed. Garret & Massie: Richmond, VA, 1939. Reprint, Fort Leavenworth, KS: U.S. Army Command and General Staff College, 1981.

Initial Impressions Report: Operation Iraqi Freedom—Stability Operations-Support Operations Center for Army Lessons Learned (CALL) Newsletter No. 04-13 (Fort Leavenworth, KS: Center for Army Lessons Learned, U.S. Army Training and Doctrine Command [TRADOC], 2004).

Initial Impressions Report: Operation Iraqi Freedom—Stability Operations-Support Operations. Center for Army Lessons Learned (CALL) Newsletter No. 03 (Fort Leavenworth, KS: Center for Army Lessons Learned, U.S. Army Training and Doctrine Command [TRADOC], 2003).

Jomini, Antoine-Henri. "The Art of War" in *Roots of Strategy: Book 2*. Harrisburg, PA: Stackpole Books, 1987.

Kemp, Anthony. *The Unknown Battle: Metz, 1944*. New York: Stein and Day, 1981.

Kennedy, John F. Message to Congress, 25 May 1961. Available at <http://www.jfklibrary.org/Historical+Resources/Archives>.

Linden, Eugene. "The Exploding Cities of the Developing World." *Foreign Affairs* (January/February 1996): 52–65.

Mabry, Robert L. et al. "United States Army Rangers in Somalia: An Analysis of Combat Casualties on an Urban Battlefield." *Journal of Special Operations Medicine* (fall 2001): 24–40.

MacArthur, Douglas. *Reminiscences*. New York: McGraw-Hill, 1964.

Manwaring, Max G. *Street Gangs: The New Urban Insurgency*. Carlisle, PA: Strategic Studies Institute, 2005.

Maxwell, Mark D. "The Law of War and Civilians on the Battlefield: Are We Undermining Civilian Protections?" *Military Review* (September/October 2004): 17–25.

McConnell, Malcolm. *Just Cause: The Real Story of America's High-Tech Invasion of Panama*. New York: St. Martin's Press, 1991.

Medby, Jamison Jo and Russell W. Glenn. *Street Smart: Intelligence Preparation of the Battlefield for Urban Operations*. Santa Monica, CA: RAND, 2002.

Mendel, William W. *Combat in Cities: The LA Riots and Operation Rio*. Fort Leavenworth, KS: Foreign Military Studies Office, 1996.

Monmonier, Mark. *How to Lie with Maps*. Chicago: University of Chicago Press, 1996.

Montgomery, Thomas M. Interview. "Ambush in Mogasdishu." *Frontline*, WGBH Boston. 29 September 1998. Available at <http://www.pbs.org/wgbh/pages/frontline/programs/1998.html>.

Moore, John C. "Sadr City: The Armor Pure Assault in Urban Terrain." *Armor* (November/December 2004): 31–37.

Natsios, Andrew S. "Commander's Guidance: A Challenge of Complex Humanitarian Emergencies." *Parameters* (summer 1996): 50–66.

______. "Food Through Force: Humanitarian Intervention and U.S. Policy." *The Washington Quarterly* (winter 1994): 136.

Nolan, Keith William. *The Battle for Saigon: Tet 1968*. New York: Pocket Books, 1996.

Oakley, Robert B. "An Envoy's Perspective." *JFQ Forum* (autumn 1993): 44–55.

Oliker, Olga. *Russia's Chechen Wars 1994–2000: Lessons from Urban Combat*. Santa Monica, CA: RAND, 2001.

Operation Restore Hope Lessons Learned Report: 3 December 1992 – 4 May 1993, Operations Other than War. Fort Leavenworth, KS: Center for Army Lessons Learned (CALL), U.S. Army Combined Arms Command (CAC), 1993. Reprint, Washington, DC: U.S. Government Printing Office, 1996.

Parkerson, John Embry, Jr. "United States Compliance with Humanitarian Law Respecting Civilians During Operation Just Cause." *Military Law Review* (1991): 31–140.

Peters, Ralph. "Our Soldiers, Their Cities." *Parameters* (spring 1996): 43–50.

Peterson, John J. *Into the Cauldron*. Clinton, MD: Clavier House, 1973.

Petraeus, David H. "Learning Counterinsurgency: Observations from Soldiering in Iraq." *Military Review* (January/February 2006): 2–12.

Ritchie, Elspeth C. and Robert Mott. "Caring for Civilians During Peace Keeping Missions: Priorities and Decisions." *Military Medicine* (Vol. 167, Supplement 3, 2002): 14–16.

Rosenberger, Leif R. "The Strategic Importance of the World Food Supply." *Parameters* (spring 1997): 84–105.

Schadlow, Nadia. "War and the Art of Governance." *Parameters* (autumn 2003): 85–94.

Shelton, H. Hugh and Timothy D. Vane. "Winning the Information War in Haiti." *Military Review* (November/December 1995): 3–9.

Shtemenko, S. M. *The Soviet General Staff at War, 1941-1945: Book One*. Moscow: Progress Publishers, 1981.

Smith, Rupert. "Wars in Our Time: A Survey of Recent and Continuing Conflicts." *World Defence Systems,* volume 3:2. London: Royal United Services Institute for Defence Studies, 2001.

Stanton, Shelby L. *Ten Corps in Korea, 1950*. Novato, CA: Presidio Press, 1989.

Taw, Jennifer Morrison. *Operation Just Cause: Lessons for Operations Other Than* War. Santa Monica, CA: RAND, 1996.

Thomas, Timothy L. "The Battle for Grozny: Deadly Classroom for Urban Combat." *Parameters* (summer 1999): 87–102.

Thomas, Timothy L. and Charles P. O'Hara. "Combat Stress in Chechnya: 'The Equal Opportunity Disorder'." *Army Medical Department Journal* (January/March 2000): 46–53.

Trinquier, Roger. *Modern Warfare: A French View of Counterinsurgency*. New York: Praeger Publisher, 1964.

United Nations Security Council Resolution 794 (1992). New York: United Nations, 1992. Available at <http://daccessdds.un.org/doc/UNDOC/GEN/N92/772/11/PDF>.

U.S. Agency for International Development, Bureau for Humanitarian Response, Office of Foreign Disaster Assistance. *Field Operations Guide for Disaster Assessment and Response, Version 3.0*. Washington, DC: U.S. Government Printing Office, 2003. Available at <http://www.usaid.gov/our_work/humanitarian_assistance/>.

U.S. Department of Transportation, Research and Special Programs Administration, Office of Hazardous Materials Initiative and Training. *2004 Emergency Response Guidebook*. Washington, DC: U.S. Government Printing Office, 2004. Available at <http://hazmat.dot.gov/pubs/erg/erg2004.pdf>.

U.S. Environmental Protection Agency, Office of Emergency and Remedial Response. *HAZMAT Team Planning Guidance*. Washington, DC: U.S. Government Printing Office, 1990.

Van Sickle, Jeffrey B. "Stability Operations in Northern Iraq: Task Force Altun Kupri." *Infantry* (January/February 2005): 25–29.

Watson, Bruce Allen. *Sieges: A Comparative Study*. Westport, CT: Praeger, 1993.

Werstein, Irving. *The Battle of Aachen*. New York: Thomas Y. Crowell Company, 1962.

Whiting, Charles. *Bloody Aachen*. New York: Stein and Day, 1976.

Wintle, Justin, ed. *The Dictionary of War Quotations*. New York: The Free Press, 1989.

Yates, Lawrence A. "Military Stability and Support Operations: Analogies, Patterns and Recurring Themes." *Military Review* (July/August 1997): 51–61.

RECOMMENDED READINGS

The following sources contain relevant supplemental information.

Alexander, John B. *Future War: Non-Lethal Weapons in Twenty-First-Century* Warfare. New York: St. Martin's Press, 1999.

Ashworth, G. J. *The City and War*. London: Routledge, 1991.

Barker, A. J. *Dunkirk: The Great Escape*. New York: David McKay, 1977.

Baumann, Robert F., Lawrence A. Yates, and Versalle F. Washington. "My Clan Against the World": *U.S. and Coalition Forces in Somalia, 1992-1994*. Fort Leavenworth, KS: Combat Studies Institute Press, 2004.

Birtle, Andrew J. *U.S. Army Counterinsurgency and Contingency Operations Doctrine: 1860-1941*. Washington, DC: Center of Military History, 2001.

Blaufarb, Douglas S. *The Counterinsurgency Era: U.S. Doctrine and Performance 1950 to Present*. New York: The Free Press, 1977.

Breuer, William B. *Hitler's Fortress Cherbourg: The Conquest of a Bastion*. New York: Stein and Day, 1984.

Brown, Stephen D. "PSYOP in Operation Uphold Democracy." *Military Review* (September/October 1996): 57–76.

Bulloch, Gavin. "Military Doctrine and Counterinsurgency: A British Perspective." *Parameters* (summer 1996): 4–16.

Urban Combat Operations: Tactics, Techniques and Procedures. Center for Army Lessons Learned (CALL) Newsletter No. 99-16 (Fort Leavenworth, KS: Center for Army Lessons Learned, U.S. Army Training and Doctrine Command [TRADOC], November 1999).

Cordesman, Anthony H. and Abraham R. Wagner. *The Lessons of Modern War—Volume II: The Iran-Iraq War*. Boulder, CO: Westview Press, 1990.

Corum, James S. and Wray R. Johnson. *Airpower in Small Wars: Fighting Insurgents and Terrorists*. Lawrence, KS: University Press of Kansas, 2003.

Desch, Michael C., ed. *Soldiers in Cities: Military Operations on Urban Terrain*. Carlisle Barracks, PA: Strategic Studies Institute, 2001.

Duffy, Christopher. *Fire & Stone: The Science of Fortress Warfare, 1660–1860*. Mechanicsburg, PA: Stackpole Books, 1996.

______. *The Fortress in the Age of Vauban and Frederick the Great, 1660–1789: Siege Warfare Volume I*. London: Routledge & Kegan Paul, 1985.

______. *Siege Warfare: The Fortress in the Early Modern World, 1494–1660*. London: Routledge & Kegan Paul, 1979.

Edwards, Sean J. A. *Mars Unmasked: The Changing Face of Urban Operations*. Santa Monica, CA: RAND, 2000.

Eikenberry, Karl W. "Take No Casualties." *Parameters* (summer 1996): 109–118.

Erickson, John. *The Road to Berlin: Continuing the History of Stalin's War with Germany*. Boulder, CO: Westview Press, 1983.

Gerwehr, Scott and Russell W. Glenn. *The Art of Darkness: Deception and Urban Operations*. Santa Monica, CA: RAND, 2000.

Glenn, Russell W. *Combat in Hell: A Consideration of Constrained Urban Warfare*. Santa Monica, CA: RAND, 1996.

______. *Marching Under Darkening Skies: The American Military and Impending Urban Operations Threat.* Santa Monica, CA: RAND, 1998.

Glenn, Russell W., ed. *Capital Preservation: Preparing for Urban Operations in the Twenty-First Century: Proceedings of the RAND Arroyo-TRADOC-MCWL-OSD Urban Operations Conference, March 22-23, 2000.* Santa Monica, CA: RAND, 2001.

Glenn, Russell W. et al. *Denying the Widow-Maker, Summary of Proceedings, RAND-DDBL Conference on Military Operations on Urbanized Terrain.* Santa Monica, CA: RAND, 1999.

______. *Getting the Musicians of Mars on the Same Sheet of Music: Army, Joint, Multinational, and Interagency C4ISR Interoperability.* Santa Monica, CA: RAND, 2000.

Glenn, Russell W. et al., eds. *The City's Many Faces: Proceedings of the RAND Arroyo-MCWL-J8 UWG Urban Operations Conference, April 13-14, 1999.* Santa Monica, CA: RAND, 2000.

Gouré, Leon. *The Siege of Leningrad.* Stanford, CA: Stanford University Press, 1962.

Grau, Lester W. and Jacob W. Kipp. "Urban Combat: Confronting the Specter." *Military Review* (July/August 1999): 9–17.

Groves, John R. "Operations in Urban Environments." *Military Review* (July/August 1998): 30–39.

Habeck, Mary R. Lecture. "Jihadist Strategies in the War on Terrorism." Washington, DC: The Heritage Foundation, 8 November 2004. Available at http://www.heritage.org/research/nationalsecurity/hl855.cfm.

Hahn, Robert F., II and Bonnie Jezior. "Urban Warfare and the Urban Warfighter of 2025." *Parameters* (summer 1999): 74–86.

Harman, Nicholas. *Dunkirk: The Patriotic Myth.* New York: Simon and Schuster, 1980.

Heckstall-Smith, Anthony. *Tobruk: The Story of a Siege.* New York: W. W. Norton, 1959.

Hibbert, Christopher. *The Battle of Arnhem.* London: Batsford, 1962.

House, John M. "The Enemy After Next." *Military Review* (March/April 1998): 22–28.

Jackson, W. G. F. *The Battle for Italy.* New York: Harper & Row, 1967.

Karslake, Basil. *1940 – The Last Act: The Story of the British Forces in France After Dunkirk.* Hamden, CT: Archon Books, 1979.

Kostof, Spiro. *The City Shaped: Urban Patterns and Meanings Through History.* London: Thames and Hudson, 1999.

______. *The City Assembled: The Elements of Urban Form Throughout History.* Boston: Little Brown and Company, 1999.

Kuehl, Daniel T. "Airpower vs. Electricity: Electric Power as a Target for Strategic Air Operations." *Journal of Strategic Studies* (March 1995): 237–266.

Larson, Eric. *Casualties and Consensus: The Historical Role of Casualties in Domestic Support of U.S. Military Operations.* Santa Monica, CA: RAND, 1996.

Lieven, Anatol. *Chechnya: Tombstone of Russian Power.* New Haven: Yale University Press, 1998.

______. "The World Turned Upside Down: Military Lessons of the Chechen War." *Armed Forces Journal International* (August 1998): 40–43.

McKee, Alexander. *Caen: Anvil of Victory.* New York: St. Martin's Press, 1964.

McLaurin, R. D. "Military Forces in Urban Antiterrorism." Aberdeen Proving Ground, MD: U.S. Army Human Engineering Laboratory, 1989.

McLaurin, R. D. and R. Miller. "Urban Counterinsurgency: Case Studies and Implications for U.S. Military Forces." Aberdeen Proving Ground, MD: U.S. Army Human Engineering Laboratory, 1989.

Mercier, Charles E. "Terrorists, WMD, and the U.S. Army Reserve." *Parameters* (autumn 1997): 98–118.

Milton, T. R. "Urban Operations: Future War." *Military Review* (February 1994): 37–46.

Moskin, J. Robert. *Among Lions: The Battle for Jerusalem, June 5–7, 1967*. New York: Ballantine Books, 1982.

Northam, Ray M. *Urban Geography*. New York: Wiley, 1975.

Peters, Ralph. "The Culture of Future Conflict." *Parameters* (winter 1995 96): 18–27.

______. "The Future of Armed Warfare." *Parameters* (autumn 1997): 50–59.

______. "The Human Terrain of Urban Operations." *Parameters* (spring 2000): 4–12.

Robertson, William G., ed. *Block by Block: The Challenges of Urban Operations*. Fort Leavenworth, KS: U.S. Army Command and Staff College Press, 2003.

Rosenau, William G. "'Every Room is a New Battle': The Lessons of Modern Urban Warfare." *Studies in Conflict and Terrorism* (October–December 1997): 371–394.

Scalard, Douglas. "People of Whom We Know Nothing: When Doctrine Isn't Enough." *Military Review* (July/August 1997): 4–11.

Scales, Robert H. "The Indirect Approach: How U.S. Military Forces Can Avoid the Pitfalls of Future Urban Warfare." *Armed Forces Journal International* (October 1998): 68–74.

Siegel, Adam B. *Eastern Exit: The Noncombatant Evacuation Operation (NEO) from Mogadishu, Somalia, in January 1991*. Alexandria, VA: Center for Naval Analyses, 1992.

Spiller, Roger J. *Sharp Corners: Urban Operations at Century's End*. Fort Leavenworth, KS: Combat Studies Institute, 2001.

Stech, Frank J. "Winning CNN Wars." *Parameters* (autumn 1994): 37–56.

Stevenson, Jonathan. *Losing Mogadishu: Testing U.S. Policy in Somalia*. Annapolis, MD: Naval Institute Press, 1995.

Taw, Jennifer Morrison and Bruce Hoffman. *The Urbanization of Insurgency: The Potential Challenge to U.S. Army Operations*. Santa Monica, CA: RAND, 1994.

Tully, Andrew. *Berlin: Story of a Battle*. Westport, CT: Greenwood Press, 1977.

United Nations Department of Economic and Social Affairs, Population Division. *World Urbanization Prospects: The 2003 Revision*. New York: United Nations, 2003.

Vick, Alan et al. *Aerospace Operations in Urban Environments: Exploring New Concepts*. Santa Monica, CA: RAND, 2000.

Wykes, Alan. *The Siege of Leningrad*. New York: Ballentine Books Inc., 1968.

Index

O

P-Q

R

S

T

U-V

W-Z

This page intentionally left blank.

FM 3-06
26 October 2006

By Order of the Secretary of the Army:

PETER J. SCHOOMAKER
General, United States Army
Chief of Staff

Official:

JOYCE E. MORROW
Administrative Assistant to the
Secretary of the Army
0627803

DISTRIBUTION:

Active Army, Army National Guard, and U. S. Army Reserve: To be distributed in accordance with initial distribution number 111233, requirements for FM 3-06.

Foreword

This manual is designed to fill a doctrinal gap. It has been 20 years since the Army published a field manual devoted exclusively to counterinsurgency operations. For the Marine Corps it has been 25 years. With our Soldiers and Marines fighting insurgents in Afghanistan and Iraq, it is essential that we give them a manual that provides principles and guidelines for counterinsurgency operations. Such guidance must be grounded in historical studies. However, it also must be informed by contemporary experiences.

This manual takes a general approach to counterinsurgency operations. The Army and Marine Corps recognize that every insurgency is contextual and presents its own set of challenges. You cannot fight former Saddamists and Islamic extremists the same way you would have fought the Viet Cong, Moros, or Tupamaros; the application of principles and fundamentals to deal with each varies considerably. Nonetheless, all insurgencies, even today's highly adaptable strains, remain wars amongst the people. They use variations of standard themes and adhere to elements of a recognizable revolutionary campaign plan. This manual therefore addresses the common characteristics of insurgencies. It strives to provide those conducting counterinsurgency campaigns with a solid foundation for understanding and addressing specific insurgencies.

A counterinsurgency campaign is, as described in this manual, a mix of offensive, defensive, and stability operations conducted along multiple lines of operations. It requires Soldiers and Marines to employ a mix of familiar combat tasks and skills more often associated with nonmilitary agencies. The balance between them depends on the local situation. Achieving this balance is not easy. It requires leaders at all levels to adjust their approach constantly. They must ensure that their Soldiers and Marines are ready to be greeted with either a handshake or a hand grenade while taking on missions only infrequently practiced until recently at our combat training centers. Soldiers and Marines are expected to be nation builders as well as warriors. They must be prepared to help reestablish institutions and local security forces and assist in rebuilding infrastructure and basic services. They must be able to facilitate establishing local governance and the rule of law. The list of such tasks is long; performing them involves extensive coordination and cooperation with many intergovernmental, host-nation, and international agencies. Indeed, the responsibilities of leaders in a counterinsurgency campaign are daunting; however, the discussions in this manual alert leaders to the challenges of such campaigns and suggest general approaches for grappling with those challenges.

Conducting a successful counterinsurgency campaign requires a flexible, adaptive force led by agile, well-informed, culturally astute leaders. It is our hope that this manual provides the guidelines needed to succeed in operations that are exceedingly difficult and complex. Our Soldiers and Marines deserve nothing less.

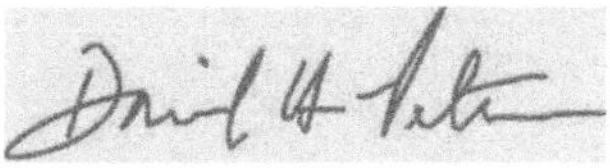

DAVID H. PETRAEUS
Lieutenant General, U.S. Army
Commander
U.S. Army Combined Arms Center

JAMES F. AMOS
Lieutenant General, U.S. Marine Corps
Deputy Commandant
Combat Development and Integration

FM 3-24
***MCWP 3-33.5**

Field Manual
No. 3-24

Headquarters
Department of the Army
Washington, DC

Marine Corps Warfighting Publication
No. 3-33.5

Headquarters
Marine Corps Combat Development Command
Department of the Navy
Headquarters
United States Marine Corps
Washington, DC

15 December 2006

COUNTERINSURGENCY

Contents

Distribution Restriction: Approved for public release; distribution is unlimited.

***This publication supersedes FMI 3-07.22, 1 October 2004, and MCWP 3-33.5, 29 January 1980.**

Marine Corps PCN: 143 000124 00

Figures

Tables

Vignettes

This publication is available at Army Knowledge Online (www.us.army.mil) and General Dennis J. Reimer Training and Doctrine Digital Library at (www.train.army.mil).

Preface

This field manual/Marine Corps warfighting publication establishes doctrine (fundamental principles) for military operations in a counterinsurgency (COIN) environment. It is based on lessons learned from previous counterinsurgencies and contemporary operations. It is also based on existing interim doctrine and doctrine recently developed.

Counterinsurgency operations generally have been neglected in broader American military doctrine and national security policies since the end of the Vietnam War over 30 years ago. This manual is designed to reverse that trend. It is also designed to merge traditional approaches to COIN with the realities of a new international arena shaped by technological advances, globalization, and the spread of extremist ideologies—some of them claiming the authority of a religious faith.

The manual begins with a description of insurgencies and counterinsurgencies. The first chapter includes a set of principles and imperatives necessary for successful COIN operations. Chapter 2 discusses nonmilitary organizations commonly involved in COIN operations and principles for integrating military and civilian activities. Chapter 3 addresses aspects of intelligence specific to COIN operations. The next two chapters discuss the design and execution of those operations. Developing host-nation security forces, an essential aspect of successful COIN operations, is the subject of chapter 6. Leadership and ethical concerns are addressed in chapter 7. Chapter 8, which concerns sustainment of COIN operations, concludes the basic manual. The appendixes contain useful supplemental information. Appendix A discusses factors to consider during the planning, preparation, execution, and assessment of a COIN operation. Appendixes B and C contain supplemental intelligence information. Appendix D addresses legal concerns. Appendix E describes the role of airpower.

Doctrine by definition is broad in scope and involves principles, tactics, techniques, and procedures applicable worldwide. Thus, this publication is not focused on any region or country and is not intended to be a stand-alone reference. Users should assess information from other sources to help them decide how to apply the doctrine in this publication to the specific circumstances facing them.

The primary audience for this manual is leaders and planners at the battalion level and above. This manual applies to the United States Marine Corps, the Active Army, the Army National Guard/Army National Guard of the United States, and the United States Army Reserve unless otherwise stated.

This publication contains copyrighted material. Copyrighted material is identified with footnotes. Other sources are identified in the source notes.

Terms that have joint, Army, or Marine Corps definitions are identified in both the glossary and the text. FM 3-24 is not the proponent field manual (the authority) for any Army term. For definitions in the text, the term is italicized and the number of the proponent manual follows the definition.

Headquarters, U.S. Army Training and Doctrine Command is the proponent for this publication. The preparing agency is the Combined Arms Doctrine Directorate, U.S. Army Combined Arms Center. Send written comments and recommendations on DA Form 2028 (Recommended Changes to Publications and Blank Forms) directly to Commander, U.S. Army Combined Arms Center and Fort Leavenworth, ATTN: ATZL-CD (FM 3-24), 201 Reynolds Avenue (Building 285), Fort Leavenworth, Kansas 66027-1352. Send comments and recommendations by e-mail to web-cadd@leavenworth.army.mil. Follow the DA Form 2028 format or submit an electronic DA Form 2028.

ACKNOWLEDGEMENTS

Counterinsurgency Warfare: Theory and Practice, David Galula. Copyright © 1964 by Frederick A. Praeger, Inc. Reproduced with permission of Greenwood Publishing Group, Inc., Westport, CT.

"Battle Lessons, What the Generals Don't Know," Dan Baum. *The New Yorker,* Jan 17, 2005. Reproduced with permission of Dan Baum.

Defeating Communist Insurgency: The Lessons of Malaya and Vietnam, Sir Robert Thompson. Copyright © 1966 by Robert Thompson. Reproduced with permission of Hailer Publishing.

Introduction

This is a game of wits and will. You've got to be learning and adapting constantly to survive.

General Peter J. Schoomaker, USA, 2004

The United States possesses overwhelming conventional military superiority. This capability has pushed its enemies to fight U.S. forces unconventionally, mixing modern technology with ancient techniques of insurgency and terrorism. Most enemies either do not try to defeat the United States with conventional operations or do not limit themselves to purely military means. They know that they cannot compete with U.S. forces on those terms. Instead, they try to exhaust U.S. national will, aiming to win by undermining and outlasting public support. Defeating such enemies presents a huge challenge to the Army and Marine Corps. Meeting it requires creative efforts by every Soldier and Marine.

Throughout its history, the U.S. military has had to relearn the principles of counterinsurgency (COIN) while conducting operations against adaptive insurgent enemies. It is time to institutionalize Army and Marine Corps knowledge of this longstanding form of conflict. This publication's purpose is to help prepare Army and Marine Corps leaders to conduct COIN operations anywhere in the world. It provides a foundation for study before deployment and the basis for operations in theater. Perhaps more importantly, it provides techniques for generating and incorporating lessons learned during those operations—an essential requirement for success against today's adaptive foes. Using these techniques and processes can keep U.S. forces more agile and adaptive than their irregular enemies. Knowledge of the history and principles of insurgency and COIN provides a solid foundation that informed leaders can use to assess insurgencies. This knowledge can also help them make appropriate decisions on employing all instruments of national power against these threats.

All insurgencies are different; however, broad historical trends underlie the factors motivating insurgents. Most insurgencies follow a similar course of development. The tactics used to successfully defeat them are likewise similar in most cases. Similarly, history shows that some tactics that are usually successful against conventional foes may fail against insurgents.

One common feature of insurgencies is that the government that is being targeted generally takes awhile to recognize that an insurgency is occurring. Insurgents take advantage of that time to build strength and gather support. Thus, counterinsurgents often have to "come from behind" when fighting an insurgency. Another common feature is that forces conducting COIN operations usually begin poorly. Western militaries too often neglect the study of insurgency. They falsely believe that armies trained to win large conventional wars are automatically prepared to win small, unconventional ones. In fact, some capabilities required for conventional success—for example, the ability to execute operational maneuver and employ massive firepower—may be of limited utility or even counterproductive in COIN operations. Nonetheless, conventional forces beginning COIN operations often try to use these capabilities to defeat insurgents; they almost always fail.

The military forces that successfully defeat insurgencies are usually those able to overcome their institutional inclination to wage conventional war against insurgents. They learn how to practice COIN and apply that knowledge. This publication can help to compress the learning curve. It is a tool for planners, trainers, and field commanders. Using it can help leaders begin the learning process sooner and build it on a larger knowledge base. Learning done before deployment results in fewer lives lost and less national treasure spent relearning past lessons in combat.

In COIN, the side that learns faster and adapts more rapidly—the better learning organization—usually wins. Counterinsurgencies have been called learning competitions. Thus, this publication identifies "Learn and Adapt" as a modern COIN imperative for U.S. forces. However, Soldiers and Marines cannot wait until they are alerted to deploy to prepare for a COIN mission. Learning to conduct complex COIN operations begins with study beforehand. This publication is a good place to start. The annotated bibliography lists a number of other sources; however, these are only a sample of the vast amount of available information on this subject. Adapting occurs as Soldiers and Marines apply what they have learned through study and experience, assess the results of their actions, and continue to learn during operations.

As learning organizations, the Army and Marine Corps encourage Soldiers and Marines to pay attention to the rapidly changing situations that characterize COIN operations. Current tactics, techniques, and procedures sometimes do not achieve the desired results. When that happens, successful leaders engage in a directed search for better ways to defeat the enemy. To win, the Army and Marine Corps must rapidly develop an institutional consensus on new doctrine, publish it, and carefully observe its impact on mission accomplishment. This learning cycle should repeat continuously as U.S. counterinsurgents seek to learn faster than the insurgent enemy. The side that learns faster and adapts more rapidly wins.

Just as there are historical principles underlying success in COIN, there are organizational traits shared by most successful learning organizations. Forces that learn COIN effectively have generally—

- Developed COIN doctrine and practices locally.
- Established local training centers during COIN operations.
- Regularly challenged their assumptions, both formally and informally.
- Learned about the broader world outside the military and requested outside assistance in understanding foreign political, cultural, social and other situations beyond their experience.
- Promoted suggestions from the field.
- Fostered open communication between senior officers and their subordinates.
- Established rapid avenues of disseminating lessons learned.
- Coordinated closely with governmental and nongovernmental partners at all command levels.
- Proved open to soliciting and evaluating advice from the local people in the conflict zone.

These are not always easy practices for an organization to establish. Adopting them is particularly challenging for a military engaged in a conflict. However, these traits are essential for any military confronting an enemy who does not fight using conventional tactics and who adapts while waging irregular warfare. Learning organizations defeat insurgencies; bureaucratic hierarchies do not.

Promoting learning is a key responsibility of commanders at all levels. The U.S. military has developed first-class lessons-learned systems that allow for collecting and rapidly disseminating information from the field. But these systems only work when commanders promote their use and create a command climate that encourages bottom-up learning. Junior leaders in the field often informally disseminate lessons based on their experiences. However, incorporating this information into institutional lessons learned, and then into doctrine, requires commanders to encourage subordinates to use institutional lessons-learned processes.

Ironically, the nature of counterinsurgency presents challenges to traditional lessons-learned systems; many nonmilitary aspects of COIN do not lend themselves to rapid tactical learning. As this publication explains, performing the many nonmilitary tasks in COIN requires knowledge of many diverse, complex subjects. These include governance, economic development, public administration, and the rule of law. Commanders with a deep-rooted knowledge of these subjects can help subordinates understand challenging, unfamiliar environments and adapt more rapidly to changing situations. Reading this publication is a first stop to developing this knowledge.

COIN campaigns are often long and difficult. Progress can be hard to measure, and the enemy may appear to have many advantages. Effective insurgents rapidly adapt to changing circumstances. They cleverly use the tools of the global information revolution to magnify the effects of their actions. The often carry out barbaric acts and do not observe accepted norms of behavior. However, by focusing on efforts to secure the safety and support of the local populace, and through a concerted effort to truly function as learning organizations, the Army and Marine Corps can defeat their insurgent enemies.

Chapter 1

Insurgency and Counterinsurgency

Counterinsurgency is not just thinking man's warfare—it is the graduate level of war.

Special Forces Officer in Iraq, 2005

This chapter provides background information on insurgency and counterinsurgency. The first half describes insurgency, while the second half examines the more complex challenge of countering it. The chapter concludes with a set of principles and imperatives that contribute to success in counterinsurgency.

OVERVIEW

1-1. Insurgency and counterinsurgency (COIN) are complex subsets of warfare. Globalization, technological advancement, urbanization, and extremists who conduct suicide attacks for their cause have certainly influenced contemporary conflict; however, warfare in the 21st century retains many of the characteristics it has exhibited since ancient times. Warfare remains a violent clash of interests between organized groups characterized by the use of force. Achieving victory still depends on a group's ability to mobilize support for its political interests (often religiously or ethnically based) and to generate enough violence to achieve political consequences. Means to achieve these goals are not limited to conventional forces employed by nation-states.

1-2. Insurgency and its tactics are as old as warfare itself. Joint doctrine defines an *insurgency* as an organized movement aimed at the overthrow of a constituted government through the use of subversion and armed conflict (JP 1-02). Stated another way, an insurgency is an organized, protracted politico-military struggle designed to weaken the control and legitimacy of an established government, occupying power, or other political authority while increasing insurgent control. *Counterinsurgency* is military, paramilitary, political, economic, psychological, and civic actions taken by a government to defeat insurgency (JP 1-02). These definitions are a good starting point, but they do not properly highlight a key paradox: though insurgency and COIN are two sides of a phenomenon that has been called revolutionary war or internal war, they are distinctly different types of operations. In addition, insurgency and COIN are included within a broad category of conflict known as irregular warfare.

1-3. Political power is the central issue in insurgencies and counterinsurgencies; each side aims to get the people to accept its governance or authority as legitimate. Insurgents use all available tools—political (including diplomatic), informational (including appeals to religious, ethnic, or ideological beliefs), military, and economic—to overthrow the existing authority. This authority may be an established government or an interim governing body. Counterinsurgents, in turn, use all instruments of national power to sustain the established or emerging government and reduce the likelihood of another crisis emerging.

1-4. Long-term success in COIN depends on the people taking charge of their own affairs and consenting to the government's rule. Achieving this condition requires the government to eliminate as many causes of the insurgency as feasible. This can include eliminating those extremists whose beliefs prevent them from ever reconciling with the government. Over time, counterinsurgents aim to enable a country or regime to provide the security and rule of law that allow establishment of social services and growth of economic activity. COIN thus involves the application of national power in the political, military, economic, social, information, and infrastructure fields and disciplines. Political and military leaders and planners should never underestimate its scale and complexity; moreover, they should recognize that the Armed Forces cannot succeed in COIN alone.

ASPECTS OF INSURGENCY

1-5. Governments can be overthrown in a number of ways. An unplanned, spontaneous explosion of popular will, for example, might result in a revolution like that in France in 1789. At another extreme is the coup d'etat, where a small group of plotters replace state leaders with little support from the people at large. Insurgencies generally fall between these two extremes. They normally seek to achieve one of two goals: to overthrow the existing social order and reallocate power within a single state, or to break away from state control and form an autonomous entity or ungoverned space that they can control. Insurgency is typically a form of internal war, one that occurs primarily within a state, not between states, and one that contains at least some elements of civil war.

1-6. The exception to this pattern of internal war involves resistance movements, where indigenous elements seek to expel or overthrow what they perceive to be a foreign or occupation government. Such a resistance movement could be mounted by a legitimate government in exile as well as by factions competing for that role.

1-7. Even in internal war, the involvement of outside actors is expected. During the Cold War, the Soviet Union and the United States participated in many such conflicts. Today, outside actors are often transnational organizations motivated by ideologies based on extremist religious or ethnic beliefs. These organizations exploit the unstable internal conditions plaguing failed and failing states. Such outside involvement, however, does not change one fact: the long-term objective for all sides remains acceptance of the legitimacy of one side's claim to political power by the people of the state or region.

1-8. The terrorist and guerrilla tactics common to insurgency have been among the most common approaches to warfare throughout history. Any combatant prefers a quick, cheap, overwhelming victory over a long, bloody, protracted struggle. But to succeed against superior resources and technology, weaker actors have had to adapt. The recent success of U.S. military forces in major combat operations undoubtedly will lead many future opponents to pursue asymmetric approaches. Because the United States retains significant advantages in fires and technical surveillance, a thinking enemy is unlikely to choose to fight U.S. forces in open battle. Some opponents have attempted to do so, such as in Panama in 1989 and Iraq in 1991 and 2003. They were defeated in conflicts measured in hours or days. Conversely, other opponents have offset America's fires and surveillance advantages by operating close to civilians, as Somali clans did in 1993 and insurgents in Iraq have done since mid-2003; these enemies have been more successful in achieving their aims. This situation does not mean that counterinsurgents do not face open warfare. Although insurgents frequently use nonviolent means like political mobilization and work stoppages (strikes), they do resort to conventional military operations when conditions seem right.

1-9. The contest of internal war is not "fair"; many of the "rules" favor insurgents. That is why insurgency has been a common approach used by the weak against the strong. At the beginning of a conflict, insurgents typically hold the strategic initiative. Though they may resort to violence because of regime changes or government actions, insurgents generally initiate the conflict. Clever insurgents strive to disguise their intentions. When these insurgents are successful at such deception, potential counterinsurgents are at a disadvantage. A coordinated reaction requires political and military leaders to recognize that an insurgency exists and to determine its makeup and characteristics. While the government prepares to respond, the insurgents gain strength and foster increasing disruption throughout the state or region. The government normally has an initial advantage in resources; however, that edge is counterbalanced by the requirement to maintain order and protect the population and critical resources. Insurgents succeed by sowing chaos and disorder anywhere; the government fails unless it maintains a degree of order everywhere.

1-10. For the reasons just mentioned, maintaining security in an unstable environment requires vast resources, whether host nation, U.S., or multinational. In contrast, a small number of highly motivated insurgents with simple weapons, good operations security, and even limited mobility can undermine security over a large area. Thus, successful COIN operations often require a high ratio of security forces to the protected population. (See paragraph 1-67.) For that reason, protracted COIN operations are hard to sustain. The effort requires a firm political will and substantial patience by the government, its people, and the countries providing support.

1-11. Revolutionary situations may result from regime changes, external interventions, or grievances carefully nurtured and manipulated by unscrupulous leaders. Sometimes societies are most prone to unrest not when conditions are the worst, but when the situation begins to improve and people's expectations rise. For example, when major combat operations conclude, people may have unrealistic expectations of the United States' capability to improve their lives. The resulting discontent can fuel unrest and insurgency. At such times, the influences of globalization and the international media may create a sense of relative deprivation, contributing to increased discontent as well.

1-12. The information environment is a critical dimension of such internal wars, and insurgents attempt to shape it to their advantage. One way they do this is by carrying out activities, such as suicide attacks, that may have little military value but create fear and uncertainty within the populace and government institutions. These actions are executed to attract high-profile media coverage or local publicity and inflate perceptions of insurgent capabilities. Resulting stories often include insurgent fabrications designed to undermine the government's legitimacy.

1-13. Insurgents have an additional advantage in shaping the information environment. Counterinsurgents seeking to preserve legitimacy must stick to the truth and make sure that words are backed up by deeds; insurgents, on the other hand, can make exorbitant promises and point out government shortcomings, many caused or aggravated by the insurgency. Ironically, as insurgents achieve more success and begin to control larger portions of the populace, many of these asymmetries diminish. That may produce new vulnerabilities that adaptive counterinsurgents can exploit.

1-14. Before most COIN operations begin, insurgents have seized and exploited the initiative, to some degree at the least. Therefore, counterinsurgents undertake offensive and defensive operations to regain the initiative and create a secure environment. However, killing insurgents—while necessary, especially with respect to extremists—by itself cannot defeat an insurgency. Gaining and retaining the initiative requires counterinsurgents to address the insurgency's causes through stability operations as well. This initially involves securing and controlling the local populace and providing for essential services. As security improves, military resources contribute to supporting government reforms and reconstruction projects. As counterinsurgents gain the initiative, offensive operations focus on eliminating the insurgent cadre, while defensive operations focus on protecting the populace and infrastructure from direct attacks. As counterinsurgents establish military ascendancy, stability operations expand across the area of operations (AO) and eventually predominate. Victory is achieved when the populace consents to the government's legitimacy and stops actively and passively supporting the insurgency.

THE EVOLUTION OF INSURGENCY

1-15. Insurgency has taken many forms over time. Past insurgencies include struggles for independence against colonial powers, the rising up of ethnic or religious groups against their rivals, and resistance to foreign invaders. Students and practitioners of COIN must begin by understanding the specific circumstances of their particular situation. The history of this form of warfare shows how varied and adaptive it can be, and why students must understand that they cannot focus on countering just one insurgent approach. This is particularly true when addressing a continually complex, changing situation like that of Iraq in 2006.

1-16. Insurgencies and counterinsurgencies have been common throughout history, but especially since the beginning of the 20th century. The United States began that century by defeating the Philippine Insurrection. The turmoil of World War I and its aftermath produced numerous internal wars. Trotsky and Lenin seized power in Russia and then defended the new regime against counterrevolutionaries. T.E. Lawrence and Arab forces used guerrilla tactics to overcome the Ottoman Turks during the Arab Revolt.

1-17. Before World War I, insurgencies were mostly conservative; insurgents were usually concerned with defending hearth, home, monarchies, and traditional religion. Governments were seldom able to completely defeat these insurgencies; violence would recur when conditions favored a rebellion. For example, the history of the British Isles includes many recurring insurgencies by subjugated peoples based on ethnic identities. Another example of a conservative insurgency is the early 19th century Spanish uprising against Napoleon that sapped French strength and contributed significantly to Napoleon's defeat.

1-18. Since World War I, insurgencies have generally had more revolutionary purposes. The Bolshevik takeover of Russia demonstrated a conspiratorial approach to overthrowing a government; it spawned a communist movement that supported further "wars of national liberation." Lawrence's experiences in the Arab Revolt made him a hero and also provide some insights for today.

1-19. The modern era of insurgencies and internal wars began after World War II. Many of the resistance movements against German and Japanese occupation continued after the Axis defeat in 1945. As nationalism rose, the imperial powers declined. Motivated by nationalism and communism, people began forming governments viewed as more responsive to their needs. The development of increasingly lethal and portable killing technologies dramatically increased the firepower available to insurgent groups. As important was the increase in the news media's ability to get close to conflicts and transmit imagery locally and globally. In 1920, T.E. Lawrence noted, "The printing press is the greatest weapon in the armory of the modern commander." Today, he might have added, "and the modern insurgent," though certainly the Internet and compact storage media like cassettes, compact disks, and digital versatile disks (DVDs) have become more important in recent years.

1-20. Thus, 20th century events transformed the purpose and character of most insurgencies. Most 19th century insurgencies were local movements to sustain the status quo. By the mid-20th century they had become national and transnational revolutionary movements. Clausewitz thought that wars by an armed populace could only serve as a strategic defense; however, theorists after World War II realized that insurgency could be a decisive form of warfare. This era spawned the Maoist, Che Guevara-type focoist, and urban approaches to insurgency.

1-21. While some Cold War insurgencies persisted after the Soviet Union's collapse, many new ones appeared. These new insurgencies typically emerged from civil wars or the collapse of states no longer propped up by Cold War rivalries. Power vacuums breed insurgencies. Similar conditions exist when regimes are changed by force or circumstances. Recently, ideologies based on extremist forms of religious or ethnic identities have replaced ideologies based on secular revolutionary ideals. These new forms of old, strongly held beliefs define the identities of the most dangerous combatants in these new internal wars. These conflicts resemble the wars of religion in Europe before and after the Reformation of the 16th century. People have replaced nonfunctioning national identities with traditional sources of unity and identity. When countering an insurgency during the Cold War, the United States normally focused on increasing a threatened but friendly government's ability to defend itself and on encouraging political and economic reforms to undercut support for the insurgency. Today, when countering an insurgency growing from state collapse or failure, counterinsurgents often face a more daunting task: helping friendly forces reestablish political order and legitimacy where these conditions may no longer exist.

1-22. Interconnectedness and information technology are new aspects of this contemporary wave of insurgencies. Using the Internet, insurgents can now link virtually with allied groups throughout a state, a region, and even the entire world. Insurgents often join loose organizations with common objectives but different motivations and no central controlling body, which makes identifying leaders difficult.

1-23. Today's operational environment also includes a new kind of insurgency, one that seeks to impose revolutionary change worldwide. Al Qaeda is a well-known example of such an insurgency. This movement seeks to transform the Islamic world and reorder its relationships with other regions and cultures. It is notable for its members' willingness to execute suicide attacks to achieve their ends. Such groups often feed on local grievances. Al Qaeda-type revolutionaries are willing to support causes they view as compatible with their own goals through the provision of funds, volunteers, and sympathetic and targeted propaganda. While the communications and technology used for this effort are often new and modern, the grievances and methods sustaining it are not. As in other insurgencies, terrorism, subversion, propaganda, and open warfare are the tools of such movements. Today, these time-tested tools have been augmented by the precision munition of extremists—suicide attacks. Defeating such enemies requires a global, strategic response—one that addresses the array of linked resources and conflicts that sustain these movements while tactically addressing the local grievances that feed them.

INSURGENTS AND THEIR MOTIVES

1-24. Each insurgency is unique, although there are often similarities among them. In all cases, insurgents aim to force political change; any military action is secondary and subordinate, a means to an end. Few insurgencies fit neatly into any rigid classification. In fact, counterinsurgent commanders may face a confusing and shifting coalition of many kinds of opponents, some of whom may be at odds with one another. Examining the specific type of insurgency they face enables commanders and staffs to build a more accurate picture of the insurgents and the thinking behind their overall approach. Such an examination identifies the following:

- Root cause or causes of the insurgency.
- Extent to which the insurgency enjoys internal and external support.
- Basis (including the ideology and narrative) on which insurgents appeal to the target population.
- Insurgents' motivation and depth of commitment.
- Likely insurgent weapons and tactics.
- Operational environment in which insurgents seek to initiate and develop their campaign and strategy.

INSURGENT APPROACHES

1-25. Counterinsurgents have to determine not only their opponents' motivation but also the approach being used to advance the insurgency. This information is essential to developing effective programs that attack the insurgency's root causes. Analysis of the insurgents' approach shapes counterinsurgent military options. Insurgent approaches include, but are not limited to, the following:

- Conspiratorial.
- Military-focused.
- Urban.
- Protracted popular war.
- Identity-focused.
- Composite and coalition.

Conspiratorial

1-26. A conspiratorial approach involves a few leaders and a militant cadre or activist party seizing control of government structures or exploiting a revolutionary situation. In 1917, Lenin used this approach in carrying out the Bolshevik Revolution. Such insurgents remain secretive as long as possible. They emerge only when success can be achieved quickly. This approach usually involves creating a small, secretive, "vanguard" party or force. Insurgents who use this approach successfully may have to create security forces and generate mass support to maintain power, as the Bolsheviks did.

Military-Focused

1-27. Users of military-focused approaches aim to create revolutionary possibilities or seize power primarily by applying military force. For example, the focoist approach, popularized by figures like Che Guevera, asserts that an insurrection itself can create the conditions needed to overthrow a government. Focoists believe that a small group of guerrillas operating in a rural environment where grievances exist can eventually gather enough support to achieve their aims. In contrast, some secessionist insurgencies have relied on major conventional forces to try to secure their independence. Military-focused insurgencies conducted by Islamic extremist groups or insurgents in Africa or Latin America have little or no political structure; they spread their control through movement of combat forces rather than political subversion.

Urban

1-28. Organizations like the Irish Republican Army, certain Latin American groups, and some Islamic extremist groups in Iraq have pursued an urban approach. This approach uses terrorist tactics in urban areas to accomplish the following:

- Sow disorder.
- Incite sectarian violence.
- Weaken the government.
- Intimidate the population.
- Kill government and opposition leaders.
- Fix and intimidate police and military forces, limiting their ability to respond to attacks.
- Create government repression.

1-29. Protracted urban terrorism waged by small, independent cells requires little or no popular support. It is difficult to counter. Historically, such activities have not generated much success without wider rural support. However, as societies have become more urbanized and insurgent networks more sophisticated, this approach has become more effective. When facing adequately run internal security forces, urban insurgencies typically assume a conspiratorial cellular structure recruited along lines of close association—family, religious affiliation, political party, or social group.

Protracted Popular War

1-30. Protracted conflicts favor insurgents, and no approach makes better use of that asymmetry than the protracted popular war. The Chinese Communists used this approach to conquer China after World War II. The North Vietnamese and Algerians adapted it to fit their respective situations. And some Al Qaeda leaders suggest it in their writings today. This approach is complex; few contemporary insurgent movements apply its full program, although many apply parts of it. It is, therefore, of more than just historical interest. Knowledge of it can be a powerful aid to understanding some insurgent movements.

Mao Zedong's Theory of Protracted War

1-31. Mao's Theory of Protracted War outlines a three-phased, politico-military approach:

- **Strategic defensive,** when the government has a stronger correlation of forces and insurgents must concentrate on survival and building support.
- **Strategic stalemate,** when force correlations approach equilibrium and guerrilla warfare becomes the most important activity.
- **Strategic counteroffensive,** when insurgents have superior strength and military forces move to conventional operations to destroy the government's military capability.

1-32. Phase I, strategic defensive, is a period of latent insurgency that allows time to wear down superior enemy strength while the insurgency gains support and establishes bases. During this phase, insurgent leaders develop the movement into an effective clandestine organization. Insurgents use a variety of subversive techniques to psychologically prepare the populace to resist the government or occupying power. These techniques may include propaganda, demonstrations, boycotts, and sabotage. In addition, movement leaders organize or develop cooperative relationships with legitimate political action groups, youth groups, trade unions, and other front organizations. Doing this develops popular support for later political and military activities. Throughout this phase, the movement leadership—

- Recruits, organizes, and trains cadre members.
- Infiltrates key government organizations and civilian groups.
- Establishes cellular intelligence, operations, and support networks.
- Solicits and obtains funds.
- Develops sources for external support.

Subversive activities are frequently executed in an organized pattern, but major combat is avoided. The primary military activity is terrorist strikes. These are executed to gain popular support, influence recalcitrant individuals, and sap enemy strength. In the advanced stages of this phase, the insurgent organization may establish a counterstate that parallels the established authority. (A *counterstate* [or shadow government] is a competing structure that a movement sets up to replace the government. It includes the administrative and bureaucratic trappings of political power and performs the normal functions of a government.)

1-33. Phase II, strategic stalemate, begins with overt guerrilla warfare as the correlation of forces approaches equilibrium. In a rural-based insurgency, guerrillas normally operate from a relatively secure base area in insurgent-controlled territory. In an urban-based insurgency, guerrillas operate clandestinely, using a cellular organization. In the political arena, the movement concentrates on undermining the people's support of the government and further expanding areas of control. Subversive activities can take the form of clandestine radio broadcasts, newspapers, and pamphlets that openly challenge the control and legitimacy of the established authority. As the populace loses faith in the established authority the people may decide to actively resist it. During this phase, a counterstate may begin to emerge to fill gaps in governance that the host-nation (HN) government is unwilling or unable to address. Two recent examples are Moqtada al Sadr's organization in Iraq and Hezbollah in Lebanon. Sadr's Mahdi Army provides security and some services in parts of southern Iraq and Baghdad under Sadr's control. (In fact, the Mahdi Army created gaps by undermining security and services; then it moved to solve the problem it created.) Hezbollah provides essential services and reconstruction assistance for its constituents as well as security. Each is an expression of Shiite identity against governments that are pluralist and relatively weak.

1-34. Phase III, strategic counteroffensive, occurs as the insurgent organization becomes stronger than the established authority. Insurgent forces transition from guerrilla warfare to conventional warfare. Military forces aim to destroy the enemy's military capability. Political actions aim to completely displace all government authorities. If successful, this phase causes the government's collapse or the occupying power's withdrawal. Without direct foreign intervention, a strategic offensive takes on the characteristics of a full-scale civil war. As it gains control of portions of the country, the insurgent movement becomes responsible for the population, resources, and territory under its control. To consolidate and preserve its gains, an effective insurgent movement continues the phase I activities listed in paragraph 1-32. In addition it—

- Establishes an effective civil administration.
- Establishes an effective military organization.
- Provides balanced social and economic development.
- Mobilizes the populace to support the insurgent organization.
- Protects the populace from hostile actions.

1-35. Effectively applying Maoist strategy does not require a sequential or complete application of all three stages. The aim is seizing political power; if the government's will and capability collapse early in the process, so much the better. If unsuccessful in a later phase, the insurgency might revert to an earlier one. Later insurgents added new twists to this strategy, to include rejecting the need to eventually switch to large-scale conventional operations. For example, the Algerian insurgents did not achieve much military success of any kind; instead they garnered decisive popular support through superior organizational skills and astute propaganda that exploited French mistakes. These and other factors, including the loss of will in France, compelled the French to withdraw.

The North Vietnamese Dau Trahn

1-36. The Vietnamese conflict offers another example of the application of Mao's strategy. The North Vietnamese developed a detailed variant of it known as dau tranh ("the struggle") that is most easily described in terms of logical lines of operations (LLOs). In this context, a *line of operations* is a logical line that connects actions on nodes and/or decisive points related in time and purpose with an objective (JP 1-02). LLOs can also be described as an operational framework/planning construct used to define the concept of multiple, and often disparate, actions arranged in a framework unified by purpose. (Chapters 4 and 5 discuss LLOs typically used in COIN operations.) Besides modifying Mao's three phases, dau tranh delineated LLOs for achieving political objectives among the enemy population, enemy soldiers, and friendly forces. The "general offensive-general uprising" envisioned in this approach did not occur during the Vietnam War; however, the approach was designed to achieve victory by whatever means were effective.

It did not attack a single enemy center of gravity; instead it put pressure on several, asserting that, over time, victory would result in one of two ways: from activities along one LLO or the combined effects of efforts along several. North Vietnamese actions after their military failure in the 1968 Tet offensive demonstrate this approach's flexibility. At that time, the North Vietnamese shifted their focus from defeating U.S. forces in Vietnam to weakening U.S. will at home. These actions expedited U.S. withdrawal and laid the groundwork for the North Vietnamese victory in 1975.

Complexity and the Shifting Mosaic

1-37. Protracted popular war approaches are conducted along multiple politico-military LLOs and are locally configured. Insurgents may use guerrilla tactics in one province while executing terrorist attacks and an urban approach in another. There may be differences in political activities between villages in the same province. The result is more than just a "three-block war": it is a shifting "mosaic war" that is difficult for counterinsurgents to envision as a coherent whole. In such situations, an effective COIN strategy must be multifaceted and flexible.

Identity-Focused

1-38. The identity-focused approach mobilizes support based on the common identity of religious affiliation, clan, tribe, or ethnic group. Some movements may be based on an appeal to a religious identity, either separately from or as part of other identities. This approach is common among contemporary insurgencies and is sometimes combined with the military-focused approach. The insurgent organization may not have the dual military/political hierarchy evident in a protracted popular war approach. Rather, communities often join the insurgent movement as a whole, bringing with them their existing social/military hierarchy. Additionally, insurgent leaders often try to mobilize the leadership of other clans and tribes to increase the movement's strength.

Composite Approaches and Coalitions

1-39. As occurred in Iraq, contemporary insurgents may use different approaches at different times, applying tactics that take best advantage of circumstances. Insurgents may also apply a composite approach that includes tactics drawn from any or all of the other approaches. In addition—and as in Iraq at present—different insurgent forces using different approaches may form loose coalitions when it serves their interests; however, these same movements may fight among themselves, even while engaging counterinsurgents. Within a single AO, there may be multiple competing entities, each seeking to maximize its survivability and influence—and this situation may be duplicated several times across a joint operations area. This reality further complicates both the mosaic that counterinsurgents must understand and the operations necessary for victory.

MOBILIZATION MEANS AND CAUSES

1-40. The primary struggle in an internal war is to mobilize people in a struggle for political control and legitimacy. Insurgents and counterinsurgents seek to mobilize popular support for their cause. Both try to sustain that struggle while discouraging support for their adversaries. Two aspects of this effort are mobilization means and causes.

Mobilization Means

1-41. There are five means to mobilize popular support:

- Persuasion.
- Coercion.
- Reaction to abuses.
- Foreign support.
- Apolitical motivations.

A mixture of them may motivate any one individual.

Persuasion

1-42. In times of turmoil, political, social, security, and economic benefits can often entice people to support one side or the other. Ideology and religion are means of persuasion, especially for the elites and leadership. In this case, legitimacy derives from the consent of the governed, though leaders and led can have very different motivations. In Iraq, for example, an issue that motivated fighters in some Baghdad neighborhoods in 2004 was lack of adequate sewer, water, electricity, and trash services. Their concerns were totally disconnected from the overall Ba'athist goal of expelling U.S. forces and retaining Sunni Arab power.

Coercion

1-43. The struggle in Iraq has produced many examples of how insurgent coercion can block government success. In the eyes of some, a government that cannot protect its people forfeits the right to rule. Legitimacy is accorded to the element that can provide security, as citizens seek to ally with groups that can guarantee their safety. In some areas of Iraq and Afghanistan, for instance, militias established themselves as extragovernmental arbiters of the populace's physical security—in some case, after first undermining that security.

1-44. Insurgents may use coercive force to provide security for people or to intimidate them and the legitimate security forces into active or passive support. Kidnapping or killing local leaders or their families is a common insurgent tactic to discourage working with the government. Militias sometimes use the promise of security, or the threat to remove it, to maintain control of cities and towns. Such militias may be sectarian or based on political parties. The HN government must recognize and remove the threat to sovereignty and legitimacy posed by extragovernmental organizations of this type. (The dangers of militias are further described in paragraphs 3-112 and 3-113.)

Reaction to Abuses

1-45. Though firmness by security forces is often necessary to establish a secure environment, a government that exceeds accepted local norms and abuses its people or is tyrannical generates resistance to its rule. People who have been maltreated or have had close friends or relatives killed by the government, particularly by its security forces, may strike back at their attackers. Security force abuses and the social upheaval caused by collateral damage from combat can be major escalating factors for insurgencies.

Foreign Support

1-46. Foreign governments can provide the expertise, international legitimacy, and money needed to start or intensify a conflict. For example, although there was little popular support for the renewal of fighting in Chechnya in 1999, the conflict resumed anyway because foreign supporters and warlords had enough money to hire a guerrilla army. Also of note, nongovernmental organizations (NGOs), even those whose stated aims are impartial and humanitarian, may wittingly or unwittingly support insurgents. For example, funds raised overseas for professed charitable purposes can be redirected to insurgent groups.

Apolitical Motivations

1-47. Insurgencies attract criminals and mercenaries. Individuals inspired by the romanticized image of the revolutionary or holy warrior and others who imagine themselves as fighters for a cause might also join. It is important to note that political solutions might not satisfy some of them enough to end their participation. Fighters who have joined for money will probably become bandits once the fighting ends unless there are jobs for them. This category also includes opportunists who exploit the absence of security to engage in economically lucrative criminal activity, such as kidnapping and theft. True extremists are unlikely to be reconciled to any other outcome than the one they seek; therefore, they must be killed or captured.

Causes

1-48. A cause is a principle or movement militantly defended or supported. Insurgent leaders often seek to adopt attractive and persuasive causes to mobilize support. These causes often stem from the unresolved contradictions existing within any society or culture. Frequently, contradictions are based on real problems. However, insurgents may create artificial contradictions using propaganda and misinformation. Insurgents can gain more support by not limiting themselves to a single cause. By selecting an assortment of causes and tailoring them for various groups within the society, insurgents increase their base of sympathetic and complicit support.

1-49. Insurgents employ deep-seated, strategic causes as well as temporary, local ones, adding or deleting them as circumstances demand. Leaders often use a bait-and-switch approach. They attract supporters by appealing to local grievances; then they lure followers into the broader movement. Without an attractive cause, an insurgency might not be able to sustain itself. But a carefully chosen cause is a formidable asset; it can provide a fledgling movement with a long-term, concrete base of support. The ideal cause attracts the most people while alienating the fewest and is one that counterinsurgents cannot co-opt.

1-50. Potential insurgents can capitalize on a number of potential causes. Any country ruled by a small group without broad, popular participation provides a political cause for insurgents. Exploited or repressed social groups—be they entire classes, ethnic or religious groups, or small elites—may support larger causes in reaction to their own narrower grievances. Economic inequities can nurture revolutionary unrest. So can real or perceived racial or ethnic persecution. For example, Islamic extremists use perceived threats to their religion by outsiders to mobilize support for their insurgency and justify terrorist tactics. As previously noted, effective insurgent propaganda can also turn an artificial problem into a real one.

1-51. Skillful counterinsurgents can deal a significant blow to an insurgency by appropriating its cause. Insurgents often exploit multiple causes, however, making counterinsurgents' challenges more difficult. In the end, any successful COIN operation must address the legitimate grievances insurgents use to generate popular support. These may be different in each local area, in which case a complex set of solutions will be needed.

Mobilizing Resources

1-52. Insurgents resort to such tactics as guerrilla warfare and terrorism for any number of reasons. These may include disadvantages in manpower or organization, relatively limited resources compared to the government, and, in some cases, a cultural predisposition to an indirect approach to conflict. To strengthen and sustain their effort once manpower is mobilized, insurgents require money, supplies, and weapons.

1-53. Weapons are especially important. In some parts of the world, lack of access to weapons may forestall insurgencies. Unfortunately, there is widespread availability of weapons in many areas, with especially large surpluses in the most violent regions of the world. Explosive hazards, such as mines and improvised explosive devices, are likely to be common weapons in insurgencies. (See FMI 3-34.119/MCIP 3-17.01 for more information on improvised explosive devices.) Insurgents can obtain weapons through legal or illegal purchases or from foreign sources. A common tactic is to capture them from government forces. Skillful counterinsurgents cut off the flow of arms into the AO and eliminate their sources.

1-54. Income is essential not only for insurgents to purchase weapons but also to pay recruits and bribe corrupt officials. Money and supplies can be obtained through many sources. Foreign support has already been mentioned. Local supporters or international front organizations may provide donations. Sometimes legitimate businesses are established to furnish funding. In areas controlled by insurgents, confiscation or taxation might be utilized. Another common source of funding is criminal activity.

INSURGENCY AND CRIME

1-55. Funding greatly influences an insurgency's character and vulnerabilities. The insurgents' approach determines the movement's requirements. Protracted popular war approaches that emphasize mobilization of the masses require the considerable resources needed to build and maintain a counterstate. In comparison, the military-focused approach, which emphasizes armed action, needs only the resources necessary to sustain a military campaign. A conspiratorial or urban approach requires even less support.

1-56. Sustainment requirements often drive insurgents into relationships with organized crime or into criminal activity themselves. Reaping windfall profits and avoiding the costs and difficulties involved in securing external support makes illegal activity attractive to insurgents. Taxing a mass base usually yields low returns. In contrast, kidnapping, extortion, bank robbery, and drug trafficking—four favorite insurgent activities—are very lucrative. The activities of the Fuerzas Armadas Revolucionarias de Colombia (FARC) illustrate this point: profits from single kidnappings often total millions of U.S. dollars. For the Maoist Communist Party of Nepal, directly taxing the mass base proved inferior to other criminal forms of "revolutionary taxation," such as extortion and kidnapping. Drugs retain the highest potential for obtaining large profits from relatively small investments. In the 1990s, insurgents in Suriname, South America, were asked why they were selling gold at half the market price; they responded that the quick profits provided seed money to invest in the drug trade, from which they "could make real money." Similarly, failed and failing states with rich natural resources like oil or poppies (which provide the basis for heroin) are particularly lucrative areas for criminal activity. State failure precipitated by violent regime change further encourages criminal activity because of the collapse of law enforcement, the courts, and penal systems.

1-57. Devoting exceptional amounts of time and effort to fund-raising requires an insurgent movement to shortchange ideological or armed action. Indeed, the method of raising funds is often at the heart of debates on characterizing movements as diverse as the Provisional Irish Republican Army in Ulster and the FARC in Colombia. The first has been involved in all sorts of criminal activity for many years; however, it remains committed to its ideological aims. The second, through its involvement in the drug trade, has become the richest self-sustaining insurgent group in history; yet it continues to claim to pursue "Bolivarian" and "socialist" or "Marxist-Leninist" ends. FARC activities, though, have increasingly been labeled "narcoterrorist" or simply criminal by a variety of critics.

1-58. Throughout history, many insurgencies have degenerated into criminality. This occurred as the primary movements disintegrated and the remaining elements were cast adrift. Such disintegration is desirable; it replaces a dangerous, ideologically inspired body of disaffiliated individuals with a less dangerous but more diverse body, normally of very uneven character. The first is a security threat, the second a law-and-order concern. This should not be interpreted, of course, as denigrating the armed capacity of a law-and-order threat. Successful counterinsurgents are prepared to address this disintegration. They also recognize that the ideal approach eliminates both the insurgency and any criminal threats its elimination produces.

ELEMENTS OF INSURGENCY

1-59. Though insurgencies take many forms, most share some common attributes. An insurgent organization normally consists of five elements:

- Movement leaders.
- Combatants (main, regional, and local forces [including militias]).
- Political cadre (also called militants or the party).
- Auxiliaries (active followers who provide important support services).
- Mass base (the bulk of the membership).

1-60. The proportion of each element relative to the larger movement depends on the strategic approach the insurgency adopts. A conspiratorial approach does not pay much attention to combatants or a mass base. Military-focused insurgencies downplay the importance of a political cadre and emphasize military action to generate popular support. The people's war approach is the most complex: if the state presence has been eliminated, the elements exist openly; if the state remains a continuous or occasional presence, the elements maintain a clandestine existence.

Movement Leaders

1-61. Movement leaders provide strategic direction to the insurgency. They are the "idea people" and the planners. They usually exercise leadership through force of personality, the power of revolutionary ideas, and personal charisma. In some insurgencies, they may hold their position through religious, clan, or tribal authority.

Combatants

1-62. Combatants (sometimes called "foot soldiers") do the actual fighting and provide security. They are often mistaken for the movement itself; however, they exist only to support the insurgency's broader political agenda and to maintain local control. Combatants protect and expand the counterstate, if the insurgency sets up such an institution. They also protect training camps and networks that facilitate the flow of money, instructions, and foreign and local fighters.

Political Cadre

1-63. The cadre forms the political core of the insurgency. They are actively engaged in the struggle to accomplish insurgent goals. They may also be designated as a formal party to signify their political importance. The cadre implement guidance and procedures provided by the movement leaders. Modern non-communist insurgencies rarely, if ever, use the term "cadre"; however these movements usually include a group that performs similar functions. Additionally, movements based on religious extremism usually include religious and spiritual advisors among their cadre.

1-64. The cadre assesses grievances in local areas and carries out activities to satisfy them. They then attribute the solutions they have provided to the insurgency. As the insurgency matures, deeds become more important to make insurgent slogans meaningful to the population. Larger societal issues, such as foreign presence, facilitate such political activism because insurgents can blame these issues for life's smaller problems. Destroying the state bureaucracy and preventing national reconstruction after a conflict (to sow disorder and sever legitimate links with the people) are also common insurgent tactics. In time, the cadre may seek to replace that bureaucracy and assume its functions in a counterstate.

Auxiliaries

1-65. Auxiliaries are active sympathizers who provide important support services. They do not participate in combat operations. Auxiliaries may do the following:

- Run safe houses.
- Store weapons and supplies.
- Act as couriers.
- Provide passive intelligence collection.
- Give early warning of counterinsurgent movements.
- Provide funding from lawful and unlawful sources.
- Provide forged or stolen documents and access or introductions to potential supporters.

Mass Base

1-66. The mass base consists of the followers of the insurgent movement—the supporting populace. Mass base members are often recruited and indoctrinated by the cadre. However, in many politically charged situations or identity-focused insurgencies, such active pursuit is not necessary. Mass base members may continue in their normal positions in society. Many, however, lead clandestine lives for the insurgent movement. They may even pursue full-time positions within the insurgency. For example, combatants normally begin as members of the mass base. In tribal- or clan-based insurgencies, such roles are particularly hard to define. There is no clear cadre in those movements, and people drift between combatant, auxiliary, and follower status as needed.

Employing the Elements

1-67. The movement leaders provide the organizational and managerial skills needed to transform mobilized individuals and communities into an effective force for armed political action. The result is a contest of resource mobilization and force deployment. No force level guarantees victory for either side. During previous conflicts, planners assumed that combatants required a 10 or 15 to 1 advantage over insurgents to win. However, no predetermined, fixed ratio of friendly troops to enemy combatants ensures success in COIN. The conditions of the operational environment and the approaches insurgents use vary too widely. A better force requirement gauge is troop density, the ratio of security forces (including the host nation's military and police forces as well as foreign counterinsurgents) to inhabitants. Most density recommendations fall within a range of 20 to 25 counterinsurgents for every 1000 residents in an AO. Twenty counterinsurgents per 1000 residents is often considered the minimum troop density required for effective COIN operations; however as with any fixed ratio, such calculations remain very dependent upon the situation.

1-68. As in any conflict, the size of the force needed to defeat an insurgency depends on the situation. However, COIN is manpower intensive because counterinsurgents must maintain widespread order and security. Moreover, counterinsurgents typically have to adopt different approaches to address each element of the insurgency. For example, auxiliaries might be co-opted by economic or political reforms, while fanatic combatants will most likely have to be killed or captured.

DYNAMICS OF AN INSURGENCY

1-69. Insurgencies are also shaped by several common dynamics:

- Leadership.
- Objectives.
- Ideology and narrative.
- Environment and geography.
- External support and sanctuaries.
- Phasing and timing.

These make up a framework that can be used to assess the insurgency's strengths and weaknesses. Although these dynamics can be examined separately, studying their interaction is necessary to fully understand an insurgency.

1-70. The interplay of these dynamics influences an insurgency's approach and organization. Effective counterinsurgents identify the organizational pattern these dynamics form and determine if it changes. For example, insurgents operating in an urban environment usually form small, cohesive, secretive organizations. In contrast, insurgents following a military-focused strategy often operate in a rural environment and exploit international support to a greater extent. A change in location or the amount of external support might lead insurgents to adjust their approach and organization.

Leadership

1-71. Leadership is critical to any insurgency. An insurgency is not simply random violence; it is directed and focused violence aimed at achieving a political objective. It requires leadership to provide vision, direction, guidance, coordination, and organizational coherence. Successful insurgent leaders make their cause known to the people and gain popular support. Their key tasks are to break the ties between the people and the government and to establish credibility for their movement. Their education, background, family and social connections, and experiences contribute to their ability to organize and inspire the people who form the insurgency.

1-72. Some insurgent movements have their roots in a clash of cultures over power and preeminence. Others begin as the tangible manifestation of some form of political estrangement. In either case, alienated elite members advance alternatives to existing conditions. As their movement grows, leaders decide which approach to adopt. The level of decentralization of responsibility and authority drives the insurgency's structure and operational procedures. Extreme decentralization results in a movement that rarely functions as a coherent body. It is, however, capable of inflicting substantial casualties and damage. Loose networks find it difficult to create a viable counterstate; they therefore have great difficulty seizing political power. However, they are also very hard to destroy and can continue to sow disorder, even when degraded. It takes very little coordination to disrupt most states.

1-73. Many contemporary insurgencies are identity-based. These insurgencies are often led by traditional authority figures, such as tribal sheikhs, local warlords, or religious leaders. As the Indonesian Dar 'ul Islam rebellions of 1948 and 1961 demonstrate, traditional authority figures often wield enough power to single-handedly drive an insurgency. This is especially true in rural areas. Identity-focused insurgencies can be defeated in some cases by co-opting the responsible traditional authority figure; in others, the authority figures have to be discredited or eliminated. Accurately determining whether a leader can be co-opted is crucial. Failed attempts to co-opt traditional leaders can backfire if those leaders choose to oppose the counterinsurgency. Their refusal to be co-opted can strengthen their standing as they gain power and influence among insurgents.

Objectives

1-74. Effective analysis of an insurgency requires identifying its strategic, operational, and tactical objectives. The strategic objective is the insurgents' desired end state. Operational objectives are those that insurgents pursue to destroy government legitimacy and progressively establish their desired end state. Tactical objectives are the immediate aims of insurgent acts. Objectives can be psychological or physical. One example of a psychological objective is discouraging support for the government by assassinating local officials. An example of a physical objective is the disruption of government services by damaging or seizing a key facility. These tactical acts are often linked to higher purposes; in fact, tactical actions by both insurgents and counterinsurgents frequently have strategic effects.

Ideology and Narrative

1-75. Ideas are a motivating factor in insurgent activities. Insurgencies can gather recruits and amass popular support through ideological appeal (including religious or other cultural identifiers). Promising potential recruits often include individuals receptive to the message that the West is dominating their region through puppet governments and local surrogates. The insurgent group channels anti-Western anger and provides members with identity, purpose, and community, in addition to physical, economic, and psychological security. The movement's ideology explains its followers' difficulties and provides a means to remedy those ills. The most powerful ideologies tap latent, emotional concerns of the populace. Examples of these concerns include religiously based objectives, a desire for justice, ethnic aspirations, and a goal of liberation from foreign occupation. Ideology provides a prism, including a vocabulary and analytical categories, through which followers perceive their situation.

1-76. The central mechanism through which ideologies are expressed and absorbed is the narrative. A narrative is an organizational scheme expressed in story form. Narratives are central to representing identity, particularly the collective identity of religious sects, ethnic groupings, and tribal elements. Stories about a community's history provide models of how actions and consequences are linked. Stories are often the basis for strategies and actions, as well as for interpreting others' intentions. Insurgent organizations like Al Qaeda use narratives very effectively in developing legitimating ideologies. In the Al Qaeda narrative, for example, Osama bin Laden depicts himself as a man purified in the mountains of Afghanistan who is gathering and inspiring followers and punishing infidels. In the collective imagination of Bin Laden and his followers, they are agents of Islamic history who will reverse the decline of the umma [Muslim community] and bring about its inevitable triumph over Western imperialism. For them, Islam can be renewed both politically and theologically only through jihad [holy war] as they define it.

1-77. Though most insurgencies have been limited to nation-states, there have been numerous transnational insurgencies. Likewise, external powers have tried to tap into or create general upheaval by coordinating national insurgencies to give them a transnational character. Al Qaeda's ongoing activities also attempt to leverage religious identity to create and support a transnational array of insurgencies. Operational-level commanders address elements of the transnational movement within their joint operations areas. Other government agencies and higher level officials deal with the national-strategic response to such threats.

1-78. As noted earlier, insurgent groups often employ religious concepts to portray their movement favorably and mobilize followers in pursuit of their political goals. For example, the Provisional Irish Republican Army frequently used Roman Catholic iconography in its publications and proclamations, although many of its members were not devout Catholics. In other cases, a religious ideology may be the source of an insurgent group's political goals. This is the case in Al Qaeda's apparent quest to "reestablish the Caliphate." For many Moslems, the Caliphate produces a positive image of the golden age of Islamic civilization. This image mobilizes support for Al Qaeda among some of the most traditional Muslims while concealing the details of the movement's goal. In fact, Al Qaeda's leaders envision the "restored Caliphate" as a totalitarian state similar to the pre-2002 Taliban regime in Afghanistan.

1-79. Religious extremist insurgents, like many secular radicals and some Marxists, frequently hold an all-encompassing worldview; they are ideologically rigid and uncompromising, seeking to control their members' private thought, expression, and behavior. Seeking power and believing themselves to be ideologically pure, violent religious extremists often brand those they consider insufficiently orthodox as enemies. For example, extreme, violent groups like Al Qaeda routinely attack Islamic sects that profess beliefs inconsistent with their religious dogma. Belief in an extremist ideology fortifies the will of believers. It confirms the idea, common among hard-core transnational terrorists, that using unlimited means is appropriate to achieve their often unlimited goals. Some ideologies, such as the one underlying the culture of martyrdom, maintain that using such means will be rewarded.

1-80. Cultural knowledge is essential to waging a successful counterinsurgency. American ideas of what is "normal" or "rational" are not universal. To the contrary, members of other societies often have different notions of rationality, appropriate behavior, level of religious devotion, and norms concerning gender. Thus, what may appear abnormal or strange to an external observer may appear as self-evidently normal to a group member. For this reason, counterinsurgents—especially commanders, planners, and small-unit leaders—should strive to avoid imposing their ideals of normalcy on a foreign cultural problem.

1-81. Many religious extremists believe that the conversion, subjugation, or destruction of their ideological opponents is inevitable. Violent extremists and terrorists are often willing to use whatever means necessary, even violence against their own followers, to meet their political goals. Nevertheless, they often pursue their ends in highly pragmatic ways based on realistic assumptions. Not all Islamic insurgents or terrorists are fighting for a global revolution. Some are pursing regional goals, such as a establishing a Sunni Arab-dominated Iraq or replacing Israel with an Arab Palestinian state. And militant groups with nationalist as well as religious agendas seek cease fires and participate in elections when such actions support their interests.

1-82. In that light, commanders must consider the presence of religious extremism in the insurgents' ideology when evaluating possible friendly and enemy courses of action. Enemy courses of action that may appear immoral or irrational to Westerners may be acceptable to extremists. Moreover, violent extremists resist changing their worldview; for them, coexistence or compromise is often unacceptable, especially when the movement is purist (like Al Qaeda), in an early stage, or small. However, some extremists are willing to overlook their worldview to achieve short-term goals. Terrorist groups, regardless of their ideology, have cooperated with seemingly incompatible groups. For example, the Palestinian group Black September used German terrorists to perform reconnaissance of the Olympic Village before its 1972 attack on Israeli athletes. Currently, the Taliban is engaged in the drug trade in South Asia. Al Qaeda cooperates with a variety of diverse groups to improve its global access as well.

1-83. The rigid worldview of such extremist groups means that friendly actions intended to create good will among the populace are unlikely to affect them. Similarly, if a group's ideology is so strong that it dominates all other issues, dialog and negotiation will probably prove unproductive. The challenge for counterinsurgents in such cases is to identify the various insurgent groups and determine their motivations. Commanders can then determine the best course of action for each group. This includes identifying the groups with goals flexible enough to allow productive negotiations and determining how to eliminate the extremists without alienating the populace.

Environment and Geography

1-84. Environment and geography, including cultural and demographic factors, affect all participants in a conflict. The manner in which insurgents and counterinsurgents adapt to these realities creates advantages and disadvantages for each. The effects of these factors are immediately visible at the tactical level. There they are perhaps the predominant influence on decisions regarding force structure and doctrine (including tactics, techniques, and procedures). Insurgencies in urban environments present different planning considerations from insurgencies in rural environments. Border areas contiguous to states that may wittingly or unwittingly provide external support and sanctuary to insurgents create a distinct vulnerability for counterinsurgents.

External Support and Sanctuaries

1-85. Access to external resources and sanctuaries has always influenced the effectiveness of insurgencies. External support can provide political, psychological, and material resources that might otherwise be limited or unavailable. Such assistance does not need to come just from neighboring states; countries from outside the region seeking political or economic influence can also support insurgencies. Insurgencies may turn to transnational criminal elements for funding or use the Internet to create a support network among NGOs. Ethnic or religious communities in other states may also provide a form of external support and sanctuary, particularly for transnational insurgencies.

1-86. The meaning of the term sanctuary is evolving. Sanctuaries traditionally were physical safe havens, such as base areas, and this form of safe haven still exists. But insurgents today can also draw on "virtual" sanctuaries in the Internet, global financial systems, and the international media. These virtual sanctuaries can be used to try to make insurgent actions seem acceptable or laudable to internal and external audiences.

1-87. Historically, sanctuaries in neighboring countries have provided insurgents places to rebuild and reorganize without fear of counterinsurgent interference. Modern target acquisition and intelligence-gathering technology make insurgents in isolation, even in neighboring states, more vulnerable than those hidden among the population. Thus, contemporary insurgencies often develop in urban environments, leveraging formal and informal networks for action. Understanding these networks is vital to defeating such insurgencies.

1-88. Insurgencies can also open up sanctuaries within a state over which the host nation's forces cannot extend control or significant influence. In these sanctuaries, nonstate actors with intentions hostile to the host nation or United States can develop unimpaired. When it is to their advantage, such elements provide support for insurgencies. The issue of sanctuaries thus cannot be ignored during planning. Effective COIN operations work to eliminate all sanctuaries.

1-89. Changes in the security environment since the end of the Cold War and the terrorist attacks of 11 September 2001 have increased concerns about the role of nonstate actors in insurgencies. Nonstate actors, such as transnational terrorist organizations, often represent a security threat beyond the areas they inhabit. Some pose a direct concern for the United States and its partners. These nonstate actors often team with insurgents and, in this sense, profit from the conflict.

1-90. A feature of today's operational environment deserving mention is the effort by Islamic extremists, including those that advocate violence, to spread their influence through the funding and use of entities that share their views or facilitate them to varying degrees. These entities may or may not be threats themselves; however, they can provide passive or active support to local or distant insurgencies. Examples include the following:

- Religious schools and mosques.
- NGOs.
- Political parties.
- Business and financial institutions.
- Militia organizations.
- Terrorist training camps and organizations.

Phasing and Timing

1-91. Insurgencies often pass through common phases of development, such as those listed in paragraph 1-31. However, not all insurgencies experience such phased development, and progression through all phases is not required for success. Moreover, a single insurgent movement may be in different phases in different parts of a country. Insurgencies under pressure can also revert to an earlier phase. They then resume development when favorable conditions return. Indeed, this flexibility is the key strength of a phased approach, which provides fallback positions for insurgents when threatened. The protracted popular war phases may not provide a complete template for understanding contemporary insurgencies; however, they do explain the shifting mosaic of activities usually present in some form.

1-92. Versions of protracted popular war have been used by movements as diverse as communist and Islamist insurgencies because the approach is sound and based on mass mobilization—which is a common requirement. Strategic movement from one phase to another does not end the operational and tactical activities typical of earlier phases; it incorporates them. The North Vietnamese explicitly recognized this fact in their doctrine, as was discussed in paragraph 1-36. Their approach emphasized that all forms of warfare occur simultaneously, even as a particular form is paramount. Debates about Vietnam that focus on whether U.S. forces should have concentrated on guerrilla or conventional operations ignore this complexity. In fact, forces that win a mosaic war are those able to respond to both types of operations, often simultaneously.

1-93. The phases of protracted popular war do not necessarily apply to the conspiratorial or military-focused approach. These approaches emphasize quick or armed action and minimize political organization. In many ways, these approaches are less difficult to counter. However, long-term political objectives, as evidenced in the protracted popular war approach, are major parts of any insurgent approach. Effective counterinsurgents understand their overall importance and address them appropriately.

INSURGENT NETWORKS

1-94. A network is a series of direct and indirect ties from one actor to a collection of others. Insurgents use technological, economic, and social means to recruit partners into their networks. Networking is a tool available to territorially rooted insurgencies, such as the FARC in Colombia. It extends the range and variety of both their military and political actions. Other groups have little physical presence in their target countries and exist almost entirely as networks. Networked organizations are difficult to destroy. In addition, they tend to heal, adapt, and learn rapidly. However, such organizations have a limited ability to attain strategic success because they cannot easily muster and focus power. The best outcome they can expect is to create a security vacuum leading to a collapse of the targeted regime's will and then to gain in the competition for the spoils. However, their enhanced abilities to sow disorder and survive present particularly difficult problems for counterinsurgents.

INSURGENT VULNERABILITIES

1-95. While this chapter so far has stressed the difficulties insurgencies present, they do have vulnerabilities that skilled counterinsurgents can exploit. Chapters 4 and 5 discuss how to do this. However, some potential vulnerabilities are worth highlighting here:

- Insurgents' need for secrecy.
- Inconsistencies in the mobilization message.
- Need to establish a base of operations.
- Reliance on external support.
- Need to obtain financial resources.
- Internal divisions.
- Need to maintain momentum.
- Informants within the insurgency.

Secrecy

1-96. Any group beginning from a position of weakness that intends to use violence to pursue its political aims must initially adopt a covert approach for its planning and activities. This practice can become counterproductive once an active insurgency begins. Excessive secrecy can limit insurgent freedom of action, reduce or distort information about insurgent goals and ideals, and restrict communication within the insurgency. Some insurgent groups try to avoid the effects of too much secrecy by splitting into political and military wings. This allows the movement to address the public (political) requirements of an insurgency while still conducting clandestine (military) actions. An example is the insurgency in Northern Ireland, comprised of Sinn Fein (its political wing) and the Irish Republican Army (its military wing). Hamas and Hezbollah also use this technique.

Mobilization and Message

1-97. In the early stages of an insurgency, a movement may be tempted to go to almost any extremes to attract followers. To mobilize their base of support, insurgent groups use a combination of propaganda and intimidation, and they may overreach in both. Effective counterinsurgents use information operations (IO) to exploit inconsistencies in the insurgents' message as well as their excessive use of force or intimidation. The insurgent cause itself may also be a vulnerability. Counterinsurgents may be able to "capture" an insurgency's cause and exploit it. For example, an insurgent ideology based on an extremist interpretation of a holy text can be countered by appealing to a moderate interpretation of the same text. When a credible religious or other respected leader passes this kind of message, the counteraction is even more effective.

Base of Operations

1-98. Insurgents can experience serious difficulties finding a viable base of operations. A base too far from the major centers of activity may be secure but risks being out of touch with the populace. It may also be vulnerable to isolation. A base too near centers of government activity risks opening the insurgency to observation and perhaps infiltration. Bases close to national borders can be attractive when they are beyond the reach of counterinsurgents yet safe enough to avoid suspicions of the neighboring authority or population. Timely, resolute counterinsurgent actions to exploit poor enemy base locations and eliminate or disrupt good ones can significantly weaken an insurgency.

External Support

1-99. Insurgent movements do not control the geographic borders of a country. In fact, insurgencies often rely heavily on freedom of movement across porous borders. Insurgencies usually cannot sustain themselves without substantial external support. An important feature of many transnational terrorist groups is the international nature of their basing. Terrorists may train in one country and fight or conduct other types of operations in another country. The movement of fighters and their support is vulnerable to intervention or attack.

Financial Weakness

1-100. All insurgencies require funding to some extent. Criminal organizations are possible funding sources; however, these groups may be unreliable. Such cooperation may attract undue attention from HN authorities and create vulnerabilities to counterinsurgent intelligence operations. In addition, cooperating with criminals may not be ideologically consistent with the movement's core beliefs, although it often does not prevent such cooperation. Funding from outside donors may come with a political price that affects the overall aim of an insurgency and weakens its popular appeal.

1-101. Counterinsurgents can exploit insurgent financial weaknesses. Controls and regulations that limit the movement and exchange of materiel and funds may compound insurgent financial vulnerabilities. These counters are especially effective when an insurgency receives funding from outside the state.

Internal Divisions

1-102. Counterinsurgents remain alert for signs of divisions within an insurgent movement. A series of successes by counterinsurgents or errors by insurgent leaders can cause some insurgents to question their cause or challenge their leaders. In addition, relations within an insurgency do not remain harmonious when factions form to vie for power. Rifts between insurgent leaders, if identified, can be exploited. Offering amnesty or a seemingly generous compromise can also cause divisions within an insurgency and present opportunities to split or weaken it.

Maintaining Momentum

1-103. Controlling the pace and timing of operations is vital to the success of any insurgency. Insurgents control when the conflict begins and have some measure of control over subsequent activity. However, many insurgencies have failed to capitalize on their initial opportunities. Others have allowed counterinsurgents to dictate the pace of events and scope of activities. If insurgents lose momentum, counterinsurgents can regain the strategic initiative.

Informants

1-104. Nothing is more demoralizing to insurgents than realizing that people inside their movement or trusted supporters among the public are deserting or providing information to government authorities. Counterinsurgents may attract deserters or informants by arousing fear of prosecution or by offering rewards. However, informers must be confident that the government can protect them and their families against retribution.

ASPECTS OF COUNTERINSURGENCY

1-105. The purpose of America's ground forces is to fight and win the Nation's wars. Throughout history, however, the Army and Marine Corps have been called on to perform many tasks beyond pure combat; this has been particularly true during the conduct of COIN operations. COIN requires Soldiers and Marines to be ready both to fight and to build—depending on the security situation and a variety of other factors. The full spectrum operations doctrine (described in FM 3-0) captures this reality.

1-106. All full spectrum operations executed overseas—including COIN operations—include offensive, defensive, and stability operations that commanders combine to achieve the desired end state. The exact mix varies depending on the situation and the mission. Commanders weight each operation based on their assessment of the campaign's phase and the situation in their AO. They shift the weight among these operations as necessary to address situations in different parts of the AO while continuing to pursue their overall objectives. (See figure 1-1.)

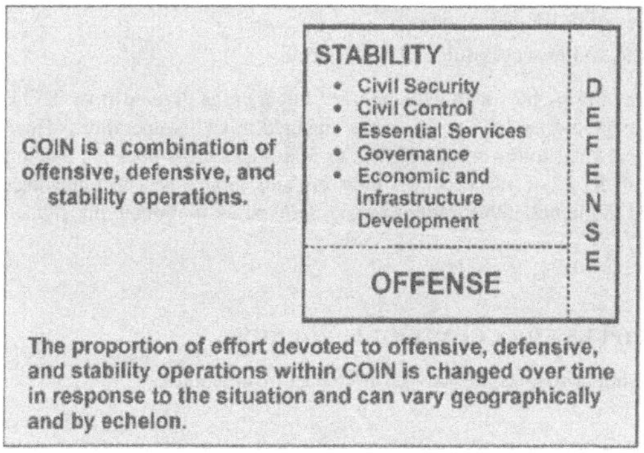

Figure 1-1. Aspects of counterinsurgency operations

1-107. Offensive and defensive operations are integral to COIN. COIN differs from peacekeeping operations in this regard; indeed, this is a key point. In peacekeeping operations, combat is not expected and the goal is an absence of violence. In COIN, such an absence may actually mask insurgent preparations for combat. This was the case, for example, in the Sadr City area of Baghdad in 2003.

1-108. In almost every case, counterinsurgents face a populace containing an active minority supporting the government and an equally small militant faction opposing it. Success requires the government to be accepted as legitimate by most of that uncommitted middle, which also includes passive supporters of both sides. (See figure 1-2.) Because of the ease of sowing disorder, it is usually not enough for counterinsurgents to get 51 percent of popular support; a solid majority is often essential. However, a passive populace may be all that is necessary for a well-supported insurgency to seize political power.

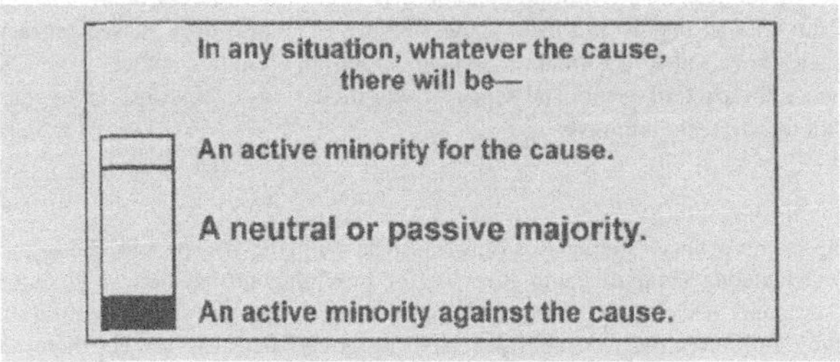

Figure 1-2. Support for an insurgency

1-109. Counterinsurgents must be prepared to identify their opponents and their approach to insurgency. Counterinsurgents must also understand the broader context within which they are operating. A mission to assist a functioning government offers different options from situations where no such viable entity exists or where a regime has been changed by conflict. The last two situations add complex sovereignty and national reconstruction issues to an already complex mission. The state of the infrastructure determines the resources required for reconstruction. The level of violence is a factor in determining how agencies outside the Department of Defense support COIN operations. An extremely violent environment may hamper their freedom of movement. These agencies may elect to operate only from secure areas within or outside of the country.

1-110. The rest of this publication describes how to conduct COIN operations. The following discussion addresses some general themes that shape the following chapters:

- Historical principles for COIN.
- Contemporary imperatives for COIN.
- Paradoxes of COIN operations.
- Successful and unsuccessful COIN practices.

1-111. The historical principles and contemporary imperatives derived from the historical record and detailed below provide some guideposts for forces engaged in COIN operations. However, COIN operations are complicated, and even following the principles and imperatives does not guarantee success. This paradox is present in all forms of warfare but is most obvious in COIN. The following principles and imperatives are presented in the belief that understanding them helps illuminate the challenges inherent in defeating an insurgency.

HISTORICAL PRINCIPLES FOR COUNTERINSURGENCY

1-112. The following principles are derived from past insurgencies.

Legitimacy Is the Main Objective

1-113. The primary objective of any COIN operation is to foster development of effective governance by a legitimate government. Counterinsurgents achieve this objective by the balanced application of both military and nonmilitary means. All governments rule through a combination of consent and coercion. Governments described as "legitimate" rule primarily with the consent of the governed; those described as "illegitimate" tend to rely mainly or entirely on coercion. Citizens of the latter obey the state for fear of the consequences of doing otherwise, rather than because they voluntarily accept its rule. A government that derives its powers from the governed tends to be accepted by its citizens as legitimate. It still uses coercion—for example, against criminals—but most of its citizens voluntarily accept its governance.

1-114. In Western liberal tradition, a government that derives its just powers from the people and responds to their desires while looking out for their welfare is accepted as legitimate. In contrast, theocratic societies fuse political and religious authority; political figures are accepted as legitimate because the populace views them as implementing the will of God. Medieval monarchies claimed "the divine right of kings." Imperial China governed with "the mandate of heaven." Since the 1979 revolution, Iran has operated under the "rule of the jurists [theocratic judges]." In other societies, "might makes right." And sometimes, the ability of a state to provide security—albeit without freedoms associated with Western democracies—can give it enough legitimacy to govern in the people's eyes, particularly if they have experienced a serious breakdown of order.

1-115. Legitimacy makes it easier for a state to carry out its key functions. These include the authority to regulate social relationships, extract resources, and take actions in the public's name. Legitimate governments can develop these capabilities more easily; this situation usually allows them to competently manage, coordinate, and sustain collective security as well as political, economic, and social development. Conversely, illegitimate states (sometimes called "police states") typically cannot regulate society or can do so only by applying overwhelming coercion. Legitimate governance is inherently stable; the societal support it engenders allows it to adequately manage the internal problems, change, and conflict that affect individual and collective well-being. Conversely, governance that is not legitimate is inherently unstable; as soon as the state's coercive power is disrupted, the populace ceases to obey it. Thus legitimate governments tend to be resilient and exercise better governance; illegitimate ones tend to be fragile and poorly administered.

1-116. Six possible indicators of legitimacy that can be used to analyze threats to stability include the following:

- The ability to provide security for the populace (including protection from internal and external threats).
- Selection of leaders at a frequency and in a manner considered just and fair by a substantial majority of the populace.
- A high level of popular participation in or support for political processes.
- A culturally acceptable level of corruption.
- A culturally acceptable level and rate of political, economic, and social development.
- A high level of regime acceptance by major social institutions.

1-117. Governments scoring high in these categories probably have the support of a sufficient majority of the population. Different cultures, however, may see acceptable levels of development, corruption, and participation differently. And for some societies, providing security and some basic services may be enough for citizens to grant a government legitimacy; indeed, the importance of security in situations where violence has escalated cannot be overemphasized. In such cases, establishing security can win the people's confidence and enable a government to develop legitimacy in other areas.

1-118. In working to understand the problem, commanders and staffs determine what the HN population defines as effective and legitimate governance. This understanding continues to evolve as information is developed. Commanders and staffs must continually diagnose what they understand legitimacy to mean to the HN population. The population's expectations will influence all ensuing operations. Additionally, planners may also consider perceptions of legitimacy held by outside supporters of the HN government and the insurgents. Differences between U.S., local, and international visions of legitimacy can further complicate operations. But the most important attitude remains that of the HN population. In the end, its members determine the ultimate victor.

1-119. The presence of the rule of law is a major factor in assuring voluntary acceptance of a government's authority and therefore its legitimacy. A government's respect for preexisting and impersonal legal rules can provide the key to gaining it widespread, enduring societal support. Such government respect for rules—ideally ones recorded in a constitution and in laws adopted through a credible, democratic process—is the essence of the rule of law. As such, it is a powerful potential tool for counterinsurgents.

1-120. Military action can address the symptoms of a loss of legitimacy. In some cases, it can eliminate substantial numbers of insurgents. However, success in the form of a durable peace requires restoring legitimacy, which, in turn, requires the use of all instruments of national power. A COIN effort cannot achieve lasting success without the HN government achieving legitimacy.

Unity of Effort Is Essential

1-121. Unity of effort must be present at every echelon of a COIN operation. Otherwise, well-intentioned but uncoordinated actions can cancel each other or provide vulnerabilities for insurgents to exploit. Ideally, a single counterinsurgent leader has authority over all government agencies involved in COIN operations. Usually, however, military commanders work to achieve unity of effort through liaison with leaders of a wide variety of nonmilitary agencies. The U.S. Ambassador and country team, along with senior HN representatives, must be key players in higher level planning; similar connections are needed throughout the chain of command.

1-122. NGOs often play an important role at the local level. Many such agencies resist being overtly involved with military forces; however, efforts to establish some kind of liaison are needed. The most important connections are those with joint, interagency, multinational, and HN organizations. The goal of these connections is to ensure that, as much as possible, objectives are shared and actions and messages synchronized. Achieving this synergy is essential.

Political Factors Are Primary

1-123. General Chang Ting-chen of Mao Zedong's central committee once stated that revolutionary war was 80 percent political action and only 20 percent military. Such an assertion is arguable and certainly depends on the insurgency's stage of development; it does, however, capture the fact that political factors have primacy in COIN. At the beginning of a COIN operation, military actions may appear predominant as security forces conduct operations to secure the populace and kill or capture insurgents; however, political objectives must guide the military's approach. Commanders must, for example, consider how operations contribute to strengthening the HN government's legitimacy and achieving U.S. political goals. This means that political and diplomatic leaders must actively participate throughout the conduct (planning, preparation, execution, and assessment) of COIN operations. The political and military aspects of insurgencies are so bound together as to be inseparable. Most insurgent approaches recognize that fact. Military actions executed without properly assessing their political effects at best result in reduced effectiveness and at worst are counterproductive. Resolving most insurgencies requires a political solution; it is thus imperative that counterinsurgent actions do not hinder achieving that political solution.

Counterinsurgents Must Understand the Environment

1-124. Successful conduct of COIN operations depends on thoroughly understanding the society and culture within which they are being conducted. Soldiers and Marines must understand the following about the population in the AO:

- Organization of key groups in the society.
- Relationships and tensions among groups.
- Ideologies and narratives that resonate with groups.
- Values of groups (including tribes), interests, and motivations.
- Means by which groups (including tribes) communicate.

- The society's leadership system.

1-125. In most COIN operations in which U.S. forces participate, insurgents hold a distinct advantage in their level of local knowledge. They speak the language, move easily within the society, and are more likely to understand the population's interests. Thus, effective COIN operations require a greater emphasis on certain skills, such as language and cultural understanding, than does conventional warfare. The interconnected, politico-military nature of insurgency and COIN requires immersion in the people and their lives to achieve victory. Specifically, successful COIN operations require Soldiers and Marines at every echelon to possess the following within the AO's cultural context:

- A clear appreciation of the essential nature and nuances of the conflict.
- An understanding of the motivation, strengths, and weaknesses of the insurgents.
- Knowledge of the roles of other actors in the AO.

Without this understanding of the environment, intelligence cannot be understood and properly applied.

Intelligence Drives Operations

1-126. Without good intelligence, counterinsurgents are like blind boxers wasting energy flailing at unseen opponents and perhaps causing unintended harm. With good intelligence, counterinsurgents are like surgeons cutting out cancerous tissue while keeping other vital organs intact. Effective operations are shaped by timely, specific, and reliable intelligence, gathered and analyzed at the lowest possible level and disseminated throughout the force.

1-127. Because of the dispersed nature of COIN operations, counterinsurgents' own actions are a key generator of intelligence. A cycle develops where operations produce intelligence that drives subsequent operations. Reporting by units, members of the country team, and associated civilian agencies is often of greater importance than reporting by specialized intelligence assets. These factors, along with the need to generate a favorable tempo (rate of military operations), drive the requirement to produce and disseminate intelligence at the lowest practical level. (Chapter 3 addresses intelligence in COIN.)

Insurgents Must be Isolated from Their Cause and Support

1-128. It is easier to separate an insurgency from its resources and let it die than to kill every insurgent. Clearly, killing or capturing insurgents will be necessary, especially when an insurgency is based in religious or ideological extremism. However, killing every insurgent is normally impossible. Attempting to do so can also be counterproductive in some cases; it risks generating popular resentment, creating martyrs that motivate new recruits, and producing cycles of revenge.

1-129. Dynamic insurgencies can replace losses quickly. Skillful counterinsurgents must thus cut off the sources of that recuperative power. Some sources can be reduced by redressing the social, political, and economic grievances that fuel the insurgency. Physical support can be cut off by population control or border security. International or local legal action might be required to limit financial support. Urban insurgents, however, are especially difficult to isolate from their cause and sources of support. They may operate in small, compartmentalized cells that are usually independent or semi-independent. These cells often have their own support mechanisms and few, if any, ties to the population that counterinsurgents can track.

1-130. As the HN government increases its legitimacy, the populace begins to assist it more actively. Eventually, the people marginalize and stigmatize insurgents to the point that the insurgency's claim to legitimacy is destroyed. However, victory is gained not when this isolation is achieved, but when the victory is permanently maintained by and with the people's active support and when insurgent forces have been defeated.

Security Under the Rule of Law is Essential

1-131. The cornerstone of any COIN effort is establishing security for the civilian populace. Without a secure environment, no permanent reforms can be implemented and disorder spreads. To establish legitimacy, commanders transition security activities from combat operations to law enforcement as quickly as feasible. When insurgents are seen as criminals, they lose public support. Using a legal system established in line with local culture and practices to deal with such criminals enhances the HN government's legitimacy. Soldiers and Marines help establish HN institutions that sustain that legal regime, including police forces, court systems, and penal facilities. It is important to remember that the violence level must be reduced enough for police forces to maintain order prior to any transition; otherwise, COIN forces will be unable to secure the populace and may lose the legitimacy gained by the transition.

1-132. Illegitimate actions are those involving the use of power without authority—whether committed by government officials, security forces, or counterinsurgents. Such actions include unjustified or excessive use of force, unlawful detention, torture, and punishment without trial. Efforts to build a legitimate government though illegitimate actions are self-defeating, even against insurgents who conceal themselves amid noncombatants and flout the law. Moreover, participation in COIN operations by U.S. forces must follow United States law, including domestic laws, treaties to which the United States is party, and certain HN laws. (See appendix D.) Any human rights abuses or legal violations committed by U.S. forces quickly become known throughout the local populace and eventually around the world. Illegitimate actions undermine both long- and short-term COIN efforts.

1-133. Every action by counterinsurgents leaves a "forensic trace" that may be required sometime later in a court of law. Counterinsurgents document all their activities to preserve, wherever possible, a chain of evidence. Accurate documentation can also be an important means to counter insurgent propaganda.

Counterinsurgents Should Prepare for a Long-Term Commitment

1-134. Insurgencies are protracted by nature. Thus, COIN operations always demand considerable expenditures of time and resources. The populace may prefer the HN government to the insurgents; however, people do not actively support a government unless they are convinced that the counterinsurgents have the means, ability, stamina, and will to win. The insurgents' primary battle is against the HN government, not the United States; however, U.S. support can be crucial to building public faith in that government's viability. The populace must have confidence in the staying power of both the counterinsurgents and the HN government. Insurgents and local populations often believe that a few casualties or a few years will cause the United States to abandon a COIN effort. Constant reaffirmations of commitment, backed by deeds, can overcome that perception and bolster faith in the steadfastness of U.S. support. But even the strongest U.S. commitment will not succeed if the populace does not perceive the HN government as having similar will and stamina. U.S. forces must help create that capacity and sustain that impression.

1-135. Preparing for a protracted COIN effort requires establishing headquarters and support structures designed for long-term operations. Planning and commitments should be based on sustainable operating tempo and personnel tempo limits for the various components of the force. (Operating tempo and personnel tempo are defined in the glossary.) Even in situations where the U.S. goal is reducing its military force levels as quickly as possible, some support for HN institutions usually remains for a long time.

1-136. At the strategic level, gaining and maintaining U.S. public support for a protracted deployment is critical. Only the most senior military officers are involved in this process at all. It is properly a political activity. However, military leaders typically take care to ensure that their actions and statements are forthright. They also ensure that the conduct of operations neither makes it harder for elected leaders to maintain public support nor undermines public confidence.

CONTEMPORARY IMPERATIVES OF COUNTERINSURGENCY

1-137. Recent COIN experiences have identified an important set of additional imperatives to keep in mind for success.

Manage Information and Expectations

1-138. Information and expectations are related; skillful counterinsurgents manage both. To limit discontent and build support, the HN government and any counterinsurgents assisting it create and maintain a realistic set of expectations among the populace, friendly military forces, and the international community. IO (including psychological operations and the related activities of public affairs and civil-military operations) are key tools to accomplish this. Achieving steady progress toward a set of reasonable expectations can increase the populace's tolerance for the inevitable inconveniences entailed by ongoing COIN operations. Where a large U.S. force is present to help establish a regime, such progress can extend the period before an army of liberation becomes perceived as an army of occupation.

1-139. U.S. forces start with a built-in challenge because of their reputation for accomplishment, what some call the "man on the moon syndrome." This refers to the expressed disbelief that a nation able to put a man on the moon cannot quickly restore basic services. U.S. agencies trying to fan enthusiasm for their efforts should avoid making unrealistic promises. In some cultures, failure to deliver promised results is automatically interpreted as deliberate deception, rather than good intentions gone awry. In other cultures, exorbitant promises are normal and people do not expect them to be kept. Effective counterinsurgents understand local norms; they use locally tailored approaches to control expectations. Managing expectations also involves demonstrating economic and political progress to show the populace how life is improving. Increasing the number of people who feel they have a stake in the success of the state and its government is a key to successful COIN operations. In the end, victory comes, in large measure, by convincing the populace that their life will be better under the HN government than under an insurgent regime.

1-140. Both counterinsurgents and the HN government ensure that their deeds match their words. They also understand that any action has an information reaction. Counterinsurgents and the HN government carefully consider that impact on the many audiences involved in the conflict and on the sidelines. They work actively to shape responses that further their ends. In particular, messages to different audiences must be consistent. In the global information environment, people in the AO can access the Internet and satellite television to determine the messages counterinsurgents are sending to the international community and the U.S. public. Any perceived inconsistency reduces credibility and undermines COIN efforts.

Use the Appropriate Level of Force

1-141. Any use of force generates a series of reactions. There may be times when an overwhelming effort is necessary to destroy or intimidate an opponent and reassure the populace. Extremist insurgent combatants often have to be killed. In any case, however, counterinsurgents should calculate carefully the type and amount of force to be applied and who wields it for any operation. An operation that kills five insurgents is counterproductive if collateral damage leads to the recruitment of fifty more insurgents.

1-142. In a COIN environment, it is vital for commanders to adopt appropriate and measured levels of force and apply that force precisely so that it accomplishes the mission without causing unnecessary loss of life or suffering. Normally, counterinsurgents can use escalation of force/force continuum procedures to minimize potential loss of life. These procedures are especially appropriate during convoy operations and at checkpoints and roadblocks. *Escalation of force* (Army)/*force continuum* (Marine Corps) refers to using lesser means of force when such use is likely to achieve the desired effects and Soldiers and Marines can do so without endangering themselves, others, or mission accomplishment. Escalation of force/force continuum procedures do not limit the right of self-defense, including the use of deadly force when such force is necessary to defend against a hostile act or demonstrated hostile intent. Commanders ensure that their Soldiers and Marines are properly trained in such procedures and, more importantly, in methods of shaping situations so that small-unit leaders have to make fewer split-second, life-or-death decisions.

1-143. Who wields force is also important. If the HN police have a reasonable reputation for competence and impartiality, it is better for them to execute urban raids; the populace is likely to view that application of force as more legitimate. This is true even if the police are not as well armed or as capable as military units. However, local circumstances affect this decision. If the police are seen as part of an ethnic or sectarian group oppressing the general population, their use may be counterproductive. Effective counterinsurgents thus understand the character of the local police and popular perceptions of both police and military units. This understanding helps ensure that the application of force is appropriate and reinforces the rule of law.

Learn and Adapt

1-144. An effective counterinsurgent force is a learning organization. Insurgents constantly shift between military and political phases and tactics. In addition, networked insurgents constantly exchange information about their enemy's vulnerabilities—even with insurgents in distant theaters. However, skillful counterinsurgents can adapt at least as fast as insurgents. Every unit needs to be able to make observations, draw and apply lessons, and assess results. Commanders must develop an effective system to circulate best practices throughout their command. Combatant commanders might also need to seek new laws or policies that authorize or resource necessary changes. Insurgents shift their AOs looking for weak links, so widespread competence is required throughout the counterinsurgent force.

Empower the Lowest Levels

1-145. *Mission command* is the conduct of military operations through decentralized execution based upon mission orders for effective mission accomplishment. Successful mission command results from subordinate leaders at all echelons exercising disciplined initiative within the commander's intent to accomplish missions. It requires an environment of trust and mutual understanding (FM 6-0). It is the Army's and Marine Corps' preferred method for commanding and controlling forces during all types of operations. Under mission command, commanders provide subordinates with a mission, their commander's intent, a concept of operations, and resources adequate to accomplish the mission. Higher commanders empower subordinates to make decisions within the commander's intent. They leave details of execution to their subordinates and expect them to use initiative and judgment to accomplish the mission.

1-146. Mission command is ideally suited to the mosaic nature of COIN operations. Local commanders have the best grasp of their situations. Under mission command, they are given access to or control of the resources needed to produce timely intelligence, conduct effective tactical operations, and manage IO and civil-military operations. Thus, effective COIN operations are decentralized, and higher commanders owe it to their subordinates to push as many capabilities as possible down to their level. Mission command encourages the initiative of subordinates and facilitates the learning that must occur at every level. It is a major characteristic of a COIN force that can adapt and react at least as quickly as the insurgents.

Support the Host Nation

1-147. U.S. forces committed to a COIN effort are there to assist a HN government. The long-term goal is to leave a government able to stand by itself. In the end, the host nation has to win on its own. Achieving this requires development of viable local leaders and institutions. U.S. forces and agencies can help, but HN elements must accept responsibilities to achieve real victory. While it may be easier for U.S. military units to conduct operations themselves, it is better to work to strengthen local forces and institutions and then assist them. HN governments have the final responsibility to solve their own problems. Eventually all foreign armies are seen as interlopers or occupiers; the sooner the main effort can transition to HN institutions, without unacceptable degradation, the better.

PARADOXES OF COUNTERINSURGENCY OPERATIONS

1-148. The principles and imperatives discussed above reveal that COIN presents a complex and often unfamiliar set of missions and considerations. In many ways, the conduct of COIN is counterintuitive to the traditional U.S. view of war—although COIN operations have actually formed a substantial part of the U.S. military experience. Some representative paradoxes of COIN are presented here as examples of the different mindset required. These paradoxes are offered to stimulate thinking, not to limit it. The applicability of the thoughts behind the paradoxes depends on a sense of the local situation and, in particular, the state of the insurgency. For example, the admonition "Sometimes, the More Force Used, the Less Effective It Is" does not apply when the enemy is "coming over the barricades"; however, that thought is applicable when increased security is achieved in an area. In short, these paradoxes should not be reduced to a checklist; rather, they should be used with considerable thought.

Sometimes, the More You Protect Your Force, the Less Secure You May Be

1-149. Ultimate success in COIN is gained by protecting the populace, not the COIN force. If military forces remain in their compounds, they lose touch with the people, appear to be running scared, and cede the initiative to the insurgents. Aggressive saturation patrolling, ambushes, and listening post operations must be conducted, risk shared with the populace, and contact maintained. The effectiveness of establishing patrol bases and operational support bases should be weighed against the effectiveness of using larger unit bases. (FM 90-8 discusses saturation patrolling and operational support bases.) These practices ensure access to the intelligence needed to drive operations. Following them reinforces the connections with the populace that help establish real legitimacy.

Sometimes, the More Force Is Used, the Less Effective It Is

1-150. Any use of force produces many effects, not all of which can be foreseen. The more force applied, the greater the chance of collateral damage and mistakes. Using substantial force also increases the opportunity for insurgent propaganda to portray lethal military activities as brutal. In contrast, using force precisely and discriminately strengthens the rule of law that needs to be established. As noted above, the key for counterinsurgents is knowing when more force is needed—and when it might be counterproductive. This judgment involves constant assessment of the security situation and a sense of timing regarding insurgents' actions.

The More Successful the Counterinsurgency Is, the Less Force Can Be Used and the More Risk Must Be Accepted

1-151. This paradox is really a corollary to the previous one. As the level of insurgent violence drops, the requirements of international law and the expectations of the populace lead to a reduction in direct military actions by counterinsurgents. More reliance is placed on police work, rules of engagement may be tightened, and troops may have to exercise increased restraint. Soldiers and Marines may also have to accept more risk to maintain involvement with the people.

Sometimes Doing Nothing Is the Best Reaction

1-152. Often insurgents carry out a terrorist act or guerrilla raid with the primary purpose of enticing counterinsurgents to overreact, or at least to react in a way that insurgents can exploit—for example, opening fire on a crowd or executing a clearing operation that creates more enemies than it takes off the streets. If an assessment of the effects of a course of action determines that more negative than positive effects may result, an alternative should be considered—potentially including not acting.

Some of the Best Weapons for Counterinsurgents Do Not Shoot

1-153. Counterinsurgents often achieve the most meaningful success in garnering public support and legitimacy for the HN government with activities that do not involve killing insurgents (though, again, killing clearly will often be necessary). Arguably, the decisive battle is for the people's minds; hence synchronizing IO with efforts along the other LLOs is critical. Every action, including uses of force, must be "wrapped in a bodyguard of information." While security is essential to setting the stage for overall progress, lasting victory comes from a vibrant economy, political participation, and restored hope. Particularly after security has been achieved, dollars and ballots will have more important effects than bombs and bullets. This is a time when "money is ammunition." Depending on the state of the insurgency, therefore, Soldiers and Marines should prepare to execute many nonmilitary missions to support COIN efforts. Everyone has a role in nation building, not just Department of State and civil affairs personnel.

The Host Nation Doing Something Tolerably Is Normally Better than Us Doing It Well

1-154. It is just as important to consider who performs an operation as to assess how well it is done. Where the United States is supporting a host nation, long-term success requires establishing viable HN leaders and institutions that can carry on without significant U.S. support. The longer that process takes, the more U.S. public support will wane and the more the local populace will question the legitimacy of their own forces and government. General Creighton Abrams, the U.S. commander in Vietnam in 1971, recognized this fact when he said, "There's very clear evidence,...in some things, that we helped too much. And we *retarded* the Vietnamese by doing *it....We* can't run this thing.... *They've* got to run it. The nearer we get to that the better off *they* are and the better off *we* are." T.E. Lawrence made a similar observation while leading the Arab Revolt against the Ottoman Empire in 1917: "Do not try to do too much with your own hands. Better the Arabs do it tolerably than that you do it perfectly. It is their war, and you are to help them, not to win it for them." However, a key word in Lawrence's advice is "tolerably." If the host nation cannot perform tolerably, counterinsurgents supporting it may have to act. Experience, knowledge of the AO, and cultural sensitivity are essential to deciding when such action is necessary.

If a Tactic Works this Week, It Might Not Work Next Week; If It Works in this Province, It Might Not Work in the Next

1-155. Competent insurgents are adaptive. They are often part of a widespread network that communicates constantly and instantly. Insurgents quickly adjust to successful COIN practices and rapidly disseminate information throughout the insurgency. Indeed, the more effective a COIN tactic is, the faster it may become out of date because insurgents have a greater need to counter it. Effective leaders at all levels avoid complacency and are at least as adaptive as their enemies. There is no "silver bullet" set of COIN procedures. Constantly developing new practices is essential.

Tactical Success Guarantees Nothing

1-156. As important as they are in achieving security, military actions by themselves cannot achieve success in COIN. Insurgents that never defeat counterinsurgents in combat still may achieve their strategic objectives. Tactical actions thus must be linked not only to strategic and operational military objectives but also to the host nation's essential political goals. Without those connections, lives and resources may be wasted for no real gain.

Many Important Decisions Are Not Made by Generals

1-157. Successful COIN operations require competence and judgment by Soldiers and Marines at all levels. Indeed, young leaders—so-called "strategic corporals"—often make decisions at the tactical level that have strategic consequences. Senior leaders set the proper direction and climate with thorough training and clear guidance; then they trust their subordinates to do the right thing. Preparation for tactical-level leaders requires more than just mastering Service doctrine; they must also be trained and educated to adapt to their local situations, understand the legal and ethical implications of their actions, and exercise initiative and sound judgment in accordance with their senior commanders' intent.

SUCCESSFUL AND UNSUCCESSFUL COUNTERINSURGENCY PRACTICES

1-158. Table 1-1 lists some practices that have contributed significantly to success or failure in past counterinsurgencies.

SUMMARY

1-159. COIN is an extremely complex form of warfare. At its core, COIN is a struggle for the population's support. The protection, welfare, and support of the people are vital to success. Gaining and maintaining that support is a formidable challenge. Achieving these aims requires synchronizing the efforts of many nonmilitary and HN agencies in a comprehensive approach.

1-160. Designing operations that achieve the desired end state requires counterinsurgents to understand the culture and the problems they face. Both insurgents and counterinsurgents are fighting for the support of the populace. However, insurgents are constrained by neither the law of war nor the bounds of human decency as Western nations understand them. In fact, some insurgents are willing to commit suicide and kill innocent civilians in carrying out their operations—and deem this a legitimate option. They also will do anything to preserve their greatest advantage, the ability to hide among the people. These amoral and often barbaric enemies survive by their wits, constantly adapting to the situation. Defeating them requires counterinsurgents to develop the ability to learn and adapt rapidly and continuously. This manual emphasizes this "Learn and Adapt" imperative as it discusses ways to gain and maintain the support of the people.

1-161. Popular support allows counterinsurgents to develop the intelligence necessary to identify and defeat insurgents. Designing and executing a comprehensive campaign to secure the populace and then gain its support requires carefully coordinating actions along several LLOs over time to produce success. One of these LLOs is developing HN security forces that can assume primary responsibility for combating the insurgency. COIN operations also place distinct burdens on leaders and logisticians. All of these aspects of COIN are described and analyzed in the chapters that follow.

Table 1-1. Successful and unsuccessful counterinsurgency operational practices

Successful practices	*Unsuccessful practices*
• Emphasize intelligence. • Focus on the population, its needs, and its security. • Establish and expand secure areas. • Isolate insurgents from the populace (population control). • Conduct effective, pervasive, and continuous information operations. • Provide amnesty and rehabilitation for those willing to support the new government. • Place host-nation police in the lead with military support as soon as the security situation permits. • Expand and diversify the host-nation police force. • Train military forces to conduct counterinsurgency operations. • Embed quality advisors and special forces with host-nation forces. • Deny sanctuary to insurgents. • Encourage strong political and military cooperation and information sharing. • Secure host-nation borders. • Protect key infrastructure.	• Overemphasize killing and capturing the enemy rather than securing and engaging the populace. • Conduct large-scale operations as the norm. • Concentrate military forces in large bases for protection. • Focus special forces primarily on raiding. • Place low priority on assigning quality advisors to host-nation forces. • Build and train host-nation security forces in the U.S. military's image. • Ignore peacetime government processes, including legal procedures. • Allow open borders, airspace, and coastlines.

This page intentionally left blank.

Chapter 2

Unity of Effort: Integrating Civilian and Military Activities

Essential though it is, the military action is secondary to the political one, its primary purpose being to afford the political power enough freedom to work safely with the population.

David Galula, *Counterinsurgency Warfare*, 1964[1]

This chapter begins with the principles involved in integrating the activities of military and civilian organizations during counterinsurgency operations. It then describes the categories of organizations usually involved. After that, it discusses assignment of responsibilities and mechanisms used to integrate civilian and military activities. It concludes by listing information commanders need to know about civilian agencies operating in their area of operations.

INTEGRATION

2-1. Military efforts are necessary and important to counterinsurgency (COIN) efforts, but they are only effective when integrated into a comprehensive strategy employing all instruments of national power. A successful COIN operation meets the contested population's needs to the extent needed to win popular support while protecting the population from the insurgents. Effective COIN operations ultimately eliminate insurgents or render them irrelevant. Success requires military forces engaged in COIN operations to—

- Know the roles and capabilities of U.S., intergovernmental, and host-nation (HN) partners.
- Include other participants, including HN partners, in planning at every level.
- Support civilian efforts, including those of nongovernmental organizations (NGOs) and intergovernmental organizations (IGOs).
- As necessary, conduct or participate in political, social, informational, and economic programs.

2-2. The integration of civilian and military efforts is crucial to successful COIN operations. All efforts focus on supporting the local populace and HN government. Political, social, and economic programs are usually more valuable than conventional military operations in addressing the root causes of conflict and undermining an insurgency. COIN participants come from many backgrounds. They may include military personnel, diplomats, police, politicians, humanitarian aid workers, contractors, and local leaders. All must make decisions and solve problems in a complex and extremely challenging environment.

2-3. Controlling the level of violence is a key aspect of the struggle. A high level of violence often benefits insurgents. The societal insecurity that violence brings discourages or precludes nonmilitary organizations, particularly external agencies, from helping the local populace. A more benign security environment allows civilian agencies greater opportunity to provide their resources and expertise. It thereby relieves military forces of this burden.

2-4. An essential COIN task for military forces is fighting insurgents; however, these forces can and should use their capabilities to meet the local populace's fundamental needs as well. Regaining the populace's active and continued support for the HN government is essential to deprive an insurgency of its power and appeal. The military forces' primary function in COIN is protecting that populace. However, employing military force is not the only way to provide civil security or defeat insurgents. Indeed, excessive use of military force can frequently undermine policy objectives at the expense of achieving the overarching political goals that define success. This dilemma places tremendous importance on the measured application of force.

2-5. Durable policy success requires balancing the measured use of force with an emphasis on nonmilitary programs. Political, social, and economic programs are most commonly and appropriately associated with civilian organizations and expertise; however, effective implementation of these programs is more important than who performs the tasks. If adequate civilian capacity is not available, military forces fill the gap. COIN programs for political, social, and economic well-being are essential to developing the local capacity that commands popular support when accurately perceived. COIN is also a battle of ideas. Insurgents seek to further their cause by creating misperceptions of COIN efforts. Comprehensive information programs are necessary to amplify the messages of positive deeds and to counter insurgent propaganda.

2-6. COIN is fought among the populace. Counterinsurgents take upon themselves responsibility for the people's well-being in all its manifestations. These include the following:

- Security from insurgent intimidation and coercion, as well as from nonpolitical violence and crime.
- Provision for basic economic needs.
- Provision of essential services, such as water, electricity, sanitation, and medical care.
- Sustainment of key social and cultural institutions.
- Other aspects that contribute to a society's basic quality of life.

Effective COIN programs address all aspects of the local populace's concerns in a unified fashion. Insurgents succeed by maintaining turbulence and highlighting local grievances the COIN effort fails to address. COIN forces succeed by eliminating turbulence and helping the host nation meet the populace's basic needs.

2-7. When the United States commits to helping a host nation defeat an insurgency, success requires applying the instruments of national power along multiple lines of operations. (Normally, these are logical lines of operations [LLOs], as described in chapter 5.) Since efforts along one LLO often affect progress in others, uncoordinated actions are frequently counterproductive.

2-8. LLOs in COIN focus primarily on the populace. Each line depends on the others. The interdependence of the lines is total: if one fails, the mission fails. Many LLOs require applying capabilities usually resident in civilian organizations, such as—

- U.S. government agencies other than the Department of Defense (DOD).
- Other nations' defense and nondefense agencies and ministries.
- IGOs, such as the United Nations and its subordinate organizations.
- NGOs.
- Private corporations.
- Other organizations that wield diplomatic, informational, and economic power.

These civilian organizations bring expertise that complements that of military forces. At the same time, civilian capabilities cannot be employed effectively without the security that military forces provide. Effective COIN leaders understand the interdependent relationship of all participants, military and civilian. COIN leaders orchestrate their efforts to achieve unity of effort and coherent results.

UNITY OF COMMAND

2-9. Unity of command is the preferred doctrinal method for achieving unity of effort by military forces. Where possible, COIN leaders achieve unity of command by establishing and maintaining the formal command or support relationships discussed in FM 3-0. Unity of command should extend to all military forces supporting a host nation. The ultimate objective of these arrangements is for military forces, police, and other security forces to establish effective control while attaining a monopoly on the legitimate use of violence within the society. Command and control of all U.S. Government organizations engaged in a COIN mission should be exercised by a single leader through a formal command and control system. (FM 6-0, chapter 5, discusses a commander's command and control system.)

2-10. The relationships and authorities between military and nonmilitary U.S. Government agencies are usually given in the document directing an agency to support the operation. (The document is usually a memorandum of agreement or understanding.) Commanders exercise only the authority those documents allow; however, the terms in those documents may form the basis for establishing some form of relationship between commanders and agency chiefs.

2-11. As important as unity of command is to military operations, it is one of the most sensitive and difficult-to-resolve issues in COIN. The participation of U.S. and multinational military forces in COIN missions is inherently problematic, as it influences perceptions of the capacity and legitimacy of local security forces. Although unity of command of military forces may be desirable, it may be impractical due to political considerations. Political sensitivities about the perceived subordination of national forces to those of other states or IGOs often preclude strong command relationships; however, the agreements that establish a multinational force provide a source for determining possible authorities and command, support, or other relationships.

2-12. The differing goals and fundamental independence of NGOs and local organizations usually prevent formal relationships governed by command authority. In the absence of such relationships, military leaders seek to persuade and influence other participants to contribute to achieving COIN objectives. Informal or less authoritative relationships include coordination and liaison. In some cases, direct interaction among various organizations may be impractical or undesirable. Basic awareness and general information sharing may be the most that can be accomplished. When unity of command with part or all of the force, including nonmilitary elements, is not possible, commanders work to achieve unity of effort through cooperation and coordination among all elements of the force—even those not part of the same command structure.

UNITY OF EFFORT

2-13. Achieving unity of effort is the goal of command and support relationships. All organizations contributing to a COIN operation should strive, or be persuaded to strive, for maximum unity of effort. Informed, strong leadership forms the foundation for achieving it. Leadership in this area focuses on the central problems affecting the local populace. A clear understanding of the desired end state should infuse all efforts, regardless of the agencies or individuals charged with their execution. Given the primacy of political considerations, military forces often support civilian efforts. However, the mosaic nature of COIN operations means that lead responsibility shifts among military, civilian, and HN authorities. Regardless, military leaders should prepare to assume local leadership for COIN efforts. The organizing imperative is focusing on what needs to be done, not on who does it.

"Hand Shake Con" in Operation Provide Comfort

Operation Provide Comfort provided relief to the Kurdish refugees from northern Iraq and protection for humanitarian relief efforts. It began on 6 April 1991 and ended 24 July 1991. General Anthony C. Zinni, USMC, the multinational force commander, relates the following conversation regarding his command and control arrangements:

"[Regarding command and control relationships with other multinational contingents, t]he Chairman of the Joint Chiefs of Staff asked me… 'The lines in your command chart, the command relationships, what are they? OpCon [operational control]? TaCon [tactical control]? Command?' 'Sir, we don't ask, because no one can sign up to any of that stuff.' 'Well, how do you do business?' 'Hand Shake Con. That's it.' No memoranda of agreement. No memoranda of understanding.… [T]he relationships are worked out on the scene, and they aren't pretty. And you don't really want to try to capture them,…distill them, and say as you go off in the future, you're going to have this sort of command relationship.… [I]t is Hand Shake Con and that's the way it works. It is consultative. It is behind-the-scene."

2-14. Countering an insurgency begins with understanding the complex environment and the numerous competing forces within it. Gaining an understanding of the environment—including the insurgents, affected populace, and different counterinsurgent organizations—is essential to an integrated COIN operation. The complexity of identifying the insurgency's causes and integrating actions addressing them across multiple, interrelated LLOs requires understanding the civilian and military capabilities, activities, and end state. Various agencies acting to reestablish stability may differ in goals and approaches, based on their experience and institutional culture. When their actions are allowed to adversely affect each other, the populace suffers and insurgents identify grievances to exploit. Integrated actions are essential to defeat the ideologies professed by insurgents. A shared understanding of the operation's purpose provides a unifying theme for COIN efforts. Through a common understanding of that purpose, the COIN leadership can design an operation that promotes effective collaboration and coordination among all agencies and the affected populace.

COORDINATION AND LIAISON

2-15. Many organizations can contribute to successful COIN operations. An insurgency's complex diplomatic, informational, military, and economic context precludes military leaders from commanding all contributing organizations—and they should not try to do so. Interagency partners, NGOs, and private organizations have many interests and agendas that military forces cannot control. Additionally, local legitimacy is frequently affected by the degree to which local institutions are perceived as independent and capable without external support. Nevertheless, military leaders should make every effort to ensure that COIN actions are as well integrated as possible. Active leadership by military leaders is imperative to effect coordination, establish liaison (formal and informal), and share information. Influencing and persuading groups outside a commander's authority requires skill and often subtlety. As actively as commanders pursue unity of effort, they should also be mindful of their prominence and recognize the wisdom of acting indirectly and in ways that allow credit for success to go to others—particularly local individuals and organizations.

2-16. Many groups often play critical roles in influencing the outcome of a COIN effort but are beyond the control of military forces or civilian governing institutions. These groups include the following:

- Local leaders.
- Informal associations.
- Religious groups.
- Families.
- Tribes.
- Some private enterprises.
- Some humanitarian groups.
- The media.

Commanders remain aware of the influence of such groups and are prepared to work with, through, or around them.

KEY COUNTERINSURGENCY PARTICIPANTS AND THEIR LIKELY ROLES

2-17. Likely participants in COIN operations include the following:

- U.S. military forces.
- Multinational (including HN) forces.
- U.S. Government agencies.
- Other governments' agencies.
- NGOs.
- IGOs.
- Multinational corporations and contractors.
- HN civil and military authorities (including local leaders).

U.S. MILITARY FORCES

2-18. Military forces play an extensive role in COIN efforts. Demanding and complex, COIN draws heavily on a broad range of the joint force's capabilities and requires a different mix of offensive, defensive, and stability operations from that expected in major combat operations. Air, land, and maritime components all contribute to successful operations and to the vital effort to separate insurgents from the people they seek to control. The Army and Marine Corps usually furnish the principal U.S. military contributions to COIN forces. Special operations forces (SOF) are particularly valuable due to their specialized capabilities:

- Civil affairs.
- Psychological operations.
- Intelligence.
- Language skills.
- Region-specific knowledge.

SOF can provide light, agile, high-capability teams able to operate discreetly in local communities. SOF can also conduct complex counterterrorist operations.

2-19. U.S. military forces are vastly capable. Designed primarily for conventional warfare, they nonetheless have the capabilities essential to successfully conduct COIN operations. The most important military assets in COIN are disciplined Soldiers and Marines with adaptive, self-aware, and intelligent leaders. Military forces also have capabilities particularly relevant to common COIN requirements. These capabilities include the following:

- Dismounted infantry.
- Human intelligence.
- Language specialists.
- Military police.
- Civil affairs.
- Engineers.
- Medical units.
- Logistic support.
- Legal affairs.
- Contracting elements.

All are found in the Army; most are found in the Marine Corps. To a limited degree, they are also found in the Air Force and Navy.

2-20. U.S. forces help HN military, paramilitary, and police forces conduct COIN operations, including area security and local security operations. U.S. forces provide advice and help find, disperse, capture, and defeat insurgent forces. Concurrently, they emphasize training HN forces to perform essential defense functions. These are the central tasks of foreign internal defense, a core SOF task. The current and more extensive national security demands for such efforts require conventional forces of all Services be prepared to contribute to establishing and training local security forces.

2-21. Land combat forces, supported by air and maritime forces, conduct full spectrum operations to disrupt or destroy insurgent military capabilities. Land forces use offensive operations to disrupt insurgent efforts to establish base areas and consolidate their forces. They conduct defensive operations to provide area and local security. They conduct stability operations to thwart insurgent efforts to control or disrupt people's lives and routine activities. In all applications of combat power, commanders first ensure that likely costs do not outweigh or undermine other more important COIN efforts.

2-22. Most valuable to long-term success in winning the support of the populace are the contributions land forces make by conducting stability operations. *Stability operations* is an overarching term encompassing various military missions, tasks, and activities conducted outside the United States in coordination with other instruments of national power to maintain or reestablish a safe and secure environment, provide essential governmental services, emergency infrastructure reconstruction, and humanitarian relief (JP 1-02).

Forces engaged in stability operations establish, safeguard, or restore basic civil services. They act directly but also support government agencies. Success in stability operations enables the local populace and HN government agencies to resume or develop the capabilities needed to conduct COIN operations and create conditions that permit U.S. military forces to disengage.

2-23. Military forces also use their capabilities to enable the efforts of nonmilitary participants. Logistics, transportation, equipment, personnel, and other assets can support interagency partners and other civilian organizations.

MULTINATIONAL (INCLUDING HOST-NATION) MILITARY FORCES

2-24. The U.S. Government prefers that U.S. military forces operate with other nations' forces and not alone. Thus, Soldiers and Marines normally function as part of a multinational force. In COIN operations, U.S. forces usually operate with the security forces of the local populace or host nation. Each multinational participant provides capabilities and strengths that U.S forces may not have. Many other countries' military forces bring cultural backgrounds, historical experiences, and other capabilities that can be particularly valuable to COIN efforts.

2-25. However, nations join coalitions for various reasons. Although the missions of multinational partners may appear similar to those of the United States, rules of engagement, home-country policies, and sensitivities may differ among partners. U.S. military leaders require a strong cultural and political awareness of HN and other multinational military partners.

NONMILITARY COUNTERINSURGENCY PARTICIPANTS

2-26. Many nonmilitary organizations may support a host nation as it confronts an insurgency. Some of these organizations are discussed below. (JP 3-08, volume II, discusses nonmilitary organizations often associated with joint operations.)

U.S. Government Organizations

2-27. Commanders' situational awareness includes being familiar with other U.S. Government organizations participating in the COIN effort and aware of their capabilities. Planning includes determining which organizations are supporting the force or operating in their area of operations (AO). Commanders and leaders of other U.S. Government organizations should collaboratively plan and coordinate actions to avoid conflict or duplication of effort. Within the U.S. Government, key organizations include—

- Department of State.
- U.S. Agency for International Development (USAID).
- Central Intelligence Agency.
- Department of Justice.
- Drug Enforcement Administration (under Department of Justice).
- Department of the Treasury.
- Department of Homeland Security.
- Department of Energy.
- Department of Agriculture.
- Department of Commerce.
- Department of Transportation.
- U.S. Coast Guard (under Department of Homeland Security).
- Federal Bureau of Investigation (under Department of Justice).
- Immigration Customs Enforcement (under Department of Homeland Security).

Other Governments' Agencies

2-28. Agencies of other national governments (such as ministries of defense, foreign affairs, development, and justice) are likely to actively participate in COIN operations. The list of possible participants from other countries is too long to list. Leaders of U.S. contingents must work closely with their multinational counterparts to become familiar with agencies that may operate in their AO. To the degree possible, military leaders should use U.S. civilian representatives to establish appropriate relationships and awareness of their multinational counterparts.

Nongovernmental Organizations

2-29. Joint doctrine defines a *nongovernmental organization* as a private, self-governing, not-for-profit organization dedicated to alleviating human suffering; and/or promoting education, health care, economic development, environmental protection, human rights, and conflict resolution; and/or encouraging the establishment of democratic institutions and civil society (JP 1-02). There are several thousand NGOs of many different types. Their activities are governed by their organizing charters and their members' motivations. Some NGOs receive at least part of their funding from national governments or IGOs. Some may become implementing partners in accordance with grants or contracts. (For example, USAID provides some NGO funding.) In these cases, the funding organization often gains oversight and authority over how the funds are used.

2-30. Some NGOs maintain strict independence from governments and belligerents and do not want to be seen directly associating with military forces. Gaining the support of and coordinating operations with these NGOs can be difficult. Establishing basic awareness of these groups and their activities may be the most commanders can achieve. NGOs play important roles in resolving insurgencies, however. Many NGOs arrive before military forces and remain afterwards. They can support lasting stability. To the greatest extent possible, commanders try to complement and not override their capabilities. Building a complementary, trust-based relationship is vital.

2-31. Examples of NGOs include—

- International Committee of the Red Cross.
- World Vision.
- Medecins sans Frontieres (Doctors Without Borders).
- Cooperative for Assistance and Relief Everywhere (CARE).
- Oxford Committee for Famine Relief (OXFAM).
- Save the Children.
- Mercy Corps.
- Academy for Educational Development.

Intergovernmental Organizations

2-32. Joint doctrine defines an *intergovernmental organization* as an organization created by a formal agreement (for example, a treaty) between two or more governments. It may be established on a global, regional, or functional basis for wide-ranging or narrowly defined purposes. IGOs are formed to protect and promote national interests shared by member states (JP 1-02). The most notable IGO is the United Nations (UN). Regional organizations like the Organization of American States and European Union may be involved in some COIN operations. The UN in particular has many subordinate and affiliated agencies active worldwide. Depending on the situation and HN needs, any number of UN organizations may be present, such as the following:

- Office of the Chief of Humanitarian Affairs.
- Department of Peacekeeping Operations.
- World Food Program.
- UN Refugee Agency (known as UNHCR, the acronym for its director, the UN High Commissioner for Refugees).
- UN High Commissioner for Human Rights.
- UN Development Program.

Multinational Corporations and Contractors

2-33. Multinational corporations often engage in reconstruction, economic development, and governance activities. At a minimum, commanders should know which companies are present in their AO and where those companies are conducting business. Such information can prevent fratricide and destruction of private property.

2-34. Recently, private contractors from firms providing military-related services have become more prominent in theaters of operations. This category includes armed contractors providing many different security services to the U.S. Government, NGOs, and private businesses. Many businesses market expertise in areas related to supporting governance, economics, education, and other aspects of civil society as well. Providing capabilities similar to some NGOs, these firms often obtain contracts through government agencies.

2-35. When contractors or other businesses are being paid to support U.S. military or other government agencies, the principle of unity of command should apply. Commanders should be able to influence contractors' performance through U.S. Government contract supervisors. When under contract to the United States, contractors should behave as an extension of the organizations or agencies for which they work. Commanders should identify contractors operating in their AO and determine the nature of their contract, existing accountability mechanisms, and appropriate coordination relationships.

Host-Nation Civil Authorities

2-36. Sovereignty issues are among the most difficult for commanders conducting COIN operations, both in regard to forces contributed by other nations and by the host nation. Often, commanders are required to lead through coordination, communication, and consensus, in addition to traditional command practices. Political sensitivities must be acknowledged. Commanders and subordinates often act as diplomats as well as warriors. Within military units, legal officers and their staffs are particularly valuable for clarifying legal arrangements with the host nation. To avoid adverse effects on operations, commanders should address all sovereignty issues through the chain of command to the U.S. Ambassador. As much as possible, sovereignty issues should be addressed before executing operations. Examples of key sovereignty issues include the following:

- Aerial ports of debarkation.
- Basing.
- Border crossings.
- Collecting and sharing information.
- Protection (tasks related to preserving the force).
- Jurisdiction over members of the U.S. and multinational forces.
- Location and access.
- Operations in the territorial waters, both sea and internal.
- Overflight rights.
- Police operations, including arrest, detention, penal, and justice authority and procedures.
- Railheads.
- Seaports of debarkation.

2-37. Commanders create coordinating mechanisms, such as committees or liaison elements, to facilitate cooperation and build trust with HN authorities. HN military or nonmilitary representatives should have leading roles in such mechanisms. These organizations facilitate operations by reducing sensitivities and misunderstandings while removing impediments. Sovereignty issues can be formally resolved with the host nation by developing appropriate technical agreements to augment existing or recently developed status of forces agreements. In many cases, security assistance organizations, NGOs, and IGOs have detailed local knowledge and reservoirs of good will that can help establish a positive, constructive relationship with the host nation.

2-38. Coordination and support should exist down to local levels (such as villages and neighborhoods). Soldiers and Marines should be aware of the political and societal structures in their AOs. Political structures usually have designated leaders responsible to the government and people. However, the societal structure may include informal leaders who operate outside the political structure. These leaders may be—

- Economic (such as businessmen).
- Theological (such as clerics and lay leaders).
- Informational (such as newspaper publishers or journalists).
- Family based (such as elders or patriarchs).

Some societal leaders may emerge due to charisma or other intangible influences. Commanders should identify the key leaders and the manner in which they are likely to influence COIN efforts.

KEY RESPONSIBILITIES IN COUNTERINSURGENCY

2-39. Participants best qualified and able to accomplish nonmilitary tasks are not always available. The realistic division of labor does not match the preferred division of labor. In those cases, military forces perform those tasks. Sometimes forces have the skills required; other times they learn them during execution.

PREFERRED DIVISION OF LABOR

2-40. In COIN it is always preferred for civilians to perform civilian tasks. Whenever possible, civilian agencies or individuals with the greatest applicable expertise should perform a task. Legitimate local authorities should receive special preference. There are many U.S. agencies and civilian IGOs with more expertise in meeting the fundamental needs of a population under assault than military forces have; however, the ability of such agencies to deploy to foreign countries in sustainable numbers and with ready access to necessary resources is usually limited. The violence level in the AO also affects civilian agencies' ability to operate. The more violent the environment, the more difficult it is for civilians to operate effectively. Hence, the preferred or ideal division of labor is frequently unattainable. The more violent the insurgency, the more unrealistic is this preferred division of labor.

REALISTIC DIVISION OF LABOR

2-41. By default, U.S. and multinational military forces often possess the only readily available capability to meet many of the local populace's fundamental needs. Human decency and the law of war require land forces to assist the populace in their AOs. Leaders at all levels prepare to address civilian needs. Commanders identify people in their units with regional and interagency expertise, civil-military competence, and other critical skills needed to support a local populace and HN government. Useful skill sets may include the following:

- Knowledge, cultural understanding, and appreciation of the host nation and region.
- Functional skills needed for interagency and HN coordination (for example, liaison, negotiation, and appropriate social or political relationships).
- Language skills needed for coordination with the host nation, NGOs, and multinational partners.
- Knowledge of basic civic functions such as governance, infrastructure, public works, economics, and emergency services.

2-42. U.S. Government agencies and IGOs rarely have the resources and capabilities needed to address all COIN tasks. Success requires adaptable leaders who prepare to perform required tasks with available resources. These leaders understand that long-term security cannot be imposed by military force alone; it requires an integrated, balanced application of effort by all participants with the goal of supporting the local populace and achieving legitimacy for the HN government. David Galula wisely notes, "To confine soldiers to purely military functions while urgent and vital tasks have to be done, and nobody else is available to undertake them, would be senseless. The soldier must then be prepared to become...a social worker, a civil engineer, a schoolteacher, a nurse, a boy scout. But only for as long as he cannot be replaced, for it is better to entrust civilian tasks to civilians." Galula's last sentence is important. Military forces can perform civilian tasks but often not as well as the civilian agencies with people trained in those skills. Further, military forces performing civilian tasks are not performing military tasks. Diverting them from those tasks should be a temporary measure, one taken to address urgent circumstances.

TRANSITIONS

2-43. Regardless of the division of labor, an important recurring feature of COIN is transitioning responsibility and participation in key LLOs. As consistently and conscientiously as possible, military leaders ensure continuity in meeting the needs of the HN government and local populace. The same general guidelines governing battle handovers apply to COIN transitions. Whether the transition is between military units or from a military unit to a civilian agency, all involved must clearly understand the tasks and responsibilities being passed. Maintaining unity of effort is particularly important during transitions, especially between organizations of different capabilities and capacities. Relationships tend to break down during transitions. A transition is not a single event where all activity happens at once. It is a rolling process of little handoffs between different actors along several streams of activities. There are usually multiple transitions for any one stream of activity over time. Using the coordination mechanisms discussed below can help create and sustain the links that support effective transitions without compromising unity of effort.

CIVILIAN AND MILITARY INTEGRATION MECHANISMS

2-44. Applying the principle of unity of effort is possible in many organizational forms. The first choice should be to identify existing coordination mechanisms and incorporate them into comprehensive COIN efforts. This includes existing U.S. Government, multinational, and HN mechanisms. Context is extremely important. Although many of these structures exist and are often employed in other types of missions (such as peacekeeping or humanitarian relief), there is an acute and fundamental difference in an insurgency environment. The nature of the conflict and its focus on the populace make civilian and military unity a critical enabling aspect of a COIN operation. The following discussion highlights some of the well-established, general mechanisms for civilian and military integration. Many civil-military organizations and mechanisms have been created for specific missions. Although the names and acronyms differ, in their general outlines they usually reflect the concepts discussed below.

2-45. The U.S. Government influences events worldwide by effectively employing the instruments of national power: diplomatic, informational, military, and economic. These instruments are coordinated by the appropriate executive branch officials, often with assistance from the National Security Council (NSC) staff.

2-46. The NSC is the President's principal forum for considering national security and foreign policy matters. It serves as the President's principal means for coordinating policy among various interagency organizations. At the strategic level, the NSC directs the creation of the interagency political-military plan for COIN. The NSC staff, guided by the deputies and principals, assists in integrating interagency processes to develop the plan for NSC approval. (See JP 1.)

JOINT INTERAGENCY COORDINATION GROUP

2-47. *Interagency coordination,* within the context of Department of Defense involvement, is the coordination that occurs between elements of Department of Defense and engaged U.S. Government agencies for the purpose of achieving an objective (JP 1-02). Joint interagency coordination groups (JIACGs) help combatant commanders conduct COIN operations by providing interagency support of plans, operations, contingencies, and initiatives. The goal of a JIACG is to provide timely, usable information and advice from an interagency perspective to the combatant commander by information sharing, integration, synchronization, training, and exercises. JIACGs may include representatives from other federal departments and agencies and state and local authorities, as well as liaison officers from other commands and DOD components. The interagency representatives and liaison officers are the subject matter experts for their respective agencies and commands. They provide the critical bridge between the combatant commander and interagency organizations. (See JP 3-08, volume I.)

COUNTRY TEAM

2-48. At the HN level, the U.S. country team is the primary interagency coordinating structure for COIN. (See figure 2-1.) The country team is the senior in-country coordinating and supervising body, headed by the U.S. chief of mission, usually the Ambassador. It is composed of the senior member of each represented department or agency. In a foreign country, the chief of mission is the highest U.S. civil authority.

The Foreign Service Act assigns the chief of mission to a foreign country responsibility for the direction, coordination, and supervision of all government executive branch employees in that country except for service members and employees under the command of a U.S. area military commander. As the senior U.S. Government official permanently assigned in the host nation, the chief of mission is responsible to the President for policy oversight of all United States government programs. The chief of mission leads the country team and is responsible for integrating U.S. efforts in support of the host nation. As permanently established interagency organizations, country teams represent a priceless COIN resource. They often provide deep reservoirs of local knowledge and interaction with the HN government and population.

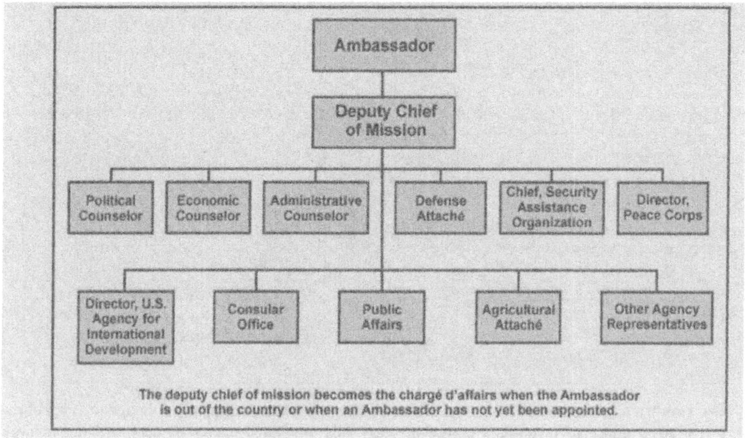

Figure 2-1. Sample country team

2-49. The more extensive the U.S. participation is in a COIN and the more dispersed U.S. forces are throughout a country, the greater the need for additional mechanisms to extend civilian oversight and assistance. However, given the limited resources of the Department of State and the other U.S. Government agencies, military forces often represent the country team in decentralized and diffuse operational environments. Operating with a clear understanding of the guiding political aims, members of the military at all levels must be prepared to exercise judgment and act without the benefit of immediate civilian oversight and control. At each subordinate political level of the HN government, military and civilian leaders should establish a coordinating structure, such as an area coordination center or civil-military operations center (CMOC), that includes representatives of the HN government and security forces, as well as U.S. and multinational forces and agencies. CMOCs facilitate the integration of military and political actions. Below the national level, additional structures where military commanders and civilian leaders can meet directly with local leaders to discuss issues may be established. Where possible, IGOs and NGOs should be encouraged to participate in coordination meetings to ensure their actions are integrated with military and HN plans. (See JP 3-07.1 for additional information about COIN planning and coordination organizations.)

2-50. In practice, the makeup of country teams varies widely, depending on the U.S. departments and agencies represented in country, the desires of the Ambassador, and the HN situation. During COIN, country team members meet regularly to coordinate U.S. Government diplomatic, informational, military, and economic activities in the host nation to ensure unity of effort. The interagency representatives usually include at least the following:

- Departments of State, Defense, Justice, and Treasury.
- USAID.
- Central Intelligence Agency.
- Drug Enforcement Administration.

Participation of other U.S. Government organizations depends on the situation.

2-51. In almost all bilateral missions, DOD is represented on the country team by the U.S. defense attache's office or the security assistance organization. They are key military sources of information for interagency coordination in foreign countries. (Security assistance organizations are called by various names, such as the office of defense cooperation, security assistance office, or military group. The choice is largely governed by HN preference.)

Provincial Reconstruction Teams in Afghanistan

A model for civil-military cooperation is the provincial reconstruction teams (PRTs) first fielded in 2003 in Afghanistan. PRTs were conceived as a means to extend the reach and enhance the legitimacy of the central government into the provinces of Afghanistan at a time when most assistance was limited to the nation's capital. Though PRTs were staffed by a number of coalition and NATO allied countries, they generally consisted of 50 to 300 troops as well as representatives from multinational development and diplomatic agencies. Within U.S. PRTs, USAID and Department of State leaders and the PRT commander formed a senior team that coordinated the policies, strategies, and activities of each agency towards a common goal. In secure areas, PRTs maintained a low profile. In areas where coalition combat operations were underway, PRTs worked closely with maneuver units and local government entities to ensure that shaping operations achieved their desired effects. Each PRT leadership team received tremendous latitude to determine its own strategy. However, each PRT used its significant funding and diverse expertise to pursue activities that fell into one of three general logical lines of operations: pursue security sector reform, build local governance, or execute reconstruction and development.

2-52. The country team determines how the United States can effectively apply interagency capabilities to assist a HN government in creating a complementary institutional capacity to deal with an insurgency. Efforts to support local officials and build HN capacity must be integrated with information operations so HN citizens are aware of their government's efforts. In addition, interagency capabilities must be applied at the tactical level to give commanders access to options such capabilities make available. The Civil Operations and Revolutionary—later Rural—Development Support approach developed during the Vietnam War provides a positive example of integrated civilian and military structures that reached every level of the COIN effort.

CORDS and Accelerated Pacification in Vietnam

During the Vietnam War, one of the most valuable and successful elements of COIN was the Civil Operations and Revolutionary—later Rural—Development Support (CORDS) program. CORDS was created in 1967 to integrate U.S. civilian and military support of the South Vietnamese government and people. CORDS achieved considerable success in supporting and protecting the South Vietnamese population and in undermining the communist insurgents' influence and appeal, particularly after implementation of accelerated pacification in 1968.

Pacification was the process by which the government asserted its influence and control in an area beset by insurgents. It included local security efforts, programs to distribute food and medical supplies, and lasting reforms (like land redistribution). In 1965, U.S. civilian contributions to pacification consisted of several civilian agencies (among them, the Central Intelligence Agency, Agency for International Development, U.S. Information Service, and Department of State). Each developed its own programs. Coordination was uneven. The U.S. military contribution to pacification consisted of thousands of advisors. By early 1966, there were military advisory teams in all of South Vietnam's 44 provinces and most of its 243 districts. But there were two separate chains of command for military and civilian pacification efforts, making it particularly difficult for the civilian-run pacification program to function.

In 1967, President Lyndon B. Johnson established CORDS within the Military Assistance Command, Vietnam (MACV), which was commanded by General William Westmoreland, USA. The purpose of CORDS was to establish closer integration of civilian and military efforts. Robert Komer was appointed to run the program, with a three-star-equivalent rank. Civilians, including an assistant chief of staff for CORDS, were integrated into military staffs at all levels. This placed civilians in charge of military personnel and resources. Komer was energetic, strong-willed, and persistent in getting the program started. Nicknamed "Blowtorch Bob" for his aggressive style, Komer was modestly successful in leading improvements in pacification before the 1968 Tet offensive.

In mid-1968, the new MACV commander, General Creighton Abrams, USA, and his new civilian deputy, William Colby, used CORDS as the implementing mechanism for an accelerated pacification program that became the priority effort for the United States. Significant allocations of personnel helped make CORDS effective. In this, the military's involvement was key. In September 1969—the high point of the pacification effort in terms of total manpower—there were 7,601 advisors assigned to province and district pacification teams. Of these 6,464 were military.

The effectiveness of CORDS was a function of integrated civilian and military teams at every level of society in Vietnam. From district to province to national level, U.S. advisors and U.S. interagency partners worked closely with their Vietnamese counterparts. The entire effort was well established under the direction of the country team, led by Ambassador Ellsworth Bunker. General Abrams and his civilian deputy were clear in their focus on pacification as the priority and ensured that military and civilian agencies worked closely together. Keen attention was given to the ultimate objective of serving the needs of the local populace. Success in meeting basic needs of the populace led, in turn, to improved intelligence that facilitated an assault on the Viet Cong political infrastructure. By early 1970, statistics indicated that 93 percent of South Vietnamese lived in "relatively secure" villages, an increase of almost 20 percent from the middle of 1968. By 1972, pacification had largely uprooted the insurgency from among the South Vietnamese population and forced the communists to rely more heavily on infiltrating conventional forces from North Vietnam and employing them in irregular and conventional operations.

In 1972, South Vietnamese forces operating with significant support from U.S. airpower defeated large-scale North Vietnamese conventional attacks. Unfortunately, a North Vietnamese conventional assault succeeded in 1975 after the withdrawal of U.S. forces, ending of U.S. air support, and curtailment of U.S. funding to South Vietnam.

Pacification, once it was integrated under CORDS, was generally led, planned, and executed well. CORDS was a successful synthesis of military and civilian efforts. It is a useful model to consider for other COIN operations.

CIVIL-MILITARY OPERATIONS CENTER

2-53. Another mechanism for bringing elements together for coordination is the CMOC. CMOCs can be established at all levels of command. CMOCs coordinate the interaction of U.S. and multinational military forces with a wide variety of civilian agencies. A CMOC is not designed, nor should it be used as, a command and control element. However, it is useful for transmitting the commander's guidance to other agencies, exchanging information, and facilitating complementary efforts.

2-54. Overall management of a CMOC may be assigned to a multinational force commander, shared by a U.S. and a multinational commander, or shared by a U.S. commander and a civilian agency head. A CMOC can be used to build on-site, interagency coordination to achieve unity of effort. There is no established CMOC structure; its size and composition depend on the situation. However, CMOCs are organic to Army civil affairs organizations, from civil affairs command to company. Senior civil affairs officers normally serve as the CMOC director and deputy director. Other military participants usually include civil affairs, legal, operations, logistic, engineering, and medical representatives of the supported headquarters. Civilian members of a CMOC may include representatives of the following:

- U.S. Government organizations.
- Multinational partners.
- IGOs.
- HN or other local organizations.
- NGOs.

(For more information on CMOCs, see FM 3-05.401/CRP 3-33.1A.)

TACTICAL-LEVEL INTERAGENCY CONSIDERATIONS

2-55. Tactical units may find interagency expertise pushed to their level when they are responsible for large AOs in a COIN environment. Tactical units down to company level must be prepared to integrate their efforts with civilian organizations.

2-56. To ensure integration of interagency capabilities, units should coordinate with all interagency representatives and organizations that enter their AO. Despite the best efforts to coordinate, the fog and friction inherent in COIN will often lead to civilian organizations entering an AO without prior coordination. (Table 2-1 is a suggested list for coordinating with interagency and other nonmilitary organizations.)

Table 2-1. Example interagency coordination checklist

• Identify organizational structures and leadership. • Identify key objectives, responsibilities, capabilities, and programs. • Develop common courses of action or options for inclusion in planning, movement coordination, and security briefings. • Determine how to ensure coordination and communications before and during the execution of the organization's activities in the unit's area of operations. • Develop relationships that enable the greatest possible integration. • Assign liaison officers to the most important civilian organizations. • Define problems in clear and unambiguous terms. • Determine the intended duration of operations. • Determine the location of bases of operations. • Determine the number, names, and descriptions of personnel. • Determine the type, color, number, and license numbers of civilian vehicles. • Identify other agency resources in the area of operations. • Identify local groups and the agencies with whom they are working. • Establish terms of reference or operating procedures, especially in the event of incidents that result in casualties. • Identify funding for interagency projects.

SUMMARY

2-57. President John F. Kennedy noted, "You [military professionals] must know something about strategy and tactics and.. .logistics, but also economics and politics and diplomacy and history. You must know everything you can know about military power, and you must also understand the limits of military power. You must understand that few of the important problems of our time have. ..been finally solved by military power alone." Nowhere is this insight more relevant than in COIN. Successful COIN efforts require unity of effort in bringing all instruments of national power to bear. Civilian agencies can contribute directly to military operations, particularly by providing information. That theme is developed further in the next chapter.

Chapter 3

Intelligence in Counterinsurgency

Everything good that happens seems to come from good intelligence.

General Creighton W. Abrams Jr., USA, 1970

Effective, accurate, and timely intelligence is essential to the conduct of any form of warfare. This maxim applies especially to counterinsurgency operations; the ultimate success or failure of the mission depends on the effectiveness of the intelligence effort. This chapter builds upon previous concepts to further describe insurgencies, requirements for intelligence preparation of the battlefield and predeployment planning, collection and analysis of intelligence in counterinsurgency, intelligence fusion, and general methodology for integrating intelligence with operations. This chapter does not supersede processes in U.S. military doctrine (see FM 2-0, FM 34-130/FMFRP 3-23-2, and FMI 2-91.4) but instead provides specific guidance for counterinsurgency.

SECTION I - INTELLIGENCE CHARACTERISTICS IN COUNTERINSURGENCY

3-1. Counterinsurgency (COIN) is an intelligence-driven endeavor. The function of intelligence in COIN is to facilitate understanding of the operational environment, with emphasis on the populace, host nation, and insurgents. Commanders require accurate intelligence about these three areas to best address the issues driving the insurgency. Both insurgents and counterinsurgents require an effective intelligence capability to be successful. Both attempt to create and maintain intelligence networks while trying to neutralize their opponent's intelligence capabilities.

3-2. Intelligence in COIN is about people. U.S. forces must understand the people of the host nation, the insurgents, and the host-nation (HN) government. Commanders and planners require insight into cultures, perceptions, values, beliefs, interests and decision-making processes of individuals and groups. These requirements are the basis for collection and analytical efforts.

3-3. Intelligence and operations feed each other. Effective intelligence drives effective operations. Effective operations produce information, which generates more intelligence. Similarly, ineffective or inaccurate intelligence produces ineffective operations, which produce the opposite results.

3-4. All operations have an intelligence component. All Soldiers and Marines collect information whenever they interact with the populace. Operations should therefore always include intelligence collection requirements.

3-5. Insurgencies are local. They vary greatly in time and space. The insurgency one battalion faces will often be different from that faced by an adjacent battalion. The mosaic nature of insurgencies, coupled with the fact that all Soldiers and Marines are potential intelligence collectors, means that all echelons both produce and consume intelligence. This situation results in a bottom-up flow of intelligence. This pattern also means that tactical units at brigade and below require a lot of support for both collection and analysis, as their organic intelligence structure is often inadequate.

3-6. COIN occurs in a joint, interagency, and multinational environment at all echelons. Commanders and staffs must coordinate intelligence collection and analysis with foreign militaries, foreign and U.S. intelligence services, and other organizations.

SECTION II - PREDEPLOYMENT PLANNING AND INTELLIGENCE PREPARATION OF THE BATTLEFIELD

3-7. *Intelligence preparation of the battlefield* is the systematic, continuous process of analyzing the threat and environment in a specific geographic area. Intelligence preparation of the battlefield (IPB) is designed to support the staff estimate and military decision-making process. Most intelligence requirements are generated as a result of the IPB process and its interrelation with the decision-making process (FM 34-130). Planning for deployment begins with a thorough mission analysis, including IPB. IPB is accomplished in four steps:

- Define the operational environment.
- Describe the effects of the operational environment.
- Evaluate the threat.
- Determine threat courses of action.

3-8. The purpose of planning and IPB before deployment is to develop an understanding of the operational environment. This understanding drives planning and predeployment training. Predeployment intelligence must be as detailed as possible. It should focus on the host nation, its people, and insurgents in the area of operations (AO). Commanders and staffs use predeployment intelligence to establish a plan for addressing the underlying causes of the insurgency and to prepare their units to interact with the populace appropriately. The goal of planning and preparation is for commanders and their subordinates not to be surprised by what they encounter in theater.

3-9. IPB in COIN operations follows the methodology described in FM 34-130/FMFRP 3-23-2. However, it places greater emphasis on civil considerations, especially people and leaders in the AO, than does IPB for conventional operations. IPB is continuous and its products are revised throughout the mission. Nonetheless, predeployment products are of particular importance for the reasons explained above. Whenever possible, planning and preparation for deployment includes a thorough and detailed IPB. IPB in COIN requires personnel to work in areas like economics, anthropology, and governance that may be outside their expertise. Therefore, integrating staffs and drawing on the knowledge of nonintelligence personnel and external experts with local and regional knowledge are critical to effective preparation.

3-10. Deployed units are the best sources of intelligence. Deploying units should make an effort to reach forward to deployed units. The Secret Internet Protocol Router Network (SIPRNET) allows deploying units to immerse themselves virtually in the situation in theater. Government agencies, such as the Department of State, U.S. Agency for International Development, and intelligence agencies, can often provide country studies and other background information as well.

3-11. *Open-source intelligence* is information of potential intelligence value that is available to the general public (JP 1-02). It is important to predeployment IPB. In many cases, background information on the populations, cultures, languages, history, and governments of states in an AO is in open sources. Open sources include books, magazines, encyclopedias, Web sites, tourist maps, and atlases. Academic sources, such as journal articles and university professors, can also be of great benefit.

DEFINE THE OPERATIONAL ENVIRONMENT

3-12. The *operational environment* is a composite of the conditions, circumstances, and influences that affect the employment of capabilities and bear on the decisions of the commander. (JP 1-02). At the tactical and operational levels, defining the operational environment involves defining a unit's AO and determining an area of interest. The *area of interest* is area of concern to the commander, including the area of influence, areas adjacent thereto, and extending into enemy territory to the objectives of current or planned operations. This area also includes areas occupied by enemy forces who could jeopardize the accomplishment of the mission (JP 1-02).

3-13. AOs may be relatively static, but people and information flow through AOs continuously. Therefore, when defining an area of interest, commanders take into account physical geography and civil considerations, particularly human factors. AOs often cut across physical lines of communications, such as roads, as well as areas that are tribally, economically, or culturally defined. For instance, tribal and family groups in Iraq and Afghanistan cross national borders into neighboring countries. The cross-border ties allow insurgents safe haven outside of their country and aid them in cross-border smuggling. The area of interest can be large relative to the AO; it must often account for various influences that affect the AO, such as—

- Family, tribal, ethnic, religious, or other links that go beyond the AO.
- Communication links to other regions.
- Economic links to other regions.
- Media influence on the local populace, U.S. public, and multinational partners.
- External financial, moral, and logistic support for the enemy.

3-14. At the combatant command level, the area of interest may be global, if, for example, enemy forces have an international financial network or are able to affect popular support within the U.S. or multinational partners. At the tactical level, commands must be aware of activities in neighboring regions and population centers that affect the population in their AO.

3-15. As explained in chapter 2, another consideration for predeployment planning and defining the operational environment is understanding the many military and nonmilitary organizations involved in the COIN effort. Intelligence planners determine the non-Department of Defense (DOD) agencies, multinational forces, nongovernmental organizations, and HN organizations in the AO. Knowledge of these organizations is needed to establish working relationships and procedures for sharing information. These relationships and procedures are critical to developing a comprehensive common operational picture and enabling unity of effort.

DESCRIBE THE EFFECTS OF THE OPERATIONAL ENVIRONMENT

3-16. This IPB step involves developing an understanding of the operational environment and is critical to the success of operations. It includes—

- Civil considerations, with emphasis on the people, history, and HN government in the AO.
- Terrain analysis (physical geography), with emphasis on the following:
  - Complex terrain.
  - Suburban and urban terrain.
  - Key infrastructure.
  - Lines of communications.
- Weather analysis, focusing on how weather affects the populace's activities, such as agriculture, smuggling, or insurgent actions.

3-17. Including all staff members in this step improves the knowledge base used to develop an understanding of the AO. For instance, civil affairs personnel receive training in analysis of populations, cultures, and economic development. These Soldiers and Marines can contribute greatly to understanding civil considerations. As another example, foreign area officers have linguistic, historical, and cultural knowledge about particular regions and have often lived there for extended periods.

3-18. The products that result from describing the effects of the operational environment influence operations at all echelons. The description informs political activities and economic policies of combatant commanders. It drives information operations (IO) and civil-military operations planning. The knowledge gained affects the way Soldiers and Marines interact with the populace.

3-19. *Civil considerations* are how the manmade infrastructure, civilian institutions, and attitudes and activities of the civilian leaders, populations, and organizations within an area of operations influence the conduct of military operations (FM 6-0). Civil considerations form one of the six categories into which relevant information is grouped for military operations. (The glossary lists the other five categories under the entry for METT-TC.) Civil considerations comprise six characteristics, expressed in the memory aid ASCOPE: areas, structures, capabilities, organizations, people, and events. (Paragraphs B-10 through B-27 discuss these characteristics.) While all characteristics of civil considerations are important, understanding the people is particularly important in COIN. In order to evaluate the people, the following six socio-cultural factors should be analyzed:

- Society.
- Social structure.
- Culture.
- Language.
- Power and authority.
- Interests.

This analysis may also identify information related to areas, structures, organizations, and events.

SOCIETY

3-20. A *society* can be defined as a population whose members are subject to the same political authority, occupy a common territory, have a common culture, and share a sense of identity. A society is not easily created or destroyed, but it is possible to do so through genocide or war.

3-21. No society is homogeneous. A society usually has a dominant culture, but can also have a vast number of secondary cultures. Different societies may share similar cultures, such as Canada and the United States do. Societies are not static, but change over time.

3-22. Understanding the societies in the AO allows counterinsurgents to achieve objectives and gain support. Commanders also consider societies outside the AO whose actions, opinions, or political influence can affect the mission.

SOCIAL STRUCTURE

3-23. Each society is composed of both social structure and culture. Social structure refers to the relations among groups of persons within a system of groups. Social structure is persistent over time. It is regular and continuous despite disturbances. The relationships among the parts hold steady, even as groups expand or contract. In a military organization, for example, the structure consists of the arrangement into groups like divisions, battalions, and companies. In a society, the social structure includes groups, institutions, organizations, and networks. Social structure involves the following:

- Arrangement of the parts that constitute society.
- Organization of social positions.
- Distribution of people within those positions.

Groups

3-24. A *group* is two or more people regularly interacting on the basis of shared expectations of others' behavior and who have interrelated statuses and roles. A social structure includes a variety of groups. These groups may be racial, ethnic, religious, or tribal. There may also be other kinship-based groups.

3-25. A *race* is a human group that defines itself or is defined by other groups as different by virtue of innate physical characteristics. Biologically, there is no such thing as race among human beings; race is a social category.

3-26. An *ethnic group* is a human community whose learned cultural practices, language, history, ancestry, or religion distinguish them from others. Members of ethnic groups see themselves as different from other groups in a society and are recognized as such by others. Religious groups may be subsets of larger ethnic groups. An ethnic group may contain members of different religions. For example, some Kurds are Muslim, while others are Christian. Other ethnic groups may be associated with a particular religion—such as Sri Lankan Sinhalese, who are almost exclusively Buddhist. Other religious groups have members of many different ethnicities. A prominent example of such a group is the Roman Catholic faith.

3-27. *Tribes* are generally defined as autonomous, genealogically structured groups in which the rights of individuals are largely determined by their ancestry and membership in a particular lineage. Tribes are essentially adaptive social networks organized by extended kinship and descent with common needs for physical and economic security.

3-28. Understanding the composition of groups in the AO is vital for effective COIN operations. This is especially true when insurgents organize around racial, ethnic, religious, or tribal identities. Furthermore, tensions or hostilities between groups may destabilize a society and provide opportunities for insurgents. Commanders should thus identify powerful groups both inside and outside their AO and obtain the following information about them:

- Formal relationships (such as treaties or alliances) between groups.
- Informal relationships (such as tolerance or friction) between groups.
- Divisions and cleavages between groups.
- Cross-cutting ties (for example, religious alignments that cut across ethnic differences) between groups.

In some cases, insurgent leaders and their followers may belong to separate groups. In others, the bulk of the population may differ from the insurgents. These characteristics may suggest courses of action aimed at reinforcing or widening seams among insurgents or between insurgents and the population.

Networks

3-29. *Networks* may be an important aspect of a social structure as well as within the insurgent organization (See paragraphs 1-94 and B-29 through B-56). Common types of networks include elite networks, prison networks, worldwide ethnic and religious communities, and neighborhood networks. Networks can have many purposes: economic, criminal, and emotional. Effective social network analysis, discussed below, considers the structure of a network and the nature of interactions between network actors.

Institutions

3-30. Groups engaged in patterned activity to complete a common task are called *institutions.* Educational institutions bring together groups and individuals whose statuses and roles concern teaching and learning. Military institutions bring together groups and individuals whose statuses and roles concern defense and security. Institutions, the basic building blocks of societies, are continuous through many generations. They continue to exist, even when the individuals who compose them are replaced.

Organizations

3-31. *Organizations,* both formal and informal, are institutions with the following characteristics:

- Bounded membership.
- Defined goals.
- Established operations.
- Fixed facilities or meeting places.
- Means of financial or logistic support.

3-32. Planners can generally group organizations into the following categories:

- **Communicating** organizations have the power to influence a population's perceptions.
- **Religious** organizations regulate norms, restrain or empower activities, reaffirm worldviews, and provide social support. A religious organization differs from a religious group. A religious group is a general category, such as Christian; a religious organization is a specific community, such as the Episcopal Church.
- **Economic** organizations provide employment, help regulate and stabilize monetary flow, assist in development, and create social networks.
- **Social** organizations provide support to the population, create social networks, and can influence ideologies. Examples include schools, civil society groups, and sports teams.

Organizations may belong to more than one category. For instance, an influential religious organization may also be a communicating organization.

3-33. Organizations may control, direct, restrain, or regulate the local populace. Thus, commanders should identify influential organizations both inside and outside of their AO. Commanders need to know which members of what groups belong to each organization and how their activities may affect military operations. The next step is to determine how these organizations affect the local populace, whose interests they fulfill, and what role they play in influencing local perceptions.

Roles and Statuses

3-34. Understanding a society requires identifying the most common roles, statuses, and institutions within the society. Individuals in a society interact as members with social positions. These positions are referred to as statuses. Most societies associate particular statuses with particular social groups, such as family, lineage, ethnicity, or religion. Statuses may be achieved by meeting certain criteria or may be ascribed by birth. Statuses are often reciprocal, such as that of husband and wife or teacher and student. Every status carries a cluster of expected behaviors known as a *role,* which includes how a person of that status is expected to think, feel, and act. A status may also include expectations about how others should treat a person of that status. Thus, in American society parents (status) have the obligation to care for their children (role) and the right to discipline them (role).

Social Norms

3-35. Violation of a role prescribed by a given status, such as failing to feed one's children, results in social disapproval. The standard of conduct for social roles is known as a *social norm.* A social norm is what people are expected to do or should do, rather than what people actually do. Norms may be either moral (incest prohibition, homicide prohibition) or customary (prayer before a meal, removing shoes before entering a house). When a person's behavior does not conform to social norms, the person may be sanctioned. Understanding the roles, statuses, and social norms of groups within an AO can clarify expected behavior and provide guidelines on how to act. Some norms that may impact military operations include the following:

- The requirement for revenge if honor is lost.
- Appropriate treatment of women and children.
- Common courtesies, such as gift giving.
- Local business practices, such as bribes and haggling.

CULTURE

3-36. Once the social structure has been thoroughly mapped out, staffs should identify and analyze the culture of the society as a whole and of each major group within the society. Social structure comprises the relationships among groups, institutions, and individuals within a society; in contrast, culture (ideas, norms, rituals, codes of behavior) provides meaning to individuals within the society. For example, families are a core institutional building block of social structure found everywhere. However, marital monogamy, expectations of a certain number of children, and willingness to live with in-laws are highly variable in different societies. They are matters of culture. Social structure can be thought of as a skeleton, with culture being the muscle on the bones. The two are mutually dependent and reinforcing. A change in one results in a change in the other.

3-37. Culture is "web of meaning" shared by members of a particular society or group within a society. (See FM 3-05.301/MCRP 3-40.6A.) Culture is—

- A system of shared beliefs, values, customs, behaviors, and artifacts that members of a society use to cope with their world and with one another.
- Learned, though a process called enculturation.
- Shared by members of a society; there is no "culture of one."
- Patterned, meaning that people in a society live and think in ways forming definite, repeating patterns.
- Changeable, through social interactions between people and groups.

- Arbitrary, meaning that Soldiers and Marines should make no assumptions regarding what a society considers right and wrong, good and bad.
- Internalized, in the sense that it is habitual, taken for granted, and perceived as "natural" by people within the society.

3-38. Culture might also be described as an "operational code" that is valid for an entire group of people. Culture conditions the individual's range of action and ideas, including what to do and not do, how to do or not do it, and whom to do it with or not to do it with. Culture also includes under what circumstances the "rules" shift and change. Culture influences how people make judgments about what is right and wrong, assess what is important and unimportant, categorize things, and deal with things that do not fit into existing categories. Cultural rules are flexible in practice. For example, the kinship system of a certain Amazonian Indian tribe requires that individuals marry a cousin. However, the definition of cousin is often changed to make people eligible for marriage.

Identity

3-39. Each individual belongs to multiple groups, through birth, assimilation, or achievement. Each group to which individuals belong influences their beliefs, values, attitudes, and perceptions. Individuals consciously or unconsciously rank their identities into primary and secondary identities. Primary identities are frequently national, racial, and religious. In contrast, secondary identities may include such things as hunter, blogger, or coffee drinker. Frequently, individuals' identities are in conflict; counterinsurgents can use these conflicts to influence key leaders' decisions.

Beliefs

3-40. Beliefs are concepts and ideas accepted as true. Beliefs can be core, intermediate, or peripheral.

3-41. *Core* beliefs are those views that are part of a person's deep identity. They are not easily changed. Examples include belief in the existence of God, the value of democratic government, the importance of individual and collective honor, and the role of the family. Core beliefs are unstated, taken for granted, resistant to change, and not consciously considered. Attempts to change the central beliefs of a culture may result in significant unintended second- and third- order consequences. Decisions to do so are made at the national-strategic level.

3-42. *Intermediate* beliefs are predicated on reference to authority figures or authoritative texts. Thus, intermediate beliefs can sometimes be influenced by co-opting opinion leaders.

3-43. From intermediate beliefs flow *peripheral* beliefs. These beliefs are open to debate, consciously considered, and easiest to change. For example, a belief about birth control may derive from an individual's beliefs about the Roman Catholic Church. Beliefs about the theory of sexual repression may come from a person's opinion of Sigmund Freud.

Values

3-44. A *value* is an enduring belief that a specific mode of conduct or end state of existence is preferable to an opposite or converse mode of conduct or end state of existence. Values include beliefs concerning such topics as toleration, stability, prosperity, social change, and self-determination. Each group to which a person belongs inculcates that person with its values and their ranking of importance. Individuals do not unquestioningly absorb all the values of the groups to which they belong; they accept some and reject others. Most individuals belong to more than one social group. The values of each group are often in conflict: religious values may conflict with generational values or gender values with organizational practices. Commanders should evaluate the values of each group in the AO. They should determine whether the values promoted by the insurgency correspond to the values of other social groups in the AO or to those of the HN government. Based on that assessment, commanders can determine whether counterinsurgents can exploit these differences in values.

Attitudes and Perceptions

3-45. *Attitudes* are affinities for and aversions to groups, persons, and objects. Attitudes affect *perception,* which is the process by which an individual selects, evaluates, and organizes information from the external environment. Commanders should consider groups' attitudes regarding the following:

- Other groups.
- Outsiders.
- HN government.
- United States.
- U.S. military.
- Globalization.

Belief Systems

3-46. The totality of the identities, beliefs, values, attitudes, and perceptions that an individual holds—and the ranking of their importance—is that person's belief system. Religions, ideologies, and all types of "isms" fall into this category. As a belief system, a religion may include such things as a concept of God, a view of the afterlife, ideas about the sacred and the profane, funeral practices, rules of conduct, and modes of worship.

3-47. A belief system acts as a filter for new information: it is the lens through which people perceive the world. What members of a particular group believe to be rational, normal, or true may appear to outsiders to be strange, irrational or illogical. Understanding the belief systems of various groups in an AO allows counterinsurgents to more effectively influence the population.

3-48. Commanders should give the belief systems of insurgents and other groups in the AO careful attention. An insurgency may frame its objectives in terms of a belief system or may use a belief system to mobilize and recruit followers. Differences between the insurgents' and civilian groups' belief systems provide opportunities for counterinsurgents to separate the insurgents from the population. If local individuals are members of more than one group, there maybe contradictions in their belief systems that can be exploited.

Cultural Forms

3-49. Cultural forms are the concrete expression of the belief systems shared by members of a particular culture. Cultural forms include rituals, symbols, ceremonies, myths, and narratives. Cultural forms are the medium for communicating ideologies, values, and norms that influence thought and behavior. Each culture constructs or invents its own cultural forms through which cultural meanings are transmitted and reproduced. A culture's belief systems can be decoded by observing and analyzing its cultural forms. Insurgent groups frequently use local cultural forms to mobilize the population. Counterinsurgents can use cultural forms to shift perceptions, gain support, or reduce support for insurgents.

3-50. The most important cultural form for counterinsurgents to understand is the narrative. A cultural narrative is a story recounted in the form of a causally linked set of events that explains an event in a group's history and expresses the values, character, or self-identity of the group. Narratives are the means through which ideologies are expressed and absorbed by members of a society. For example, at the Boston Tea Party in 1773, Samuel Adams and the Sons of Liberty dumped five tons of tea into the Boston Harbor to protest what they considered unfair British taxation. This narrative explains in part why the Revolutionary War began. However, it also tells Americans something about themselves each time they hear the story: that fairness, independence, and justice are worth fighting for. As this example indicates, narratives may not conform to historical facts or they may drastically simplify facts to more clearly express basic cultural values. (For example, Americans in 1773 were taxed less than their British counterparts and most British attempts to raise revenues from the colonies were designed to help reduce the crushing national debt incurred in their defense.) By listening to narratives, counterinsurgents can identify a society's core values. Commanders should pay particular attention to cultural narratives of the HN population pertaining to outlaws, revolutionary heroes, and historical resistance figures. Insurgents may use these narratives to mobilize the population.

3-51. Other cultural forms include ritual and symbols. A *ritual* is a stereotyped sequence of activities involving gestures, words, and objects performed to influence supernatural entities or forces on behalf of the actors' goals and interests. Rituals can be either sacred or secular. A *symbol* is the smallest unit of cultural meaning. Symbols are filled with a vast amount of information that can be decoded by a knowledgeable observer. Symbols can be objects, activities, words, relationships, events, or gestures. Institutions and organizations often use cultural symbols to amass political power or generate resistance against external groups. Commanders should pay careful attention to the meaning of common symbols in the AO and how various groups use them.

LANGUAGE

3-52. *Language* is a system of symbols that people use to communicate with one another. It is a learned element of culture. Successful communication requires more than just grammatical knowledge; it also requires understanding the social setting, appropriate behaviors towards people of different statuses, and nonverbal cues, among other things. An understanding of the social environment can facilitate effective communication, even if counterinsurgents do not speak the local language and must work through translators or interpreters.

3-53. The languages used in an AO have a major impact on operations. Languages must be identified to facilitate language training, communication aids such as phrase cards, and requisitioning of translators. Translators are critical for collecting intelligence, interacting with local citizens and community leaders, and developing IO products. (Appendix C addresses linguist support.)

3-54. The transliteration of names not normally written using the English alphabet affects all intelligence operations, especially collection, analysis, and fusion. Unfamiliar and similar place names can make it hard find places on a map and cause targeting errors. In addition, detained insurgents may be released if their name is misidentified. In countries that do not use the English alphabet, a combatant-command-wide standard for spelling names agreed upon by non-DOD agencies should be set.

POWER AND AUTHORITY

3-55. Once they have mapped the social structure and understand the culture, staffs must determine how power is apportioned and used within a society. *Power* is the probability that one actor within a social relationship will be in a position to carry out his or her own will despite resistance. Understanding power is the key to manipulating the interests of groups within a society.

3-56. There may be many formal and informal power holders in a society. The formal political system includes the following organizations:

- Central governments.
- Local governments.
- Political interest groups.
- Political parties.
- Unions.
- Government agencies.
- Regional and international political bodies.

Understanding the formal political system is necessary but not sufficient for COIN operations. Informal power holders are often more important. They may include ethno-religious groups, social elites, and religious figures.

3-57. For each group in an AO, counterinsurgents should identify the type of power the group has, what it uses the power for (such as amassing resources and protecting followers), and how it acquires and maintains power. Commanders should also determine the same information about leaders within particular groups. There are four major forms of power in a society:

- Coercive force.
- Social capital.
- Economic resources.
- Authority.

Coercive Force

3-58. *Coercion* is the ability to compel a person to act through threat of harm or by use of physical force. Coercive force can be positive or negative. An example of coercion used positively is a group providing security for its members (as in policing and defending of territory). An example of coercion used negatively is a group intimidating or threatening group members or outsiders.

3-59. One essential role of government is providing physical security for its citizens by monopolizing the use of coercive force within its territory. When a government fails to provide security to its citizens or becomes a threat to them, citizens may seek alternative security guarantees. Ethnic, political, religious, or tribal groups in the AO may provide such guarantees.

3-60. Insurgents and other nongovernmental groups may possess considerable means of coercive force. Such groups may use coercion to gain power over the population. Examples of organizations providing such force are paramilitary units, tribal militias, gangs, and organizational security personnel. Groups may use their coercive means for a variety of purposes unrelated to the insurgency. Protecting their community members, carrying out vendettas, and engaging in criminal activities are examples. What may appear to be insurgent violence against innocent civilians could in fact be related to a tribal blood feud rather than the insurgency.

Social Capital

3-61. *Social capital* refers to the power of individuals and groups to use social networks of reciprocity and exchange to accomplish their goals. In many non-Western societies, patron-client relationships are an important form of social capital. In a system based on patron-client relationships, an individual in a powerful position provides goods, services, security, or other resources to followers in exchange for political support or loyalty, thereby amassing power. Counterinsurgents must identify, where possible, groups and individuals with social capital and how they attract and maintain followers.

Economic Power

3-62. *Economic power* is the power of groups or individuals to use economic incentives and disincentives to change people's behavior. Economic systems can be formal, informal, or a mixture of both. In weak or failed states, the formal economy may not function well. In such cases, the informal economy plays a central role in people's daily lives. The informal economy refers to such activities as smuggling, black market activities, barter, and exchange. For example, in many societies, monies and other economic goods are distributed though the tribal or clan networks and are connected to indigenous patronage systems. Those groups able to provide their members with economic resources through an informal economy gain followers and may amass considerable political power. Therefore, counterinsurgents must monitor the local informal economy and evaluate the role played by various groups and individuals within it. Insurgent organizations may also attract followers through criminal activities that provide income.

Authority

3-63. *Authority* is legitimate power associated with social positions. It is justified by the beliefs of the obedient. There are three primary types of authority:

- **Rational-legal authority,** which is grounded in law and contract, codified in impersonal rules, and most commonly found in developed, Western societies.
- **Charismatic authority,** which is exercised by leaders who develop allegiance among their followers because of their unique, individual charismatic appeal, whether ideological, religious, political, or social.

- **Traditional authority,** which is usually invested in a hereditary line or particular office by a higher power.

3-64. Traditional authority relies on the precedent of history. It is a common type of authority in non-Western societies. In particular, tribal and religious forms of organization rely heavily on traditional authority. Traditional authority figures often wield enough power, especially in rural areas, to single-handedly drive an insurgency. Understanding the types of authority at work in the formal and informal political systems of the AO helps counterinsurgents identify agents of influence who can help or hinder achieving objectives.

INTERESTS

3-65. Commanders and staffs analyze the culture of the society as a whole and that of each group within the society. They identify who holds formal and informal power and why. Then they consider ways to reduce support for insurgents and gain support for the HN government.

3-66. Accomplishing these tasks requires commanders and staffs to understand the population's interests. *Interests* refer to the core motivations that drive behavior. These include physical security, basic necessities, economic well-being, political participation, and social identity. During times of instability, when the government cannot function, the groups and organizations to which people belong satisfy some or all of their interests. Understanding a group's interests allows commanders to identify opportunities to meet or frustrate those interests. A group's interests may become grievances if the HN government does not satisfy them.

Physical Security

3-67. During any period of instability, people's primary interest is physical security for themselves and their families. When HN forces fail to provide security or threaten the security of civilians, the population is likely to seek security guarantees from insurgents, militias, or other armed groups. This situation can feed support for an insurgency. However, when HN forces provide physical security, people are more likely to support the government. Commanders therefore identify the following:

- Whether the population is safe from harm.
- Whether there is a functioning police and judiciary system.
- Whether the police and courts are fair and nondiscriminatory.
- Who provides security for each group when no effective, fair government security apparatus exists.

The provision of security by the HN government must occur in conjunction with political and economic reform.

Essential Services

3-68. *Essential services* provide those things needed to sustain life. Examples of these essential needs are food, water, clothing, shelter, and medical treatment. Stabilizing a population requires meeting these needs. People pursue essential needs until they are met, at any cost and from any source. People support the source that meets their needs. If it is an insurgent source, the population is likely to support the insurgency. If the HN government provides reliable essential services, the population is more likely to support it. Commanders therefore identify who provides essential services to each group within the population.

Economy

3-69. A society's individuals and groups satisfy their economic interests by producing, distributing, and consuming goods and services. How individuals satisfy their economic needs depends on the society's level and type of economic development. For instance, in a rural-based society, land ownership may be a major part of any economic development plan. For a more urban society, public- and private-sector jobs may be the greatest concern.

3-70. Sometimes economic disparities between groups contribute to political instability. Insurgent leadership or traditional authority figures often use real or perceived injustices to drive an insurgency. Perceived injustices may include the following:

- Economic disenfranchisement.
- Exploitative economic arrangements.
- Significant income disparity that creates (or allows for) intractable class distinctions.

3-71. Military operations or insurgent actions can adversely affect the economy. Such disruption can generate resentment against the HN government. Conversely, restoring production and distribution systems can energize the economy, create jobs and growth, and positively influence local perceptions. To determine how to reduce support for insurgency and increase support for the government, commanders determine the following:

- Whether the society has a functioning economy.
- Whether people have fair access to land and property.
- How to minimize the economic grievances of the civilian population.

Political Participation

3-72. Another interest of the population is political participation. Many insurgencies begin because groups within a society believe that they have been denied political rights. Groups may use preexisting cultural narratives and symbols to mobilize for political action. Very often, they rally around traditional or charismatic authority figures. Commanders should investigate whether—

- All members of the civilian population have a guarantee of political participation.
- Ethnic, religious, or other forms of discrimination exist.
- Legal, social, or other policies are creating grievances that contribute to the insurgency.

Commanders should also identify traditional or charismatic authority figures and what narratives mobilize political action.

Grievances

3-73. Unsatisfied interests may become grievances. Table 3-1 lists factors to consider when an interest has become a grievance.

Table 3-1. Factors to consider when addressing grievances

• What are the insurgents' grievances? • What are the population's grievances? • Would a reasonable person consider the population's grievances valid? The validity of a grievance should be assessed using both subjective and objective criteria. • Are the grievances of the population and those of the insurgency the same? • What does the host-nation government believe to be the population's grievances? Does it consider those grievances valid? • Are the population's grievances the same as those perceived by the host-nation government? • Has the host-nation government made genuine efforts to address these grievances? • Are there practical actions the host-nation government can take to address these grievances? Or is addressing the source of the grievances beyond the host-nation government's capability? (For example, major social and economic dislocations may be caused by globalization.) • Can U.S. forces act to address these interests or grievances?

Evaluate the Threat

3-74. The purpose of evaluating the insurgency and related threats is to understand the enemy, enemy capabilities, and enemy vulnerabilities. This evaluation also identifies opportunities commanders may exploit.

3-75. Evaluating an insurgency is difficult. Neatly arrayed enemy orders of battle are neither available nor what commanders need to know. Insurgent organizational structures are functionally based and continually adaptive. Attempts to apply traditional order of battle factors and templates can produce oversimplified, misleading conclusions. Commanders require knowledge of difficult-to-measure characteristics. These may include the following:

- Insurgent goals.
- Grievances insurgents exploit.
- Means insurgents use to generate support.
- Organization of insurgent forces.
- Accurate locations of key insurgent leaders.

However, insurgents usually look no different from the general populace and do their best to blend with noncombatants. Insurgents may publicly claim motivations and goals different from what is truly driving their actions. Further complicating matters, insurgent organizations are often rooted in ethnic or tribal groups. They often take part in criminal activities or link themselves to political parties, charities, or religious organizations as well. These conditions and practices make it difficult to determine what and who constitutes the threat. Table 3-2 lists characteristics of an insurgency that can provide a basis for evaluating a threat. Table 3-2 also lists the conventional order of battle element as a supplement for analysts.

Table 3-2. Insurgency characteristics and order of battle factors

Insurgency characteristics	*Conventional order of battle factors*
• Insurgent objectives • Insurgent motivations • Popular support or tolerance • Support activities, capabilities, and vulnerabilities • Information activities, capabilities, and vulnerabilities • Political activities, capabilities, and vulnerabilities • Violent activities, capabilities, and vulnerabilities • Organization • Key leaders and personalities	• Composition • Disposition • Strength • Tactics and operations • Training • Logistics • Operational effectiveness • Electronic technical data • Personalities • Miscellaneous data • Other factors

3-76. The following insurgency characteristics are often the most important intelligence requirements and the most difficult to ascertain:

- Objectives.
- Motivations.
- Means of generating popular support or tolerance.

In particular, the ability to generate and sustain popular support, or at least acquiescence and tolerance, often has the greatest impact on the insurgency's long-term effectiveness. This ability is usually the insurgency's center of gravity. Support or tolerance, provided either willingly or unwillingly, provides the following for an insurgency:

- Safe havens.
- Freedom of movement.
- Logistic support.
- Financial support.
- Intelligence.
- New recruits.

Support or tolerance is often generated using violent coercion and intimidation of the populace. In these cases, even if people do not favor the insurgent cause, they are forced to tolerate the insurgents or provide them material support.

3-77. Understanding the attitudes and perceptions of the society's groups is very important to understanding the threat. It is important to know how the population perceives the insurgents, the host nation, and U.S. forces. In addition, HN and insurgent perceptions of one another and of U.S. forces are also very important. Attitudes and perceptions of different groups and organizations inform decision-making processes and shape popular thinking on the legitimacy of the actors in the conflict.

3-78. As analysts perform IPB, they should focus on insurgent vulnerabilities to exploit and strengths to mitigate. (See chapter 1.) Evaluating threats from insurgents and other armed groups and learning the people's interests and attitudes lets analysts identify divisions between the insurgents and the populace. This analysis also identifies divisions between the HN government and the people. For instance, if the insurgent ideology is unpopular, insurgents may use intimidation to generate support. Another example is discovering that insurgents gain support by providing social services that the HN government neglects or cannot provide. Determining such divisions identifies opportunities to conduct operations that expand splits between the insurgents and the populace or lessen divides between the HN government and the people.

OBJECTIVE AND MOTIVATION IDENTIFICATION

3-79. Insurgents have political objectives and are motivated by an ideology or grievances. The grievances may be real or perceived. Identifying insurgent objectives and motivations lets counterinsurgents address the conflict's underlying causes. Broadly speaking, there are two kinds of insurgencies: national insurgencies and resistance movements. Both can be further classified according to the five approaches explained in paragraphs 1-25 through 1-39.

3-80. In a *national insurgency*, the conflict is between the government and one or more segments of the population. In this type of insurgency, insurgents seek to change the political system, take control of the government, or secede from the country. A national insurgency polarizes the population and is generally a struggle between the government and insurgents for legitimacy and popular support.

3-81. In contrast, a *resistance movement* (sometimes called a liberation insurgency) occurs when insurgents seek to expel or overthrow what they consider a foreign or occupation government. The grievance is foreign rule or foreign intervention. Resistance movements tend to unite insurgents with different objectives and motivations. However, such an insurgency can split into competing factions when foreign forces leave and the focus of resistance is gone. That situation may result in a civil war.

3-82. Identification of insurgent goals and motivations can be difficult for a number of reasons:

- There may be multiple insurgent groups with differing goals and motivations. This case requires separately monitoring each group's goals and motivations.
- Insurgent leaders may change and the movement's goals change with them.
- Movement leaders may have different motivations from their followers. For instance, a leader may want to become a new dictator; followers may be motivated by a combination of political ideology and money.
- Insurgents may hide their true motivations and make false claims. For instance, the differences between insurgents and outsiders generally make resistance movements easier to unify and mobilize. Thus, insurgents may try to portray a national insurgency as a resistance movement.
- The goals of an insurgency may change due to changes in the operational environment. Foreign forces joining a COIN effort can transform a national insurgency into resistance movement. The reverse may happen when foreign forces depart.

For all these reasons, analysts continuously track insurgent actions, internal communications, and public rhetoric to determine insurgent goals and motivations.

Popular Support or Tolerance

3-83. Developing passive support (tolerance or acquiescence) early in an insurgency is often critical to an insurgent organization's survival and growth. Such support often has a great effect on the insurgency's long-term effectiveness. As an insurgent group gains support, its capabilities grow. New capabilities enable the group to gain more support. Insurgents generally view popular support as a zero-sum commodity; that is, a gain for the insurgency is a loss for the government, and a loss for the government is a gain for the insurgency.

Forms of Popular Support

3-84. Popular support comes in many forms. It can originate internally or externally, and it is either active or passive. There are four forms of popular support:

- Active external.
- Passive external.
- Active internal.
- Passive internal.

The relative importance of each form of support varies depending on the circumstances. However, all forms benefit an insurgency.

3-85. *Active external support* includes finance, logistics, training, fighters, and safe havens. These forms of support may be provided by a foreign government or by nongovernmental organizations, such as charities.

3-86. *Passive external support* occurs when a foreign government supports an insurgency through inaction. Forms of passive support include the following:

- Not curtailing the activities of insurgents living or operating within the state's borders.
- Recognizing the legitimacy of an insurgent group.
- Denying the legitimacy of the HN government.

3-87. *Active internal support* is usually the most important to an insurgent group. Forms of active support include the following:

- Individuals or groups joining the insurgency.
- Providing logistic or financial support.
- Providing intelligence.
- Providing safe havens.
- Providing medical assistance.
- Providing transportation.
- Carrying out actions on behalf of the insurgents.

3-88. *Passive internal support* is also beneficial. Passive supporters do not provide material support; however, they do allow insurgents to operate and do not provide information to counterinsurgents. This form of support is often referred to as tolerance or acquiescence.

Methods of Generating Popular Support

3-89. Insurgents use numerous methods to generate popular support. These include the following:

- Persuasion.
- Coercion.
- Encouraging overreaction.
- Apolitical fighters.

3-90. *Persuasion* can be used to obtain either internal or external support. Forms of persuasion include—

- Charismatic attraction to a leader or group.
- Appeal to an ideology.

- Promises to address grievances.
- Demonstrations of potency, such as large-scale attacks or social programs for the poor.

Persuasion through demonstrations of potency can be the most effective technique because it can create the perception that the insurgency has momentum and will succeed.

3-91. Insurgents use *coercion* to force people to support or tolerate insurgent activities. Means of coercion include terrorist tactics, violence, and the threat of violence. Coercion may be used to alter the behavior of individuals, groups, organizations, or governments. Coercion is often very effective in the short term, particularly at the community level. However, terrorism against the general populace and popular leaders or attacks that negatively affect people's way of life can undermine insurgent popularity. Coercion is an easy way for insurgents to generate passive support; however, this support exists only as long as insurgents are able to intimidate.

3-92. *Encouraging overreaction* refers to enticing counterinsurgents to use repressive tactics that alienate the populace and bring scrutiny upon the government. It is also referred to as provocation of a government response.

3-93. *Apolitical fighters* may be attracted by many nonideological means. These means include monetary incentives, the promise of revenge, and the romance of fighting a revolutionary war.

3-94. Although difficult to quantify, analysts evaluate the popular support an insurgent group receives and its ability to generate more support. Open sources and intelligence reporting provide data to support this analysis. Polling data can be a valuable, though imprecise, means of gauging support for the HN government and support for the insurgency. Media and other open-source publications are important at all echelons. Assessing community attitudes, by gauging such things as the reactions of local populace to the presence of troops or government leaders, can be used to estimate popular support at the tactical level. At a minimum, the information in table 3-3 should be known.

Table 3-3. Critical information regarding popular support

• Overall level of popular support to the insurgency relative to that for the government. • Forms of popular support the insurgents receive (active, passive, internal, external) and their relative importance. • Foreign government support. • Support from nongovernmental organizations, including charities and transnational terrorist organizations.	• Criminal network support. • Segments of the populace supporting the insurgency. • Methods used to generate popular support and their relative effectiveness. • Grievances (real or perceived) exploited by insurgents. • Capabilities and vulnerabilities in generating popular support.

Support Activities

3-95. Although noticeable, violence may be only a small part of overall insurgent activity. Unseen insurgent activities include training and logistic actions. These are the support activities that sustain insurgencies. They come from an insurgency's ability to generate popular support. Like conventional military forces, insurgencies usually require more sustainers than fighters. Insurgent support networks may be large, even when violence levels are low. For this reason, it is easy to overlook them early in the development of an insurgency.

3-96. Undermining an insurgency's popular support is the most effective way to reduce insurgent support capabilities. However, identifying support capabilities and vulnerabilities is still important. Doing this lets analysts evaluate potential threat courses of action. Such analysis also lets commanders target vulnerable parts of the insurgents' support network. Table 3-4 lists support activities and capabilities to evaluate.

Table 3-4. Insurgent support activities and capabilities

• Locations of safe havens • Freedom of movement • Logistic support • Financial support • Means of communication	• Ability to train recruits in military or terrorist tactics and techniques • Means of collecting intelligence • Means of maintaining operations security

Information and Media Activities

3-97. Information and media activities can be an insurgency's main effort, with violence used in support. Insurgents use information activities to accomplish the following:

- Undermine HN government legitimacy.
- Undermine COIN forces.
- Excuse insurgent transgressions of national and international laws and norms.
- Generate popular support.

To achieve these effects, insurgents broadcast their successes, counterinsurgent failures, HN government failures, and illegal or immoral actions by counterinsurgents or the HN government. Insurgent broadcasts need not be factual; they need only appeal to the populace. Table 3-5 lists media forms that insurgents commonly use.

Table 3-5. Media forms insurgents use

• Word of mouth • Speeches by elites and key leaders • Flyers and handouts • Newspapers • Journals or magazines • Books • Audio recordings	• Video recordings • Radio broadcasts • Television broadcasts • Web sites • E-mail • Internet • Cellular telephones • Text messaging

3-98. To supplement their own media activities, insurgents take advantage of existing private and public media companies through press releases and interviews. These efforts, in addition to using the Internet, broadcast insurgent messages worldwide. By broadcasting to a global audience, insurgents directly attack public support for the COIN effort. Information and media activities to evaluate include the following:

- Commitment of assets and personnel to information activities.
- Types of media employed.
- Professionalism of products, such as newspaper articles or videos.
- Effectiveness and reach of information activities.

Political Activities

3-99. Insurgents use political activities to achieve their goals and enhance their cause's legitimacy. Political activities are tightly linked to information activities and violent acts. Political parties affiliated with an insurgent organization may negotiate or communicate on behalf of the insurgency, thereby serving as its public face. Insurgencies may grow out of political parties, or political parties may grow out of insurgencies. However, links between insurgents and political parties may be weak or easily broken by disputes between insurgents and politicians. In such cases, political parties may not be able to keep promises to end violent conflict. It is important to understand not only the links between insurgent groups and political organizations but also the amount of control each exerts over the other.

3-100. Understanding insurgent political activities enables effective political engagement of insurgents. Without this knowledge, the wrong political party may be engaged, the wrong messages may be used, or the government may make deals with political parties that cannot deliver on their promises. Political activities to be evaluated include the following:

- Links, if any, between the insurgency and political parties.
- Influence of political parties over the insurgency and vice versa.
- Political indoctrination and recruiting by insurgent groups.

violent Activities

3-101. Violent actions by insurgents include three major types, which may occur simultaneously:

- Terrorist.
- Guerrilla.
- Conventional.

3-102. Insurgents are by nature an asymmetric threat. They do not use terrorist and guerrilla tactics because they are cowards afraid of a "fair fight"; insurgents use these tactics because they are the best means available to achieve the insurgency's goals. Terrorist and guerrilla attacks are usually planned to achieve the greatest political and informational impact with the lowest amount of risk to insurgents. Thus, commanders need to understand insurgent tactics and targeting as well as how the insurgent organization uses violence to achieve its goals and how violent actions are linked to political and informational actions.

Asymmetric Tactics in Ireland

In 1847, Irish insurgents were advised to engage the British Army in the following way:

"The force of England is entrenched and fortified. You must draw it out of position; break up its mass; break its trained line of march and manoeuvre, its equal step and serried array... nullify its tactic and strategy, as well as its discipline; decompose the science and system of war, and resolve them into their first elements."

3-103. Terrorist tactics employ violence primarily against noncombatants. Terror attacks generally require fewer personnel than guerrilla warfare or conventional warfare. They allow insurgents greater security and have relatively low support requirements. Insurgencies often rely on terrorist tactics early in their formation due to these factors. Terrorist tactics do not involve mindless destruction nor are they employed randomly. Insurgents choose targets that produce the maximum informational and political effects. Terrorist tactics can be effective for generating popular support and altering the behavior of governments.

3-104. Guerrilla tactics, in contrast, feature hit-and-run attacks by lightly armed groups. The primarily targets are HN government activities, security forces, and other COIN elements. Insurgents using guerrilla tactics usually avoid decisive confrontations unless they know they can win. Instead, they focus on harassing counterinsurgents. As with terrorist tactics, guerrilla tactics are neither mindless nor random. Insurgents choose targets that produce maximum informational and political effects. The goal is not to militarily defeat COIN forces but to outlast them while building popular support for the insurgency. Terrorist and guerrilla tactics are not mutually exclusive. An insurgent group may employ both forms of violent action simultaneously.

3-105. Insurgents rarely use conventional tactics. Conventional operations are not always necessary for an insurgency's success. However, insurgents may engage in conventional operations after the insurgency develops extensive popular support and sustainment capabilities. The insurgents can then generate a conventional military force that can engage HN government forces.

3-106. Knowledge of violent capabilities is used to evaluate insurgent courses of action. Commanders use this knowledge to determine appropriate protection measures and tactics to counter insurgent actions. In addition, knowledge of how insurgents conduct attacks provides a baseline that helps determine the effectiveness of COIN operations. The following should be evaluated to determine insurgents' capabilities for violent action:

- Forms of violent action used.
- Weapons available and their capabilities.
- Training.
- Known methods of operating.
  - Frequency of attacks.
  - Timing of attacks.
  - Targets of attacks.
  - Tactics and techniques.
- Known linkages between violent, political, and information actions. How do the insurgents use violence to increase their popular support and undermine counterinsurgents?
- Means of command and control during attacks (including communications means used).

Insurgent Organizational Structure and Key Personalities

3-107. Conducting the preceding activities requires some form of organizational structure and leadership. Insurgencies can be organized in several ways. Each structure has its own strengths and limitations. The structure used balances the following:

- Security.
- Efficiency and speed of action.
- Unity of effort.
- Survivability.
- Geography.
- Social structures and cultures of the society.

Organizations also vary greatly by region and time. Insurgent organizations are often based on existing social networks—familial, tribal, ethnic, religious, professional, or others. Analysts can use social network analysis to determine organizational structure. (See paragraphs B-15 through B-18.)

3-108. An insurgency's structure often determines whether it is more effective to target enemy forces or enemy leaders. For instance, if an insurgent organization is hierarchical with few leaders, removing the leaders may greatly degrade the organization's capabilities. However, if the insurgent organization is nonhierarchical, targeting the leadership may not have much effect. Understanding an insurgent organization's structure requires answers to the following questions:

- Is the organization hierarchical or nonhierarchical?
- Is the organization highly structured or unsystematic?
- Are movement members specialists or generalists?
- Do leaders exercise centralized control or do they allow autonomous action and initiative?
- Are there a few leaders (promotes rapid decision making) or is there redundant leadership (promotes survivability)?
- Does the movement operate independently or does it have links to other organizations and networks (such as criminal, religious, and political organizations)?
- Does the movement place more weight on political action or violent action?

3-109. As explained in paragraphs 1-58 through 1-66, insurgents fall into five overlapping categories: movement leaders, combatants, political cadre, auxiliaries, and the mass base. Movement leaders are important because they choose the insurgency's organization, approach, and tactics. The movement leaders' personalities and decisions often determine whether the insurgency succeeds. Therefore, the movement leaders must be identified and their basic beliefs, intentions, capabilities, and vulnerabilities understood. Important leader characteristics include the following:

- Role in the organization.
- Known activities.
- Known associates.
- Background and personal history.
- Beliefs, motivations, and ideology.
- Education and training.
- Temperament (for example, careful, impulsive, thoughtful, or violent).
- Importance of the organization.
- Popularity outside the organization.

ASSOCIATED THREATS

3-110. When an insurgency has widespread support, it usually means the HN government is weak and losing control. In such situations, other armed groups—particularly criminal organizations, militias, and terrorist groups—can be significant players. Moreover, these groups can support each other's operations.

Criminal Networks

3-111. Criminal networks may not be a part of an insurgency. However, their activities—for example, banditry, hijackings, kidnappings, and smuggling—can further undermine the HN government's authority. Insurgent organizations often link themselves to criminal networks to obtain funding and logistic support. In some cases, insurgent networks and criminal networks become indistinguishable. As commanders work to reassert government control, they need to know the following:

- Which criminal networks are present.
- What their activities are.
- How they interact with insurgents.

Nongovernment Militias

3-112. As the HN government weakens and violence increases, people look for ways to protect themselves. If the government cannot provide protection, people may organize into armed militias to provide that essential service. Examples of this sort of militia include the following:

- Loyalist militias formed in Northern Ireland.
- Right-wing paramilitary organizations formed in Colombia to counter the FARC.
- Militias of various ethnic and political groups formed in Iraq during Operation Iraqi Freedom.

If militias are outside the HN government's control, they can often be obstacles to ending an insurgency. Militias may become more powerful than the HN government, particularly at the local level. They may also fuel the insurgency and a precipitate a downward spiral into full-scale civil war.

3-113. Militias may or may not be an immediate threat to U.S. forces; however, they constitute a long-term threat to law and order. The intelligence staff should track them just like insurgent and other armed groups. Commanders need to understand the role militias play in the insurgency, the role they play in politics, and how they can be disarmed.

DETERMINE THREAT COURSES OF ACTION

3-114. The purpose of this IPB step is to understand insurgent approaches and tactics so they can be effectively countered. The initial determination of threat courses of action focuses on two levels of analysis. The first is determining the overall approach, or combination of approaches, the movement leaders have selected to achieve their goals. The second is determining tactical courses of action used to execute that approach.

3-115. The insurgents' approach is based on their objectives, desired end state, and requirements of the operational environment. The approach and the tactics used to execute it set the conditions for the insurgents to achieve their desired end state. Insurgents can accomplish this goal by maintaining preexisting adverse conditions or by creating those conditions.

INSURGENT APPROACHES

3-116. As indicated in paragraphs 1-25 through 1-39, there are six approaches insurgents may follow:

- Conspiratorial.
- Military-focused.
- Urban.
- Protracted popular war.
- Identity-focused.
- Composite and coalition.

3-117. These approaches may be combined with one another. They may also occur in parallel as different insurgent groups follow different paths, even within a single AO. In addition, insurgents may change approaches over time. The approach pursued affects the insurgents' organization, types of activities, and emphasis placed on different activities.

3-118. Table 3-6 (page 3-22) lists potential indicators of different insurgent approaches. The conspiratorial and identity-focused approaches present distinct collection challenges. Insurgents using a conspiratorial approach execute few overt acts until the conditions appear ripe to seize power. Thus, this approach is difficult to identify without sources within the insurgent organization. Similarly, members of an identity-focused insurgency strongly identify with the insurgent organization. It is difficult to get such people to provide useful information.

Table 3-6. Potential indicators of insurgent approaches

- **Conspiratorial**
  - Absence of overt violent or informational actions.
  - Large cadre relative to the number of combatants in the organization.
  - Small mass base or no mass base at all.
- **Military-focused**
  - Presence of leaders and combatants, but little, if any, cadre or mass base.
- **Urban**
  - Terrorist attacks in urban areas.
  - Infiltration and subversion of host-nation government and security forces in urban areas.
  - Organization composed of small, compartmentalized cells.
  - Cadre and mass base small relative to the number of combatants.
- **Protracted popular war**
  - A large mass base.
  - Overt violence.
  - Heavy use of informational and political activities.
  - Focus on building popular support for the insurgency.
- **Identity-focused**
  - Presence of a resistance movement.
  - Presence of an "us-and-them" gap between the government and one or more ethnic, tribal, or religious groups.
  - Large mass base of passive and active supporters built around preexisting social networks.
  - Many auxiliaries.
  - Small cadre composed primarily of traditional authority figures.
  - Large numbers of part-time combatants.

3-119. It should be noted that insurgents may be inept at the use of a given approach. Alternatively, they may misread the operational environment and use an inappropriate approach. Knowledge of misapplication of approach or the use of different approaches by different insurgent groups may provide opportunities for counterinsurgents to exploit. It is imperative not only to identify insurgent approaches but also to understand their strengths and weaknesses in the context of the operational environment.

TACTICAL COURSES OF ACTION

3-120. Insurgents base their tactical courses of action on their capabilities and intentions. Evaluating the support, information, political, and violent capabilities of insurgent organizations was discussed in paragraphs 3-95 through 3-106. The intentions come from goals, motivations, approach, culture, perceptions, and leadership personalities. Insurgents may pursue many different courses of action in an AO at any time. Their tactical courses of action change with both time and location. People and their attitudes, both within the nation and often outside it, are the ultimate targets of the insurgents. Therefore, commanders pay special attention to the effects insurgent actions have on the populace and how the insurgents achieve those effects. Finally, tactical actions can have strategic effects. This is because insurgent propaganda and media reporting can reach a global audience, multiplying the effects of insurgent tactical actions. Insurgents can employ a wide variety of tactics. (See table 3-7.)

Table 3-7. Examples of insurgent tactics

- **Ambushes.** Guerrilla-style attacks to kill or intimidate counterinsurgents.
- **Assassination.** A term generally applied to the killing of prominent persons and symbolic personnel. It may also pertain to "traitors" who defect from the insurgency, human intelligence sources, and others who work with or for the host-nation government or U.S. forces.
- **Arson.** Less dramatic than most tactics, arson has the advantage of low risk to the perpetrator. It requires only a low level of technical knowledge.
- **Bombing and high explosives.** The improvised explosive device is often the insurgent's weapon of choice. Improvised explosive devices can be inexpensive to produce and, because of the various detonation techniques available, may be a low risk to the perpetrator. However, suicidal bombing cannot be overlooked as an employment method. Another advantage is the publicity that such attacks produce. Yet another is the insurgents' ability to control casualties through timed detonation and careful placement of the device. Attacks are also easily deniable should the action produce undesirable results. From 1983 through 1996, approximately half of all terrorist incidents worldwide involved the use of explosives.
- **Chemical, biological, radiological, or nuclear weapons.** There is a potential for use of both chemical and biological weapons in the future. These types of weapons are relatively cheap and easy to make. They may be used in place of conventional explosives in many situations. The potential for mass destruction and the deep-seated fear most people have for these weapons make them attractive to groups wishing to attract international attention. Although an explosive nuclear device is acknowledged to be beyond the financial and technical reach of most terrorist groups, a chemical or biological weapon, or a radiological dispersion device using nuclear contaminants, is not. The technology is simple and the payoff is potentially higher than that of conventional explosives.
- **Demonstrations.** Demonstrations can be used to incite violent responses by counter-insurgents and also to display the popularity of the insurgent cause.
- **Denial and deception.**
  - *Denial* consists of measures taken by the threat to block, prevent, or impair U.S. intelligence collection. Examples include killing human intelligence sources.
  - *Deception* involves deliberately manipulating information and perceptions in order to mislead. Examples include providing false intelligence.

- **Hijacking and skyjacking.** Sometimes insurgents employ hijacking as a means of escape. However, hijackings are normally executed to produce a spectacular hostage situation. Although trains, buses, and ships have been hijacked, aircraft are the preferred target because of their greater mobility and because they are difficult to enter once they have been seized.
- **Hoaxes.** Any credible insurgent or terrorist group can employ a hoax with considerable success. A threat against a person's life causes that person and those associated with that individual to devote time and efforts to security measures. A bomb threat can close a commercial building, empty a theater, or delay an aircraft flight at no cost to insurgents. False alarms dull the analytical and operational efficiency of key security personnel, thus degrading readiness.
- **Hostage taking.** Hostage taking is an overt seizure of one or more people to gain publicity or other concessions in return for releasing the hostage. While dramatic, hostage and hostage barricade situations are risky for perpetrators.
- **Indirect fire.** Insurgents may use indirect fire, such as mortars, to harass counterinsurgents or cause them to commit forces that can then be attacked by secondary ambushes.
- **Infiltration and subversion.** Insurgents use these tactics to gain intelligence and degrade the effectiveness of host-nation government organizations. These tactics involve getting host-nation government agencies to hire insurgent agents or by convincing members of the government to support the insurgency. Subversion may be achieved through intimidation, indoctrination of sympathetic individuals, or bribes.
- **Kidnapping.** While similar to hostage taking, kidnapping has significant differences. Kidnapping is usually a covert seizure of one or more persons to obtain specific results. It is normally very difficult to execute. Perpetrators may or may not be known for a long time. Media attention is initially intense but decreases overtime. Because of the time involved, successful kidnapping requires elaborate planning and logistics. The risk to the perpetrators may be less than in the hostage situation.
- **Propaganda.** Insurgents may disseminate propaganda using any form of media, including face-to-face talks.
- **Raids or attacks on facilities.** Armed attacks on facilities are usually undertaken to—
  - Demonstrate the host-nation government's inability to secure critical facilities or national symbols.
  - Acquire resources (for example, robbery of a bank or armory).
  - Kill host-nation government personnel.
  - Intimidate the host-nation government and the populace.
- **Sabotage.** The objective in most sabotage incidents is to demonstrate how vulnerable the society or host-nation government is to terrorist actions. Industrialized areas are more vulnerable to sabotage than less developed areas. Utility, communication, and transportation systems are interdependent; a serious disruption of any affects them all and gains immediate public attention. Sabotage of industrial and commercial facilities can create significant disruption while making a statement of future intent. Military facilities and installations, information systems, and information infrastructures are possible targets.
- **Seizure.** Seizure usuaiiy involves a building or object of vaiue in the eyes of the audience. There is some risk to perpetrators because security forces have time to react.

SECTION III - INTELLIGENCE, SURVEILLANCE, AND RECONNAISSANCE OPERATIONS

3-121. The purpose of intelligence, surveillance, and reconnaissance (ISR) operations during a COIN is to develop the intelligence needed to address the issues driving the insurgency. Several factors are particularly important for ISR operations in COIN environments. These include the following:

- A focus on the local populace.
- Collection occurring at all echelons.
- Localized nature of insurgencies.

- All Soldiers and Marines functioning as potential collectors.
- Insurgent use of complex terrain.

3-122. Intelligence gaps and information requirements determined during IPB may range from insurgent leaders' locations, to the populace's perceptions of insurgents, to HN political parties' status. In general, collection focuses on the populace, insurgents, and host nation.

3-123. The fact that all units collect and report information, combined with the mosaic nature of insurgencies, means that the intelligence flow in COIN is more bottom up than top down. Conducting aggressive ISR operations and pushing intelligence collection assets and analysts to the tactical level, sometimes as far as company level, therefore benefits all echelons. It strengthens local intelligence, enhances regional and national reporting, and bolsters operations at all levels. Two techniques— either attaching a basic intelligence analytical capability down to battalion or company level, or forming a company information management capability from assigned personnel—can help commanders handle the tactical information flow better.

3-124. Collection may occur in any unit and collectors may be pushed to the lowest levels; nonetheless, the overall intelligence synchronization plan (formerly the collection plan) must remain synchronized so that all echelons receive the intelligence they require. There are several means of ensuring this happens. One is to ensure that priority intelligence requirements (PIRs) are "nested" at all echelons. They may be tailored to local or regional circumstances, but tactical and operational collection efforts should support one another. Headquarters monitor requests for information from lower echelons and taskings from higher echelons to get information to requestors when they need it. Commanders ensure their subordinates understand the PIRs. Such understanding helps Soldiers and Marines know when to report something and what they should report.

3-125. Feedback from analysts and intelligence consumers to collectors is important to synchronizing the ISR effort. Responses tell collectors that a report is of interest and that they should follow it up. Such feedback may come from any unit at any echelon.

3-126. Also affecting intelligence synchronization is the requirement to work closely with U.S. Government agencies, HN security and intelligence organizations, and multinational intelligence organizations. Operational-level ISR planning drives the synchronization of these agencies' and organizations' efforts; however, coordination occurs at all echelons. Communication among collection managers and collectors down to the battalion level is important; it can eliminate circular reporting and unnecessary duplicate work. (See section IV.)

3-127. Insurgents often try to use complex terrain and seams between maneuver units to their advantage. (Seams are boundaries between units not adequately covered by any unit.) Collection managers do not ignore areas of complex terrain. They monitor seams to ensure insurgents do not establish undetected bases of operation.

THE INTELLIGENCE-OPERATIONS DYNAMIC

3-128. Intelligence and operations have a dynamic relationship. Even in permissive environments where a great deal is known about the enemy, there is an intelligence aspect to all operations. Intelligence drives operations and successful operations generate additional intelligence. For instance, an operation increasing the security and general happiness of a town often increases the amount of information its inhabitants offer. This information is processed into more intelligence, which results in more effective operations. The reverse is also true. Operations conducted without accurate intelligence may upset the populace and lead them to offer less information. In many cases, newly arrived units have little intelligence on their AO. They have to conduct operations to generate intelligence.

3-129. Because intelligence and operations are so closely related, it is important for collectors to be linked directly to the analysts and operators they support. Analysts must remain responsive to their supported units' intelligence requirements. Further, collectors should not passively wait for operators to submit requirements; rather, they should closely monitor the operational environment and recommend requirements based on their understanding of operators' needs.

HUMAN INTELLIGENCE AND OPERATIONAL REPORTING

3-130. *Human intelligence* (HUMINT) is the collection of information by a trained human intelligence collector from people and their associated documents and media sources to identify elements, intentions, composition, strength, dispositions, tactics, equipment, personnel, and capabilities (FM 2-22.3). (Trained HUMINT collectors are Soldiers holding military occupational specialties 97E, 351Y [formerly 351C], 351M [formerly 351E], 35E, and 35F, and Marines holding the specialty 0251.) HUMINT uses human sources as tools and a variety of collection methods, both passive and active, to gather information to satisfy intelligence requirements and cross-cue other intelligence disciplines. Interrogation is just one of the HUMINT tasks. (FM 2-22.3 provides the authoritative doctrine for HUMINT operations. It also contains policy for interrogation.) HUMINT operations often collect information that is difficult or sometimes impossible to obtain by other, more technical, means. During COIN operations, much intelligence is based on information gathered from people.

3-131. Operational reporting may also have information of intelligence value that originates from the local populace. People may approach Soldiers and Marines during the course of their day-to-day operations and offer information. Soldiers and Marines should take information and report it to the intelligence section. Doing so allows for verification of the information and establishes a means for HUMINT collectors to contact individuals offering information of value.

3-132. The lives of people offering information on insurgents are often in danger. Insurgents continuously try to defeat collection operations. Careless handling of human sources by untrained personnel can result in murder or intimidation of these sources. When this occurs, HUMINT can be dramatically reduced due to the word spreading that U.S. forces are careless or callous about protecting their sources. HUMINT collectors are trained in procedures that limit the risk to sources and handlers.

3-133. Counterinsurgents should not expect people to willingly provide information if insurgents have the ability to violently intimidate sources. HUMINT reporting increases if counterinsurgents protect the populace from insurgents and people begin to believe the insurgency will be defeated.

3-134. People often provide inaccurate and conflicting information to counterinsurgents. They may be spreading rumors or providing inaccurate information purposefully for their own reasons. Examples of reasons include accomplishing the following:

- Using counterinsurgents to settle tribal, ethnic, or business disputes.
- Leading counterinsurgents into ambushes.
- Enticing counterinsurgents into executing operations that upset the populace.
- Learning about U.S. planning time and tactics.
- Stretching COIN forces thin by causing them to react to false reports.

The accuracy of information obtained by Soldiers and Marines should be verified before being used to support operations. This means that information reported to patrols should be verified with all-source intelligence.

MILITARY SOURCE OPERATIONS

3-135. Because of their continuous contact with the populace, Soldiers and Marines regularly identify potential sources for HUMINT personnel to develop. It is therefore imperative that all counterinsurgents know the PIRs and that every patrol is debriefed. These debriefings should be as detailed as possible. Analysts and HUMINT collectors should work closely with operations staffs and other personnel to ensure new sources are properly developed. (Table 3-8) lists some potential HUMINT sources.)

3-136. Establishing a reliable source network is an effective collection method. Military source operations provide the COIN equivalent of the reconnaissance and surveillance conducted by scouts in conventional operations. HUMINT sources serve as "eyes and ears" on the street and provide an early warning system for tracking insurgent activity. Although counterinsurgents regularly get information from "walk-in" or "walk-up" sources, only HUMINT personnel are trained and authorized to work with HUMINT sources. Due to legal considerations, the potential danger sources face if identified, and the potential danger to troops involved, only HUMINT personnel may conduct military source operations. All Soldiers and Marines may record information given to them by walk-up contacts, including liaison relationships, but they may not develop HUMINT sources or networks. (Refer to FM 2-22.3 for more information on military source operations.)

INTERROGATION OF DETAINEES AND DEBRIEFING OF DEFECTORS

3-137. Detainees and insurgent defectors are important HUMINT sources. The information they provide about the internal workings of an insurgency may be better than any other HUMINT source can provide. In addition, defectors can provide otherwise unobtainable insights into an insurgent organization's perceptions, motivations, goals, morale, organization, and tactics. Both detainees and defectors should be thoroughly questioned on all aspects of an insurgency discussed in section II. Their answers should be considered along with information obtained from captured equipment, pocket litter, and documents to build a better understanding of the insurgency. Properly trained Soldiers and Marines can conduct immediate tactical questioning of detainees or defectors. However, only trained HUMINT personnel are legally authorized to conduct interrogations. A trained debriefer should be used for questioning a defector. All questioning of detainees is conducted to comply with U.S. law and regulation, international law, execution orders and other operationally specific guidelines. (FM 2-22.3 provides the authoritative doctrine and policy for interrogation. Chapter 7 and appendix D of this manual also address this subject.)

Table 3-8. Potential sources of human intelligence

- **Patrol debriefings and after-action reviews**. Patrols regularly encounter individuals offering information and observe new enemy tactics and techniques. Patrol debriefings are especially important to units at brigade level and below; however, the information collected can be of higher echelon significance.
- **Civil affairs reports**. These reports are especially useful for gathering information about politics, economy, and infrastructure. Civil affairs personnel also regularly come into contact with individuals offering information.
- **Psychological operations (PSYOP) reports**. PSYOP personnel conduct opinion polls and gather information on community attitudes, perceptions, interests, and grievances. PSYOP personnel also regularly encounter individuals offering information.
- **Special operations forces reporting**. Special operations forces often work closely with local nationals and produce valuable human intelligence reports.
- **Leadership liaison**. Commanders and leaders regularly meet with their counterparts in the host-nation security forces and with community leaders. These meetings produce information or tips.
- **Contracting**. Contracting officers work with theater contractors, both host nation and external, performing support functions or building infrastructure. Contractors may offer information to contracting officers.
- **Multinational operations centers**. These provide a place to share information between host-nation and U.S. personnel
- **Tips hotlines**. Telephone or e-mail hotlines provide a safe means for people to provide information. They are especially useful for obtaining time-sensitive intelligence, such as warning of an attack or the current location of an insurgent.
- **U.S. persons**. There will be times when U.S. civilians, such as contractors or journalists, offer information to counterinsurgents. For legal reasons, it is important to understand regulations regarding intelligence-related information collected on U.S. persons. (See FM 2-22.3.)

SURVEILLANCE AND RECONNAISSANCE CONSIDERATIONS

3-138. Because all Soldiers and Marines are potential collectors, the ISR plan addresses all day-to-day tactical operations. This means every patrol or mission should be given intelligence collection requirements as well as operations requirements.

3-139. Overt area and zone reconnaissances are excellent means for tactical units to learn more about their AO, especially the terrain, infrastructure, people, government, local leaders, and insurgents. Overt reconnaissance by patrols allows commanders to fill intelligence gaps and develop relationships with local leaders, while simultaneously providing security to the populace.

3-140. Covert reconnaissance and surveillance operations employing scouts or concealed observation posts are often ineffective in places where the populace is alert and suspicious of outsiders. Such places include urban areas, suburban areas, and close-knit communities. In those places, it is very difficult for scouts to conduct reconnaissance or surveillance without being observed by insurgents or people who may tip off insurgents. Reconnaissance of a target may be noticed and cause insurgents to leave the area. Likewise, small groups of scouts may be attractive targets for insurgent attacks if the scouts' location is known. For these reasons, using a HUMINT network or aerial imagery platforms is often preferable to ground reconnaissance and surveillance. Successful ground reconnaissance in populated areas requires leaders to be creative in how they establish observation posts. One technique is for dismounted night patrols to leave a small "stay behind" observation post while the rest of a patrol moves on. Another effective technique is secretly photographing a place of interest while driving by it. However, commanders must weigh the benefits of these operations with the potential cost of insurgents receiving early warning of counterinsurgent intentions.

CONSIDERATIONS FOR OTHER INTELLIGENCE DISCIPLINES

3-141. An *intelligence discipline* is a well-defined area of intelligence collection, processing, exploitation, and reporting using a specific category of technical or human resources (JP 1-02). HUMINT is one of these disciplines. The following discussion addresses COIN-specific considerations for other selected intelligence disciplines and information types. Because of their importance to COIN, counterintelligence is covered separately in section IV and all-source analysis in section V.

SIGNALS INTELLIGENCE

3-142. In conventional environments, signals intelligence (SIGINT) collection is a good source for determining enemy locations, intentions, capabilities, and morale. The same applies in COIN operations. SIGINT is often helpful for confirming or denying HUMINT reporting and may be the primary source of intelligence in areas under insurgent control. Pushing SIGINT collection platforms down to tactical units can therefore improve intelligence collection.

OPEN-SOURCE INTELLIGENCE

3-143. Open-source intelligence (OSINT) is valuable for understanding the operational environment. It is often more useful than any other discipline for understanding public attitudes and public support for insurgents and counterinsurgents. OSINT is also an important means of determining the effectiveness of IO. Monitoring a wide variety of media in multiple languages benefits the COIN effort. If possible, monitoring should occur at every echelon with collection requirements. Each echelon should monitor the media that contain information relevant to operations at that echelon. For instance, reporting by major news networks often matters a lot at the combatant command level; in contrast, local newspapers or radio stations may be more important to tactical units.

IMAGERY INTELLIGENCE

3-144. In COIN operations, imagery intelligence (IMINT) platforms may be used for surveillance of likely insurgent safe houses and other facilities. Further, aerial IMINT platforms are also effective at detecting unusual personnel and supply movements. This information can help commanders determine where best to interdict insurgent lines of communications.

3-145. Static imagery, such as aerial photos of facilities, is useful for detecting long-term changes in structures or activities.

3-146. Real-time video, often from aerial surveillance platforms, is critical to assessing whether particular locations are likely sites of insurgent activity. This capability may also be used to track insurgents during operations. If flown high enough that insurgents cannot hear the platform, real-time video provides surveillance in areas where it is difficult or impossible to use observation posts.

TECHNICAL INTELLIGENCE

3-147. Insurgents often adapt their tactics, techniques, and procedures rapidly. Technical intelligence on insurgent equipment can help understand insurgent capabilities. These may include how insurgents are using improvised explosive devices, homemade mortars, and other pieces of customized military equipment.

MEASUREMENT AND SIGNATURES INTELLIGENCE

3-148. Measurement and signature intelligence (MASINT) sensors can provide remote monitoring of avenues of approach or border regions for smugglers or insurgents. They can also be used to locate insurgent safe havens and cache sites and determining insurgent activities and capabilities. MASINT can also contribute to targeting.

GEOSPATIAL INTELLIGENCE

3-149. Geospatial intelligence (GEOINT) is the exploitation and analysis of imagery and geospatial information to describe, assess, and visually depict physical features and geographically referenced activities on the Earth. GEOINT consists of imagery, IMINT, and geospatial information. GEOINT may have some benefit for identifying smuggling routes and safe havens. Imagery can be very beneficial to operations in urban areas as well. It can help identify structures of interest and aid urban terrain navigation. (Paragraphs B-5 through B-9 contain more GEOINT-related information.)

INTELLIGENCE-RELATED ACTIVITIES

3-150. There are several activities and information sources important to COIN that are not intelligence disciplines but are related to them. Chief among these are target exploitation (TAREX), document exploitation (DOCEX), property ownership records, and financial records.

Target Exploitation and Document Exploitation

3-151. Documents and pocket litter, as well as information found in computers and cell phones, can provide critical information that analysts need to evaluate insurgent organizations, capabilities, and intentions. TAREX and DOCEX are also of great benefit to HUMINT collectors in substantiating what detainees know and whether they are telling the truth.

3-152. TAREX in a COIN environment is like evidence collection in a law enforcement environment. Procedures that ensure captured equipment and documents are tracked accurately and attached to the correct insurgents is necessary. Evidence needs to be enough to justify using operational resources to apprehend the individuals in question; however, it does not necessarily need be enough to convict in a court of law. Pushing HUMINT or law enforcement personnel to the battalion level and below can improve TAREX and DOCEX by tactical units. Procedures for ensuring that tactical units get the results of higher level TAREX and DOCEX are also important. Units must be able to receive intelligence collected from the documents, equipment, and personnel they capture in enough time to exploit it.

Property Ownership Records

3-153. Property ownership records include census records, deeds, and other means of determining ownership of land and buildings. They help counterinsurgents to determine who should or should not be living in a specific area and help them secure the populace. In some cases, it may be necessary for Soldiers and Marines to go door to door and collect census data themselves.

Financial Records

3-154. Information gathered on sources of insurgent funding can be very helpful to the COIN effort. Collection of financial records often requires help from agencies like the Department of the Treasury and financial institutions. It may also require analyzing criminal activities or traditional means of currency transfer.

SECTION IV - COUNTERINTELLIGENCE AND COUNTERRECONNAISSANCE

3-155. Counterintelligence counters or neutralizes intelligence collection efforts through collection, counterintelligence investigations, operations, analysis and production, and functional and technical services. Counterintelligence includes all actions taken to detect, identify, exploit, and neutralize the multidiscipline intelligence activities of friends, competitors, opponents, adversaries, and enemies.

3-156. Insurgents place heavy emphasis on gathering intelligence. They use informants, double agents, reconnaissance, surveillance, open-source media, and open-source imagery. Insurgents can potentially use any person interacting with U.S. or multinational personnel as an informant. These include the same people that U.S. forces use as potential HUMINT sources. Operations security is thus very important; U.S. personnel must carefully screen the contractors, informants, translators, and other personnel working with them. Failure to do so can result in infiltration of U.S. facilities and deaths of U.S. personnel and their partners.

3-157. Background screenings should include collection of personal and biometric data and a search through available reporting databases to determine whether the person is an insurgent. (Biometrics concerns the measurement and analysis of unique physical or behavioral characteristics [as fingerprint or voice patterns].) Identification badges may be useful for providing security and personnel accountability for local people working on U.S. and HN government facilities. Biometric data is preferable, when available, because identification badges may be forged or stolen and insurgents can use them to identify people working with the HN government.

3-158. Insurgents have their own reconnaissance and surveillance networks. Because they usually blend well with the populace, insurgents can execute reconnaissance without easily being identified. They also have an early warning system composed of citizens who inform them of counterinsurgent movements. Identifying the techniques and weaknesses of enemy reconnaissance and surveillance enables commanders to detect signs of insurgent preparations and to surprise insurgents by neutralizing their early warning systems.

3-159. Insurgents may also have a SIGINT capability based on commercially available scanners and radios, wiretaps, or captured counterinsurgent equipment. Counterinsurgents should not use commercial radios or phones because insurgents can collect information from them. If Soldiers and Marines must use commercial equipment or unencrypted communications, they should employ authorized brevity codes to reduce insurgents' ability to collect on them.

SECTION V - ALL-SOURCE INTELLIGENCE

3-160. Joint doctrine defines *all-source intelligence* as products and/or organizations and activities that incorporate all sources of information, most frequently including human resources intelligence, imagery intelligence, measurement and signature intelligence, signals intelligence, and open-source data in the production of finished intelligence (JP 1-02). Intelligence organizations fuse data and information into all-source intelligence products to support COIN operations. Analysis for COIN operations is very challenging, due in part to the—

- Need to understand perceptions and culture.
- Need to track hundreds or thousands of personalities.
- Local nature of insurgencies.
- Tendency of insurgencies to change over time.

3-161. Databases are very important for analyzing insurgent activities and personalities. At a minimum, there should be common searchable combatant command databases of insurgent actions and personnel, as well as another database of all intelligence reporting. These should be accessible by analysts in and out of theater. The common operational picture should include reporting from all units and organizations involved in the effort.

3-162. Because all echelons collect and use intelligence, all staffs are heavily involved in analysis. Units are simultaneously intelligence producers and consumers. This situation is normal at brigade and above; however, battalion staffs often do not have the personnel to collect patrol debriefs, analyze incoming information from multiple sources, produce finished intelligence products, and disseminate products to appropriate consumers. In many cases brigade intelligence sections may also be inadequate for a COIN environment.

3-163. COIN requirements may require pushing analysts to battalion and brigade staffs to give those echelons the required analytical support. There are also instances when analysts can be beneficial at the company level. This is the case when a maneuver company must collect large amounts of information on the local populace and insurgents. An analyst can help collect and process this information and develop an operational picture of the AO. Pushing analysts to brigade level and below places analysts closer to collectors, improves the common operational picture, and helps higher echelon staffs receive answers to their PIRs. Commanders may need to be creative in developing analytical capabilities within their units. Though it is not ideal, commanders have assigned nonintelligence personnel to work in the intelligence section.

3-164. Analysis at brigade and below is the basis for operational-level intelligence. This is due to the bottom-up flow of intelligence in COIN. Battalions and brigades develop intelligence for their AOs; higher echelons fuse it into combatant-command-wide intelligence of the insurgency. Operational-level intelligence adds information about national and international politics and their effects on the operational environment.

3-165. Analysis of enemy actions and comprehensive insurgency analysis are done at battalion level and above. These processes build on IPB and use the tools discussed in appendix B. Analysis of enemy actions, commonly called current operations, focuses on what the enemy is doing now. Comprehensive insurgency analysis focuses on the people in the AO. It develops information about relationships among them and the ideas and beliefs driving their actions. Comprehensive insurgency analysis brings together all other forms of analysis.

CURRENT OPERATIONS

3-166. Current operations intelligence supports a commander's understanding of what insurgents are currently doing. The basic tasks of analysts working in current operations are to—

- Analyze past and current enemy actions (event analysis and pattern analysis) to look for changes in the insurgents' approach or tactics.
- Track the effects of friendly operations on the populace and insurgents.
- Provide intelligence support to ongoing operations.
- Disseminate immediate threat warnings to appropriate consumers.

3-167. Intelligence for current operations comes from a variety of sources, but operations reports are particularly important. This is because current enemy activities are more often reported by patrols, units conducting raids, or observation posts than they are by dedicated intelligence collectors. OSINT is important for tracking IO effects. Current operations analysis depends on the insurgent actions database for determining changes in insurgent tactics and techniques.

COMPREHENSIVE INSURGENCY ANALYSIS

3-168. Accurate and thorough intelligence on insurgent organizations, leadership, financial support networks, and the operational environment contribute to more effective friendly operations. Comprehensive insurgency analysis integrates a range of analytic tools to develop this intelligence. (These tools include social network analysis and socio-cultural factors analysis.) Comprehensive insurgency analysis provides information upon which commanders and staffs base their understanding of the following:

- Insurgent organization.
- Insurgent leadership.
- Key nodes in the insurgent organization.
- Insurgents' approach, capabilities, and motivations.
- Insurgents' support base.
- Insurgent links to the community.

Effectively developing and integrating information from a range of intelligence and operations sources provides the detailed knowledge and insights required to exploit insurgents' vulnerabilities and mitigate their strengths. Table 3-9 lists key tasks associated with comprehensive insurgency analysis.

Table 3-9. Comprehensive insurgency analysis tasks

• Identify insurgent strategic, operational, and tactical goals, objectives, and imperatives.
• Identify motivations, fears, concerns, and perceptions that shape the actions of insurgents and their supporters.
• Identify grievances, fears, and concerns that the insurgents exploit.
• Determine how culture, interests, and history inform insurgent and host-nation decision making.
• Understand links among political, religious, tribal, criminal, and other social networks.
• Determine how social networks, key leaders, and groups interact with insurgent networks.
• Determine the structure and function of insurgent organizations.
• Identify key insurgent activities and leaders.
• Understand popular and insurgent perceptions of the host-nation, insurgency, and counterinsurgents—and how these affect the insurgency.

3-169. Developing knowledge and using network analytic tools requires an unusually large investment of time compared to conventional analytic problem-solving methods. Comprehensive insurgency analysis may not provide immediate usable intelligence. Analysts may have to spend weeks or months analyzing numerous all-source intelligence reports before providing an accurate picture of insurgent groups, leaders and activities. It is essential that commanders designate a group of analysts to perform comprehensive insurgency analysis. This team must be insulated from the short-term demands of current operations and day-to-day intelligence demands. These analysts focus on long-term intelligence development. It is ultimately the commander's responsibility to ensure that comprehensive and basic insurgent network analysis still occurs despite high-profile demands and time sensitive requirements.

3-170. Comprehensive insurgency analysis examines interactions among individuals, groups, and beliefs within the operational environment's historic and cultural context. One of the more important products of this analysis is an understanding of how local people think. This knowledge allows predictive analysis of enemy actions. It also contributes to the ability to develop effective IO and civil-military operations.

REACHBACK

3-171. *Reachback* refers to the process of obtaining products, services, and applications, or forces, or equipment, or material from organizations that are not forward deployed (JP 1-02). Deployed or deploying units should use reach capabilities to "outsource" time-intensive aspects of analysis. Reachback is particularly useful when deployments occur with little warning and when organizations used for reach have a great deal of expertise available on a given subject. Analysts may receive reach assistance from higher echelons or external sources. Most organizations affiliated with DOD regard assisting field commanders as one of their primary missions.

ANALYTIC CONTINUITY

3-172. The complexity and difficulty of analyzing an insurgency means it often takes analysts months to understand the operational environment and the insurgency. The most productive analysts and action officers generally have more than a year focused on an aspect of the insurgency problem; therefore, commanders should try to maintain continuity among their analysts. Intelligence and other staff sections should track operations from their home station and immerse themselves in related intelligence before deploying. This flattens the learning curve of units rotating in to an AO and increases their effectiveness during the deployment.

3-173. Part of unit transition should include exchange of relevant databases and knowledge of the AO. Effective intelligence handover saves time and effort for an incoming unit and ensures consistency of approach.

SECTION VI - INTELLIGENCE COLLABORATION

3-174. Effective intelligence collaboration organizes the collection and analysis actions of various units and organizations into a coherent, mutually supportive intelligence effort. Collaboration and synchronization of the effort between lower and higher echelon units and organizations reduces the likelihood of gaps in the intelligence effort. Some of the important operational- and strategic-level analytic support is called "tactical overwatch." The intelligence portion of the common operational picture and other supporting intelligence for COIN operations is complex. Insurgencies do not normally lend themselves to generalizations like "if this leader is removed, the insurgency is over" or "this group drives the movement." It is important not to oversimplify an insurgency. However, analysts and commanders still require a commonly understood means of defining and describing the enemy. One such means is using the following categories to track and report the insurgency:

- Region.
- Insurgent organization.
- Key personalities.
- Insurgent goals and motivations.

The mutual support that various intelligence units and organizations provide across all echelons facilitates timely and relevant intelligence.

3-175. Insurgencies are often localized; however, most have national or international aspects to them. This characteristic complicates intelligence collaboration between adjacent units and among various echelons. For instance, if numerous insurgent groups operate in one country, adjacent battalions within the country may face very different threats. Higher echelon analysts must then understand multiple insurgent organizations and determine the links, if any, among them. Usually, battalions focus on the population and insurgents in their AO. Higher echelon analysts determine links and interactions among the populace and insurgents across unit boundaries. Combatant-command-level analysts determine the major linkages within the area of responsibility and internationally. Based on these requirements, a common database based on intelligence reporting is a prerequisite for effective intelligence fusion.

3-176. Also complicating collaboration is the fact that COIN operations involve many government agencies and foreign security forces. Analysts must establish good working relationships with various agencies and elements to ensure they can fuse intelligence.

INTELLIGENCE CELLS AND WORKING GROUPS

3-177. Intelligence community assets operating in an AO work in or coordinate with the intelligence cell in one of the unit's command posts. They are under the staff supervision of the unit intelligence officer. Table 3-10 (page 3-34) lists examples of intelligence community assets that may operate in a division AO.

3-178. As necessary, intelligence officers form working groups or boards to synchronize collection, analysis, and targeting efforts. Cells and working groups conduct regular meetings to accomplish the following:

- Establish and maintain shared situational awareness.
- Share collection priorities.
- Deconflict activities and operations.

- Discuss target development.
- Share results of operations.

These meetings build mutual trust and understanding of each member's mission, capabilities, and limitations. Meetings should be coordinated with meetings of other staff cells, working groups, and boards (for example, the targeting board) as part of the command post's battle rhythm.

3-179. An effective intelligence cell enhances the commander's knowledge of the enemy, local populace, and friendly forces and agencies operating in the unit's AO. Incorporating HN representatives (for example, intelligence services, military forces, and local government officials) and multinational partners into the intelligence cell should also be considered to foster teamwork, gain insight into the local society, and prepare the host nation to assume the COIN mission when multinational forces depart.

Table 3-10. Possible intelligence community assets in a division area of operations

• Defense Intelligence Agency ▪ Defense human intelligence case officers ▪ Reports officers ▪ Document exploitation teams • Central Intelligence Agency ▪ Chief of base ▪ Case officers ▪ Reports officers • National intelligence support team (from the U.S. Army Intelligence and Security Command —INSCOM) • National Geospatial-Intelligence Agency • Department of the Treasury analytic support teams	• U.S. Special Operations Command ▪ Special forces operational detachment-alpha teams ▪ Special mission units ▪ Civil affairs teams ▪ Psychological operations teams • Immigration and Customs Enforcement/ Department of Homeland Security agents • Air Force Office of Special Investigations special agents • Federal Bureau of Investigation agents • Department of State political advisor • National Security Agency

Protecting Sources

3-180. Protecting sources is another important consideration when sharing intelligence. Organizations may sometimes choose not to share information because acting on intelligence can compromise its sources. Using the targeting process to synchronize targeting decisions is usually a good way to protect sources. (See paragraphs 5-100 through 5-112.)

HOST-NATION INTEGRATION

3-181. COIN operations require U.S. personnel to work closely with the host nation. Sharing intelligence with HN security forces and government personnel is an important and effective means of supporting their COIN efforts. However, HN intelligence services may not be well developed and their personnel may not be well trained. Thus, HN intelligence should be considered useful but definitely not the only intelligence available. Usually, HN services are not the most important intelligence source. It is essential for U.S. personnel to evaluate HN intelligence capabilities and offer training as required.

3-182. In addition, infiltration of HN security forces by insurgents or foreign intelligence services can create drawbacks to intelligence sharing. Insurgents may learn what is known about them, gain insight into COIN intelligence sources and capabilities, and get early warning of targeting efforts.

3-183. When sharing intelligence with the host nation, it is important to understand the level of infiltration by insurgents or foreign intelligence services. Insofar as possible, intelligence should be tailored so required intelligence still gets to HN consumers but does not give away information about sources and capabilities. In addition, care is needed when providing targeting information; it should be done such that insurgents do not receive early warning of an upcoming operation. As trust develops between HN and U.S. personnel, the amount of intelligence shared should grow. This will make the COIN effort more effective.

SECTION VII - SUMMARY

3-184. What makes intelligence analysis for COIN so distinct and so challenging is the amount of socio-cultural information that must be gathered and understood. However, truly grasping the operational environment requires commanders and staffs to devote at least as much effort to understanding the people they support as they do to understanding the enemy. All this information is essential to get at the root causes of the insurgency and to determine the best ways to combat it. Identifying the real problem and developing solutions is the essence of operational design, which is discussed in the next chapter.

This page intentionally left blank.

Chapter 4

Designing Counterinsurgency Campaigns and Operations

The first, the supreme, the most far-reaching act of judgment that the statesman and commander have to make is to establish...the kind of war on which they are embarking; neither mistaking it for, nor trying to turn it into, something that is alien to its nature. This is the first of all strategic questions and the most comprehensive.

Carl von Clausewitz, *On War*

This chapter describes considerations for designing counterinsurgency campaigns and operations. For Army forces, this chapter applies aspects of command and control doctrine and planning doctrine to counterinsurgency campaign planning. While campaign design is most often associated with a joint force command, all commanders and staffs need to understand it.

THE IMPORTANCE OF CAMPAIGN DESIGN

4-1. In chapter 1, insurgency is described as an organized, protracted politico-military struggle designed to weaken government control and legitimacy while increasing insurgent control. Ultimately, the long-term objective for both sides in that struggle remains acceptance by the people of the state or region of the legitimacy of one side's claim to political power. The reason an insurgency forms to challenge the existing order is different in each case. The complexity of insurgency presents problems that have incomplete, contradictory, and changing requirements. The solutions to these intensely challenging and complex problems are often difficult to recognize as such because of complex interdependencies. While attempting to solve an intensely complex problem, the solution of one of its aspects may reveal or create another, even more complex, problem. The purpose of design is to achieve a greater understanding, a proposed solution based on that understanding, and a means to learn and adapt. For a U.S. military commander directed to counter an insurgency, knowing why an insurgent movement has gained support and the purpose of American involvement is essential in designing a counterinsurgency (COIN) campaign. Failure to understand both factors can have disastrous consequences, as illustrated by Napoleon's experience in Spain.

Campaign Assessment and Reassessment

During Napoleon's occupation of Spain in 1808, it seems little thought was given to the potential challenges of subduing the Spanish populace. Conditioned by the decisive victories at Austerlitz and Jena, Napoleon believed the conquest of Spain would be little more than a "military promenade." Napoleon's campaign included a rapid conventional military victory but ignored the immediate requirement to provide a stable environment for the populace.

The French failed to analyze the Spanish people, their history, culture, motivations, and potential to support or hinder the achievement of French political objectives. The Spanish people were accustomed to hardship, suspicious of foreigners and constantly involved in skirmishes with security forces. Napoleon's cultural miscalculation resulted in a protracted occupation struggle that lasted nearly six years and ultimately required approximately three-fifths of the Empire's total armed strength, almost four times the force of 80,000 Napoleon originally designated.

> The Spanish resistance drained the resources of the French Empire. It was the beginning of the end for Napoleon. At the theater level, a complete understanding of the problem and a campaign design that allowed the counterinsurgency force to learn and adapt was lacking.

4-2. Design and planning are qualitatively different yet interrelated activities essential for solving complex problems. While planning activities receive consistent emphasis in both doctrine and practice, discussion of design remains largely abstract and is rarely practiced. Presented a problem, staffs often rush directly into planning without clearly understanding the complex environment of the situation, purpose of military involvement, and approach required to address the core issues. This situation is particularly problematic with insurgencies. Campaign design informs and is informed by planning and operations. It has an intellectual foundation that aids continuous assessment of operations and the operational environment. Commanders should lead the design process and communicate the resulting framework to other commanders for planning, preparation, and execution.

THE RELATIONSHIP BETWEEN DESIGN AND PLANNING

4-3. It is important to understand the distinction between design and planning. (See figure 4-1.) While both activities seek to formulate ways to bring about preferable futures, they are cognitively different. Planning applies established procedures to solve a largely understood problem within an accepted framework. Design inquires into the nature of a problem to conceive a framework for solving that problem. In general, planning is problem solving, while design is problem setting. Where planning focuses on generating a plan—a series of executable actions—design focuses on learning about the nature of an unfamiliar problem.

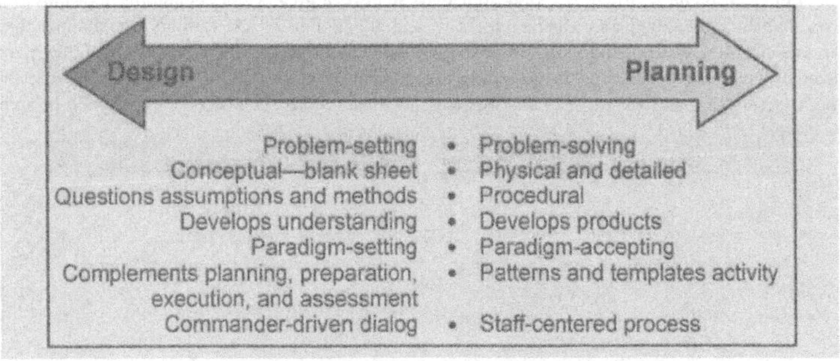

Figure 4-1. Design and planning continuum

4-4. When situations do not conform to established frames of reference—when the hardest part of the problem is figuring out what the problem is—planning alone is inadequate and design becomes essential. In these situations, absent a design process to engage the problem's essential nature, planners default to doctrinal norms; they develop plans based on the familiar rather than an understanding of the real situation. Design provides a means to conceptualize and hypothesize about the underlying causes and dynamics that explain an unfamiliar problem. Design provides a means to gain understanding of a complex problem and insights towards achieving a workable solution.

4-5. This description of design at the tactical level is a form of what Army doctrine calls commander's visualization. Commanders begin developing their design upon receipt of a mission. Design precedes and forms the foundation for staff planning. However, design is also continuous throughout the operation. As part of assessment, commanders continuously test and refine their design to ensure the relevance of military action to the situation. In this sense, design guides and informs planning, preparation, execution, and assessment. However, a plan is necessary to translate a design into execution. (FM 6-0, paragraphs 4-17 through 4-25, discusses commander's visualization.)

4-6. Planning focuses on the physical actions intended to directly affect the enemy or environment. Planners typically are assigned a mission and a set of resources; they devise a plan to use those resources to accomplish that mission. Planners start with a design (whether explicit or implicit) and focus on generating a plan—a series of executable actions and control measures. Planning generally is analytic and reductionist. It breaks the design into manageable pieces assignable as tasks, which is essential to transforming the design into an executable plan. Planning implies a stepwise process in which each step produces an output that is the necessary input for the next step. (FM 5-0 contains Army planning doctrine. MCDP 5 contains Marine Corps planning doctrine.)

THE NATURE OF DESIGN

4-7. Given the difficult and multifaceted problems of insurgencies, dialog among the commander, principal planners, members of the interagency team, and host-nation (HN) representatives helps develop a coherent design. This involvement of all participants is essential. The object of this dialog is to achieve a level of situational understanding at which the approach to the problem's solution becomes clear. The underlying premise is this: when participants achieve a level of understanding such that the situation no longer appears complex, they can exercise logic and intuition effectively. As a result, design focuses on framing the problem rather than developing courses of action.

4-8. COIN design must be iterative. By their nature, COIN efforts require repeated assessments from different perspectives to see the various factors and relationships required for adequate understanding. Assessment and learning enable incremental improvements to the design. The aim is to rationalize the problem—to construct a logical explanation of observed events and subsequently construct the guiding logic that unravels the problem. The essence of this is the mechanism necessary to achieve success. This mechanism may not be a military activity—or it may involve military actions in support of nonmilitary activities. Once commanders understand the problem and what needs to be accomplished to succeed, they identify the means to assess effectiveness and the related information requirements that support assessment. This feedback becomes the basis for learning, adaptation, and subsequent design adjustment.

CONSIDERATIONS FOR DESIGN

4-9. Key design considerations include the following:

- Critical discussion.
- Systems thinking.
- Model making.
- Intuitive decision making.
- Continuous assessment.
- Structured learning.

4-10. Rigorous and structured *critical discussion* provides an opportunity for interactive learning. It deepens shared understanding and leverages the collective intelligence and experiences of many people.

4-11. *Systems thinking* involves developing an understanding of the relationships within the insurgency and the environment. It also concerns the relationships of actions within the various logical lines of operations (LLOs). This element is based on the perspective of the systems sciences that seeks to understand the interconnectedness, complexity, and wholeness of the elements of systems in relation to one another.

4-12. In *model making,* the model describes an approach to the COIN campaign, initially as a hypothesis. The model includes operational terms of reference and concepts that shape the language governing the conduct (planning, preparation, execution, and assessment) of the operation. It addresses questions like these: Will planning, preparation, execution, and assessment activities use traditional constructs like center of gravity, decisive points, and LLOs? Or are other constructs—such as leverage points, fault lines, or critical variables—more appropriate to the situation?

4-13. The Army and Marine Corps define *intuitive decision making* as the act of reaching a conclusion which emphasizes pattern recognition based on knowledge, judgment, experience, education, intelligence, boldness, perception, and character. This approach focuses on assessment of the situation vice comparison of multiple options (FM 6-0/MCRP 5-12A). An operational design emerges intuitively as understanding of the insurgency deepens.

4-14. *Continuous assessment* is essential as an operation unfolds because of the inherent complexity of COIN operations. No design or model completely matches reality. The object of continuous assessment is to identify where and how the design is working or failing and to consider adjustments to the design and operation.

4-15. The objective of *structured learning* is to develop a reasonable initial design and then learn, adapt, and iteratively and continuously improve that design as more about the dynamics of the COIN problem become evident.

DESIGN FOR COUNTERINSURGENCY

4-16. Through design commanders gain an understanding of the problem and the COIN operation's purpose within the strategic context. Communicating this understanding of the problem, purpose, and context to subordinates allows them to exercise subordinates' initiative. *Subordinates' initiative* is assumption of responsibility for deciding and initiating independent actions when the concept of operations or order no longer applies or when an unanticipated opportunity leading to the accomplishment of the commander's intent presents itself (FM 6-0). (Subordinates' initiative is discussed in FM 6-0, paragraphs 2-83 through 2-92.) It facilitates decentralized execution and continuous assessment of operations at all levels throughout the campaign. While traditional aspects of campaign design as expressed in joint and Service doctrine remain relevant, they are not adequate for a discussion of the broader design construct for a COIN environment. Inherent in this construct is the tension created by understanding that military capabilities provide only one component of an overall approach to a COIN campaign. Design of a COIN campaign must be viewed holistically. Only a comprehensive approach employing all relevant design components, including the other instruments of national power, is likely to reach the desired end state.

4-17. As noted above, this description of campaign design is a form that Army doctrine calls commander's visualization. Design begins with identification of the end state, as derived from the policy aim. (Joint doctrine defines the *end state* as the set of required conditions that defines achievement of the commander's objectives [JP 1-02]). The end state provides context and logic for operational and tactical decision making. Consequently, strategic goals must be communicated clearly to commanders at every level. While strategy drives design, which in turn drives tactical actions, the reverse is also true. The observations of tactical actions result in learning and greater understanding that may generate modifications to the design, which in turn may have strategic implications. The COIN imperative to "Learn and Adapt" is essential in making the design process work correctly. Figure 4-2 illustrates the iterative nature of COIN campaign design and the large number of factors involved.

COMMANDER'S INTENT AND VISION OF RESOLUTION

4-18. Guided by the campaign's purpose, commanders articulate an operational logic for the campaign that expresses in clear, concise, conceptual language a broad vision of what they plan to accomplish. The operational logic is the commander's assessment of the problem and approach toward solving it. Commanders express it as the commander's intent. Ideally, the operational logic is expressed clearly and simply but in comprehensive terms, such as what the commander envisions achieving with various components or particular LLOs. This short statement of the operational logic helps subordinate commanders and planners, as well as members of other agencies and organizations, see the campaign's direction. It provides a unifying theme for interagency planning.

4-19. In addition, commanders also issue a form of planning guidance called the vision of resolution. The vision of resolution is usually expressed in the form of LLOs. LLOs for a counterinsurgency may include the following:

- Conduct information operations.
- Conduct combat operations/civil security operations.
- Train and employ HN security forces.

- Establish or restore essential services.
- Support development of better governance.
- Support economic development.

This list is an example only. Commanders determine the LLOs appropriate to the situation based on their assessment and their dialog with the leaders of other participating organizations.

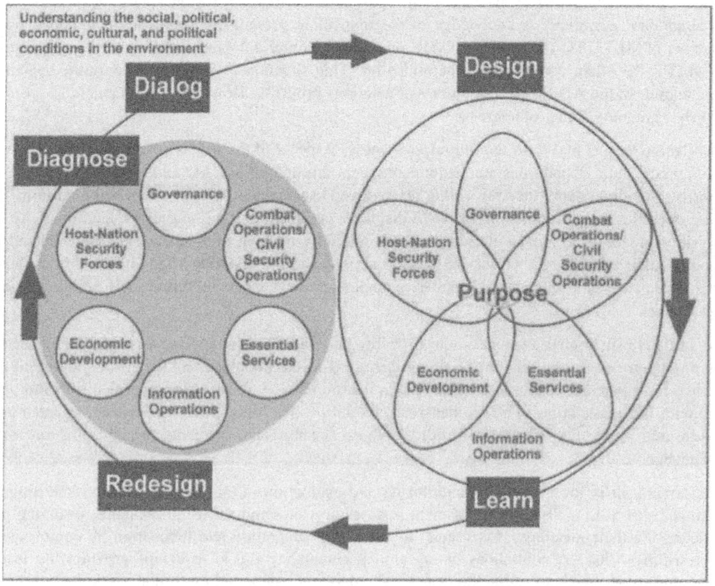

Figure 4-2. Iterative counterinsurgency campaign design

4-20. LLOs like those listed in paragraph 4-19 are not intended as a "success template." Selecting and applying them requires judgment. The mosaic nature of insurgencies and the shifting circumstances within each area of operations (AO) requires a different emphasis on and interrelationship among the various lines. The situation may also require that military forces closely support, or temporarily assume responsibility for, tasks normally accomplished by other government agencies and private organizations. By broadly describing how the LLOs interact to achieve the end state, commanders provide the operational logic to link the various components in a comprehensive framework. This framework guides the initiative of subordinate commanders as they establish local conditions that support achieving the overall end state. It also promotes unity of effort among joint, interagency, multinational, and HN partners.

LOCAL RELEVANCE

4-21. Informed by the commander's intent—including the end state and vision of resolution—subordinate commanders tailor and prioritize their actions within the LLOs based on the distinct and evolving circumstances within their respective AOs. Military forces are accustomed to unity of command; however, the interagency and multinational nature of COIN operations usually makes such arrangements unlikely. All participating organizations do share attitudes and goals. General cooperation on matters of mutual concern, established through informal agreements, may be the most practicable arrangement. Therefore, effective commanders empower subordinate leaders to perform the coordination, cooperation, and innovation needed to achieve unity of effort and execute operations in the manner best suited to local conditions. The design—consisting of the commander's intent, vision of resolution and other guidance issued as the campaign unfolds, and end state—provides the framework within which subordinates exercise this form of initiative.

LEARNING IN EXECUTION

4-22. Before commanders deploy their units, they make every effort to mentally prepare their Soldiers or Marines for the anticipated challenges, with a particular focus on situational awareness of the anticipated AO. *Situational awareness* is knowledge of the immediate present environment, including knowledge of the factors of METT-TC (FMI 5-0.1). COIN operations require a greater focus on civil considerations—the C in METT-TC—than conventional operations do. This situational awareness is only the beginning of an understanding of the AO that will mature as operations progress. However, commanders use it to begin to establish a common frame of reference.

4-23. Design begins based on this initial awareness. Aspects of the problem and means of resolving them do not remain static. Conditions are seldom consistent throughout any AO and continue to change based on actions by friendly, enemy, neutral, and other involved organizations. Rather than being uniform in character, the operational environment is likely to display a complex, shifting mosaic of conditions. To be effective, commanders—and indeed all personnel—continually develop and enhance their understanding of the mosaic peculiar to their AO. Observing tactical actions and the resulting changing conditions deepens understanding of the environment and enables commanders to relearn and refine their design and implementation actions.

4-24. Initially, situational awareness will probably be relatively low and the design will, by necessity, require a number of assumptions, especially with respect to the populace and the force's ability to positively influence their perception of events. The design can be viewed as an experiment that tests the operational logic, with the expectation of a less-than-perfect solution. As the experiment unfolds, interaction with the populace and insurgents reveals the validity of those assumptions, revealing the strengths and weaknesses of the design.

4-25. *Assessment* is the continuous monitoring and evaluation of the current situation and progress of an operation (FMI 5-0.1). Effective assessment is necessary for commanders to recognize changing conditions and determine their meaning. It is crucial to successful adaptation and innovation by commanders within their respective AOs. A continuous dialog among commanders at all echelons provides the feedback the senior commander needs to refine the design. The dialog is supported by formal assessment techniques and red-teaming to ensure commanders are fully cognizant of the causal relationships between their actions and the insurgents' adaptations. Accordingly, assessment is a learning activity and a critical aspect of design. This learning leads to redesign. Therefore, design can be viewed as a perpetual design-learn-redesign activity, with the commander's intent, vision of resolution, and end state providing the unifying themes.

4-26. The critical role of assessment necessitates establishing measures of effectiveness during planning. Commanders should choose these carefully so that they align with the design and reflect the emphasis on and interrelationship among the LLOs. Commanders and staffs revise their assessment and measures of effectiveness during the operation in order to facilitate redesign and stay abreast of the current situation. Sound assessment blends qualitative and quantitative analysis with the judgment and intuition of all leaders. Great care must be applied here, as COIN operations often involve complex societal issues that may not lend themselves to quantifiable measures of effectiveness. Moreover, bad assumptions and false data can undermine the validity of both assessments and the conclusions drawn from them. Data and metrics can inform a commander's assessment. However they must not be allowed to dominate it in uncertain situations. Subjective and intuitive assessment must not be replaced by an exclusive focus on data or metrics. Commanders must exercise their professional judgment in determining the proper balance.

GOALS IN COUNTERINSURGENCY

4-27. In an ideal world, the commander of military forces engaged in COIN operations would enjoy clear and well-defined goals for the campaign from the very beginning. However, the reality is that many goals emerge only as the campaign develops. For this reason, counterinsurgents usually have a combination of defined and emerging goals toward which to work. Likewise, the complex problems encountered during COIN operations can be so difficult to understand that a clear design cannot be developed initially. Often, the best choice is to create iterative solutions to better understand the problem. In this case, these iterative solutions allow the initiation of intelligent interaction with the environment. The experiences of the 1st Marine Division during Operation Iraqi Freedom II illustrate this situation.

Iterative Design During Operation Iraqi Freedom II

During Operation Iraqi Freedom II (2004-2005), the 1st Marine Division employed an operational design similar to that used during the Philippine Insurrection (circa 1902). The commanding general, Major General James N. Mattis, USMC, began with an assessment of the people that the Marines, Soldiers, and Sailors would encounter within the division's area of operations. The area of operations was in western Iraq/Al Anbar Province, which had a considerably different demographic than the imam-led Shia areas in which the division had operated during Operation Iraqi Freedom I.

Major General Mattis classified provincial constituents into three basic groups: the tribes, former regime elements, and foreign fighters. The tribes constituted the primary identity group in western Iraq/Al Anbar Province. They had various internal tribal affiliations and looked to a diverse array of sheiks and elders for leadership. The former regime elements were a minority that included individuals with personal, political, business, and professional ties to the Ba'ath Party. These included civil servants and career military personnel with the skills needed to run government institutions. Initially, they saw little gain from a democratic Iraq. The foreign fighters were a small but dangerous minority of transnational Islamic subversives.

To be successful, U.S. forces had to apply a different approach to each of these groups within the framework of an overarching plan. As in any society, some portion of each group included a criminal element, further complicating planning and interaction. Major General Mattis's vision of resolution comprised two major elements encompassed in an overarching "bodyguard" of information operations. (See figure 4-3, page 4-8.)

The first element and main effort was diminishing support for insurgency. Guided by the maxims of "first do no harm" and "no better friend–no worse enemy," the objective was to establish a secure local environment for the indigenous population so they could pursue their economic, social, cultural, and political well-being and achieve some degree of local normalcy. Establishing a secure environment involved both offensive and defensive combat operations with a heavy emphasis on training and advising the security forces of the fledgling Iraqi government. It also included putting the populace to work. Simply put, an Iraqi with a job was less likely to succumb to ideological or economic pressure to support the insurgency. Other tasks included the delivery of essential services, economic development, and the promotion of governance. All were geared towards increasing employment opportunities and furthering the establishment of local normalcy. Essentially, diminishing support for insurgency entailed gaining and maintaining the support of the tribes, as well as converting as many of the former regime members as possible. "Fence-sitters" were considered a winnable constituency and addressed as such.

The second element involved neutralizing the bad actors, a combination of irreconcilable former regime elements and foreign fighters. Offensive combat operations were conducted to defeat recalcitrant former regime members. The task was to make those who were not killed outright see the futility of resistance and give up the fight. With respect to the hard-core extremists, who would never give up, the task was more straightforward: their complete and utter destruction. Neutralizing the bad actors supported the main effort by improving the local security environment. Neutralization had to be accomplished in a discrete and discriminate manner, however, in order to avoid unintentionally increasing support for insurgency.

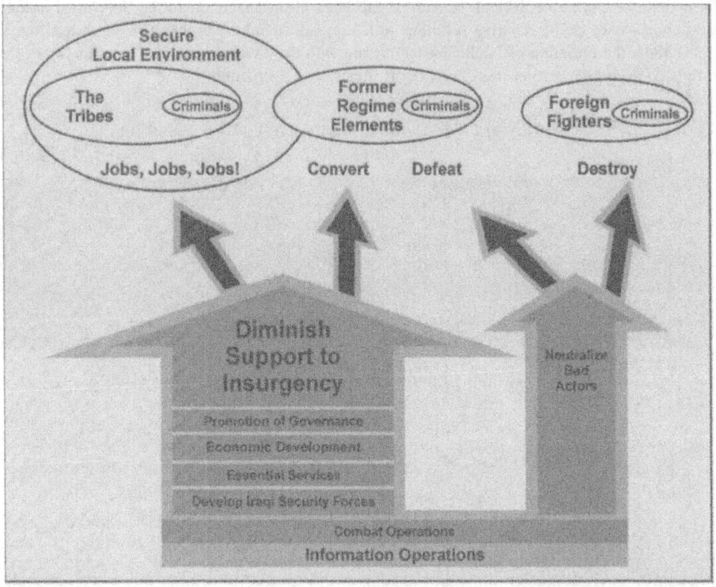

Figure 4-3. 1st Marine Division's operational design for Operation Iraqi Freedom II

Both elements described above were wrapped in an overarching "bodyguard" of information operations. Information operations, both proactive and responsive, were aggressively employed to favorably influence the populace's perception of all coalition actions while simultaneously discrediting the insurgents. These tasks were incredibly difficult for a number of reasons. Corruption had historically been prevalent among Iraqi officials, generating cynicism toward any government. Additionally, decades of Arab media mischaracterization of U.S. actions had instilled distrust of American motives. The magnitude of that cynicism and distrust highlighted the critical importance of using information operations to influence every situation.

In pursuing this vision of resolution, the 1st Marine Division faced an adaptive enemy. Persistent American presence and interaction with the populace threatened the insurgents and caused them to employ more open violence in selected areas of Al Anbar province. This response resulted in learning and adaptation within the 1st Marine Division. The design enabled 1st Marine Division to adjust the blend of "diminishing support for insurgents" and "neutralizing bad actors" to meet the local challenges. Throughout the operation, 1st Marine Division continued learning and adapting with the vision of resolution providing a constant guide to direct and unify the effort.

SUMMARY

4-28. Campaign design may very well be the most important aspect of countering an insurgency. It is certainly the area in which the commander and staff can have the most influence. Design is not a function to be accomplished, but rather a living process. It should reflect ongoing learning and adaptation and the growing appreciation counterinsurgents share for the environment and all actors within it, especially the insurgents, populace, and HN government. Though design precedes planning, it continues throughout planning, preparation, and execution. It is dynamic, even as the environment and the counterinsurgents' understanding of the environment is dynamic. The resulting growth in understanding requires integrated assessment and a rich dialog among leaders at various levels to determine the need for adaptation throughout the COIN force. Design should reflect a comprehensive approach that works across all LLOs in a manner applicable to the stage of the campaign. There should only be one campaign and therefore one design. This single campaign should bring in all players, with particular attention placed on the HN participants. Design and operations are integral to the COIN imperative to "Learn and Adapt," enabling a continuous cycle of design-learn-redesign to achieve the end state.

This page intentionally left blank.

Chapter 5

Executing Counterinsurgency Operations

It is a persistently methodical approach and steady pressure which will gradually wear the insurgent down. The government must not allow itself to be diverted either by countermoves on the part of the insurgent or by the critics on its own side who will be seeking a simpler and quicker solution. There are no short-cuts and no gimmicks.

Sir Robert Thompson,
Defeating Communist Insurgency: The Lessons of Malaya and Vietnam, 1966 [1]

This chapter addresses principles and tactics for executing counterinsurgency (COIN) operations. It begins by describing the different stages of a COIN operation and logical lines of operations that commanders can use to design one. It continues with discussions of three COIN approaches and how to continuously assess a COIN operation. The chapter concludes by describing lethal and nonlethal targeting in a COIN environment.

THE NATURE OF COUNTERINSURGENCY OPERATIONS

5-1. Counterinsurgency (COIN) operations require synchronized application of military, paramilitary, political, economic, psychological, and civic actions. Successful counterinsurgents support or develop local institutions with legitimacy and the ability to provide basic services, economic opportunity, public order, and security. The political issues at stake are often rooted in culture, ideology, societal tensions, and injustice. As such, they defy nonviolent solutions. Military forces can compel obedience and secure areas; however, they cannot by themselves achieve the political settlement needed to resolve the situation. Successful COIN efforts include civilian agencies, U.S. military forces, and multinational forces. These efforts purposefully attack the basis for the insurgency rather than just its fighters and comprehensively address the host nation's core problems. Host-nation (HN) leaders must be purposefully engaged in this effort and ultimately must take lead responsibility for it.

5-2. There are five overarching requirements for successful COIN operations:

- U.S. and HN military commanders and the HN government together must devise the plan for attacking the insurgents' strategy and focusing the collective effort to bolster or restore government legitimacy.
- HN forces and other counterinsurgents must establish control of one or more areas from which to operate. HN forces must secure the people continuously within these areas.
- Operations should be initiated from the HN government's areas of strength against areas under insurgent control. The host nation must retain or regain control of the major population centers to stabilize the situation, secure the government's support base, and maintain the government's legitimacy.
- Regaining control of insurgent areas requires the HN government to expand operations to secure and support the population. If the insurgents have established firm control of a region, their military apparatus there must be eliminated and their politico-administrative apparatus rooted out.
- Information operations (IO) must be aggressively employed to accomplish the following:

- Favorably influence perceptions of HN legitimacy and capabilities.
- Obtain local, regional, and international support for COIN operations.
- Publicize insurgent violence.
- Discredit insurgent propaganda and provide a more compelling alternative to the insurgent ideology and narrative.

5-3. COIN operations combine offensive, defensive, and stability operations to achieve the stable and secure environment needed for effective governance, essential services, and economic development. The focus of COIN operations generally progresses through three indistinct stages that can be envisioned with a medical analogy:

- Stop the bleeding.
- Inpatient care—recovery.
- Outpatient care—movement to self-sufficiency.

Understanding this evolution and recognizing the relative maturity of the operational environment are important to the conduct (planning, preparation, execution, and assessment) of COIN operations. This knowledge allows commanders to ensure that their activities are appropriate to the current situation.

INITIAL STAGE: "STOP THE BLEEDING"

5-4. Initially, COIN operations are similar to emergency first aid for the patient. The goal is to protect the population, break the insurgents' initiative and momentum, and set the conditions for further engagement. Limited offensive operations may be undertaken, but are complemented by stability operations focused on civil security. During this stage, friendly and enemy information needed to complete the common operational picture is collected and initial running estimates are developed. Counterinsurgents also begin shaping the information environment, including the expectations of the local populace.

MIDDLE STAGE: "INPATIENT CARE—RECOVERY"

5-5. The middle stage is characterized by efforts aimed at assisting the patient through long-term recovery or restoration of health—which in this case means achieving stability. Counterinsurgents are most active here, working aggressively along all logical lines of operations (LLOs). The desire in this stage is to develop and build resident capability and capacity in the HN government and security forces. As civil security is assured, focus expands to include governance, provision of essential services, and stimulation of economic development. Relationships with HN counterparts in the government and security forces and with the local populace are developed and strengthened. These relationships increase the flow of human and other types of intelligence. This intelligence facilitates measured offensive operations in conjunction with the HN security forces. The host nation increases its legitimacy through providing security, expanding effective governance, providing essential services, and achieving incremental success in meeting public expectations.

LATE STAGE: "OUTPATIENT CARE—MOVEMENT TO SELF-SUFFICIENCY"

5-6. Stage three is characterized by the expansion of stability operations across contested regions, ideally using HN forces. The main goal for this stage is to transition responsibility for COIN operations to HN leadership. In this stage, the multinational force works with the host nation in an increasingly supporting role, turning over responsibility wherever and whenever appropriate. Quick reaction forces and fire support capabilities may still be needed in some areas, but more functions along all LLOs are performed by HN forces with the low-key assistance of multinational advisors. As the security, governing, and economic capacity of the host nation increases, the need for foreign assistance is reduced. At this stage, the host nation has established or reestablished the systems needed to provide effective and stable government that sustains the rule of law. The government secures its citizens continuously, sustains and builds legitimacy through effective governance, has effectively isolated the insurgency, and can manage and meet the expectations of the nation's entire population.

LOGICAL LINES OF OPERATIONS IN COUNTERINSURGENCY

5-7. Commanders use LLOs to visualize, describe, and direct operations when positional reference to enemy forces has little relevance. (See figure 5-1.) LLOs are appropriate for synchronizing operations against enemies that hide among the populace. A plan based on LLOs unifies the efforts of joint, interagency, multinational, and HN forces toward a common purpose. Each LLO represents a conceptual category along which the HN government and COIN force commander intend to attack the insurgent strategy and establish HN government legitimacy. LLOs are closely related. Successful achievement of the end state requires careful coordination of actions undertaken along all LLOs.

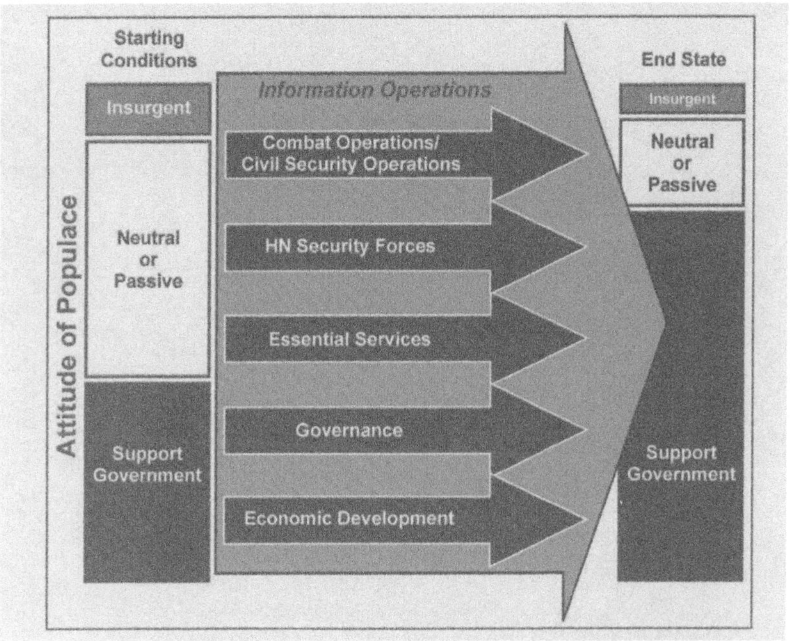

Figure 5-1. Example logical lines of operations for a counterinsurgency

5-8. Success in one LLO reinforces successes in the others. Progress along each LLO contributes to attaining a stable and secure environment for the host nation. Stability is reinforced by popular recognition of the HN government's legitimacy, improved governance, and progressive, substantive reduction of the root causes of the insurgency. There is no list of LLOs that applies in all cases. Commanders select LLOs based on their understanding of the nature of the insurgency and what the COIN force must do to counter it. Commanders designate LLOs that best focus counterinsurgent efforts against the insurgents' subversive strategy.

5-9. Commanders at all echelons can use LLOs. Lower echelon operations are nested within the higher echelon's operational design and LLOs; however, lower echelon operations are conducted based on the operational environment in each unit's area of operations (AO).

5-10. The commander's intent and vision of resolution, expressed as LLOs, describe the design for a COIN operation. Commanders and staffs synchronize activities along all LLOs to gain unity of effort. This approach ensures the LLOs converge on a well-defined, commonly understood end state.

5-11. LLOs are directly related to one another. They connect objectives that, when accomplished, support achieving the end state. Operations designed using LLOs typically employ an extended, event-driven timeline with short-, mid-, and long-term goals. These operations combine the effects of long-term operations, such as neutralizing the insurgent infrastructure, with cyclic and short-term events, like regular trash collection and attacks against insurgent bases. (See figure 5-2.)

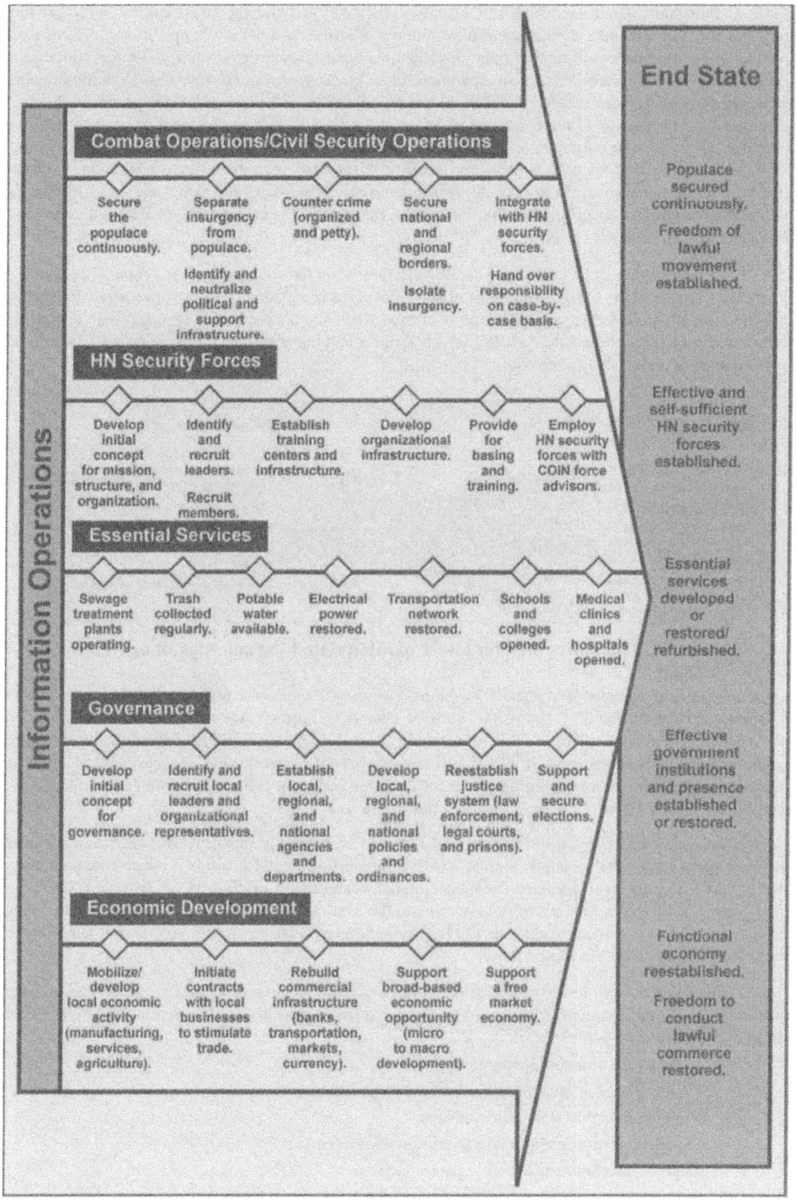

Figure 5-2. Example goals and objectives along logical lines of operations

5-12. Commanders determine which LLOs apply to their AO and how the LLOs connect with and support one another. For example, commanders may conduct offensive and defensive operations to form a shield behind which simultaneous stability operations can maintain a secure environment for the populace. Accomplishing the objectives of combat operations/civil security operations sets the conditions needed to achieve essential services and economic development objectives. When the populace perceives that the environment is safe enough to leave families at home, workers will seek employment or conduct public economic activity. Popular participation in civil and economic life facilitates further provision of essential services and development of greater economic activity. Over time such activities establish an environment that attracts outside capital for further development. Neglecting objectives along one LLO risks creating vulnerable conditions along another that insurgents can exploit. Achieving the desired end state requires linked successes along all LLOs.

5-13. The relationship of LLOs to the overall operation is similar to the stands of a rope. (See figure 5-3.) Each LLO is a separate string. Operations along it cannot accomplish all objectives required for success in a COIN operation. However, a strong rope is created when strands are woven together. The overall COIN effort is further strengthened through IO, which support and enhance operations along all LLOs by highlighting the successes along each one.

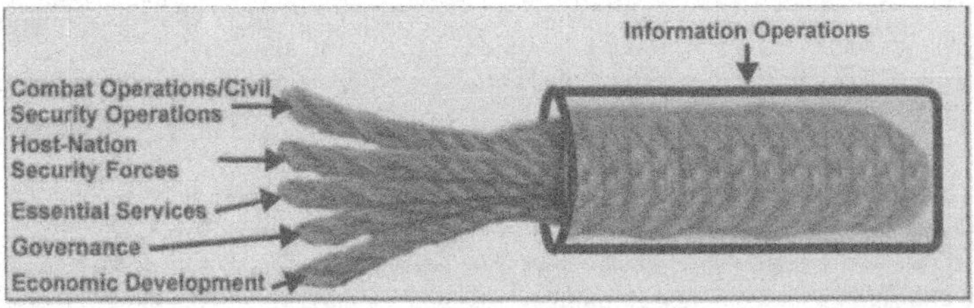

Figure 5-3. The strengthening effect of interrelated logical lines of operations

5-14. LLOs help commanders identify missions, assign tasks, allocate resources, and assess operations. Commanders specify the LLO that is the decisive operation; others shape the operational environment for the decisive operation's success. This prioritization usually changes as COIN operations create insurgent vulnerabilities, insurgents react or adjust their activities, or the environment changes. In this sense, commanders adapt their operations not only to the state of the insurgency, but also to the environment's overall condition. Greater stability indicates progress toward the end state.

5-15. Well-designed operations are based on LLOs that are mutually supportive between echelons and adjacent organizations. For example, similar LLOs among brigade combat teams produce complementary effects, while brigade-level accomplishments reinforce achievement of division objectives. LLOs are normally used at brigade and higher levels, where the staff and unit resources needed to use them are available; however, battalions can use LLOs. Commanders at various levels may expect subordinates to describe their operations in these terms.

5-16. Commanders at all levels should select the LLOs that relate best to achieving the desired end state in accordance with the commander's intent. The following list of possible LLOs is not all inclusive. However, it gives commanders a place to start:

- Conduct information operations.
- Conduct combat operations/civil security operations.
- Train and employ HN security forces.
- Establish or restore essential services.
- Support development of better governance.
- Support economic development.

The Importance of Multiple Lines of Operations in COIN

The Chinese Civil War illustrates the importance of pursuing and linking multiple logical lines of operations. Chiang Kai-shek's defeat in 1949 resulted from his failure to properly establish security, good governance, the rule of law, essential services, and economic stability. Failures in each undermined his government's position in the others.

In China during 1945, Chiang Kai-shek adopted a strategy to secure and defend the coastal financial and industrial centers from the communist insurgency led by Mao Zedong. These areas had been the Chinese government's prewar core support areas. Although a logical plan, this strategy suffered in implementation. Republic of China administration and military forces were often corrupt. They neither provided good governance and security nor facilitated provision of essential services. Furthermore, the government, because of an insufficient number of soldiers, relied on warlord forces, which lacked quality and discipline. Their actions undermined the legitimacy of and popular support for the government within the very core areas vital to government rejuvenation. Likewise, when government forces attempted to reestablish their presence in the Chinese countryside, their undisciplined conduct towards the rural populace further undermined the legitimacy of the Chinese government. Because of these actions, Chiang's forces were unable to secure or expand their support base.

As a result, there was increasing lack of material and political support for the government, whose legitimacy was undercut. The government's inability to enforce ethical adherence to the rule of law by its officials and forces, combined with widespread corruption and economic collapse, served to move millions from being supporters into the undecided middle. When economic chaos eliminated any government ability to fund even proper and justified efforts, an insurgent victory led by the Chinese Communist Party became inevitable.

As government defeat followed defeat, a collapse of morale magnified the impact of material shortages. Chiang's defeat in Manchuria, in particular, created a psychological loss of support within China. It caused economic dislocation due to substantial price inflation of foodstuffs and sowed discord and dissension among government allies. As the regime lost moral authority, it also faced a decreasing ability to govern. All these factors served to create a mythical yet very powerful psychological impression that the success of the Chinese Communist Party was historically inevitable. The failure of the leaders of the Republic of China to address the requirements of logical lines of operations like good governance, economic development, and essential services magnified their military shortcomings and forced their abandonment of the Chinese mainland.

5-17. These lines can be customized, renamed, changed altogether, or simply not used. Commanders may combine two or more of the listed LLOs or split one LLO into several. For example, IO are integrated into all LLOs; however, commanders may designate a separate LLO for IO if necessary to better describe their intent. Likewise, some commanders may designate separate LLOs for combat operations and civil security operations.

5-18. LLOs should be used to isolate the insurgents from the population, address and correct the root causes of the insurgency, and create or reinforce the societal systems required to sustain the legitimacy of the HN government. The following discussion addresses six LLOs common during COIN operations. The IO LLO may be the most important one. However, IO are interwoven throughout all LLOs and shape the information environment in which COIN operations are executed.

CONDUCT INFORMATION OPERATIONS

5-19. The IO LLO may often be the decisive LLO. By shaping the information environment, IO make significant contributions to setting conditions for the success of all other LLOs. (See JP 3-13 and FM 3-13 for IO doctrine. IO include elements not addressed here.) By publicizing government policies, the actual situation, and counterinsurgent accomplishments, IO, synchronized with public affairs, can neutralize insurgent propaganda and false claims. Major IO task categories include the following:

- Ensure that IO are synchronized at all levels and nested within the interagency strategic communications operation.
- Identify all the audiences (local, regional, and international), the various news cycles, and how to reach them with the HN government's message.
- Manage the local populace's expectations regarding what counterinsurgents can achieve.
- Develop common, multiechelon themes based on and consistent with HN government policies and the operation's objectives. Sustain unity of the message.
- Coordinate and provide a comprehensive assessment of the information environment, incorporating the activities of all other LLOs.
- Remember actions always speak louder than words—every Soldier and Marine is an integral part of IO communications. IO are executed every day through the actions of firm, fair, professional, and alert Soldiers and Marines on the streets among the populace.
- Work to establish and sustain transparency that helps maintain HN government legitimacy.

5-20. Commanders and staffs synchronize IO with operations along all other LLOs. IO address and manage the public's expectations by reporting and explaining HN government and counterinsurgent actions. When effectively used, IO address the subject of root causes that insurgents use to gain support. (Table 5-1) lists considerations for developing the IO LLO.)

5-21. IO are tailored to address the concerns of the populace of specific areas. IO should inform the public of successfully completed projects and improvements, including accomplishments in security, infrastructure, essential services, and economic development. This publicity furthers popular acceptance of the HN government's legitimacy.

5-22. Effective IO use consistent themes based on policy, facts, and deeds—not claims or future plans, because these can be thwarted. Themes must be reinforced by actions along all LLOs. Making unsubstantiated claims can undermine the long-term credibility and legitimacy of the HN government. Counterinsurgents should never knowingly commit themselves to an action that cannot be completed. However, to reduce the negative effects of a broken promise, counterinsurgents should publicly address the reasons expectations cannot be met before insurgents can take advantage of them.

5-23. Command themes and messages based on policy should be distributed simultaneously or as soon as possible using all available media. Radio, television, newspapers, flyers, billboards, and the Internet are all useful dissemination means. Polling and analysis should be conducted to determine which media allow the widest dissemination of themes to the desired audiences at the local, regional, national, and international levels.

5-24. Insurgents are not constrained by truth; they create propaganda that furthers their aims. Insurgent propaganda may include lying, deception, and creating false causes. Historically, as the environment changes, insurgents change their message to address the issues that gain them support. IO should point out the insurgency's propaganda and lies to the local populace. Doing so creates doubt regarding the viability of the insurgents' short- and long-term intentions among the uncommitted public and the insurgency's supporters.

Table 5-1. Considerations for developing the information operations LLO

- Consider word choices carefully. Words are important—they have specific meanings and describe policy. For example, are counterinsurgents liberators or occupiers? Occupiers generate a "resistance," whereas liberators may be welcomed for a time. Soldiers and Marines can be influenced likewise. In a conflict among the people, terms like "battlefield" influence perceptions and confuse the critical nature of a synchronized approach. Refrain from referring to and considering the area of operations as a "battlefield" or it may continue to be one.
- Publicize insurgent violence and use of terror to discredit the insurgency. Identify barbaric actions by extremists and the insurgents' disregard for civilian losses.
- Admit mistakes (or actions perceived as mistakes) quickly. Explain these mistakes and actions as fully as possible—including mistakes committed by U.S. military forces. However, do not attempt to explain actions by the host-nation government. Instead encourage host-nation officials to handle such information themselves. They know the cultural implications of their actions better, and honesty should help to build legitimacy.
- Highlight successes of the host-nation government and counterinsurgents promptly. Positive results speak loudly and resonate with people. Do not delay announcements while waiting for all results. Initiate communications immediately to let people know what counterinsurgents are doing and why. Delaying announcements creates "old news" and misses news cycles.
- Respond quickly to insurgent propaganda. Delaying responses can let the insurgent story dominate several news cycles. That situation can lead to the insurgents' version of events becoming widespread and accepted. This consideration may require giving increased information assets and responsibilities to lower level leaders.
- Shape the populace's expectations. People generally expect too much too soon. When the host-nation government or counterinsurgency force is slow to deliver, people become easily and perhaps unfairly disgruntled.
- Give the populace some way to voice their opinions and grievances, even if that activity appears at first to cause friction. Such opportunities are important to both the formal political process and to informal, local issues (where government touches people directly). Develop a feedback mechanism from the populace to the local government to identify needs and align perceptions.
- Keep Soldiers and Marines engaged with the populace. Presence patrols facilitate Soldiers and Marines mingling with the people. As the populace and counterinsurgents learn to know each other better, two-way communication develops, building trust and producing intelligence.
- Conduct ongoing perception assessments. Identify leaders who influence the people at the local, regional, and national levels. Determine a population's relevant lines of loyalty as accurately as possible.
- Treat detainees professionally and publicize their treatment. Arrange for host-nation leaders to visit and tour your detention facility. Consider allowing them to speak to detainees and eat the same food detainees receive. If news media or host-nation government representatives visit your detention facility, allow them as much access as prudent. Provide a guided tour and explain your procedures.
- Consider encouraging host-nation leaders to provide a forum for initiating a dialog with the opposition. This does not equate to "negotiating with terrorists." It is an attempt to open the door to mutual understanding. There may be no common ground and the animosity may be such that nothing specifically or directly comes of the dialog. However, if counterinsurgents are talking with their adversaries, they are using a positive approach and may learn something useful. If the host nation is reluctant to communicate with insurgents, other counterinsurgents may have to initiate contact. Consider adopting a "We understand why you fight" attitude and stating this position to the insurgents.

- Work to convince insurgent leaders that the time for resistance has ended and that other ways to accomplish what they desire exist.
- Turn the insurgents' demands on the insurgents. Examine the disputed issues objectively; then work with host-nation leaders to resolve them where possible. Portray any success as a sign of responsiveness and improvement.
- Portray the counterinsurgency force as robust, persistent, and willing to help the population through the present difficulty.
- Learn the insurgents' messages or narratives. Develop countermessages and counter-narratives to attack the insurgents' ideology. Understanding the local culture is required to do this. Host-nation personnel can play a key role.
- Remember that the media's responsibility is to report the news. The standard against which the media should be judged is accuracy complemented by the provision of context and proper characterization of overall trends, not whether it portrays the actions of counterinsurgents, host-nation forces, and host-nation officials positively or negatively.
- Conduct town meetings to assess and address areas where counterinsurgents can make things better.
- When insurgents follow an ideology based on religious extremism, information operations should encourage, strengthen, and protect the society's moderating elements. Command themes need to portray a credible, publicly attractive vision that resonates with local culture. At the same time, commanders should avoid the appearance of interfering in the society's internal religious affairs.

5-25. Impartiality is a common theme for information activities when there are political, social, and sectarian divisions in the host nation. Counterinsurgents should avoid taking sides, when possible. Perceived favoritism can exacerbate civil strife and make counterinsurgents more desirable targets for sectarian violence.

5-26. Effective commanders directly engage in a dialog with the media and communicate command themes and messages personally. The worldwide proliferation of sophisticated communication technologies means that media coverage significantly affects COIN operations at all echelons. Civilian and military media coverage influences the perceptions of the political leaders and public of the host nation, United States, and international community. The media directly influence the attitude of key audiences toward counterinsurgents, their operations, and the opposing insurgency. This situation creates a war of perceptions between insurgents and counterinsurgents conducted continuously using the news media.

5-27. Commanders often directly engage the local populace and stakeholders through face-to-face meetings, town meetings, and community events highlighting counterinsurgent community improvements. These engagements give commanders additional opportunities to assess their efforts' effects, address community issues and concerns, and personally dispel misinformation. These events often occur in the civil-military operations center.

5-28. The media are a permanent part of the information environment. Effective media/public affairs operations are critical to successful military operations. All aspects of military operations are subject to immediate scrutiny. Well-planned, properly coordinated, and clearly expressed themes and messages can significantly clarify confusing situations. Clear, accurate portrayals can improve the effectiveness and morale of counterinsurgents, reinforce the will of the U.S. public, and increase popular support for the HN government. The right messages can reduce misinformation, distractions, confusion, uncertainty, and other factors that cause public distress and undermine the COIN effort. Constructive and transparent information enhances understanding and support for continuing operations against the insurgency.

5-29. There are several methods for working with the media to facilitate accurate and timely information flow. These include the following:

- Embedded media.
- Press conferences.
- Applying resources.
- Network with media outlets.

5-30. Embedded media representatives experience Soldiers' and Marines' perspectives of operations in the COIN environment. Media representatives should be embedded for as long as practicable. Representatives embedded for weeks become better prepared to present informed reports. Embedding for days rather than weeks risks media representatives not gaining a real understanding of the context of operations. Such short exposure may actually lead to unintended misinformation. The media should be given access to Soldiers and Marines in the field. These young people nearly always do a superb job of articulating the important issues for a broad audience. Given a chance, they can share their courage and sense of purpose with the American people and the world.

5-31. Commanders may hold weekly press conferences to explain operations and provide transparency to the people most affected by COIN efforts. Ideally, these sessions should include the HN media and HN officials. Such events provide opportunities to highlight the accomplishments of counterinsurgents and the HN government.

5-32. Commanders should apply time, effort, and money to establish the proper combination of media outlets and communications to transmit the repetitive themes of HN government accomplishments and insurgent violence against the populace. This might require counterinsurgents to be proactive, alerting the media to news opportunities and perhaps providing transportation or other services to ensure proper coverage. Helping establish effective HN media is another important COIN requirement. A word of caution: the populace and HN media must never perceive that counterinsurgents and HN forces are manipulating the media. Even the slightest appearance of impropriety can undermine the credibility of the COIN force and the host nation.

5-33. Good working relationships between counterinsurgent leaders and members of the U.S. media are in the Nation's interest. Similar relationships can be established with international media sources. When they do not understand COIN efforts, U.S. media representatives portray the situation to the American public based on what they do know. Such reports can be incomplete, if not incorrect. Through professional relationships, military leaders can ensure U.S. citizens better understand what their military is doing in support of the Nation's interests.

5-34. The media are ever present and influence perceptions of the COIN environment. Therefore, successful leaders engage the media, create positive relationships, and help the media tell the story. Operations security must always be maintained; however, security should not be used as an excuse to create a media blackout. In the absence of official information, some media representatives develop stories on their own that may be inaccurate and may not include the COIN force perspective. (See JP 3-61, FM 46-1, FM 3-61.1 for public affairs doctrine.)

CONDUCT COMBAT OPERATIONS/CIVIL SECURITY OPERATIONS

5-35. This LLO is the most familiar to military forces. Care must be taken not to apply too many resources to this LLO at the expense of other LLOs that facilitate the development or reinforcement of the HN government's legitimacy. Commanders may describe actions related to combat operations and civil security operations as a single LLO or as multiple LLOs. Commanders base their decision on conditions in the AO and their objectives. (Table 5-2 [page 5-12] lists considerations for developing the combat operations/civil security LLO.)

5-36. Under full spectrum operations, forces conduct simultaneous offensive, defensive, and stability operations. Offensive and defensive operations focus on defeating enemy forces. Security operations, including area security, pertain to actions taken to protect the force. They are associated with offensive and defensive operations. In contrast, stability operations focus on security and control of areas, resources, and populations. Civil security and civil control are types of stability operations. Army commanders expect a mission of protecting and providing security for a population to be expressed in terms of civil security or civil control.

5-37. Within the COIN context, Marine Corps doctrine does not draw a distinction in this manner; rather, it places tasks related to civil security and area security under combat operations. A Marine force assigned an area security mission during a COIN operation executes it as a combat operation. The force establishes and maintains measures to protect people and infrastructure from hostile acts or influences while actively seeking out and engaging insurgent forces.

Table 5-2. Considerations for developing the combat operations/civil security operations LLO

- Develop cultural intelligence, which assumes a prominent role. Make every effort to learn as much about the environment as possible. Human dynamics tend to matter the most.
- Ensure that rules of engagement adequately guide Soldiers and Marines engaged in combat while encouraging the prudent use of force commensurate with mission accomplishment and self-defense.
- Consider how the populace might react when planning tactical situations, even for something as simple as a traffic control point. Anticipate how people might respond to each operation.
- Identify tasks the host-nation government and populace generally perceive to be productive and appropriate for an outside force. Focus counterinsurgents on them.
- Win over, exhaust, divide, capture, or eliminate the senior- and mid-level insurgent leaders as well as network links. (Appendix B discusses networks and links.)
- Frustrate insurgent recruiting.
- Disrupt base areas and sanctuaries.
- Deny outside patronage (external support). Make every effort to stop insurgents from bringing materiel support across international and territorial borders.
- When patrolling in or occupying an area, clear only what the unit intends to hold. Otherwise, the effort will be wasted as the insurgents reoccupy the area. An exception to this policy is when commanders deem disruption of enemy strongholds necessary.
- When Soldiers and Marines interact with the populace, encourage them to treat people with respect to avoid alienating anyone.
- Support efforts to disarm, demobilize, and reintegrate into society members of armed groups outside of government control, such as militias and paramilitary organizations. Also include insurgents who are captured, surrender, or accept amnesty.
- Take a census as soon as is practicable. Help the host-nation government do this. This information can be helpful for learning about the people and meeting their needs. The census also establishes a necessary database for civil security operations.
- Help the host-nation government produce and distribute identification cards. Register all citizens—or at least those nearing a predetermined, adult age. Identification cards may help to track people's movements. This information is useful in identifying illicit activity and also contributes to civil security.

5-38. Insurgents use unlawful violence to weaken the HN government, intimidate people into passive or active support, and murder those who oppose the insurgency. Measured combat operations are always required to address insurgents who cannot be co-opted into operating inside the rule of law. These operations may sometimes require overwhelming force and the killing of fanatic insurgents. However, COIN is "war amongst the people." Combat operations must therefore be executed with an appropriate level of restraint to minimize or avoid injuring innocent people. Not only is there a moral basis for the use of restraint or measured force; there are practical reasons as well. Needlessly harming innocents can turn the populace against the COIN effort. Discriminating use of fires and calculated, disciplined response should characterize COIN operations. Kindness and compassion can often be as important as killing and capturing insurgents.

5-39. Battalion-sized and smaller unit operations are often most effective for countering insurgent activities. Counterinsurgents need to get as close as possible to the people to secure them and glean the maximum amount of quality information. Doing this helps counterinsurgents gain a fluidity of action equal or superior to that of the enemy. This does not mean larger unit operations are not required. Brigades are usually synchronizing headquarters. Divisions shape the environment to set conditions and facilitate brigade and battalion success. The sooner counterinsurgents can execute small-unit operations effectively, the better.

TRAIN AND EMPLOY HOST-NATION SECURITY FORCES

5-40. Most societal and government functions require a secure environment. Although U.S. and multinational forces can provide direct assistance to establish and maintain security, this situation is at best a provisional solution. Ultimately, the host nation must secure its own people. (Table 5-3 lists considerations for developing the HN security forces LLO. Chapter 6 addresses this LLO in detail.)

Table 5-3. Considerations for developing the host-nation security forces LLO

- Understand the security problem. The function, capabilities, and capacities required for host-nation security (military and police) forces should align with the theater strategy, host-nation culture, and the threat these forces face.
- Take a comprehensive approach, beginning with planning. Consult local representatives to determine local needs. Include host-nation government and military authorities as partners. Consult with multinational partners and intergovernmental organizations that may be involved. Share leadership with local authorities to achieve legitimacy in the eyes of the populace.
- Avoid mirror-imaging (trying to make host-nation forces look like the U.S. military). That solution fits few cultures or situations.
- Establish separate training academies for military and police forces. Staff them with multinational personnel. (Tap into the talents of as many nations as possible for this.) Maintaining public order, enforcing laws, and preventing and detecting crime requires a trained and capable host-nation police force. Likewise, an operational penal system providing adequate and humane confinement is needed to support host-nation police and judicial processes.
- Establish mobile training teams.
- Train the host-nation cadre first. Focus on identifying leaders. Use trained leaders to establish new units, staff training academies, and staff mobile training teams (where appropriate).
- Create special-purpose forces based on threats facing the host nation.
  - For police, consider special reaction teams, a counterespionage organization, port security, and public figure security.
  - For the host-nation military, consider riverine operations forces, explosive ordinance disposal, and other specialized units.
- Put host-nation personnel in charge of as much as possible as soon as possible.
- Conduct operations with host-nation forces. Show that you respect their partnership. Once host-nation forces are ready to work with the counterinsurgency force, include them in planning. Encourage host-nation leaders to take ownership of plans and operations as they move toward self-sufficiency.
- Respect host-nation security force leaders in public and private. Show the populace that their security forces have earned counterinsurgents' respect. However, do not tolerate abuses. Base respect on generally upright comportment by host-nation security forces.
- Provide advisors for host-nation units under development. Place liaison officers with trained host-nation units. Make sure all involved understand how an advisor differs from a liaison officer.
- Establish competent military and police administrative structures early. Provision and pay host-nation forces on time. Pay should come from the host-nation organization, not the counterinsurgency force.
- Encourage insurgents to change sides—welcome them in with an "open-arms" policy. However, identify insurgents seeking to join the security forces under false pretext. Vetting repatriated insurgents is a task for the host-nation government in partnership with the country team.
- Encourage the host nation to establish a repatriation or amnesty program to allow insurgents an alternative to the insurgency.

5-41. The U.S. military can help the host nation develop the forces required to establish and sustain stability within its borders. This task usually involves other government agencies and multinational partners. This assistance can include developing, equipping, training, and employing HN security forces. It may extend to operations in which multinational military units fight alongside the newly formed, expanded, or reformed HN forces.

ESTABLISH OR RESTORE ESSENTIAL SERVICES

5-42. Essential services address the life support needs of the HN population. The U.S. military's primary task is normally to provide a safe and secure environment. HN or interagency organizations can then develop the services or infrastructure needed. In an unstable environment, the military may initially have the leading role. Other agencies may not be present or might not have enough capability or capacity to meet HN needs. Therefore, COIN military planning includes preparing to perform these tasks for an extended period. (Table 5-4 lists considerations for developing the essential services LLO.)

Table 5-4. Considerations for developing the essential services LLO

- Make this effort a genuine partnership between counterinsurgents and host-nation authorities. Use as much local leadership, talent, and labor as soon as possible.
- Plan for a macro and a micro assessment effort. Acknowledge early what is known and not known—and honestly appraise what needs to be accomplished. The macro assessment concerns national-level needs; it is long term in focus. The micro assessment focuses on the local level; it determines, by region, specific short-term needs.
- Appreciate local preferences. An accurate needs assessment reflects cultural sensitivity; otherwise, great time and expense can be wasted on something the populace considers of little value. Ask, How do I know this effort matters to the local populace? If there is no answer, the effort may not be important. Host-nation authorities are a good place to start with this question. A local perception assessment may also be useful. (See appendix B.)
- Establish realistic, measurable goals. Establish ways to assess their achievement.
- Form interagency planning teams to discuss design, assessment, and redesign. Recognize and understand other agencies' institutional cultures.
- Meet with representatives from organizations beyond the host-nation government team. Many nongovernmental organizations do not want to appear closely aligned with the counterinsurgency effort. Encourage their participation in planning, even if it means holding meetings in neutral areas. When meeting with these organizations, help them understand mutual interests in achieving local security, stability, and relief objectives.
- Be as transparent as possible with the local populace. Do your best to help people understand what counterinsurgents are doing and why they are doing it.
- Consider the role of women in the society and how this cultural factor may influence these activities.
- Discourage the attitude that counterinsurgents have arrived to "save the day"—or that their arrival will only cause greater problems. Helping the populace understand what is possible avoids frustrations based on unrealized high expectations.

5-43. Counterinsurgents should work closely with the host nation in establishing achievable goals. If lofty goals are set and not achieved, both counterinsurgents and the HN government can lose the populace's respect. The long-term objective is for the host nation to assume full responsibility and accountability for these services. Establishing activities that the HN government is unable to sustain may be counterproductive. IO nested within this LLO manage expectations and ensure that the public understands the problems involved in providing these services, for example, infrastructure sabotage by insurgents. Figure 5-4 shows an example of common essential services categories. Accomplishing objectives in each category contributes to achieving the higher commander's desired end state.

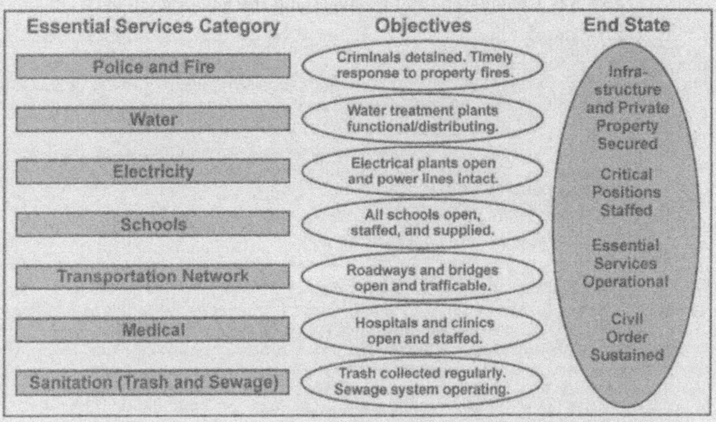

Figure 5-4. Example of essential services categories and objectives

SUPPORT DEVELOPMENT OF BETTER GOVERNANCE

5-44. This LLO relates to the HN government's ability to gather and distribute resources while providing direction and control for society. These include regulation of public activity; taxation; maintenance of security, control, and essential services; and normalizing the means of succession of power. Good governance is normally a key requirement to achieve legitimacy for the HN government. Activities related to it include the following:

- Controlling military and police activities.
- Establishing and enforcing the rule of law.
- Public administration.
- Justice (a judiciary system, prosecutor/defense representation, and corrections).
- Property records and control.
- Public finance.
- Civil information.
- Historical, cultural, and recreational services.
- An electoral process for representative government.
- Disaster preparedness and response.

5-45. Sometimes no HN government exists or the government is unable or unwilling to assume full responsibility for governance. In those cases, this LLO may involve establishing and maintaining a military government or a civil administration while creating and organizing a HN capability to govern. In the long run, developing better governance will probably affect the lives of the populace more than any other COIN activities. When well executed, these actions may eliminate the root causes of the insurgency. Governance activities are among the most important of all in establishing lasting stability for a region or nation. (Table 5-5 (page 5-16) lists considerations for developing the governance LLO.)

Table 5-5. Considerations for developing the governance LLO

• Encourage community leaders to participate in local governance. If no local council exists, encourage the populace to create one. Ask teachers, businesspeople, and others whom the community respects to form a temporary council until a more permanent organization can be formed.
• Help (or encourage) the host-nation government to remove or reduce genuine grievances, expose imaginary ones, and resolve contradictions, immediately where possible. Accomplishing these tasks may be difficult because—
  ▪ Genuine grievances may be hard to separate from unreasonable complaints.
  ▪ Host-nation leaders may be unable or unwilling to give up the necessary power to local governments.
• Make only commitments that can be fulfilled in the foreseeable future.
• Help the host nation develop and empower competent and responsive leaders and strengthen their civil service and security forces. Doing this is often difficult; however, backing an incompetent (or worse) host-nation leader can backfire. Do not be afraid to step in and make a bold change where necessary. A corrupt official, such as a chief of police who is working for both sides, can be doing more harm than good. You may be forced to replace him. If so, move decisively. Arrange the removal of all officials necessary to solve the problem. The pain of the affair may be acute, but it will be brief and final. Wherever possible, have host-nation authorities conduct the actual removal.
• Be accessible to the populace to facilitate two-way communication. Establish rapport for the host-nation government and counterinsurgents.
• Encourage the host nation to grant local demands and meet acceptable aspirations. Some of these might be driving the insurgency.
• Emphasize the national perspective in all host-nation government activities. Downplay sectarian divides.
• Provide liaison officers to host-nation government ministries or agencies. When possible, use an interagency team approach. Structure teams based on function.
• Once the legal system is functioning, send someone to observe firsthand a person moving through the legal process (arrest by police, trial, and punishment by confinement to a correctional facility). Ask to see the docket of the judges at the provincial courthouse. If there is no one on the docket or if it is full and there are no proceedings, there may be a problem.
• Create a system for citizens to pursue redress for perceived wrongs by authorities. Rule of law includes means for a citizen to petition the government for redress of government wrongs. The ability to petition the counterinsurgency force for redress of wrongs perpetrated by that force (intentionally or otherwise) is also required.
• Build on existing capabilities wherever possible. Host nations often have some capability; counterinsurgents may only need to help develop greater capacity.

SUPPORT ECONOMIC DEVELOPMENT

5-46. The economic development LLO includes both short- and long-term aspects. The short-term aspect concerns immediate problems, such as large-scale unemployment and reestablishing an economy at all levels. The long-term aspect involves stimulating indigenous, robust, and broad economic activity. The stability a nation enjoys is often related to its people's economic situation and its adherence to the rule of law. However, a nation's economic health also depends on its government's ability to continuously secure its population.

5-47. Planning economic development requires understanding the society, culture, and operational environment. For example, in a rural society, land ownership and the availability of agricultural equipment, seed, and fertilizer may be the chief parts of any economic development plan. In an urban, diversified society, the availability of jobs and the infrastructure to support commercial activities may be more important. Except for completely socialist economies, governments do not create jobs other than in the public bureaucracy. However, the micro economy can be positively stimulated by encouraging small businesses development. Jump-starting small businesses requires micro finance in the form of some sort of banking activities. So then, supporting economic development requires attention to both the macro economy and the micro economy.

5-48. Without a viable economy and employment opportunities, the public is likely to pursue false promises offered by insurgents. Sometimes insurgents foster the conditions keeping the economy stagnant. Insurgencies attempt to exploit a lack of employment or job opportunities to gain active and passive support for their cause and ultimately undermine the government's legitimacy. Unemployed males of military age may join the insurgency to provide for their families. Hiring these people for public works projects or a local civil defense corps can remove the economic incentive to join the insurgency. The major categories of economic activity include the following:

- Fossil fuels, mining, and related refining infrastructure.
- Generation and transmission of power and energy.
- Transportation and movement networks.
- Stock and commodities exchange.
- Banking.
- Manufacturing and warehousing.
- Building trades and services.
- Agriculture, food processing, fisheries, and stockyard processing.
- Labor relations.
- Education and training.

5-49. Table 5-6 lists considerations for developing the economic development LLO.

Table 5-6. Considerations for the economic development LLO

• Work with the host-nation government to strengthen the economy and quality of life. In the long run, success depends on supporting people's livelihoods. • Create an environment where business can thrive. In every state (except perhaps a completely socialist one), business drives the economy. To strengthen the economy, find ways to encourage and support legitimate business activities. Even providing security is part of a positive business environment. • Work with the host-nation government to reduce unemployment to a manageable level. • Seek to understand the effects of military operations on business activities and vice versa. Understand the effects of outsourcing and military support on the local economy and employment level. • Use economic leverage to enter new areas and reach new people. Remember that in many societies, monies are distributed though tribal or clan networks. For instance, giving a clan leader a contracting job may lead to employing many local men. Those employees are less likely to join the insurgency. It may be necessary to pay more than seems fair for a job; however, the money is well spent if it keeps people from supporting the insurgency. • Ensure that noncompliance with government policies has an economic price. Likewise, show that compliance with those policies is profitable. In the broadest sense, counterinsurgency operations should reflect that "peace pays." • Program funds for commanders to use for economic projects in their area of operations from the beginning of any operation. No one appreciates the situation better than those "on the ground." Creating these funds may require congressional action. (Appendix D contains a description of relevant funding sources.)

COUNTERINSURGENCY APPROACHES

5-50. There are many approaches to achieving success in a COIN effort. The components of each approach are not mutually exclusive. Several are shared by multiple approaches. The approaches described below are not the only choices available and are neither discrete nor exclusive. They may be combined, depending on the environment and available resources. The following methods and their components have proven effective. However, they must be adapted to the demands of the local environment. Three examples of approaches are—

- Clear-hold-build.
- Combined action.
- Limited support.

CLEAR-HOLD-BUILD

5-51. A clear-hold-build operation is executed in a specific, high-priority area experiencing overt insurgent operations. It has the following objectives:

- Create a secure physical and psychological environment.
- Establish firm government control of the populace and area.
- Gain the populace's support.

Popular support can be measured in terms of local participation in HN programs to counter the insurgency and whether people give counterinsurgents usable information about insurgent locations and activities.

5-52. COIN efforts should begin by controlling key areas. Security and influence then spread out from secured areas. The pattern of this approach is to clear, hold, and build one village, area, or city—and then reinforce success by expanding to other areas. This approach aims to develop a long-term, effective HN government framework and presence that secures the people and facilitates meeting their basic needs. Success reinforces the HN government's legitimacy. The primary tasks to accomplish during clear-hold-build are—

- Provide continuous security for the local populace.
- Eliminate insurgent presence.
- Reinforce political primacy.
- Enforce the rule of law.
- Rebuild local HN institutions.

5-53. To create success that can spread, a clear-hold-build operation should not begin by assaulting the main insurgent stronghold. However, some cases may require attacks to disrupt such strongholds, even if counterinsurgents cannot clear and hold the area. "Disrupt and leave" may be needed to degrade the insurgents' ability to mount attacks against cleared areas. Clear-hold-build objectives require lots of resources and time. U.S. and HN commanders should prepare for a long-term effort. All operations require unity of effort by civil authorities, intelligence agencies, and security forces. Coherent IO are also needed.

5-54. Clear-hold-build operations should expand outward from a secure base. An example is an urban industrial complex whose population supports the government effort and where security forces are in firm control. No population subjected to the intense organizational efforts of an insurgent organization can be won back until certain conditions are created:

- The counterinsurgent forces are clearly superior to forces available to the insurgents.
- Enough nonmilitary resources are available to effectively carry out all essential improvements needed to provide basic services and control the population.
- The insurgents are cleared from the area.
- The insurgent organizational infrastructure and its support have been neutralized or eliminated.
- A HN government presence is established to replace the insurgents' presence, and the local populace willingly supports this HN presence.

5-55. The following discussion describes some examples of activities involved in the clear-hold-build approach. Its execution involves activities across all LLOs. There can be overlap between steps—especially between hold and build, where relevant activities are often conducted simultaneously.

Clearing the Area

5-56. *Clear* is a tactical mission task that requires the commander to remove all enemy forces and eliminate organized resistance in an assigned area (FM 3-90). The force does this by destroying, capturing, or forcing the withdrawal of insurgent combatants. This task is most effectively initiated by a clear-in-zone or cordon-and-search operation. This operation's purpose is to disrupt insurgent forces and force a reaction by major insurgent elements in the area. Commanders employ a combination of offensive small-unit operations. These may include area saturation patrolling that enables the force to defeat insurgents in the area, interdiction ambushes, and targeted raids.

5-57. These offensive operations are only the beginning, not the end state. Eliminating insurgent forces does not remove the entrenched insurgent infrastructure. While their infrastructure exists, insurgents continue to recruit among the population, attempt to undermine the HN government, and try to coerce the populace through intimidation and violence. After insurgent forces have been eliminated, removing the insurgent infrastructure begins. This should be done so as to minimize the impact on the local populace. Rooting out such infrastructure is essentially a police action that relies heavily on military and intelligence forces until HN police, courts, and legal processes can assume responsibility for law enforcement within the cleared area.

5-58. If insurgent forces are not eliminated but instead are expelled or have broken into smaller groups, they must be prevented from reentering the area or reestablishing an organizational structure inside the area. Once counterinsurgents have established their support bases, security elements cannot remain static. They should be mobile and establish a constant presence throughout the area. Use of special funds should be readily available for all units to pay compensation for damages that occur while clearing the area of insurgent forces. Offensive and stability operations are continued to maintain gains and set the conditions for future activities. These include—

- Isolating the area to cut off external support and to kill or capture escaping insurgents.
- Conducting periodic patrols to identify, disrupt, eliminate, or expel insurgents.
- Employing security forces and government representatives throughout the area to secure the populace and facilitate follow-on stages of development.

5-59. Operations to clear an area are supplemented by IO focused on two key audiences: the local populace and the insurgents. The message to the populace focuses on gaining and maintaining their overt support for the COIN effort. This command theme is that the continuous security provided by U.S. and HN forces is enough to protect the people from insurgent reprisals for their cooperation. Conversely, the populace should understand that actively supporting the insurgency will prolong combat operations, creating a risk to themselves and their neighbors. The command message to the insurgents focuses on convincing them that they cannot win and that the most constructive alternatives are to surrender or cease their activities.

Holding with Security Forces

5-60. Ideally HN forces execute this part of the clear-hold-build approach. Establishment of HN security forces in bases among the population furthers the continued disruption, identification, and elimination of the local insurgent leadership and infrastructure. The success or failure of the effort depends, first, on effectively and continuously securing the populace and, second, on effectively reestablishing a HN government presence at the local level. Measured offensive operations continue against insurgents as opportunities arise, but the main effort is focused on the population.

5-61. Key infrastructure must be secured. Since resources are always limited, parts of the infrastructure vital for stability and vulnerable to attack receive priority for protection. These critical assets should be identified during planning. For instance, a glassmaking factory may be important for economic recovery, but it may not be at risk of insurgent attack and therefore may not require security.

5-62. There are four key target audiences during the hold stage:

- Population.
- Insurgents.
- COIN force.
- Regional and international audiences.

5-63. Command themes and messages to the population should affirm that security forces supporting the HN government are in the area to accomplish the following:

- Protect the population from insurgent intimidation, coercion, and reprisals.
- Eliminate insurgent leaders and infrastructure.
- Improve essential services where possible.
- Reinstate HN government presence.

IO should also emphasize that U.S. and HN security forces will remain until the current situation is resolved or stated objectives are attained. This message of a persistent presence can be reinforced by making long-term contracts with local people for supply or construction requirements.

5-64. The commander's message to the insurgents is to surrender or leave the area. It emphasizes the permanent nature of the government victory and presence. The HN government might try to exploit success by offering a local amnesty. Insurgent forces will probably not surrender, but they may cease hostile actions against the HN government agencies in the area.

5-65. The commander's message to the COIN force should explain changes in missions and responsibilities associated with creating or reinforcing the HN government's legitimacy. The importance of protecting the populace, gaining people's support by assisting them, and using measured force when fighting insurgents should be reinforced and understood.

5-66. Operations during this stage are designed to—

- Continuously secure the people and separate them from the insurgents.
- Establish a firm government presence and control over the area and populace.
- Recruit, organize, equip, and train local security forces.
- Establish a government political apparatus to replace the insurgent apparatus.
- Develop a dependable network of sources by authorized intelligence agents.

5-67. Major actions occurring during this stage include—

- Designating and allocating area-oriented counterinsurgent forces to continue offensive operations. Other forces that participated in clearing actions are released or assigned to other tasks.
- A thorough population screening to identify and eliminate remaining insurgents and to identify any lingering insurgent support structures.
- Conducting area surveys to determine available resources and the populace's needs. Local leaders should be involved.
- Environmental improvements designed to convince the populace to support the HN government, participate in securing their area, and contribute to the reconstruction effort.
- Training of local paramilitary security forces, including arming them and integrating them into successful operations against the insurgents.
- Establishing a communications system that integrates the area into the HN communications grid and system.

Building Support and Protecting the Population

5-68. Progress in building support for the HN government requires protecting the local populace. People who do not believe they are secure from insurgent intimidation, coercion, and reprisals will not risk overtly supporting COIN efforts. The populace decides when it feels secure enough to support COIN efforts.

5-69. To protect the populace, HN security forces continuously conduct patrols and use measured force against insurgent targets of opportunity. Contact with the people is critical to the local COIN effort's success. Actions to eliminate the remaining covert insurgent political infrastructure must be continued; an insurgent presence will continue to threaten and influence people.

5-70. Tasks that provide an overt and direct benefit for the community are key, initial priorities. Special funds (or other available resources) should be available to pay wages to local people to do such beneficial work. Accomplishing these tasks can begin the process of establishing HN government legitimacy. Sample tasks include—

- Collecting and clearing trash from the streets.
- Removing or painting over insurgent symbols or colors.
- Building and improving roads.
- Digging wells.
- Preparing and building an indigenous local security force.
- Securing, moving, and distributing supplies.
- Providing guides, sentries, and translators.
- Building and improving schools and similar facilities.

Population Control Measures

5-71. Population control includes determining who lives in an area and what they do. This task requires determining societal relationships—family, clan, tribe, interpersonal, and professional. Establishing control normally begins with conducting a census and issuing identification cards. A census must be advertised and executed systematically. Census tasks include establishing who resides in which building and each household's family head. Those heads of households are required to report any changes to the appropriate agencies. Census records provide information regarding real property ownership, relationships, and business associations.

5-72. Insurgents may try to force people to destroy their identification cards. The benefits of retaining identification cards must be enough to motivate people to resist losing them. Insurgents may participate in the census to obtain valid identification cards. Requiring applicants to bring two men from outside their family to swear to their identity can reduce this probability. Those who affirm the status of an applicant are accountable for their official statements made on behalf of the applicant. Identification cards should have a code that indicates where the holders live.

5-73. Other population control measures include—

- Curfews.
- A pass system (for example, one using travel permits or registration cards) administered by security forces or civil authorities.
- Limits on the length of time people can travel.
- Limits on the number of visitors from outside the area combined with a requirement to register them with local security forces or civil authorities.
- Checkpoints along major routes to monitor and enforce compliance with population control measures.

5-74. The HN government should explain and justify new control measures to the affected population. People need to understand what is necessary to protect them from insurgent intimidation, coercion, and reprisals. Once control measures are in place, the HN government should have an established system of punishments for offenses related to them. These should be announced and enforced. The host nation should establish this system to ensure uniform enforcement and conformity with the rule of law throughout its territory. The HN government must be able to impose fines and other punishments for such civil infractions.

Increasing Popular Support

5-75. Counterinsurgents should use every opportunity to help the populace and meet its needs and expectations. Projects to improve economic, social, cultural, and medical needs can begin immediately. Actions speak louder than words. Once the insurgent political infrastructure is destroyed and local leaders begin to establish themselves, necessary political reforms can be implemented. Other important tasks include the following:

- Establishing HN government agencies to perform routine administrative functions and begin improvement programs.
- Providing HN government support to those willing to participate in reconstruction. Selection for participation should be based on need and ability to help. People should also be willing to secure what they create.
- Beginning efforts to develop regional and national consciousness and rapport between the population and its government. Efforts may include participating in local elections, making community improvements, forming youth clubs, and executing other projects.

5-76. Commanders can use IO to increase popular support. Command messages are addressed to the populace, insurgents, and counterinsurgents.

5-77. The IO message to the population has three facets:

- Obtaining the understanding or approval of security force actions that affect the populace, such as control measures or a census. Tell the people what forces are doing and why they are doing it.
- Establishing human intelligence sources that lead to identification and destruction of any remaining insurgent infrastructure in the area.
- Winning over passive or neutral people by demonstrating how the HN government is going to make their life better.

5-78. The IO message to insurgents should aim to create divisions between the movement leaders and the mass base by emphasizing failures of the insurgency and successes of the government. Success is indicated when insurgents abandon the movement and return to work with the HN government.

5-79. Commanders should emphasize that counterinsurgents must remain friendly towards the populace while staying vigilant against insurgent actions. Commanders must ensure Soldiers and Marines understand the rules of engagement, which become more restrictive as peace and stability return.

5-80. The most important activities during the build stage are conducted by nonmilitary agencies. HN government representatives reestablish political offices and normal administrative procedures. National and international development agencies rebuild infrastructure and key facilities. Local leaders are developed and given authority. Life for the area's inhabitants begins to return to normal. Activities along the combat operations/civil security operations LLO and HN security force LLO become secondary to those involved in essential services, good governance, and essential services LLOs.

Clear-Hold-Build in Tal Afar

In early 2005, the city of Tal Afar in northern Iraq had become a focal point for Iraqi insurgent efforts. The insurgents tried to assert control over the population. They used violence and intimidation to inflame ethnic and sectarian tensions. They took control of all schools and mosques, while destroying police stations. There were frequent abductions and executions. The insurgents achieved some success as the populace divided into communities defined by sectarian boundaries. Additionally, Tal Afar became an insurgent support base and sanctuary for launching attacks in the major regional city of Mosul and throughout Nineveh province.

During the summer of 2005, the 3d Armored Cavalry Regiment (ACR) assumed the lead for military efforts in and around Tal Afar. In the months that followed, the 3d ACR applied a clear-hold-build approach to reclaim Tal Afar from the insurgents.

Destruction or Expulsion of Insurgent Forces (Clear)

In August 2005, the 3d ACR and Iraqi forces began the process of destroying the insurgency in Tal Afar. Their first step was to conduct reconnaissance to understand the enemy situation; understand the ethnic, tribal, and sectarian dynamics; and set the conditions for effective operations. Iraqi security forces and U.S. Soldiers isolated the insurgents from external support by controlling nearby border areas and creating an eight-foot-high berm around the city. The berm's purpose was to deny the enemy freedom of movement and safe haven in outlying communities. The berm prevented free movement of fighters and weapons and forced all traffic to go through security checkpoints manned by U.S. and Iraqi forces. Multinational checkpoints frequently included informants who could identify insurgents. Multinational forces supervised the movement of civilians out of contentious areas. Forces conducted house-to-house searches. When they met violent resistance, they used precision fires from artillery and aviation. Targets were chosen through area reconnaissance operations, interaction with the local populace, and information from U.S. and Iraqi sources. Hundreds of insurgents were killed or captured during the encirclement and clearing of the city. Carefully controlled application of violence limited the cost to residents.

Deployment of Security Forces (Hold)

Following the defeat of enemy fighters, U.S. and Iraqi forces established security inside Tal Afar. The security forces immediately enhanced personnel screening at checkpoints based on information from the local population. To enhance police legitimacy in the people's eyes, multinational forces began recruiting Iraqi police from a more diverse, representative mix comprising city residents and residents of surrounding communities. Police recruits received extensive training in a police academy. U.S. forces and the Iraqi Army also trained Iraqi police in military skills. Concurrently, the local and provincial government dismissed or prosecuted Iraqi police involved in offenses against the populace. The government assigned new police leaders to the city from Mosul and other locations. U.S. forces assisted to ensure Iraqi Army, police, and their own forces shared common boundaries and were positioned to provide mutual support to one another. At the same time, U.S. forces continued to equip and train a border defense brigade, which increased the capability to interdict the insurgents' external support. Among its successes, the multinational force destroyed an insurgent network that included a chain of safe houses between Syria and Tal Afar.

Improving Living Conditions and Restoring Normalcy (Build)

With insurgents driven out of their city, the local population accepted guidance and projects to reestablish control by the Iraqi government. The 3d ACR commander noted, "The people of Tal Afar understood that this was an operation for them—an operation to bring back security to the city."

With the assistance of the Department of State and the U.S. Agency for International Development's Office of Transition Initiatives, efforts to reestablish municipal and economic systems began in earnest. These initiatives included providing essential services (water, electricity, sewage, and trash collection), education projects, police stations, parks, and reconstruction efforts. A legal claims process and compensation program to address local grievances for damages was also established.

As security and living conditions in Tal Afar improved, citizens began providing information that helped eliminate the insurgency's infrastructure. In addition to information received on the streets, multinational forces established joint coordination centers in Tal Afar and nearby communities that became multinational command posts and intelligence-sharing facilities with the Iraqi Army and the Iraqi police.

Unity of effort by local Iraqi leaders, Iraqi security forces, and U.S. forces was critical to success. Success became evident when many families who had fled the area returned to the secured city.

COMBINED ACTION

5-81. Combined action is a technique that involves joining U.S. and HN troops in a single organization, usually a platoon or company, to conduct COIN operations. This technique is appropriate in environments where large insurgent forces do not exist or where insurgents lack resources and freedom of maneuver. Combined action normally involves joining a U.S. rifle squad or platoon with a HN platoon or company, respectively. Commanders use this approach to hold and build while providing a persistent counterinsurgent presence among the populace. This approach attempts to first achieve security and stability in a local area, followed by offensive operations against insurgent forces now denied access or support. Combined action units are not designed for offensive operations themselves and rely on more robust combat units to perform this task. Combined action units can also establish mutual support among villages to secure a wider area.

5-82. A combined action program can work only in areas with limited insurgent activity. The technique should not be used to isolate or expel a well-established and supported insurgent force. Combined action is most effective after an area has been cleared of armed insurgents.

5-83. The following geographic and demographic factors can also influence the likelihood of success:

- Towns relatively isolated from other population centers are simpler to secure continuously.
- Towns and villages with a limited number of roads passing through them are easier to secure than those with many routes in and out. All approaches must be guarded.
- Existing avenues of approach into a town should be observable from the town. Keeping these areas under observation facilitates interdiction of insurgents and control of population movements.
- The local populace should be small and constant. People should know one another and be able to easily identify outsiders. In towns or small cities where this is not the case, a census is the most effective tool to establish initial accountability for everyone.
- Combined action or local defense forces must establish mutual support with forces operating in nearby towns. Larger reaction or reserve forces as well as close air support, attack aviation, and air assault support should be quickly available. Engineer and explosive ordnance disposal assets should also be available.

5-84. Combined action unit members must develop and build positive relationships with their associated HN security forces and with the town leadership. By living among the people, combined action units serve an important purpose. They demonstrate the commitment and competence of counterinsurgents while sharing experiences and relationships with local people. These working relationships build trust and enhance the HN government's legitimacy. To build trust further, U.S. members should ask HN security forces for training on local customs, key terrain, possible insurgent hideouts, and relevant cultural dynamics. HN forces should also be asked to describe recent local events.

5-85. Combined action units are integrated into a regional scheme of mutually supporting security and influence; however, they should remain organic to their parent unit. Positioning reinforced squad-sized units (13 to 15 Soldiers or Marines) among HN citizens creates a dispersal risk. Parent units can mitigate this risk with on-call reserve and reaction forces along with mutual support from adjacent villages and towns.

5-86. Thoroughly integrating U.S. and HN combined action personnel supports the effective teamwork critical to the success of each team and the overall program. U.S. members should be drawn from some of the parent unit's best personnel. Designating potential members before deployment facilitates the training and team building needed for combined action unit success in theater. Preferably, team members should have had prior experience in the host nation. Other desirable characteristics include—

- The ability to operate effectively as part of a team.
- Strong leadership qualities, among them—
  - Communicating clearly.
  - Maturity.
  - Leading by example.
  - Making good decisions.
- The ability to apply the commander's intent in the absence of orders.
- Possession of cultural awareness and understanding of the HN environment.
- The absence of obvious prejudices.

- Mutual respect when operating with HN personnel.
- Experience with the HN language, the ability to learn languages, or support of reliable translators.
- Patience and tolerance when dealing with language and translation barriers.

5-87. Appropriate tasks for combined action units include, but are not limited to, the following:

- Helping HN security forces maintain entry control points.
- Providing reaction force capabilities through the parent unit.
- Conducting multinational, coordinated day and night patrols to secure the town and area.
- Facilitating local contacts to gather information in conjunction with local HN security force representatives. (Ensure information gathered is made available promptly and on a regular basis to the parent unit for timely fusion and action.)
- Training HN security forces in leadership and general military subjects so they can secure the town or area on their own.
- Conducting operations with other multinational forces and HN units, if required.
- Operating as a team with HN security forces to instill pride, leadership, and patriotism.
- Assisting HN government representatives with civic action programs to establish an environment where the people have a stake in the future of their town and nation.
- Protecting HN judicial and government representatives and helping them establish the rule of law.

Combined Action Program

Building on their early 20th-century counterinsurgency experiences in Haiti and Nicaragua, the Marine Corps implemented an innovative program in South Vietnam in 1965 called the Combined Action Program. This program paired teams of about 15 Marines led by a noncommissioned officer with approximately 20 host-nation security personnel. These combined action platoons operated in the hamlets and villages in the northern two provinces of South Vietnam adjacent to the demilitarized zone. These Marines earned the trust of villagers by living among them while helping villagers defend themselves. Marines trained and led the local defense forces and learned the villagers' customs and language. The Marines were very successful in denying the Viet Cong access to areas under their control. The Combined Action Program became a model for countering insurgencies. Many lessons learned from it were used in various peace enforcement and humanitarian assistance operations that Marines conducted during the 1990s. These operations included Operations Provide Comfort in northern Iraq (1991) and Restore Hope in Somalia (1992 through 1993).

LIMITED SUPPORT

5-88. Not all COIN efforts require large combat formations. In many cases, U.S. support is limited, focused on missions like advising security forces and providing fire support or sustainment. The longstanding U.S. support to the Philippines is an example of such limited support. The limited support approach focuses on building HN capability and capacity. Under this approach, HN security forces are expected to conduct combat operations, including any clearing and holding missions.

PATTERN OF TRANSITION

5-89. COIN efforts may require Soldiers and Marines to create the initial secure environment for the populace. Ideally HN forces hold cleared areas. As HN military and civil capabilities are further strengthened, U.S. military activity may shift toward combined action and limited support. As HN forces assume internal and external security requirements, U.S. forces can redeploy to support bases, reduce force strength, and eventually withdraw. Special operations forces and conventional forces continue to provide support as needed to achieve internal defense and development objectives.

ASSESSMENT OF COUNTERINSURGENCY OPERATIONS

The two best guides, which can not be readily reduced to statistics or processed through a computer, are an improvement in intelligence voluntarily given by the population and a decrease in the insurgents' recruiting rate. Much can be learnt merely from the faces of the population in villages that are subject to clear-and-hold operations, if these are visited at regular intervals. Faces which at first are resigned and apathetic, or even sullen, six months or a year later are full of cheerful welcoming smiles. The people know who is winning.

Sir Robert Thompson
Defeating Communist Insurgency: The Lessons of Malaya and Vietnam, 1966[1]

5-90. *Assessment* is the continuous monitoring and evaluation of the current situation and progress of an operation (FMI 5-0.1). Assessment precedes and is integrated into every operations-process activity and entails two tasks:

- Continuously monitoring the current situation (including the environment) and progress of the operation.
- Evaluating the operation against established criteria.

Commanders, assisted by the staff, continuously compare the operation's progress with their commander's visualization and intent. Based on their assessments, commanders adjust the operation and associated activities to better achieve the desired end state. (See FM 6-0, paragraphs 6-90 through 6-92 and 6-110 through 6-121.)

DEVELOPING MEASUREMENT CRITERIA

5-91. Assessment requires determining why and when progress is being achieved along each LLO. Traditionally, commanders use discrete quantitative and qualitative measurements to evaluate progress. However, the complex nature of COIN operations makes progress difficult to measure. Subjective assessment at all levels is essential to understand the diverse and complex nature of COIN problems. It is also needed to measure local success or failure against the overall operation's end state. Additionally, commanders need to know how actions along different LLOs complement each other; therefore, planners evaluate not only progress along each LLO but also interactions among LLOs.

ASSESSMENT TOOLS

5-92. Assessment tools help commanders and staffs determine—

- Completion of tasks and their impact.
- Level of achievement of objectives.
- Whether a condition of success has been established.
- Whether the operation's end state has been attained.
- Whether the commander's intent was achieved.

For example, planning for transition of responsibility to the host nation is an integral part of COIN operational design and planning. Assessment tools may be used to assess the geographic and administrative transfer of control and responsibility to the HN government as it develops its capabilities. Assessments differ for every mission, task, and LLO, and for different phases of an operation. Leaders adjust assessment methods as insurgents adapt to counterinsurgent tactics and the environment changes.

5-93. The two most common types of assessment measures are measures of effectiveness (MOEs) and measures of performance (MOPs).

5-94. A *measure of effectiveness* is a criterion used to assess changes in system behavior, capability, or operational environment that is tied to measuring the attainment of an end state, achievement of an objective, or creation of an effect (JP 1-02). MOEs focus on the results or consequences of actions. MOEs answer the question, Are we achieving results that move us towards the desired end state, or are additional or alternative actions required?

5-95. A *measure of performance* is a criterion to assess friendly actions that is tied to measuring task accomplishment (JP 1-02). MOPs answers the question, Was the task or action performed as the commander intended?

5-96. Leaders may use observable, quantifiable, objective data as well as subjective indicators to assess progress measured against expectations. A combination of both types of indicators is recommended to reduce the chance of misconstruing trends.

5-97. All MOEs and MOPs for assessing COIN operations should be designed with the same characteristics. These four characteristics are—

- **Measurable.** MOEs and MOPs should have quantitative or qualitative standards against which they can be measured. The most effective measurement would be a combination of quantitative and qualitative measures to guard against an inaccurate view of results.
- **Discrete.** Each MOE and MOP must measure a separate, distinct aspect of the task, purpose, or condition.
- **Relevant.** MOEs and MOPs must be relevant to the measured task, outcome, and condition. HN local, regional, and national leaders, and nongovernmental organization personnel, may provide practical, astute, and professional ideas and feedback to craft relevant MOPs and MOEs.
- **Responsive.** Assessment tools must detect environmental and situational changes quickly and accurately enough to facilitate developing an effective response or counter.

BROAD INDICATORS OF PROGRESS

5-98. Numerical and statistical indicators have limits when measuring social environments. For example, in South Vietnam U.S. forces used the body count to evaluate success or failure of combat operations. Yet, the body count only communicated a small part of the information commanders needed to assess their operations. It was therefore misleading. Body count can be a partial, effective indicator only when adversaries and their identities can be verified. (Normally, this identification is determined through a uniform or possession of an insurgent identification card.) Additionally, an accurate appreciation of what insurgent casualty numbers might indicate regarding enemy strength or capability requires knowing the exact number of insurgent armed fighters initially present. In addition, this indicator does not measure several important factors: for example, which side the local populace blames for collateral damage, whether this fighting and resultant casualties damaged the insurgent infrastructure and affected the insurgency strategy in that area, and where families of dead insurgents reside and how they might react. For another example, within the essential services LLO the number of schools built or renovated does not equate to the effective operation of an educational system.

5-99. Planners should start with broad measures of social and economic health or weakness when assessing environmental conditions. (Table 5-7 [page 5-28] lists possible examples of useful indicators in COIN.)

Table 5-7. Example progress indicators

- **Acts of violence** (numbers of attacks, friendly/host-nation casualties).
- **Dislocated civilians.** The number, population, and demographics of dislocated civilian camps or the lack thereof are a resultant indicator of overall security and stability. A drop in the number of people in the camps indicates an increasing return to normalcy. People and families exiled from or fleeing their homes and property and people returning to them are measurable and revealing.[1]
- **Human movement and religious attendance**. In societies where the culture is dominated by religion, activities related to the predominant faith may indicate the ease of movement and confidence in security, people's use of free will and volition, and the presence of freedom of religion. Possible indicators include the following:
  - Flow of religious pilgrims or lack thereof.
  - Development and active use of places of worship.
  - Number of temples and churches closed by a government.
- **Presence and activity of small- and medium-sized businesses.** When danger or insecure conditions exist, these businesses close. Patrols can report on the number of businesses that are open and how many customers they have. Tax collections may indicate the overall amount of sales activity.
- **Level of agricultural activity**.
  - Is a region or nation self-sustaining, or must life-support type foodstuffs be imported?
  - How many acres are in cultivation? Are the fields well maintained and watered?
  - Are agricultural goods getting to market? Has the annual need increased or decreased?
- **Presence or absence of associations.** The formation and presence of multiple political parties indicates more involvement of the people in government. Meetings of independent professional associations demonstrate the viability of the middle class and professions. Trade union activity indicates worker involvement in the economy and politics.
- **Participation in elections**, especially when insurgents publicly threaten violence against participants.
- **Government services available**. Examples include the following:
  - Police stations operational and police officers present throughout the area.
  - Clinics and hospitals in full operation, and whether new facilities sponsored by the private sector are open and operational.
  - Schools and universities open and functioning.
- **Freedom of movement of people, goods, and communications**. This is a classic measure to determine if an insurgency has denied areas in the physical, electronic, or print domains.
- **Tax Revenue**. If people are paying taxes, this can be an indicator of host-nation government influence and subsequent civil stability.
- **Industry exports**.
- **Employment/unemployment rate**.
- **Availability of electricity**.
- **Specific attacks on infrastructure**.

[1] *Dislocated civilian* is a broad term that includes a displaced person, an evacuee, an expellee, an internally displaced person, a migrant, a refugee, or a stateless person (JP 1-02). Dislocated civilians are a product of the deliberate violence associated with insurgencies and their counteraction. (See FM 3-05.40/MCRP 3-33.1A for additional information on dislocated civilians.)

TARGETING

5-100. The targeting process focuses operations and the use of limited assets and time. Commanders and staffs use the targeting process to achieve effects that support the LLOs in a COIN campaign plan. It is important to understand that targeting is done for all operations, not just attacks against insurgents. The targeting process can support IO, civil-military operations (CMO), and even meetings between commanders and HN leaders, based on the commander's desires. The targeting process occurs in the targeting cell of the appropriate command post. (See JP 3-60, FM 3-09.31/MCRP 3-16C, and FM 6-20-10 for joint and Army targeting doctrine. FM 3-13, appendix E, describes how to apply the targeting process to IO-related targets.)

5-101. Targeting in a COIN environment requires creating a targeting board or working group at all echelons. The intelligence cell provides representatives to the targeting board or working group to synchronize targeting with intelligence sharing and intelligence, surveillance, and reconnaissance operations. The goal is to prioritize targets and determine the means of engaging them that best supports the commander's intent and the operation plan.

5-102. The focus for targeting is on people, both insurgents and noncombatants. There are several different approaches to targeting in COIN. For example, all of the following are potential targets that can link objectives with effects:

- Insurgents (leaders, combatants, political cadre, auxiliaries, and the mass base).
- Insurgent internal support structure (bases of operations, finance base, lines of communications, and population).
- Insurgent external support systems (sanctuaries, media, and lines of communications).
- Legitimate government and functions (essential services, promotion of governance, development of security forces, and institutions).

5-103. Effective targeting identifies the targeting options, both lethal and nonlethal, to achieve effects that support the commander's objectives. Lethal targets are best addressed with operations to capture or kill; nonlethal targets are best engaged with CMO, IO, negotiation, political programs, economic programs, social programs and other noncombat methods. Nonlethal targets are usually more important than lethal targets in COIN; they are never less important. (Table 5-8 [page 5-30] lists examples of lethal and nonlethal targets.)

5-104. The targeting process comprises the following four activities:

- Decide which targets to engage.
- Detect the targets.
- Deliver (conduct the operation).
- Assess the effects of the operation.

5-105. Commanders issue targeting guidance during the "decide" activity. The commander's guidance drives subsequent targeting-process activities. Actions during the "detect" activity may give commanders the intelligence needed to refine the guidance. It may be difficult to identify targets when a COIN campaign begins. The focus during the "decide" activity should be on decisive points commanders can engage.

DECIDE

5-106. The decide activity draws on a detailed intelligence preparation of the battlefield and continuous assessment of the situation. Intelligence personnel, with the commander and other staff members, decide when a target is developed well enough to engage. Continuous staff integration and regular meetings of the intelligence cell and targeting board enable this activity. Staff members consider finished intelligence products in light of their understanding of the AO and advise commanders on targeting decisions. Intelligence personnel provide information on the relative importance of different target personalities and areas and the projected effects of lethal or nonlethal engagement. Specifically, the intelligence analysts need to identify individuals and groups to engage as potential COIN supporters, targets to isolate from the population, and targets to eliminate.

5-107. During the decide activity, the targeting board produces a prioritized list of targets and a recommended course of action associated with each. Executing targeting decisions may require the operations section to issue fragmentary orders. Each of these orders is a task that should be nested within the higher headquarters' plan and the commander's intent. Targeting decisions may require changing the intelligence synchronization plan.

Table 5-8. Examples of lethal and nonlethal targets

Personality targets
• **Lethal** ▪ Insurgent leaders to be captured or killed. • **Nonlethal** ▪ People like community leaders and those insurgents who should be engaged through outreach, negotiation, meetings, and other interaction. ▪ Corrupt host-nation leaders who may have to be replaced.
Area targets
• **Lethal** ▪ Insurgent bases and logistic depots or caches. ▪ Smuggling routes. • **Lethal and nonlethal mix** ▪ Populated areas where insurgents commonly operate. ▪ Populated areas controlled by insurgents where the presence of U.S. or host-nation personnel providing security could undermine support to insurgents. • **Nonlethal** ▪ Populations potentially receptive to civil-military operations or information operations.

DETECT

5-108. The detect activity is performed continuously. It requires much analytical work by intelligence personnel. They analyze large quantities of all-source intelligence reporting to determine the following:

- Threat validity.
- Actual importance of potential targets.
- Best means to engage the target.
- Expected effects of engaging the targets (which will guide actions to mitigate negative effects).
- Any changes required to the exploitation plan.

As mentioned in paragraph 3-152 target exploitation in a COIN environment is similar to that in law enforcement. An exploitation plan not only facilitates gathering evidence but also may lead to follow-on targets after successful exploitation. This requires a detailed understanding of social networks, insurgent networks, insurgent actions, and the community's attitude toward counterinsurgents. (See appendix B.)

5-109. Intelligence regarding the perceptions and interests of the populace requires particular attention. This intelligence is crucial to IO and CMO targeting. It is also important for developing political, social, and economic programs.

DELIVER

5-110. The deliver activity involves executing the missions decided upon by the commander.

ASSESS

5-111. The assess activity occurs continuously throughout an operation. During assessment, collectors and analysts evaluate the operation's progress. They adjust the intelligence synchronization plan and analyses based on this evaluation. In addition to assessing changes to their own operations, intelligence personnel look for reports indicating effects on all aspects of the operational environment, including insurgents and civilians. Relevant reporting can come from any intelligence discipline, open sources, or operational reporting. Commanders adjust an operation based on its effects. They may expand the operation, continue it as is, halt it, execute a branch or sequel, or take steps to correct a mistake's damage. Therefore, an accurate after-action assessment is very important. Metrics often include the following:

- Changes in local attitudes (friendliness towards U.S. and HN personnel).
- Changes in public perceptions.
- Changes in the quality or quantity of information provided by individuals or groups.
- Changes in the economic or political situation of an area.
- Changes in insurgent patterns.
- Captured and killed insurgents.
- Captured equipment and documents.

5-112. As indicated in chapter 3, detainees, captured documents, and captured equipment may yield a lot of information. Its exploitation and processing into intelligence often adds to the overall understanding of the enemy. This understanding can lead to more targeting decisions. In addition, the assessment of the operation should be fed back to collectors. This allows them to see if their sources are credible. In addition, effective operations often cause the local populace to provide more information, which drives future operations.

LEARNING AND ADAPTING

5-113. When an operation is executed, commanders may develop the situation to gain a more thorough situational understanding. This increased environmental understanding represents a form of operational learning and applies across all LLOs. Commanders and staffs adjust the operation's design and plan based on what they learn. The result is an ongoing design-learn-redesign cycle.

5-114. COIN operations involve complex, changing relations among all the direct and peripheral participants. These participants adapt and respond to each other throughout an operation. A cycle of adaptation usually develops between insurgents and counterinsurgents; both sides continually adapt to neutralize existing adversary advantages and develop new (usually short-lived) advantages of their own. Victory is gained through a tempo or rhythm of adaptation that is beyond the other side's ability to achieve or sustain. Therefore, counterinsurgents should seek to gain and sustain advantages over insurgents by emphasizing the learning and adaptation that this manual stresses throughout.

5-115. Learning and adapting in COIN is very difficult due to the complexity of the problems commanders must solve. Generally, there is not a single adversary that can be singularly classified as the enemy. Many insurgencies include multiple competing groups. Success requires the HN government and counter-insurgents to adapt based on understanding this very intricate environment. But the key to effective COIN design and execution remains the ability to adjust better and faster than the insurgents.

SUMMARY

5-116. Executing COIN operations is complex, demanding, and tedious. There are no simple, quick solutions. Success often seems elusive. However, contributing to the complexity of the problem is the manner in which counterinsurgents view the environment and how they define success. The specific design of the COIN operation and the manner in which it is executed must be based on a holistic treatment of the environment and remain focused on the commander's intent and end state. Success requires unity of effort across all LLOs to achieve objectives that contribute to the desired end state—establishing legitimacy and gaining popular support for the HN government. Operational design and execution cannot really be separated. They are both part of the same whole.

This page intentionally left blank.

Chapter 6

Developing Host-Nation Security Forces

[H]elping others to help themselves is critical to winning the long war.

Quadrennial Defense Review Report, 2006

This chapter addresses aspects of developing host-nation security forces. It begins with a discussion of challenges involved and resources required. It provides a framework for organizing the development effort. It concludes with a discussion of the role of police in counterinsurgency operations.

OVERVIEW

6-1. Success in counterinsurgency (COIN) operations requires establishing a legitimate government supported by the people and able to address the fundamental causes that insurgents use to gain support. Achieving these goals requires the host nation to defeat insurgents or render them irrelevant, uphold the rule of law, and provide a basic level of essential services and security for the populace. Key to all these tasks is developing an effective host-nation (HN) security force. In some cases, U.S. forces might be actively engaged in fighting insurgents while simultaneously helping the host nation build its own security forces.

6-2. Just as insurgency and COIN are defined by a complex array of factors, training HN security forces is also affected by a variety of determinants. These include whether sovereignty in the host nation is being exercised by an indigenous government or by a U.S. or multinational element. The second gives counterinsurgents more freedom of maneuver, but the first is important for legitimate governance, a key goal of any COIN effort. If the host nation is sovereign, the quality of its governance also has an impact. The scale of the effort is another factor; what works in a small country might not work in a large one. Terrain and civil considerations are also important. A nation compartmentalized by mountains, rivers, or ethnicity presents different challenges for the COIN effort. A large "occupying" force or international COIN effort can facilitate success in training HN security forces; however, it also complicates the situation. Other factors to consider include the following:

- Type of security forces that previously existed.
- Whether the effort involves creating a completely new security force or changing an existing one.
- Existence of sectarian divisions within the forces.
- Resources available.
- Popular support.

Commanders must adapt these doctrinal foundations to the situation in the area of operations (AO).

6-3. The term "security forces" includes all HN forces with the mission of protecting against internal and external threats. Elements of the security forces include, but are not limited to, the following:

- Military forces.
- Police.
- Corrections personnel.
- Border guards (including the coast guard).

Elements of the HN security forces exist at the local through national levels. Only in unusual cases will COIN forces experience a situation where the host nation has no security force. With this in mind, this chapter addresses methods to develop security forces, realizing that the range of assistance varies depending on the situation.

6-4. JP 3-07.1 contains foreign internal defense (FID) doctrine. JP 3-07.1 addresses the legal and fiscal regulations and responsibilities concerning the planning, development, and administration of FID programs. It also discusses command and supervisory relationships of U.S. diplomatic missions, geographic combatant commands, and joint task forces in applying military aid, support, and advisory missions. The tenets presented in this chapter reinforce and supplement those in JP 3-07.1.

CHALLENGES, RESOURCES, AND END STATE

6-5. Each instance of developing security forces is as unique as each insurgency. In Vietnam, the United States committed thousands of advisors for South Vietnamese units and hundreds of thousands of combat troops but ultimately failed to achieve its strategic objectives. In El Salvador, a relative handful of American advisors were enough enable the HN government to execute a successful counterinsurgency, even though that situation had evolved into a recognized, full-blown civil war. Many factors influence the amount and type of aid required. These are discussed in more detail later, but include the following:

- Existing HN security force capabilities.
- Character of the insurgency.
- Population and culture.
- Level of commitment and sovereignty of the host nation.
- Level of commitment from the United States and other nations.

6-6. U.S. and multinational forces may need to help the host nation in improve security; however, insurgents can use the presence of foreign forces as a reason to question the HN government's legitimacy. A government reliant on foreign forces for internal security risks not being recognized as legitimate. While combat operations with significant U.S. and multinational participation may be necessary, U.S. combat operations are secondary to enabling the host nation's ability to provide for its own security.

CHALLENGES TO DEVELOPING EFFECTIVE SECURITY FORCES

6-7. Many common problems and issues arose in training missions U.S. forces undertook after World War II. These problems generally fall under differing national perspectives in one of four broad categories:

- Resources.
- Leadership.
- Exercising power.
- Organizational structures.

6-8. Governments must properly balance national resources to meet the people's expectations. Funding for services, education, and health care can limit resources available for security forces. The result HN spending priorities may be a security force capable of protecting only the capital and key government facilities, leaving the rest of the country unsecured. Undeveloped countries often lack resources to maintain logistic units. This situation results in chronic sustainment problems. Conducting effective COIN operations requires allocating resources to ensure integration of efforts to develop all aspects of the security force. Recognizing the interrelationship of security and governance, the HN government must devote adequate resources to meeting basic needs like health care, clean water, and electricity.

6-9. Counterinsurgents may need to adjust the existing HN approach to leadership. HN leaders may be appointed and promoted based on family ties or membership in a party or faction, rather than on demonstrated competence or performance. Leaders may not seek to develop subordinates. The need to ensure the welfare of subordinates may not be a commonly shared trait. In some cases, leaders enforce the subordinates' obedience by fear and use their leadership position to exploit them. Positions of power can lead to corruption, which can also be affected by local culture.

6-10. The behavior of HN security force personnel is often a primary cause of public dissatisfaction. Corrupting influences of power must be guarded against. Cultural and ethnic differences within a population may lead to significant discrimination within the security forces and by security forces against minority groups. In more ideological struggles, discrimination may be against members of other political parties, whether in a minority cultural group or not. Security forces that abuse civilians do not win the populace's trust and confidence; they may even be a cause of the insurgency. A comprehensive security force development program identifies and addresses biases as well as improper or corrupt practices.

6-11. Perhaps the biggest hurdle for U.S. forces is accepting that the host nation can ensure security using practices that differ from U.S. practices. Commanders must recognize and continuously address that this "The American way is best" bias is unhelpful. While relationships among U.S. police, customs, and military organizations works for the United States, those relationships may not exist in other nations that have developed differently.

RESOURCES

6-12. For Soldiers and Marines, the mission of developing HN security forces goes beyond a task assigned to a few specialists. The scope and scale of training programs today and the scale of programs likely to be required in the future have grown. While FID has been traditionally the primary responsibility of the special operations forces (SOF), training foreign forces is now a core competency of regular and reserve units of all Services. Multinational partners are often willing to help a nation against an insurgency by helping to train HN forces. Partner nations may develop joint training teams or assign teams to a specific element of the security force or a particular specialty. Training resources may be received from the following organizations and programs:

- Special operations forces.
- Ground forces.
- Joint forces.
- Interagency resources.
- Multinational resources.
- International Military Education and Training (IMET) Program.
- Contractor support.

Special Operations Forces

6-13. SOF focus on specific regions of the world and the study of languages and cultures. SOF have long been the lead organization in training and advising foreign armed forces. (FM 31-20-3 outlines Army special forces training programs and tactics, techniques, and procedures.) While SOF personnel may be ideal for some training and advisory roles, their limited numbers restrict their ability to carry out large-scale missions to develop HN security forces. In a low-level COIN, SOF personnel may be the only forces assigned; at the higher end of the spectrum, SOF may train only their counterparts in the HN forces.

Ground Forces

6-14. Large-scale training and advisory missions need to use large numbers of Soldiers and Marines who may not have language training or regional expertise to levels common in SOF. However, such conventional forces may have some advantages in training HN counterparts with similar missions. SOF and conventional ground forces need linguist augmentation and additional cultural training. (Appendix C discusses linguist support.) Commanders must assign the best qualified Soldiers and Marines to training and advisory missions. Those personnel normally come from active-duty forces, but large-scale efforts require using Reserve Component personnel. All land forces assigned to this high-priority mission need thorough training, both before deploying and in theater.

Joint Forces

6-15. Although other Services often play smaller roles, they can still make significant contributions because of their considerable experience in training foreign forces. For example, the Navy and Air Force can train their HN counterparts. The Coast Guard may also be of value, since its coastal patrol, fisheries oversight, and port security missions correlate with the responsibilities of navies in developing countries. To minimize the burden on land forces, specialists—such as lawyers and medical personnel—from other Services participate in HN training wherever possible.

Interagency Resources

6-16. Interagency resources can support training HN security forces. Perhaps most important is training nonmilitary security forces. The Departments of Justice and State can send law enforcement specialists overseas to train and advise HN police forces. Police are best trained by other police. The quick reaction capability of these agencies is limited, although they can attain necessary levels when given time. Such forces are also expensive. During intensive counterinsurgencies, the environment's high-threat nature limits the effectiveness of civilian police advisors and trainers. These forces work more effectively when operating in a benign environment or when security is provided separately. Many legal restrictions about training nonmilitary forces exist. Normally the Department of State takes the lead in such efforts. However, the President occasionally assigns military forces to these missions.

Multinational Resources

6-17. Although their support frequently plays more of a legitimizing role, multinational partners also assist materially in training HN security forces. Some nations more willingly train HN forces, especially police forces, than provide troops for combat operations. Some multinational forces come with significant employment restrictions. Each international contribution is considered on its own merits, but such assistance is rarely declined. Good faith efforts to integrate multinational partners and achieve optimum effectiveness are required.

International Military Education and Training Program

6-18. For more than 50 years, the U.S. military has run the IMET program to provide opportunities for foreign personnel to attend U.S. military schools and courses. Most of these commissioned officers and noncommissioned officers (NCOs) receive English language training before attending the U.S. courses. In the case of Latin American armed forces, the United States operates courses in Spanish.

contractor support

6-19. In some cases, additional training support from contractors enables commanders to use Soldiers and Marines more efficiently. Contractor support can provide HN training and education, including the following:

- Institutional training.
- Developing security ministries and headquarters.
- Establishing administrative and logistic systems.

Contracted police development capabilities through the Department of State's Bureau of International Narcotics and Law Enforcement Affairs can provide expertise not resident in the uniformed military.

Organizing U.S. Forces to Develop Host-Nation Security Forces

6-20. Developing HN security forces is a complex and challenging mission. The United States and multinational partners can only succeed if they approach the mission with the same deliberate planning and preparation, energetic execution, and appropriate resourcing as the combat aspects of the COIN operation. Accordingly, COIN force commanders and staffs need to consider the task of developing HN security forces during their initial mission analysis. They must make that task an integral part of all assessments, planning, coordination, and preparation.

6-21. As planning unfolds, mission requirements should drive the initial organization for the unit charged with developing security forces. To achieve unity of effort, a single organization should receive this responsibility.

6-22. For small-scale COIN efforts, SOF may be the only forces used. SOF organizations may be ideally suited for developing security forces through the FID portion of their doctrinal mission.

6-23. If only a single component (land, maritime, air, or special operations) is being developed, commanders can assign the mission to a single-Service task force. For example, if the host nation requires a maritime capability to guard oil distribution platforms, a Navy task force may receive the mission.

6-24. In an area in which COIN operations are already underway, developing security forces can be assigned to a specific unit, such as a brigade combat team, division, or Marine air-ground task force.

6-25. For large, multi-Service, long-duration missions, a separate organization with the sole responsibility of developing security forces and subordinate to the COIN force commander may be required. Such an organization may be multi-Service, multinational, and interagency.

Table 6-1. Staff functions required when training host-nation security forces

- **Financial manager** for managing the significant monetary resources required for training, equipping, and building security forces. A separate internal auditor may be required as a check to ensure host-nation resources are safeguarded and effectively managed.
- **Staff judge advocate** with specific specialties and a robust capability for contract law, military justice, and the law of land warfare.
- **Construction engineer** management to oversee and manage the construction of security forces infrastructure, such as the following:
  - Bases, ranges, and training areas.
  - Depots and logistic facilities.
  - Police stations.
- **Political-military advisors** to ensure development of security forces is integrated with development of civilian ministries and capabilities.
- **Public affairs**, with a focused capability to build the populace's confidence in the host-nation security forces and to develop the host-nation forces' public affairs capability.
- **Force protection and focused intelligence staff** to address the challenge of and threats to the relatively small teams that may be embedded with host-nation security forces and not co-located with U.S. or multinational forces.
- **Materiel management** until such a capability is developed in the host-nation forces. The equipping and supplying of new security forces is critical to their development and employment. It may not be able to wait until the host-nation develops that capability.
- **Health affairs**, since most developing countries have poor health care systems. Host-nation personnel are more likely to stay in new units and fight when they believe that they will be properly treated if wounded. Additionally, disease is a significant threat that must be addressed with preventive medicine and robust care.
- **Security assistance** (IMET) to manage the external training efforts and foreign military sales, and to employ well-developed procedures for purchasing weapons, equipment, goods, and services. In counterinsurgencies, these functions are probably performed by higher headquarters staff elements rather than a stand-alone office (such as an office of military cooperation). U.S. security assistance programs normally try to sell U.S.-manufactured equipment. The organization responsible for equipping host-nation forces should not be constrained to purchase U.S. equipment. It requires the flexibility to procure equipment where time, cost, and quality are appropriate for host-nation needs.
- **Civilian law enforcement**. Staff officers with a civilian law enforcement background or actual civilian law enforcement personnel can play a vitally important role in advising the commander. Traditionally officers from the Reserve Components have done this.

6-26. The internal structure of the organization charged with developing security forces must reflect the desired end state of those security forces. For example, if army, police, air, naval, and special operations capabilities are being developed, the organization in charge of those programs requires teams charged specifically with each of those tasks. If civilian security components, such as a ministry of defense or interior, are being developed, then ministerial teams are needed. Developing security forces in terms of professionalism and ethics is important; a separate element focused on training those values may be needed.

6-27. The U.S. or multinational force responsible for these programs requires a headquarters and staff task-organized for the functions required to support all aspects of developing the security forces. (See paragraph 6-31.) In addition to traditional staff functions, some or all of functions listed in table 6-1 (page 6-5) may require augmentation.

6-28. An effective security force development organization is flexible and adaptive. Requirements for developing the type, character, composition, and quantity of security forces change as security forces grow and the COIN operation matures. The organization must anticipate such changes, since joint manning document procedures and requests for forces have limited responsiveness. Temporary duty and contract personnel may provide support to fill gaps until more permanent individuals or units arrive.

DESIRED END STATE

6-29. Training HN security forces is a slow and painstaking process. It does not lend itself to a "quick fix." Real success does not appear as a single decisive victory. To ensure long-term success, commanders clarify their desired end state for training programs early. This end state consists of a set of military characteristics common to all militaries. (See table 6-2.) Those characteristics have nuances in different countries, but well-trained HN security forces should—

- Provide reasonable levels of security from external threats while not threatening regional security.
- Provide reasonable levels of internal security without infringing upon the populace's civil liberties or posing a coup threat.
- Be founded upon the rule of law.
- Be sustainable by the host nation after U.S. and multinational forces depart.

6-30. When dealing with insurgents, HN military and police forces may perform functions not normally considered conventional. The military may fill an internal security role usually reserved for the police. Police may have forces so heavily armed that they would normally be part of the military. In the near term, the HN security forces should—

- Focus on COIN operations, integrating military capabilities with those of local, regional, and national police.
- Maintain the flexibility to transition to more conventional roles of external and internal defense, based on long-term requirements.

To meet both near- and long-term objectives, trainers remember the cumulative effects of training. Effective training programs have short-, mid-, and long-term effects.

6-31. To achieve this end state and intermediate objectives, the host nation should develop a plan—with multinational assistance when necessary. The plan should address all aspects of force development. U.S. doctrine divides force development into domains: doctrine, organization, training, materiel, leadership and education, personnel and facilities (DOTMLPF). Doctrine is listed first. However, these elements are tightly linked, simultaneously pursued, and difficult to prioritize. Commanders monitor progress in all domains. There is always a temptation for Soldiers and Marines involved in such programs to impose their own doctrine and judgment on the host nation. The first U.S. advisors and trainers working with the South Vietnamese Army aimed to create a conventional force to fight another Korean War. They did not recognize their allies' abilities or the real nature of the threat. The organization and doctrine adopted did not suit the South Vietnamese situation and proved vulnerable to North Vietnamese guerrilla tactics. HN security force doctrine, like the remaining DOTMLPF domains discussed in this chapter, must be appropriate to HN capabilities and requirements.

FRAMEWORK FOR DEVELOPMENT

6-32. The mission to develop HN security forces can be organized around these tasks:

- Assess.
- Organize.
- Build or rebuild facilities.

- Train.
- Equip.
- Advise.

These incorporate all DOTMLPF requirements. Although described sequentially, these tasks are normally performed concurrently. For example, training and equipping operations must be integrated and, as the operation progresses, assessments will lead to changes. If U.S. forces are directly involved in operations against insurgents, the development program requires a transition period during which major COIN operations are handed over to HN security forces.

Table 6-2. Characteristics of effective host-nation security forces

- **Flexible**. Forces capable of accomplishing the broad missions required by the host nation—not only to defeat insurgents or defend against outside aggression but also to increase security in all areas. This requires an effective command and organizational structure that makes sense for the host nation.
- **Proficient**.
  - Security forces capable of working effectively in close coordination with each other to suppress lawlessness and insurgency.
  - Military units tactically and technically proficient, capable of ensuring their aspect of national security and capable of integrating their operations with those of multinational partners.
  - Nonmilitary security forces competent in maintaining civil order, enforcing laws, controlling borders, securing key infrastructure (such as power plants), and detaining criminal suspects.
  - Nonmilitary security forces thoroughly trained in modern police ethos and procedures, and who understand the basics of investigation, evidence collection, and proper court and legal procedures.
- **Self-sustained**. Forces capable of managing their own equipment throughout its life cycle (procurement to disposal) and performing administrative support.
- **Well led**. Leaders at all levels who possess sound professional standards and appropriate military values, and are selected and promoted based on competence and merit.
- **Professional**.
  - Security forces that are honest, impartial, and committed to protecting and serving the entire population, operating under the rule of law, and respecting human rights.
  - Security forces that are loyal to the central government and serving national interests, recognizing their role as the people's servants and not their masters.
- **Integrated into society**. Forces that represent the host nation's major ethnic groups and are not seen as instruments of just one faction. Cultural sensitivities toward the incorporation of women must be observed, but efforts should also be made to include women in police and military organizations.

ASSESS

6-33. Commanders assess the situation at the start of every major military operation. The assessment is one part of the comprehensive program of analyzing the insurgency. It includes looking at the society and the economy. The analysis is performed in close collaboration with the U.S. country team, the host nation, and multinational partners. These partners continually assess the security situation and its influence on other logical lines of operations. From the assessment, planners develop short-, mid-, and long-range goals and programs. As circumstances change, so do the goals and programs. A raging insurgency might require the early employment of HN forces at various stages of development. Some existing security forces may be so dysfunctional or corrupt that the organizations must be disbanded rather than rehabilitated. In some cases, commanders will need to replace some HN leaders before their units will become functional.

6-34. While every situation is different, leaders of the development program should assess the following factors throughout planning, preparation, and execution of the operation:

- Social structure, organization, demographics, interrelationships, and education level of security force elements.
- Methods, successes, and failures of HN COIN efforts.
- State of training at all levels, and the specialties and education of leaders.
- Equipment and priority placed on maintenance.
- Logistic and support structure, and its ability to meet the force's requirements.
- Level of sovereignty of the HN government.
- Extent of acceptance of ethnic and religious minorities.
- Laws and regulations governing the security forces and their relationship to national leaders.

6-35. A comprehensive mission analysis should provide a basis for establishing the scope of effort required. This analysis includes a troop-to-task analysis that determines the type and size of forces needed. The HN security forces may require complete reestablishment, or they may only require assistance to increase capacity. They may completely lack a capability (for example, internal affairs, federal investigative department, corrections, logistics for military forces, formal schools for leaders), or they may only require temporary reinforcement. As with other military operations, efforts to assist security forces should reinforce success. For example, instead of building new police stations in every town, improve the good stations and use them as models for weaker organizations.

6-36. Leaders need decisions on what shortfalls to address first. The extent of the insurgency combined with resource limitations inevitably forces commanders to set priorities. Follow-on assessments should start by reviewing areas with restricted resources, determining where resources should be committed or redirected, and deciding whether to request additional resources. If the U.S. or another multinational partner or international entity exercises sovereignty, such as during an occupation or regime change, decisions about security force actions can be imposed on a host nation; however, it is always better to take efforts to legitimize the HN leaders by including them in decisions.

6-37. Developing a strategic analysis and outlining a strategic plan for training the forces of a country facing an insurgency is not necessarily a long process. In fact, situations that include a security vacuum or very active insurgency often require starting programs as soon as possible. Assessment is continuous; initial assessments and the programs they inspire must be adjusted as more experience and information are gained. In 1981, when El Salvador faced a major insurgency, a team of ten U.S. officers visited there. The team consulted with the HN command and the U.S. military assistance advisory group (referred to as the MILGRP) for ten days. In that time, the team outlined a five-year comprehensive plan to rebuild, reorganize, train, and reequip the Salvadoran armed forces to counter the insurgency. The U.S. plan became part of the foundation of a successful national COIN strategy. A team with a similar mission today should include specialists from the Departments of State, Justice, and Homeland Security (in particular, border security and customs experts) to assess the security force requirements.

ORGANIZE

6-38. Organizing HN forces depends on the host nation's social and economic conditions, cultural and historical factors, and security threat. The development program's aim is to create an efficient organization with a command, intelligence, logistic, and operations structure that makes sense for the host nation. Conventional forces with limited special purpose teams (such as explosive ordnance disposal and special weapons and tactics [SWAT]) are preferred. Elite units tend to divert a large share of the best leadership and remove critical talent from the regular forces. Doctrine should be standard across the force, as should unit structures. The organization must facilitate the collection, processing, and dissemination of intelligence across and throughout all security forces.

6-39. Another organizational approach is establishing home guard units. In many COIN operations, these units have effectively provided increased security to the populace. Home guards are part-time, lightly armed, local security forces under HN government control. Often, career military and police officers supervise home guards at the provincial and national levels. Home guards provide point security. They guard vital installations that insurgents will likely target, such as government buildings and businesses. Home guards can also provide security for small villages and man gates and checkpoints. While home guards are not trained to conduct offensive operations, their constant presence reminds the populace that the HN government can provide security. Effective home guards can free police and military forces from stationary guard duties.

General Considerations

6-40. As much as possible, the host nation should determine the security force organization's structure. The host nation may be open to proposals from U.S. and multinational forces but should at least approve all organizational designs. As the HN government strengthens, U.S. leaders and trainers should expect increasingly independent organizational decisions. These may include changing the numbers of forces, types of units, and internal organizational designs. Culture and conditions might result in security forces given what U.S. experience considers nontraditional roles and missions. HN police may be more paramilitary than their U.S. counterparts, and the military may have a role in internal security. Eventually, police and military roles should clearly differ. Police should counter crime while the military should address external threats. However, the exact nature of these missions depends on the HN situation. In any event, police and military roles should be clearly delineated.

6-41. Organized units should include all appropriate warfighting functions (formerly, battlefield operating systems) or some adaptation for police forces. Some systems may be initially deployed in limited numbers or excluded for various reasons, for example, cost, relevance, or training requirements. However, organizational plans should include eventually establishing all appropriate capabilities.

6-42. Organization should address all security force elements, from the ministerial level to the patrolling police officer and soldier. Figure 6-1 (page 6-10) illustrates the complex matrix of simultaneous development programs. Building a competent HN civilian infrastructure—including civilian command and control systems—is critical for success in COIN. The COIN force commander works with HN ministries responsible for national and internal security, including the ministry of national defense, the interior ministry, and the justice ministry. The commander assesses strengths and weaknesses of the ministerial organization as well as training requirements of civilian defense officials and employees. The U.S. and multinational advisory team at the ministry level help the host nation develop a procurement and management system that effectively meets its requirements.

6-43. A thorough review of HN military and police doctrine is a necessary first step in setting up a training program. Advisors review security force regulations to ensure they provide clear, complete instructions for discipline, acquisitions, and support activities. Advisors review and refine doctrine (including tactics, techniques, and procedures) to address COIN operations. Regulations should fit security force personnel's level of education and sophistication. Treatment of prisoners, detainees, and suspected persons should be clear and consistent with the norms of international and military law.

personnel Considerations

6-44. Organizing a security force requires resolving issues related to the following areas:

- Recruiting.
- Promotion screening and selection.
- Pay and benefits.
- Leader recruiting and selection.
- Personnel accountability.
- Demobilization of security force personnel.

Recruiting

6-45. Recruiting is critical when establishing security forces. The host nation designs the recruiting program, considering local culture and themes that resonate with the populace. The program should ensure that security forces include members from all major demographic groups. U.S. and multinational partners should encourage and support HN efforts to recruit from minority populations. A mobile recruiting capability should be established to target specific areas, ethnic groups, or tribes to ensure demographic representation within the security forces. Moderate groups and factions within hostile or potentially hostile ethnic groups should be encouraged to join the HN security forces. Most HN governments will likely resist recruiting disaffected ethnic groups into their security forces. However, even moderate success in recruiting from these groups provides enormous payoffs. It builds the security forces' legitimacy and often quiets legitimate fears of such groups regarding their relationship to the HN government. Effectively disarming, demobilizing, and reintegrating former insurgents and their groups must be part of the overall COIN plan. It must be included in the recruiting effort.

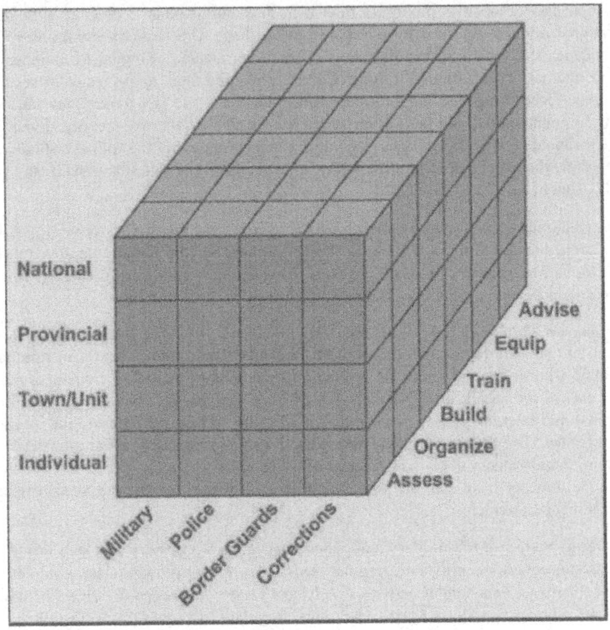

Figure 6-1. Factors affecting security force development

6-46. A proper recruiting program requires a clear set of appropriate mental, physical, and moral standards. Ideally, recruits are centrally screened and inducted. Recruiting centers need to be located in areas safe from insurgent attacks; these centers are attractive targets. All recruits should undergo a basic security check and be vetted against lists of suspected insurgents. As much as possible, HN agencies and personnel should perform this screening. Membership in illegal organizations needs to be carefully monitored. Past membership need not preclude a person from joining the security forces; however, any ongoing relationship with an illegal organization requires constant monitoring. HN personnel need to ensure that no single military or police unit contains too many prior members of an illegal unit, tribal militia, or other militant faction.

Promotion Screening and Selection

6-47. Selection for promotion must stem from proven performance and aptitude for increased responsibility. Objective evaluations ensure promotion is by merit and not through influence or family ties. Two methods for selecting leaders may be worth considering. One method identifies the most competent performers, trains them, and recommends them for promotion. The second method identifies those with social or professional status within the training group, then trains and recommends them for promotion. The first method may lead to more competent leaders but could be resisted for cultural reasons. The second method ensures the new leader will be accepted culturally but may sacrifice competence. The most effective solution comes from combining the two methods.

Pay and Benefits

6-48. Appropriate compensation levels help prevent a culture of corruption in the security forces. It is cheaper to spend the money needed for adequate wages and produce effective security forces than to pay less and end up with corrupt and abusive forces that alienate the populace. Paying the police adequately is especially important; the nature of their duties and contact with the civilian community often expose them to opportunities for corruption. (Table 6-3 lists some important considerations concerning security force pay.)

Table 6-3. Security force pay considerations

- Pay for commissioned officers, noncommissioned officers, and technical specialists should be competitive with that of other host-nation professionals. Police officers need to be paid enough that they do not have to supplement their income with part-time jobs or resort to illegal behavior.
- Pay should be disbursed through host-nation government channels, not U.S. channels.
- Cultural norms should be addressed to ensure that any questionable practices, such as "taxing" subordinates, are minimized.
- Good pay and attractive benefits must be combined with a strict code of conduct that allows for the immediate dismissal of corrupt security personnel.
- Pensions should be available to compensate families of security force members in the event of a service-related death.
- Pay for military forces should come from central government budgets. Military forces must not be paid from kickbacks or locally procured revenue. This practice results in the populace questioning the military's loyalty, and corruption will be likely.
- Pay for military and police personnel should be roughly equivalent. General pay parity helps ensure that no single force attracts all the best qualified personnel.

6-49. Effective security forces can help improve HN social and economic development through the benefits each member receives. Every recruit should receive a basic education, job training, and morals and values inculcation.

Leader Recruiting and Selection

6-50. Officer candidate standards should be high. Candidates should be in good health and pass an academic test with a higher standard than the test for enlisted recruits. Officer candidates should be carefully vetted to ensure that they do not have close ties to any radical or insurgent organization.

6-51. NCOs should be selected from the best enlisted security force members. Objective standards, including proficiency tests, should be established and enforced to ensure that promotion to the NCO ranks comes from merit, not through influence or family ties. Many armies lack a professional NCO corps; establishing one for a host nation may be difficult. In the meantime, adjustments will have to be made, placing more responsibility on commissioned officers.

Personnel Accountability

6-52. HN leaders must carefully track and account for security force personnel. Proper personnel accountability reduces corruption, particularly in countries with manual banking systems where security force personnel are paid in cash. In addition, large numbers of personnel failing to report for duty can indicate possible attacks, low unit morale, or insurgent and militia influences on the security forces.

Demobilization of Security Force Personnel

6-53. Host nations should develop programs to keep a class of impoverished and disgruntled former officers and soldiers from forming. As the security forces mature, commissioned officers and NCOs who perform poorly or fail to meet the new, higher standards of the force will need to be removed. Providing some form of government-provided education grants or low-interest business loans enables discharged personnel to earn a living outside the military. Commissioned officers and NCOs who serve for several years and are then removed should receive a lump-sum payment or small pension to ease their transition to civilian life. These programs should not apply to those guilty of major human rights abuses or corruption. Demobilization planning should start as soon as commanders anticipate the need. (It may not be required in all cases.) Any plan should evolve with HN security force development to ensure its feasibility. Similar programs may be required when demobilizing nongovernment militias.

6-54. As a conflict ends, some security forces may need to be disarmed, demobilized, and reintegrated into civil society. To avoid producing a pool of recruits for the insurgency, the host nation should establish programs to keep large numbers of demobilized security force members from becoming immediately unemployed. Civil service departments should provide a hiring preference to people completing an honorable term of service. Government-financed education programs for demobilized members are another possibility.

BUILD OR REBUILD FACILITIES

6-55. HN security forces need infrastructure support. People need buildings for storage, training, and shelter. Often requirements include barracks, ranges, motor pools, and other military facilities. Construction takes time; the host nation needs to invest early in such facilities if they are to be available when needed. Protection must be considered in any infrastructure design, including headquarters facilities, as infrastructure provides attractive targets for insurgents. (See FM 5-104 for information on hardening measures to increase infrastructure survivability and improve protection.)

6-56. During an insurgency, HN military and police forces often operate from local bases. Building training centers and unit garrisons requires a long-term force-basing plan. If possible, garrisons should include housing for the commissioned officers, NCOs, and families; government-provided medical care for the families; and other benefits that make national service attractive.

6-57. The host nation may need to make large investments in time and resources to restore or create the nationwide infrastructure necessary to effectively command and control HN security forces. Communications facilities are especially important. Besides building local bases and police stations, the host nation will need functional regional and national headquarters and ministries.

TRAIN

6-58. U.S. and multinational training assistance should address shortfalls at every level with the purpose of establishing self-sustaining training systems.

Training the U.S. Trainers

6-59. Soldiers and Marines assigned training missions should receive training on the specific requirements of developing HN forces. The course should emphasize the host nation's cultural background, introduce its language, and provide cultural tips for developing a good rapport with HN personnel. The course should also include protection training for troops working with HN forces. U.S. trainees must become familiar with the HN organization and equipment, especially weapons not in the U.S. inventory. This training must emphasize the following:

- Sustaining training and reinforcing individual and team skills.
- Using the smallest possible student-to-instructor ratio.
- Developing HN trainers.
- Training to standards—not to time.
- Providing immediate feedback; using after-action reviews.

- Respecting the HN culture, but learning to distinguish between cultural practices and excuses.
- Learning the HN language.
- Working with interpreters.

6-60. U.S. forces should show respect for local religions and traditions. Soldiers and Marines should willingly accept many aspects of the local and national culture, including food (if sanitation standards permit). U.S. forces must make clear that they do not intend to undermine or change the local religion or traditions. However, Soldiers and Marines have a mission to reduce the effects of dysfunctional social practices that affect the ability to conduct effective security operations. U.S. trainers and advisors must have enough awareness to identify and stop inappropriate behavior, or at least report it to the multinational and HN chains of command.

Establishing Training Standards

6-61. Insurgent approaches and their corresponding responses from targeted governments vary widely; however, the host nation and trainers can still establish clear measures for evaluating the training of individuals, leaders, and units. COIN operations require many of the same individual and collective skills performed in conventional military operations but also include additional requirements for COIN. Small units execute most COIN operations; therefore, effective COIN forces require strong junior leaders. All levels of training for all components should include values training. Metrics for evaluating units should include subjective measures, such as loyalty to the HN government, as well as competence in military tasks. Soldiers and Marines know how to evaluate military training. However, the acceptance of values, such as ethnic equality or the rejection of corruption, may be a better measure of training effectiveness in some COIN situations. Gauging this acceptance is far more difficult than evaluating task performance.

6-62. Effective training programs require clear, detailed individual, leader, and unit performance standards. These standards take into account cultural factors that directly affect the ability of the individual or unit to operate. For example, training a unit to conduct effective operations requires more time in countries where the average soldier is illiterate. Similarly, staff training proves more difficult in countries with a low educational level. Building a security force from the ground up takes far more time than creating one around a trained cadre of HN personnel. With this in mind, it is usually better to use existing military personnel to form units and cadres for units, rather than creating novice security forces. Vetting may be required to determine loyalties and validate the abilities of existing security forces.

6-63. Poorly trained leaders and units are more prone to committing human rights violations than well-trained, well-led units. Leaders and units unprepared for the pressure of active operations tend to use indiscriminate force, target civilians, and abuse prisoners. These actions can threaten the popular support and government legitimacy essential for COIN success. Badly disciplined and poorly led security forces become effective recruiters and propagandists for insurgents.

6-64. Setting realistic metrics, both objective and subjective, for HN security forces and following through on training plans consume time. The pressure is strong to find training shortcuts, employ "quick fixes," or to train personnel on the job. Trainers should resist such approaches. In the long term, such approaches create more problems than they solve. However, trainers should also avoid the temptation to create long, complex training programs based on unrealistic standards. Effective programs account for the host nation's culture, resources, and short-term security needs. No firm rules exist on how long particular training programs should take, but trainers can use existing and historical U.S. or multinational training programs as starting points for planning. To a certain extent, the insurgent threat dictates how long training can take. As security improves, training programs can expand to facilitate achievement of the long-term end state.

Training Methods

6-65. Training programs should be designed to prepare HN personnel to train themselves. HN trainers are the best trainers and should be used as much as possible. Many training methods have proven successful; some also enhance developing HN training capability. (Table 6-4 [page 6-14] lists several successful training methods.)

Table 6-4. Methods of training host-nation security forces

- **Formal schools** initially run by U.S. forces with selected graduates returning as instructors. This includes entry-level individual training.
- **Mobile training teams** to reinforce individual and collective training as needed.
- **Partnership training** with U.S. forces tasked to train and advise host-nation units with whom they are partnered. U.S. forces support host-nation units. As training progresses, host-nation squads, platoons, and companies may work with their U.S. partners in security or combat operations. In this manner, the whole U.S. unit mentors their partners. Habitual training relationships should be maintained between partners until host-nation units meet standards for full capability.
- **Advisor teams** detailed to assist host-nation units with minimal segregation between U.S. and host-nation personnel. Advisor teams work especially well in training senior ministry personnel.
- **U.S. personnel embedded in key positions in host-nation units.** This may be required where host-nation security forces are needed but leader training is still in its early stages. This approach increases dependency on U.S. forces and should be used only in extreme circumstances. As host-nation capabilities improve, host-nation personnel should replace the embedded U.S. or multinational personnel.
- **Contractors** can also be used to assist with training, though care is required to ensure the training is closely supervised and meets standards.

Training Soldiers

6-66. Security force members must be developed through a systematic training program. The program first builds their basic skills, then teaches them to work together as a team, and finally allows them to function as a unit. Basic military training should focus first on COIN-related skills, such as first aid, marksmanship, and fire discipline. Leaders must be trained in tactics, including patrolling and urban operations. Everyone must master rules of engagement and the law of armed conflict. HN units should train to standard for conducting the major COIN missions they will face. Required skills include the following:

- Manage their security.
- Provide effective personnel management.
- Conduct logistic (planning, maintenance, sustainment, and movement) operations.
- Conduct basic intelligence functions.
- Coordinate indirect fires.
- Provide for effective medical support.

Training Leaders

6-67. The effectiveness of the HN security forces directly relates to the quality of their leadership. Building effective leaders requires a comprehensive program of officer, staff, and specialized training. The ultimate success of any U.S. involvement in a COIN effort depends on creating viable HN leaders able to carry on the fight at all levels and build their nation on their own.

Leader Training Standards

6-68. The leader training methodology must reinforce the different levels of authority within the HN security force. The roles and responsibilities of each commissioned officer and NCO rank must be firmly established so recruits understand what is expected of them. Their subordinate relationship to civilian authorities must also be reinforced to ensure civilian control. In addition, training should establish team dynamics. In some cultures, security forces may need training to understand the vital role of members not in primary leadership positions.

6-69. In addition to tactical skills, commissioned officers should be trained in accountability, decision making, delegating authority, values, and ethics. Special requirements for COIN should be the primary focus of the initial curriculum. These subjects include the following:

- Intelligence collection.
- Day and night patrolling.
- Point security.
- Cordon and search operations.
- Operations with police.
- Treatment of detainees and prisoners.
- Psychological operations.
- Civic action.

As the insurgency declines, the curriculum can be adjusted for a long-term focus.

6-70. In addition, leader training should be conducted in a way that shows—

- How to work as a team.
- How to develop and take advantage of subordinates' skills.
- How to train subordinates.
- How to maintain discipline and assume responsibility for one's own and subordinates' actions.
- How to understand and enforce the rules of engagement.

Basic Commissioned Officer Education

6-71. Various models for basic officer education exist. These include the following:

- One-year military college.
- Four-year military college.
- Officer candidate school (OCS).
- Military training at civilian universities.

6-72. Time available may determine which model to use as the primary commissioning source. If the situation allows, four-year programs at military or civilian institutions may be the best choice. If not, the OCS and one-year college models may be better. Theoretically, having a few high-quality officers may be better than many adequate ones, but the insurgents may not allow this luxury. Citizens under attack would rather have an adequate officer and unit now than a better leader and organization years later.

6-73. The British Army uses the one-year military college model. Prospective officers attend Sandhurst, an intensive course that includes a rigorous program of basic training and a thorough grounding in the British Army's history and culture. It also emphasizes developing each future officer as a leader. At the end of the year, each new officer attends a shorter specialty branch course.

6-74. The four-year military and civilian college models provide the best overall education, while preparing officers for work at the tactical and operational levels. The longer programs also are good for inculcating values. However, they require significant time and resources.

6-75. Under an OCS-style program, outstanding individuals come from the enlisted ranks or society. They receive intensive training in the military skills junior officers require. OCS programs often last less than a year. An OCS-style course should be followed by specialized branch training.

Intermediate and Advanced Commissioned Officer Education

6-76. Military units only become effective when their commanders and staffs can effectively plan, prepare, execute, and assess operations. Initial intermediate-level commissioned officer training should focus on building effective commanders and staffs for small units, then progressively move to higher echelons. Thus, initial intermediate-level officer training focuses on the company and battalion levels (or police station level). Later courses address higher echelons, depending on the size of the overall force to be developed.

6-77. A cadre of carefully selected low- and mid-level commissioned officers can receive an advanced education at existing formal schools in the United States or other partner nations through IMET-like programs. This type of program builds a qualified leadership cadre. This cadre can, upon their return home, assume leadership positions and become the faculty for HN schools. These officers should have increased credibility when they return to their country. Officer students usually make and maintain strong personal connections with their foreign hosts during and after their stay abroad. One of the key goals for developing HN security forces is to professionalize them. The first-hand experience of their officers training at foreign military schools, living abroad, and seeing military professional standards practiced proves invaluable. As with officer commissioning programs, time is a key consideration. IMET-like programs are expensive and time consuming. The best officers—those normally selected for such training—may be needed more in the country's combat forces fighting the insurgency.

Operational Employment of Newly Trained Forces

6-78. Building the morale and confidence of security forces should be a primary strategic objective. Committing poorly trained and badly led forces results in high casualties and invites tactical defeats. While defeat in a small operation may have little strategic consequence in a conventional war, even a small tactical defeat of HN forces can have serious strategic consequences in a COIN. Insurgent warfare is largely about perceptions. Effective insurgent leaders can quickly turn minor wins into major propaganda victories. Defeat of one government force can quickly degrade the morale of others. If a HN force fails, the local populace may begin to lose confidence in the government's ability to protect them. A string of relatively minor insurgent victories can cause widespread loss of morale in the HN forces and encourage the "neutral majority" to side with the insurgents. In short, the HN security forces must be prepared for operations so that they have every possible advantage. The decision to commit units to their first actions and their employment method requires careful consideration. As much as possible, HN forces should begin with simpler missions. As their confidence and competence grows, these forces can assume more complex assignments. Collaborating with U.S. or multinational units can help new HN units to become accustomed to the stresses of combat.

6-79. Newly trained units should enter their first combat operation in support of more experienced HN, U.S., or multinational forces. Operational performance of such inexperienced organizations should be carefully monitored and evaluated so that weaknesses can be quickly corrected. The employment plan for HN security units should allow enough time for additional training after each operation. Introducing units into combat gradually allows identification of poor leaders to the HN leadership for retraining or other action. Competent leaders are also identified and given greater authority and responsibility.

Training Defense Ministry Civilians

6-80. U.S. forces tasked with training HN personnel must also ensure that the military and security forces have capable management in the top ranks. Combatant commanders place experienced U.S. officers and Department of Defense personnel inside the HN defense and interior ministries as trainers and advisors for HN managers and leaders. U.S. forces should also develop a training program for civilian personnel of the ministry of defense. Personnel training should address the following:

- Equipment acquisition.
- Departmental administration.
- Personnel management.
- Financial management.
- Maintenance and inventory controls.
- Strategic (or national) level operations.

Selected ministry of defense personnel may receive specialized training in defense management through U.S. or multinational partners' schools (for example, the National Defense University) or in civilian institutions that specialize in graduate programs for security studies.

EQUIP

6-81. The strategic plan for security force development should outline HN equipment requirements. Equipment should meet the host nation's requirements. Appropriate equipment is affordable and suitable against the threat. The host nation must also be able to train on the equipment. Interoperability may be desired in some cases. A central consideration includes the host nation's long-term ability to support and maintain the equipment.

6-82. The initial development plan should include phases with goals for HN forces to meet over a period of three to four years. Due to the highly adaptive nature of insurgents and the often rapidly changing situation on the ground, commanders must continually asses the direction and progress of developing HN security forces.

6-83. The requirement to provide equipment may be as simple as assisting with maintenance of existing formations or as extensive as providing everything from shoes and clothing to vehicles, communications, and investigation kits. If insurgents use heavy machine guns and rocket-propelled grenades, HN security forces need comparable or better equipment. This especially applies to police forces, which are often lightly armed and vulnerable to well-armed insurgents.

6-84. Primary considerations should include maintainability, ease of operation, and long-term sustainment costs. Few developing nations can support highly complex equipment. In COIN operations, having many versatile vehicles that require simple maintenance is often better than having a few highly capable armored vehicles or combat systems that require extensive maintenance. Developing an effective HN maintenance system often begins with major maintenance performed by contractors. The program then progresses to partnership arrangements with U.S. forces as HN personnel are trained to perform the support mission.

6-85. Sources for HN materiel include U.S. foreign military sales, multinational or third-nation resale of property, HN contracts with internal suppliers, or HN purchases on the international market. (Paragraphs D-27 through D-34 discusses relevant legal considerations.) The HN, U.S., and multinational organizations responsible for equipping HN forces should have the flexibility necessary to obtain equipment that meets the HN force needs for quality, timeliness and cost. As part of their training, HN security forces also need to learn the practices and importance of property accountability to reduce corruption and ensure proper equipment usage. Part of equipping HN forces includes training them in the practices and importance of property accountability. HN forces are expected to provide equipment the same level of control and protection that U.S. forces provide for similar equipment. (See AR 12-1 and DODD 5105.38M.)

ADVISE

6-86. Advisors are the most prominent group of U.S. personnel that serve with HN units. Advisors live, work, and (when authorized) fight with their HN units. Segregation is kept at an absolute minimum. The relationship between advisors and HN forces is vital. U.S. commanders must remember that advisors are not liaison officers, nor do they command HN units.

6-87. Effective advisors are an enormous force enhancer. The importance of the job means that the most capable individuals should fill these positions. Advisors should be Soldiers and Marines known to take the initiative and who set the standards for others. (FM 31-20-3 provides additional information and guidelines for advisors.)

6-88. More than anything else, professional knowledge and competence win the respect of HN troops. Effective advisors develop a healthy rapport with HN personnel but avoid the temptation to adopt HN positions contrary to U.S. or multinational values or policy.

6-89. Advisors who understand the HN military culture understand that local politics have national effects. Effective advisors recognize and use cultural factors that support HN commitment and teamwork. A good advisor uses the culture's positive aspects to get the best performance from each security force member and leader. Table 6-5 (page 6-18) lists important guidelines for advisors.

Table 6-5. Guidelines for advisors

- Try to learn enough of the language for simple conversation.
- Be patient. Be subtle. In guiding host-nation counterparts, explain the benefits of an action and convince them to accept the idea as their own. Respect the rank and positions of host-nation counterparts.
- Be diplomatic in correcting host-nation forces. Praise each success and work to instill pride in the unit.
- Understand that an advisor is not the unit commander but an enabler. The host-nation commander makes decisions and commands the unit. Advisors help with this task.
- Keep host-nation counterparts informed; try not to hide agendas.
- Work to continually train and improve the unit, even in the combat zone. Help the commander develop unit standing operating procedures.
- Be prepared to act as a liaison to multinational assets, especially air support and logistics. Maintain liaison with civil affairs and humanitarian teams in the area of operations.
- Be ready to advise on the maintenance of equipment and supplies.
- Have a thorough knowledge of light infantry tactics and unit security procedures.
- Use "confidence" missions to validate training.
- Stay integrated with the unit. Eat their food. Do not become isolated from them.
- Be aware of the operations in the immediate area to prevent fratricide.
- Insist on host-nation adherence to the recognized human rights standards concerning treatment of civilians, detainees, and captured insurgents. Report any violations to the chain of command.
- Be objective in reports on host-nation unit and leader proficiency. Report gross corruption or incompetence.
- Train host-nation units to standard and fight alongside them. Consider host-nation limitations and adjust. Flexibility is key. It is impossible to plan completely for everything in this type of operation. Therefore, constantly look forward to the next issue and be ready to develop solutions to problems that cannot be answered with a doctrinal solution.
- Remember that most actions have long-term strategic implications.
- Maintain a proper military bearing and professional manner.

Multinational Security Transition Command–Iraq

The experience of Multinational Security Transition Command–Iraq (MNSTC-I) in 2005 demonstrates the challenges of developing and assisting host-nation security forces facing an intensive insurgency. MNSTC-I programs were built on three pillars:

- Training and equipping the Iraqi security forces to standard.
- Using transition teams to guide the development of leaders and staffs.
- Partnerships between U.S. and multinational forces on the ground and developing Iraqi forces.

Initial plans called for only a small army (to deal with external threats) supplemented by conventional police forces (to maintain internal law and order). As the insurgency matured, the decision was made to develop a larger Iraqi Army and to focus it on the internal threat. A more robust police force was also developed. Training programs matured, becoming longer and more focused on counterinsurgency tasks. Decisions by the Iraqi government also necessitated training and organizational changes. Training programs were adjusted, based on the experience of recruits, and eventually lengthened and changed in response to the increasing lethality of the insurgency and lessons learned. Advisors assigned to Iraqi units were termed "transition teams"

> and instructed to focus on the development of leaders and staffs at battalion level and above.
>
> As security forces grew significantly, it became evident that civilian ministry infrastructure development was not keeping pace. So MNSTC-I assumed the additional mission of creating ministerial-level organizations, finding resources, and changing culture. Another significant challenge involved selecting and training competent host-nation leadership—commissioned and noncommissioned officers. This grew more challenging as the host nation regained sovereignty and asserted its authority, which led to a greater focus on developing staffs by transition teams.
>
> The MNSTC-I experience shows the advantage of doing a prompt initial assessment and then adjusting as conditions change and lessons are learned. Developing HN security forces from scratch in an active insurgency environment is often more about overcoming friction than about perfect planning. MNSTC-I also demanded robust interagency and multinational participation in the training effort and fought for the funding necessary to make it effective. MNSTC-I found that easily measured measures of effectiveness, such as soldiers equipped or battalions fielded, were not as useful as more complicated and more subjective metrics, such as the training level of fielded units and their loyalty to the national government.

POLICE IN COUNTERINSURGENCY

6-90. The primary frontline COIN force is often the police—not the military. The primary COIN objective is to enable local institutions. Therefore, supporting the police is essential. But the police are only a part of the rule of law. Police require support from a law code, judicial courts, and a penal system. Such support provides a coherent and transparent system that imparts justice. Upholding the rule of law also requires other civil institutions and a HN ability to support the legal system. Commanders should ensure that robust coordination mechanisms linking their efforts with the larger developmental process exist. If parts of the rule of law do not work, then commanders must be prepared to meet detention requirements.

6-91. Few military units can match a good police unit in developing an accurate human intelligence picture of their AO. Because of their frequent contact with populace, police often are the best force for countering small insurgent bands supported by the local populace. In COIN operations, special police strike units may move to different AOs, while patrol police remain in the local area on a daily basis and build a detailed intelligence picture of the insurgent strength, organization, and support.

ORGANIZING THE POLICE

6-92. Police often consist of several independent but mutually supporting forces. These may include—

- Criminal and traffic police.
- Border police.
- Transport police for security of rail lines and public transport.
- Specialized paramilitary strike forces.

In addition, a host nation may establish various reserve police units or home guards to provide local security. The force may include paramilitary units. Police might be organized on a national or local basis. Whatever police organization is established, Soldiers and Marines must understand it and help the host nation effectively organize and use it. This often means dealing with several police organizations and developing plans for training and advising each one.

6-93. A formal link or liaison channel must exist between the HN police and military forces. This channel for coordination, deconfliction, and information sharing enables successful COIN operations.

6-94. Military forces might have to perform police duties at the start of an insurgency; however, it is best to establish police forces to assume these duties as soon as possible. U.S., multinational, and HN partners should institute a comprehensive program of police training. Moreover, plans for police training need to envision a several-year program to systematically build institutions and leadership.

6-95. Although roles of the police and military forces in COIN operations may blur, important distinctions between the two forces exist. If security forces treat insurgents as criminals, the police may retain the primary responsibility for their arrest, detention, and prosecution.

6-96. Countering an insurgency requires a police force that is visible day and night. The host nation will not gain legitimacy if the populace believes that insurgents and criminals control the streets. Well-sited and protected police stations can establish a presence in communities as long as the police do not hide in those stations. Police presence provides security to communities and builds support for the HN government. When police have daily contact with the local populace, they can collect information for counterinsurgents.

6-97. Good pay and attractive benefits must be combined with a strict code of conduct that follows the rule of law and allows for the immediate dismissal of police officers for gross corruption. Good planning ensures that police pay, housing, benefits, and work conditions attract a high quality of police recruit as well as discourage petty corruption. Such corruption undermines the populace's confidence in the police and government. An important step in organizing a police force involves setting up an independent review board composed of experts, government officials, or nongovernmental organization members. It should not be under the direct command of the police force. This board should have the authority to investigate charges of police abuse and corruption, oversee the complaints process, and dismiss and fine police found guilty of misconduct.

TRAINING THE POLICE IN COUNTERINSURGENCY

6-98. Police training is best conducted as an interagency and multinational operation. In a multinational effort, a separate multinational police training and advisory command could work with the military training command. Ideally, leaders for police training are civilian police officers from the Departments of Justice and State along with senior police officers from multinational partners. Civilian police forces have personnel with extensive experience in large city operations. Department of Justice and multinational police organizations have extensive experience operating against organized crime groups. Experience countering organized crime is especially relevant to COIN; most insurgent groups are more similar to organized crime in their organizational structure and relations with the populace than they are to military units. U.S. military police units serve best when operating as a support force for the professional civilian police trainers. However, military forces may be assigned the primary responsibility for police training; they must be prepared to assume that role if required. (See paragraph D-3 for legal considerations associated with this mission.)

6-99. Military police can provide much of the initial police training. They are especially suited to teach the HN police forces the following skills:

- Weapons handling.
- Small-unit tactics.
- Special weapons employment.
- Convoy escort.
- Riot control.
- Traffic control.
- Prisoner and detainee handling and processing.
- Police intelligence.
- Criminal intelligence.
- Criminal handling.
- Stations management.

Higher level police skills—such as civilian criminal investigation procedures, antiorganized crime operations, and police intelligence operations—are best taught by civilian experts.

6-100. Military police or corrections personnel can also provide training for detention and corrections operations. HN personnel should be trained to handle and interrogate detainees and prisoners according to internationally recognized human rights norms. Prisoner and detainee management procedures should provide for the security and the fair and efficient processing of those detained.

6-101. Police forces, just like military forces, need quality support personnel to be effective. This requires training teams to ensure that training in support functions is established. Specially trained personnel required by police forces include the following:

- Armorers.
- Supply specialists.
- Communications personnel.
- Administrative personnel.
- Vehicle mechanics.

6-102. Effective policing also requires an effective justice system that can process arrests, detentions, warrants, and other judicial records. Such a system includes trained judges, prosecutors, defense counsels, prison officials, and court personnel. These people are important to establishing the rule of law.

6-103. Advisors should help the host nation establish and enforce police roles and authority. The authority to detain and interrogate, procedures for detention facilities, and human rights standards are important considerations.

POLICE/MILITARY OPERATIONS

6-104. In COIN operations, police forces from the host nation, United States, and multinational partners often conduct operations together. To work effectively together, the police and military coordinate rules of engagement. These forces also—

- Establish common standing operating procedures.
- Conduct supporting information operations.
- Perform combined planning.
- Ensure command and control interoperability.

6-105. Military forces can support the police in carrying out numerous COIN functions in accordance with U.S. law. This support can include the following:

- Assisting in the arrest of war criminals.
- Supporting police presence and search patrols.
- Providing logistic support.
- Controlling crowds and urban unrest.
- Detaining suspected felons.
- Securing key facilities.
- Providing advisors to the police.

6-106. The Departments of Justice and State normally take the lead in helping the host nation develop a workable judicial system through a ministerial-level advisory program. The military's staff judge advocate and civil affairs personnel may help develop the HN judicial system. As this system is developed and reformed, the military commander's legal and political advisors should ensure that the military's concerns are addressed.

Developing a Police Force in Malaya

In 1948, the Malayan Communist Party, whose members were primarily ethnic Chinese, began an insurgency against the British colonial government. The British first responded by dramatically expanding the Malayan security forces. The police, not the army, served as the lead counterinsurgency force. Between 1948 and 1950, the number of Malayan police expanded fivefold to 50,000, while the British army garrison expanded to 40,000. However, there was only time to provide a few weeks of rudimentary training to the new police officers before throwing them into operations.

> Police with little training and little competent leadership were ineffective in conducting operations. They also abused the civilian population and fell into corrupt practices. The population largely regarded the police as hostile; they were reluctant to give them information on the insurgents.
>
> By 1952, the insurgency had reached a stalemate. The British then established a new strategy. The strategy included reforming and retraining the entire Malaya Police Force. First, 10,000 corrupt or incompetent police officers were removed from the force. Then, police officers who had proven the most competent in operations were made instructors in new police schools. During 1952 and 1953, every police officer attended a four-month basic training course. Police commissioned and noncommissioned officers were sent to three- to four-month advanced courses. All senior Malayan police officers were required to attend the police intelligence school. There they learned the latest criminal investigation techniques. Teams of Britain's top police officers taught them intelligence collection and analysis methods as well. Dozens of the most promising Malayan officers attended the full yearlong course in advanced police operations in Britain.
>
> To win the ethnic Chinese away from the insurgents, the British worked closely with ethnic Chinese organizations to recruit Chinese for the Malaya Police Force. In 1952, the number of ethnic Chinese in the force more than doubled. Although the percentage of ethnic Chinese in the police force did not equal their percentage in the population, the ethnic Chinese saw this reaching out as a sign that the government was addressing their interests. At the same time, some Chinese and Malay political groups were building a coalition to establish an independent Malaya in which all the major ethnic groups would participate. The two efforts complemented each other.
>
> Better trained police officers and soldiers led by fully trained commissioned and noncommissioned officers dramatically improved the Malayan security forces' discipline. Better relations between the population and security forces resulted, and the people began to provide information on the insurgents. Thanks to their intelligence training, the security forces could develop intelligence from that information and act on it. They begin to break the insurgent organization. In 1953, the government gained the initiative. After that, the insurgent forces and support structure declined rapidly. In late 1953, the British began withdrawing forces. They progressively turned the war over to the Malayans, who were fully prepared to conduct counterinsurgency operations without a drop in efficiency.
>
> The Malaya insurgency provides lessons applicable to combating any insurgency. Manpower is not enough; well-trained and well-disciplined forces are required. The Malayan example also illustrates the central role that police play in counterinsurgency operations. British leaders concentrated on training the Malayan leadership. The British insisted that chosen personnel receive the full British Army and police officer courses. These actions built the Malayan security forces on a sound foundation. By taking a comprehensive approach to security force training and reform, the British commanders transformed a demoralized organization into a winning force. This transformation required only 15 months.

SUMMARY

6-107. A successful COIN effort establishes HN institutions that can sustain government legitimacy. Developing effective HN security forces—including military, police, and paramilitary forces—is one of the highest priority COIN tasks. Soldiers and Marines can make vital contributions to this mission by training and advising the HN security forces. Effective commanders understand the importance of this mission and select the right personnel as trainers and advisors. Developing all necessary HN security forces require a considerable interagency effort and normally includes a significant multinational involvement as well.

Chapter 7

Leadership and Ethics for Counterinsurgency

Leaders must have a strong sense of the great responsibility of their office; the resources they will expend in war are human lives.

Marine Corps Doctrinal Publication 1, 1997

There are leadership and ethical imperatives that are prominent and, in some cases, unique to counterinsurgency. The dynamic and ambiguous environment of modern counterinsurgency places a premium on leadership at every level, from sergeant to general. Combat in counterinsurgency is frequently a small-unit leader's fight; however, commanders' actions at brigade and division levels can be more significant. Senior leaders set the conditions and the tone for all actions by subordinates. Today's Soldiers and Marines are required to be competent in a broad array of tasks. They must also rapidly adapt cognitively and emotionally to the perplexing challenges of counterinsurgency and master new competencies as well as new contexts. Those in leadership positions must provide the moral compass for their subordinates as they navigate this complex environment. Underscoring these imperatives is the fact that exercising leadership in the midst of ambiguity requires intense, discriminating professional judgment.

LEADERSHIP IN COUNTERINSURGENCY

7-1. Army and Marine Corps leaders are expected to act ethically and in accordance with shared national values and Constitutional principles, which are reflected in the law and military oaths of service. These leaders have the unique professional responsibility of exercising military judgment on behalf of the American people they serve. They continually reconcile mission effectiveness, ethical standards, and thoughtful stewardship of the Nation's precious resources—human and material—in the pursuit of national aims.

7-2. Army and Marine Corps leaders work proactively to establish and maintain the proper ethical climate of their organizations. They serve as visible examples for every subordinate, demonstrating cherished values and military virtues in their decisions and actions. Leaders must ensure that the trying counterinsurgency (COIN) environment does not undermine the values of their Soldiers and Marines. Under all conditions, they must remain faithful to basic American, Army, and Marine Corps standards of proper behavior and respect for the sanctity of life.

7-3. Leaders educate and train their subordinates. They create standing operating procedures and other internal systems to prevent violations of legal and ethical rules. They check routinely on what Soldiers and Marines are doing. Effective leaders respond quickly and aggressively to signs of illegal or unethical behavior. The Nation's and the profession's values are not negotiable. Violations of them are not just mistakes; they are failures in meeting the fundamental standards of the profession of arms.

LARGE- AND SMALL-UNIT LEADERSHIP TENETS

7-4. There are basic leadership tenets that apply to all levels of command and leadership in COIN, though their application and importance may vary.

7-5. Effective leaders ensure that Soldiers and Marines are properly trained and educated. Such training includes cultural preparation for the operational environment. In a COIN environment, it is often counterproductive to use troops that are poorly trained or unfamiliar with operating close to the local populace. COIN forces aim to mobilize the good will of the people against the insurgents. Therefore, the populace must feel protected, not threatened, by COIN forces' actions and operations.

7-6. Proper training addresses many possible scenarios of the COIN environment. Education should prepare Soldiers and Marines to deal with the unexpected and unknown. Senior commanders should, at a minimum, ensure that their small-unit leaders are inculcated with tactical cunning and mature judgment. Tactical cunning is the art of employing fundamental skills of the profession in shrewd and crafty ways to out-think and out-adapt enemies. Developing mature judgment and cunning requires a rigorous regimen of preparation that begins before deployment and continues throughout. Junior leaders especially need these skills in a COIN environment because of the decentralized nature of operations.

7-7. Senior leaders must determine the purpose of their operations. This entails, as discussed in chapter 4, a design process that focuses on learning about the nature of unfamiliar problems. Effective commanders know the people, topography, economy, history, and culture of their area of operations (AO). They know every village, road, field, population group, tribal leader, and ancient grievance within it. The COIN environment changes continually; good leaders appreciate that state of flux and constantly assess their situation.

7-8. Another part of analyzing a COIN mission involves assuming responsibility for everyone in the AO. This means that leaders feel the pulse of the local populace, understand their motivations, and care about what they want and need. Genuine compassion and empathy for the populace provide an effective weapon against insurgents.

7-9. Senior leaders exercise a leadership role throughout their AO. Leaders directly influence those in the chain of command while indirectly leading everyone else within their AO. Elements engaged in COIN efforts often look to the military for leadership. Therefore, military actions and words must be beyond reproach. The greatest challenge for leaders may be in setting an example for the local populace. Effective senior and junior leaders embrace this role and understand its significance. It involves more than just killing insurgents; it includes the responsibility to serve as a moral compass that extends beyond the COIN force and into the community. It is that moral compass that distinguishes Soldiers and Marines from the insurgents.

7-10. Senior commanders must maintain the "moral high ground" in all their units' deeds and words. Information operations complement and reinforce actions, and actions reinforce the operational narrative. All COIN force activity is wrapped in a blanket of truth. Maintaining credibility requires commanders to immediately investigate all allegations of immoral or unethical behavior and provide a prudent degree of transparency.

7-11. Army and Marine Corps leaders emphasize that on the battlefield the principles of honor and morality are inextricably linked. Leaders do not allow subordinates to fall victim to the enormous pressures associated with prolonged combat against elusive, unethical, and indiscriminate foes. The environment that fosters insurgency is characterized by violence, immorality, distrust, and deceit; nonetheless, Army and Marine Corps leaders continue to demand and embrace honor, courage, and commitment to the highest standards. They know when to inspire and embolden their Soldiers and Marines and when to enforce restraint and discipline. Effective leaders at all levels get out and around their units, and out among the populace. Such leaders get a true sense of the complex situation in their AO by seeing what subordinates are actually doing, exchanging information with military and interagency leaders, and—most importantly—listening.

7-12. Leaders at every level establish an ethical tone and climate that guards against the moral complacency and frustrations that build up in protracted COIN operations. Leaders remain aware of the emotional toll that constant combat takes on their subordinates and the potential for injuries resulting from combat stress. Such injuries can result from cumulative stress over a prolonged period, witnessing the death of a comrade, or killing other human beings. Caring leaders recognize these pressures and provide emotional "shock absorbers" for their subordinates. Soldiers and Marines must have outlets to share their feelings and reach closure on traumatic experiences. These psychological burdens may be carried for a long time. Leaders watch for signs of possible combat stress within individuals and units. These signs include—

- Physical and mental fatigue.
- Lack of respect for human life.
- Loss of appetite, trouble with sleep, and no interest in physical hygiene.
- Lack of unit cohesion and discipline.
- Depression and fatalism.

(See FM 6-22.5/MCRP 6-11C for techniques first-line leaders can use to prevent, identify, and treat combat stress reactions.)

7-13. Combat requires commanders to be prepared to take some risk, especially at the tactical level. Though this tenet is true for the entire spectrum of conflict, it is particularly important during COIN operations, where insurgents seek to hide among the local populace. Risk takes many forms. Sometimes accepting it is necessary to generate overwhelming force. However, in COIN operations, commanders may need to accept substantial risk to de-escalate a dangerous situation. The following vignette illustrates such a case.

Defusing a Confrontation

[On 3 April 2005, a] small unit of American soldiers was walking along a street in Najaf [en route to a meeting with a religious leader] when hundreds of Iraqis poured out of the buildings on either side. Fists waving, throats taut, they pressed in on the Americans, who glanced at one another in terror. ...The Iraqis were shrieking, frantic with rage.... [It appeared that a shot would] come from somewhere, the Americans [would] open fire, and the world [would] witness the My Lai massacre of the Iraq war. At that moment, an American officer stepped through the crowd holding his rifle high over his head with the barrel pointed to the ground. Against the backdrop of the seething crowd, it was a striking gesture.... "Take a knee," the officer said.... The Soldiers looked at him as if he were crazy. Then, one after another, swaying in their bulky body armor and gear, they knelt before the boiling crowd and pointed their guns at the ground. The Iraqis fell silent, and their anger subsided. The officer ordered his men to withdraw [and continue on their patrol].

© Dan Baum, "Battle Lessons, What the Generals Don't Know," *The New Yorker*, Jan 17, 2005.

7-14. Leaders prepare to indirectly inflict suffering on their Soldiers and Marines by sending them into harm's way to accomplish the mission. At the same time, leaders attempt to avoid, at great length, injury and death to innocents. This requirement gets to the very essence of what some describe as "the burden of command." The fortitude to see Soldiers and Marines closing with the enemy and sustaining casualties day in and day out requires resolve and mental toughness in commanders and units. Leaders must develop these characteristics in peacetime through study and hard training. They must maintain them in combat.

7-15. Success in COIN operations requires small-unit leaders agile enough to transition among many types of missions and able to adapt to change. They must be able to shift through a number of activities from nation building to combat and back again in days, or even hours. Alert junior leaders recognize the dynamic context of a tactical situation and can apply informed judgment to achieve the commander's intent in a stressful and ambiguous environment. COIN operations are characterized by rapid changes in tactical and operational environments. The presence of the local populace within which insurgents may disappear creates a high degree of ambiguity. Adaptable leaders observe the rapidly changing situation, identify its key characteristics, ascertain what has to be done in consultation with subordinates, and determine the best method to accomplish the mission.

7-16. Cultural awareness has become an increasingly important competency for small-unit leaders. Perceptive junior leaders learn how cultures affect military operations. They study major world cultures and put a priority on learning the details of the new operational environment when deployed. Different solutions are required in different cultural contexts. Effective small-unit leaders adapt to new situations, realizing their words and actions may be interpreted differently in different cultures. Like all other competencies, cultural awareness requires self-awareness, self-directed learning, and adaptability.

7-17. Self-aware leaders understand the need to assess their capabilities and limitations continually. They are humble, self-confident, and brave enough to admit their faults and shortcomings. More important, self-aware leaders work to improve and grow. After-action reviews, exchanging information with subordinate and interagency leaders, and open discussions throughout a COIN force are essential to achieve understanding and improvement. Soldiers and Marines can become better, stronger leaders through a similar habit of self-examination, awareness, and focused corrective effort.

7-18. Commanders exercise initiative as leaders and fighters. Learning and adapting, with appropriate decision-making authority, are critical to gaining an advantage over insurgents. Effective senior leaders establish a climate that promotes decentralized modes of command and control—what the Army calls mission command and the Marine Corps calls mission command and control. Under mission command, commanders create the conditions for subordinates' success. These leaders provide general guidance and the commander's intent and assign small-unit leaders authority commensurate with their responsibilities. Commanders establish control measures to monitor subordinates' actions and keep them within the bounds established by commander's intent without micromanaging. At the same time, Soldiers and Marines must feel the commander's presence throughout the AO, especially at decisive points. The operation's purpose and commander's intent must be clearly understood throughout the force.

7-19. The practice of leaders sharing hardship and danger with subordinates builds confidence and esprit. Soldiers and Marines are more confident in their chances of success when they know that their leaders are involved. They understand their leaders are committing them to courses of action based on firsthand knowledge. However, this concept of leaders being fighters does not absolve leaders from remembering their position and avoiding needless risk.

7-20. COIN operations require leaders to exhibit patience, persistence, and presence. While leading Soldiers and Marines, commanders cooperate with, and leverage the capabilities of, multinational partners, U.S. Government agencies, and nongovernmental organizations. Commanders also gain the confidence of the local populace while defeating and discrediting the insurgents.

Patience, Presence, and Courage

For the first two months of 2006, the Marine platoon of the 22d Marine Expeditionary Unit had walked the streets in Iraq on foot without serious incident. Their patrols had moved fearlessly around lines of cars and through packed markets. For the most part, their house calls began with knocks, not kicks. It was their aim to win the respect of the city's Sunni Arab population.

Suddenly things changed. An armored HMMWV on night patrol hit an improvised explosive device. The bomb destroyed the vehicle. Five Marines were wounded and two died shortly thereafter. A third Marine, a popular noncommissioned officer, later died of his wounds as well.

The platoon was stunned. Some of the more veteran noncommissioned officers shrugged it off, but the younger Marines were keyed up and wanted to make the elusive enemy pay a price. A squad leader stood up in the squad bay asserted that there would be a pile of dead Arabs on the street when the platoon went out the next day.

Just then, the company commander walked in. He was widely respected and generally short on words. He quickly sensed the unit's mood and recognized the potential danger in their dark attitude. Speaking directly to his Marines, the commander urged them to remember why they were there. He reminded them that a very small percentage of the populace was out to create problems. It was that minority that benefited from creating chaos. The enemy would love to see an overreaction to the attack, and they would benefit from any actions that detracted from the Marines' honor or purpose. The commander urged his Marines not to get caught up in the anger of the moment and do something they all would regret for a long time. Rather, they needed to focus on what the force was trying to accomplish and keep their minds on the mission. They had taken some hits and lost some good men, the commander said, but escalating the violence would not help them win. It would fall for the insurgents' strategy instead of sticking to the Marines' game plan of winning the respect of the populace.

The commander knew his Marines and understood the operational environment. He assessed the situation and acted aggressively to counter a dangerous situation that threatened mission accomplishment. By his actions, the commander demonstrated patience, presence, and courage.

ETHICS

7-21. Article VI of the U.S. Constitution and the Army Values, Soldier's Creed, and Core Values of U.S. Marines all require obedience to the law of armed conflict. They hold Soldiers and Marines to the highest standards of moral and ethical conduct. Conflict brings to bear enormous moral challenges, as well as the burden of life-and-death decisions with profound ethical considerations. Combat, including counterinsurgency and other forms of unconventional warfare, often obligates Soldiers and Marines to accept some risk to minimize harm to noncombatants. This risk taking is an essential part of the Warrior Ethos. In conventional conflicts, balancing competing responsibilities of mission accomplishment with protection of noncombatants is difficult enough. Complex COIN operations place the toughest of ethical demands on Soldiers, Marines, and their leaders.

7-22. Even in conventional combat operations, Soldiers and Marines are not permitted to use force disproportionately or indiscriminately. Typically, more force reduces risk in the short term. But American military values obligate Soldiers and Marines to accomplish their missions while taking measures to limit the destruction caused during military operations, particularly in terms of collateral harm to noncombatants. It is wrong to harm innocents, regardless of their citizenship.

7-23. Limiting the misery caused by war requires combatants to consider certain rules, principles, and consequences that restrain the amount of force they may apply. At the same time, combatants are not required to take so much risk that they fail in their mission or forfeit their lives. As long as their use of force is proportional to the gain to be achieved and discriminates in distinguishing between combatants and noncombatants. Soldiers and Marines may take actions where they knowingly risk, but do not intend, harm to noncombatants.

7-24. Ethically speaking, COIN environments can be much more complex than conventional ones. Insurgency is more than combat between armed groups; it is a political struggle with a high level of violence. Insurgents try to use this violence to destabilize and ultimately overthrow a government. Counterinsurgents that use excessive force to limit short-term risk alienate the local populace. They deprive themselves of the support or tolerance of the people. This situation is what insurgents want. It increases the threat they pose. Sometimes lethal responses are counterproductive. At other times, they are essential. The art of command includes knowing the difference and directing the appropriate action.

7-25. A key part of any insurgent's strategy is to attack the will of the domestic and international opposition. One of the insurgents' most effective ways to undermine and erode political will is to portray their opposition as untrustworthy or illegitimate. These attacks work especially well when insurgents can portray their opposition as unethical by the opposition's own standards. To combat these efforts, Soldiers and Marines treat noncombatants and detainees humanely, according to American values and internationally recognized human rights standards. In COIN, preserving noncombatant lives and dignity is central to mission accomplishment. This imperative creates a complex ethical environment.

WARFIGHTING VERSUS POLICING

7-26. In counterinsurgencies, warfighting and policing are dynamically linked. The moral purpose of combat operations is to secure peace. The moral purpose of policing is to maintain the peace. In COIN operations, military forces defeat enemies to establish civil security; then, having done so, these same forces preserve it until host-nation (HN) police forces can assume responsibility for maintaining the civil order.

When combatants conduct stability operations in a way that undermines civil security, they undermine the moral and practical purposes they serve. There is a clear difference between warfighting and policing. COIN operations require that every unit be adept at both and capable of moving rapidly between one and the other.

7-27. The COIN environment frequently and rapidly shifts from warfighting to policing and back again. There are many examples from Iraq and Afghanistan where U.S. forces drove insurgents out of urban areas only to have the insurgents later return and reestablish operations. Insurgents were able to return because U.S. forces had difficulty maintaining civil security. U.S. forces then had to deal with insurgents as an organized combatant force all over again. To prevent such situations, counterinsurgents that establish civil security need to be prepared to maintain it. Maintaining civil security entails very different ethical obligations than establishing it.

7-28. Civil security holds when institutions, civil law, courts, prisons, and effective police are in place and can protect the recognized rights of individuals. Typically this requires that—

- The enemy is defeated or transformed into a threat not capable of challenging a government's sovereignty.
- Institutions necessary for law enforcement—including police, courts, and prisons—are functioning.
- These institutions are credible, and people trust them to resolve disputes.

7-29. Where a functioning civil authority does not exist, COIN forces must work to establish it. Where U.S. forces are trying to build a HN government, the interim government should transition to HN authority as soon as possible. Counterinsurgents must work within the framework of the institutions established to maintain order and security. In these conditions, COIN operations more closely resemble police work than combat operations.

PROPORTIONALITY AND DISCRIMINATION

7-30. The principle of proportionality requires that the anticipated loss of life and damage to property incidental to attacks must not be excessive in relation to the concrete and direct military advantage expected to be gained. Proportionality and discrimination require combatants not only to minimize the harm to noncombatants but also to make positive commitments to—

- Preserve noncombatant lives by limiting the damage they do.
- Assume additional risk to minimize potential harm.

7-31. Proportionality requires that the advantage gained by a military operation not be exceeded by the collateral harm. The law of war principle of proportionality requires collateral damage to civilians and civilian property not be excessive in relation to the military advantage expected to be gained by executing the operation. Soldiers and Marines must take all feasible precautions when choosing means and methods of attack to avoid and minimize loss of civilian life, injury to civilians, and damage to civilian objects.

7-32. In conventional operations, proportionality is usually calculated in simple utilitarian terms: civilian lives and property lost versus enemy destroyed and military advantage gained. But in COIN operations, advantage is best calculated not in terms of how many insurgents are killed or detained, but rather which enemies are killed or detained. If certain key insurgent leaders are essential to the insurgents' ability to conduct operations, then military leaders need to consider their relative importance when determining how best to pursue them. In COIN environments, the number of civilian lives lost and property destroyed needs to be measured against how much harm the targeted insurgent could do if allowed to escape. If the target in question is relatively inconsequential, then proportionality requires combatants to forego severe action, or seek noncombative means of engagement.

7-33. When conditions of civil security exist, Soldiers and Marines may not take any actions that might knowingly harm noncombatants. This does not mean they cannot take risks that might put the populace in danger. But those risks are subject to the same rules of proportionality. The benefit anticipated must outweigh the risk taken.

7-34. Discrimination requires combatants to differentiate between enemy combatants, who represent a threat, and noncombatants, who do not. In conventional operations, this restriction means that combatants cannot intend to harm noncombatants, though proportionality permits them to act, knowing some noncombatants may be harmed.

7-35. In COIN operations, it is difficult to distinguish insurgents from noncombatants. It is also difficult to determine whether the situation permits harm to noncombatants. Two levels of discrimination are necessary:

- Deciding between targets.
- Determining an acceptable risk to noncombatants and bystanders.

7-36. Discrimination applies to the means by which combatants engage the enemy. The COIN environment requires counterinsurgents to not only determine the kinds of weapons to use and how to employ them but also establish whether lethal means are desired—or even permitted. (FM 27-10 discusses forbidden means of waging war.) Soldiers and Marines require an innate understanding of the effects of their actions and weapons on all aspects of the operational environment. Leaders must consider not only the first-order, desired effects of a munition or action but also possible second- and third-order effects—including undesired ones. For example, bombs delivered by fixed-wing close air support may effectively destroy the source of small arms fire from a building in an urban area; however, direct-fire weapons may be more appropriate due to the risk of collateral damage to nearby buildings and noncombatants. The leader at the scene assesses the risks and makes the decision. Achieving the desired effects requires employing tactics and weapons appropriate to the situation. In some cases, this means avoiding the use of area munitions to minimize the potential harm inflicted on noncombatants located nearby. In situations where civil security exists, even tenuously, Soldiers and Marines should pursue nonlethal means first, using lethal force only when necessary.

7-37. The principles of discrimination in the use of force and proportionality in actions are important to counterinsurgents for practical reasons as well as for their ethical or moral implications. Fires that cause unnecessary harm or death to noncombatants may create more resistance and increase the insurgency's appeal—especially if the populace perceives a lack of discrimination in their use. The use of discriminating, proportionate force as a mindset goes beyond the adherence to the rules of engagement. Proportionality and discrimination applied in COIN require leaders to ensure that their units employ the right tools correctly with mature discernment, good judgment and moral resolve.

DETENTION AND INTERROGATION

7-38. Detentions and interrogations are critical components to any military operation. The nature of COIN operations sometimes makes it difficult to separate potential detainees from innocent bystanders, since insurgents lack distinctive uniforms and deliberately mingle with the local populace. Interrogators are often under extreme pressure to get information that can lead to follow-on operations or save the lives of noncombatants, Soldiers, or Marines. While enemy prisoners in conventional war are considered moral and legal equals, the moral and legal status of insurgents is ambiguous and often contested. What is not ambiguous is the legal obligation of Soldiers and Marines to treat all prisoners and detainees according to the law. All captured or detained personnel, regardless of status, shall be treated humanely, and in accordance with the Detainee Treatment Act of 2005 and DODD 2310.01E. No person in the custody or under the control of DOD, regardless of nationality or physical location, shall be subject to torture or cruel, inhuman, or degrading treatment or punishment, in accordance with, and as defined in, U.S. law. (Appendix D provides more guidance on the legal issues concerning detention and interrogation.)

LIMITS ON DETENTION

7-39. Mistreatment of noncombatants, including prisoners and detainees is illegal and immoral. It will not be condoned. The Detainee Treatment Act of 2005 makes the standard clear:

> *No person in the custody or under the effective control of the Department of Defense or under detention in a Department of Defense facility shall be subject to any treatment or technique of interrogation not authorized by and listed in the United States Army Field Manual on Intelligence Interrogation* [FM 2-22.3].

No individual in the custody or under the physical control of the United States Government, regardless of nationality or physical location, shall be subject to cruel, inhuman, or degrading treatment or punishment.

7-40. In COIN environments, distinguishing an insurgent from a civilian is difficult and often impossible. Treating a civilian like an insurgent, however, is a sure recipe for failure. Individuals suspected of insurgent or terrorist activity may be detained for two reasons:

- To prevent them from conducting further attacks.
- To gather information to prevent other insurgents and terrorists from conducting attacks.

These reasons allow for two classes of persons to be detained and interrogated:

- Persons who have engaged in, or assisted those who engage in, terrorist or insurgent activities.
- Persons who have incidentally obtained knowledge regarding insurgent and terrorist activity, but who are not guilty of associating with such groups.

People engaging in insurgent activities may be detained as enemies. Persons not guilty of associating with insurgent or terrorist groups may be detained and questioned for specific information. However, since these people have not—by virtue of their activities—represented a threat, they may be detained only long enough to obtain the relevant information. Since persons in the second category have not engaged in criminal or insurgent activities, they must be released, even if they refuse to provide information.

7-41. At no time can Soldiers and Marines detain family members or close associates to compel suspected insurgents to surrender or provide information. This kind of hostage taking is both unethical and illegal.

LIMITS ON INTERROGATION

7-42. Abuse of detained persons is immoral, illegal, and unprofessional. Those who engage in cruel or inhuman treatment of prisoners betray the standards of the profession of arms and U.S. laws. They are subject to punishment under the Uniform Code of Military Justice. The Geneva Conventions, as well as the Convention against Torture and Other Cruel, Inhuman or Degrading Treatment or Punishment, agree on unacceptable interrogating techniques. Torture and cruel, inhuman, and degrading treatment is never a morally permissible option, even if lives depend on gaining information. No exceptional circumstances permit the use of torture and other cruel, inhuman, or degrading treatment. Only personnel trained and certified to interrogate can conduct interrogations. They use legal, approved methods of convincing enemy prisoners of war and detainees to give their cooperation. Interrogation sources are detainees, including enemy prisoners of war. (FM 2-22.3 provides the authoritative doctrine and policy for interrogation. Chapter 3 and appendix D of this manual also address this subject.)

7-43. The ethical challenges posed in COIN operations require commanders' attention and action. Proactive commanders establish procedures and checks to ensure proper handling of detainees. Commanders verify that subordinate leaders do not allow apparent urgent requirements to result in violations of these procedures. Prohibitions against mistreatment may sometimes clash with leaders' moral imperative to accomplish their mission with minimum losses. Such situations place leaders in difficult situations, where they must choose between obedience to the law and the lives of their Soldiers and Marines. U.S. law and professional values compel commanders to forbid mistreatment of noncombatants, including captured enemies. Senior commanders clearly define the limits of acceptable behavior to their subordinates and take positive measures to ensure their standards are met.

7-44. To the extent that the work of interrogators is indispensable to fulfilling the state's obligation to secure its citizens' lives and liberties, conducting interrogations is a moral obligation. The methods used, however, must reflect the Nation's commitment to human dignity and international humanitarian law. A commander's need for information remains valid and can be met while observing relevant regulations and ethical standards. Acting morally does not necessarily mean that leaders give up obtaining critical information. Acting morally does mean that leaders must relinquish certain methods of obtaining information, even if that decision requires Soldiers and Marines to take greater risk.

Lose Moral Legitimacy, Lose the War

During the Algerian war of independence between 1954 and 1962, French leaders decided to permit torture against suspected insurgents. Though they were aware that it was against the law and morality of war, they argued that—

- This was a new form of war and these rules did not apply.
- The threat the enemy represented, communism, was a great evil that justified extraordinary means.
- The application of torture against insurgents was measured and nongratuitous.

This official condoning of torture on the part of French Army leadership had several negative consequences. It empowered the moral legitimacy of the opposition, undermined the French moral legitimacy, and caused internal fragmentation among serving officers that led to an unsuccessful coup attempt in 1962. In the end, failure to comply with moral and legal restrictions against torture severely undermined French efforts and contributed to their loss despite several significant military victories. Illegal and immoral activities made the counterinsurgents extremely vulnerable to enemy propaganda inside Algeria among the Muslim population, as well as in the United Nations and the French media. These actions also degraded the ethical climate throughout the French Army. France eventually recognized Algerian independence in July 1963.

THE LEARNING IMPERATIVE

7-45. Today's operational environment requires military organizations at all echelons to prepare for a broader range of missions than ever before. The Services are preparing for stability operations and post-conflict reconstruction tasks with the same degree of professionalism and study given to the conduct of combat operations. Similarly, COIN operations are receiving the attention and study merited by their frequency and potential impact. This broader mission set has significant leader development, education, and training implications, especially for land forces.

7-46. Army and Marine Corps leaders need to visualize the operational and informational impact of many tactical actions and relate their operations to larger strategic purposes. Effectively blending traditional military operations with other forms of influence is necessary. Effective leaders place a stronger emphasis on organizational change, develop subordinates, and empower them to execute critical tasks in consonance with broad guidance. Commanders must influence directly and indirectly the behavior of others outside their chain of command. Leaders are increasingly responsible for creating environments in which individuals and organizations learn from their experiences and for establishing climates that tap the full ingenuity of subordinates. Open channels of discussion and debate are needed to encourage growth of a learning environment in which experience is rapidly shared and lessons adapted for new challenges. The speed with which leaders adapt the organization must outpace insurgents' efforts to identify and exploit weaknesses or develop countermeasures.

7-47. Effective individual professional development programs develop and reward initiative and adaptability in junior leaders. Self-development, life-long learning, and reflection on experience should be encouraged and rewarded. Cultural sensitivity, development of nonauthoritarian interpersonal skills, and foreign language ability must be encouraged. Institutional professional development programs must develop leaders' judgment to help them recognize when situations change from combat to policing. Effective leaders are as skilled at limiting lethal force as they are in concentrating it. Indeed, they must learn that nonlethal solutions may often be preferable.

SUMMARY

7-48. Senior leaders must model and transmit to their subordinates the appropriate respect for professional standards of self-discipline and adherence to ethical values. Effective leaders create command climates that reward professional conduct and punish unethical behavior. They also are comfortable delegating authority. However, as always, accountability for the overall behavior and performance of a command cannot be delegated. Commanders remain accountable for the attainment of objectives and the manner in which they are attained.

Chapter 8
Sustainment

In my experience in previous wars, the logistic soldier was generally regarded as a rear area soldier.... Over here in Vietnam, that is completely changed....There is no rear area soldier, as such. Because of this, more than ever before, the man in logistics has to be first, a soldier, in the full sense of the word, and yet at the same time he has to know his MOS so that he can do his logistics job.

Major General James M. Heiser, USA, 1969

This chapter begins with a general discussion and analysis of how logistics in counterinsurgency (COIN) operations differ from logistics in conventional operations. This is followed by a survey of COIN-specific factors that affect how commanders can leverage available logistic assets and assign logisticians to meet special requirements needed to support different COIN logical lines of operations. Discussions that follow acknowledge that COIN operations may be entered into from various military conditions ranging from unstable peace to general war. The chapter concludes with a discussion of contracting support to COIN operations.

LOGISTIC CONSIDERATIONS IN COUNTERINSURGENCY

8-1. In counterinsurgency (COIN), the support provided by sustainment units often extends beyond sustaining operations; support provided to the population may become an important shaping operation or even the decisive operation. Logistic providers are often no longer the tail but the nose of a COIN force. Some of the most valuable services that military logisticians can provide to COIN operations include the means and knowledge for setting up or restarting self-perpetuating sustainment designs. The development of effective sustainment designs gives the populace a stake in stability and hope for the future. One COIN paradox is that many of the logisticians' best weapons for countering an insurgency do not shoot. Logistic units provide some of the most versatile and effective nonlethal resources available to Soldiers and Marines. Logisticians prepare to provide support across all logical lines of operations (LLOs) visualized and articulated by the commander. Often, logisticians already supporting COIN combat operations may be the only available source of prompt, essential knowledge, capabilities, and materials. This chapter focuses on capabilities and responsibilities of logistic units and logisticians. Commanders of all types of units at all levels must also be aware of the characteristics of COIN support.

WHAT IS DIFFERENT ABOUT LOGISTICS IN COUNTERINSURGENCY

8-2. In COIN operations, logistic units and other logistic providers perform all the functions they do in conventional operations as well as some different ones. Conventional operations usually involve two recognizable military organizations engaging each other in contiguous areas of operations. In COIN operations, the usual logistic functions—as well as COIN-specific activities—are performed in a frequently disorienting environment complicated by important social, political, and economic implications. Security conditions in these environments can rapidly change from moment to moment and every few hundred yards over various terrain conditions. (Table 8-1 [page 8-2] lists differences between the characteristics of logistic support to conventional operations and to COIN operations.)

Table 8-1. Conventional and counterinsurgency operations contrasted

	Conventional operations	*Counterinsurgency operations*
Mission	• Support combat unit missions. • Sustain and build combat power. • Support a mobile force with clear organization and structure. • Typically in direct support. • Logistic units and assets conduct only sustaining operations (focused on the force).	• Same as conventional operations plus support of logical lines of operations specific to counterinsurgency. • Support both a static force and mobile force. • Increased requirements for area support operations. • Logistic units and assets can be assigned as decisive and shaping operations (focused on the environment).
Enemy	• Enemy forces have supply trains and support echelons. • Friendly operational surprise (masking possible). • Difficult for enemy to perform pattern analysis. • Targeting logistic units is the enemy's shaping effort and considered a second front.	• Insurgents use nonstandard, covert supply methods that are difficult to template. • Limited operational surprise. • Easy for enemy to observe patterns in friendly logistic operations. • Insurgents place a high value on attacking logistic units and other less formidable, soft, high-payoff targets.
Terrain	• Fought in a definable area of operations. • Focus on destruction of enemy combat forces. • Few constraints. • Echeloned formations and discernable, hierarchical logistic organizations supporting well-defined, contiguous areas of operations. • Relatively secure lines of communications facilitate distribution operations from theater to corps to division to brigade.	• Operational environment poorly defined with multiple dimensions. • Support of the host nation population is the key objective. • Constrained time to achieve results, yet many counterinsurgency tasks are inherently time consuming. • Noncontiguous areas of operations and wide dispersion of units. • No front; everything is potentially close, yet far. • Need to maximize multiple lines of communications capacity/greater complexity. • Potentially decreased throughput capabilities. • Increased area support requirements. • Lines of communications vulnerable.
Troops and support available	• Uniformed personnel always suitable. • Contractor personnel suitable for secure areas only.	• Uniformed personnel usually suitable. • Suitability of contractor personnel judged case by case. • Task and location dependent; must be part of economic pluralism promotion plan.
Time available	• Tempo quicker. • Geared toward decisive major combat.	• Long-duration operations. • Continuity/logistics hand-off planning often required.
Civil considerations	• Secondary to considerations of how to defeat the enemy.	• May be the primary determinant of victory. • May figure prominently in logistic planning.

8-3. Differences in COIN logistics fit into the following major considerations:

- Logistic units are an essential part of COIN operations.
- Logistic units are perceived by insurgents as high-payoff targets and potential sources of supplies; thus, lines of communications (LOCs) are a main battle area for insurgents.

What is Different: Insurgent Perceptions of Military Logistics

Insurgents have a long history of exploiting their enemies' lines of communications as sources of supply. During the Revolutionary War, American forces significantly provisioned themselves from the British Army's overindulgent and carelessly defended logistic tail. In the 1930s, Mao Zedong codified a doctrine for insurgency logistics during the fight against the Japanese occupation of China. Without exaggerating, Mao stated, "We have a claim on the output of the arsenals of [our enemies],...and, what is more, it is delivered to us by the enemy's transport corps. This is the sober truth, it is not a jest." For Mao's forces, his enemy's supply trains provided a valuable source of supply. Mao believed the enemy's rear was the guerrillas' front; the guerrillas' advantage was that they had no discernable logistic rear.

This relative lack of logistic capacity was not an insurmountable problem for Mao or one of his logistic theorists, Ming Fan. According to Ming, "Weapons are not difficult to obtain. They can be purchased from the people's 'self-preservation corps.' Almost every home has some sort of weapon that can be put to use.... Ammunition can be obtained in the following ways: (1) From supplies given by friendly troops and headquarters on higher echelons. (2) Purchased or appropriated from the people. (3) Captured by ambushing enemy supply columns. (4) Purchased undercover from the enemy army. (5) From salvage in combat areas. (6) From the field of battle. (7) Self-made. (8) Manufactured by guerrilla organizations. (Such items as hand grenades, ammunition, etc.)" Beyond these specifics, this doctrine prescribes a mindset of actively seeking parasitic logistic relationships with not only the conventional enemy forces that the insurgents seek to co-opt and defeat but also active linkages to local black market activities and the cultivation of host-nation sympathizers.

For these reasons, forces conducting counterinsurgency operations must protect all potential supplies. Forces must also vigorously protect their lines of communications, scrupulously collect and positively control dud munitions and access to other convertible materiel, and actively seek ways to separate insurgents from black market activities.

In one moment in time, our service members will be feeding and clothing displaced refugees, providing humanitarian assistance. In the next moment, they will be holding two warring tribes apart—conducting peacekeeping operations—and, finally, they will be fighting a highly lethal mid-intensity battle—all in the same day...all within three city blocks. It will be what we call the "three block war."

General Charles C. Krulak, USMC, 1997

8-4. In a COIN environment, units providing logistic support potentially can be involved in more complicated tasks than even those of the three block war metaphor. COIN operations significantly differ in that logistic units must prepare to provide conventional logistic support to highly lethal, mid-intensity combat operations while supporting humanitarian operations. Logistic units may be required to maintain this support until conditions stabilize locally and civilian organizations can assume those duties.

8-5. The COIN environment requires logisticians to seek distribution efficiencies wherever possible. Logisticians must strive to eliminate backtracking and unnecessary distribution traffic. Ideally, logisticians maximize throughput methods that bypass—either on the ground or by air—population centers and heavily used civilian transportation nets. These practices are especially valuable in COIN operations. They improve logistic security, speed delivery, and minimize adverse effects and stress on the local populace.

8-6. Because of the diverse requirements, logisticians stay involved from the beginning of the planning process. They begin planning in detail as early as possible. Because of the complex logistic requirements and conditions under which COIN operations are pursued, commanders must ensure a careful logistic preparation of the area of operations (AO).

LOGISTIC PREPARATION OF THE COUNTERINSURGENCY AREA OF OPERATIONS

8-7. Logistic preparation of the AO relates to and can be treated as a COIN-specific logistic preparation of the theater. (FM 4-0, paragraphs 5-34 through 5-57, discusses logistics preparation of the theater.) In COIN operations, detailed analysis of civil logistic and economic assets takes on great importance. These assets can potentially support insurgents as well as the development and sustainment of host-nation (HN) security forces and the restoration of other essential services. Some examples of essential information for COIN logistic planning include the following:

- Analysis of the HN conventional force's existing logistic resources as a source of supply for developing HN security forces as well as the potential for insurgent black market activity.
- Effects of requirements generated by combat operations and collateral damage.
- Effects of multinational distribution requirements on HN lines of commerce.
- HN economic base (such as industry, manufacturing, and agriculture).
- HN lines of commerce (such as main supply routes, industrial cities, technical cities, pipelines, rail lines, and air and maritime ports).
- HN public works, utilities, and health, transportation, legal, and justice systems.
- Potential or existing dislocated civilian requirements.

ANALYSIS OF INSURGENT LOGISTIC CAPABILITIES

8-8. In COIN operations, analysis of the insurgents' logistic capabilities and shortfalls is especially significant. Logisticians and intelligence personnel perform what was formerly known as reverse-BOS (battlefield operating systems) analysis. This analysis does not just target enemy logistic capabilities and LOCs; it also assesses the suitability of supply sources for developing and sustaining insurgent forces. Effective analysis includes assessment of black market material, including salvage goods that insurgents might use to improvise equipment.

LOGISTIC SUPPORT TO LOGICAL LINES OF OPERATIONS

8-9. Although logisticians support all LLOs, logistic support during COIN focuses on the following LLOs:

- Conduct combat operations/civil security operations.
- Train and employ HN security forces.
- Establish or restore essential services.
- Support development of better governance.
- Support economic development.

SUPPORT TO COMBAT OPERATIONS/CIVIL SECURITY OPERATIONS

8-10. Most logisticians and nonlogisticians are familiar with the combat operations/civil security operations LLO. The paramount role of logistic units remains to support Soldiers and Marines in accomplishing the mission and meeting other Title 10 responsibilities. Using logistic units to augment civil programs supporting other LLOs must not detract from the logistic system's capability to support combined arms forces engaged in combat operations.

Support to and from Operating Bases

8-11. Logistic support to COIN combat operations is often accomplished from bases (see FM 3-90, paragraph E-19 through E-29) or forward operating bases (see FM 3-05). Operating bases provide combined arms units with relatively secure locations from which to plan and prepare for operations. As a result, these bases take on new significance in operations executed in a noncontiguous COIN environment.

Considerations for Situating Operating Bases

8-12. In COIN operations, base site selection becomes extremely important for more reasons than providing optimal support to combat operations. Under certain geographic conditions, such as in rugged mountains with few passes or desolate desert terrain, placing secure operating bases astride or near the insurgents' LOCs can improve counterinsurgents' interdiction and disruption capabilities. In urban areas and jungles, insurgents may negate advantages of such a position by rerouting their LOCs around the base. This happened when U.S. forces tried to interdict insurgent supplies on the Ho Chi Minh Trail in Vietnam.

8-13. Other reasons for carefully considering base placements involve sensitivities and concerns of the local populace. The potential for ill-considered bases to substantially disrupt the local populace's daily lives and produce other unintentionally negative effects is significant, even if counterinsurgents arrive with positive intentions. Bases must be set up so that they do not project an image of undue permanency or a posture suggesting a long-term foreign occupation. Similarly, logistic postures that project an image of unduly luxurious living by foreign forces while HN civilians suffer in poverty should be avoided. Such postures undermine the COIN message and mission. Insurgent propaganda can twist such images into evidence of bad intentions by counterinsurgents. While these considerations take on special significance in COIN operations, none of them override the primary concern that operating bases be securable and defendable.

8-14. Selecting and developing operating base sites requires the additional consideration and balance of other factors. In COIN operations, logisticians must provide support through a careful mix of supply-based or supply-point ("just-in-case") practices with distribution-based or unit distribution ("just-in-time") logistic methods. Situations can swiftly develop that require equally rapid logistic responses to prevent further deterioration of security conditions. Under these COIN-specific circumstances, just-in-time practices may still not be quick enough; using just-in-case capabilities may be more appropriate, effective, and timely, while conserving resources. A fire-fighting analogy best illustrates this dilemma of COIN logistics. A small fire confined to a pan on the stove can be put out easily with the five-pound extinguisher. But when this extinguisher is not immediately available to put out the fire, half the house may burn before fire fighters arrive. Then extinguishing the fire requires trucks, hoses, and thousands of gallons of water. The house also needs construction materials and time to be restored to its former state. Commanders and logisticians supporting COIN operations must correctly identify which materials equate to "five-pound fire extinguishers" for counterinsurgents. Logisticians must then ensure that items are available at the most appropriate location.

8-15. This carefully considered balance between distribution- and supply-based methods supports the goal of minimizing the size of operating bases. When required, Soldiers and Marines can relocate smaller bases more easily. Such bases are also less intrusive and antagonizing to the local populace.

8-16. Planners must consider an operating base's purpose when selecting its location. If planners anticipate extensive logistic throughput, they pay close attention to entry and exit points. Ideally, more than one entry and exit point should exist. (FM 5-104 discusses the construction of entry control points and facilities.) Where possible, at least one control point should not require convoys to travel though a populated area. In addition, at least one entry point requires a staging area for convoys and should be located to avoid having to transit the base to form up.

8-17. Due to the noncontiguous nature of COIN operations, logisticians develop weblike LOCs and main supply routes between operating bases and logistic bases. Weblike links between bases have two advantages. By dispersing logistic operations, weblike LOCs minimize intrusive effects of these operations on the populace. They also provide redundancy in distribution capabilities, making the system more robust and limiting the effects of any one LOC's interdiction. In addition, more ground LOC routes provide more opportunities to observe the populace and gather information from them. Wherever possible in COIN operations, planners should identify multiple LOCs between bases.

Protection

8-18. Protection of logistic activities takes on greater significance during COIN operations. Historically, insurgents have deliberately sought out and engaged logistic units, particularly poorly defended, easy targets. COIN operations have intensive manpower requirements and a dispersed nature. Logistic units cannot assume that combined arms units will be available to assist them with protection. For this reason, logistic units play a larger role in defending bases and LOCs. These units must assume responsibility to protect civilian logistic augmenters, whether Department of Defense (DOD) civilians or contractors, working in their AOs.

Vietnam: Meeting the Enemy and Convoy Security

The year 1968 proved a turning point for units of the 48th Transportation Group stationed in the Republic of Vietnam. The group was assigned to transport supplies to 25th Infantry Division units operating in Vietnam's Cu Chi province. In August, North Vietnamese Army and Vietcong units ambushed a supply convoy from the group. The maneuver brigade responsible for clearing that part of the main supply route had been recently assigned other missions, and its resources were spread thin. The ambushers chose their moment to attack well. Monsoon conditions prevailed, and the site was outside the range of supporting indirect fire. In addition, the supplies were destined for the unit tasked with responding to such attacks. Under dangerous weather conditions, two UH-1C "Huey" gun ships arrived first to assist the beleaguered convoy. From the air, aviators witnessed enemy soldiers unloading supplies from U.S. vehicles onto trucks hidden in the tree line off the road. Almost three hours later, the first relief force arrived on the ground. This force barely had enough capability to continue a minimal defense of the remaining convoy assets and surviving personnel. Finally, seven hours later, a U.S. armored cavalry force arrived and forced the attackers to withdraw.

Thirty Soldiers were killed, 45 were wounded, and 2 were taken prisoner. This event forced the 48th Group and 25th Infantry Division to rethink their convoy tactics. The two units started to hold detailed convoy planning meetings and renewed their enforcement of Soldier discipline. They placed security guards on every vehicle, hardened cabs of supply trucks with steel plates, and mounted M-60 machine guns on every vehicle possible. The greatest improvement was the clarification of command and support relationships and responsibilities between the 48th Group and 25th Infantry Division. This included publishing common convoy standing operating procedures. With these new practices in place, convoy ambushes soon had different endings. A change in thinking about a logistic problem converted convoy operations from unglamorous defensive activities into valuable opportunities to engage insurgents offensively.

Combat Logistic Convoys

8-19. During COIN operations, every logistic package or resupply operation becomes a mounted combat operation, or combat logistic convoy. Insurgents see attacks on resupply operations as a potential source of dramatic propaganda as well as a source of supplies and materiel. For this reason, combat logistic convoys should project a resolute ("hard and prickly") image that suggests that they will not be an easy ("soft and chewy") target. Combat logistic convoys project their available combat power to the maximum extent possible, as would any other combat convoy or patrol. Under these conditions, logistic units—or anyone else involved in resupply operations—perform a detailed intelligence preparation of the battlefield and prepare a fire support plan. These units also identify usable intelligence, surveillance, and reconnaissance assets. Additionally, combat logistic convoys should gather information, report on road statuses, and contribute to intelligence collection plans. Logisticians must remember that while the materiel is in transit, it is not only unusable but also vulnerable to insurgent attacks. In a COIN environment, distribution-based practices may actually provide insurgents with more opportunities to target resupply activities. These opportunities stem from the large blocks of time the materiel is in ground transit. Likewise, logisticians must remember that insurgents constantly and deliberately seek out adaptive countermeasures to logistic activities. For example, the development and proliferation of improvised explosive devices were a natural counter to U.S. distribution-based doctrine. As a result, engineer assets and other combined arms elements should prepare to execute periodic route clearance operations. Logisticians must carefully analyze conditions and perform thorough combat preparations before launching combat logistic convoys.

Unit Equipment for Counterinsurgency

8-20. COIN operations shift quite rapidly. Logistic units and supported units often deploy with some equipment unsuitable for prevailing operational and tactical trends when they arrive. This dynamic often forces logisticians to seek equipment modifications and new items. For this reason, COIN operations especially benefit from new procurement programs, such as rapid fielding initiatives and the purchase of commercial off-the-shelf (COTS) items. These approaches make sense for counterinsurgents when insurgents' capabilities also come from their creative exploitation of commercially available technologies and materials as well as their lack of bureaucratic encumbrance.

8-21. Counterinsurgents benefit from using more streamlined materiel procurement procedures. They receive what they need when they need it. Specific and localized environmental and cultural conditions create unseen needs. Streamlined procedures often meet these needs more quickly. Examples of COIN requirements that can be met by COTS procurement are—

- Public address systems.
- Language translation devices.
- Nonlethal weapons.
- Backpack drinking water systems.
- Cargo all-terrain vehicles ("Gators").

8-22. Examples of rapid fielding initiatives are—

- Up-armoring kits for light wheeled vehicles.
- Body armoring improvements.
- Improved explosive detectors.
- Improvised explosive device signal jammers.

8-23. A potential drawback when adopting COTS equipment concerns maintenance support packages. Repair parts may be inadequate or difficult to obtain in theater. Many commercial manufacturers lack the experience or infrastructure needed to support their equipment under military conditions and in quantity. Often, they are unaccustomed to operating in hostile austere theaters, far from their regular markets and customer base. It may take time to get needed parts into normal supply channels and trained personnel in theater to fix COTS equipment. Logisticians must consider other measures to assure the continuous operation of this type of new, vital equipment. They should consider establishing pools of low-density COTS items that can provide exchanges through procedures similar to those used for operational readiness floats. These procedures provide time to evacuate exchanged equipment to locations where it can be maintained and repaired and where the required commercial parts are readily available.

8-24. During COIN operations, units may temporarily have to draw additional or specialized equipment in theater. For long-term COIN operations, leaders may have to establish theater property books. These procedures help maintain and account for rotationally issued additions to standard equipment as well as specialized or specially modified equipment. In-theater special issues and fieldings may include materiel and equipment procured through military channels, rapid fielding initiatives, or COTS sources. Units might draw supplementary or modified equipment to a greater or lesser degree depending on the unit's non-COIN primary function. For example, an artillery unit normally equipped with self-propelled howitzers may draw numerous hardened HMMWVs to conduct security missions. This unit would probably leave many of its howitzers at home station. Conversely, a military police unit might already be well equipped to conduct security activities and might only have to draw a few pieces of the latest specialized equipment. Examples of other items that counterinsurgents might need include—

- Up-armored vehicles.
- Cargo trucks.
- All-terrain cargo vehicles ("Gators").
- Improvised explosive device jammers.
- Body armor.
- Specialized mine-clearing equipment ("Buffaloes").

8-25. Units conducting COIN operations can expect somewhat different maintenance requirements than in conventional operations. Units that put high mileage on their wheeled vehicles need more frequent servicing. Armor packages may wear out shock absorbers and springs much faster; these would require replacement sooner than in conventional operations. COIN missions and the remoteness of many operating bases may compel maintenance sections to perform higher echelons of maintenance than normal and encourage greater organic capability.

Unit Basic Loads and Operational Reach

8-26. In places like Somalia and Sadr City during Operation Iraqi Freedom, Soldiers and Marines conducting COIN operations risked being cut off from bases and forward operating bases due to weather changes, enemy action, or civil protests. Such rapidly developing situations can deny units access to resupply for extended periods. Units operating away from supporting bases carry the maximum amount of basic supplies—water, food, ammunition, first aid, and equipment batteries—on their vehicles. Additionally, some COIN operations can consume surprisingly high quantities of ammunition (specifically small arms) because of combined defensive and offensive actions. Logisticians supporting these operations adjust stockage levels for unit basic loads and other sustainment commodities. In turn, logisticians and their supported units should rethink how to best configure their supporting vehicles as supply platforms to meet these COIN-specific needs. Competent authorities should validate successful solutions. This process ensures standardization across formations to ensure safety and to support planning for effective employment.

8-27. When developing their operating base requirements, commanders and logisticians must plan for areas to store ammunition and explosives. Ideally, units are issued ammunition and explosives, anticipating that it may be some time before resupply is available. Units normally carry only their basic load. The rest should be staged appropriately.

Aerial Distribution

8-28. During COIN operations, logisticians should maximize intratheater aerial resupply. This practice reduces the vulnerability of resupply activities to ground-based attacks by insurgents. Its also minimizes negative effects of COIN logistic activities on public roadways and reduces the potential for alienating the populace. Site selection for bases and forward operating bases includes assessing aircraft support capabilities. Site selection also considers maximizing the possible options for aerial delivery by rotary- and fixed-wing aircraft, airdrops, and landings. Additionally bases with medical capability (level II or higher) require a helicopter pad near the medical area. (FM 4-02, paragraph 2-4, and NAVMED P-117, article 19-24, discuss levels of medical care.)

Air Delivery in Iraq: Maximizing Counterinsurgency Potential

For almost five months in 2004, two Marine battalions with attached units operating in remote areas of Iraq were resupplied by airdrops from rotary- and fixed-wing aircraft. Until then, cargo airdrops from helicopters had been suspended since the mid-1990s due to technical difficulties that posed unacceptable peacetime risks. The high tempo in Iraq, coupled with severe challenges in using ground supply methods to reach remote locations, made circumstances opportune to reexamine helicopter airdrop possibilities.

A careful analysis of earlier challenges resulted in a clarification of flying procedures during drops. Furthermore, technological advances made it possible to drop bundles by parachute with some directional adjustment after release. This new ability greatly increased the accuracy and utility of helicopter airdrops. These improved procedures replaced dangerous nine-hour convoys with quicker and more secure flights covering three-quarters the ground distance. During this period, Marines received more than 103 tons of supplies from airdrops. A logistic method that previously had been set aside as unworkable found new utility when operational conditions changed and planners addressed technological shortcomings.

With the need to deliver many supplies to diverse locations quickly, counterinsurgents discovered that adopting various air delivery procedures significantly improved their logistic posture under challenging conditions. Such procedures also minimized the risk of negative encounters with the populace and insurgents.

SUPPORT TO TRAINING AND EMPLOYING HOST-NATION SECURITY FORCES

8-29. One of the most important LLOs for U.S. forces engaged in COIN operations is training and employing HN security forces. Various support and training activities contribute to security sector reform. Each activity can substantially involve military logisticians. Usually, developing HN police forces and sustaining their training falls under the auspices of non-DOD agencies. These agencies include the Department of State, the Department of Justice, or United Nations mandated missions. The development and support of HN military forces is a COIN mission that military logisticians must prepare to support, from planning at the strategic level to practical implementation on the ground. (Chapter 6 covers the support of HN security forces in more detail.)

8-30. Some tasks required to establish HN security forces might initially fall to military logistic units until other government agencies' programs start, logistic support can be contracted, or HN logistic organizations begin to function. Examples of HN security force support tasks include the following:

- Providing operating base space or establishing other supportable, secure locations to recruit, receive, and train HN security forces.
- Providing initial logistic support to forming HN security forces, possibly to include equipping, arming, feeding, billeting, fueling, and providing medical support.
- Providing logistic training to newly formed HN security force logistic organizations.

Equipping and sustaining Host-Nation security Forces

8-31. Logisticians can expect to help develop plans and programs for sustaining HN security forces. They must ensure that equipment selected is suitable to and sustainable through the host nation's capabilities. Equipment and support programs must fall within the host nation's resources, including budget and technological capabilities. In many cases "good enough to meet standards" equipment that is indigenously sustainable is preferable to "high-technology, best available" equipment that requires substantial foreign assistance for long-term maintenance. Foreign high-technology equipment can provide the insurgent movement with a valuable propaganda point that could negate any potential technological advantages.

8-32. One acknowledged difficulty in establishing HN security forces is identifying sources of suitable materiel and equipment. Often, multinational partners develop plans to equip the host nation from multiple donor nations and agency sources. Logisticians may need to become familiar with these agencies' capabilities and donor nations' supply and maintenance systems, even though support packages may not be included with the donation.

Host-Nation security Forces Logistics

8-33. Logisticians involved in training logistic personnel to support HN security forces need to be aware of several special challenges. Part of the problem with previously dysfunctional military cultures in many unstable countries is a pervasive climate of corruption and graft that can cripple attempts to develop effective support services. Logisticians conducting such training should expect to stress repeatedly the long-term benefits of supply discipline and materiel accountability. Logisticians must also emphasize how these practices affect HN security and development. For this reason, emphasis should be placed on inventory procedures. Logisticians should monitor the black market to check for pilfered military equipment and to evaluate the effectiveness of logistic procedures and accountability training. Of all capabilities, logistic functions may take the longest for HN forces to develop. These functions are rife with inherent complexity and potential cultural challenges. Hence, HN forces may take a long time to operate independently of U.S. or multinational logistic support. (AR 12-1 and DODD 5105-38M discuss control and accountability requirements for property transferred to a host nation [end-use monitoring].)

8-34. HN-produced materiel should be procured and used to support HN security forces whenever it can meet requirements and is reasonably and reliably available. This practice helps stimulate the HN economic base and promotes an attitude of self-sufficiency in HN forces. It reinforces the important political message that HN security forces are of the people, not agents of foreign powers. When promoting these practices, logisticians may find themselves outside the normal scope of their duties when they assess the suitability of locally available materials and advise how to make such materials suitable for self-sustainment. The most valuable lesson logisticians may give to HN security forces and those supporting them is not "what to do" but "how to think about the problem of sustainment" and its link to security effectiveness.

Building a Military: Sustainment Failure

By 1969, pressure was on for U.S. forces in Vietnam to turn the war over to the host nation in a process now known as Vietnamization. While assisting South Vietnamese military forces, the United States armed and equipped them with modern small arms, communications, and transportation equipment—all items produced by and sustained from the U.S. industrial base. This modern equipment required an equally sophisticated maintenance and supply system to sustain it. Sustaining this equipment challenged the South Vietnamese economically and culturally, despite the training of several thousand South Vietnamese in American supply and maintenance practices. In short, the American way of war was not indigenously sustainable and was incompatible with the Vietnamese material culture and economic capabilities. South Vietnam's predominately agrarian-based economy could not sustain the high-technology equipment and computer-based systems established by U.S. forces and contractors. Consequently, the South Vietnamese military transformation was artificial and superficial. Many South Vietnamese involved in running the sustainment systems had little faith in them. Such attitudes encouraged poor administration and rampant corruption. After U.S. forces left and most U.S. support ended, the logistic shortcomings of the supposedly modern South Vietnamese military contributed to its rapid disintegration when the North Vietnamese advanced in 1975.

SUPPORT TO ESTABLISHING OR RESTORING ESSENTIAL SERVICES

8-35. According to existing U.S. military logistic doctrine, no provision exists for U.S. forces to become decisively or exclusively engaged in providing essential services to the populace. However, this doctrinal position does not prohibit units from applying skills and expertise to help assess HN essential services needs. Along with these assessments, logistic and other units may be used to meet immediate needs, where possible and in the commander's interest. These units can also assist in the handoff of essential service functions to appropriate U.S. agencies, HN agencies, and other civilian organizations.

Assessing Essential services Requirements

8-36. Logistic preparation of the theater and detailed assessments of COIN-specific issues should give logisticians good insights into HN capabilities, requirements, and shortfalls. (See paragraph 8-7.) Logistic assessments should be combined with information from civil affairs area assessments. Logisticians working closely with other branches can contribute to these area assessments. (Table 8-2 [page 8-12] identifies the areas that concern civil affairs. Logisticians can help assess these areas.)

Table 8-2. Civil affairs capabilities used by logisticians

Service	Branch	Capability
Police and Fire	Military police	• Assessment of police capabilities (law enforcement, crowd control, traffic control, crime trends, criminal investigations) and detention and correction operations.
	Military firefighter	• Assessment of firefighting capabilities and equipment.
Water	Quartermaster	• Water purification.
	Medical service	• Preventive medicine and sanitation specialists.
	Engineer	• Water pipelines, facilities, and drilling support.
	Ordnance	• Pumps and mechanicals repair.
Electricity	Engineer	• Power generation and facility repair and construction specialists.
Schools	Engineer	• Vertical construction specialists.
	All	• Training and education.
Transportation network	Engineer	• Assessment of rail, bus, and port facilities; roadway and bridge assessments; and airfield capabilities and upgrades.
	Transportation	• Assessment of rail, bus, and ferry capacities.
	Ordnance	• Assessment of mechanical maintenance of rail, bus, truck, and ferry operating equipment.
	Military police	• Roadway and traffic flow assessments.
Medical	Medical service and medical corps	• Assessment of local health care capabilities and needs.
Sanitation (trash and sewage)	Medical service	• Preventive medicine and sanitation specialists.
	Quartermaster	• Water specialists.
	Engineer	• Earthmoving specialists, plumbing construction, soil analysis, concrete casting, and sanitary landfill management.
	Ordnance	• Heavy wheeled vehicle, pump, and mechanical repair.
Food supply	Veterinary corps	• Food source and quality; vector control.
	Quartermaster	• Food packaging and distribution.
Fuel	Quartermaster	• Fuel specialists; testing of locally procured fuel supplies.
Financial	Finance	• Identify requirements to reestablish accountability and security of host-nation and captured funds. • Assess financial support requirements and banking and currency needs.

Time as a Logistic commodity

8-37. The speed with which COIN operations are executed may determine their success and whether the populace supports them. This is especially true for operations that involve restoring essential services. Planners must strive to have the smallest possible gap of time between when they assess essential services and when U.S. forces begin remediation efforts. To keep this time gap as small as possible and manage the development of popular expectations, logistic units may need to initiate remedial services until HN authorities and agencies can assume these functions. For example, think of the populace as a patient destabilized by the trauma of insurgency. In COIN operations, logistic and other units may need to function much like the first-responder medic on the scene. The medic conducts initial assessments of a patient's needs, provides lifesaving first aid, and lets the hospital know what more specific higher level care the patient requires. Medical and COIN first responders work best when they can assess and initiate life support treatment immediately. In COIN operations, "treatment" is harder to determine without obvious calibrated vital signs—such as blood pressure, pulse, and temperature—and "immediately" may last weeks or months. For this reason, COIN logistic units may have to take whatever measures they can for immediately stabilizing essential services and preventing deteriorating conditions. (Table 8-3 [page 8-37] lists some examples of how military logistic assets and capabilities can be used to bridge the essential services gap and meet immediate and essential service needs.)

Handoff of Essential Services

8-38. Frequently, logisticians who have provided stopgap essential services may be the only ones with accurate knowledge of essential services needs and priorities. Logisticians providing these services may stay actively involved in the handoff to other government agencies and their designated civil organizations. Their involvement should continue until HN agencies and activities can function and meet essential services needs. A poor handoff can provide insurgents with propaganda opportunities and evidence of the "insincerity" of COIN efforts.

8-39. When U.S. forces restore and transition essential services to the HN government, they remove one of the principal causes insurgents exploit. This action greatly assists the HN government in its struggle for legitimacy. Competent leaders can expect insurgents to conduct attacks against restored services. During this handoff, multinational logistic assets may need to maintain a logistic quick reaction force. This force ensures the continuity of services and marginalizes counteractions and messages by insurgents.

Public Transportation, Population Movement, and Life Support to Internally Displaced persons and Refugees

8-40. Under conditions of national crisis or insurgency, public transportation systems often fall into disarray. Counterinsurgents may have to recover stolen or misappropriated buses, trucks, cars, and other government vehicles (including former military equipment) and restore them to public service. This action helps alleviate urgent requirements for public transportation. It also sends an unmistakable message of resumption of governmental authority and can substantially reduce the amount of replacement equipment counterinsurgents must procure. Logistic units and personnel can expect to assist in this process—from reestablishing accountability procedures to assessing repair and maintenance needs—until competent HN public or government authorities can resume these duties.

8-41. An insurgency often creates many groups of internally displaced persons and refugees on short notice. Attending to internally displaced person and refugee needs can quickly become an urgent logistic requirement. Planners draw on all essential services to provide secure emergency shelter, camps for internally displaced persons and refugees, and life support (food, water, and medical care). Nongovernmental organizations and other civilian agencies normally furnish this support to internally displaced persons and refugees. However, conditions may prevent these agencies from providing these services quickly. Furthermore, in COIN operations, internally displaced person and refugee security may take on heightened military importance. Traumatized and dislocated persons may become vulnerable to insurgent threats and recruitment. The restoration and maintenance of public transportation services can help internally displaced persons and refugees. Figure 8-1 (page 8-15) shows that as essential services projects take root, they start to provide tangible benefits for the populace. Progress in these individual endeavors may experience individual setbacks as planners calibrate programs and projects to localized needs. Before implanting agency transfers, planners should measure overall progress. As essential services become more effective, insurgent activities lose influence and popular support. As a general rule, it is best to provide essential services to people in their native areas and thereby discourage their displacement.

Table 8-3. Logistic units and their capabilities

Logistics activities	*Capabilities*
Contingency contracting officers	• Procure commercial public utility equipment with Title 22 funds. • Employ theater contractors and external theater subject matter experts and trainers to maintain assets.
Reverse osmosis water purification units	• Provide immediate sources of potable water until water pumps at purification plants are restored.
Distribution companies and supply support activities	• Provide temporary storage and distribution of foreign humanitarian assistance.
Explosive ordnance disposal units	• Dispose of munitions and other unexploded explosive ordnance in populated areas. • May require transportation or engineer support.
Combat logistic convoys	• Provide security for nongovernmental organization (NGO) transportation of critical humanitarian assistance. (Not all NGOs agree to this.)
Medical units	• Provide medical civic action program teams to conduct visiting clinics at small or remote villages. • May augment NGOs, such as Doctors Without Borders.
Brigade surgeons and brigade engineers	• Work with contracting officer representatives to restore clinic or hospital services.
Medical personnel and units	• Assist with upgrading and restoring host-nation medical training programs to meet civil health care providers' critical shortfalls.
Senior power generation technicians	• Provide advice concerning and troubleshoot municipal power sources.
Class I (rations) sections	• Account for, preserve, and distribute humanitarian daily rations (a Department of State-controlled item).
Medium truck companies	• Move internally displaced persons.
Logistic units	• Provide life support to internally displaced person (or refugee) camps—billeting, food service, personnel (biometric), accountability, and work placement.
Preventive medicine teams	• In conjunction with veterinary support, conduct vector and parasite analysis on farm livestock (host-nation food source).
Engineers	• Repair critical highways, renovate bridges, or build buildings (such as clinics or schools).

SUPPORT TO DEVELOPING BETTER GOVERNANCE

8-42. COIN operations strive to restore order, the rule of law, and civil procedures to the authority of the HN government. All counterinsurgent actions must be those of agents of a legitimate and law-abiding HN government. Multinational and U.S. forces brought in to support this objective must remember that the populace will scrutinize their actions. People will watch to see if Soldiers and Marines stay consistent with this avowed purpose. Inconsistent actions furnish insurgents with valuable issues for manipulation in propaganda.

Legal Support to Operations

8-43. Legal support to COIN operations can cover many areas. (See FM 4-0, chapter 12; MCWP 4-12, page A-4; and appendix D of this manual.) It includes continuously monitoring and evaluating rules of engagement. This legal status may affect the conduct of contractors and their requirements for protection. Legal support also includes status of forces agreements. These agreements need to be negotiated and revised as the HN government becomes able to responsibly assume and exercise sovereignty. Status of forces agreements affect how legal disputes between U.S. forces and local nationals are handled, including those disputes emerging from contracting and other commercial activities. Contracts and claims require sensitive and fair construction and execution so perceptions of exploitation and favoritism do not undermine overall COIN initiatives. COIN operations often depend on many funding sources. Leaders get judge advocate legal advice on fiscal law to ensure compliance with domestic statutes governing the funding of military and nonmilitary activities. Leaders may ask judge advocates to advise the HN government at all political levels about how to establish and administer appropriate legal safeguards.

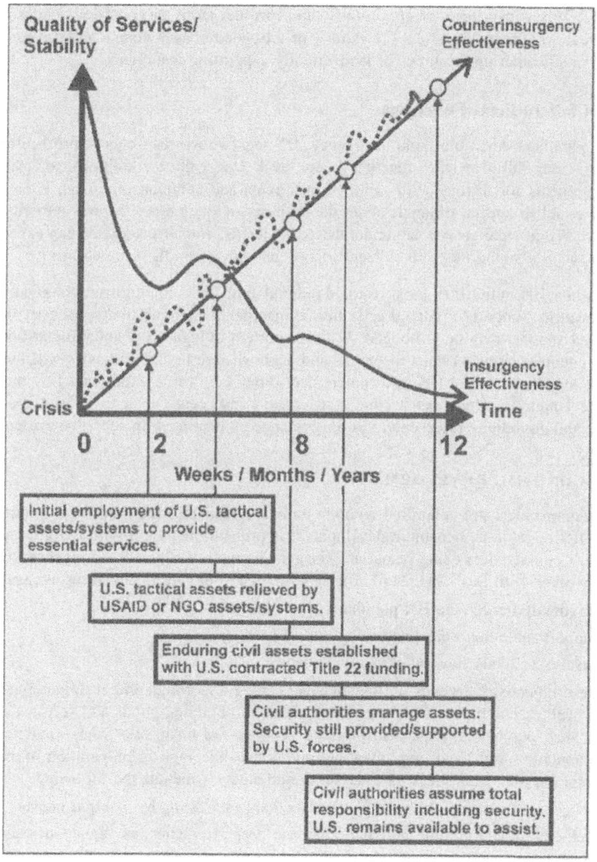

Figure 8-1. Comparison of essential services availability to insurgency effectiveness

Legal Aspects of contracting and claims

8-44. In COIN operations, two circumstances may require extensive civil law support. The first situation is when counterinsurgents engage in commercial contracts with people for provision of goods and or services. The second occasion is when people seek compensation for damages, injuries, or deaths that they or their relatives claim to have suffered due to actions by counterinsurgents.

8-45. Legally reviewing COIN contracts negotiated with HN contractors establishes several important conditions. First, the process makes clear to HN contractors that established procedures rooted in law govern such transactions. It sends the message that favoritism and partisanship are not part of the process in a legitimate government. Second, this review potentially forestalls contracts going to individuals who may be part of the insurgency and may already be named or identified as subjects of other ongoing investigations or legal actions.

8-46. In the case of claims for damages allegedly caused by counterinsurgents, legal reviews show genuinely wronged people that the HN government takes their grievances seriously. (See paragraphs D-35 through D-37.) When insurgents or opportunists misrepresent terms or conditions under which "damages" occur, legal reviews effectively assess the validity or falsehood of such claims and thereby prevent counterinsurgents from squandering resources or inadvertently supporting insurgents.

Restoration of Civil Judicial Functions

8-47. In periods of extreme unrest and insurgency, HN legal structures—courts, prosecutors, defense assistance, prisons—may fail to exist or function at any level. Under these conditions, to establish legal procedures and precedents for dealing with captured insurgents and common criminals, U.S. forces may make provisions to establish special tribunals under the auspices of either a provisional authority or a United Nations mandate. While legal actions fall under these provisions, counterinsurgents can expect to provide sustainment and security support as well as legal support and advice on these functions.

8-48. Even when HN authorities have restored judicial functions, counterinsurgents may have to provide logistic and security support to judicial activities. If insurgents continue to disrupt activities supporting the rule of law, the support may be prolonged. With restoration of legislative and judicial functions to the HN government, counterinsurgents must recognize and acknowledge that not all laws will look familiar. Laws passed by the legislature of the HN government may differ from those experienced by multinational forces in their home countries. Under such conditions, counterinsurgents need to consult their legal advisors, commanders, and diplomatic representatives for guidance on dealing with sensitive matters.

SUPPORT TO ECONOMIC DEVELOPMENT

8-49. Many commanders are unfamiliar with the tools and resources required for promoting economic pluralism. In COIN operations, economic development is probably the LLO with the greatest logistic significance. Military commanders using resource managers (comptrollers) and contingency contracting officers are usually involved with this LLO. Staff officers will need to deliver financial resources that—

- Maximally benefit the HN population.
- Support achieving objectives along other LLOs.
- Ensure the funds stay out of insurgent hands.

Achieving these objectives depends on logisticians keeping a thorough and accurate logistic preparation of the theater as well as commanders and contracting officers obtaining goods and services consistent with its assessments. Such purchasing must also promote vendors and businesses whose practices support widespread job stimulation and local investment. In addition to the logistic preparation of the AO issues discussed in paragraph 8-7, other areas for assessment and analysis include the following:

- HN economic capabilities and shortfalls suitable for filling by external means.
- Methods of determining land and other real property ownership, means of transfers, and dispute resolution.
- Methods for promoting and protecting property and asset rights as well as open access to trade goods and services.

- Prevailing wage rate standards and correlation to occupational category (unskilled, skilled, and professional labor).
- Historic market demographics.
- Identification of potential vendors with local sources of supply in the AO.

8-50. Various funding sources usually support COIN operations. (See paragraphs D-27 through D-31.) U.S. forces most commonly operate under two types of funds:

- Title 10 funds, which are strictly for the supply, support, and sustainment of DOD service members and employees.
- Title 22 funds, which are appropriated for foreign relations purposes and used solely for the benefit and support of the HN government and population.

8-51. Other sources of funding in COIN operations may include those provided by other government agencies. Such agencies can include the U.S. Agency for International Development, the Department of State, other donor nations and agencies, the United Nations, or even the host nation. In some cases, counterinsurgents may seize or capture misappropriated or illicit funds. Confiscated funds may be redistributed to fund COIN activities. Under these complex fiscal circumstances, resource managers and staff judge advocates provide the best guidance on the legal use of different types of funds. (Appendix D covers this subject in more detail.)

CONTRACTED LOGISTIC SUPPORT

8-52. In COIN operations, logistic contractors might support U.S. forces. Contractor activities fall into three different categories:

- Theater support contractors.
- External support contractors.
- System contractors.

(See FM 4-0, paragraphs 5-92 through 5-95; MCWP 4-12, page 4-8.)

8-53. Theater support contractors can make significant contributions to promoting economic pluralism because they rely the most on HN employees and vendors. External support contractors and the Logistic Civilian Augmentation Program (LOGCAP) provide logistic services, usually through large-scale prearranged contracts with major contractors, who may in turn subcontract various components of their large contracts to smaller theater-based providers. Systems contracts are designed by systems program managers to support special or complex equipment; they generally have little influence on promoting economic pluralism.

THEATER SUPPORT CONTRACTORS

8-54. Theater support contractors can be obtained either under prearranged contracts or by contracting officers serving under the direct authority of the theater principle assistant responsible for contracting. Theater support contractors usually obtain most of their materials, goods, and labor from the local manufacturing and vendor base. Some examples of goods and services that theater support contractors provide include—

- Construction, delivery, and installation of concrete security barriers for the defense of counterinsurgent bases and HN public buildings.
- Construction of security fencing.
- Public building construction and renovations (such as site preparation, structure construction, electrical and plumbing installation, and roofing).
- Sanitation services.
- Maintenance augmentation in motor pools.
- Road construction and repair.
- Trucking and cartage.

- Manual labor details (such as grounds maintenance and sandbag filling).
- Housekeeping (such as warehouses).

COUNTERINSURGENCY CONTRACTOR CONSIDERATIONS

8-55. In a COIN environment, commanders carefully consider when to use theater support contractors and local hires. Commanders also supervise contracted personnel to ensure they do not undermine achieving COIN objectives. Due to the subversive nature of many insurgent activities, all contractors and their employees require vetting through the intelligence section. All contractors and their employees require tamperproof, photograph, biometric-tagged identification. This identification needs to be coded to indicate access areas, security level, and supervision required. In the case of HN employees, "badging" can also be an accountability tool if U.S. forces issue and receive badges at entry control points daily. Contractor security breaches are one concern; another is the security and safety of the contractor's employees. Though insurgents may target logistic contractors and their employees, the employees are not combatants. They are classified as "civilians accompanying the force." This status must not be jeopardized and the military units with which they work must keep them secure in the workplace. Units employing HN contractors and employees must watch for signs of exploitive or corrupt business practices that may alienate segments of the local populace and inadvertently undermine COIN objectives. Treated fairly and respectfully, HN employees can provide good insights into the local language, culture, and perceptions of COIN activities as well as other issues affecting communities in the AO.

Host-Nation Contracting: A Potential Double-Edged Sword

Early in Operation Iraqi Freedom, a brigade from the 101st Airborne Division was assigned a large area of operations near Tal Afar, in northern Iraq. The terrain the unit was required to cover and support exceeded the distribution capabilities of its ground transportation assets. Logistic officers supporting the brigade sought out and found a local business leader with a family-owned transportation company. He was positive towards U.S. aims for improving Iraq and willing to work with U.S. forces by providing various truck and bus services.

After two months of ad hoc daily arrangements for services at the U.S. forces' compound entry point, the unit established a six-month contract to make this transportation support more regular. As the working relationship became more solid, the contractor and his employees also furnished insights into the effectiveness of U.S. information operations as well as information on the presence and activities of suspicious persons possibly affiliated with the insurgency. The arrangement worked exceptionally well, effectively supported counterinsurgency activities, and maintained peace and security—as long as the original unit that established the services was stationed in the area.

Eventually, a smaller task force replaced the first unit that established the contract, and the security situation in the area began to deteriorate. Upon detecting this change in security posture, insurgents quickly found the contractor and killed him. No doubt their intent was to degrade the U.S. forces' logistic posture and to send the message to other local vendors that doing business with the Americans was costly. Eventually the contactor's brother took over operations, but understandably support deteriorated.

When setting up logistic contracting arrangements with HN contractors in a COIN environment, U.S. logisticians and contracting officers must remember the grave risks people take by accepting these jobs. Insurgents are exceptionally adept at finding ways to attack logistics. When insurgents attack people branded as traitors, there is an added terror or political message benefit. Inadequate or shifting U.S. security arrangements can provide openings for insurgents to more easily attack host-nation contractors and logistic providers.

EXTERNAL SUPPORT CONTRACTORS

8-56. Many of the same considerations that apply to theater contractors apply to external support contractors and their subcontracted employees, particularly if they are local or in-theater hires who are not U.S. or multinational partner citizens.

SYSTEM CONTRACTORS

8-57. System contractors generally work on technologically complicated military systems, such as vehicles, weapons systems, aircraft, and information systems. This support is provided under contracts negotiated by program executive officers and program managers. These contractors provide systems expertise. Most are U.S. citizens, and many of them are former U.S. military members. These contractors generally meet deployment and security requirements similar to DOD civilian requirements.

CONTINGENCY CONTRACTING OFFICERS AND OTHER AGENTS

8-58. In COIN operations, the timely and well-placed distribution of funds at the local level can serve as an invaluable force multiplier. Many challenges to accomplishing payments and purchases in the COIN environment exist. These challenges can include—

- Problems with the security of financial institutions, agents, and instruments.
- Potential for sudden volatility in the HN economy.
- Reliability issues with local supplier and vendors.
- Peculiarities of local business cultures.

8-59. Contracting officers and other agents authorized to make payments to support COIN activities often find it difficult to obtain reliable information upon which to make decisions and conduct negotiations. Military means to accomplish this type of purchasing are found at two levels:

- Contingency contracting officer, who acts on unit-generated purchasing requests and committals.
- Ordering officer for smaller purchases.

8-60. Legal requirements keep U.S. funds for different purposes separate and distinct. (See paragraphs D-27 through D-31.) COIN units must maintain two types of purchasing officer teams:

- Field ordering officer teams for Title 10 funds.
- Project ordering officer teams for Title 22 funds.

(Figure 8-2 [page 8-20] portrays the relationships and roles among different contracting and ordering officers, and the types of funds that they manage.)

CONTINGENCY CONTRACTING

8-61. Because contingency contracting officers can set contracts for larger sums than ordering officers can, they normally place purchasing request and committal requirements out for bid to local vendors. During COIN operations, contacting officers must spread contracts across different vendors to forestall any appearance of partiality.

FIELD ORDERING AND PROJECT ORDERING OFFICER TEAMS

8-62. Both field ordering and project ordering officer teams consist of the respective contracting officer agent, a paying agent, and a security detail. Both teams operate under similar regulatory constraints. These officers' duties differ not only with respect to the type of funds they disburse but also in the increment caps applied. Field ordering officers with Title 10 funds are limited to individual payments not to exceed $2,500. Project ordering officers can make individual contract payments of up to $20,000, since projects are associated with higher costs and scales. In both cases, these teams are an invaluable asset for reaching into HN communities and promoting economic pluralism while assessing the economic effects of purchasing activities and economic stability initiatives.

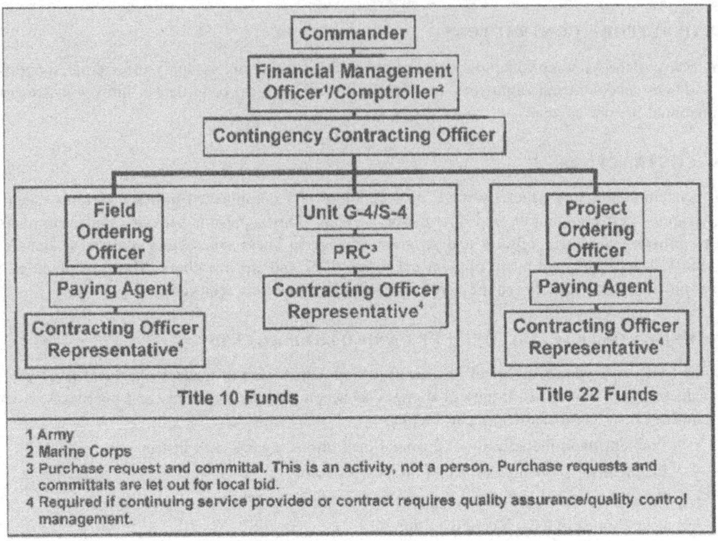

Figure 8-2. Tactical financial management organizations

8-63. Due to their activities throughout HN communities and their cultivation of local business connections, field ordering and project ordering officer teams should be viewed as potential means of gathering and distributing information. Vendor observations and actions may also reveal much about the real status of the COIN effort. Vendors may notice commonplace events that, taken together, are significant. For example, vendors may notice outsiders moving into the area, contractors failing to deliver goods, and HN employees asking to leave early. Together these events might indicate an impending attack. By doing business with counterinsurgents, local contractors and vendors may put themselves at great risk. Protection of their activities can pose a great challenge to U.S. forces and must be seriously considered when doing business with them.

SUMMARY

8-64. Logistic activities are an integral part of COIN operations. These activities take the traditional form of support to combat and security forces as well as the unconventional form of providing mixes of essential and timely support to many HN security and stability-enhancing activities that may seem purely civil in character. Initially, uniformed military logistic providers may have to provide this support. However, COIN logistic objectives should include encouraging and promoting HN providers as soon as security conditions make this feasible. This transition is a delicate one. Logistic providers must constantly determine whether their practices are contributing to achieving the end state and adjust their methods if necessary. If there is a final paradox in counterinsurgency, it is that logistic postures and practices are a major part of the effort and may well determine the operation's success.

Appendix A
A Guide for Action

Translating lessons of this manual into practice begins with planning and preparation for deployment. Successful counterinsurgents execute wisely and continually assess their area of operations, the impact of their own operations, and their enemy's strategy and tactics to adapt and win. This appendix discusses several techniques that have proven successful during counterinsurgency operations. They are discussed within the framework of the operations process. However, this does not limit their use to any operations process activity. Successful counterinsurgents assess the operational environment continuously and apply appropriate techniques when they are needed.

PLAN

A-1. *Planning* is the process by which commanders (and staffs, if available) translate the commander's visualization into a specific course of action for preparation and execution, focusing on the expected results (FMI 5-0.1). Planning for counterinsurgency (COIN) operations is no different from planning for conventional operations. However, effective COIN planning requires paying at least as much attention to aspects of the environment as to the enemy force.

ASSESS DURING PLANNING: PERFORM MISSION ANALYSIS

A-2. Learn about the people, topography, economy, history, religion, and culture of the area of operations (AOs). Know every village, road, field, population group, tribal leader, and ancient grievance. Become the expert on these topics. If the precise destination is unknown, study the general area. Focus on the precise destination when it is determined. Ensure leaders and staffs use the Secret Internet Protocol Router Network (SIPRNET) to immerse themselves virtually in the AO into which the unit is deploying. Understand factors in adjacent AOs and the information environment that can influence AOs. These can be many, particularly when insurgents draw on global grievances.

A-3. Read the map like a book. Study it every night before sleep and redraw it from memory every morning. Do this until its patterns become second nature. Develop a mental model of the AO. Use it as a frame into which to fit every new piece of knowledge.

A-4. Study handover notes from predecessors. Better still, get in touch with personnel from the unit in theater and pick their brains. In an ideal world, intelligence officers and area experts provide briefings. This may not occur. Even if it does, there is no substitute for personal mastery.

A-5. Require each subordinate leader, including noncommissioned officers, to specialize on some aspect of the AO and brief the others.

ANALYZE THE PROBLEM

A-6. Mastery of the AO provides a foundation for analyzing the problem. Who are the insurgents? What drives them? What makes local leaders tick? An insurgency is basically a competition among many groups, each seeking to mobilize the populace in support of its agenda. Thus, COIN operations are always more than two sided.

A-7. Understand what motivates the people and how to mobilize them. Knowing why and how the insurgents are getting followers is essential. This requires knowing the real enemy, not a cardboard cutout. Insurgents are adaptive, resourceful, and probably from the area. The local populace has known them since they were young. U.S. forces are the outsiders. The worst opponents are not the psychopathic terrorists of the movies; rather, they are charismatic warriors who would excel in any armed force. Insurgents are not necessarily misled or naive. Much of their success may stem from bad government policies or security forces that alienate the local populace.

A-8. Work the problem collectively with subordinate leaders. Discuss ideas and explore possible solutions. Once leaders understand the situation, seek a consensus on how to address it. If this sounds unmilitary, get over it. Such discussions help subordinates understand the commander's intent. Once in theater, situations requiring immediate action will arise too quickly for orders. Subordinates will need to exercise subordinates' initiative and act based on the commander's intent informed by whatever knowledge they have developed. Corporals and privates will have to make quick decisions that may result in actions with strategic implications. Such circumstances require a shared situational understanding. They also require a command climate that encourages subordinates to assess the situation, act on it, and accept responsibility for their actions. Employing mission command is essential in this environment. (Mission command, subordinates' initiative and commander's intent are defined in the glossary. See FM 6-0, paragraphs 1-67 through 1-80, 2-83 through 2-92, and 4-26 through 4-31 for discussions of the principles involved.)

PREPARE

A-9. *Preparation* consists of activities by the unit before execution to improve its ability to conduct the operation, including, but not limited to, the following: plan refinement, rehearsals, reconnaissance, coordination, inspection, and movement (FM 3-0). Compared with conventional operations, preparing for COIN operations requires greater emphasis on organizing for intelligence and for working with nonmilitary organizations. These operations also require more emphasis on preparing small-unit leaders for increased responsibility and maintaining flexibility.

ORGANIZE FOR INTELLIGENCE

A-10. Intelligence and operations are always complementary, especially in COIN operations. COIN operations are intelligence driven, and units often develop much of their own intelligence. Commanders must organize their assets to do that.

A-11. Each company may require an intelligence section, including analysts and an individual designated as the "S-2." Platoon leaders may also have to designate individuals to perform intelligence and operations functions. A reconnaissance and surveillance element is also essential. Augmentation for these positions is normally not available, but companies still must perform the tasks. Put the smartest Soldiers and Marines in the intelligence section and the reconnaissance and surveillance element. This placement results in one less rifle squad, but an intelligence section pays for itself in lives and effort saved.

A-12. There are never enough linguists. Commanders consider with care where best to use them. Linguists are a battle-winning asset, but like any other scarce resource, commanders must allocate them carefully. During predeployment, the best use of linguists may be to train Soldiers and Marines in basic language skills.

ORGANIZE FOR INTERAGENCY OPERATIONS

A-13. Almost everything in COIN is interagency. Everything from policing to intelligence to civil-military operations (CMO) to trash collection involves working with interagency and host-nation (HN) partners. These agencies are not under military control, but their success is essential to accomplishing the mission. Train Soldiers and Marines in conducting interagency operations. Get a briefing from the Department of State, aid agencies, and the local police or fire departments. Designate interagency subject matter experts in each subordinate element and train them. Look at the situation through the eyes of a civilian who knows nothing about the military. Many civilians find rifles, helmets, and body armor intimidating. Learn how not to scare them. Seek advice from those who come from that nation or culture. Most importantly, know that military operations create temporary breathing space. But to prevail, civilian agencies need long-term development and stabilization.

TRAVEL LIGHT AND HARDEN YOUR SUSTAINMENT ASSETS

A-14. A normal combat load for Soldiers and Marines includes body armor, rations, extra ammunition, communications gear, and many other things—all of which are heavy. Insurgents may carry a rifle or rocket-propelled grenade, a headscarf, and a water bottle. Without the extra weight, insurgents can run and maneuver easily. U.S. forces must lighten their combat loads and enforce a habit of speed and mobility. Otherwise, insurgents consistently outrun and outmaneuver them. However, make sure Soldiers and Marines can always reach back for fires or other support.

A-15. Remember to harden sustainment bases. Insurgents often consider them weak points and attack there. Most attacks on coalition forces in Iraq in 2004 and 2005, other than combat actions, were against sustainment installations and convoys. Ensure sustainment assets are hardened and have communications. Make sure to prepare Soldiers and Marines whose primary task is providing logistic support to fight ground combat operations. While executing their sustaining operations, they may do more fighting than some rifle squads.

FIND A POLITICAL AND CULTURAL ADVISOR

A-16. A force optimized for COIN operations would have political and cultural advisors at company level. The current force structure gives corps and division commanders a political advisor. Lower echelon commanders must improvise. They select a political and cultural advisor from among their troops. This person may be a commissioned officer, but may not. The position requires someone with "people skills" and a feel for the environment. Commanders should not try to be their own cultural advisor. They must be fully aware of the political and cultural dimension, but this is a different role. In addition, this position is not suitable for intelligence professionals. They can help, but their task is to understand the environment. The political advisor's job is to help shape the environment.

TRAIN THE SQUAD LEADERS—THEN TRUST THEM

A-17. Squads and platoons execute mostly COIN operations. Small-unit actions in a COIN environment often have more impact than similar actions during major combat operations. Engagements are often won or lost in moments; whoever can bring combat power to bear in seconds wins. The on-scene leader controls the fight. This situation requires mission command and subordinates' initiative. Train leaders at the lowest echelons to act intelligently and independently.

A-18. Training should focus on basic skills: marksmanship, patrolling, security on the move and at the halt, and basic drills. When in doubt, spend less time on company and platoon training and more time on squads. Ruthlessly replace ineffective leaders. Once trained, give Soldiers and Marines a clear commander's intent and trust them to exercise subordinates' initiative within it. This allows subordinates to execute COIN operations at the level at which they are won.

IDENTIFY AND USE TALENT

A-19. Not everyone is good at counterinsurgency. Many leaders do not understand it, and some who do cannot execute it. COIN operations are difficult and anyone can learn the basics. However, people able to intuitively grasp, master, and execute COIN techniques are rare. Learn how to spot these people and put them into positions where they can make a difference. Rank may not indicate the required talent. In COIN operations, a few good Soldiers and Marines under a smart junior noncommissioned officer doing the right things can succeed, while a large force doing the wrong things will fail.

CONTINUE TO ASSESS AND PLAN DURING PREPARATION: BE FLEXIBLE

A-20. *Commander's visualization* is the mental process of developing situational understanding, determining a desired end state, and envisioning how the force will achieve that end state (FMI 5-0.1). It begins with mission receipt and continues throughout any operation. The commander's visualization forms the basis for conducting (planning, preparing for, executing, and assessing) an operation.

A-21. Commanders continually refine their visualization based on their assessment of the operational environment. They describe and direct any changes they want made as the changes are needed. They do not wait for a set point in any process. This flexibility is essential during preparation for COIN operations. Some are tempted to try to finalize a plan too early. They then prepare to execute the plan rather than what changes in the operational environment require. However, as commanders gain knowledge, their situational understanding improves. They get a better idea of what to do and of their own limitations. This lets them refine their visualization and direct changes to the plan and their preparations. Even with this, any plan will change once operations begin. If there is a major shift in the environment, commanders may need to scrap the plan. However, a plan is still needed. Developing it gives leaders a simple robust idea of what to achieve, even if the methods change. Directing changes to it based on continuous assessment is one aspect of the art of command.

A-22. One planning approach is to identify phases of the operation in terms of major objectives to achieve such as establishing dominance, building local networks, and marginalizing the enemy. Make sure forces can easily transition between phases, both forward to exploit successes and backward to recover from setbacks. Insurgents can adapt their activity to friendly tactics. The plan must be simple enough to survive setbacks without collapsing. This plan is the solution that began with the shared analysis and consensus that began preparation. It must be simple and known to everyone.

EXECUTE

A-23. *Execute* means to put a plan into action by applying combat power to accomplish the mission and using situational understanding to assess progress and make execution and adjustment decisions (FM 6-0). The execution of COIN operations demands all the skills required to execute conventional operations. In addition, it also requires mastery of building alliances and personal relationships, attention to the local and global media, and additional skills that are not as heavily tasked in conventional operations.

ESTABLISH AND MAINTAIN A PRESENCE

A-24. The first rule of COIN operations is to establish the force's presence in the AO. If Soldiers and Marines are not present when an incident happens, they usually cannot do much about it. The force cannot be everywhere at once. The more time Soldiers and Marines spend in the AO, the more likely they are where the action is. If the force is not large enough to establish a presence throughout the AO, then determine the most important places and focus on them. This requires living in the AO close to the populace. Raiding from remote, secure bases does not work. Movement on foot, sleeping in villages, and night patrolling all seem more dangerous than they are—and they are what ground forces are trained to do. Being on the ground establishes links with the local people. They begin to see Soldiers and Marines as real people they can trust and do business with, rather than as aliens who descended from armored boxes. Driving around in an armored convoy actually degrades situational awareness. It makes Soldiers and Marines targets and is ultimately more dangerous than moving on foot and remaining close to the populace.

ASSESS DURING EXECUTION: AVOID HASTY ACTIONS

A-25. Do not act rashly; get the facts first. Continuous assessment, important during all operations, is vital during COIN operations. Violence can indicate several things. It may be part of the insurgent strategy, interest groups fighting among themselves, or individuals settling vendettas. Or, it may just be daily life. Take the time to learn what normalcy looks like. Insurgents may try to goad Soldiers and Marines into lashing out at the local populace or making a similar mistake. Unless leaders are on the spot when an incident occurs, they receive only second-hand reports and may misunderstand the local context or interpretation. This means that first impressions are often highly misleading, particularly in urban areas. Of course, leaders cannot avoid making judgments. When there is time, ask an older hand or trusted local people for their opinions. If possible, keep one or two officers from your predecessor unit for the first part of the tour. Avoid rushing to judgment.

BUILD TRUSTED NETWORKS

A-26. Once the unit settles into the AO, its next task is to build trusted networks. This is the true meaning of the phrase "hearts and minds," which comprises two separate components. "Hearts" means persuading people that their best interests are served by COIN success. "Minds" means convincing them that the force can protect them and that resisting it is pointless. Note that neither concerns whether people like Soldiers and Marines. Calculated self-interest, not emotion, is what counts. Over time, successful trusted networks grow like roots into the populace. They displace enemy networks, which forces enemies into the open, letting military forces seize the initiative and destroy the insurgents.

A-27. Trusted networks are diverse. They include local allies, community leaders, and local security forces. Networks should also include nongovernmental organizations (NGOs), other friendly or neutral nonstate actors in the AO, and the media.

A-28. Building trusted networks begins with conducting village and neighborhood surveys to identify community needs. Then follow through to meet them, build common interests, and mobilize popular support. This is the true main effort; everything else is secondary. Actions that help build trusted networks support the COIN effort. Actions that undermine trust or disrupt these networks—even those that provide a short-term military advantage—help the enemy.

GO WITH THE GRAIN AND SEEK EARLY VICTORIES

A-29. Do not try to crack the hardest nut first. Do not go straight for the main insurgent stronghold or try to take on villages that support insurgents. Instead, start from secure areas and work gradually outwards. Extend influence through the local people's networks. Go with, not against, the grain of the local populace. First, win the confidence of a few villages, and then work with those with whom they trade, intermarry, or do business. This tactic develops local allies, a mobilized populace, and trusted networks.

A-30. Seek a victory early in the operation to demonstrate dominance of the AO. This may not be a combat victory. Early combat without an accurate situational understanding may create unnecessary collateral damage and ill will. Instead, victories may involve resolving a long-standing issue or co-opting a key local leader. Achieving even a small early victory can set the tone for the tour and help commanders seize the initiative.

PRACTICE DETERRENT PATROLLING

A-31. Establish patrolling tactics that deter enemy attacks. An approach using combat patrols to provoke, then defeat, enemy attacks is counterproductive. It leads to a raiding mindset, or worse, a bunker mentality. Deterrent patrolling is a better approach. It keeps the enemy off balance and the local populace reassured. Constant, unpredictable activity over time deters attacks and creates a more secure environment. Accomplishing this requires one- to two-thirds of the force to be on patrol at any time, day or night.

BE PREPARED FOR SETBACKS

A-32. Setbacks are normal in counterinsurgencies, as in all operations. Leaders make mistakes and lose people. Soldiers and Marines occasionally kill or detain the wrong person. It may not be possible to build or expand trusted networks. If this happens, drop back to the previous phase of the plan, recover, and resume operations. It is normal in company-level COIN operations for some platoons to do well while others do badly. This situation is not necessarily evidence of failure. Give subordinate leaders the freedom to adjust their posture to local conditions. This creates flexibility that helps survive setbacks.

REMEMBER THE GLOBAL AUDIENCE

A-33. The omnipresence and global reach of today's news media affects the conduct of military operations more than ever before. Satellite receivers are common, even in developing countries. Bloggers and print, radio, and television reporters monitor and comment on everything military forces do. Insurgents use terrorist tactics to produce graphics that they hope will influence public opinion—both locally and globally.

A-34. Train Soldiers and Marines to consider how the global audience might perceive their actions. Soldiers and Marines should assume that the media will publicize everything they say or do. Also, treat the media as an ally. Help reporters get their story. That helps them portray military actions favorably. Trade information with media representatives. Good relationships with nonembedded media, especially HN media, can dramatically increase situational awareness.

ENGAGE THE WOMEN; BE CAUTIOUS AROUND THE CHILDREN

A-35. Most insurgent fighters are men. However, in traditional societies, women are hugely influential in forming the social networks that insurgents use for support. When women support COIN efforts, families support COIN efforts. Getting the support of families is a big step toward mobilizing the local populace against the insurgency. Co-opting neutral or friendly women through targeted social and economic programs builds networks of enlightened self-interest that eventually undermine insurgents. Female counterinsurgents, including interagency people, are required to do this effectively.

A-36. Conversely, be cautious about allowing Soldiers and Marines to fraternize with local children. Homesick troops want to drop their guard with kids. But insurgents are watching. They notice any friendships between troops and children. They may either harm the children as punishment or use them as agents. It requires discipline to keep the children at arm's length while maintaining the empathy needed to win local support.

ASSESS DURING EXECUTION

A-37. Develop measures of effectiveness early and continuously refine them as the operation progresses. These measures should cover a range of social, informational, military, and economic issues. Use them to develop an in-depth operational picture. See how the operation is changing, not just that it is starting or ending. Typical measures of effectiveness include the following:

- Percentage of engagements initiated by friendly forces versus those initiated by insurgents.
- Longevity of friendly local leaders in positions of authority.
- Number and quality of tips on insurgent activity that originate spontaneously.
- Economic activity at markets and shops.

These mean virtually nothing as a snapshot; trends over time indicate the true progress.

A-38. Avoid using body counts as a measure of effectiveness. They actually measure very little and may provide misleading numbers. Using body counts to measure effectiveness accurately requires answers to the following questions:

- How many insurgents were there at the start?
- How many insurgents have moved into the area?
- How many insurgents have transferred from supporter to combatant status?
- How many new fighters has the conflict created?

Accurate information of this sort is usually not available.

MAINTAIN MISSION FOCUS THROUGHOUT

A-39. Once a unit is established in its AO, Soldiers and Marines settle into a routine. A routine is good as long as the mission is being accomplished. However, leaders should be alert for the complacency that often accompanies routines.

A-40. It often takes Soldiers and Marines at least one-third of the tour to become effective. Toward the tour's end, leaders struggle against the "short-timer" mentality. Thus, the middle part of the tour is often the most productive. However, leaders must work to keep Soldiers and Marines focused on the mission and attentive to the environment.

EXPLOIT A SINGLE NARRATIVE

A-41. Since counterinsurgency is a competition to mobilize popular support, it pays to know how people are mobilized. Most societies include opinion-makers—local leaders, religious figures, media personalities, and others who set trends and influence public perceptions. This influence often follows a single narrative—a simple, unifying, easily expressed story or explanation that organizes people's experience—and provides a framework for understanding events. Nationalist and ethnic historical myths and sectarian creeds are examples of such narratives. Insurgents often try to use the local narrative to support their cause. Undercutting their influence requires exploiting an alternative narrative. An even better approach is tapping into an existing narrative that excludes insurgents.

A-42. Higher headquarters usually establishes the COIN narrative. However, only leaders, Soldiers, and Marines at the lowest levels know the details needed to tailor it to local conditions and generate leverage from it. For example, a nationalist narrative can be used to marginalize foreign fighters. A narrative of national redemption can undermine former regime elements seeking to regain power. Company-level leaders apply the narrative gradually. They get to know local opinion makers, win their trust, and learn what motivates them. Then they build on this knowledge to find a single narrative that emphasizes the inevitability and rightness of the COIN operation's success. This is art, not science.

HAVE LOCAL FORCES MIRROR THE ENEMY, NOT U.S. FORCES

A-43. By mid-tour, U.S. forces should be working closely with local forces, training or supporting them and building an indigenous security capability. The natural tendency is to create forces in a U.S. image. This is a mistake. Instead, local HN forces need to mirror the enemy's capabilities and seek to supplant the insurgent's role. This does not mean they should be irregular in the sense of being brutal or outside proper control. Rather, they should move, equip, and organize like insurgents but have access to U.S. support and be under the firm control of their parent societies. Combined with a mobilized populace and trusted networks, these characteristics allow HN forces to separate the insurgents from the population.

A-44. U.S. forces should support HN forces. At the company level, this means raising, training, and employing local HN auxiliary forces (police and military). These tasks require high-level clearance, but if permission is given, companies should each establish a training cell. Platoons should aim to train one local squad and then use that squad as a nucleus for a partner platoon. The company headquarters should train an HN leadership team. This process mirrors the development of trusted networks. It tends to emerge naturally with the emergence of local allies willing to take up arms to defend themselves.

CONDUCT CIVIL-MILITARY OPERATIONS

A-45. COIN operations can be characterized as armed social work. It includes attempts to redress basic social and political problems while being shot at. This makes CMO a central COIN activity, not an afterthought. Civil-military operations are one means of restructuring the environment to displace the enemy from it. They must focus on meeting basic needs first. A series of village or neighborhood surveys, regularly updated, are invaluable to understanding what the populace needs and tracking progress in meeting them.

A-46. Effective CMO require close cooperation with national, international, and local interagency partners. These partners are not under military control. Many NGOs, for example, do not want to be too closely associated with military forces because they need to preserve their perceived neutrality. Interagency cooperation may involve a shared analysis of the problem, building a consensus that allows synchronization of military and interagency efforts. The military's role is to provide protection, identify needs, facilitate CMO, and use improvements in social conditions as leverage to build networks and mobilize the populace.

A-47. There is no such thing as impartial humanitarian assistance or CMO in COIN. Whenever someone is helped, someone else is hurt, not least the insurgents. So civil and humanitarian assistance personnel often become targets. Protecting them is a matter not only of providing a close-in defense, but also of creating a secure environment by co-opting local beneficiaries of aid and their leaders.

REMEMBER SMALL IS BEAUTIFUL

A-48. Another tendency is to attempt large-scale, mass programs. In particular, Soldiers and Marines tend to apply ideas that succeed in one area to another area. They also try to take successful small programs and replicate them on a larger scale. This usually does not work. Often small-scale programs succeed because of local conditions or because their size kept them below the enemy's notice and helped them flourish unharmed. Company-level programs that succeed in one AO often succeed in another; however, small-scale projects rarely proceed smoothly into large programs. Keep programs small. This makes them cheap, sustainable, low-key, and (importantly) recoverable if they fail. Leaders can add new programs—also small, cheap, and tailored to local conditions—as the situation allows.

FIGHT THE ENEMY'S STRATEGY

A-49. When COIN efforts succeed, insurgents often transition to the offensive. COIN successes create a situation that threatens to separate insurgents from the populace. Insurgents attack military forces and the local populace to reassert their presence and continue the insurgency. This activity does not necessarily indicate an error in COIN tactics (though it may, depending on whether insurgents successfully mobilized the population). It is normal, even in the most successful operations, to have spikes of offensive insurgent activity.

A-50. The obvious military response is a counteroffensive to destroy enemy's forces. This is rarely the best choice at company level. Only attack insurgents when they get in the way. Try not to be distracted or forced into a series of reactive moves by a desire to kill or capture them. Provoking combat usually plays into the enemy's hands by undermining the population's confidence. Instead, attack the enemy's strategy. If insurgents are seeking to recapture a community's allegiance, co-opt that group against them. If they are trying to provoke a sectarian conflict, transition to peace enforcement operations. The possible situations are endless, but the same principle governs the response: fight the enemy's strategy, not enemy forces.

ASSESS DURING EXECUTION: RECOGNIZE AND EXPLOIT SUCCESS

A-51. Implement the plan developed early in the campaign and refined through interaction with local partners. Focus on the environment, not the enemy. Aim at dominating the whole district and implementing solutions to its systemic problems. Continuously assess results and adjust as needed.

A-52. Achieving success means that, particularly late in the campaign, it may be necessary to negotiate with the enemy. Local people supporting the COIN operation know the enemy's leaders. They even may have grown up together. Valid negotiating partners sometimes emerge as the campaign progresses. Again, use close interagency relationships to exploit opportunities to co-opt segments of the enemy. This helps wind down the insurgency without alienating potential local allies who have relatives or friends among insurgents. As an insurgency ends, a defection is better than a surrender, a surrender better than a capture, and a capture better than a kill.

PREPARE DURING EXECUTION: GET READY FOR HANDOVER FROM DAY ONE

A-53. It is unlikely the insurgency will end during a troop's tour. There will be a relief in place, and the relieving unit will need as much knowledge as can be passed to them. Start handover folders in every platoon and specialist squad immediately upon arrival, if they are not available from the unit being relieved. The folders should include lessons learned, details about the populace, village and patrol reports, updated maps, and photographs—anything that will help newcomers master the environment. Computerized databases are fine. Keep good back-ups and ensure a hard copy of key artifacts and documents exists. Developing and keeping this information current is boring, tedious work. But it is essential to both short- and long-term success. The corporate memory this develops gives Soldiers and Marines the knowledge they need to stay alive. Passing it on to the relieving unit does the same for them. It also reduces the loss of momentum that occurs during any handover.

ENDING THE TOUR

A-54. As the end of the tour approaches, the key leadership challenge becomes keeping the Soldiers and Marines focused. They must not drop their guard. They must continue to monitor and execute the many programs, projects, and operations.

A-55. The tactics discussed above remain applicable as the end-of-tour transition approaches. However, there is an important new one: keep the transition plan secret. The temptation to talk about home becomes almost unbearable toward the end of a tour. The local people know that Soldiers and Marines are leaving and probably have a good idea of the generic transition plan. They have seen units come and go. But details of the transition plan must be protected; otherwise, the enemy might use the handover to undermine any progress made during the tour. Insurgents may stage a high-profile attack. They may try to recapture the populace's allegiance by scare tactics. Insurgents will try to convince the local populace that military forces will not protect them after the transition. Insurgents may try to persuade the local populace that the successor unit will be oppressive or incompetent. Set the follow-on unit up for success. Keep the transition plan details secret within a tightly controlled compartment in the headquarters. Tell the Soldiers and Marines to resist the temptation to say goodbye to local allies. They can always send a postcard from home.

THREE "WHAT IFS"

A-56. The discussion above describes what should happen, but things do go wrong. Here are some "what ifs" to consider.

WHAT IF YOU GET MOVED TO A DIFFERENT AREA?

A-57. Efforts made preparing for operations in one AO are not wasted if a unit is moved to another area. In mastering the first area, Soldiers and Marines learned techniques applicable to the new one. For example, they know how to analyze an AO and decide what matters in the local society. The experience provides a mental structure for analyzing the new AO. Soldiers and Marines can focus on what is different, making the process easier and faster. They need to apply this same skill when they are moved within battalion or brigade AOs.

WHAT IF YOU HAVE NO RESOURCES?

A-58. Things can be things done in a low-priority AO. However, commanders need to focus on self-reliance, keeping things small and sustainable and ruthlessly prioritizing efforts. Local leaders can help. They know what matters to them. Commanders should be honest with them, discuss possible projects and options, and ask them to recommend priorities. Often commanders can find translators, building supplies, or expertise. They may only expect support and protection in making their projects work. Negotiation and consultation can help mobilize their support and strengthen social cohesion. Setting achievable goals is key to making the situation work.

WHAT IF THE THEATER SITUATION SHIFTS?

A-59. Sometimes everything goes well at the tactical level, but the theater situation changes and invalidates those efforts. When that happens, drop back a stage, consolidate, regain balance, and prepare to expand again when the situation allows. A flexible, adaptive plan helps in such situations. Friendly forces may have to cede the initiative for a time; however, they must regain it as soon as the situation allows.

SUMMARY

A-60. This appendix has summarized one set of tactics for conducting COIN operations. Like all tactics, they need interpretation. Constant study of the AO is needed to apply them to the specific circumstances a unit faces. Observations and experience helps Soldiers and Marines apply them better. Whatever else is done, the focus must remain on gaining and maintaining the support of the population. With their support, victory is assured; without it, COIN efforts cannot succeed.

This page intentionally left blank.

Appendix B

Social Network Analysis and Other Analytical Tools

Situational understanding involves determining the relationships among the factors of METT-TC. This appendix discusses several tools used to describe the effects of the operational environment and evaluate the threat. One of the most important of these is social network analysis, a powerful threat evaluation tool. Commanders and staffs use these tools to help them understand the operational environment. This understanding facilitates making decisions and developing plans and orders.

DESCRIBE THE EFFECTS OF THE OPERATIONAL ENVIRONMENT

B-1. Describing the effects of the operational environment requires an analysis of the terrain, weather, and civil considerations. This discussion addresses terrain and civil considerations at length because of their importance in a counterinsurgency (COIN) environment. Terrain and civil considerations have distinct aspects in COIN that analysts must understand to effectively describe the operational environment. Imagery, geospatial analysis tools, overlays, and graphics can help depict these aspects.

DESCRIBE TERRAIN EFFECTS

B-2. As in conventional operations, terrain analysis in COIN includes examining the terrain's effects on the movement of military units and enemy personnel. However, because COIN focuses on people, terrain analysis usually centers on populated areas and the effects of terrain on the local populace. During COIN operations, Soldiers and Marines spend a lot of time in suburban and urban areas interacting with the populace. This battlefield is three dimensional. Multistory buildings and underground lines of communications, such as tunnels and sewers, can be very important. Insurgents also use complex natural terrain to their advantage as well. Mountains, caves, jungles, forests, swamps, and other complex terrain are potential bases of operations for insurgents. (See FM 34-130 for additional information on terrain analysis. See FMI 2-91.4 for terrain analysis in urban operations.)

B-3. An important terrain consideration in COIN is urban and suburban land navigation. This can be difficult in areas without an address system and in cities where 10-digit grids may not be accurate enough to locate a specific apartment. Knowledge of how local people find one another's houses and what type of address system they use are beneficial. Recent, accurate maps that use overhead imagery are also helpful. In addition, tourist maps and locally produced maps facilitate understanding the names local people use to describe places.

Military Aspects of Terrain for Counterinsurgency

B-4. At the tactical level, Soldiers and Marines consider different details of the military aspects of terrain to describe the operational environment.

- **Observation and fields of fire.** In COIN operations, Soldiers and Marines look for areas with good fields of fire that may serve as ambush points. In addition, they also consider different ways insurgents might observe them. These ways include surveillance, the use of spies and infiltrators, and locations with line of sight on counterinsurgent positions.
- **Avenues of approach.** Insurgents use any means possible to get into counterinsurgent installations. Possible entry points include sewers, rooftops, roads, and sidewalks. Insurgents exploit their ability to blend with the populace. They may try to infiltrate by posing as contractors working for counterinsurgents or the host-nation (HN) government. Along border regions, insurgents

may use smuggling routes to move people and materiel in and out of the country. An additional avenue of approach to consider is how insurgents influence public opinion.

- **Key terrain.** Tactically, key terrain may be important structures, economically and politically important areas, areas of religious or cultural significance, access control points, and lines of communications.
- **Obstacles.** In addition to terrain obstacles, obstacles in a COIN environment include anything that hinders insurgent freedom of operation or counterinsurgent operations. Traffic control points, electronic security systems, and guard plans are examples of obstacles to insurgents. Use of places protected under rules of engagement, translators, the ability to communicate with the populace, culture, and politics may all be obstacles for U.S. and HN government forces.
- **Cover and concealment.** In COIN, cover and concealment includes the means by which insurgents hide themselves and their activities. These include using disguises and false identification badges, and hiding supplies underground or in buildings.

Geospatial Intelligence

B-5. Geospatial intelligence (GEOINT) is the intelligence derived from the exploitation, analysis, and fusion of imagery with geospatial information to describe, assess, and visually depict physical features and geographically referenced activities in an area of operations (AO). GEOINT consists of imagery, imagery intelligence, and geospatial information. Geospatial information and services remains a core mission of the engineer branch and provides the foundation for GEOINT. Imagery intelligence remains a core mission of the military intelligence branch and provides the intelligence layers and analytic fusion for GEOINT. The result is digitally integrated intelligence products that support all-source analysis, planning, decision making and support to current operations.

Geospatial Tools

B-6. Geospatial products (tools) that can be provided by the geospatial information and services team include the following:

- Terrain databases.
- Special terrain studies and products prepared by U.S. or HN agencies, and special maps, charts, and geodetic studies.
- Current photography.
- Real-time terrain reconnaissance.
- Terrain factor matrices.

Imagery

B-7. Imagery products include both aerial photography and satellite imagery. In many cases, aerial reconnaissance platforms, such as unmanned aircraft systems, respond directly to commanders. This practice aids timely, focused data collection. Each collection system has its own capabilities. The situation determines whether black and white or infrared imagery offers the better view of a target. (Figure B-1 shows an example of an imagery product.)

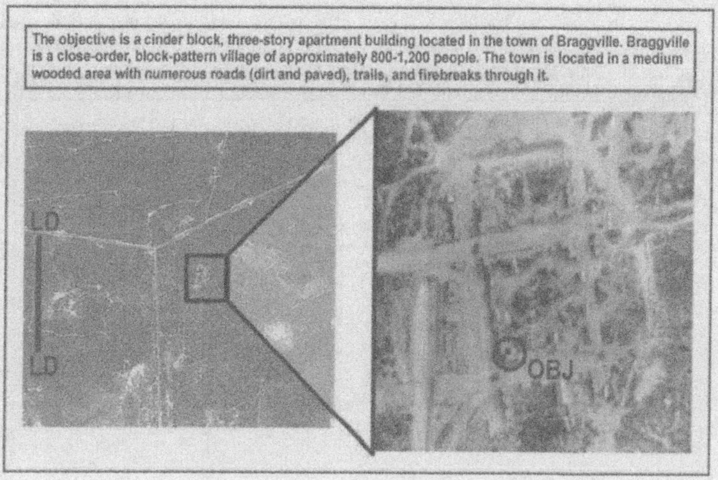

Figure B-1. Example imagery photograph

B-8. A key element in future operations may be the imagery downlink capabilities of space-based intelligence collection platforms. Space-based systems use state-of-the-art spectral, infrared, electro-optical, and synthetic aperture radar imaging. They can provide important information. Data collected from such sources is transferred in a digital format that can be manipulated to address specific requirements. Intelligence staffs remain aware of the capabilities and limitations of these systems and the procedures for requesting this support.

B-9. Advanced GEOINT products are produced using any combination of imaging platforms—visible, infrared, radar, or spectral—depending on requestor needs. These products have many applications. Presenting imagery in an oblique perspective by combining it with digital terrain elevation data provides a perspective view. Spectral imagery uses heat distribution patterns and changes in a scene imaged at various times to discover and distinguish manmade from indigenous activity. Other uses include facility analysis, structural analysis, target detection, soil analysis, and damage assessment.

DESCRIBE CIVIL CONSIDERATIONS (ASCOPE)

B-10. *Civil considerations* concern the manmade infrastructure, civilian institutions, and attitudes and activities of the civilian leaders, populations, and organizations within an area of operations influence the conduct of military operations (FM 6-0). Because the purpose of COIN is to support a HN government in gaining legitimacy and the support of the populace, civil considerations are often the most important factors to consider during mission analysis.

B-11. Civil considerations generally focus on the immediate impact of civilians on operations in progress. However, at higher levels, they also include larger, long-term diplomatic, informational, and economic issues. At the tactical level, civil considerations directly relate to key civilian areas, structures, capabilities, organizations, people, and events within the AO. These characteristics are represented by the memory aid ASCOPE. Socio-cultural factors analysis, discussed in paragraphs 3-19 through 3-73, provides a more in-depth evaluation of civil considerations.

Areas

B-12. Key civilian areas are localities or aspects of the terrain within an AO that have significance to the local populace. This characteristic approaches terrain analysis from a civilian perspective. Commanders analyze key civilian areas in terms of how they affect the missions of individual units as well as how military operations affect these areas. (Table B-1 lists examples of key civilian areas.)

Table B-1. Examples of key civilian areas

- Areas defined by political boundaries, such as—
  - Districts or neighborhoods within a city.
  - Municipalities within a region.
  - Provinces within a country.
- Areas of high economic value, such as industrial centers, farming regions, and mines.
- Centers of government and politics.
- Culturally important areas.
- Social, ethnic, tribal, political, religious, criminal, or other important enclaves.
- Trade routes and smuggling routes.
- Possible sites for the temporary settlement of dislocated civilians or other civil functions.

Structures

B-13. Analyzing a structure involves determining how its location, functions, and capabilities can support operations. Commanders also consider the consequences of using it. Using a structure for military purposes often competes with civilian requirements. Commanders carefully weigh the expected military benefits against costs to the community that will have to be addressed in the future. (Table B-2 lists examples of important structures in an AO.)

Capabilities

B-14. Capabilities can refer to the ability of local authorities—those of the host nation or some other body—to provide a populace with key functions or services. Commanders and staffs analyze capabilities from different perspectives but generally put priority on understanding the capability of the HN government to support the mission. The most essential capabilities are those required to save, sustain, or enhance life, in that order. Some of the more important capabilities are—

- Public administration—effectiveness of bureaucracy, courts, and other parts of the HN government.
- Public safety—provided by the security forces and military, police, and intelligence organizations.
- Emergency services—such agencies as fire departments and ambulance services.
- Public health—clinics and hospitals.
- Food.
- Water.
- Sanitation.

Organizations

B-15. Organizations are nonmilitary groups or institutions in the AO. They influence and interact with the populace, counterinsurgents, and each other. They generally have a hierarchical structure, defined goals, established operations, fixed facilities or meeting places, and a means of financial or logistic support. Some organizations may be indigenous to the area. These may include—

- Religious organizations.
- Political parties.
- Patriotic or service organizations.

- Labor unions.
- Criminal organizations.
- Community organizations.

B-16. Other organizations may come from outside the AO. Examples of these include—

- Multinational corporations.
- Intergovernmental organizations (IGOs), such as United Nations agencies.
- Nongovernmental organizations (NGOs), such as the International Red Cross.

B-17. Operations often require commanders to coordinate with IGOs and NGOs. Information required for evaluation includes these groups' activities, capabilities, and limitations. Situational understanding includes knowing how the activities of different organizations may affect military operations and how military operations may affect these organizations' activities. From this analysis, commanders can determine how organizations and military forces can work together toward common goals.

Table B-2. Examples of important structures

- **Government centers**—necessary for the government to function.
- **Headquarters and bases for security forces**—necessary for security forces to function.
- **Police stations, courthouses, and jails**—necessary for countering crime and beneficial for counterinsurgency operations.
- **Communications and media infrastructure**—important to information flow and the opinions of the populace. These include the following:
  - **Radio towers.**
  - **Television stations.**
  - **Cellular towers.**
  - **Newspaper offices.**
  - **Printing presses.**
- **Roads and bridges**—allow for movement of people, goods, insurgents, and counterinsurgents.
- **Ports of entry, such as airports and seaports**—allow for movement of people, goods, insurgents, and counterinsurgents.
- **Dams**—provide electric power, drinking water, and flood control.
- **Electrical power stations and substations**—enable the economy's functioning and often important for day-to-day life of the populace.
- **Refineries and other sources of fuel**—enable the economy's functioning and often important for day-to-day life of the populace.
- **Sources of potable water**—important for public health.
- **Sewage systems**—important for public health.
- **Clinics and hospitals**—important for the health of the populace; these often are protected sites.
- **Schools and universities**—affect the opinions of the populace; these often are protected sites.
- **Places of religious worship**—affect opinions of the populace; often of great cultural importance; these often are protected sites.

B-18. In almost every case, military forces have more resources than civilian organizations. However, some civilian organizations possess specialized capabilities that they may be willing to share. Commanders do not command civilian organizations in their AOs. However some operations require achieving unity of effort with these groups. These situations require commanders to influence the leaders of these organizations through persuasion, relying on the force of argument and the example of actions.

People

B-19. *People* refers to nonmilitary personnel encountered by military forces. The term includes all civilians within an AO (the populace) as well as those outside the AO whose actions, opinions, or political influence can affect the mission. To display different aspects of the populace, analysts can use population support overlays and religion, race, and ethnicity overlays. (FMI 2-91.4 contains information about these overlays.) Perception is another significant people factor in COIN. The perception assessment matrix is a tool that compares the intent of friendly operations to the populace's perception of those operations.

Population Support Overlay

B-20. The population support overlay can graphically depict the sectors of the populace that are progovernment, antigovernment, proinsurgent, anti-insurgent, uncommitted, and neutral. (See figure B-2.) These overlays are important because they help analysts determine whether the local populace is likely support the HN government or support the insurgency.

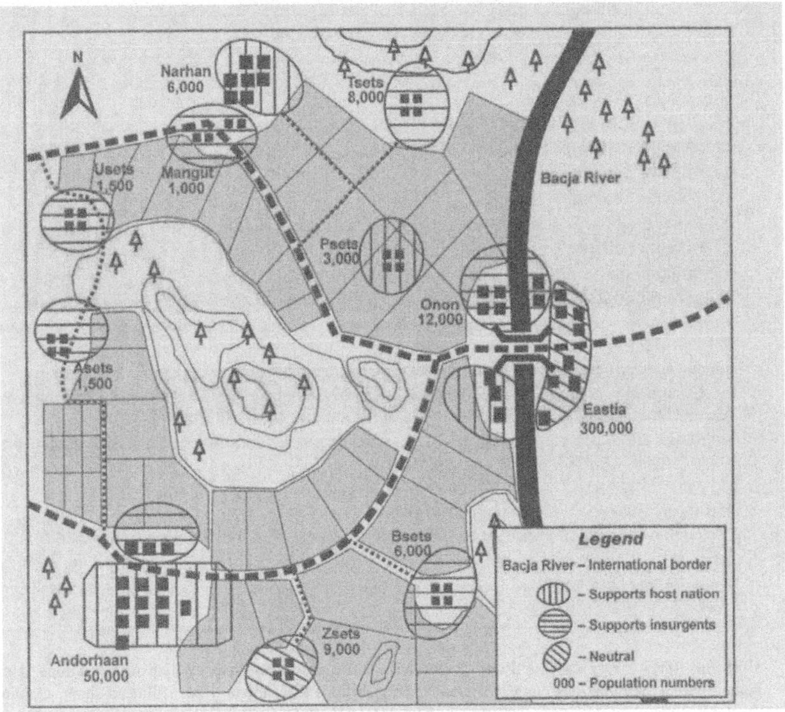

Figure B-2. Example population support overlay

Religion, Race, and Ethnicity Overlay

B-21. Religion, race, and ethnicity issues often contribute to conflicts. Religious, race, and ethnicity overlays depict the current ethnic and religious make-up of an AO. These overlays can also display any specific religious-, racial-, or ethnicity-specific areas and any zones of separation agreed upon by peace accords. These three overlays may be separate or combined. (Figure B-3 shows an example of an ethnicity overlay.)

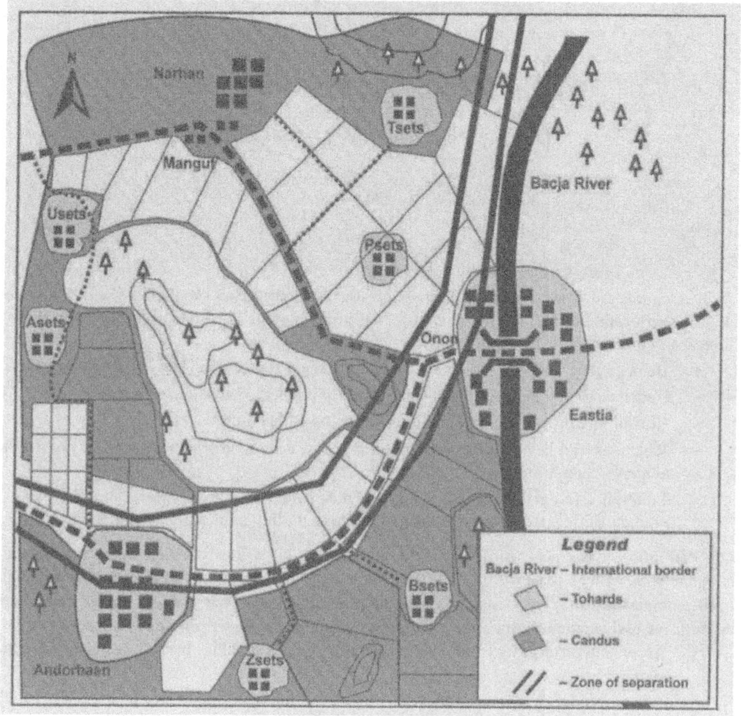

Figure B-3. Example ethnicity overlay

Perception Assessment Matrix

B-22. Perceptions influence how insurgents are targeted and engaged. Important considerations include how insurgents perceive counterinsurgents, themselves, their environment, the nature and reasons for the conflict, and their success criteria. Perception is complicated but key to successfully targeting, engaging, and evaluating success. In-depth knowledge and understanding of the national, regional, and local cultures, norms, moralities, and taboos are needed to understand the operational environment and reactions of the insurgents and populace.

B-23. Perception assessment matrices are often used by psychological operations personnel and other staff elements and can be a valuable tool for intelligence analysts. (See figure B-4 [page B-9].) Counterinsurgent activities intended to be benign or benevolent might have negative results if the populace's perceptions are not considered, and then evaluated or measured. This is true because perceptions—more than reality—drive a commander's decision making and can influence the populace's reactions. A perception assessment matrix displays how well counterinsurgents are able to achieve an effect during an operation. In this sense, the matrix can be used to directly display the effectiveness of the unit's civil affairs, public affairs, and

The Importance of Perceptions in Operation Uphold Democracy

One proposed psychological operations action developed for Operation Uphold Democracy in Haiti in 1994 illustrates why perception assessment is necessary. Before deployment, leaflets were prepared informing the Haitian populace of U.S. intentions. The original leaflet was printed in Dutch, the language of the Haitian elite. However, the one actually used was published in Creole, the official language of Haiti, because an astute team member realized the need to publish to the wider audience.

If a flier in Dutch had been dropped, it could have undermined the American mission to the country in several ways. The majority of the population would have been unable to read the flier. The subsequent deployment of U.S. forces into the country, therefore, could have been perceived as hostile. The mission, which was intended in part to restore equity within Haiti's social structure, could have backfired if the Haitians viewed the Dutch flier as an indication of U.S. favoritism toward the Haitian elite.

B-24. Perception can work against operational objectives. Perceptions should therefore be assessed before and during an operation. It is not possible to read the minds of the local populace; however, there are several means of measuring its perceptions. These include the following:

- Demographic analysis and cultural intelligence are key components of perception analysis.
- Understanding a population's history can help predict expectations and reactions.
- Human intelligence can provide information on perceptions.
- Reactions and key activities can be observed to determine whether people act based on real or perceived conditions.
- Editorial and opinion pieces of relevant newspapers can be monitored for changes in tone or opinion shifts that can steer, or may be reacting to, the opinions of a population group.

B-25. Perception assessment matrices aim to measure the disparities between friendly force actions and what population groups perceive. In addition to assessing the perceptions of the population groups within an AO, commanders may also want to assess the perceptions that their Soldiers and Marines have of unit activities. Assessing counterinsurgents' perceptions can begin to answer the following questions:

- Are counterinsurgents exhibiting Western or American values that the populace does not appreciate?
- Are embedded American beliefs preventing the unit from understanding the HN population or its multinational partners?
- Is what the intelligence and command staff perceives really what is happening?
- Does the populace believe what the unit believes?
- Is there something that is part of the populace's (or a subgroup's) perception that can be detrimental to the unit?

Condition	Cultural norm	Alternative proposed by COIN force	Population's perception	Acceptable difference in perception?	Root of difference	Possible to change perception?	Proposed solution
Food shortages	Rice	Meat and potatoes	Inadequate/ inconsiderate	No	Culturally accepted norms/ standards; no known physically detrimental effects	No; logistically restricted	Offer potatoes; seek exchange for rice
Use of guns	Criminal elements carry weapons openly	Confiscate all weapons	Unfair/ population not protected by traditional means	No	Culture/ criminal element provides a measure of security for the local populace	No; population and friendly forces at risk	Psychological operations campaign; weapons turn-in program
Government structure	Tribal	Hierarchical	Tolerable as long as needs are fulfilled by group in charge	Yes	History	No	Bargain
Language	Dual languages: Creole and Dutch	Respect all languages	Unfair/show of favoritism	Yes	History/ national language	Yes	Communicate in all languages when possible

Events

B-26. Events are routine, cyclical, planned, or spontaneous activities that significantly affect organizations, people, and military operations. They are often symbols, as described in paragraph 3-51. Examples include the following:

- National and religious holidays.
- Agricultural crop, livestock, and market cycles.
- Elections.
- Civil disturbances.
- Celebrations.

B-27. Other events include disasters from natural, manmade, or technological sources. These create hardships and require emergency responses. Examples of events precipitated by military forces include combat operations, deployments, redeployments, and paydays. Once significant events are determined, it is important to template the events and analyze them for their political, economic, psychological, environmental, and legal implications.

EVALUATE THE THREAT

B-28. Evaluating the threat involves analyzing insurgent organizations, capabilities, and tactics to identify vulnerabilities to exploit. Tools like social network analysis, link diagrams, and association matrices help analysts do this. Other tools such as historical time lines and pattern analysis tools assist in developing event and doctrinal templates to depict enemy tactics.

SOCIAL NETWORK ANALYSIS

B-29. Social network analysis (SNA) is a tool for understanding the organizational dynamics of an insurgency and how best to attack or exploit it. It allows analysts to identify and portray the details of a network structure. Its shows how an insurgency's networked organization behaves and how that connectivity affects its behavior. SNA allows analysts to assess the network's design, how its member may or may not act autonomously, where the leadership resides or how it is distributed among members, and how hierarchical dynamics may mix or not mix with network dynamics.

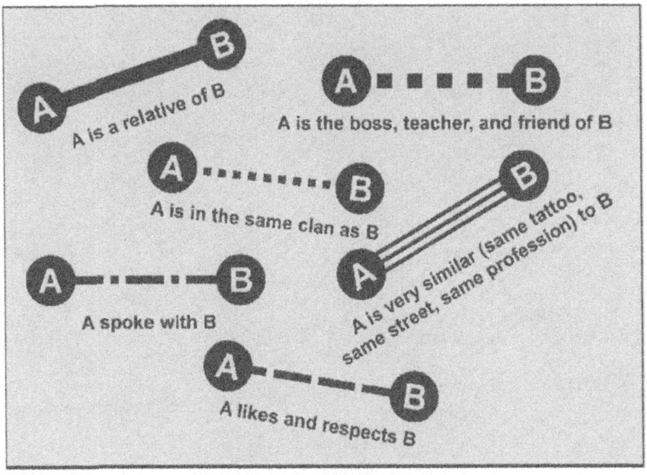

Figure B-5. Examples of dyads

B-30. SNA supports a commander's requirement to describe, estimate, and predict the dynamic structure of an insurgent organization. It also provides commanders a useful tool to gauge their operations' effectiveness. SNA allows analysts assess the insurgency's adaptation to the operational environment and friendly operations.

Social Network Analysis—Terms and Concepts

B-31. The social network graph is the building block of social network analysis. A social network graph consists of individuals and connections between them. Individuals in a network are called *actors* or *nodes*. (Actor and node are often used interchangeably.) The contacts between nodes are called *links*. The basic element of a social network graph is the *dyad*. A dyad consists of two nodes and a single link. In the simplest form of a network, the two nodes represent people and the link represents a relationship between them. (See figure B-5.)

B-32. *Social network measures* allow units to analyze and describe networks. They fall into two categories: organizational-level and individual-level.

Organizational-Level Analysis

B-33. *Organizational-level analysis* provides insight about the insurgent organization's form, efficiency, and cohesion. A regional insurgency may consist of large numbers of disconnected subinsurgencies. As a result, each group should be analyzed based on its capacities as compared to the other groups. Organizational-level capacities can be described in terms of network density, cohesion, efficiency, and core-periphery. Each measure describes a characteristic of a networked organization's structure. Different network structures can support or hinder an organization's capabilities. Therefore, each organizational measure supports the analyst's assessment of subgroup capabilities.

B-34. *Network density* is a general indicator of how connected people are in the network. Network or global-level density is the proportion of ties in a network relative to the total number possible. Comparing network densities between insurgent subgroups provides commanders with an indication of which group is most capable of a coordinated attack and which group is the most difficult to disrupt. (Figure B-6 shows three networks with different densities.)

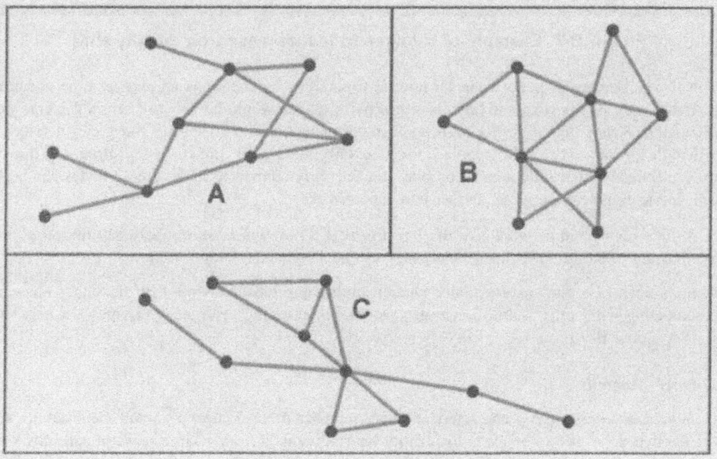

Figure B-6. Comparison of network densities

B-35. Most network measures, including network density, can be mapped out to evaluate performance over time. Based on changes in network density over time, a commander can—

- Monitor enemy capabilities.
- Monitor the effects of recent operations.
- Develop tactics to further fragment the insurgency.

B-36. An increase in network density indicates the likelihood that the insurgent group can conduct coordinated attacks. A decrease in network density means the group is reduced to fragmented or individual-level attacks. (Figure B-7 illustrates an example of how tactics and activities can change based on network density.) A well-executed COIN eventually faces only low-network-density subgroups. This is because high-network-density subgroups require only the capture of one highly connected insurgent to lead counterinsurgents to the rest of the group. So while high-network-density groups are the most dangerous, they are also the easiest to defeat and disrupt.

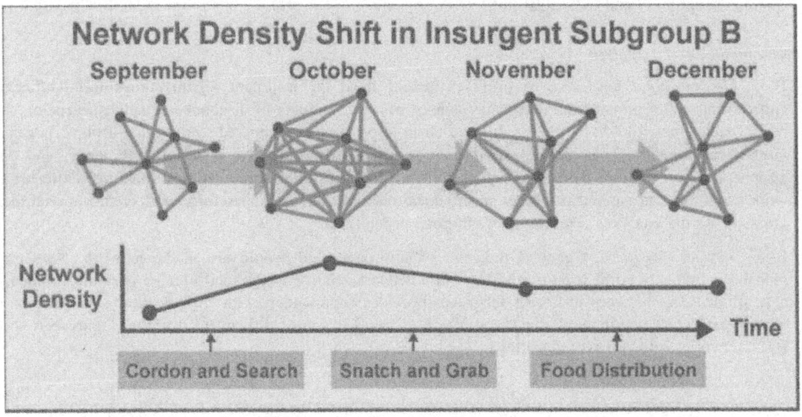

Figure B-7. Example of changes to tactics based on density shift

B-37. Network density does not consider how distributed the connections are between the nodes in a network. Better metrics of group and organizational performance would be network centrality, core-periphery, and diameter. A few nodes with a high number of connections can push up the group network density, even though the majority of the people nodes are only marginally linked to the group. In the case of a highly centralized network dominated by one or a few very connected nodes, these nodes can be removed or damaged to fragment the group further into subnetworks.

B-38. A fully connected network like the one figure B-8 portrays is an unlikely description of the enemy insurgent order of battle. A regional insurgency can be fragmented within itself.

B-39. Sometimes a region may actually contain multiple subinsurgencies that are either unaware of, or even competing with, other subinsurgent groups. In this case, the insurgency resembles a fragmented network. (See figure B-9.)

Individual-Level Analysis

B-40. *Individual-level analysis* characterizes every member of an organization and identifies its key members. Effective SNA allows analysts to identify key individuals from a large mass of data. SNA describes individuals based on their network position in relation to the network position of every other individual in the network. Descriptions are in terms of the following individual-level measures: degree centrality, betweenness centrality, and diameter. Individual network centralities provide insight into an individual's location in the network. The relationship between the centralities of all nodes can reveal much about the overall network structure.

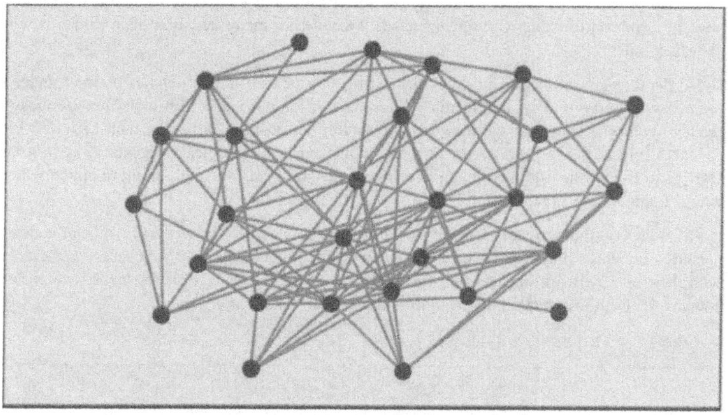

Figure B-8. Networked organization with high connections

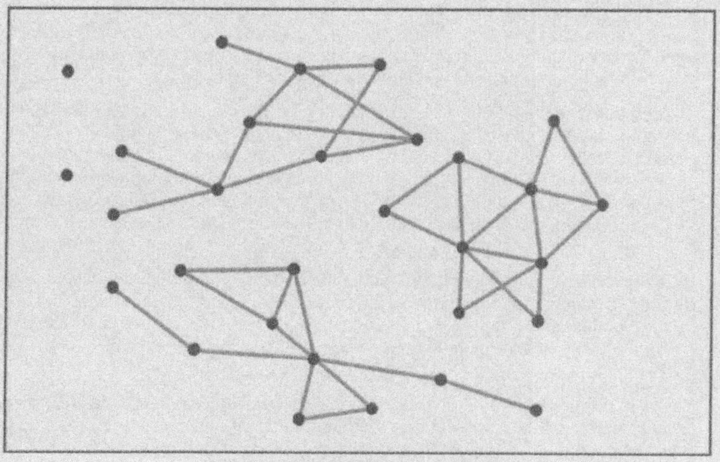

Figure B-9. Fragmented network

B-41. One node or a very few central nodes dominate a very centralized network. If these nodes are removed or damaged, the network may quickly fragment into unconnected subnetworks. *Hubs* are nodes with a very high degree of centrality. A network centralized around a well-connected hub can fail abruptly if that hub is disabled or removed.

B-42. A less centralized network has no single points of failure. It is resilient in the face of many intentional attacks or random failures. Many nodes or links can fail while allowing the remaining nodes to still reach each other over other, redundant network paths.

B-43. *Degree centrality* describes how active an individual is in the network. Network activity for a node is measured using the concept of degrees—the number of direct connections a node has. Nodes with the most direct connections are the most active in their networks. Common wisdom in organizations is "the more connections, the better." This is not always so. What really matters is where those connections lead and how they connect the otherwise unconnected. If a node has many ties, it is often said to be either prominent or influential.

B-44. *Betweenness centrality* indicates the extent to which an individual lies between other individuals in the network, serving as an intermediary, liaison, or bridge. A node with high "betweenness" has great influence over what flows in the network. Depending on position, a person with high betweenness plays a "broker" role in the network. A major opportunity exists for counterinsurgents if, as in group C of figure B-6 (page B-11), the high betweenness centrality person is also a single point of failure which, if removed, would fragment the organization.

B-45. Nodes on the periphery receive very low centrality scores. However, peripheral nodes are often connected to networks that are not currently mapped. The outer nodes may be resource gatherers or individuals with their own network outside their insurgent group. These characteristics make them very important resources for fresh information not available inside their insurgent group.

The Capture of Saddam Hussein

The capture of Saddam Hussein in December 2003 was the result of hard work along with continuous intelligence gathering and analysis. Each day another piece of the puzzle fell into place. Each led to coalition forces identifying and locating more of the key players in the insurgent network—both highly visible ones like Saddam Hussein and the lesser ones who sustained and supported the insurgency. This process produced detailed diagrams that showed the structure of Hussein's personal security apparatus and the relationships among the persons identified.

The intelligence analysts and commanders in the 4th Infantry Division spent the summer of 2003 building link diagrams showing everyone related to Hussein by blood or tribe. Those family diagrams led counterinsurgents to the lower level, but nonetheless highly trusted, relatives and clan members harboring Hussein and helping him move around the countryside. The circle of bodyguards and mid-level military officers, drivers, and gardeners protecting Hussein was described as a "Mafia organization," where access to Hussein controlled relative power within the network.

Over days and months, coalition forces tracked how the enemy operated. Analysts traced trends and patterns, examined enemy tactics, and related enemy tendencies to the names and groups on the tracking charts. This process involved making continual adjustments to the network template and constantly determining which critical data points were missing.

Late in the year, a series of operations produced an abundance of new intelligence about the insurgency and Hussein's whereabouts. Commanders then designed a series of raids to capture key individuals and leaders of the former regime who could lead counterinsurgents to him. Each mission gained additional information, which shaped the next raid. This cycle continued as a number of mid-level leaders of the former regime were caught, eventually leading coalition forces into Hussein's most trusted inner circle and finally to Hussein's capture.

Social Network Analysis and Counterinsurgency

B-46. Figure B-10 shows a simple, social network of key individuals and relationships. The nodes in this data set are from a modified, subnetwork of the link diagram representing Saddam Hussein and his connections to various family members, former regime members, friends, and associates. The original diagram contained hundreds of names and took shape on a large 36-by-36-inch board. Each "box" in the network contained personal information on a particular individual. This information included roles and positions of certain people within the network—for example, chief of staff, chief of operations, and personal secretary. These were not necessarily positions the individuals occupied before the fall of Hussein; rather they were based on an understanding of the role they were filling in the insurgency or Saddam's underground operations. Analysts assigned these roles based on an assessment of various personalities and recent reports. Such a process helped coalition forces focus their efforts in determining those who were closest to Hussein and their importance.

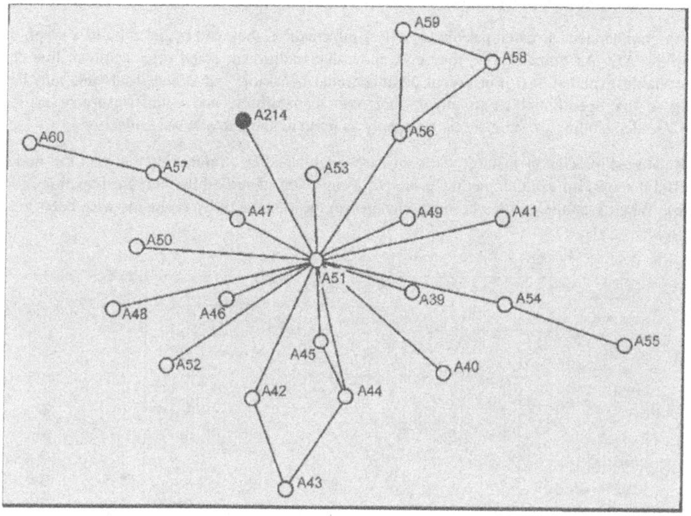

Figure B-10. Simple network

B-47. For an insurgency, a social network is not just a description of who is in the insurgent organization; it is a picture of the population, how it is put together and how members interact with one another. A tribal society already has affiliated social, economic, and military networks easily adapted to warfighting. The ways in which insurgents exploit a tribal network does not represent an evolved form of insurgency but the expression of inherent cultural and social customs. The social dynamic that sustains ongoing fighting is best understood when considered in tribal terms—in particular, from the perspective of a traditionally networked society. It is the traditional tribal network that offers rebels and insurgents a ready-made insurrectionary infrastructure on which to draw.

B-48. The full functioning of a network depends on how well, and in what ways, its members are personally known and connected to one another. This is the classic level of SNA, where strong personal ties, often ones that rest on friendship and bonding experiences, ensure high degrees of trust and loyalty. To function well, networks may require higher degrees of interpersonal trust than do other approaches to organization, like hierarchies. Kinship ties, be they of blood or brotherhood, are a fundamental aspect of many terrorist, criminal, and gang organizations. For example, news about Osama bin Laden and the Al Qaeda network reveal his, and its, dependence on personal relationships formed over years with "Afghan Arabs" from Egypt and elsewhere. These people are committed to anti-United States terrorism and Islamic fundamentalism.

B-49. To draw an accurate picture of a network, units need to identify ties among its members. Strong bonds formed over time by family, friendship, or organizational association characterize these ties. Units gather information on these ties by analyzing historical documents and records, interviewing individuals, and studying photos and books. It is painstaking work, but there is really no alternative when trying to piece together a network that does not want to be identified. Charts and diagrams lead to understanding the insurgents' means of operations. These same diagrams are also useful for understanding tribal, family, NGO, and transnational terrorist elements. Each diagram and chart may have links to another or several others, but they are not created overnight. It takes time, patience, detailed patrolling, and reporting and recording of efforts.

B-50. As a unit builds its situational awareness, it must create easy-to-understand, adaptable, and accurate diagrams and information sheets. These products feed one another and allow units to maintain and contribute to their understanding of the situation.

B-51. As commanders dispatch patrols to collect information, they can begin to build a graph of the population in the AO. As graphs grow, they may show that traditional, static organizational line charts do not produce viable explanations of insurgent organizational behavior. Individual insurgents may be constantly adapting to the operational environment, their own capabilities, and counterinsurgent tactics. A commander's understanding of the insurgency is only as good as the patrol's last collection.

B-52. Relationships (links) in large data sets are established by similarities between the nodes (people). Figure B-11 shows an example *activities matrix*. People are identified by their participation in independent activities. When graphed, pairs who have engaged in the same activity (columns with dots) are designated with a link.

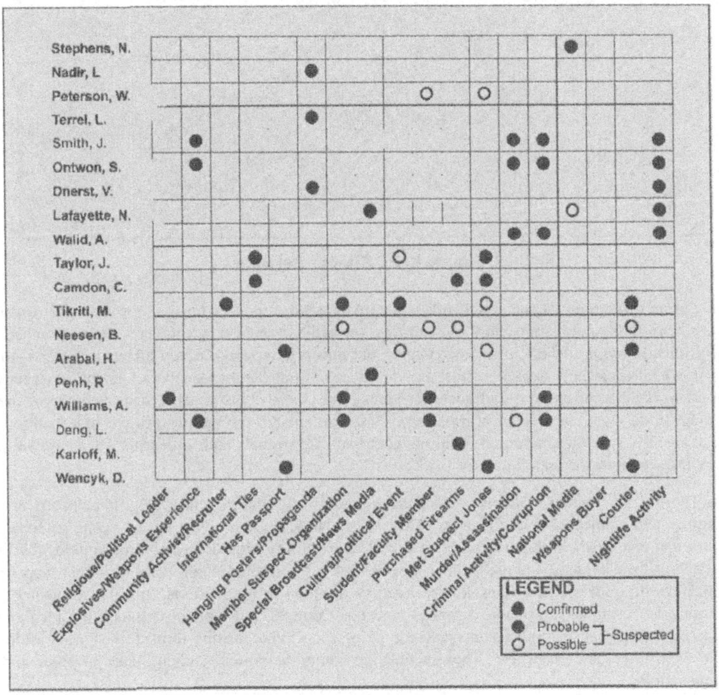

Figure B-11. Example activities matrix

B-53. An *association matrix* portrays the existence of an association, known or suspected, between individuals. (See figure B-12.) Direct connections include such things as face-to-face meetings and confirmed telephonic conversations. Association matrices provide a one-dimensional view of the relationships and tend to focus on the immediate AO. Analysts can use association matrices to identify those personalities and associations needing a more in-depth analysis to determine the degree of relationship, contacts, or knowledge between the individuals. The structure of the insurgent organization is identified as connections between personalities are made.

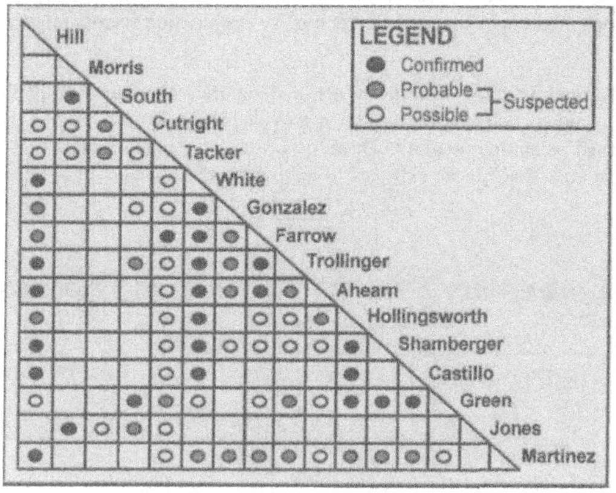

Figure B-12. Example association matrix

Social Network Analysis Summary

B-54. Insurgents often form a networked organization embedded in a sympathetic population. Differentiating between insurgents, insurgent supporters, neutrals, and the HN government supporters is difficult. With every counterinsurgent success, the insurgent organization becomes further fragmented but remains dangerous.

B-55. SNA helps units formalize the informality of insurgent networks by portraying the structure of something not readily observed. Network concepts let commanders highlight the structure of a previously unobserved association by focusing on the preexisting relationships and ties that bind together such groups. By focusing on roles, organizational positions, and prominent or influential actors, commanders may get a sense of how the organization is structured and thus how the group functions, how members are influenced and power exerted, and how resources are exchanged.

B-56. COIN operations require assessing the political and social architecture of the operational environment, from both friendly and enemy perspectives. SNA can help commanders understand how an insurgent organization operates. Insurgent networks often do not behave like normal social networks. However, SNA can help commanders determine what kind of social network an insurgent organization is. That knowledge helps commanders understand what the network looks like, how it is connected, and how best to defeat it.

HISTORICAL TIME LINE

B-57. A *time line* is list of significant dates along with relevant information and analysis. Time lines seek to provide a context to operational conditions. (See figure B-13.) Time lines often contain information related to areas and people as well as events. Some time lines describe population movements (areas) and political shifts (power and authority) that are relevant to the AO. Time lines can also include a brief historical record of the population or area, highlighting the activities (events) of a certain population sector. As analytic tools, time lines might help analysts predict how key population sectors might react to certain circumstances.

B-58. Key local national holidays, historic events, and significant cultural and political events can be extremely important. Soldiers and Marines are often provided with a list of these key dates to identify potential dates of increased or unusual activity. These lists, however, rarely include a description of why these dates are significant and what can be expected to happen on the holiday. In some cases, days of the week are significant.

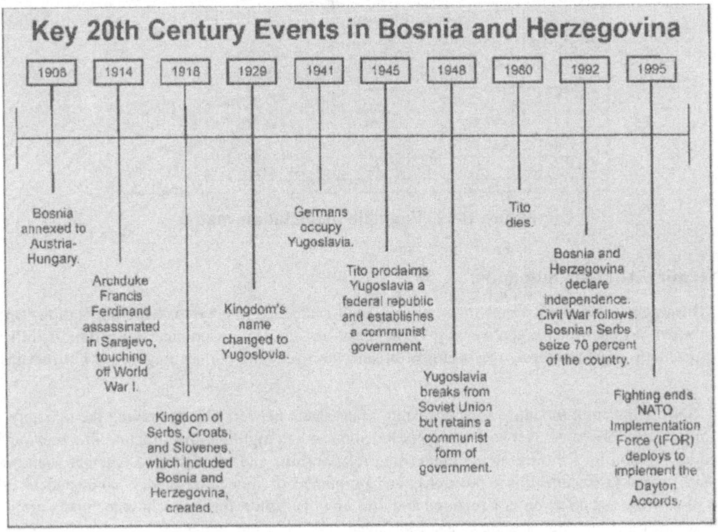

Figure B-13. Example historical time line

PATTERN ANALYSIS

B-59. Pattern analysis plot sheets, time-event charts, and coordinates registers are pattern analysis tools used to evaluate a threat and determine threat courses of action. (FM 2-22.3 discusses how use these tools.)

Pattern Analysis Plot Sheet

B-60. Pattern analysis plot sheets focus on the time and date of each serious incident that occurs within the AO. (See figure B-14.) The rings depict days of the month; the segments depict the hours of the day. As shown in the plot sheet's legend, the chart depicts the actual events and identifies each by using an alphanumeric designation that corresponds to the legend used on the coordinates register. (See paragraph B-61.) Another type of pattern analysis plot sheet helps distinguish patterns in activities associated with particular days, dates, or times. When used in conjunction with the coordinates register and doctrinal templates, a pattern analysis plot sheet supplies most of the data needed for an event template. Analysts may choose to modify this product to track shorter periods to avoid clutter and confusion.

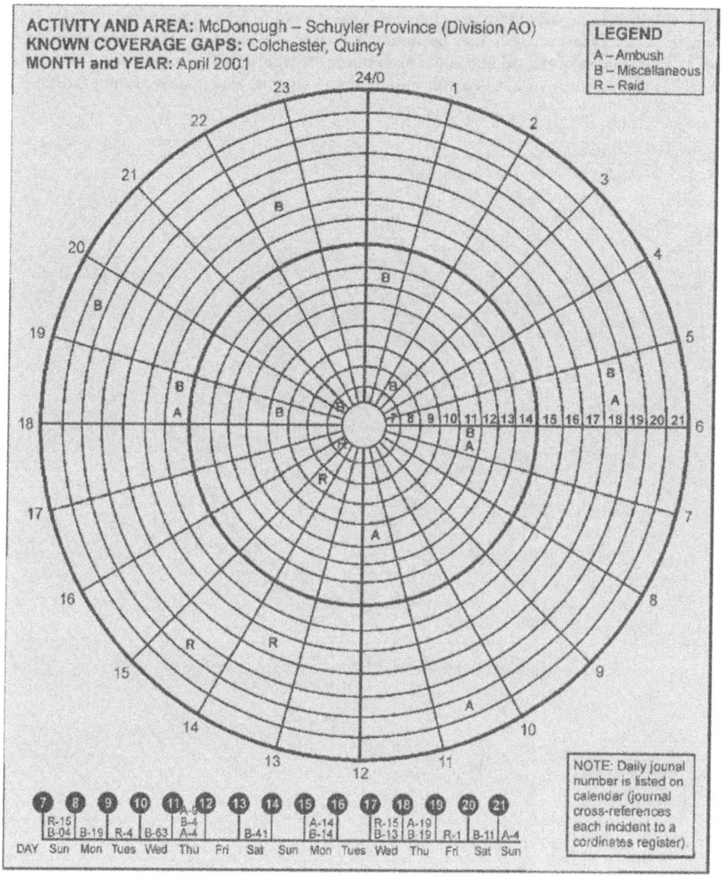

Figure B-14. Example pattern analysis plot sheet

Coordinates Register

B-61. Another pattern analysis tool is the coordinates register, also known as an incident map. (See figure B-15.) A coordinates register illustrates cumulative events that have occurred within an AO. It focuses on the "where" of an event. Analysts may use multiple coordinates registers, each focusing on an individual subject or a blend of subjects. Additionally, a coordinates register includes information like notes or graphics. Analysts should always use the coordinates register in with the pattern analysis plot sheet.

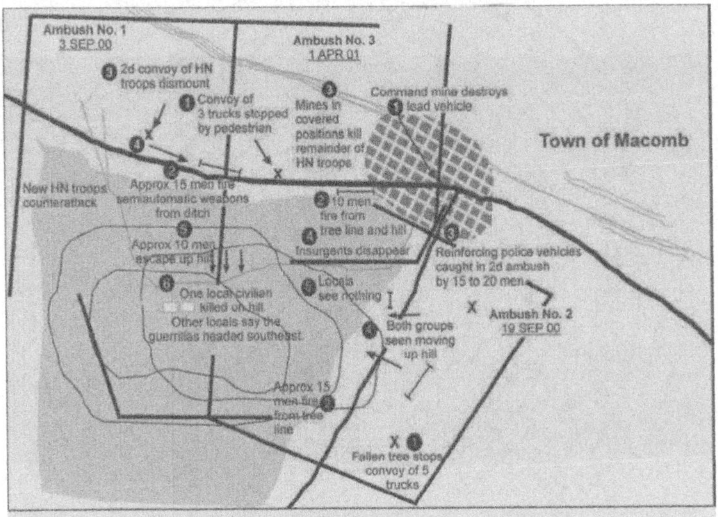

Figure B-15. Example coordinates register

Link Diagrams

B-62. The link diagram graphically depicts relationships between people, events, locations, or other factors deemed significant in any given situation. (See Figure B-16.) Link diagrams help analysts better understand how people and factors are interrelated in order to determine key links. (For more information on link diagrams, see FM 2-22.3.)

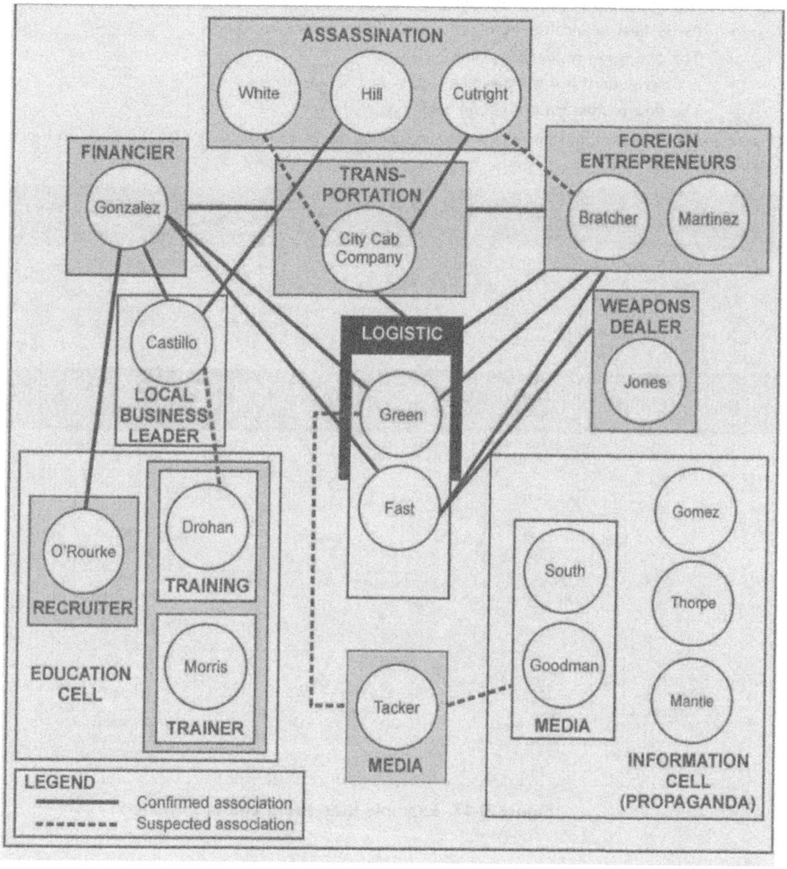

Figure B-16. Example link diagram

Time-Event Charts

B-63. Time-event charts are chronological records of individual or group activities. They are designed to store and display large amounts of information in a small space. Analysts can use time-event charts to help analyze larger scale patterns of such things as activities and relationships. There is great latitude in preparing time-event charts. Some of their common characteristics are as follows:

- The beginning and ends of the chart are shown with triangles.
- Other events are shown with squares.
- Particularly noteworthy events have an X drawn across the square.
- The date is always on the symbol.
- A description is below the symbol.
- The flow is from left to right for each row.

(Figure B-17 is an example showing events surrounding the plot to attack several landmarks in New York City in the early 1990s.)

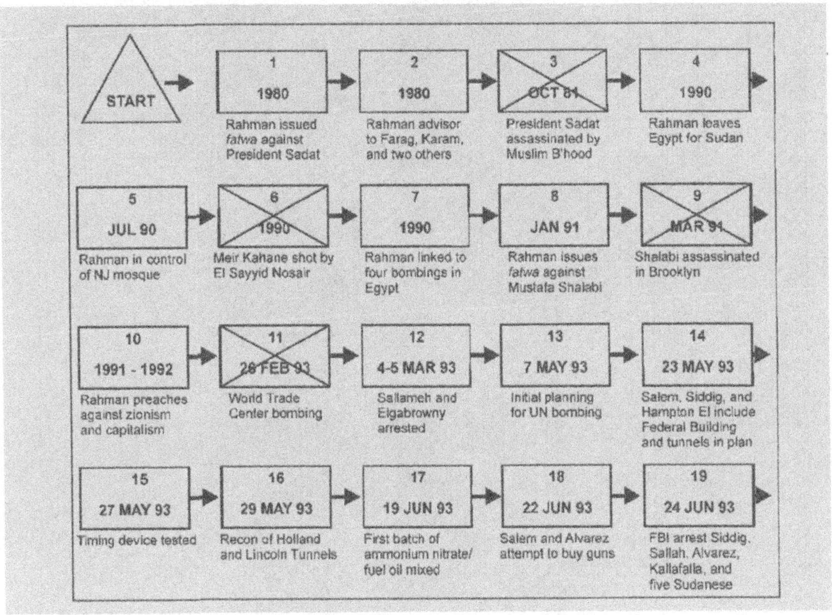

Figure B-17. Example time-event chart

Appendix C
Linguist Support

U.S. forces conducting counterinsurgency operations in foreign nations require linguist support. Military intelligence units assigned to brigade and higher level commands have organic interpreters (linguists) to perform human intelligence and signals intelligence functions. However, the need for interpreters usually exceeds organic capabilities, and commanders should obtain external interpreter support early.

LINGUIST SUPPORT CATEGORIES

C-1. When possible, interpreters should be U.S. military personnel or category II or III linguists. Unit intelligence officers should maintain language rosters at home station to track assigned personnel with linguistic capabilities before deployment. When requirements exceed organic capabilities, unit commanders can hire host-nation (HN) personnel to support their operations. Contracted linguists can provide interpreter support and perform intelligence functions. They fall into three categories.

C-2. *Category I linguists* usually are hired locally and require vetting. They do not have a security clearance. They are the most abundant resource pool; however, their skill level is limited. Category I linguists should be used for basic interpretation for activities such as patrols, base entrance coverage, open-source intelligence collection, and civil-military operations. Commanders should plan for 30 to 40 linguists from category I for an infantry battalion. Brigade headquarters should maintain roughly 15 category I linguists for surge operations.

C-3. *Category II linguists* are U.S. citizens with a secret clearance. Often they possess good oral and written communication skills. They should be managed carefully due to limited availability. Category II linguists interpret for battalion and higher level commanders or tactical human intelligence teams. Brigade commanders should plan for 10 to 15 linguists from category II. That breaks down to one linguist for the brigade commander, one for each infantry battalion commander, and approximately 10 linguists for the supporting military intelligence company. Of those 10, three translate for each tactical human intelligence team or operations management team, and two translate for each signals intelligence collection platform.

C-4. *Category III linguists* are U.S. citizens with a top secret clearance. They are a scarce commodity and often retained at division and higher levels of command. They have excellent oral and written communications skills.

C-5. Some private companies provide linguist support through contracts. The required statement of work or contract should define the linguist's requirements and the unit's responsibilities. Contracted category II and III linguists should provide their own equipment, such as flak vests, Kevlar, and uniforms. (Category I linguists normally do not.) The unit designates a linguist manager to identify language requirements and manage assets. Site managers for the contractor are located at the division level to manage personnel issues such as leave, vacation, pay, and equipment.

C-6. When hiring HN personnel to perform category I linguist requirements as interpreters, units should consider the guidelines under the following categories:

- Selecting interpreters.
- Establishing rapport.
- Orienting interpreters.
- Preparing for presentations.

- Conducting presentations.
- Speaking techniques.

SELECTING INTERPRETERS

C-7. Soldiers and Marines must try to vet interpreters before hiring them. All interpreters must meet a basic set of criteria. They should be native speakers. The target audience should willingly accept their social status. All interpreters should speak English fluently. They should be able to translate correctly. Intelligent translators are mandatory; those with technical knowledge are desired. Interpreters should be reliable, loyal, and compatible with the military personnel. Their gender, age, race, and ethnicity must be compatible with the target audience.

NATIVE SPEAKER

C-8. Interpreters should be native speakers of the socially or geographically determined dialect. Their speech, background, and mannerisms should be completely acceptable to the target audience. The interpreters should not distract the interviewees. The target audience should give no attention to the way interpreters talk, only to what they say. Native speakers can better distinguish dialects of different regions and provinces. This knowledge can help identify interviewees from other countries or from outside the local area.

SOCIAL STATUS AND ETHNO-RELIGIOUS IDENTITY

C-9. If their social standing is considerably lower than that of the audience, interpreters may be limited in their effectiveness. Examples include significant differences in military rank or membership in a shunned ethnic or religious group. Soldiers and Marines must communicate with the local population. They must be tolerant of local prejudices and choose an interpreter who is least likely to cause suspicion or miscommunication. Interpreters should also have a good reputation in the community and not be intimidated when dealing with important audiences.

ENGLISH FLUENCY

C-10. An often overlooked consideration is how well the interpreter speaks English. If the interpreter understands the speaker and the speaker understands the interpreter, then the interpreter's command of English is satisfactory. Soldiers and Marines can check that understanding by speaking something to the interpreter in English and asking the interpreter to paraphrase it.

UNDERSTANDING OF THE AUDIENCE

C-11. Interpreting goes both ways. Interpreters should accurately convey information expressed by interviewees or the target audience. This is especially important when commanders speak with HN civilian leaders and military personnel. Linguists involved in military discussions should understand military terms and doctrine.

INTELLECTUAL CAPABILITIES

C-12. Interpreters should be quick and alert, able to respond to changing conditions and situations. They should be able to grasp complex concepts and discuss them clearly and logically. Although education does not equate to intelligence, it does expose students to diverse and complex topics. As a result, the better educated the interpreters, the better they perform.

TECHNICAL ABILITY

C-13. Sometimes Soldiers and Marines need interpreters with technical training or experience in special subject areas. Such interpreters can translate the meaning as well as the words. For instance, if the subject is nuclear physics, background knowledge is useful.

RELIABILITY

C-14. Soldiers and Marines should avoid a potential interpreter who arrives late for the vetting interview. Throughout the world, the concept of time varies widely. In many countries, timeliness is relatively unimportant. Soldiers and Marines should stress the importance of punctuality with interpreters.

LOYALTY

C-15. If interpreters are local nationals, their first loyalty is probably to the host nation or ethnic group, not to the U.S. military. The security implications are clear. Soldiers should be cautious when they explain concepts. They should limit what information interpreters can overhear. Some interpreters, for political or personal reasons, may have ulterior motives or a hidden agenda. Soldiers and Marines who detect or suspect such motives should tell the commander or security manager.

GENDER, AGE, RACE, AND ETHNICITY

C-16. Gender, age, and race can seriously affect the mission effectiveness of interpreters. In predominantly Muslim countries, cultural prohibitions may cause difficulties with gender. A female interpreter may be ineffective in communicating with males, while a female interpreter may be needed to communicate with females. In regions featuring ethnic strife, such as the Balkans, ethnic divisions may limit the effectiveness of an interpreter from outside the target audience. Since traditions, values, and biases vary from country to country, Soldiers and Marines must thoroughly study the culture to determine the most favorable characteristics for interpreters.

COMPATIBILITY

C-17. The target audience quickly recognizes personality conflicts between Soldiers and Marines and their interpreters. Such friction can undermine the effectiveness of the communication. When selecting interpreters, Soldiers and Marines should look for compatible traits and strive for a harmonious working relationship.

EMPLOYING LINGUISTS

C-18. If several qualified interpreters are available, Soldiers and Marines should select at least two. This is particularly important if the interpreter works during long conferences or courses of instruction. With two interpreters available, they should each work for thirty-minute periods. Due to the mental strain associated with this task, four hours of active interpreting a day is usually the most that interpreters can work before effectiveness declines. During short meetings and conversations with two or more available interpreters, one can provide quality control and assistance for the one translating. This technique is useful when conducting coordination or negotiation meetings, as one interpreter can actively interpret while the other pays attention to the body language and side conversations of the audience. Many times, Soldiers and Marines can learn important auxiliary information from listening to what others are saying among themselves. This information can help in later negotiations.

C-19. Commanders must protect their interpreters. They should emplace security measures to keep interpreters and their families safe. Insurgents know the value of good interpreters and will often try to intimidate or kill interpreters and their family members. Insurgents may also coerce interpreters to gather information on U.S. operations. Soldiers and Marines must actively protect against subversion and espionage, to include using a polygraph.

C-20. Certain tactical situations may require using uncleared HN personnel as interpreters. Commanders should recognize the increased security risk when using such personnel and carefully weigh the risk versus potential gain. If uncleared interpreters are used, Soldiers and Marines must limit discussing sensitive information.

ESTABLISHING RAPPORT

C-21. Interpreters are a vital link among Soldiers, Marines, and the target audience. Without supportive, cooperative interpreters, the mission is jeopardized. Mutual respect and understanding are essential to effective teamwork. Soldiers and Marines should establish rapport early and maintain it throughout the operation. Problems establishing rapport stem mostly from a lack of personal communication skills and misunderstandings regarding culture.

C-22. Before they meet interpreters, Soldiers and Marines study the area of operations and its inhabitants. This process is discussed in chapter 3. Many foreigners have some knowledge about the United States. Unfortunately, much of this comes from commercial movies and television shows. Soldiers and Marines may need to teach the interpreter something realistic about the United States as well.

C-23. Soldiers and Marines working with an interpreter should research and verify the interpreter's background. They should be genuinely interested in the interpreter and the interpreter's family, aspirations, career, and education. Many cultures emphasize family roles differently from the United States, so Soldiers and Marines should first understand the interpreter's home life. Though Soldiers and Marines should gain as much cultural information as possible before deploying, their interpreters can be valuable sources for filling gaps. However, information from interpreters will likely represent the views of the group to which they belong. Members of opposing groups almost certainly see things differently and often view culture and history differently.

C-24. Soldiers and Marines should gain an interpreter's trust and confidence before discussing sensitive issues. These issues include religion, likes, dislikes, and prejudices. Soldiers and Marines should approach these topics carefully. Although deeply held personal beliefs may be revealing and useful in a professional relationship, Soldiers and Marines should draw these out of their interpreters gently and tactfully.

C-25. One way to reinforce the bond between military personnel and their interpreter is to make sure the interpreter has every available comfort. This includes providing personal protection equipment—boots, helmets, and body armor—that the interpreter (especially a category I interpreter) may not already have. Soldiers and Marines must give interpreters the same comforts that military personnel enjoy. Interpreters need the same base comforts—shelter, air conditioning, and heat—as military personnel. If and when an interpreter is assigned to a specific unit, the interpreter ought to live with that organization to develop a bond. If there are several interpreters, it may be more effective for the interpreters to live together on the unit compound.

ORIENTING INTERPRETERS

C-26. Early in the relationship, Soldiers and Marines must explain to interpreters their duties, expected standards of conduct, interview techniques, and any other requirements and expectations. (Table C-1 lists some information to include when orienting interpreters.)

PREPARING FOR PRESENTATIONS

C-27. Sites for interviews, meetings, or classes should be carefully selected and arranged. The physical arrangement can be especially significant for certain groups or cultures.

C-28. Speakers should understand unique cultural practices before interviewing, instructing, or talking with foreign nationals. For example, speakers and interpreters should know when to stand, sit, or cross one's legs. Gestures are a learned behavior and vary from culture to culture. If properly selected, interpreters should be helpful in this regard.

C-29. Interpreters should mirror the speaker's tone and personality. They must not add their own questions or emotions. Speakers should instruct interpreters to inform them discreetly if they notice inconsistencies or peculiarities of speech, dress, and behavior.

Table C-1. Orientation for interpreters

Background for the interpreter includes—
• The current tactical situation. • Information on the source, interviewee, or target audience. • Specific objectives for the interview, meeting, or interrogation. • The conduct of the interview, lesson, or interrogation. • The physical arrangements of interviewing site, if applicable.
Duties of the interpreter include—
• Informing the interviewer about inconsistencies the interviewee uses in language. An example would be someone who claims to be a college professor, yet speaks like an uneducated person. • Possibly assisting in after-action reviews or assessments.
Interview techniques for the interpreter include—
• Simultaneous—when the interpreter listens and translates at the same time (not recommended). • Consecutive—when the interpreter listens to an entire phrase, sentence, or paragraph, and then translates during natural pauses.
Standards of conduct for the interpreter include—
• Being careful not to inject their personality, ideas, or questions. • Mirroring the speaker's tone and personality. • Translating the exact meaning without adding or deleting information.

C-30. Soldiers and Marines must carefully analyze the target audience. This analysis goes beyond the scope of this appendix. Mature judgment, thoughtful consideration of the target audience as individuals, and a genuine concern for their receiving accurate information helps Soldiers and Marines accomplish the mission. Soldiers and Marines should remember, for example, that a farmer from a small village has markedly different expectations and requirements than a city executive.

C-31. Soldiers and Marines who work through an interpreter may take double or triple the time normally required for an event. They may save time if they give the interpreter pertinent information beforehand. This information may include briefing slides, questions to ask, a lesson plan, copies of any handouts, or a glossary of difficult terms.

CONDUCTING PRESENTATIONS

C-32. As part of the initial training for interpreters, Soldiers and Marines emphasize that interpreters follow their speaker's lead. They become a vital communication link between the speaker and target audience. Soldiers and Marines should appeal to the interpreter's professional pride. They clarify how the quality and quantity of the information sent and received directly depends on linguistic skills. Although interpreters perform some editing as a function of the interpreting process, they must transmit the exact meaning without additions or deletions.

C-33. Speakers should avoid simultaneous translations—the speaker and interpreter talking at the same time—when conducting an interview or presenting a lesson. They should talk directly to the individual or audience for a minute or less in a neutral, relaxed manner. The interpreter should watch the speaker carefully. While translating, the interpreter should mimic the speaker's body language as well as interpret verbal meaning. Speakers should observe interpreters closely to detect any inconsistent behaviors. After speakers present one major thought in its entirety, interpreters then reconstruct it in their language. One way to ensure that the interpreter is communicating exactly what the speaker means is to have a senior interpreter observe several conversations. The senior interpreter can provide feedback along with further training.

C-34. Soldiers and Marines should be aware that some interpreters might attempt to save face or to protect themselves by concealing their lack of understanding. They may translate what they believe the speaker or audience said or meant without asking for clarification. This situation can result in misinformation and confusion. It can also impact the speaker's credibility. Interpreters must know that when in doubt they should always ask for clarification.

C-35. During an interview or lesson, if the interviewee asks questions, interpreters should immediately relay them to the speaker for an answer. Interpreters should never attempt to answer questions, even if they know the correct answers. Neither speakers nor interpreters should correct each other in front of an interviewee or class. They should settle all differences away from the subject or audience.

C-36. Establishing rapport with the interpreter is vital; establishing rapport with interviewees or the target audience is equally important. Speakers and their interpreters should concentrate on this task. To establish rapport, interviewees and target audiences should be treated as mature, important people who are worthy and capable.

C-37. Several methods ensure that the speaker communicates directly to the target audience using the interpreter as only a mechanism for that communication. One technique is to have the interpreter stand to the side of and just behind the speaker. This position lets the speaker stand face-to-face with the target audience. The speaker should always look at and talk directly to the target audience, rather than to the interpreter. This method allows the speaker and the target audience to establish a personal relationship.

SPEAKING TECHNIQUES

C-38. An important first step for Soldiers and Marines communicating in a foreign language is to reinforce and polish their English language skills. These skills are important, even when no attempt has been made to learn the HN language. They should use correct words, without idioms or slang. The more clearly Soldiers and Marines speak in English, the easier it is for interpreters to translate exactly. For instance, speakers may want to add words usually left out in colloquial English, such as the "air" in airplane. This ensures they are not misinterpreted as referring to the Great Plains or a carpenter's plane.

C-39. Speakers should not use profanity at all and should avoid slang. Many times, interpreters cannot translate such expressions. Even those they can translate might lose the desired meaning. Terms of surprise or reaction such as "gee whiz" and "golly" are difficult to translate.

C-40. Speakers should avoid using acronyms. While these are part of everyday military language, most interpreters and target audiences do not know them. The interpreter may have to interrupt the interview for clarification. This can disrupt the rhythm of the interview or lesson. If interpreters constantly interrupt the speaker for explanation, they could lose credibility in the eyes of the target audience. Such a reaction could jeopardize the interview or lesson. If speakers use technical terms or expressions, they should be sure interpreters convey the proper meaning. This preparation is best done in advance.

C-41. Before speaking impulsively, Soldiers and Marines should consider what they wish to say. They should break their thoughts into logical bits and articulate them one at a time. Using short, simple words and sentences helps the interpreter to translate quickly and easily. Speakers should never say more in one sentence than they can easily repeat immediately after saying it. Each sentence should contain a complete thought without the extra words.

C-42. Speakers should avoid American "folk" and culture-specific references. Target audiences may have no idea what is being talked about. Even when interpreters understand the reference, they may find it difficult to quickly identify an appropriate equivalent in the target audience's cultural frame of reference.

C-43. Transitional phrases and qualifiers may confuse nonnative speakers and waste valuable time. Examples include "for example," "in most cases," "maybe," and "perhaps."

C-44. Speakers should avoid American humor. Humor is culturally specific and does not translate well. Cultural and language differences can lead to misinterpretations by foreigners.

C-45. Speakers must consider some "dos" and "don'ts" for when working with interpreters. Table C-2 clarifies what speakers should and should not do.

Table C-2. Good and bad practices for speakers

Speakers should—
• Position the interpreter by their side (or a step back). This keeps the subject or target audience from shifting their attention or fixating on the interpreter rather than on the leader. • Always look at and talk directly to the subject or target audience. Guard against the tendency to talk to the interpreter. • Speak slowly and clearly. Repeat as often as necessary. • Speak to the individual or group as if they understand English. Be enthusiastic. Use gestures, movements, voice intonations, and inflections normally used with an English-speaking group. Remember that considerable nonverbal meaning is conveyed through voice and body movements. Encourage interpreters to mimic the same delivery. • Periodically check an interpreter's accuracy, consistency, and clarity. Request a U.S. citizen fluent in the language to check that the interpreter is not distorting the translation, intentionally or unintentionally. Learn some of the language. • Check with the audience whenever a misunderstanding is suspected and clarify immediately. Using the interpreter, ask questions to elicit answers to indicate whether the point is clear. If it is not, rephrase the instruction and illustrate the point differently. Use repetition and examples whenever necessary to facilitate learning. If the interviewees ask few questions, it may mean they have not understood the instruction or the message is unclear. • Ensure interpreters understand they are valuable team members. Recognize them based on the importance of their contributions. Protect interpreters; the insurgents and criminal elements may target them.
Speakers should not—
• Address the subject or audience in the third person through the interpreter. For example, avoid saying, "Tell them I'm glad to be their instructor." Instead, directly address the subject or audience saying, "I am glad to be your instructor." Make continual eye contact with the audience. Watch them, not the interpreter. • Make side comments to the interpreter that are not interpreted. This action is rude, discourteous, and creates the wrong atmosphere. • Distract the audience while the interpreter is translating. Avoid pacing, writing on the blackboard, teetering on the lectern, drinking beverages, or doing any other distracting activity while the interpreter is translating.

This page intentionally left blank.

Appendix D
Legal Considerations

Law and policy govern the actions of the U.S. forces in all military operations, including counterinsurgency. For U.S. forces to conduct operations, a legal basis must exist. This legal basis profoundly influences many aspects of the operation. It affects the rules of engagement, how U.S. forces organize and train foreign forces, the authority to spend funds to benefit the host nation, and the authority of U.S. forces to detain and interrogate. Under the Constitution, the President is commander in chief of the U.S. forces. Therefore, orders issued by the President or the Secretary of Defense to a combatant commander provide the starting point in determining the legal basis. This appendix summarizes some of the laws and policies that bear upon U.S. military operations in support of foreign counterinsurgencies. Laws are legislation passed by Congress and signed into law by the President, as well as treaties to which the United States is party. Policies are executive orders, departmental directives and regulations, and other authoritative statements issued by government officials. No summary provided here can replace a consultation with the unit's supporting staff judge advocate.

AUTHORITY TO ASSIST A FOREIGN GOVERNMENT

D-1. U.S. forces have limited authority to provide assistance to foreign governments. For foreign internal defense, U.S. forces may be authorized to make limited contributions. Assistance to police by U.S. forces is permitted, but not with the Department of Defense (DOD) as the lead governmental department.

AUTHORITY FOR FOREIGN INTERNAL DEFENSE

D-2. The President or Secretary of Defense give the deployment and execution order. Without receiving a deployment or execution order, U.S. forces may be authorized to make only limited contributions in support of a host nation's counterinsurgency (COIN) effort. If the Secretary of State requests and the Secretary of Defense approves, U.S. forces can participate in this action. The request and approval go through standing statutory authorities in Title 22, United States Code. Title 22 contains the Foreign Assistance Act, the Arms Export Control Act, and other laws. It authorizes security assistance, developmental assistance, and other forms of bilateral aid. The request and approval might also occur under various provisions in Title 10, United States Code. Title 10 authorizes certain types of military-to-military contacts, exchanges, exercises, and limited forms of humanitarian and civic assistance in coordination with the U.S. Ambassador for the host nation. In such situations, U.S. military personnel work as administrative and technical personnel. They are part of the U.S. diplomatic mission, pursuant to a status of forces agreement, or pursuant to an exchange of letters. This cooperation and assistance is limited to liaison, contacts, training, equipping, and providing defense articles and services. It does not include direct involvement in operations.

DOD USUALLY NOT LEAD—GENERAL PROHIBITION ON ASSISTANCE TO POLICE

D-3. DOD is usually not the lead governmental department for assisting foreign governments, even for the provision of security assistance—that is, military training, equipment, and defense articles and services—to the host nation's military forces. DOD contribution may be large, but the legal authority is typically one exercised by the Department of State. With regard to provision of training to a foreign government's police or other civil interior forces, the U.S. military typically has no authorized role. The Foreign Assistance Act specifically prohibits assistance to foreign police forces except within carefully circumscribed exceptions, and under a Presidential directive, and the lead role in providing police assistance within those exceptions has been normally delegated to the Department of State's Bureau of International Narcotics and Law Enforcement Affairs. However, the President did sign a decision directive in 2004 granting authority to train and equip Iraqi police to the Commander, U.S. Central Command.

AUTHORIZATION TO USE MILITARY FORCE

D-4. Two types of resolutions authorize involvement of U. S. forces: a Congressional resolution and the 1973 War Powers Resolution. Congressional support is necessary if U.S. forces will be involved in actual operations overseas. The 1973 Resolution lets the President authorize military forces for a limited time.

CONGRESSIONAL RESOLUTION

D-5. Congressional support is needed for any prolonged involvement of U.S. forces in actual operations overseas. Often a Congressional resolution provides the central legal basis for such involvement within domestic law. This is especially likely if U.S. forces are anticipated, at least initially, to be engaged in combat operations against an identified hostile force.

STANDING WAR POWERS RESOLUTION

D-6. The 1973 War Powers Resolution requires the President to consult and report to Congress when introducing U.S. forces into certain situations. There are times though when a specific Congressional authorization for use of force is absent. In the absence of this authorization, the President—without conceding that the 1973 Resolution binds the President's own constitutional authority—makes a report to Congress. The President must make the report within 48 hours of introducing substantial U.S. forces into the host nation. This report details the circumstances necessitating introduction or enlargement of troops. The President bases his or her action on the Constitutional or legislative authority and the estimated scope and duration of the deployment or combat action. The 1973 Resolution states that if Congress does not declare war or specifically authorize the deployment or combat action within 60 days of the report, the President must terminate U.S. military involvement and redeploy U.S. forces.

RULES OF ENGAGEMENT

D-7. *Rules of engagement* (ROE) are directives issued by competent military authority that delineate the circumstances and limitations under which United States forces will initiate and/or continue combat engagement with other forces encountered (JP 1-02). Often these directives are specific to the operation. If there are no operation-specific ROE, U.S. forces apply standing rules of engagement (SROE). When working with a multinational force, commanders must coordinate the ROE thoroughly.

OPERATION-SPECIFIC RULES OF ENGAGEMENT

D-8. In a large-scale deployment, the Secretary of Defense may issue ROE specific to the operation to a combatant commander. The combatant commander and subordinate commanders then issue ROE consistent with the ROE received from the Secretary of Defense. ROE state the circumstances under which Soldiers or Marines may open fire. They may fire when they positively identify a member of a hostile force or they have clear indications of hostile intent. ROE may include rules concerning when civilians may be detained, specify levels of approval authority for using heavy weapons, or identify facilities that may be protected with deadly force. All ROE comply with the law of war. ROE in COIN are dynamic. Commanders must regularly review ROE for their effectiveness in the complex COIN environment. Training counterinsurgents in ROE should be reinforced regularly.

CHAIRMAN OF THE JOINT CHIEFS OF STAFF STANDING RULES OF ENGAGEMENT

D-9. In the absence of operation-specific ROE, U.S. forces apply CJCSI 3121.01B. This instruction refers to SROE for U.S. forces. The SROE establish fundamental policies and procedures governing the actions of U.S. force commanders in certain events. These events include military attacks against the United States and during all military operations, contingencies, terrorist attacks, or prolonged conflicts outside the territorial jurisdiction of the United States. The SROE do not limit a commander's inherent authority and obligation to use all necessary means available. They also do not limit the commander's authority and obligation to take all appropriate action in self-defense of the commander's unit and other U.S. forces in the vicinity. The SROE prescribe how supplemental ROE for specific operations are provided as well as the format by which subordinate commanders may request ROE.

MULTINATIONAL RULES OF ENGAGEMENT

D-10. When U.S. forces, under U.S. operational or tactical control, operate with a multinational force, reasonable efforts are made to effect common ROE. If such ROE cannot be established, U.S. forces operate under the SROE or operation-specific ROE provided by U.S. authorities. To avoid misunderstanding, commanders thoroughly discuss among multinational forces any differences in ROE or ROE interpretation. They disseminate any differences in ROE to the units involved.

THE LAW OF WAR

D-11. COIN and international armed conflicts often overlap. COIN may take place before, after, or simultaneously with a war occurring between nations. U.S. forces obey the law of war. The law of war is a body of international treaties and customs, recognized by the United States as binding. It regulates the conduct of hostilities and protects noncombatants. The main law of war protections come from the Hague and Geneva Conventions. They apply at the tactical and operational levels and are summarized in ten rules:

- Soldiers and Marines fight only enemy combatants.
- Soldiers and Marines do not harm enemies who surrender. They disarm them and turn them over to their superiors.
- Soldiers and Marines do not kill or torture enemy prisoners of war.
- Soldiers and Marines collect and care for the wounded, whether friend or foe.
- Soldiers and Marines do not attack medical personnel, facilities, or equipment.
- Soldiers and Marines destroy no more than the mission requires.
- Soldiers and Marines treat all civilians humanely.
- Soldiers and Marines do not steal. They respect private property and possessions.
- Soldiers and Marines do their best to prevent violations of the law of war.
- Soldiers and Marines report all violations of the law of war to their superior.

D-12. When insurgency occurs during occupation, the law of war includes rules governing situations in which the military forces of one state occupy the territory of another. Occupation is not a transfer of sovereignty. It does however grant the occupying power the authority and responsibility to restore and maintain public order and safety. The occupying power must respect, as much as possible, the laws in force in the host nation. One of the four Geneva Conventions of 1949—the Geneva Convention Relative to the Protection of Civilian Persons in Time of War—becomes a prominent source of law during occupation.

INTERNAL ARMED CONFLICT

D-13. During COIN operations, commanders must be aware of Common Article 3 of the Geneva Conventions and the status of insurgents under the laws of the host nation.

GENEVA CONVENTION, COMMON ARTICLE 3

D-14. Although insurgencies can occur simultaneously with a legal state of war between two nations, they are classically conflicts internal to a single nation, between uniformed government forces and armed elements that do not wear uniforms with fixed distinctive insignia, carry arms openly, or otherwise obey the laws and customs of war. As such, the main body of the law of war does not strictly apply to these conflicts—a legal fact that can be a source of confusion to commanders and Soldiers. It bears emphasis, however, that one article contained in all four of the Geneva Conventions—Common Article 3—is specifically intended to apply to internal armed conflicts:

In the case of armed conflict not of an international character occurring in the territory of one of the High Contracting Parties, each Party to the conflict shall be bound to apply, as a minimum, the following provisions:

(1) Persons taking no active part in the hostilities, including members of armed forces who have laid down their arms and those placed "hors de combat" by sickness, wounds, detention, or any other cause, shall in all circumstances be treated humanely, without any adverse distinction founded on race, colour, religion or faith, sex, birth or wealth, or any other similar criteria.

To this end, the following acts are and shall remain prohibited at any time and in any place whatsoever with respect to the above-mentioned persons:

(a) Violence to life and person, in particular murder of all kinds, mutilation, cruel treatment and torture;

(b) Taking of hostages;

(c) Outrages upon personal dignity, in particular humiliating and degrading treatment;

(d) The passing of sentences and the carrying out of executions without previous judgment pronounced by a regularly constituted court, affording all the judicial guarantees which are recognized as indispensable by civilized peoples.

(2) The wounded and sick shall be collected and cared for.

An impartial humanitarian body, such as the International Committee of the Red Cross, may offer its services to the Parties to the conflict.

The Parties to the conflict should further endeavor to bring into force, by means of special agreements, all or part of the other provisions of the present Convention.

The application of the preceding provisions shall not affect the legal status of the Parties to the conflict.

APPLICATION OF CRIMINAL LAWS OF THE HOST NATION

D-15. The final sentence of Common Article 3 makes clear that insurgents have no special status under international law. They are not, when captured, prisoners of war. Insurgents may be prosecuted legally as criminals for bearing arms against the government and for other offenses, so long as they are accorded the minimum protections described in Common Article 3. U.S. forces conducting COIN should remember that the insurgents are, as a legal matter, criminal suspects within the legal system of the host nation. Counterinsurgents must carefully preserve weapons, witness statements, photographs, and other evidence collected at the scene. This evidence will be used to process the insurgents into the legal system and thus hold them accountable for their crimes while still promoting the rule of law.

D-16. Status of forces agreements establish the legal status of military personnel in foreign countries. Criminal and civil jurisdiction, taxation, and claims for damages and injuries are some of the topics usually covered in a status of forces agreement. In the absence of an agreement or some other arrangement with the host nation, DOD personnel in foreign countries may be subject to its laws.

DETENTION AND INTERROGATION

D-17. Chapters 3, 5, and 7 indicate the need for human intelligence in COIN operations. This need can create great pressure to obtain time-sensitive information from detained individuals. The Detainee Treatment Act of 2005, FM 2-22.3, and other specific standards were created to guide U.S. forces working with detainees.

DETAINEE TREATMENT ACT OF 2005

D-18. U.S. law clearly prohibits U.S. forces, including officials from other government agencies, from using certain methods to obtain information. Instances of detainee abuse, including maltreatment involving interrogation, were documented. In response, Congress passed, and the President signed into law, the Detainee Treatment Act of 2005. (See Table D-1.)

Table D-1. Extract of the Detainee Treatment Act of 2005

Section 1002: Uniform Standards for the Interrogation of Persons Under the Detention of the Department of Defense
(a) In General.—No person in the custody or under the effective control of the Department of Defense or under detention in a Department of Defense facility shall be subject to any treatment or technique of interrogation not authorized by and listed in the United States Army Field Manual on Intelligence Interrogation [FM 2-22.3]. (b) Applicability.—Subsection (a) shall not apply with respect to any person in the custody or under the effective control of the Department of Defense pursuant to a criminal law or immigration law of the United States. (c) Construction.—Nothing in this section shall be construed to affect the rights under the United States Constitution of any person in the custody or under the physical jurisdiction of the United States.
Section 1003: Prohibition on Cruel, Inhuman, or Degrading Treatment or Punishment of Persons Under Custody or Control of the United States Government
(a) In General.—No individual in the custody or under the physical control of the United States Government, regardless of nationality or physical location, shall be subject to cruel, inhuman, or degrading treatment or punishment. (b) Construction.—Nothing in this section shall be construed to impose any geographical limitation on the applicability of the prohibition against cruel, inhuman, or degrading treatment or punishment under this section. (c) Limitation on Supersedure.—The provisions of this section shall not be superseded, except by a provision of law enacted after the date of the enactment of this Act which specifically repeals, modifies, or supersedes the provisions of this section. (d) Cruel, Inhuman, or Degrading Treatment or Punishment defined.—In this section, the term "cruel, inhuman, or degrading treatment or punishment" means the cruel, unusual, and inhumane treatment or punishment prohibited by the Fifth, Eighth, and Fourteenth Amendments to the Constitution of the United States, as defined in the United States Reservations, Declarations and Understandings to the United Nations Convention Against Torture and Other Forms of Cruel, Inhuman, or Degrading Treatment or Punishment done at New York, December 10, 1984.

INTERROGATION FIELD MANUAL

D-19. The Detainee Treatment Act established FM 2-22.3 as the legal standard. No techniques other than those prescribed by the field manual are authorized by U.S. forces. Commanders must ensure that interrogators receive proper training and supervision.

STANDARDS FOR DETENTION AND INTERNMENT

D-20. Regardless of the precise legal status of those persons captured, detained, or otherwise held in custody by U.S. forces, they must receive humane treatment until properly released. They also must be provided the minimum protections of the Geneva Conventions. Specially trained, organized, and equipped military police units in adequately designed and resourced facilities should accomplish prolonged detention. Such detention must follow the detailed standards contained in AR 190-8/MCO 3461.1. The military police personnel operating such facilities shall not be used to assist in or "set the conditions for" interrogation.

TRANSFER OF DETAINEES TO THE HOST NATION

D-21. There are certain conditions under which U.S. forces may not transfer the custody of detainees to the host nation or any other foreign government. U.S. forces retain custody if they have substantial grounds to believe that the detainees would be in danger in the custody of others. Such danger could include being subjected to torture or inhumane treatment. (For more information on transferring detainees, see DODD 2310.01E and consult the legal advisor or staffjudge advocate.)

ENFORCING DISCIPLINE OF U.S. FORCES

D-22. Despite rigorous selection and training, some personnel require discipline. The Uniform Code of Military Justice is the criminal code of military justice applicable to all military members. Commanders and general officers are responsible for their subordinates and their behavior. Commanders must give clear guidance and ensure compliance. All civilians working for the U.S. Government also must comply with the laws.

UNIFORM CODE OF MILITARY JUSTICE

D-23. Although the vast majority of well-led and well-trained U.S. military personnel perform their duties honorably and lawfully, history records that some commit crimes amidst the decentralized command and control, the strains of opposing a treacherous and hidden enemy, and the often complex ROE that characterize the COIN environment. Uniformed personnel remain subject at all times to the Uniform Code of Military Justice and must be investigated and prosecuted, as appropriate, for violations of orders, maltreatment of detainees, assaults, thefts, sexual offenses, destruction of property, and other crimes, including homicides, that they may commit during COIN.

COMMAND RESPONSIBILITY

D-24. In some cases, military commanders may be deemed responsible for crimes committed by subordinates or others subject to their control. This situation arises when the criminal acts are committed pursuant to the commander's order. Commanders are also responsible if they have actual knowledge, or should have knowledge, through reports received or through other means, that troops or other persons subject to their control are about to commit or have committed a crime, and they fail to take the necessary and reasonable steps to ensure compliance with the law or to punish violators.

GENERAL ORDERS

D-25. Orders issued by general officers in command during COIN likely include provisions, such as a prohibition against drinking alcohol or against entering places of religious worship, important to maintaining discipline of the force, to safeguarding the image of U.S. forces, and to promoting the legitimacy of the host government. These orders are readily enforceable under the Uniform Code of Military Justice.

CIVILIAN PERSONNEL AND CONTRACTORS

D-26. Modern COIN operations involve many DOD civilians as well as civilian personnel employed by government contractors. The means of disciplining such persons for violations differ from the means of disciplining uniformed personnel. These civilians may be made subject to general orders. They are also subject to U.S. laws and to the laws of the host nation. These civilians may be prosecuted or receive adverse administrative action by the United States or contract employers. DOD directives contain further policy and guidance pertaining to U.S. civilians accompanying our forces in COIN.

HUMANITARIAN RELIEF AND RECONSTRUCTION

D-27. In COIN, like all operations, commands require specific authority to expend funds. That authority is normally found in the DOD Appropriations Act, specifically, operation and maintenance funds. In recent COIN operations, Congress appropriated additional funds to commanders for the specific purpose of dealing with COIN. Recent examples include the commander's emergency response program (CERP), the Iraq Relief and Reconstruction Fund, Iraq Freedom Fund, and Commander's Humanitarian Relief and Reconstruction Program funds.

DOD FUNDS GENERALLY NOT EXPENDABLE BY COMMANDERS FOR THIS PURPOSE

D-28. Congress specifically appropriates funds for foreign assistance. The United States Agency for International Development expends such funds under the legal authorities in Title 22, United States Code. Provisions of Title 10 authorize small amounts of money. These funds are appropriated annually for commanders to provide humanitarian relief, disaster relief, or civic assistance in conjunction with military operations. These standing authorities are narrowly defined and generally require significant advance coordination within the DOD and the Department of State. As such, they are of limited value to ongoing COIN operations.

COMMANDERS EMERGENCY RESPONSE PROGRAM

D-29. Beginning in November of 2003, Congress authorized use of a specific amount of operations and maintenance funds for a CERP in Iraq and Afghanistan. The legislation was renewed in successive appropriations and authorization acts. It specified that commanders could spend the funds for urgent humanitarian relief and reconstruction projects. These projects had to immediately assist the Iraqi and Afghan peoples within a commander's area of operations. Congress did not intend the funds to be used as—

- Security assistance such as weapons, ammunition, and supplies for security forces.
- Salaries for Iraqi or Afghan forces or employees.
- Rewards for information.
- Payments in satisfaction of claims made by Iraqis or Afghanis against the United States (specific legislation must authorize such payments).

D-30. The CERP provided tactical commanders a ready source of cash for small-scale projects. They could repair public buildings, clear debris from roadways, provide supplies to hospitals and schools, and meet other local needs. Because Congress had provided special authority for the program, normal federal acquisition laws and regulations did not apply. The reporting requirements were minimal.

D-31. The CERP is not a standing program. Any similar future program should be governed by whatever specific legislative provision Congress chooses to enact. In any program similar to CERP, commanders and staffs must make sound, well-coordinated decisions on how to spend the funds. They must ensure that maximum goodwill is created. Commanders must verify that the extra cash does not create harmful effects in the local economy. One such side effect would be creating unsustainable wages that divert skilled labor from a host-nation (HN) program essential to its legitimacy. Commanders must also ensure that projects can be responsibly administered to achieve the desired objective and that they avoid financing insurgents inadvertently.

TRAINING AND EQUIPPING FOREIGN FORCES

D-32. Effective foreign forces need training and equipment. U.S. laws require Congress to authorize such expenditures. U.S. laws also require the Department of State to verify that the host nation receiving the assistance is not in violation of human rights.

NEED FOR SPECIFIC AUTHORITY

D-33. All training and equipping of foreign security forces must be specifically authorized. Usually, DOD involvement is limited to a precise level of man-hours and materiel requested from the Department of State under the Foreign Assistance Act. The President may authorize deployed U.S. forces to train or advise HN security forces as part of the operational mission. In this case, DOD personnel, operations, and maintenance appropriations provide an incidental benefit to those security forces. All other weapons, training, equipment, logistic support, supplies, and services provided to foreign forces must be paid for with funds appropriated by Congress for that purpose. Examples include the Iraq Security Forces Fund and the Afghan Security Forces Fund of fiscal year 2005. Moreover, the President must give specific authority to the DOD for its role in such "train and equip" efforts. In May of 2004, the President signed a decision directive that made the commander, U.S. Central Command, under policy guidance from the chief of mission, responsible for coordinating all U.S. Government efforts to organize, train, and equip Iraqi Security Forces, including police. Absent such a directive, DOD lacks authority to take the lead in assisting a host nation to train and equip its security forces.

HUMAN RIGHTS VETTING

D-34. Congress typically limits when it will fund training or equipment for foreign security forces. If the Department of State has credible information that the foreign security force unit identified to receive the training or equipment has committed a gross violation of human rights, Congress prohibits funding. Such prohibitions impose a requirement upon Department of State and DOD. These departments must vet the proposed recipient units against a database of credible reports of human rights violations.

CLAIMS AND SOLATIA

D-35. Under certain conditions, the U.S. Government will make payments to HN civilians. The Foreign Claims Act permits claims to be filed against the U.S. Government. In some countries, solatia payments are made.

FOREIGN CLAIMS ACT

D-36. Under the Foreign Claims Act, claims by HN civilians for property losses, injury, or death caused by service members or the civilian component of the U.S. forces may be paid to promote and maintain friendly relations with the host nation. Claims that result from noncombat activities or negligent or wrongful acts or omissions are also payable. Claims that are not payable under the Foreign Claims Act include losses from combat, contractual matters, domestic obligations, and claims which are either not in the best interest of the United States to pay, or which are contrary to public policy. Because payment of claims is specifically governed by law and because many claims prove, upon investigation, to be not payable, U.S. forces must be careful not to raise expectations by promising payment.

SOLATIA

D-37. If U.S. forces are conducting COIN in a country where payments in sympathy or recognition of loss are common, solatia payments to accident victims may be legally payable. Solatia payments are not claims payments. They are payments in money or in kind to a victim or to a victim's family as an expression of sympathy or condolence. The payments are customarily immediate and generally nominal. The individual or unit involved in the damage has no legal obligation to pay; compensation is simply offered as an expression of remorse in accordance with local custom. Solatia payments should not be made without prior coordination with the combatant command.

ESTABLISHING THE RULE OF LAW

D-38. Establishing the rule of law is a key goal and end state in COIN. Defining that end state requires extensive coordination between the instruments of U.S. power, the host nation, and multinational partners. Additionally, attaining that end state is usually the province of HN authorities, international and intergovernmental organizations, the Department of State, and other U.S. Government agencies, with support from U.S. forces in some cases. Some key aspects of the rule of law include:

- **A government that derives its powers from the governed** and competently manages, coordinates, and sustains collective security, as well as political, social, and economic development. This includes local, regional, and national government.
- **Sustainable security institutions.** These include a civilian-controlled military as well as police, court, and penal institutions. The latter should be perceived by the local populace as fair, just, and transparent.
- **Fundamental human rights.** The United Nations Declaration on Human Rights and the International Convention for Civil and Political Rights provide a guide for applicable human rights. The latter provides for derogation from certain rights, however, during a state of emergency. Respect for the full panoply of human rights should be the goal of the host nation; derogation and

violation of these rights by HN security forces, in particular, often provides an excuse for insurgent activities.

D-39. In periods of extreme unrest and insurgency, HN legal structures—courts, prosecutors, defense assistance, and prisons—may cease to exist or function at any level. Under these conditions, counterinsurgents may need to undertake a significant role in the reconstruction of the HN judicial system in order to establish legal procedures and systems to deal with captured insurgents and common criminals. During judicial reconstruction, counterinsurgents can expect to be involved in providing sustainment and security support. They can also expect to provide legal support and advice to the HN judicial entities. Even when judicial functions are restored, counterinsurgents may still have to provide logistic and security support to judicial activities for a prolonged period. This support continues as long as insurgents continue to disrupt activities that support the legitimate rule of law.

This page intentionally left blank.

Appendix E

Airpower in Counterinsurgency

Counterinsurgency operations are, by their nature, joint operations—and airpower and landpower are interdependent elements of such operations. As this appendix explains, airpower and spacepower are important force multipliers for U.S., multinational, and host-nation forces fighting an insurgency.

OVERVIEW

E-1. Airpower can contribute significant support to land forces conducting counterinsurgency (COIN) operations. Aircraft can, for example, strike insurgents, and that can be enormously important in many situations. However, given the nature of the COIN environment, airpower will most often transport troops, equipment, and supplies and perform intelligence, surveillance, and reconnaissance missions. Rough terrain and poor transportation networks can create serious obstacles for COIN forces while giving advantages to insurgents. Airpower helps counterinsurgents overcome these obstacles. Thus, airpower both serves as a significant force multiplier and enables counterinsurgents to operate more effectively.

E-2. Airpower provides considerable asymmetric advantages to counterinsurgents. If insurgents assemble a conventional force, air assets can respond quickly with precision fires. In a sudden crisis, air mobility can immediately move land forces where they are needed. In numerous COIN operations, airpower has demonstrated a vital supporting role. In Malaya (1948 through 1960) and El Salvador (1980 through 1992), as well as more recently in Colombia and Afghanistan, airpower contributed significantly to successful COIN operations. In these cases, the ability to airlift British and U.S. Army and police units to remote locations proved important in tracking down and eliminating insurgents. Airpower enables counterinsurgents to operate in rough and remote terrain, areas that insurgents traditionally have used as safe havens.

E-3. Effective leaders also use airpower in roles other than delivering ordnance. In Colombia, aerial crop dusters sprayed and eradicated coca fields that provided drug income for insurgents. During the El Salvador insurgency, medical evacuation (MEDEVAC) helicopters provided to the Salvadoran forces played a central role in improving the Salvadorans' fighting capabilities. Salvadoran morale improved noticeably when soldiers knew that, if they were wounded, MEDEVAC helicopters would get them to a hospital in minutes. With this air support, the Salvadoran Army became much more aggressive in tracking down and engaging insurgents.

E-4. Air transport can also quickly deliver humanitarian assistance. In isolated regions, using air transport to airlift or airdrop food and medical supplies to civilians can help win the populace's support. Air transport is also important for COIN logistics. In areas where ground convoys are vulnerable, U.S. forces can airlift supplies, enabling commanders to maintain forces in remote but strategically important locations.

AIRPOWER IN THE STRIKE ROLE

E-5. Precision air attacks can be of enormous value in COIN operations; however, commanders exercise exceptional care when using airpower in the strike role. Bombing, even with the most precise weapons, can cause unintended civilian casualties. Effective leaders weigh the benefits of every air strike against its risks. An air strike can cause collateral damage that turns people against the host-nation (HN) government and provides insurgents with a major propaganda victory. Even when justified under the law of war, bombings that result in civilian casualties can bring media coverage that works to the insurgents' benefit. For example, some Palestinian militants have fired rockets or artillery from near a school or village to draw a retaliatory air strike that kills or wounds civilians. If that occurs, the insurgents display those killed and wounded to the media as victims of aggression.

E-6. Even when destroying an obvious insurgent headquarters or command center, counterinsurgents must take care to minimize civilian casualties. New, precise munitions with smaller blast effects can limit collateral damage. When considering the risk of civilian casualties, commanders must weigh collateral damage against the unintended consequences of taking no action. Avoiding all risk may embolden insurgents while providing them sanctuary. The proper and well-executed use of aerial attack can conserve resources, increase effectiveness, and reduce risk to U.S. forces. Given timely, accurate intelligence, precisely delivered weapons with a demonstrated low failure rate, appropriate yield, and proper fuse can achieve desired effects while mitigating adverse effects. However, inappropriate or indiscriminate use of air strikes can erode popular support and fuel insurgent propaganda. For these reasons, commanders should consider the use of air strikes carefully during COIN operations, neither disregarding them outright nor employing them excessively.

AIRPOWER IN INTELLIGENCE COLLECTION

E-7. Given the challenges faced by human intelligence (HUMINT) assets in finding and penetrating insurgent networks, counterinsurgents must effectively employ all available intelligence collection capabilities. A combination of unmanned aircraft systems, manned aircraft, and space-based platforms can provide counterinsurgents with many collection capabilities.

E-8. When insurgents operate in rural or remote areas, aerial reconnaissance and surveillance proves useful. Working with signals intelligence (SIGINT), aerial reconnaissance and surveillance uses imagery and infrared systems to find hidden base camps and insurgent defensive positions. Persistent aerial surveillance can often identify people, vehicles, and buildings—even when they are hidden under heavy growth. Manned and unmanned aircraft can patrol roads to locate insurgent ambushes and improvised explosive devices. Air-mounted SIGINT collection platforms can detect insurgent communications and locate their points of origin.

E-9. Air assets have proven important in tactical operations and in convoy and route protection. Helicopters have been especially useful in providing overwatch, fire support, alternate communications, and MEDEVAC support. At the tactical level, air support requires a decentralized command and control system that gives supported units immediate access to available combat air assets and to information collected by air reconnaissance and support assets.

E-10. However, intelligence obtained through air and space platforms works best when it is quickly and efficiently routed to a joint intelligence center. This center fuses HUMINT information with that collected by other intelligence disciplines. To provide a complete picture, air and space intelligence must be combined with HUMINT. For example, while SIGINT and aerial surveillance and reconnaissance assets can determine that people are evacuating a village, they cannot explain why the people are leaving.

E-11. HUMINT is also a key enabler of airpower in the strike role. Commanders require the best possible intelligence about a target and its surrounding area when considering an air strike. With proper placement and access to a target, a HUMINT source can often provide the most accurate target data. Details might include optimum strike times, detailed descriptions of the surrounding area, and the presence of sensitive sites like hospitals, churches, and mosques. Target data can include other important factors for collateral damage considerations. Poststrike HUMINT sources equipped with a cell phone, radio, or camera can provide an initial battle damage assessment in near real time. With a thorough debriefing, the HUMINT source can provide an accurate assessment of the functional and psychological effects achieved on the target. Commanders can use this information to assess restrike options.

AIR AND SPACE INFORMATION OPERATIONS

E-12. Air and space forces have information operations (IO) capabilities that include collecting, controlling, exploiting, and protecting information. To make IO most effective, commanders should seamlessly integrate it among all Service components. Air and space forces contribute to the execution of three IO missions:

- Influence operations.
- Electronic warfare.
- Network operations.

E-13. Air and space forces conduct and support many influence operations. These operations include the following:

- Counterpropaganda.
- Psychological operations.
- Military deception.
- Operations security.
- Counterintelligence.
- Public affairs (a related activity of IO).

Commanders must preplan and deconflict these activities to ensure success.

E-14. Airpower and spacepower also contribute to information superiority through electronic warfare operations. Air and space assets are critical in the effort to shape, exploit, and degrade the enemy's electronic devices while protecting and enhancing those of counterinsurgents. The electronic warfare spectrum is not limited to radio frequencies; it includes the optical and infrared regions as well.

E-15. In this context, *network operations* are activities conducted to operate and defend the Global Information Grid (JP 1-02). Commanders enhance these operations by using air and space systems. Such tools help achieve desired effects across the interconnected analog and digital network portions of the Global Information Grid.

HIGH-TECHNOLOGY ASSETS

E-16. Today's high-technology air and space systems have proven their worth in COIN operations. Unmanned aircraft systems, such as the Predator, give counterinsurgents unprecedented capabilities in surveillance and target acquisition. Aerial surveillance platforms with long loiter times can place an entire region under constant surveillance. Tactical air control parties now provide ground commanders beyond-line-of-sight awareness with ROVER (remote operations video enhanced receiver), which links to aircraft targeting pods and unmanned aircraft systems. Predators have been equipped with precision munitions and successfully employed in the strike role against senior terrorist leaders. Air- and space-based SIGINT platforms give U.S. forces and multinational partners important information collection capabilities. Modern munitions, such as the joint direct attack munition, can guide accurately through clouds and bad weather to destroy insurgent targets under adverse conditions.

LOW-TECHNOLOGY ASSETS

E-17. Today's low-technology aspects of airpower have also proven effective in COIN operations. Light, slow, inexpensive civilian aircraft often have successfully patrolled border areas. In the 1980s, Guatemala mobilized its civilian light aircraft, formed them into an air force reserve, and used them to patrol main roads to report suspected ambushes. This successfully deterred insurgent attacks along Guatemala's major routes. In Africa in the 1980s, South African forces used light aircraft to locate small groups of insurgents trying to infiltrate Namibia from Angola. In Iraq, light aircraft fly patrols to spot insurgents crossing the border. Israel and the United States have even used stationary balloons equipped with video cameras and infrared sensors to watch for border incursions. These unmanned balloons are a simple, inexpensive, and effective means to monitor activity in remote areas.

E-18. The United States and many small nations have effectively used aerial gunships as close air support weapons in COIN operations. A gunship is a transport aircraft modified to carry and fire heavy guns and light artillery from fixed mounts. Many gunship models exist. They range from the Air Force's AC-130 to smaller transports modified to carry weapons ranging from .50-caliber machine guns to 40-millimeter rapid-fire cannons. The gunship's major limitation is its vulnerability to antiaircraft weapons and missiles. Gunships require a relatively benign environment to operate.

AIRLIFT

E-19. Airlift provides a significant asymmetric advantage to COIN forces, enabling commanders to rapidly deploy, reposition, sustain, and redeploy land forces. While land forces can execute these basic missions alone, airlift bypasses weaknesses insurgents have traditionally exploited. For example, airlift enables land forces to operate in rough and remote terrain and to avoid lines of communications (LOCs) targeted by insurgents. During Operation Iraqi Freedom, airlift has provided protected LOCs through convoy mitigation flights. These flights rerouted typical convoy supplies and vehicles. Since insurgents frequently attacked ground convoys, convoy mitigation flights saved lives.

E-20. Sources of airlift include multinational and HN rotary- and fixed-wing assets. Special operations forces provide specialized airlift capabilities for inserting and extracting troops. Strategic intertheater airlift platforms can provide a logistic pipeline. This pipeline moves large quantities of time-critical equipment, supplies, and personnel into and out of a theater. Modern strategic airlift can often provide a direct delivery capability, landing at relatively short, austere fields formerly serviced only by intratheater airlift.

E-21. Modes of airlift include airland and airdrop. Each mode provides advantages and disadvantages, depending on the environment. Airland missions carry greater payloads, resulting in less potential for damage. They also provide backhaul capability (critical for MEDEVAC), troop rotation, equipment repair, and repositioning and redeployment of COIN forces. Fixed-wing assets on airland missions require longer and better prepared landing surfaces. Vertical-lift assets on airland missions can operate from much smaller, more austere fields; however, they fly at slower speeds and often have smaller payloads and shorter ranges. Airdrop missions require the least amount of infrastructure at the receiving end and allow for rapid buildup of forces—up to brigade size. Equally important, airdrop can provide precision insertion and sustainment of numerous small units. Advances in precision-guided, steerable parachutes increase the capability of high-value airdrop missions.

E-22. Airlift is more costly than surface transportation. It is usually a small percentage of the overall transportation network during major combat operations; however, in particularly challenging situations, airlift may become the primary transportation mode for sustainment and repositioning.

E-23. Airlift supports every logical line of operations. For example, it supports IO when COIN forces provide humanitarian airlift to a battered populace. It clearly supports combat operations. Likewise, airlift supports the essential services, governance, and economic development logical lines of operations. HN security forces thus should include airlift development as the host nation's first component of airpower and spacepower.

THE AIRPOWER COMMAND STRUCTURE

E-24. COIN operations require a joint, multinational command and control architecture for air and space that is effective and responsive. The joint structure applies to more than just U.S. forces; it involves coordinating air assets of multinational partners and the host nation. COIN planning must thus establish a joint and multinational airpower command and control system and policies on the rules and conditions for employing airpower in the theater.

E-25. During COIN operations, most planning occurs at lower echelons. Ideally, components at the operational level fully coordinate these plans. Air and space planners require visibility of actions planned at all echelons to provide the most effective air and space support. Furthermore, COIN planning is often fluid and develops along short planning and execution timelines, necessitating informal and formal coordination and integration for safety and efficiency.

E-26. U.S. and multinational air units, along with HN forces, will likely use expeditionary airfields. COIN planners must consider where to locate airfields, including those intended for use as aerial ports of debarkation and other air operations. Factors to consider include—

- Projected near-, mid- and long-term uses of the airfield.
- Types and ranges of aircraft to be operated.
- Shoulder-launched, surface-to-air-missile threats to aircraft.

- Stand-off threats to airfields.
- Proximity to other threats.
- Proximity to land LOCs.
- Availability of fuels.

Airpower operating from remote or dispersed airfields may present a smaller signature than large numbers of land forces, possibly lessening HN sensitivities to foreign military presence. Commanders must properly protect their bases and coordinate their defense among all counterinsurgents.

BUILDING HOST-NATION AIRPOWER CAPABILITY

E-27. U.S. and multinational operations strive to enable the host nation to provide its own internal and external defense. Planners therefore need to establish a long-term program to develop a HN airpower capability. The HN air force should be appropriate for that nation's requirements. For conducting effective COIN operations, a HN air force requires the following basic capabilities:

- Aerial reconnaissance and surveillance.
- Air transport.
- Close air support for land forces.
- Helicopter troop lift.
- MEDEVAC.
- Counterair.
- Interdiction.

E-28. The first step in developing HN airpower is developing the right organizational model for a HN air force. Planning should identify gaps in the host nation's ability to command, control, and employ airpower in COIN operations.

E-29. The next step is to help the host nation develop its aviation infrastructure under a long-term plan. Most developing nations need considerable assistance to develop an appropriate organization, a suitable force structure, and basing plans. As airpower assets represent a large cost for a small nation, an effective airfield security program is also necessary.

E-30. An important training asset is the U.S. Air Force Special Operations Command. This command has teams qualified to operate the most common equipment used in developing nations. These teams also have the language and cultural training to effectively support aircrew and personnel training. The Air Force can also train HN pilots and aircrews through the International Military Education and Training Program.

E-31. Planners should consider HN economic and technological resources when selecting equipment. In most cases, the host nation acquires, or the U.S. and multinational partners provide, a small air force. Although this air force often has limited resources, the host nation still should effectively operate and maintain its aircraft and supporting systems. Multinational support in training and equipping the HN air force can be very important. U.S. aircraft have tremendous capabilities, but they can be too expensive and too complex for some developing nations to operate and maintain. Multinational partners with capable, but less expensive and less sophisticated, aircraft can often help equip the host nation.

E-32. Training and developing a capable HN air force takes considerable time due to the requirements to qualify aircrews, maintenance personnel, and other specialists. Working effectively in joint operations and coordinating support to land forces requires a high skill level. Even when the HN army and police are trained, U.S. personnel will likely stay with HN forces to perform liaison for supporting U.S. air assets and to advise HN forces in the use of their own airpower.

E-33. Developing capable air forces usually takes longer than developing land forces. As a result, Air Force units, advisors, and trainers will likely remain after land force trainers and advisors have completed their mission. Effective air and land operations are complex and require many resources. Often host nations continue to rely on U.S. air liaison personnel, land controllers, and aircraft for an extended period. Thus, COIN planners must consider the long-term U.S. air support requirements in comprehensive COIN planning.

This page intentionally left blank.

Source Notes

These are the sources used, quoted, or paraphrased in this publication. They are listed by page number. Where material appears in a paragraph, both page and paragraph number are listed. Boldface indicates the titles of historical vignettes. Web sites were accessed during December 2006.

ix "This is a game of wits and will...": Peter J. Schoomaker, quoted in "Serving a Nation at War: A Campaign Quality Army with Joint and Expeditionary Capabilities," (Statement before the House Armed Services Committee, 21 Jul 2004), 17. United States House of Representatives Web site
< http://www.house.gov/hasc/openingstatementsandpressreleases/108thcongress/ >

1-1 "Counterinsurgency is not just thinking man's warfare.": This sentence appeared in an electronic mail message from a special forces officer serving in Iraq to the military assistant to the deputy secretary of defense in 2005.

1-4 para 1-19. "The printing press is...": T.E. Lawrence, "Evolution of a Revolt," *The Army Quarterly* (Devon, United Kingdom) 1, 1 (Oct 1920; reprint, Fort Leavenworth, KS: U.S. Army Command and General Staff College, Combat Studies Institute, 1990): 11. Combined Arms Research Library Web site < http://cgsc.leavenworth.army.mil/carl/download/csipubs >.

1-12 1-67. .planners assumed that combatants required a 10 or 15 to 1 advantage.: Andrew F. Krepinevich, *The Army and Vietnam* (Baltimore: Johns Hopkins University Press, 1986), 157; A better force requirement gauge..." John J. McGrath, *Boots on the Ground: Troop Density in Contingency Operations,* Global War on Terrorism Occasional Paper 16 (Fort Leavenworth, KS: Combat Studies Institute Press, 2006), 1, 6n1.

1-22 para 1-123. ...revolutionary war was 80 percent political action and only 20 percent military: cited in David Galula, *Counterinsurgency Warfare: Theory and Practice* (1964; reprint New York: Praeger, 2005), 89 (hereafter cited as Galula); ...he was involved with establishing special schools...: Walter Sullivan, "China's Communists Train Political Corps to Aid Army," *The New York Times,* 4 Jul 1949.

1-27 1-154. "There's very clear evidence,...": Creighton W. Abrams Jr., Commander's Weekly Intelligence Estimate (Headquarters, Military Assistance Command, Vietnam, 21 Nov 1970), audiotape, U.S. Army Military History Institute, Carlisle Barracks, PA; Lewis Sorley, editor, *Vietnam Chronicles: The Abrams Tapes 1968-1972* (Lubbock: Texas Tech University Press, 2004), 633 (hereafter cited as Sorley); "Do not try and do too much.": T.E. Lawrence, "Twenty-seven Articles," *Arab Bulletin* (20 August 1917), in T.E. Lawrence, *Secret Despatches* [sic] *from Arabia,* A. W. Lawrence, ed. (London, Golden Cockerel Press, 1939), 126-133. T.E. Lawrence.net Web site < http://telawrence.net/telawrencenet/works/ >.

2-1 "Essential though it is,.": Galula, 89.

2-3 **"Hand Shake Con" in Operation Provide Comfort:** Anthony C. Zinni, "Non-Traditional Military Missions: Their Nature, and the Need for Cultural Awareness and Flexible Thinking," chapter 4 in *Capital "W" War: A Case for Strategic Principles of War (Because Wars Are Conflicts of Societies, Not Tactical Exercises Writ Large),* by Joe Strange, Perspectives on Warfighting 6 (Quantico, VA: Marine Corps University, 1998), 265-266. (Operation Provide Comfort began on April 6, 1991 and ended July 24, 1991. Operation Provide Comfort II was a show of force to deter new Iraqi attacks on the Kurds and had only limited humanitarian aspects. Provide Comfort II began 24 July 1991 and ended 31 December 1996.)

2-9 para 2-42. "To confine soldiers...": Galula, 88.

2-10 para 2-48. The Foreign Service Act assigns...: Foreign Service Act of 1980, Public Law 96-465, 96th Congress, 2d session (17 Oct 1980), sections 102(3) and 207a(1).

2-12 **Provincial Reconstruction Teams in Afghanistan:** Prepared based on the experiences of several participants in operations in Afghanistan between 2003 and 2005.

2-12 **civil operations and Rural Development support and Accelerated pacification in Vietnam:** Richard A. Hunt, *Pacification: the American Struggle for Vietnam's Hearts and Minds* (Boulder, CO: Westview Press, 1995), 86-279; Thomas W. Scoville, *Reorganizing for Pacification Support* (Washington, DC: Center of Military History, 1991), 43-83.

2-14 para 2-57. "You [military professionals] must know...": John F. Kennedy, *Public Papers of the Presidents of the United States: John F. Kennedy, Containing the Public Messages, Speeches, and Statements of the President, January 20 to December 31,1961* (Washington, DC: Government Printing Office, 1962), 448. (The remarks were made at Annapolis, MD, to the graduating class of the U.S. Naval Academy on 7 June 1961.)

3-1 "Everything *good* that happens.": Sorley, 506.

3-18 **Asymmetric Tactics in ireland:** James Fintan Lalor to d'Arcy McGee, 30 March 1847, *Collected Writings of James Fintan Lalor,* edited by L. Fogarty (Dublin, 1918), 83. Quoted in *Handbook for Volunteers of the Irish Republican Army: Notes on Guerrilla Warfare* (Boulder, CO: Paladin Press, 1985), 3; Charles Townshend, *Political Violence in Ireland: Government and Resistance Since 1848* (New York: Oxford University Press, 1983), 32.

4-1 "The first, the supreme,." Carl von Clausewitz, *On War,* ed. and trans. Michael Howard and Peter Paret (Princeton, NJ: Princeton University Press, 1976; reprint 1984), 88-89.

4-1 **Campaign Assessment and Reassessment:** George W. Smith, Jr., "Avoiding a Napoleonic Ulcer: Bridging the Gap of Cultural Intelligence (Or, Have We Focused on the Wrong Transformation?)" (Quantico, VA: Marine Corps War College, 2004). < http://www.mcu.usmc.mil/MCWAR/IRP/default.htm >

4-7 **iterative Design During operation iraqi Freedom ii:** Lieutenant General James N. Mattis, USMC, interviewed by Colonel Douglas M. King, USMC, Marine Corps Combat Development Command, Feb-Mar, 2006.

5-1 "It is a persistently methodical approach.": Sir Robert Thompson, *Defeating Communist Insurgency: The Lessons of Malaya and Vietnam* ([New York: Praeger, 1966]; reprint with a foreword by Robert Bowie, St. Petersburg, FL: Hailer Publishing, 2005), 171 (hereafter cited as Thompson).

5-6 **The importance of Multiple Lines of operations in coiN:** Thomas A Marks, *Counterrevolution in China: Wang Sheng and the Kuomintang* (London: Frank Cass, 1998), 77-117.

5-21 **Clear-Hold-Build in Tal Afar, 2005-2006:** Office of the Assistant Secretary of Defense (Public Affairs), News Transcript, News Briefing with COL H.R. McMaster, 27 Jan 2006, DOD News: News Briefing with COL H.R. McMaster Web site. < http://www.defenselink.mil /transcripts/2006/tr20060127-12385.html >

5-24 **Combined Action Program:** Matthew Danner, "The Combined Action Platoon: Seeds of Success in Iraq," United States Marine Corps Combined Action Platoon (USMC CAP) Web site, < http://capmarine.com/personal/iraq/danner.htm >.

5-25 "The two best guides,.": Thompson, 170.

6-1 "[H]elping others to help themselves.": "Quadrennial Defense Review Report," (Washington, DC: Office of the Secretary of Defense, 6 Feb 2006), 11.

6-18 **Multinational Security Transition Command-Iraq:** John R. Martin, "Training Indigenous Security Forces at the Upper End of the Counterinsurgency Spectrum," *Military Review* 86, 6 (Nov-Dec 2006): 58-64.

6-21 **Developing a Police Force in Malaya:** James S. Corum, *Training Indigenous Forces in Counterinsurgency: A Tale of Two Insurgencies* (Carlisle, PA: Strategic Studies Institute, U.S. Army War College, Mar, 2006), 4-24 < http://www.StrategicStudiesInstitute.army.mil >.

7-1 "Leaders must have a strong sense." MCDP 1, *Warfighting* (20 June 1997), 57.

7-3 **Defusing a Confrontation:** Excerpted from Dan Baum, "Battle Lessons, What the Generals Don't Know," *The New Yorker,* Jan 17, 2005. The New Yorker : fact: content Web site < http://www.newyorker.com/fact/content/7050117fa_fact >.

7-4 **Patience, Presence, and Courage:** Adapted from Nelson Hernandez, "Attacks Rock Foundation That Marines Built in Anbar," *Washington Post,* Feb. 7, 2006: 1.

7-7 para 7-39. "No person in the custody...": Detainee Treatment Act of 2005, Public Law 109-148, 109th Congress, 1st Session, (30 Dec 2005), sections 1002(a), 1003(a).

7-8 **Lose Moral Legitimacy, Lose the War:** Lou DiMarco, "Losing the Moral Compass: Torture and *Guerre Revolutionnaire* in the Algerian War," *Parameters* 36, 2 (Summer, 2006): 63-76.

8-1 "In my experience in previous wars,...: Senior Officer Debriefing Report: MG J.M. Heiser, Jr., Commanding General, 1st Logistical Command, Period 2 August 1968 to 23 August 1969 (Department of the Army: Office of the Adjutant General, 23 Sep 1969), 343.

8-3 **What is Different: Insurgent Perceptions of Military Logistics:** "We have a claim..." Mao Tse-Tung, "Problems of Strategy in China's Revolutionary War," *Selected Military Writings of Mao Tse-Tung* (Peking: Foreign Language Press, 1966), 147; "Weapons are not difficult to obtain..": Ming Fan, "A Textbook on Guerrilla Warfare," translated by and reprinted in Gene Z. Hanrahan, ed. *Chinese Communist Guerrilla Warfare Tactics* (Boulder, CO: Paladin Press, 1974), 76-77.

8-3 "In one moment in time,.": Charles C. Krulak, "The Three Block War: Fighting in Urban Areas," *Vital Speeches of the Day* 64, 5 (New York: December 15, 1997): 139-142. (Speech given before the National Press Club, October 10, 1997. Quote appears on page 139.)

8-6 **Vietnam: Meeting the Enemy and Convoy Security:** John H. Hay, *Vietnam Studies: Tactical and Materiel Innovations* (Washington, DC: Department of the Army, 1989), 154-155.

8-8 **Air Delivery in Iraq: Maximizing Counterinsurgency Potential:** Marine Corps Center for Lessons Learned, "Initial Observations Report-Operation Iraqi Freedom II-1, Subject: Air Delivery Operations," (March, 2005).

8-10 **Building a Military: Sustainment Failure:** Julian Thompson, *The Lifeblood of War: Logistics in Armed Conflict* (New York: Brassey's, 1991), 217-218; Dong Van Khuyen, *RVNAF Logistics* (Washington, DC: U.S. Army Center of Military History, 1980).

8-17 **Host-Nation Contracting: A Potential Double-Edged Sword:** Interview between Major Sean Davis, USA, and Lieutenant Colonel Marian Vlasak, USA (Fort Leavenworth, KS, 1 May 2006).

A-1 Appendix A is based on David Kilcullen, "Countering Global Insurgency," *Journal of Strategic Studies* 28, 4 (August 2005): 597-617.

B-14 **The Capture of Saddam Hussein:** Brian J. Reed, "Formalizing the Informal: A Network Analysis of an Insurgency" (Ph.D. dissertation, Department of Sociology, University of Maryland, College Park, MD, 2006), 81-88.

D-3 para D-14. "In the case of armed conflict not.": Geneva Convention Relative to the Treatment of Prisoners of War, Article 3, August 12, 1949, 6 United States Treaties 3114, 75 United Nations Treaty Series 31.

D-5 table D-1. Detainee Treatment Act of 2005, Public Law 109-148, 109th Congress, 1st Session, (30 Dec 2005), sections 1002, 1003.

This page intentionally left blank.

Glossary

The glossary lists acronyms and terms with Army, multi-Service, or joint definitions, and other selected terms. Where Army and joint definitions are different, *(Army)* follows the term. The proponent manual for other terms is listed in parentheses after the definition. Terms for which the Army and Marine Corps have agreed on a common definition are followed by *(Army-Marine Corps).*

SECTION I - ACRONYMS AND ABBREVIATIONS

AcR	armored cavalry regiment
Ao	area of operations
AR	Army regulation
AscopE	A memory aid for the characteristics of civil considerations: area, structures, capabilities, organizations, people, and events. *See also* METT-TC.
cERp	commander's emergency response program
CJCSI	Chairman of the Joint Chiefs of Staff instruction
CMO	civil-military operations
CMOC	civil-military operations center
coiN	counterinsurgency
coRDs	civil operations and revolutionary (rural) development support
coTs	commercial off-the-shelf
DA	Department of the Army
DOCEX	document exploitation
DoD	Department of Defense
DoDD	Department of Defense Directive
DOTMLPF	Memory aid for the force development domains: doctrine, organization, training, materiel, leadership and education, personnel, and facilities.
FARc	Fuerzas Armadas Revolucionarias de Colombia (Revolutionary Armed Forces of Columbia)
FBi	Federal Bureau of Investigation
FiD	foreign internal defense
FM	field manual
FMFRP	Fleet Marine Force reference publication
FMI	field manual interim
G-4	assistant chief of staff, logistics
GEoiNT	geospatial intelligence
HMMWV	high-mobility, multipurpose, wheeled vehicle
HN	host-nation

HUMINT	human intelligence
iGo	intergovernmental organization
iMET	international military education and training
iMiNT	imagery intelligence
io	information operations
ipB	intelligence preparation of the battelfield
isR	intelligence, surveillance, and reconnaissance
JiAcG	joint interagency coordination group
Jp	joint publication
LLo	logical line of operations
Loc	line of communications
MAcv	Military Assistance Command, Vietnam
MAsiNT	measurement and signature intelligence
McDp	Marine Corps doctrinal publication
Mcip	Marine Corps interim publication
Mco	Marine Corps order
McRp	Marine Corps reference publication
Mcwp	Marine Corps warfighting publication
MEDEvAc	medical evacuation
METT-Tc	A memory aid for mission, enemy, terrain and weather, troops and support available, time available, civil considerations used in two contexts: (1) In the context of information management, the major subject categories into which relevant information is grouped for military operations (FM 6-0); (2) In the context of tactics, the major factors considered during mission analysis (FM 3-90). [Note: The Marine Corps uses METT-T: mission, enemy, terrain and weather, troops and support available, time available.]
MNsTc-i	Multinational Security Transition Command-Iraq
MoE	measure of effectiveness
Mop	measure of performance
NATo	North Atlantic Treaty Organization
Nco	noncommissioned officer
NGo	nongovernmental organization
Nsc	National Security Council
ocs	officer candidate school
osiNT	open-source intelligence
piR	priority intelligence requirement
pRc	purchase request and committal
pRT	provincial reconstruction team
RoE	rules of engagement
S-2	intelligence staff officer
S-4	logistics staff officer
siGiNT	signals intelligence

SNA	social network analysis
SOF	special operations forces
SROE	standing rules of engagement
TAREX	target exploitation
UN	United Nations
U.S.	United States
USA	United States Army
USAID	United States Agency for International Development
USMC	United States Marine Corps

SECTION II - TERMS AND DEFINITIONS

all-source intelligence

(joint) Intelligence products and/or organizations and activities that incorporate all sources of information, most frequently including human resources intelligence, imagery intelligence, measurement and signature intelligence, signals intelligence, and open-source data in the production of finished intelligence. (JP 1-02)

area of interest

(joint) That area of concern to the commander, including the area of influence, areas adjacent thereto, and extending into enemy territory to the objectives of current or planned operations. This area also includes areas occupied by enemy forces who could jeopardize the accomplishment of the mission. (JP 1-02)

area of operations

(joint) An operational area defined by the joint force commander for land and maritime forces. Areas of operations do not typically encompass the entire operational area of the joint force commander, but should be large enough for component commanders to accomplish their missions and protect their forces. (JP 1-02)

area security

A form of security operations conducted to protect friendly forces, installation routes, and actions within a specific area. (FM 3-90)

assessment

(Army) The continuous monitoring and evaluation of the current situation and progress of an operation. (FMI 5-0.1)

board

A temporary grouping of selected staff representatives delegated decision authority for a particular purpose or function. (FMI 5-0.1)

center of gravity

(joint) The source of power that provides moral or physical strength, freedom of action, or will to act. (JP 1-02)

civil considerations

How the manmade infrastructure, civilian institutions, and attitudes and activities of the civilian leaders, populations, and organizations within an area of operations influence the conduct of military operations. (FM 6-0) *See also* METT-TC.

clear

(Army) A tactical mission task that requires the commander to remove all enemy forces and eliminate organized resistance in an assigned area. (FM 3-90)

coalition

(joint) An ad hoc arrangement between two or more nations for common action. (JP 1-02)

combatant commander

(joint) A commander of one of the unified or specified combatant commands established by the President. (JP 1-02)

command and control system

(joint) The facilities, equipment, communications, procedures, and personnel essential to a commander for planning, directing, and controlling operations of assigned forces pursuant to the missions assigned. (JP 1-02) (Army) The arrangement of personnel, information management, procedures, and equipment and facilities essential for the commander to conduct operations. (FM 6-0)

commander's intent

(Army) A clear, concise statement of what the force must do and the conditions the force must meet to succeed with respect to the enemy, terrain, and civil considerations that represent the operation's desired end state. (FMI 5-0.1) (Marine Corps) A commander's clear, concise articulation of the purpose(s) behind one or more tasks assigned to a subordinate. It is one of two parts of every mission statement which guides the exercise of initiative in the absence of instructions. (MCRP 5-12A)

commander's visualization

The mental process of developing situational understanding, determining a desired end state, and envisioning how the force will achieve that end state. (FMI 5-0.1)

command post cell

A grouping of personnel and equipment by warfighting function or purpose to facilitate command and control during operations. (FMI 5-0.1)

common operational picture

(joint) A single identical display of relevant information shared by more than one command. A common operational picture facilitates collaborative planning and assists all echelons to achieve situational awareness. (JP 1-02) (Army) An operational picture tailored to the user's requirements, based on common data and information shared by more than one command. (FM 3-0)

counterinsurgency

(joint) Those military, paramilitary, political, economic, psychological, and civic actions taken by a government to defeat insurgency. (JP 1-02)

counterintelligence

(joint) Information gathered and activities conducted to protect against espionage, other intelligence activities, sabotage, or assassinations conducted by or on behalf of foreign governments or elements thereof, foreign organizations, or foreign persons, or international terrorist activities. (JP 1-02) (Army) Counterintelligence counters or neutralizes intelligence collection efforts through collection, counterintelligence investigations, operations, analysis and production, and functional and technical services. Counterintelligence includes all actions taken to detect, identify, exploit, and neutralize the multidiscipline intelligence activities of friends, competitors, opponents, adversaries, and enemies; and is the key intelligence community contributor to protect United States interests and equities. (FM 2-0)

counterterrorism

(joint) Operations that include the offensive measures taken to prevent, deter, preempt, and respond to terrorism. (JP 1-02)

decisive point

(joint) A geographic place, specific key event, critical system or function that, when acted upon, allows commanders to gain a marked advantage over an enemy or contribute materially to achieving success. (JP 1-02)

dislocated civilian

(joint) A broad term that includes a displaced person, an evacuee, an expellee, an internally displaced person, a migrant, a refugee, or a stateless person. (JP 1-02)

end state

(joint) The set of required conditions that defines achievement of the commander's objectives. (JP 1-02)

execute

To put a plan into action by applying combat power to accomplish the mission and using situational understanding to assess progress and make execution and adjustment decisions. (FM 6-0)

foreign internal defense

(joint) Participation by civilian and military agencies of a government in any of the action programs taken by another government or other designated organization to free and protect its society from subversion, lawlessness, and insurgency. (JP 1-02)

forward operations base

(joint) In special operations, a base usually located in friendly territory or afloat that is established to extend command and control or communications or to provide support for training and tactical operations. Facilities may be established for temporary or longer duration operations and may include an airfield or an unimproved airstrip, an anchorage, or a pier. A forward operations base may be the location of special operations component headquarters or a smaller unit that is controlled and/or supported by a main operations base. (JP 1-02) [Note: Army special operations forces term is "forward operational base."]

full spectrum operations

The range of operations Army forces conduct in war and military operations other than war. (FM 3-0) [Note: A new definition for this term is being staffed for the revision of FM 3-0. Upon publication of FM 3-0, the definition it contains will replace this definition.]

host nation

(joint) A nation that receives the forces and/or supplies of allied nations, coalition partners, and/or NATO organizations to be located on, to operate in, or to transit through its territory. (JP 1-02)

human intelligence

(Army) The collection of information by a trained human intelligence collector from people and their associated documents and media sources to identify elements, intentions, composition, strength, dispositions, tactics, equipment, personnel, and capabilities (FM 2-22.3). [Note: Trained HUMINT collectors are Soldiers holding military occupational specialties 97E, 351Y {formerly 351C}, 351M {formerly 351E}, 35E, and 35F, and Marines holding the specialty 0251.]

information environment

(joint) The aggregate of individuals, organizations or systems that collect, process, or disseminate or act on information. (JP 1-02)

information operations

(joint) The integrated employment of the core capabilities of eletronic warfare, computer network operations, psychological operations, military deception, and operations security, in concert with specified supporting and relted capabilities, to influence, disrupt, corrupt, or usurp adversarial human and automated decision making while protecting our own. (JP 1-02) (Army) The employment of the core capabilities of electronic warfare, computer network operations, psychological operations, military deception, and operations security, in concert with specified supporting and related capabilities, to affect and defend information and information systems and to influence decisionmaking. (FM 3-13)

insurgency

(joint) An organized movement aimed at the overthrow of a constituted government through the use of subversion and armed conflict. (JP 1-02)

intelligence discipline

(joint) A well-defined area of intelligence collection, processing, exploitation, and reporting using a specific category of technical or human resources. There are seven major disciplines: human intelligence, imagery intelligence, measurement and signature intelligence, signals intelligence, open-source intelligence, technical intelligence, and counterintelligence. [Note: The Army definition replaces "all-source analysis and production" with "open-source intelligence."] (JP 1-02)

intelligence preparation of the battlefield

The systematic, continuous process of analyzing the threat and environment in a specific geographic area. Intelligence preparation of the battlefield (IPB) is designed to support the staff estimate and military decision-making process. Most intelligence requirements are generated as a result of the IPB process and its interrelation with the decision-making process. (FM 34-130)

interagency coordination

(joint) Within the context of Department of Defense involvement, the coordination that occurs between elements of Department of Defense and engaged U.S. Government agencies for the purpose of achieving an objective. (JP 1-02)

intergovernmental organization

(joint) An organization created by a formal agreement (e.g. a treaty) between two or more governments. It may be established on a global, regional, or functional basis for wide-ranging or narrowly defined purposes. Formed to protect and promote national interests shared by member states. Examples include the United Nations, North Atlantic Treaty Organization, and the African Union. (JP 1-02)

intuitive decisionmaking

(Army-Marine Corps) The act of reaching a conclusion which emphasizes pattern recognition based on knowledge, judgment, experience, education, intelligence, boldness, perception, and character. This approach focuses on assessment of the situation vice comparison of multiple options. (FM 6-0; MCRP 5-12A)

line of communications

(joint) A route, either land, water, and/or air, that connects an operating military force with a base of operations and along which supplies and military forces move. (JP 1-02)

line of operations

(joint) 1. A logical line that connects actions on nodes and/or decisive points related in time and purpose with an objective(s). 2. A physical line that defines the interior or exterior orientation of the force in relation to the enemy or that connects actions on nodes and/or decisive points related in time and space to an objective(s). (JP 1-02)

measure of effectiveness

(joint) A criterion used to assess changes in system behavior, capability, or operational environment that is tied to measuring the attainment of an end state, achievement of an objective, or creation of an effect. (JP 1-02)

measure of performance

(joint) A criterion to assess friendly actions that is tied to measuring task accomplishment. (JP 1-02)

mission command

The conduct of military operations through decentralized execution based upon mission orders for effective mission accomplishment. Successful mission command results from subordinate leaders at all echelons exercising disciplined initiative within the commander's intent to accomplish missions. It requires an environment of trust and mutual understanding. (FM 6-0)

narrative

The central mechanism, expressed in story form, through which ideologies are expressed and absorbed.

nongovernmental organization

(joint) A private, self-governing, not-for-profit organization dedicated to alleviating human suffering; and/or promoting education, health care, economic development, environmental protection, human rights, and conflict resolution; and/or encouraging the establishment of democratic institutions and civil society. (JP 1-02)

open-source intelligence

(joint) Information of potential intelligence value that is available to the general public. (JP 1-02)

operating tempo

The annual operating miles or hours for the major equipment system in a battalion-level or equivalent organization. Commanders use operating tempo to forecast and allocate funds for fuel and repair parts for training events and programs. (FM 7-0) [Usually OPTEMPO.]

operational environment

(joint) A composite of the conditions, circumstances, and influences that affect the employment of capabilities and bear on the decisions of the commander. (JP 1-02)

operational picture

A single display of relevant information within a commander's area of interest. (FM 3-0)

personnel tempo

The time a service member is deployed. [Usually PERSTEMPO.]

planning

The process by which commanders (and staffs, if available) translate the commander's visualization into a specific course of action for preparation and execution, focusing on the expected results. (FMI 5-0.1)

preparation

Activities by the unit before execution to improve its ability to conduct the operation, including, but not limited to, the following: plan refinement, rehearsals, reconnaissance, coordination, inspection, and movement. (FM 3-0)

reachback

(joint) The process of obtaining products, services, and applications, or forces, or equipment, or material from organizations that are not forward deployed. (JP 1-02)

riverine area

(joint) An inland or coastal area comprising both land and water, characterized by limited land lines of communications, with extensive water surface and/or inland waterways that provide natural routes for surface transportation and communications. (JP 1-02)

rules of engagement

(joint) Directives issued by competent military authority that delineate the circumstances and limitations under which United States forces will initiate and/or continue combat engagement with other forces encountered. (JP 1-02)

running estimate

A staff section's continuous assessment of current and future operations to determine if the current operation is proceeding according to the commander's intent and if future operations are supportable. (FMI 5-0.1)

security

(joint) 1. Measures taken by a military unit, an activity or installation to protect itself against all acts designed to, or which may, impair its effectiveness. 2. A condition that results from the establishment and maintenance of protective measures that ensure a state of inviolability from hostile acts or influences. (JP 1-02)

situational awareness

Knowledge of the immediate present environment, including knowledge of the factors of METT-TC. (FMI 5-0.1)

situational understanding

(Army) The product of applying analysis and judgment to the common operational picture to determine the relationship among the factors of METT-TC. (FM 3-0) (Marine Corps) Knowledge and understanding of the current situation which promotes timely, relevant, and accurate assessment of friendly, enemy, and other operations within the battlespace in order to facilitate decisionmaking. An informational perspective and skill that foster an ability to determine quickly the context and relevance of events that are unfolding. (MCRP 5-12A)

stability operations

(joint) An overarching term encompassing various military missions, tasks, and activities conducted outside the United States in coordination with other instruments of national power to maintain or reestablish a safe and secure environment, provide essential governmental services, emergency infrastructure reconstruction, and humanitarian relief. (JP 1-02)

staff estimate

See running estimate.

status of forces agreement

(joint) An agreement that defines the legal position of a visiting military force deployed in the territory of a friendly state. Agreements delineating the status of visiting military forces may be bilateral or multilateral. Provisions pertaining to the status of visiting forces may be set forth in a separate agreement, or they may form a part of a more comprehensive agreement. These provisions describe how the authorities of a visiting force may control members of that force and the amenability of the force or its members to the local law or to the authority of local officials. To the extent that agreements delineate matters affecting the relations between a military force and civilian authorities and population, they may be considered as civil affairs agreements. (JP 1-02)

strike

(joint) An attack to damage or destroy an objective or capability. (JP 1-02)

subordinates' initiative

The assumption of responsibility for deciding and initiating independent actions when the concept of operations or order no longer applies or when an unanticipated opportunity leading to the accomplishment of the commander's intent presents itself. (FM 6-0)

tempo

(Army) The rate of military action. (FM 3-0) (Marine Corps) The relative speed and rhythm of military operations over time with respect to the enemy. (MCRP 5-12A)

theater of war

(joint) Defined by the Secretary of Defense or the geographic combatant commander, the area of air, land, and water that is, or may become, directly involved in the conduct of the war. A theater of war does not normally encompass the geographic combatant commander's entire area of responsibility and may contain more than one theater of operations. (JP 1-02)

warfighting function

A group of tasks and systems (people, organizations, information, and processes) united by a common purpose that commanders use to accomplish missions and training objectives. (FMI 5-0.1)

working group

A temporary grouping of predetermined staff representatives who meet to coordinate and provide recommendations for a particular purpose or function. (FMI 5-0.1)

Annotated Bibliography

This bibliography is a tool for Army and Marine Corps leaders to help them increase their knowledge of insurgency and counterinsurgency. Reading what others have written provides a foundation that leaders can use to assess counterinsurgency situations and make appropriate decisions. The books and articles that follow are not the only good ones on this subject. The field is vast and rich. They are, however, some of the more useful for Soldiers and Marines. (Web sites were accessed during December 2006.)

THE CLASSICS

Calwell, Charles E. *Small Wars: Their Principles and Practice.* Lincoln, NE: University of Nebraska Press, 1996. (Reprint of *Small Wars: A Tactical Textbook for Imperial Soldiers* [London: Greenhill Books, 1890]. A British major general who fought in small wars in Afghanistan and the Boer War provides lessons learned that remain applicable today.)

Galula, David. *Counterinsurgency Warfare: Theory and Practice.* London: Praeger, 1964. (Lessons derived from the author's observation of insurgency and counterinsurgency in Greece, China, and Algeria.)

Gurr, Ted Robert. *Why Men Rebel.* Princeton, NJ: Princeton University Press, 1971. (Describes the relative deprivation theory, which states that unmet expectations motivate those who join rebel movements.)

Hoffer, Eric. *The True Believer: Thoughts on the Nature of Mass Movements.* New York: Harper Perennial Modern Classics, 2002. (This book, originally published in 1951, explains why people become members of cults and similar groups.)

Horne, Alistair. *A Savage War of Peace.* New York: Viking, 1977. (One of the best analyses of the approaches and problems on both sides during the war in Algeria. For more on this conflict, see *The Battle of Algiers,* a troubling and instructive 1966 movie.)

Jeapes, Tony. *SAS Secret War.* London: Greenhill Books, 2005. (How the British Special Air Service raised and employed irregular tribal forces to counter a communist insurgency in Oman during the 1960s and 1970s.)

Kitson, Frank. *Low Intensity Operations: Subversion, Insurgency and Peacekeeping.* London: Faber and Faber, 1971. (Explanation of the British school of counterinsurgency from one of its best practitioners.)

Komer, Robert. *Bureaucracy Does Its Thing: Institutional Constraints on U.S.-GVN Performance in Vietnam.* Washington, D.C.: RAND, 1972. Rand Corporation Web site < http://www.rand.org/pubs/reports/R967/ > (Bureaucracies do what they do—even if they lose the war.)

Larteguy, Jean. *The Centurions.* New York: Dutton, 1962. (A fact-based novel about the French experience in Vietnam and Algeria that depicts the leadership and ethical dilemmas involved in counterinsurgency. The sequel *The Praetorians* is also a classic depiction of the impact of ethical erosion on a military organization.)

Lawrence, T.E. *Seven Pillars of Wisdom: A Triumph.* New York: Anchor, 1991. (Reprint of 1917 book published in London by George Doran. Autobiographical account of Lawrence of Arabia's attempts to organize Arab nationalism during World War I.)

———. "The 27 Articles of T.E. Lawrence." *The Arab Bulletin* (20 Aug 1917). Defense and the National Interest Web site < http://www.d-n-i.net/fcs/lawrence_27_articles.htm > (Much of the best of *Seven Pillars of Wisdom* in easily digestible bullet points.)

Linn, Brian McAllister. *The Philippine War, 1899-1902.* Lawrence, KS: University Press of Kansas, 2002. (The definitive treatment of successful U.S. counterinsurgency operations in the Philippines.)

Mao Zedong. *On Guerrilla Warfare.* London: Cassell, 1965. (Mao describes the principles which he used so well in seizing power in China and which have inspired many imitators.)

McCuen, John J. *The Art of Counter-Revolutionary War.* St. Petersburg, FL: Hailer Publishing, 2005. (Originally published by Harrisburg, PA: Stackpole Books, 1966. Discusses theory, practice, and historical keys to victory.)

Race, Jeffrey. *War Comes to Long An: Revolutionary Conflict in a Vietnamese Province.* Berkeley, CA: University of California Press, 1972. (Counterinsurgency is scalable. Depicts the evolution of insurgency in one province in Vietnam.)

Thompson, Robert. *Defeating Communist Insurgency.* St. Petersburg, FL: Hailer Publishing, 2005. (Written in 1966. Provides lessons from the author's counterinsurgency experience in Malaya and Vietnam.)

Trinquier, Roger. *Modern Warfare: A French View of Counterinsurgency.* New York: Praeger, 1964. (The French school of counterinsurgency with a focus on "whatever means necessary.")

United States Marine Corps. *Small Wars Manual.* Washington, D.C.: Government Printing Office, 1987. Air War College Gateway to the Internet Web site < http://www.au.af.mil/au/ > (This book, originally published in 1940, covers lessons learned from the Corps' experience in the interwar years.)

West, Bing. *The Village.* New York: Pocket Books, 1972. (A first-person account of military advisors embedded with Vietnamese units.)

OVERVIEWS AND SPECIAL SUBJECTS IN COUNTERINSURGENCY

Asprey, Robert. *War in the Shadows: The Guerrilla in History.* 2 vols. New York: William Morrow, 1994. (First published in 1975. Presents the history of guerrilla war from ancient Persia to modern Afghanistan.)

Baker, Ralph O. "The Decisive Weapon: A Brigade Combat Team Commander's Perspective on Information Operations." *Military Review* 86, 3 (May-Jun 2006), 13-32. (A brigade combat team commander in Iraq in 2003-2004 gives his perspective on information operations.)

Corum, James and Wray Johnson. *Airpower in Small Wars: Fighting Insurgents and Terrorists.* Lawrence, KS: University Press of Kansas, 2003. (Depicts uses and limits of airpower and technology in counterinsurgency.)

Davidson, Phillip. *Secrets of the Vietnam War.* Novato, CA: Presidio Press, 1990. (MACV commander General Westmoreland's intelligence officer provides an insightful analysis of the intricacies of the North Vietnamese strategy of dau tranh ["the struggle"].)

Ellis, John. *From the Barrel of a Gun: A History of Guerrilla, Revolutionary, and Counter-insurgency Warfare from the Romans to the Present.* London: Greenhill, 1995. (A comprehensive short overview of counterinsurgency.)

Hammes, T.X. *The Sling and the Stone: On War in the 21st Century.* Osceola, WI: Zenith Press, 2004. (The future of warfare for the West is insurgency and terror according to a Marine with Operation Iraqi Freedom experience.)

Krepinevich, Andrew Jr. *The Army and Vietnam.* Baltimore: Johns Hopkins University Press, 1986. (Argues that the Army never adapted to the insurgency in Vietnam, preferring to fight the war as a conventional conflict with an emphasis on firepower.)

Merom, Gil. *How Democracies Lose Small Wars: State, Society, and the Failures of France in Algeria, Israel in Lebanon, and the United States in Vietnam.* New York: Cambridge University Press, 2003. (Examines the cases of Algeria, Lebanon, and Vietnam. Determines that great powers lose small wars when they lose public support at home.)

Nagl, John A. *Learning to Eat Soup with a Knife: Counterinsurgency Lessons from Malaya and Vietnam.* Chicago: University of Chicago Press, 2005. (How to learn to defeat an insurgency. Foreword by Peter J. Schoomaker.)

O'Neill, Bard E. *Insurgency and Terrorism: From Revolution to Apocalypse.* Dulles, VA: Potomac Books, 2005. (A framework for analyzing insurgency operations and a good first book in insurgency studies.)

Sepp, Kalev I. "Best Practices in Counterinsurgency." *Military Review* 85, 3 (May-Jun 2005), 8-12. (Historical best practices for success in counterinsurgency.)

Shy, John and Thomas W. Collier. "Revolutionary War" in Peter Paret, ed. *Makers of Modern Strategy: From Machiavelli to the Nuclear Age.* Princeton, NJ: Princeton Univ. Press, 1986. (One of the best overview of the various counterinsurgency schools, discussing both the writings and the contexts in which they were developed.)

Sorley, Lewis. *A Better War: The Unexamined Victories and Final Tragedy of America's Last Years in Vietnam.* New York: Harvest/HBJ, 2000. (Describes the impact of General Creighton Abrams on the conduct of the war in South Vietnam. While he improved unity of effort in counterinsurgency, the North Vietnamese were successfully focusing on facilitating American withdrawal by targeting will in the United States.)

Taber, Robert. *War of the Flea: The Classic Study of Guerrilla Warfare.* Dulles, VA: Potomac Books, 2002. (Explains the advantages of the insurgent and how to overcome them.)

CONTEMPORARY EXPERIENCES AND THE WAR ON TERRORISM

Alwin-Foster, Nigel R.F. "Changing the Army for Counterinsurgency Operations." *Military Review* 85, 6 (Nov-Dec 2005), 2-15. (A provocative look at U.S. counterinsurgency operations in Iraq in 2003-2004 from a British practitioner.)

Barno, David W. "Challenges in Fighting a Global Insurgency." *Parameters* 36, 2 (Summer 2006), 15-29. (Observations from a three-star commander in Afghanistan.)

Chiarelli, Peter W. and Patrick R. Michaelis. "Winning the Peace: The Requirement for Full-Spectrum Operations," *Military Review* 85, 4 (Jul-Aug 2005), 4-17. (The commander of Task Force Baghdad in 2004 describes his lessons learned.)

Collins, Joseph J. "Afghanistan: Winning a Three Block War." *The Journal of Conflict Studies* 24, 2 (Winter 2004), 61-77. (The former deputy assistant secretary of defense for stability operations provides his views on achieving success in Afghanistan.)

Crane, Conrad and W. Andrew Terrill. *Reconstructing Iraq: Insights, Challenges, and Missions for Military Forces in a Post-conflict Scenario.* Carlisle Barracks, PA: U.S. Army War College, 2003. < http://www.strategicstudiesinstitute.army.mil/pubs > (Prescient look at the demands of rebuilding a state after changing a regime.)

Filkins, Dexter. "What the War Did to Colonel Sassaman." *The New York Times Magazine* (23 Oct 2005), 92. (Case study of a talented 4th Infantry Division battalion commander in Iraq in 2003-2004 who made some questionable ethical decisions that ended his career.)

Gunaratna, Rohan. *Inside Al Qaeda: Global Network of Terror.* Berkeley, CA: University of Berkeley Press, 2003. (The story behind the rise of the transnational insurgency.)

Hoffman, Bruce. *Insurgency and Counterinsurgency in Iraq.* Santa Monica, CA: RAND, 2004. Rand Corporation Web site < http://www.rand.org/pubs/occasional_papers/OP127/ > (Analysis of America's efforts in Iraq in 2003 informed by good history and theory.)

Kepel, Gilles. *The War for Muslim Minds: Islam and the West.* Cambridge, MA: Belknap Press, 2004. (A French explanation for the rise of Islamic extremism with suggestions for defeating it.)

Kilcullen, David. "Countering Global Insurgency: A Strategy for the War on Terrorism." *Journal of Strategic Studies* 28, 4 (Aug 2005), 597-617. (Describes the war on terrorism as a counterinsurgency campaign.)

———. "'Twenty-Eight Articles': Fundamentals of Company-level Counterinsurgency." *Military Review* 86, 3 (May-Jun 2006), 103-108. (Australian counterinsurgent prescribes actions for captains in counterinsurgency campaigns.)

———. "Counterinsurgency *Redux,*" *Survival* 48, 4 (Winter 2006-2007), 111-130. (Discusses insurgency's evolution from the classic Maoist form to the modern transnational, shifting coalitions that challenge the United States today.)

Lewis, Bernard. *The Crisis of Islam: Holy War and Unholy Terror.* New York: Modern Library, 2003. (A controversial but important analysis of the philosophical origins of transnational insurgency.)

McFate, Montgomery. "Iraq: The Social Context of IEDs." *Military Review* 85, 3 (May-Jun 2005), 37-40. (The insurgents' best weapon doesn't grow next to roads—it's constructed and planted there. Understanding who does that, and why, helps defeat improvised explosive devices.)

Metz, Steven and Raymond Millen, *Insurgency and Counterinsurgency in the 21st Century: Reconceptualizing Threat and Response.* Carlisle Barracks, PA: U.S. Army War College, 2004. (Longtime scholars of counterinsurgency put the war on terrorism in historical context.)

Multi-national Force-Iraq. "Counterinsurgency Handbook," 1st ed. Camp Taji, Iraq: Counterinsurgency Center for Excellence, May, 2006. (Designed to help leaders at all levels conduct counterinsurgency operations but focused at the company, platoon, and squad levels. Contains a variety of principles, considerations, and checklists.)

Packer, George. *The Assassins' Gate: America in Iraq.* New York: Farrar, Straus and Giroux, 2005. (A journalist for *The New Yorker* talks to Iraqis and Americans about Operation Iraqi Freedom.)

———. "The Lesson of Tal Afar: Is It Too Late for the Administration to Correct Its Course in Iraq?" *The New Yorker* (10 Apr 2006), 48-65. (The 2005 success of the 3d Armored Cavalry Regiment with the clear-hold-build tactic in Tal Afar.)

Petraeus, David. "Learning Counterinsurgency: Observations from Soldiering in Iraq." *Military Review* 86, 1 (Jan-Feb 2006), 2-12. (Commander of the 101st and Multinational Security Transition Command-Iraq passes on his personal lessons learned from two years in Iraq.)

Sageman, Marc. *Understanding Terror Networks.* Philadelphia, PA: University of Pennsylvania Press, 2004. (A former foreign service officer with Afghanistan experience explains the motivation of terrorists—not deprivation, but the need to belong.)

Military References

Army publications that are assigned a Marine Corps number are indicated with an asterisk.

REQUIRED PUBLICATIONS

These documents must be available to intended users of this publication.

*FM 1-02/MCRP 5-12A. *Operational Terms and Graphics.* 21 Sep 2004.

JP 1-02. *Department of Defense Dictionary of Military and Associated Terms.* 4 Dec 2001. (DOD Dictionary of Military Terms Web site < http://www.dtic.mil/doctrine/jel/doddict/ >)

RELATED PUBLICATIONS

These sources contain relevant supplemental information.

JOINT AND DEPARTMENT OF DEFENSE PUBLICATIONS

CJCSI 3121.01B. *Standing Rules of Engagement for U.S. Forces.* 15 Jan 2000.

DODD 2310.01E. *The Department of Defense Detainee Program.* 5 Sep 2006.

DODD 5105.38M. *Security Assistance Management Manual.* 3 Oct 2003. (Published by the Defense Security Cooperation Agency. Chapter 8 addresses end-use monitoring. AR 12-1 implements for the Army.)

JP 1. *Joint Warfare of the Armed Forces of the United States.* 14 Nov 2000.

JP 3-0. *Joint Operations.* 17 Sep 2006.

JP 3-07.1. *Joint Tactics, Techniques, and Procedures for Foreign Internal Defense.* 30 Apr 2004.

JP 3-08. *Interagency, Intergovernmental Organization, and Nongovernmental Organization Coordination During Joint Operations.* 2 vols. 17 Mar 2006.

JP 3-60. *Joint Doctrine for Targeting.* 17 Jan 2002.

JP 3-13. *Information Operations.* 13 Feb 2006.

JP 3-61. *Public Affairs.* 9 May 2005.

SERVICE PUBLICATIONS

AR 12-1. *Security Assistance, International Logistics, Training, and Technical Assistance Support Policy and Responsibilities.* 24 Jan 2000. (Marine Corps follows DODD 5105.38M.)

*AR 190-8/MCO 3461.1. *Enemy Prisoners of War, Retained Personnel, Civilian Internees and Other Detainees.* 1 Oct 1997.

FM 2-0 (34-1). *Intelligence.* 17 May 2004.

FM 2-22.3 (34-52). *Human Intelligence Collector Operations.* 6 Sep 2006. (See MarAdmin 458.06 for Marine Corps policy and guidance on intelligence interrogations.)

FM 3-0. *Operations.* 14 Jun 2001. (Under revision. Projected for republication during fiscal year 2007.)

FM 3-05 (100-25), *Army Special Operations Forces.* 20 Sep 2006.

FM 3-05.40 (41-10). *Civil Affairs Operations.* 29 Sep 2006. (MCRP 3-33.1 contains Marine Corps civil affairs doctrine.)

*FM 3-05.301/MCRP 3-40.6A. *Psychological Operations Tactics, Techniques, and Procedures.* 31 Dec 2003. (Distribution limited to government agencies only.)

*FM 3-05.401/MCRP 3-33.1A. *Civil Affairs Tactics, Techniques, and Procedures.* 23 Sep 2003.

*FM 3-09.31 (6-71)/MCRP 3-16C. *Tactics, Techniques, and Procedures for Fire Supportfor the Combined Arms Commander.* 1 Oct 2002.

FM 3-13 (100-6). *Information Operations: Doctrine, Tactics, Techniques, and Procedures.* 28 Nov 2003. (Appendix E addresses information operations targeting.)

FM 3-61.1. *Public Affairs Tactics, Techniques, and Procedures.* 1 Oct 2000.

FM 3-90. *Tactics.* 4 Jul 2001.

FM 4-0 (100-10). *Combat Service Support.* 29 Aug 2003.

FM 4-02 (8-10). *Force Health Protection in a Global Environment.* 13 Feb 2003. (NAVMED P-117, chapter 19, contains corresponding Marine Corps doctrine.)

FM 5-0 (101-5). *Army Planning and Orders Production.* 20 Jan 2005. (MCDP 5 contains Marine Corps planning doctrine.)

FM 5-104. *General Engineering.* 12 Nov 1986. (Will be republished as FM 3-34.300.)

FM 5-250. *Explosives and Demolitions. 30 Jul 1998.* (Will be republished as FM 3-34.214.)

FM 6-0. *Mission Command: Command and Control of Army Forces.* 11 Aug 2003.

FM 6-20-10. *Tactics, Techniques, and Procedures for the Targeting Process.* 8 May 1996.

FM 6-22 (22-100). *Army Leadership.*12 Oct 2006.

*FM 6-22.5 (22-9)/MCRP 6-11C. *Combat Stress.* 23 Jun 2000.

FM 7-98. *Operations in a Low-Intensity Conflict.* 19 Oct 1992. (Contains tactical-level guidance for brigade and battalion operations in an irregular warfare and peace operations environment.)

FM 20-32. *Mine/Countermine Operations.* 29 May 1998. (Will be republished as FM 3-34.210, *Explosive Hazards Operations.)*

FM 27-10. *The Law of Land Warfare.* 18 Jul 1956.

FM 31-20-3. *Foreign Internal Defense: Tactics, Techniques, and Procedures for Special Forces.* 20 Sep 1994. (Will be republished as FM 3-05.202.)

*FM 34-130/FMFRP 3-23-2. *Intelligence Preparation of the Battlefield.* 8 Jul 1994. (Will be republished as FM 2-01.3/MRCP 2-3A.)

FM 46-1. *Public Affairs Operations.* 30 May 1997.

FM 90-8. *Counterguerrilla Operations.* 29 Aug 1986.

FMI 2-91.4. *Intelligence Support to Operations in the Urban Environment.* 30 Jun 2005. (Expires 30 Jun 2007. Distribution limited to government agencies only. Available in electronic media only. Army Doctrine and Training Digital Library Web site < www.adtdl.army.mil >).

*FMI 3-34.119/MCIP 3-17.01. *Improved Explosive Device Defeat.* 21 Sep 2005. (Expires 21 Sep 2007. Distribution limited to government agencies only. Available in electronic media only. Army Doctrine and Training Digital Library Web site < www.adtdl.army.mil >.)

FMI 5-0.1. *The Operations Process.* 31 Mar 2006. (Expires 31 Mar 2008. When FM 3-0 is republished, it will address the material in FMI 5-0.1 that is relevant to this publication.)

MarAdmin (Marine Administrative Message) 458/06. "USMC Interim Policy and Guidance for Intelligence Interrogations." 22 Sep 2006. (States that FM 2-22.3 provides DOD-wide doctrine on intelligence interrogations. Lists sections of FM 2-22.33 that apply. Marine Corps Publications Web site < http://www.usmc.mil/maradmins/ >.)

MCDP 1. *Warfighting.* 20 Jun 1997.

MCDP 4. *Logistics.* 21 Feb 1997.

MCDP 5. *Planning.* 21 July 1997.

MCDP 6. *Command and Control.* 4 Oct 1996.

MCRP 3-33.1A. *Civil Affairs Operations.* 14 Feb 2000. (FM 3-05.40 contains Army civil affairs doctrine.)

MCWP 4-12. *Operational-Level Logistics.* 30 Jan 2002.

NAVMED P-117. *Manual of the Medical Department, U.S. Navy.* Chapter 19, "Fleet Marine Force," change 117. 21 Jun 2001. (Article 19-24 discusses levels of care. FM 4-02 contains the corresponding Army doctrine. When published, MCRP 4-11.1G will supersede this publication.)

PRESCRIBED FORMS

None

REFERENCED FORMS

None

This page intentionally left blank.

Index

Entries are by paragraph number unless a page is specified.

Entries are by paragraph number unless a page is specified.

Entries are by paragraph number unless a page is specified.

Entries are by paragraph number unless a page is specified.

Entries are by paragraph number unless a page is specified.

Entries are by paragraph number unless a page is specified

Entries are by paragraph number unless a page is specified.

Entries are by paragraph number unless a page is specified.

Entries are by paragraph number unless a page is specified.

Entries are by paragraph number unless a page is specified.

Entries are by paragraph number unless a page is specified.

Entries are by paragraph number unless a page is specified.

Entries are by paragraph number unless a page is specified.

Entries are by paragraph number unless a page is specified.

T

Entries are by paragraph number unless a page is specified.

U-V

Entries are by paragraph number unless a page is specified.

W-X-Y-Z

FM 3
MCWP 3-3
15 December 2

By Order of the Secretary of the Army:

PETER J. SCHOOMAK
General, United States A
Chief of Staff

Official:

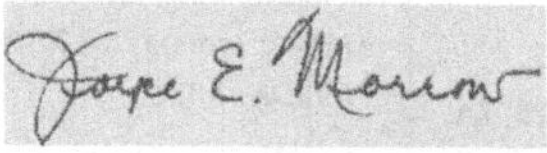

JOYCE E. MORROW
Administrative Assistant to the
Secretary of the Army
0634603

By Direction of the Commandant of the Marine Corps:

JAMES F. AMOS
Lieutenant General, U.S. Marine C
Deputy Commandant
Combat Development and Integra

DISTRIBUTION:

Active Army, Army National Guard, and U.S. Army Reserve: To be distributed in accordance with initial distribution number 121724, requirements for FM 3-24.

Made in United States
Orlando, FL
04 November 2026

90206690R00331